Turbulent Flows

This a graduate text on turbulent flows, an important topic in fluid dynamics. It is up to date, comprehensive, designed for teaching, and based on a course taught by the author at Cornell University for a number of years.

The book consists of two parts followed by a number of appendices. Part I provides a general introduction to turbulent flows, how they behave, how they can be described quantitatively, and the fundamental physical processes involved. The topics covered include: the Navier–Stokes equations; the statistical representation of turbulent flelds; mean-flow equations; the behavior of simple free shear and wall-bounded flows; the energy cascade; turbulence spectra; and the Kolmogorov hypotheses. Part II is concerned with various approaches for modelling or simulating turbulent flows. The approaches described are: direct numerical simulation (DNS); turbulent-viscosity models (e.g., the k–ε model); Reynolds-stress models; probability-density-function (PDF) methods; and large-eddy simulation (LES). There are numerous appendices in which the necessary mathematical techniques are presented.

This book is primarily intended as a graduate-level text in turbulent flows for engineering students, but it may also be valuable to students in applied mathematics, physics, oceanography, and atmospheric sciences, as well as researchers, and practicing engineers.

STEPHEN B. POPE is the Sibley College Professor in the Sibley School of Mechanical and Aerospace Engineering at Cornell University. Over the past 25 years he has performed research on turbulent flows and turbulent combustion, covering a broad range of approaches. He pioneered the application of PDF methods to turbulent flows; and his work on DNS won First Prize in the 1990 IBM Supercomputing Competition. Prior to joining Cornell University in 1982, Pope held positions at the Massachussetts Institute of Technology, the California Institute of Technology and Imperial College, London, from which he received his Ph.D. and D.Sc. He is a Fellow of the Royal Society, Member of the National Academy of Engineering, and Fellow of the American Academy of Arts and Sciences; and he is the recipient of the Fluid Dynamics Prize of the American Physical Society and the Zeldovich Gold Medal of the Combustion Institute. He is a member of the editorial boards of several journals, including *Physics of Fluids, Flow, Turbulence and Combustion*, and *Combustion Theory and Modelling*.

Turbulent Flows

Stephen B. Pope

Cornell University

CAMBRIDGE
UNIVERSITY PRESS

University Printing House, Cambridge CB2 8BS, United Kingdom

One Liberty Plaza, 20th Floor, New York, NY 10006, USA

477 Williamstown Road, Port Melbourne, VIC 3207, Australia

314-321, 3rd Floor, Plot 3, Splendor Forum, Jasola District Centre, New Delhi - 110025, India

79 Anson Road, #06-04/06, Singapore 079906

Cambridge University Press is part of the University of Cambridge.

It furthers the University's mission by disseminating knowledge in the pursuit of education, learning and research at the highest international levels of excellence.

www.cambridge.org
Information on this title: www.cambridge.org/9780521598866

First published 2000
4th printing 2018

A catalogue record for this publication is available from the British Library

Library of Congress Cataloging in Publication data
Pope, S. B.
Turbulent flows / S. B. Pope.
p.   cm.
ISBN 0 521 59125 2 (hc.)—ISBN 0 521 59886 9 (pbk.)
1.   Turbulence.   I.   Title.
QA913.P64   2000
532'.05217-dc21   99-044583   CIP

ISBN 978-0-521-59125-6 Hardback
ISBN 978-0-521-59886-6 Paperback

To my wife Linda
and to our children
Sarah and Sam

Contents

List of tables

Preface

This book is primarily intended as a graduate text on turbulent flows for engineering students, but it may also be valuable to students in atmospheric sciences, applied mathematics, and physics, as well as to researchers and practicing engineers.

The principal questions addressed are the following.

(i) How do turbulent flows behave?
(ii) How can they be described quantitatively?
(iii) What are the fundamental physical processes involved?
(iv) How can equations be constructed to simulate or model the behavior of turbulent flows?

In 1972 Tennekes and Lumley produced a textbook that admirably addresses the first three of these questions. In the intervening years, due in part to advances in computing, great strides have been made toward providing answers to the fourth question. Approaches such as Reynolds-stress modelling, probability-density-function (PDF) methods, and large-eddy simulation (LES) have been developed that, to an extent, provide quantitative models for turbulent flows. Accordingly, here (in Part II) an emphasis is placed on understanding how model equations can be constructed to describe turbulent flows; and this objective provides focus to the first three questions mentioned above (which are addressed in Part I). However, in contrast to the book by Wilcox (1993), this text is not intended to be a practical guide to turbulence modelling. Rather, it explains the concepts and develops the mathematical tools that underlie a broad range of approaches.

There is a vast literature on turbulence and turbulent flows, with many worthwhile questions addressed by many different approaches. In a one-semester course, or in a book of reasonable length, it is possible to cover only a fraction of the topics, and then with only a few of the possible approaches. The present selection of topics and approaches has evolved over the 20 years I have been teaching graduate courses on turbulence at MIT and Cornell. The emphasis on turbulent flows – rather than on the theory of homogeneous turbulence – is appropriate to applications in engineering, atmospheric sciences, and elsewhere. The emphasis on quantitative theories and models is consistent with the scientific objective – of developing a tractable, quantitatively accurate theory of the phenomenon – and is ideal for providing a solid understanding of computational approaches to turbulent flows, e.g., turbulence models and LES.

With the exceptions of LES and direct numerical simulation (DNS), the theories and models presented stem from the *statistical* approach, pioneered by Osborne Reynolds, G. I. Taylor, Prandtl, von Kármán, and Kolmogorov. A sizable fraction of the academic research work in the last 25 years has emphasized a more *deterministic* viewpoint: for example experiments on coherent structures, and models based on low-dimensional dynamical systems (e.g., Holmes, Lumley, and Berkooz (1996)). At this stage, this alternative approach has not led to a generally applicable quantitative model, neither – for better or for worse – has it had a major impact on the statistical approaches. Consequently, the deterministic viewpoint is neither emphasized nor systematically presented.

The book consists of two parts followed by a number of appendices. Part I provides a general introduction to turbulent flows, including the Navier–Stokes equations, the statistical representation of turbulent fields, mean-flow equations, the behavior of simple free-shear and wall-bounded flows, the energy cascade, turbulence spectra, and the Kolmogorov hypotheses. In the first five chapters, the focus is first on the mean velocity fields, and how they are affected by the Reynolds stresses. The concept of 'turbulent viscosity' is introduced with a thorough discussion of its deficiencies. The focus then shifts to the turbulence itself, in particular to the production and dissipation of turbulent kinetic energy. This sets the stage for a description (in Chapter 6) of the energy cascade and the Kolmogorov hypotheses. The spectral description of homogeneous turbulence in terms of Fourier modes in wavenumber space is developed in some detail. This provides an alternative perspective on the energy cascade; and it is also used in subsequent chapters in the descriptions of DNS, LES, and rapid distortion theory (RDT).

Simple wall-bounded flows are described in Chapter 7, starting with the mean velocity fields and proceeding to the Reynolds stresses. The exact transport equations for the Reynolds stresses are introduced, and their balances in turbulent boundary layers are examined.

The simulation and modelling approaches described in Part II are: DNS, turbulent viscosity models (e.g., the k–ε model), Reynolds-stress models, PDF methods, and LES. It is natural to consider DNS first (in Chapter 9) since it is conceptually the most straightforward approach. However, its restriction to simple, low-Reynolds-number flows motivates the consideration of other approaches. The most widely used turbulence models are the turbulent-viscosity models described in Chapter 10. Reynolds-stress models (Chapter 11) provide a more satisfactory connection to the physics of turbulence. The Reynolds-stress balance equations can be obtained from the Navier–Stokes equations, and the various contributions to this balance have been measured in experiments and simulations. Rapid-distortion theory is introduced to shed light on the effects that mean velocity gradients have on the Reynolds stresses. In developing and presenting modelled Reynolds-stress equations, the emphasis is on the fundamental concepts and principles, rather than on the detailed forms of particular models.

Chapter 12 deals with PDF methods. The primary object of study is the (one-point, one-time, Eulerian) joint probability density function (PDF) of velocity. The first moments of this PDF are the mean velocities; the second moments are the Reynolds stresses. For several reasons it is both natural and advantageous to proceed from the Reynolds stress to the PDF level of description: in the PDF equation, convection (by both mean and fluctuating velocity) appears in closed form, and hence does not have to be modelled; the effect of rapid distortions on turbulence can (in a limited sense) be treated exactly; and PDF methods are becoming widely used for turbulent reactive flows (e.g., turbulent combustion) because they are able to treat reaction exactly – without modelling assumptions.

Essential ingredients in PDF methods are stochastic Lagrangian models, such as the Langevin model for the velocity following a fluid particle. These models are also described in the context of turbulent dispersion (where they originated with G. I. Taylor's 1921 classic paper).

The final chapter describes LES, in which the large-scale turbulent motions are directly represented, while the effects of the smaller, subgrid-scale motions are modelled. Many of the concepts and techniques developed in Chapters 9–12 find application in the modelling of the subgrid-scale processes.

I use this book in a one-semester course, taught to students who previously have taken one or more graduate courses in fluid mechanics and applied mathematics. For most students, there is a good deal of new material, but I find that they can successfully master it, provided that it is clearly and fully explained. Accordingly there are many appendixes that provide the necessary development and explanation of mathematical techniques and results used in the text. In my experience, it is best not to rely upon the students' prior knowledge of probability theory, and consequently the necessary material is provided in the text (e.g., Sections 3.2–3.5).

For a less demanding pace, Parts I and II can be covered in two semesters – there is ample material. Alternatively, if a coverage of modelling is not required, Part I by itself provides a reasonably complete introduction to turbulent flows.

Many of the exercises ask the reader to 'show that …,' and thereby introduce additional results and observations. Consequently, it is recommended that all the exercises be read, even if they are not performed. The book is designed to be a self-contained text, but sufficient references are given to provide an entry into the research literature.

However much care is taken in the preparation of a book of this nature, it is inevitable that there will be errors in the first printing. A list of known corrections is given at `http://pope.mae.cornell.edu/TurbulentFlows .html`. The reader is asked to report any further corrections to the author at `s.b.pope@cornell.edu`.

I am profoundly grateful to many people for their help in the preparation of this work. For their support and technical input I thank my colleagues at Cornell, David Caughey, Sidney Leibovich, John Lumley, Dietmar Rempfer, and Zellman Warhaft. For their valuable suggestions based on reading draft chapters, I am grateful to Peter Bradshaw, Paul Durbin, Rodney Fox, Kemo Hanjalić, Charles Meneveau, Robert Moser, Blair Perot, Ugo Piomelli, P. K. Yeung, and Norman Zabusky. Similarly, I am grateful to the following Cornell graduates for their feedback on drafts of the book: Bertrand Delarue, Thomas Dreeben, Matthew Overholt, Paul Van Slooten, Jun Xu, Cem Albukrek, Dawn Chamberlain, Timothy Fisher, Laurent Mydlarski, Gad Reinhorn, Shankar Subramaniam, and Walter Welton. The first five mentioned are also thanked for their assistance in producing the figures. Most of the typescript was prepared by June Meyermann, whose patience, accuracy, and enthusiasm are greatly appreciated. The accuracy of the bibliography has been much improved by the careful checking performed by Sarah Pope. Above all, I wish to thank my wife, Linda, for her patience, support, and encouragement during this project and over the years.

Nomenclature

The notation used is given here in the following order: upper-case Roman, lower-case Roman, upper-case Greek, lower-case Greek, superscripts, subscripts, symbols, and abbreviations. Then the symbols $\mathcal{O}(\ )$, $o(\ )$, and $\sim$ that are used to denote the order of a quantity are explained.

Upper-case Roman

A^+	van Driest constant (Eq. (7.145))
$\mathcal{A}$	control surface bounding $\mathcal{V}$
B	log-law constant (Eq. (7.43))
B_1	constant in the velocity-defect law (Eq. (7.50))
B_2	Loitsyanskii integral (Eq. (6.92))
B_2	log-law constant for fully-rough walls (Eq. (7.120))
$\tilde{B}(s/\delta_v)$	log-law constant for rough walls (Eq. (7.121))
C	Kolmogorov constant related to $E(\kappa)$ (Eq. (6.16))
C_0	coefficient in the Langevin equation (Eqs. (12.26) and (12.100))
C_1	Kolmogorov constant related to $E_{11}(\kappa_1)$ (Eq. (6.228))
C_1'	Kolmogorov constant related to $E_{22}(\kappa_1)$ (Eq. (6.231))
C_2	Kolmogorov constant related to D_{LL} (Eq. (6.30))
C_2	constant in the IP model (Eq. (11.129))
C_3	constant in the model equation for ω^* (Eq. (12.194))
C_E	LES dissipation coefficient (Eq. (13.285))
C_f	skin-friction coefficient ($\tau_w/(\frac{1}{2}\rho \bar{U}^2)$)
C_R	Rotta constant (Eq. (11.24))
C_S	Smagorinsky coefficient (Eq. (13.128))
C_s	constant in Reynolds-stress transport models (Eq. (11.147))

C_ε	constant in the model equation for ε (Eq. (11.150))
$C_{\varepsilon 1}, C_{\varepsilon 2}$	constants in the model equation for ε (Eq. (10.53))
C_μ	turbulent-viscosity constant in the k–ε model (Eq. (10.47))
C_ν	LES eddy-viscosity coefficient (Eq. (13.286))
C_ϕ	constant in the IEM mixing model (Eq. (12.326))
C_Ω	constant in the definition of Ω (Eq. (12.193))
$C_{\omega 1}, C_{\omega 2}$	constants in the model equation for ω (Eq. (10.93))
C_0	Kolmogorov constant (Eq. (12.96))
C_{ij}^o	cross stress (Eq. (13.101))
D	pipe diameter
D_{ij}	second-order velocity structure function (Eq. (6.23))
$D_L(s)$	second-order Lagrangian structure function (Eq. (12.95))
D_{LL}	longitudinal second-order velocity structure function
D_{LLL}	longitudinal third-order velocity structure function (Eq. (6.86))
D_{NN}	transverse second-order velocity structure function
$D_n(r)$	nth-order longitudinal velocity structure function (Eq. (6.304))
$\mathrm{D}/\mathrm{D}t$	substantial derivative ($\partial/\partial t + \boldsymbol{U} \cdot \nabla$)
$\bar{\mathrm{D}}/\bar{\mathrm{D}}t$	mean substantial derivative ($\partial/\partial t + \langle \boldsymbol{U} \rangle \cdot \nabla$)
$\overline{\mathrm{D}}/\overline{\mathrm{D}}t$	substantial derivative based on filtered velocity
E	Cartesian coordinate system with basis vectors $\boldsymbol{e}_i$
$\overline{E}$	Cartesian coordinate system with basis vectors $\bar{\boldsymbol{e}}_i$
$E(\boldsymbol{x}, t)$	kinetic energy ($\frac{1}{2}\boldsymbol{U} \cdot \boldsymbol{U}$)
$\bar{E}(\boldsymbol{x}, t)$	kinetic energy of the mean flow ($\frac{1}{2}\langle \boldsymbol{U} \rangle \cdot \langle \boldsymbol{U} \rangle$)
$\dot{E}(x)$	kinetic energy flow rate of the mean flow
$E(\kappa)$	energy-spectrum function (Eq. (3.166))
$E_{ij}(\kappa_1)$	one-dimensional energy spectrum (Eq. (6.206))
$\overline{E}(\kappa)$	energy-spectrum function of filtered velocity (Eq. (13.62))
$E(\omega)$	frequency spectrum (defined for positive frequencies, Eq. (3.140))
$\check{E}(\omega)$	frequency spectrum (defined for positive and negative frequencies, Eq. (E.31))
F	determinant of the normalized Reynolds stress (Eq. (11.52))
$F(V)$	cumulative distribution function (CDF) of U (Eq. (3.7))
$F_{\mathrm{D}}(y/\delta)$	velocity-defect law (Eq. (7.46))
$\mathcal{F}$	Fourier transform (Eq. (D.1))
$\mathcal{F}^{-1}$	inverse Fourier transform (Eq. (D.2))

$\mathcal{F}_\kappa$	Fourier integral operator (Eq. (6.116))
G_{ij}	coefficient in the GLM (Eqs. (12.26) and (12.110))
$G(r)$	LES filter function
$\widehat{G}(\kappa)$	LES filter transfer function
H	shape factor (δ^*/θ)
$H(x)$	Heaviside function (Eq. (C.33))
$\mathbf{I}$	identity matrix
$I(\boldsymbol{x},t)$	indicator function for intermittency (Eq. (5.299))
$\mathrm{I}_s,\ \mathrm{II}_s,\ \mathrm{III}_s$	principal invariants of the second-order tensor s (Eqs. (B.31)–(B.33))
K	kurtosis of the longitudinal velocity derivative
K_ϕ	kurtosis of ϕ
Kn	Knudsen number
$K_\nu(z)$	modified Bessel function of the second kind
L	lengthscale $(k^{\frac{3}{2}}/\varepsilon)$
$\bar{L}$	lengthscale (u'^3/ε)
L_{11}	longitudinal integral lengthscale (Eq. (3.161))
L_{22}	lateral integral lengthscale (Eq. (6.48))
$\mathcal{L}$	charactcristic lengthscale of the flow
$\mathcal{L}$	length of side of cube in physical space
$\mathcal{L}_{ij}$	resolved stress (Eq. (13.252))
$\mathcal{L}_{ij}^\circ$	Leonard stress (Eq. (13.100))
$\dot{M}(x)$	momentum flow rate of the mean flow
M_{ij}	scaled composite rate-of-strain tensor (Eq. (13.255))
M_n	normalized nth moment of the longitudinal velocity derivative (Eq. (6.303))
Ma	Mach number
$\mathcal{N}(\mu,\sigma^2)$	normal distribution with mean μ and variance σ^2
$\mathcal{O}(h)$	quantity of big order h
$o(h)$	quantity of little order h
P	pressure (Eq. (2.32))
$P(A)$	probability of event A
$P(\boldsymbol{x},t)$	particle pressure (Eq. (12.225))
$P_{jk}(\boldsymbol{\kappa})$	projection tensor (Eq. (6.133))
$\mathcal{P}$	production: rate of production of turbulent kinetic energy (Eq. (5.133))
$\mathcal{P}_{ij}$	rate of production of Reynolds stress (Eq. (7.179))
$\mathcal{P}_r$	rate of production of residual kinetic energy (Eq. (13.123))
$\mathcal{P}_\phi$	rate of production of scalar variance (Eq. (5.282))

R	pipe radius
$R(s)$	autocovariance (Eq. (3.134))
$R_{ij}(\boldsymbol{r}, \boldsymbol{x}; t)$	two-point velocity correlation (Eq. (3.160))
$\hat{R}_{ij}(\boldsymbol{\kappa})$	Fourier coefficient of two-point velocity correlation (Eq. (6.152))
R_T	turbulent Reynolds number (Eq. (5.85))
R_λ	Taylor-scale Reynolds number (Eq. (6.63))
Re	Reynolds number
Re	Reynolds number ($2\bar{U}\delta/v$)
Re_0	Reynolds number ($U_0\delta/v$)
Re_L	turbulence Reynolds number ($k^{1/2}L/v = k^2/(\varepsilon v)$)
Re_T	turbulence Reynolds number ($u'L_{11}/v$)
Re_x	Reynolds number ($U_0 x/v$)
Re_δ	Reynolds number ($U_0\delta/v$)
Re_{δ^*}	Reynolds number ($U_0\delta^*/v$)
Re_θ	Reynolds number ($U_0\theta/v$)
Re_τ	Reynolds number based on friction velocity ($u_\tau\delta/v$)
$\mathcal{R}_{ij}$	pressure–rate-of-strain tensor (Eq. (7.187))
$\mathcal{R}_{ij}^o$	SGS Reynolds stress (Eq. (13.102))
$\mathcal{R}_{ij}^{(a)}$	redistribution term (anisotropic part of Π_{ij}, Eq. (11.6))
$\mathcal{R}_{ij}^c(\boldsymbol{v}, \boldsymbol{x}, t)$	conditional pressure–rate-of-strain tensor (Eq. (12.20))
$\mathcal{R}_{ij}^{(e)}$	redistribution term used in elliptic-relaxation model (Eq. (11.198))
$\mathcal{R}_{ij}^{(r)}$	rapid pressure–rate-of-strain tensor (Eq. (11.13))
$\mathcal{R}_{ij}^{(s)}$	slow pressure–rate-of-strain tensor
S	spreading rate of a free shear flow
S	velocity-derivative skewness (Eq. (6.85))
$S(\phi)$	chemical source term (Eq. (12.321))
S'	velocity structure function skewness (Eq. (6.89))
S_{ij}	rate-of-strain tensor ($\frac{1}{2}(\partial U_i/\partial x_j + \partial U_j/\partial x_i)$)
$\bar{S}_{ij}$	mean rate-of-strain tensor ($\frac{1}{2}(\partial\langle U_i\rangle/\partial x_j + \partial\langle U_j\rangle/\partial x_i)$)
$\hat{S}_{ij}$	normalized mean rate-of-strain tensor ($(k/\varepsilon)\bar{S}_{ij}$)
$\overline{S}_{ij}$	filtered rate-of-strain tensor (Eq. (13.73))
$\widetilde{\overline{S}}_{ij}$	doubly filtered rate-of-strain tensor
$\bar{S}_{ijk}(\boldsymbol{r}, t)$	two-point triple velocity correlation (Eq. (6.72))
S_ϕ	skewness of ϕ
S_ω	mean source of turbulence frequency (Eq. (12.184))
$\mathcal{S}$	characteristic mean strain rate ($2\bar{S}_{ij}\bar{S}_{ij})^{\frac{1}{2}}$ ($\mathcal{S} = \partial\langle U_1\rangle/\partial x_2$ in simple shear flow)

$\bar{S}$	filtered rate-of-strain invariant $(2\bar{S}_{ij}\bar{S}_{ij})^{\frac{1}{2}}$
$\tilde{\bar{S}}$	doubly filtered rate-of-strain invariant $(2\tilde{\bar{S}}_{ij}\tilde{\bar{S}}_{ij})^{\frac{1}{2}}$
$\mathcal{S}(\kappa)$	sphere in wavenumber space of radius κ
$\mathcal{S}_{\lambda}$	principal mean strain rate: largest eigenvalue of $\bar{S}_{ij}$
T	time interval
T	turbulent timescale defined by Eq. (11.163)
$\hat{T}(\kappa)$	rate of energy transfer to Fourier mode of wavenumber κ from other modes (Eq. 6.162)
T_{kij}	flux of Reynolds stress (Eq. (7.195))
$T_{kij}^{(p)}$	flux of Reynolds stress due to fluctuating pressure (Eq. (7.193))
$T_{kij}^{(p')}$	isotropic flux of Reynolds stress due to fluctuating pressure (Eq. (11.140))
$T_{kij}^{(u)}$	flux of Reynolds stress due to turbulent convection $(\langle u_k u_i u_j \rangle)$
$T_{kij}^{(v)}$	diffusive flux of Reynolds stress (Eq. (7.196))
T_{L}	Lagrangian integral timescale (Eq. (12.93))
$\mathcal{T}(\ell)$	rate of transfer of energy from eddies larger than ℓ to those smaller than ℓ
$\mathcal{T}_{\mathrm{EI}}$	rate of transfer of energy from large eddies to small eddies
$\mathcal{T}_{\mathrm{DI}}$	rate of transfer of energy into the dissipation range $(\ell < \ell_{\mathrm{DI}})$ from larger scales
$U(t)$	random process
$U(x,t)$	Eulerian velocity
$U(x,y,z)$	x component of velocity
$U(x,r,\theta)$	x component of velocity
$\bar{U}$	bulk velocity in channel (Eq. (7.3)) and pipe flow (Eq. (7.94))
$U^{+}(t)$	fluid-particle velocity
$U^{*}(t)$	model for the fluid-particle velocity
$\overline{U}(x,t)$	filtered (resolved) velocity field
U_0	mean centerline velocity in channel and pipe flow
$U_0(x)$	mean centerline velocity in a jet
$U_0(x)$	freestream velocity
$U_{\mathrm{c}}(x)$	characteristic convective velocity
U_{J}	jet-nozzle velocity
U_{h}	velocity of high-speed stream in a mixing layer
U_{l}	velocity of low-speed stream in a mixing layer
$U_s(x)$	characteristic velocity difference

$\mathcal{U}$	characteristic velocity scale of the flow
V	sample space variable corresponding to U
$\mathbf{V}$	sample space variable corresponding to velocity $\mathbf{U}$
$V(x,r,\theta)$	r component of velocity
$V(x,y,z)$	y component of velocity
$\mathcal{V}$	control volume in physical space bounded by $\mathcal{A}$
$W(t)$	Wiener process
$\mathbf{W}(t)$	vector-valued Wiener process
$W(x,r,\theta)$	θ component of velocity
$W(x,y,z)$	z component of velocity
$\mathbf{X}^{+}(t,\mathbf{Y})$	fluid-particle position: position at time t of fluid particle that is at $\mathbf{Y}$ at the reference time t_0
$\mathbf{X}^{*}(t)$	model for fluid-particle position (Eq. (12.108))
$\mathbf{Y}$	fluid particle position at the reference time t_0

Lower-case Roman

a	drift coefficient of a diffusion process (Eq. (J.27))
a_{ij}	anisotropic Reynolds stresses ($\langle u_i u_j \rangle - \frac{2}{3}k\delta_{ij}$)
a_{ij}	direction cosines (Eq. (A.11))
a_{f}	LES filter constant (Eq. (13.77))
b^2	diffusion coefficient of a diffusion process (Eq. (J.27))
b_{ij}	normalized Reynolds-stress anisotropy ($a_{ij}/(2k)$)
c_{f}	skin-friction coefficient ($\tau_{\mathrm{w}}/(\frac{1}{2}\rho U_0^2)$)
c_{S}	Smagorinsky coefficient (Eq. (13.253))
d	jet-nozzle diameter
$\hat{\mathbf{e}}(t)$	unit wavevector (Eq. (11.84))
$\mathbf{e}_i$	unit vector in the i-coordinate direction
f	friction factor (Eq. (7.97))
$f, \bar{f}$	self-similar mean axial velocity profile
$f(r,t)$	longitudinal velocity autocorrelation function (Eq. (6.45))
$f(V)$	probability density function (PDF) of U (Eq. (3.14))
$f(\mathbf{V};\mathbf{x},t)$	Eulerian PDF of velocity (Eq. (3.153))
$f'(\mathbf{V};\mathbf{x},t)$	fine-grained Eulerian PDF of velocity (Eq. (H.1))
$f^{*}(\mathbf{V};\mathbf{x},t)$	modelled Eulerian PDF of velocity (Eq. (12.116))
$f^{*}(\mathbf{V}|\mathbf{x};t)$	conditional PDF of particle velocity (Eq. (12.205))
$\overline{f}(\mathbf{V};\mathbf{x},t)$	filtered density function (Eq. (13.287))
$\tilde{f}(\mathbf{V},\theta;\mathbf{x},t)$	joint PDF of velocity and turbulence frequency
$\hat{f}(\mathbf{V},\boldsymbol{\psi};\mathbf{x},t)$	velocity–composition joint PDF
$f_{2|1}(V_2|V_1)$	PDF of U_2 conditional on $U_1 = V_1$ (Eq. (3.95))

$f_L(V, x; t \mid Y)$	Lagrangian velocity–position joint PDF (Eq. (12.76))
$f_L^*(V, x; t)$	joint PDF of $U^*(t)$ and $X^*(t)$
$f_N(\psi; x, t)$	non-turbulent conditional PDF of scalar $\phi(x, t)$
$f_T(\psi; x, t)$	turbulent conditional PDF of scalar $\phi(x, t)$
$f_w(y^+)$	law of the wall (Eq. (7.37))
$f_X(x; t \mid Y)$	PDF of fluid-particle position
$f_X^*(x; t)$	PDF of $X^*(t)$
f_μ	damping function in k–ε model (Eq. (11.155))
$f_\phi(\psi; x, t)$	PDF of scalar $\phi(x, t)$
$f_\omega(\theta; x, t)$	PDF of turbulence frequency
$g, \bar{g}$	self-similar shear-stress profile in a free shear flow
g	gravitational acceleration
$\boldsymbol{g}$	gravitational force per unit mass
$g(r, t)$	transverse velocity autocorrelation function (Eq. (6.45))
$g(v; x, t)$	Eulerian PDF of the fluctuating velocity
$h, \bar{h}$	self-similar mean lateral velocity profile
h	grid spacing
k	turbulent kinetic energy ($\frac{1}{2}\langle \boldsymbol{u} \cdot \boldsymbol{u} \rangle$)
$\bar{k}(r, t)$	longitudinal two-point triple correlation (Eq. (6.73))
k_r	residual kinetic energy (Eq. (13.92))
$k_{(\kappa_a, \kappa_b)}$	turbulent kinetic energy in the wavenumber range (κ_a, κ_b)
l	lengthscale defined as ν_T / u'
ℓ	lengthscale
ℓ	characteristic eddy size
ℓ_0	lengthscale of the largest eddies
ℓ_{DI}	demarcation lengthscale between the dissipation range ($\ell < \ell_{DI}$) and the inertial subrange ($\ell > \ell_{DI}$)
ℓ_{EI}	demarcation lengthscale between the energy-containing range of eddies ($\ell > \ell_{EI}$) and smaller eddies ($\ell < \ell_{EI}$)
ℓ_m	mixing length (Eq. (7.91))
ℓ_m^+	mixing length in wall units (ℓ_m / δ_ν)
ℓ_S	Smagorinsky lengthscale (Eq. (13.128))
$\ell_w(x)$	distance between x and the nearest solid surface
$\dot{m}(x)$	mass flow rate of the mean flow
$\boldsymbol{n}$	unit normal vector
$o(h)$	small order h (Eq. (J.34))
p	exponent in power-law spectrum (Eq. (G.5))
$p(x, t)$	modified pressure
$p'(x, t)$	fluctuating (modified) pressure

$p^{(h)}(x, t)$	harmonic pressure (Eq. (2.49))
$p^{(r)}(x, t)$	rapid pressure (Eq. (11.11))
$p^{(s)}(x, t)$	slow pressure (Eq. (11.12))
$p_0(x)$	freestream pressure
$p_w(x)$	wall pressure
q	exponent in power-law structure function (Eq. (G.6))
r	radial coordinate
$r_{1/2}(x)$	half-width of jet or wake
s	time interval
s	lengthscale of wall roughness
s_{ij}	fluctuating rate-of-strain tensor ($\frac{1}{2}(\partial u_i/\partial x_j + \partial u_j/\partial x_i)$)
t	time
u	x component of fluctuating velocity
$u(\ell)$	characteristic velocity of an eddy of size ℓ
$u(x, t)$	fluctuating velocity
$\hat{u}(\kappa, t)$	Fourier coefficient of velocity (Eq. (6.102))
u'	r.m.s. velocity
$u^*(t)$	fluctuating component of particle velocity (Eq. (12.207))
u^+	mean velocity normalized by the friction velocity
$u'(x, t)$	residual (SGS) velocity field (Eq. (13.3))
$u'_0(x)$	r.m.s. axial velocity
u_0	velocity scale of the largest eddies
u_e	propagation velocity of the viscous superlayer
u_η	Kolmogorov velocity (Eq. (5.151))
u_τ	friction velocity ($\sqrt{\tau_w/\rho}$)
v	y or r component of fluctuating velocity
$\mathbf{v}$	sample space variable corresponding to u
w	z or θ component of fluctuating velocity
$w(y/\delta)$	law of the wake function (Eq. (7.149))
x	position
x	Cartesian or polar cylindrical coordinate
x_0	virtual origin
y	Cartesian coordinate
y^+	distance from the wall normalized by the viscous lengthscale, δ_v
$y_{0.1}(x)$	cross-stream location in mixing layer (also $y_{0.9}(x)$ etc., see Eq. (5.203))
$y_{1/2}(x)$	half-width of jet or wake
y_p	distance from the wall at which wall functions are applied

y_ϕ	half-width of scalar profile
z	Cartesian coordinate

Upper-case Greek

Γ	molecular diffusivity
$\Gamma(z)$	gamma function (Eq. (3.67))
Γ_{eff}	effective diffusivity ($\Gamma_T + \Gamma$)
Γ_T	turbulent diffusivity (Eq. (4.42))
$\Delta, \bar{\Delta}$	filter width
$\bar{\Delta}$	grid filter width in the dynamic model
$\hat{\Delta}$	test filter width in the dynamic model
$\widehat{\bar{\Delta}}$	effective width of combined test and grid filters (Eq. (13.247))
Δ_{h}	temporal increment operator (Eq. (J.4))
Δ_i	filter width in direction i
$\Delta_r u$	longitudinal velocity increment (Eq. (6.305))
Π	wake-strength parameter (Eq. (7.148))
Π_{ij}	velocity–pressure-gradient tensor (Eq. (7.180))
$\Phi\left(\dfrac{y}{\delta_v}, \dfrac{y}{\delta}\right)$	universal velocity-gradient function for channel flow (Eq. (7.31))
$\Phi(\boldsymbol{\kappa}, t)$	kinetic energy of Fourier mode with wavenumber $\boldsymbol{\kappa}$ (Eq. (6.103))
$\Phi_{ij}(\boldsymbol{\kappa})$	velocity-spectrum tensor (Eq. (3.163))
Ψ	gravitational potential ($\boldsymbol{g} = -\nabla\Psi$)
$\Psi(s)$	characteristic function (Eq. (I.1))
Ω	characteristic mean rotation rate $(2\bar{\Omega}_{ij}\bar{\Omega}_{ij})^{1/2}$
$\Omega(\boldsymbol{x}, t)$	conditional mean turbulence frequency (Eq. (12.193))
Ω_{ij}	rate-of-rotation tensor ($\frac{1}{2}(\partial U_i/\partial x_j - \partial U_j/\partial x_i)$)
$\bar{\Omega}_{ij}$	mean rate-of-rotation tensor ($\frac{1}{2}(\partial \langle U_i \rangle/\partial x_j - \partial \langle U_j \rangle/\partial x_i)$)
$\hat{\Omega}_{ij}$	normalized mean rate-of-rotation tensor ($(k/\varepsilon)\bar{\Omega}_{ij}$)
$\tilde{\Omega}_{ij}$	rate of rotation of coordinate axes (Eq. (2.97))

Lower-case Greek

β_o	constant in the exponential spectrum (Eq. (6.253))
$\gamma(\boldsymbol{x}, t)$	intermittency factor (Eq. (5.300))
δ	half-height of channel
$\delta(x)$	Dirac delta function
$\delta(x)$	characteristic flow width

$\delta(x)$	boundary-layer thickness
$\delta^*(x)$	displacement thickness
δ_{ij}	Kronecker delta (Eq. (A.1))
$\delta_{\kappa,\kappa'}$	Kronecker delta defined by Eq. (6.111)
δ_v	viscous lengthscale (Eq. (7.26))
ϵ	error
ε	rate of dissipation of turbulent kinetic energy $(2v\langle s_{ij}s_{ij}\rangle)$
$\tilde{\varepsilon}$	pseudo-dissipation $\left(v\langle\dfrac{\partial u_i}{\partial x_j}\dfrac{\partial u_i}{\partial x_j}\rangle\right)$
$\varepsilon_{(\kappa_a,\kappa_b)}$	dissipation in the wavenumber range (κ_a,κ_b)
ε_0	instantaneous dissipation rate $(2vs_{ij}s_{ij})$
$\hat{\varepsilon}_0$	one-dimensional surrogate for ε_0 (Eq. (6.314))
ε_{ij}	dissipation tensor $\left(2v\langle\dfrac{\partial u_i}{\partial x_k}\dfrac{\partial u_j}{\partial x_k}\rangle\right)$
ε_{ijk}	alternating symbol (Eq. (A.56))
$\varepsilon_{ij}^c(v,x,t)$	conditional dissipation tensor (Eq. (12.21))
$\varepsilon_\phi^c(\psi,x,t)$	conditional scalar dissipation rate (Eq. (12.346))
ε_r	average of ε_0 over volume of radius r (Eq. (6.313))
$\hat{\varepsilon}_r$	one-dimensional surrogate for ε_r (Eq. (6.315))
ε_ϕ	scalar dissipation rate (Eq. (5.283))
ζ_n	nth-order structure function exponent (Eq. (6.307))
η	Kolmogorov lengthscale (Eq. (5.149))
η	normalized lateral coordinate in free shear flows
η	invariant of the Reynolds-stress anisotropy tensor (Eq. (11.28))
θ	circumferential coordinate
θ	sample-space variable corresponding to ω^*
$\theta(x)$	displacement thickness (Eq. (7.127))
ϑ	specific volume $(\vartheta=1/\rho)$
κ	von Kármán constant (Eq. (7.43))
κ	wavenumber
$\boldsymbol{\kappa}$	wavenumber vector
$\hat{\boldsymbol{\kappa}}(t)$	time-dependent wavenumber vector (Eq. (11.80))
κ_0	lowest wavenumber
κ_c	filter cutoff wavenumber $(\kappa_c=\pi/\Delta)$
κ_{DI}	demarcation wavenumber between the dissipation range $(\kappa>\kappa_{DI})$ and the inertial subrange $(\kappa<\kappa_{DI})$
κ_{EI}	demarcation wavenumber between the energy-containing range $(\kappa<\kappa_{EI})$ and the inertial subrange $(\kappa>\kappa_{EI})$
λ	mean free path

λ_f	longitudinal Taylor microscale (Eq. (6.53))
λ_g	transverse Taylor microscale (Eq. (6.57))
μ	viscosity
μ	internal intermittency exponent (Eq. (6.317))
μ	mean of a distribution
μ_n	nth central moment (Eq. (3.25))
$\hat{\mu}_n$	standardized nth central moment (Eq. (3.37))
ν	kinematic viscosity ($\nu = \mu/\rho$)
ν_{eff}	effective viscosity ($\nu_{\text{T}} + \nu$)
ν_r	residual (SGS) eddy viscosity (Eq. (13.127))
ν_{T}	turbulent viscosity (Eq. (4.45))
ξ	normalized lateral coordinate in free shear flows
ξ	invariant of the Reynolds-stress anisotropy tensor (Eq. (11.29))
ρ	density
$\rho(s)$	autocorrelation function (Eq. (3.135))
ρ_{12}	correlation coefficient between u_1 and u_2 (Eq. (3.93))
ρ_{uv}	correlation coefficient between u and v (Eq. (3.93))
σ	standard deviation
σ	Prandtl number
σ_k	turbulent Prandtl number for kinetic energy (Eq. (10.41))
σ_{T}	turbulent Prandtl number ($\nu_{\text{T}}/\Gamma_{\text{T}}$)
σ_X	r.m.s. fluid-particle dispersion (Eq. (12.149))
σ_ε	turbulent Prandtl number for dissipation (Eq. (10.53))
τ	turbulence timescale (k/ε)
$\bar{\tau}$	integral timescale (Eq. (3.139))
$\tau(\ell)$	characteristic timescale of an eddy of size ℓ
$\tau(y)$	total shear stress in simple shear flow (Eq. (7.10))
τ_0	timescale of largest eddies (u_0/ℓ_0)
τ_{ij}	stress tensor (Eq. (2.32))
τ_{ij}^{R}	residual (SGS) stress tensor (Eq. (13.90))
τ_{ij}^r	deviatoric residual (SGS) stress tensor (Eq. (13.93))
τ_{w}	wall shear stress
τ_η	Kolmogorov timescale (Eq. (5.150))
τ_ϕ	scalar timescale ($\langle\phi'^2\rangle/\varepsilon_\phi$)
$\phi(\boldsymbol{x}, t)$	conserved passive scalar
φ	self-similar profile of a conserved passive scalar
ψ	sample-space variable corresponding to ϕ
$\psi(x, r)$	Stokes stream function (Eq. (5.86))
ω	frequency

ω	turbulence frequency ε/k
$\boldsymbol{\omega}(\boldsymbol{x},t)$	vorticity ($\boldsymbol{\omega} = \nabla \times \boldsymbol{U}$)
ω^2	enstrophy ($\omega^2 = \boldsymbol{\omega}\cdot\boldsymbol{\omega}$)
$\omega^*(t)$	model for turbulence frequency

Superscripts

ϕ^*	complex conjugate of ϕ
ϕ^+	indicates Lagrangian variable (Eq. (2.9))
$\hat{\phi}(\boldsymbol{\kappa})$	indicates Fourier coefficient at wavenumber $\boldsymbol{\kappa}$ of function $\phi(\boldsymbol{x})$ (Eq. (6.113))
$\hat{\phi}$	indicates standardized random variable or PDF
$\dot{\phi}$	rate of change of ϕ ($\dot{\phi} = \mathrm{d}\phi/\mathrm{d}t$)
ϕ'	fluctuating component ($\phi' = \phi - \langle\phi\rangle$)
ϕ'_{T}	conditional turbulent r.m.s. of ϕ (Eq. (5.304))
ϕ'_{N}	conditional non-turbulent r.m.s. of ϕ
f'	derivative ($f'(x) = \mathrm{d}f(x)/\mathrm{d}x$)
u'	residual (from filtering, Eq. (13.3))
$v^{\parallel}(\boldsymbol{\kappa})$	component of $v(\boldsymbol{\kappa})$ parallel to $\boldsymbol{\kappa}$ (Eq. (6.129))
$v^{\perp}(\boldsymbol{\kappa})$	component of $v(\boldsymbol{\kappa})$ perpendicular to $\boldsymbol{\kappa}$ (Eq. (6.131))
A^{T}	transpose of A
$\overline{U}$	filtered quantity (filter width Δ or $\overline{\Delta}$)
$\widetilde{U}$	filtered quantity (filter width $\widetilde{\Delta}$)
$\widetilde{\overline{U}}$	filtered quantity (filter width $\widetilde{\overline{\Delta}}$)

Subscripts

$\langle Q\rangle_{\mathcal{L}}$	volume average of $Q(\boldsymbol{x})$ over a cube of side $\mathcal{L}$ (Eq. (3.175))
$\langle Q\rangle_N$	mean of Q over an ensemble of N samples (Eq. (3.108))
$\langle Q\rangle_{\mathrm{N}}$	non-turbulent conditional mean of Q
Q_{p}	quantity evaluated at y_{p} in wall functions
$\langle Q\rangle_T$	time average of $Q(t)$ over time interval T (Eq. (3.173))
$\langle Q\rangle_{\mathrm{T}}$	turbulent conditional mean of Q
$(Q)_y$	quantity evaluated at y

Symbols

$\det(A)$	determinant of A
$\Im(z)$	imaginary part of z

$\lim\limits_{h\downarrow 0}$	the limit as the positive quantity h tends to zero
$\max(a, b)$	the greater of a and b
$\min(a, b)$	the lesser of a and b
$\Re(z)$	real part of z
$\mathrm{sdev}(U)$	standard deviation of U (Eq. (3.24))
$\mathrm{trace}(A)$	trace of tensor A (Eq. (B.3))
$\mathrm{var}(U)$	variance of U (Eq. (3.23))
∇	gradient operator (Eq. (A.48))
$\nabla\cdot$	divergence operator (Eq. (A.52))
$\nabla\times$	curl operator (Eq. (A.60))
∇^2	Laplacian operator (Eq. (A.53))
$\times$	vector cross product (Eq. (A.57))
$\oint(\ )\,\mathrm{d}\mathcal{S}(r)$	integral over the surface of the sphere of radius r
$\langle Q \rangle$	mean or expectation of Q
$\langle Q \vert V \rangle$	mean of Q conditional on $U = V$ (Eq. (3.97))
$U \stackrel{\mathrm{D}}{=} f$	the random variable U has the distribution f
$f \sim g$	f varies as (or scales with) g

Abbreviations

ASM	algebraic stress model
CFD	computational fluid dynamics
DFT	discrete Fourier transform
DNS	direct numerical simulation
FFT	fast Fourier transform
GLM	generalized Langevin model
IEM	interaction by exchange with the mean
i.i.d.	independent and identically distributed
IP	isotropization of production
LES	large-eddy simulation
LES-NWM	LES with near-wall modelling
LES-NWR	LES with near-wall resolution
LIPM	Lagrangian isotropization of production model
LMSE	linear mean-square estimation
LRR	Reynolds-stress model of Launder, Reece, and Rodi (1975)
PDF	probability density function
POD	proper orthogonal decomposition
RANS	Reynolds-averaged Navier–Stokes

RDT	rapid distortion theory
r.m.s.	root-mean square
SGS	subgrid scale
SLM	simplified Langevin model
SSG	Reynolds-stress model of Speziale, Sarkar, and Gatski (1991)
VLES	very-large-eddy simulation

Use of symbols for order and scaling

The statement that 'the variable f is of order g' has different meanings depending on the context and the type of 'order' implied. The symbols $\mathcal{O}(h)$ (read 'big order h' or 'big O of h') and $o(h)$ (read 'little order h' or 'little O of h') indicate quantities, dependent on h, such that

$$\lim_{h \to 0} \frac{\mathcal{O}(h)}{h} = A, \quad \text{for } |A| < \infty,$$

$$\lim_{h \to 0} \frac{o(h)}{h} = 0.$$

Thus, for example, the Taylor series for a function $f(x)$ can be written

$$f(x + h) = f(x) + h f'(x) + \mathcal{O}(h^2)$$
$$= f(x) + h f'(x) + o(h).$$

In the expression $f(h) \sim h^p$, the symbol $\sim$ can be read 'varies as' or 'scales with', and it indicates that the quantity $f(h)/h^p$ is approximately constant (possibly over a limited range of h). In some contexts this type of relation is also stated as '$f(h)$ is of order h^p': for example, the FFT of N data points can be computed in of order $N \log N$ operations.

A statement such as 'f is of order 100' is used to indicate the approximate magnitude of f to the nearest power of ten. Thus, in this case, the value of f is roughly between 30 and 300.

Part one
Fundamentals

1

Introduction

1.1 The nature of turbulent flows

There are many opportunities to observe turbulent flows in our everyday surroundings, whether it be smoke from a chimney, water in a river or waterfall, or the buffeting of a strong wind. In observing a waterfall, we immediately see that the flow is unsteady, irregular, seemingly random and chaotic, and surely the motion of every eddy or droplet is unpredictable. In the plume formed by a solid rocket motor (see Fig. 1.1), turbulent motions of many scales can be observed, from eddies and bulges comparable in size to the width of the plume, to the smallest scales the camera can resolve. The features mentioned in these two examples are common to all turbulent flows.

More detailed and careful observations can be made in laboratory experiments. Figure 1.2 shows planar images of a turbulent jet at two different Reynolds numbers. Again, the concentration fields are irregular, and a large range of length scales can be observed.

As implied by the above discussion, an essential feature of turbulent flows is that the fluid velocity field varies significantly and irregularly in both position and time. The velocity field (which is properly introduced in Section 2.1) is denoted by $U(x,t)$, where x is the position and t is time.

Figure 1.3 shows the time history $U_1(t)$ of the axial component of velocity measured on the centerline of a turbulent jet (similar to that shown in Fig. 1.2). The horizontal line (in Fig. 1.3) shows the mean velocity denoted by $\langle U_1 \rangle$, and defined in Section 3.2. It may be observed that the velocity $U_1(t)$ displays significant fluctuations (about 25% of $\langle U_1 \rangle$), and that, far from being periodic, the time history exhibits variations on a wide range of timescales. Very importantly, we observe that $U_1(t)$ and its mean $\langle U_1 \rangle$ are in some sense 'stable': huge variations in $U_1(t)$ are not observed; neither does $U_1(t)$ spend long periods of time near values different than $\langle U_1 \rangle$.

3

Fig. 1.1. A photograph of the turbulent plume from the ground test of a Titan IV rocket motor. The nozzle's exit diameter is 3 m, the estimated plume height is 1,500 m, and the estimated Reynolds number is 200×10^6. For more details see Mungal and Hollingsworth (1989). With permission of San Jose Mercury & News.

Figure 1.4 shows the profile of the mean velocity $\langle U_1 \rangle$ measured in a similar turbulent jet as a function of the cross-stream coordinate x_2. In marked contrast to the velocity U_1, the mean velocity $\langle U_1 \rangle$ has a smooth profile, with no fine structure. Indeed, the shape of the profile is little different than that of a laminar jet.

In engineering applications turbulent flows are prevalent, but less easily seen. In the processing of liquids or gases with pumps, compressors, pipe lines, etc., the flows are generally turbulent. Similarly the flows around vehicles – e.g., airplanes, automobiles, ships, and submarines – are turbulent. The mixing of fuel and air in engines, boilers, and furnaces, and the mixing of the reactants in chemical reactors take place in turbulent flows.

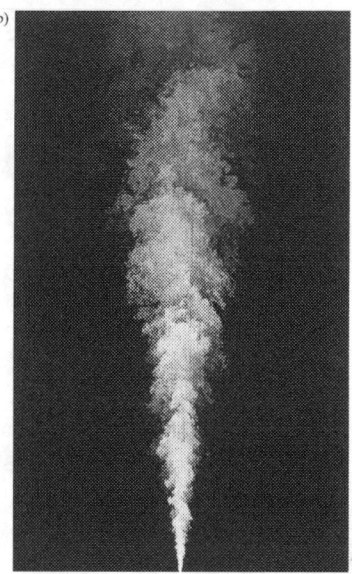

Fig. 1.2. Planar images of concentration in a turbulent jet: (a) Re = 5,000 and (b) Re = 20,000. From Dahm and Dimotakis (1990).

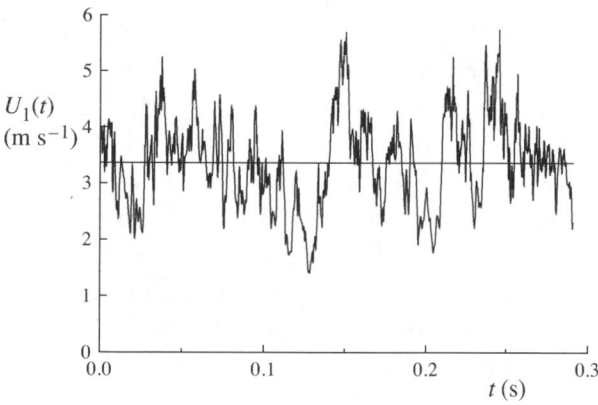

Fig. 1.3. The time history of the axial component of velocity $U_1(t)$ on the centerline of a turbulent jet. From the experiment of Tong and Warhaft (1995).

An important characteristic of turbulence is its ability to transport and mix fluid much more effectively than a comparable laminar flow. This is well demonstrated by an experiment first reported by Osborne Reynolds (1883). Dye is steadily injected on the centerline of a long pipe in which water is flowing. As Reynolds (1894) later established, this flow is characterized by a single non-dimensional parameter, now known as the Reynolds number Re. In general, it is defined by Re $= \mathcal{U}\mathcal{L}/\nu$, where $\mathcal{U}$ and $\mathcal{L}$ are characteristic

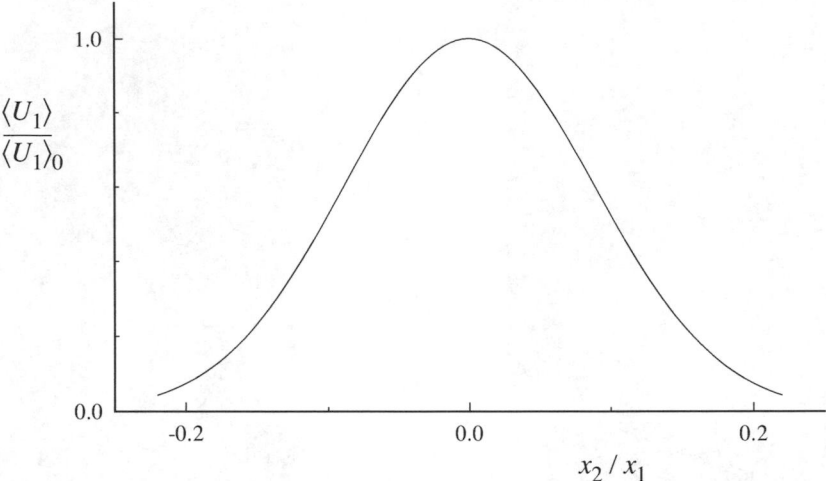

Fig. 1.4. The mean axial velocity profile in a turbulent jet. The mean velocity $\langle U_1 \rangle$ is normalized by its value on the centerline, $\langle U_1 \rangle_0$; and the cross-stream (radial) coordinate x_2 is normalized by the distance from the nozzle x_1. The Reynolds number is 95,500. Adapted from Hussein, Capp, and George (1994).

velocity and length scales of the flow, and v is the kinematic viscosity of the fluid. (For pipe flow, $\mathcal{U}$ and $\mathcal{L}$ are taken to be the area-averaged axial velocity and the pipe diameter, respectively.) In Reynolds' pipe-flow experiment, if Re is less than about 2,300, the flow is laminar – the fluid velocity does not change with time, and all streamlines are parallel to the axis of the pipe. In this (laminar) case, the dye injected on the centerline forms a long streak that increases in diameter only slightly with downstream distance. If, on the other hand, Re exceeds about 4,000, then the flow is turbulent.[1] Close to the injector, the dye streak is jiggled about by the turbulent motion; it becomes progressively less distinct with downstream distance; and eventually mixing with the surrounding water reduces the peak dye concentration to the extent that it is no longer visible.

(Visualizations from a reproduction of Reynolds' experiment, and from other canonical turbulent flows, are contained in Van Dyke (1982). There is great educational value in studying this collection of photographs.)

The effectiveness of turbulence for transporting and mixing fluids is of prime importance in many applications. When different fluid streams are brought together to mix, it is generally desirable for this mixing to take place as rapidly as possible. This is certainly the case for pollutant streams

---

[1] As the Reynolds number is increased, the transition from laminar to turbulent flow occurs over a range of Re, and this range depends on the details of the experiment.

released into the atmosphere or into bodies of water, and for the mixing of different reactants in combustion devices and chemical reactors.

Turbulence is also effective at 'mixing' the momentum of the fluid. As a consequence, on aircraft's wings and ships' hulls the wall shear stress (and hence the drag) is much larger than it would be if the flow were laminar. Similarly, compared with laminar flow, rates of heat and mass transfer at solid–fluid and liquid–gas interfaces are much enhanced in turbulent flows.

The major motivation for the study of turbulent flows is the combination of the three preceding observations: the vast majority of flows is turbulent; the transport and mixing of matter, momentum, and heat in flows is of great practical importance; and turbulence greatly enhances the rates of these processes.

1.2 The study of turbulent flows

Many different techniques have been used to address many different questions concerning turbulence and turbulent flows. The first step toward providing a categorization of these studies is to distinguish between small-scale turbulence and the large-scale motions in turbulent flows.

As is discussed in detail in Chapter 6, at high Reynolds number there is a separation of scales. The large-scale motions are strongly influenced by the geometry of the flow (i.e., by the boundary conditions), and they control the transport and mixing. The behavior of the small-scale motions, on the other hand, is determined almost entirely by the rate at which they receive energy from the large scales, and by the viscosity. Hence these small-scale motions have a universal character, independent of the flow geometry. It is natural to ask what the characteristics of the small-scale motions are. Can they be predicted from the equations governing fluid motion? These are questions of *turbulence theory*, which are addressed in the books of Batchelor (1953), Monin and Yaglom (1975), Panchev (1971), Lesieur (1990), McComb (1990), and others, and that are touched on only slightly in this book (in Chapter 6).

The focus of this book is on *turbulent flows*, studies of which can be divided into three categories.

(i) Discovery: experimental (or simulation) studies aimed at providing qualitative or quantitative information about particular flows.
(ii) Modelling: theoretical (or modelling) studies, aimed at developing tractable mathematical models that can accurately predict properties of turbulent flows.

(iii) Control: studies (usually involving both experimental and theoretical components) aimed at manipulating or controlling the flow or the turbulence in a beneficial way – for example, changing the boundary geometry to enhance mixing; or using active control to reduce drag.

The remainder of Part I of this book is based primarily on studies in the first category, the objective being to develop in the reader an understanding for the important characteristics of simple turbulent flows, of the dominant physical processes, and how they are related to the equations of fluid motion. The description of turbulent flows contained in Part I is not comprehensive: additional material can be found in the books of Monin and Yaglom (1971), Townsend (1976), Hinze (1975), and Schlichting (1979).

For studies in the second category, that aim at developing tractable mathematical models, the word 'tractable' is crucial. For fluid flows, be they laminar or turbulent, the governing laws are embodied in the Navier–Stokes equations, which have been known for over a century. (These equations are reviewed in Chapter 2.) Considering the diversity and complexity of fluid flows, it is quite remarkable that the relatively simple Navier–Stokes equations describe them accurately and in complete detail. However, in the context of turbulent flows, their power is also their weakness: the equations describe every detail of the turbulent velocity field from the largest to the smallest length and time scales. The amount of information contained in the velocity field is vast, and as a consequence (in general) the direct approach of solving the Navier–Stokes equations is impossible. So, while the Navier–Stokes equations accurately describe turbulent flows, they do not provide a *tractable* model for them.

The direct approach of solving the Navier–Stokes equations for turbulent flows is called direct numerical simulation (DNS), and is described in Chapter 9. While DNS is intractable for the high-Reynolds-number flows of practical interest, it is nevertheless a powerful research tool for investigating simple turbulent flows at moderate Reynolds numbers. In the description of wall-bounded flows in Chapter 7, DNS results are used extensively to investigate the physical processes involved.

For the high-Reynolds-number flows that are prevalent in applications, the natural alternative is to pursue a statistical approach. That is, to describe the turbulent flow, not in terms of the velocity $U(x, t)$, but in terms of some statistics, the simplest being the mean velocity field $\langle U(x, t) \rangle$. A model based on such statistics can lead to a tractable set of equations, because statistical fields vary smoothly (if at all) in position and time. In Chapter 3 we present the concepts and techniques used in the statistical representation of turbulent flow fields; while in Part II we describe statistical models that can be used to calculate the properties of turbulent flows. The approaches described include: turbulent viscosity models, e.g., the k–ε model (Chapter 10); Reynolds-stress models (Chapter 11); models based on the probability density function (PDF) of velocity (Chapter 12); and large-eddy-simulations (LES) (Chapter 13).

The statistical models described in Part II can be used in some studies in the third category mentioned above – that is, studies aimed at manipulating or controlling the flow or the turbulence. However, such studies are not explicitly discussed here.

A broad range of mathematical techniques is used to describe and model turbulent flows. Appendices on several of these techniques are provided to serve as brief tutorials and summaries. The first of these is on Cartesian tensors, which are used extensively. The reader may wish to review this material (Appendix A) before proceeding. There are exercises throughout the book, which provide the reader with the opportunity to practice the mathematical techniques employed. Most of these exercises also contain additional results and observations. A list of nomenclature and abbreviations is provided on page xxi.

2

The equations of fluid motion

In this chapter we briefly review the Navier–Stokes equations which govern the flow of constant-property Newtonian fluids. More comprehensive accounts can be found in the texts of Batchelor (1967), Panton (1984), and Tritton (1988). Two topics that are important in the study of turbulent flows, that are not extensively discussed in these texts, are the Poisson equation for pressure (Section 2.5), and the transformation properties of the Navier–Stokes equations (Section 2.9). The equations of fluid motion are expressed either in vector notation or in Cartesian tensor notation, which is reviewed in Appendix A.

2.1 Continuum fluid properties

The idea of treating fluids as continuous media is both natural and familiar. It is, however, worthwhile to review the *continuum hypothesis* – that reconciles the discrete molecular nature of fluids with the continuum view – so as to avoid confusion when quantities such as 'fluid particles' and 'infinitesimal material elements' are introduced.

The length and time scales of molecular motion are extremely small compared with human scales. Taking air under atmospheric conditions as an example, the average spacing between molecules is 3×10^{-9} m, the mean free path, λ, is 6×10^{-8} m, and the mean time between successive collision of a molecule is 10^{-10} s. In comparison, the smallest geometric length scale in a flow, ℓ, is seldom less than 0.1 mm $= 10^{-4}$ m, which, for flow velocities up to 100 m s^{-1}, yields a flow timescale larger than 10^{-6} s. Thus, even for this example of a flow with small length and time scales, these flow scales exceed the molecular scales by three or more orders of magnitude.

10

The separation of the length scales is quantified by the Knudsen number

$$\text{Kn} \equiv \lambda/\ell. \tag{2.1}$$

In the above example, Kn is less than 10^{-3}, while in general the continuum approach is appropriate for $\text{Kn} \ll 1$.

For very small Kn, because of the separation of scales, there exist intermediate length scales ℓ^*, such that ℓ^* is large compared with molecular scales, yet small compared with flow scales (i.e., $\lambda \ll \ell^* \ll \ell$). Roughly speaking, the continuum fluid properties can be thought of as the molecular properties averaged over a volume of size $V = \ell^{*3}$. Let $\mathcal{V}_x$ denote a spherical region of volume V centered on the point x. Then, at time t, the fluid's density $\rho(x, t)$ is the mass of molecules in $\mathcal{V}_x$, divided by V.

Similarly the fluid's velocity $U(x, t)$ is the average velocity of the molecules within $\mathcal{V}_x$. Because of the separation of scales, the dependence of the continuum properties on the choice of ℓ^* is negligible.

(While the approach presented in the previous paragraph is standard (see, e.g., Batchelor (1967) and Panton (1984)), as Exercise 2.1 illustrates, more care is needed to provide a proper definition of the continuum properties in terms of averaging over a scale ℓ^*. In fact, continuum fields are best defined as *expectations* of molecular properties, see Chapter 12.)

It is important to appreciate that, once we invoke the continuum hypothesis to obtain continuous fields, such as $\rho(x, t)$ and $U(x, t)$, we can leave behind all notions of the discrete molecular nature of the fluid, and molecular scales cease to be relevant. We can talk meaningfully of 'the density at x, t,' even though (in the microscopic view) in all likelihood there is no matter at (x, t). Similarly, we can consider differences in properties over distances smaller than molecular scales: indeed we do so when we define gradients,

$$\frac{\partial \rho}{\partial x_1} \equiv \lim_{h \to 0} \left(\frac{1}{h} \left[\rho(x_1 + h, x_2, x_3, t) - \rho(x_1, x_2, x_3, t) \right] \right). \tag{2.2}$$

EXERCISE

2.1   In the flow of an ideal gas, let $m^{(i)}, x^{(i)}(t)$ and $u^{(i)}(t)$ be the mass, position, and velocity of the ith molecule. As a generalization of the standard continuum hypothesis, consider the definition

$$U(x, t) \equiv \frac{\sum_i m^{(i)} u^{(i)} K(|r^{(i)}|)}{\sum_j m^{(j)} K(|r^{(j)}|)}, \tag{2.3}$$

where $r^{(i)} \equiv x^{(i)} - x$, and $K(r)$ is a smooth kernel such as

$$K(r) = \exp(-\tfrac{1}{2}r^2/\ell^{*2}), \tag{2.4}$$

with ℓ^* being a specified length scale. Show that the velocity gradients are

$$\frac{\partial U_k}{\partial x_\ell} = \frac{\sum_i m^{(i)}(U_k - u_k^{(i)})K'(|\boldsymbol{r}^{(i)}|)r_\ell^{(i)}/|\boldsymbol{r}^{(i)}|}{\sum_j m^{(j)}K(|\boldsymbol{r}^{(j)}|)}, \qquad (2.5)$$

where $K'(r) = \mathrm{d}K(r)/\mathrm{d}r$.

(Evidently, the continuum field defined by Eq. (2.3) inherits the mathematical continuity properties of the kernel. If, as in the standard treatment, $K(r)$ is piecewise constant, i.e.,

$$K(r) = \begin{cases} 1, & \text{if } r \leq \ell^*, \\ 0, & \text{if } r > \ell^*, \end{cases} \qquad (2.6)$$

then $\boldsymbol{U}(\boldsymbol{x},t)$ is piecewise constant, and hence not continuously differentiable.)

2.2 Eulerian and Lagrangian fields

The continuum density and velocity fields, $\rho(\boldsymbol{x},t)$ and $\boldsymbol{U}(\boldsymbol{x},t)$, are *Eulerian fields* in that they are indexed by the position $\boldsymbol{x}$ in an inertial frame. The starting point for the alternative Lagrangian description is the definition of a *fluid particle* – which is a continuum concept. By definition, a fluid particle is a point that moves with the local fluid velocity: $\boldsymbol{X}^+(t,\boldsymbol{Y})$ denotes the position at time t of the fluid particle that is located at $\boldsymbol{Y}$ at the specified fixed reference time t_0, see Fig. 2.1. Mathematically, the fluid particle position

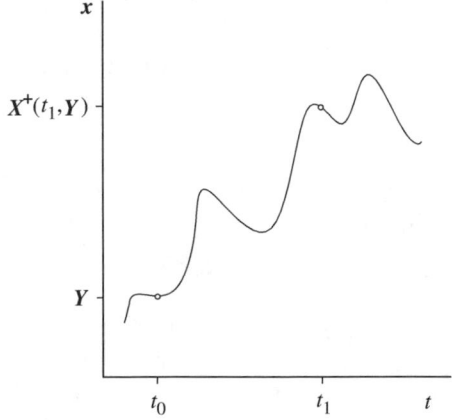

Fig. 2.1. A sketch of the trajectory $\boldsymbol{X}^+(t,\boldsymbol{Y})$ of a fluid particle in x–t space, showing its position $\boldsymbol{Y}$ at the reference time t_0, and at a later time t_1.

$X^+(t, Y)$ is defined by two equations. First, the position at the reference time t_0 is defined to be

$$X^+(t_0, Y) = Y. \tag{2.7}$$

Second, the equation

$$\frac{\partial}{\partial t} X^+(t, Y) = U(X^+(t, Y), t), \tag{2.8}$$

expresses the fact that the fluid particle moves with the local fluid velocity. Given the Eulerian velocity field $U(x, t)$, then, for any Y, Eq. (2.8) can be integrated backward and forward in time to obtain $X^+(t, Y)$ for all t.

Lagrangian fields of density and velocity, for example, are defined in terms of their Eulerian counterparts by

$$\rho^+(t, Y) \equiv \rho(X^+(t, Y), t), \tag{2.9}$$

$$U^+(t, Y) \equiv U(X^+(t, Y), t). \tag{2.10}$$

Note that the Lagrangian fields ρ^+ and U^+ are indexed not by the current position of the fluid particle, but by its position Y at the reference time t_0. Hence, Y is called the *Lagrangian coordinate* or the *material coordinate*.

For fixed Y, $X^+(t, Y)$ defines a trajectory (in x–t space) that is the *fluid-particle path*, and similarly $\rho^+(t, Y)$ is the *fluid-particle density*. The partial derivative $\partial \rho^+(t, Y)/\partial t$ is the rate of change of density at fixed Y, i.e., following a fluid particle. From Eq. (2.9) we obtain

$$\begin{aligned}
\frac{\partial}{\partial t} \rho^+(t, Y) &= \frac{\partial}{\partial t} \rho(X^+(t, Y), t) \\
&= \left(\frac{\partial}{\partial t} \rho(x, t) \right)_{x = X^+(t,Y)} + \frac{\partial}{\partial t} X_i^+(t, Y) \left(\frac{\partial}{\partial x_i} \rho(x, t) \right)_{x = X^+(t,Y)} \\
&= \left(\frac{\partial}{\partial t} \rho(x, t) + U_i(x, t) \frac{\partial}{\partial x_i} \rho(x, t) \right)_{x = X^+(t,Y)} \\
&= \left(\frac{D}{Dt} \rho(x, t) \right)_{x = X^+(t,Y)}, \tag{2.11}
\end{aligned}$$

where the *material derivative*, or *substantial derivative*, is defined by

$$\frac{D}{Dt} \equiv \frac{\partial}{\partial t} + U_i \frac{\partial}{\partial x_i} = \frac{\partial}{\partial t} + U \cdot \nabla. \tag{2.12}$$

Thus the rate of change of density following a fluid particle is given by the partial derivative of the Lagrangian field (i.e., $\partial \rho^+/\partial t$) and by the substantial derivative of the Eulerian field (i.e., $D\rho/Dt$).

Similarly, for fixed Y, $U^+(t, Y)$ is the fluid particle velocity, and

$$\frac{\partial}{\partial t} U^+(t, Y) = \left(\frac{D}{Dt} U(x, t) \right)_{x = X^+(t, Y)} \tag{2.13}$$

is the rate of change of fluid particle velocity, i.e., the fluid particle acceleration.

A fluid particle is also called a *material point* and we have seen that it is defined by its position Y at time t_0 and by its movement with the local fluid velocity (Eq. (2.8)). Material lines, surfaces, and volumes are defined similarly. For example, consider at time t_0 a simple closed surface $\mathcal{S}_0$ that encloses the volume $\mathcal{V}_0$. The corresponding material surface $\mathcal{S}(t)$ is defined to be coincident with $\mathcal{S}_0$ at time t_0, and by the property that every point of $\mathcal{S}(t)$ moves with the local fluid velocity. Thus $\mathcal{S}(t)$ is composed of the fluid particles $X^+(t, Y)$, which at t_0 compose the surface $\mathcal{S}_0$:

$$\mathcal{S}(t) \equiv \{X^+(t, Y) : Y \in \mathcal{S}_0\}. \tag{2.14}$$

Because a material surface moves with the fluid, the relative velocity between the surface and the fluid is zero. Consequently a fluid particle cannot cross a material surface; neither is there a mass flux across a material surface.

EXERCISE

2.2   Consider two fluid particles that, at the reference time t_0 are located at Y and $Y + dY$, where dY is an infinitesimal displacement. At time t, the line between the two particles forms the *infinitesimal line element*

$$s(t) \equiv X^+(t, Y + dY) - X^+(t, Y). \tag{2.15}$$

Show that $s(t)$ evolves by

$$\frac{ds}{dt} = s \cdot (\nabla U)_{x = X^+(t, Y)}. \tag{2.16}$$

(Hint: expand $U^+(t, Y + dY) = U(X^+(t, Y) + s(t), t)$ in a Taylor series. Since s is infinitesimal, only the leading-order terms need be retained.)

2.3 The continuity equation

The *mass-conservation* or *continuity* equation is

$$\frac{\partial \rho}{\partial t} + \nabla \cdot (\rho U) = 0. \tag{2.17}$$

The derivation and interpretation of this equation in terms of control volumes and material volumes should be familiar to the reader and are not repeated here. A further useful interpretation is in terms of the specific volume of the fluid $\vartheta(x, t) = 1/\rho(x, t)$. Manipulation of Eq. (2.17) yields

$$\frac{D \ln \vartheta}{Dt} = \nabla \cdot U. \tag{2.18}$$

The left-hand side is the logarithmic rate of increase of the specific volume, while (as Exercises 2.3 and 2.4 show) the *dilatation* $\nabla \cdot U$ gives the logarithmic rate of increase of the volume of an infinitesimal material volume. Hence the continuity equation can be viewed as a consistency condition between the change of the specific volume following a fluid particle, and the change in the volume of an infinitesimal material volume element.

In this book we consider constant-density flows (i.e., flows in which ρ is independent both of x and of t). In this case the evolution equation Eq. (2.17) degenerates to the kinematic condition that the velocity field be *solenoidal* or *divergence-free*:

$$\nabla \cdot U = 0. \tag{2.19}$$

EXERCISES _____

2.3   Let $\mathcal{V}(t)$ be a material volume bounded by the material surface $\mathcal{S}(t)$. Show from geometry that the volume of fluid $V(t)$ within $\mathcal{V}(t)$ evolves by

$$\frac{dV(t)}{dt} = \iint_{\mathcal{S}(t)} U \cdot n \, dA, \tag{2.20}$$

where dA is an area element on $\mathcal{S}(t)$, and n is the outward pointing normal. Use the divergence theorem to obtain

$$\frac{dV(t)}{dt} = \iiint_{\mathcal{V}(t)} \nabla \cdot U \, dx. \tag{2.21}$$

Show that, for the infinitesimal volume $dV(t)$ of an infinitesimal material volume $d\mathcal{V}$,

$$\frac{d}{dt} \ln dV(t) = \nabla \cdot U. \tag{2.22}$$

2.4   The determinant of the Jacobian

$$J(t, Y) \equiv \det \left(\frac{\partial X_i^+(t, Y)}{\partial Y_j} \right) \tag{2.23}$$

gives the volume ratio between an infinitesimal material volume $dV(t)$ at time t, and its volume $dV(t_0)$ at time t_0. To first order in the infinitesimal dt, show that

$$X_i^+(t_0 + dt, Y) = Y_i + U_i(Y, t_0) \, dt, \tag{2.24}$$

$$\frac{\partial X_i^+(t_0 + dt, \boldsymbol{Y})}{\partial Y_j} = \delta_{ij} + \left(\frac{\partial U_i}{\partial x_j}\right)_{\boldsymbol{Y}, t_0} dt, \tag{2.25}$$

$$J(t_0 + dt, \boldsymbol{Y}) = 1 + (\nabla \cdot \boldsymbol{U})_{\boldsymbol{Y}, t_0} dt. \tag{2.26}$$

Hence show that

$$\left(\frac{\partial}{\partial t} \ln J(t, \boldsymbol{Y})\right)_{t=t_0} = (\nabla \cdot \boldsymbol{U})_{\boldsymbol{Y}, t_0}. \tag{2.27}$$

2.5   The volume $V(t)$ defined in Exercise 2.3 can be written

$$V(t) = \iiint_{\mathcal{V}(t)} d\boldsymbol{x} = \iiint_{\mathcal{V}(t_0)} J(t, \boldsymbol{Y}) d\boldsymbol{Y}. \tag{2.28}$$

Differentiate the first and last expressions in this equation with respect to time, and compare the result with Eq. (2.21) to obtain

$$\frac{\partial}{\partial t} \ln J(t, \boldsymbol{Y}) = (\nabla \cdot \boldsymbol{U})_{X^+(t, \boldsymbol{Y})}. \tag{2.29}$$

Hence argue that, in constant-density flows, $J(t, \boldsymbol{Y})$ is unity.

2.4 The momentum equation

The momentum equation, based on Newton's second law, relates the fluid particle acceleration $D\boldsymbol{U}/Dt$ to the *surface forces* and *body forces* experienced by the fluid. In general, the surface forces, which are of molecular origin, are described by the *stress tensor* $\tau_{ij}(\boldsymbol{x}, t)$ – which is symmetric, i.e., $\tau_{ij} = \tau_{ji}$. The body force of interest is gravity. With Ψ being the *gravitational potential* (i.e., the potential energy per unit mass associated with gravity), the body force per unit mass is

$$\boldsymbol{g} = -\nabla\Psi. \tag{2.30}$$

(For a constant gravitational field the potential is $\Psi = gz$, where g is the gravitational acceleration, and z is the vertical coordinate.) These forces cause the fluid to accelerate according to the momentum equation

$$\rho \frac{DU_j}{Dt} = \frac{\partial \tau_{ij}}{\partial x_i} - \rho \frac{\partial \Psi}{\partial x_j}. \tag{2.31}$$

We now specialize the momentum equation to flows of *constant-property Newtonian fluids* – the fundamental class of flows considered in this book. In this case, the stress tensor is

$$\tau_{ij} = -P\delta_{ij} + \mu\left(\frac{\partial U_i}{\partial x_j} + \frac{\partial U_j}{\partial x_i}\right), \tag{2.32}$$

where P is the pressure, and μ is the (constant) coefficient of viscosity. Recalling that (for the constant-density flows considered) the velocity field is solenoidal (i.e., $\partial U_i / \partial x_i = 0$), we observe that Eq. (2.32) expresses the stress as the sum of isotropic ($-P \delta_{ij}$) and deviatoric contributions.

By substituting this expression for the stress tensor (Eq. (2.32)) into the general momentum equation Eq. (2.31) (and exploiting the facts that ρ and μ are uniform and that $\nabla \cdot U = 0$), we obtain the *Navier–Stokes equations*

$$\rho \frac{DU_j}{Dt} = \mu \frac{\partial^2 U_j}{\partial x_i \, \partial x_i} - \frac{\partial P}{\partial x_j} - \rho \frac{\partial \Psi}{\partial x_j}. \tag{2.33}$$

Further, defining the *modified pressure*, p, by

$$p = P + \rho \Psi, \tag{2.34}$$

this equation simplifies to

$$\frac{DU}{Dt} = -\frac{1}{\rho} \nabla p + \nu \nabla^2 U, \tag{2.35}$$

where $\nu \equiv \mu / \rho$ is the kinematic viscosity. In summary: the flow of constant-property Newtonian fluids is governed by the Navier–Stokes equations Eq. (2.35) together with the solenoidal condition $\nabla \cdot U = 0$ stemming from mass conservation.

At a stationary solid wall with unit normal n, the boundary conditions satisfied by the velocity are the impermeability condition

$$n \cdot U = 0, \tag{2.36}$$

and the no-slip condition

$$U - n (n \cdot U) = 0, \tag{2.37}$$

(which together yield $U = 0$).

It is sometimes useful to consider the hypothetical case of an ideal (inviscid) fluid, which is defined to have the isotropic stress tensor

$$\tau_{ij} = -P \delta_{ij}. \tag{2.38}$$

The conservation of momentum is given by the *Euler equations*

$$\frac{DU}{Dt} = -\frac{1}{\rho} \nabla p, \tag{2.39}$$

which follow from Eqs. (2.31), (2.34), and (2.38). Because the Euler equations do not contain second spatial derivatives of velocity, they require different boundary conditions than those of the Navier–Stokes equations. At a stationary solid wall, for example, only the impermeability condition can be applied, and in general the tangential components of velocity are non-zero.

While it is preferable to obtain the Euler equations (and other equations derived from them) directly from the definition of τ_{ij} being isotropic (Eq. (2.38)), it may nevertheless be observed that the Euler equations can be obtained from the Navier–Stokes equations by setting v to zero. It is important to appreciate, however, that $v = 0$ is a singular limit: solutions to the Navier–Stokes equations in the limit of vanishing viscosity ($v \to 0$) are different than solutions to the Euler equations. For one thing, even in this limit, the equations require different boundary conditions.

2.5 The role of pressure

The role of pressure in the (constant-density) Navier–Stokes equations requires further comment. First we observe that isotropic stresses and conservative body forces have the same effect, which is expressed by the modified pressure gradient. Hence the body force has no effect on the velocity field and on the modified pressure field. (This is, of course, in contrast to variable-density flows, in which buoyancy forces can be important.) Henceforth, we refer to p simply as 'pressure'.

We may be accustomed to thinking of pressure as a thermodynamic variable, related to density and temperature by an equation of state. However, for constant-density flows, there is no connection between pressure and density, and a different understanding of pressure is required.

To this end, we take the divergence of the Navier–Stokes equations Eq. (2.35), without assuming the velocity field to be solenoidal, but instead writing Δ for the dilatation rate (i.e., $\Delta = \nabla \cdot U$). The result is

$$\left(\frac{\mathrm{D}}{\mathrm{D}t} - v\,\nabla^2\right)\Delta = R, \tag{2.40}$$

where

$$R \equiv -\frac{1}{\rho}\nabla^2 p - \frac{\partial U_i}{\partial x_j}\frac{\partial U_j}{\partial x_i}. \tag{2.41}$$

Consider the solution to Eq. (2.40) with initial and boundary conditions $\Delta = 0$. The solution is $\Delta = 0$ if, and only if, R is zero everywhere, which in turn implies (from Eq. (2.41)) that p satisfies the Poisson equation

$$\nabla^2 p = S \equiv -\rho\frac{\partial U_i}{\partial x_j}\frac{\partial U_j}{\partial x_i}. \tag{2.42}$$

Thus, we conclude that *the satisfaction of this Poisson equation is a necessary and sufficient condition for a solenoidal velocity field to remain solenoidal.*

At a stationary, plane solid surface, the Navier–Stokes equations Eq. (2.35) reduce to

$$\frac{\partial p}{\partial n} = \mu \frac{\partial^2 U_n}{\partial n^2}, \tag{2.43}$$

where n is a coordinate in the wall-normal direction, and U_n is the velocity component normal to the wall. This equation provides a Neumann boundary condition for the Poisson equation, Eq. (2.42). Given Neumann conditions of this form, the Poisson equation Eq. (2.42) determines the pressure field $p(x, t)$ (to within a constant) in terms of the velocity field at the same instant of time. Thus, ∇p is uniquely determined by the current velocity field, independent of the flow's history.

The solution to the Poisson equation Eq. (2.42) can be written explicitly in terms of Green's functions. Consider the Poisson equation

$$\nabla^2 f(x) = S(x). \tag{2.44}$$

in a domain V. The source $S(x)$ can be written

$$S(x) = \iiint_V S(y)\delta(x - y)\,dy, \tag{2.45}$$

where y is a point in V, and $\delta(x - y)$ is the three-dimensional Dirac delta function[1] at y. A solution to the Poisson equation

$$\nabla^2 g(x|y) = \delta(x - y) \tag{2.46}$$

is

$$g(x|y) = \frac{-1}{4\pi |x - y|}. \tag{2.47}$$

(As implied by the notation, the solution depends both on x and on the location of the delta function, y.) When it is multiplied by $S(y)$ and integrated over V, Eq. (2.46) becomes $\nabla^2 f = S$ (i.e., Eq. (2.44)), and hence Eq. (2.47) becomes a solution:

$$f(x) = \iiint_V g(x|y)S(y)\,dy = \frac{-1}{4\pi}\iiint_V \frac{S(y)}{|x - y|}\,dy. \tag{2.48}$$

The solution to the Poisson equation for pressure Eq. (2.42) is, therefore,

$$p(x, t) = p^{(h)}(x, t) + \frac{\rho}{4\pi}\iiint_V \left(\frac{\partial U_i}{\partial x_j}\frac{\partial U_j}{\partial x_i}\right)_{y,t}\frac{dy}{|x - y|}, \tag{2.49}$$

---

[1] The properties of Dirac delta functions are reviewed in Appendix C.

where $p^{(h)}$ is a harmonic function ($\nabla^2 p^{(h)} = 0$) dependent on the boundary conditions. (It is possible to express $p^{(h)}$ in terms of surface integrals over the boundary of $\mathcal{V}$, see e.g., Kellogg (1967).)

EXERCISES

2.6   Show that (away from the origin)

$$\nabla^2 |x|^{-1} = \frac{\partial^2}{\partial x_i \, \partial x_i} (x_j x_j)^{-1/2} = 0. \tag{2.50}$$

2.7   A simple numerical method for solving the Navier–Stokes equations for constant-property flow advances the solution in small time steps Δt, starting from the initial condition $U(x, 0)$. On the nth step the numerical solution is denoted by $U^{(n)}(x)$, which is an approximation to $U(x, n\,\Delta t)$. Each time step consists of two sub-steps, the first of which yields an intermediate result $\widehat{U}^{(n+1)}(x)$ defined by

$$\widehat{U}_j^{(n+1)} \equiv U_j^{(n)} + \Delta t \left(v \frac{\partial^2 U_j^{(n)}}{\partial x_i \, \partial x_i} - U_k^{(n)} \frac{\partial U_j^{(n)}}{\partial x_k} \right). \tag{2.51}$$

The second sub-step is

$$U_j^{(n+1)} = \widehat{U}_j^{(n+1)} - \Delta t \frac{\partial \phi^{(n)}}{\partial x_j}, \tag{2.52}$$

where $\phi^{(n)}(x)$ is a scalar field.

(a) Comment on the connection between Eq. (2.51) and the Navier–Stokes equations.

(b) Assuming that $U^{(n)}$ is divergence-free, obtain from Eq. (2.51) an expression (in terms of $U^{(n)}$) for the divergence of $\widehat{U}^{(n+1)}$.

(c) Obtain from Eq. (2.52) an expression for the divergence of $U^{(n+1)}$.

(d) Hence show that the requirement $\nabla \cdot U^{(n+1)} = 0$ is satisfied if, and only if, $\phi^{(n)}(x)$ satisfies the Poisson equation

$$\nabla^2 \phi^{(n)} = -\frac{\partial U_k^{(n)}}{\partial x_j} \frac{\partial U_j^{(n)}}{\partial x_k}. \tag{2.53}$$

(e) What is the connection between $\phi^{(n)}(x)$ and the pressure?

2.6 Conserved passive scalars

In addition to the velocity $U(x, t)$, we consider a *conserved passive scalar* denoted by $\phi(x, t)$. In a constant-property flow, the conservation equation for ϕ is

$$\frac{D\phi}{Dt} = \Gamma \nabla^2 \phi, \qquad (2.54)$$

where Γ is the (constant and uniform) diffusivity. The scalar ϕ is *conserved*, because there is no source or sink term in Eq. (2.54). It is *passive* because (by assumption) its value has no effect on material properties (i.e., ρ, v, and Γ), and hence it has no effect on the flow.

The scalar ϕ can represent various physical properties. It can be a small excess in temperature – sufficiently small that its effect on material properties is negligible. In this case Γ is the thermal diffusivity, and the ratio v/Γ is the Prandtl number, Pr. Alternatively, ϕ can be the concentration of a trace species, in which case Γ is the molecular diffusivity, and v/Γ is the Schmidt number, Sc.

An important property of the scalar is its *boundedness*. If the initial and boundary values of ϕ lie within a given range

$$\phi_{min} \le \phi \le \phi_{max}, \qquad (2.55)$$

then $\phi(x, t)$ for all (x, t) also lies in this range: values of ϕ greater than ϕ_{max} or less that ϕ_{min} cannot occur.

To show this result we examine local maxima in the scalar field. Suppose that there is a local maximum at $\bar{x}$ at time $\bar{t}$, and we choose a coordinate system such that $\partial^2 \phi / (\partial x_i \, \partial x_j)$ is in principal axes there. The mathematical properties of a maximum imply that

$$(\nabla \phi)_{\bar{x}, \bar{t}} = 0, \qquad (2.56)$$

and that the second derivatives $\partial^2 \phi / \partial x_1^2$, $\partial^2 \phi / \partial x_2^2$, and $\partial^2 \phi / \partial x_3^2$ are negative or zero. Consequently, for their sum, the Laplacian, we have

$$(\nabla^2 \phi)_{\bar{x}, \bar{t}} \le 0. \qquad (2.57)$$

Then, from the conservation equation for ϕ (Eq. (2.54)), we obtain

$$\left(\frac{D\phi}{Dt}\right)_{\bar{x}, \bar{t}} = \left(\frac{\partial \phi}{\partial t} + V \cdot \nabla \phi\right)_{\bar{x}, \bar{t}} = \Gamma (\nabla^2 \phi)_{\bar{x}, \bar{t}} \le 0, \qquad (2.58)$$

for every vector V; showing that, following any trajectory from the local maximum, the value of ϕ does not increase. Consequently, there is no way in which ϕ can increase beyond the upper bound ϕ_{max} imposed by the initial and boundary conditions. Obviously, a similar argument applies to the lower bound, $\phi_{\min}$.

2.7 The vorticity equation

An essential feature of turbulent flows is that they are rotational: that is, they have non-zero vorticity. The vorticity $\omega(x, t)$ is the curl of the velocity

$$\omega = \nabla \times U, \tag{2.59}$$

and it equals twice the rate of rotation of the fluid at (x, t).

The equation for the evolution of the vorticity is obtained by taking the curl of the Navier–Stokes equations Eq. (2.35):

$$\frac{D\omega}{Dt} = \nu \nabla^2 \omega + \omega \cdot \nabla U. \tag{2.60}$$

The pressure term $(-\nabla \times \nabla p/\rho)$ vanishes for constant-density flows.

The equation for the evolution of an infinitesimal line element of material $s(t)$ (see Eq. (2.16)) is

$$\frac{ds}{dt} = s \cdot \nabla U, \tag{2.61}$$

which, apart from the viscous term, is identical to the vorticity equation. Hence, in inviscid flow, the vorticity vector behaves in the same way as an infinitesimal material line element (Helmholtz theorem). If the strain rate produced by the velocity gradients acts to stretch the material line element aligned with ω, then the magnitude of ω increases correspondingly. This is the phenomenon of *vortex stretching*, which is an important process in turbulent flows, and $\omega \cdot \nabla U$ is referred to as the vortex-stretching term.

For two-dimensional flows, the vortex-stretching term vanishes, and the one non-zero component of vorticity evolves as a conserved scalar. Because of the absence of vortex stretching, two-dimensional turbulence (which can occur in special circumstances) is qualitatively different than three-dimensional turbulence.

EXERCISES _____

2.8   Use suffix notation to verify the relations:

$$\nabla \cdot \omega = 0, \tag{2.62}$$

$$\nabla \times \nabla \phi = 0, \tag{2.63}$$

$$\nabla \times (\nabla \times U) = \nabla(\nabla \cdot U) - \nabla^2 U, \qquad (2.64)$$

$$U \times \omega = \tfrac{1}{2}\nabla(U \cdot U) - U \cdot \nabla U. \qquad (2.65)$$

Are the expressions in Eqs. (2.64) and (2.65) tensors?

2.9 Show that the Navier–Stokes equations (Eq. (2.35)) can be written in the *Stokes form*

$$\frac{\partial U}{\partial t} - U \times \omega + \nabla\left(\tfrac{1}{2}U \cdot U + \frac{p}{\rho}\right) = v\,\nabla^2 U. \qquad (2.66)$$

Hence obtain *Bernoulli's theorem*: for a steady, inviscid, constant-density flow, the Bernoulli integral,

$$H \equiv \tfrac{1}{2}U \cdot U + \frac{p}{\rho}, \qquad (2.67)$$

is constant

(a) along streamlines,
(b) along vortex lines (i.e., lines parallel to ω), and
(c) everywhere in irrotational flow ($\omega = 0$).

2.10 Show that the vorticity squared – or *enstrophy* – $\omega^2 = \omega \cdot \omega$ evolves by

$$\frac{D\omega^2}{Dt} = v\,\nabla^2\omega^2 + 2\omega_i\omega_j\frac{\partial U_i}{\partial x_j} - 2v\frac{\partial \omega_i}{\partial x_j}\frac{\partial \omega_i}{\partial x_j}. \qquad (2.68)$$

2.8 Rates of strain and rotation

The velocity gradients $\partial U_i/\partial x_j$ are the components of a second-order tensor, the general properties of which are described in Appendix B. The decomposition of $\partial U_i/\partial x_j$ into isotropic, symmetric-deviatoric, and antisymmetric parts is

$$\frac{\partial U_i}{\partial x_j} = \tfrac{1}{3}\Delta\delta_{ij} + S_{ij} + \Omega_{ij}, \qquad (2.69)$$

where the dilatation $\Delta = \nabla \cdot U$ is zero for constant-density flow, S_{ij} is the symmetric, deviatoric *rate-of-strain tensor*

$$S_{ij} \equiv \frac{1}{2}\left(\frac{\partial U_i}{\partial x_j} + \frac{\partial U_j}{\partial x_i}\right), \qquad (2.70)$$

and Ω_{ij} is the antisymmetric *rate-of-rotation tensor*

$$\Omega_{ij} \equiv \frac{1}{2}\left(\frac{\partial U_i}{\partial x_j} - \frac{\partial U_j}{\partial x_i}\right). \qquad (2.71)$$

(For variable-density flow, S_{ij} is defined as $S_{ij} \equiv \tfrac{1}{2}(\partial U_i/\partial x_j + \partial U_j/\partial x_i) - \tfrac{1}{3}\Delta\delta_{ij}$.)

It may be observed that the Newtonian stress law (Eq. (2.32)) can be re-expressed as

$$\tau_{ij} = -P\delta_{ij} + 2\mu S_{ij}, \tag{2.72}$$

showing that the viscous stress depends linearly on the rate of strain, independent of the rate of rotation.

The vorticity and the rate of rotation are related by

$$\omega_i = -\varepsilon_{ijk}\Omega_{jk}, \tag{2.73}$$

$$\Omega_{ij} = -\tfrac{1}{2}\varepsilon_{ijk}\omega_k, \tag{2.74}$$

where ε_{ijk} is the alternating symbol. Thus Ω_{ij} and ω_i contain the same information, but (as discussed in Appendix A) Ω_{ij} is a tensor whereas ω_i is not.

EXERCISES

2.11   From Eq. (2.16), derive an equation for the evolution of the length of an infinitesimal material line element. Show that the rate of growth of the line depends linearly on the rate of strain, and is independent of the rate of rotation.

2.12   Show that the vorticity equation (Eq. (2.60)) can alternatively be written

$$\frac{D\omega_i}{Dt} = \nu \frac{\partial^2 \omega_i}{\partial x_j \partial x_j} + S_{ij}\omega_j. \tag{2.75}$$

Re-express the source in the Poisson equation for pressure (Eq. (2.42)) in terms of S_{ij} and Ω_{ij}.

2.13   In a simple shear flow, all the velocity gradients are zero except for $\partial U_1/\partial x_2$. For this case write down the components of S_{ij} and Ω_{ij} (as matrices) and of ω.

2.9 Transformation properties

By studying the behavior of the Navier–Stokes equations when they are subjected to various transformations, we are able to deduce important properties of the fluid flows that they describe. The most important of these properties are Reynolds number similarity, invariance under fixed rotations and reflections of the coordinate axes, Galilean invariance, and the lack of invariance under frame rotations.

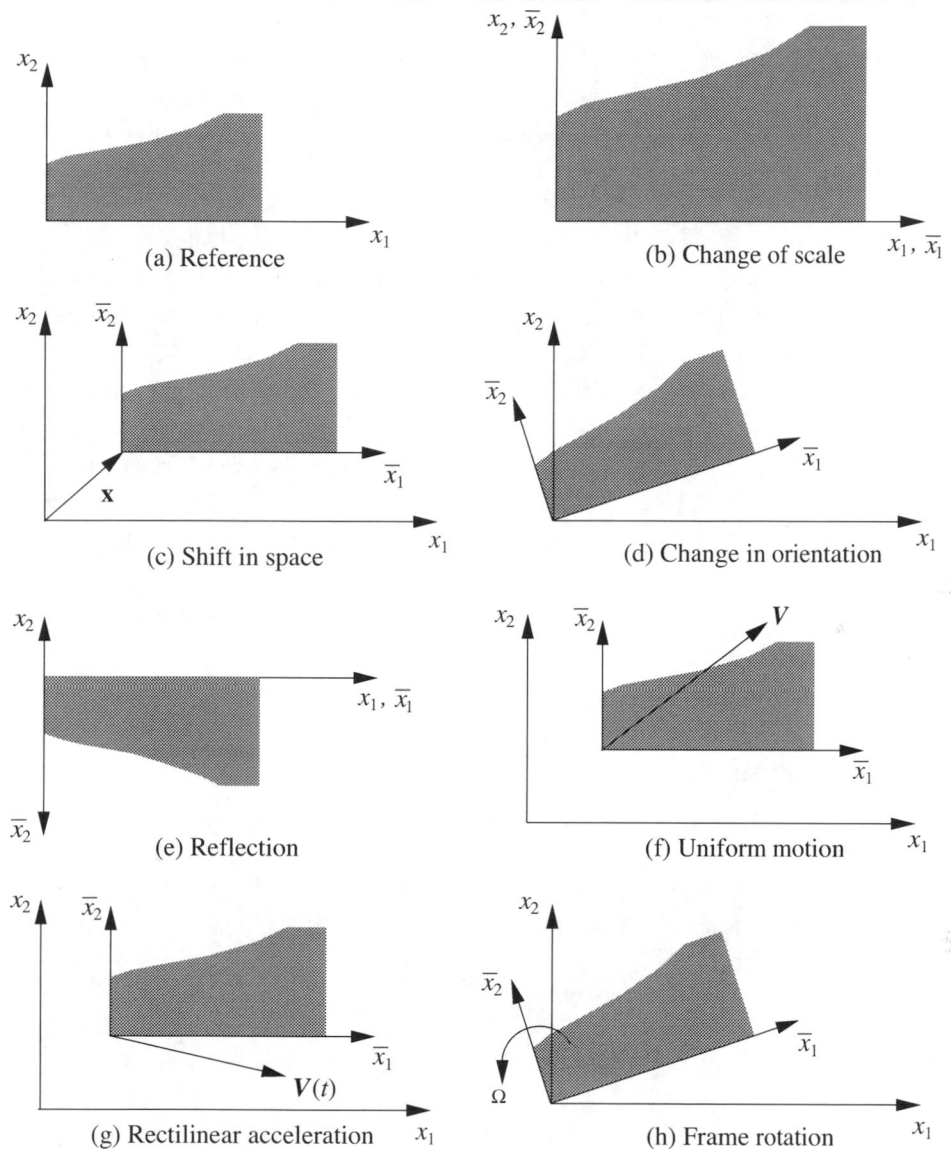

Fig. 2.2. Sketches of experiments used to study the transformation properties of the Navier–Stokes equations: (a) reference experiment (referred to the E coordinate system); (b)–(h) other experiments (referred to the $\bar{E}$ coordinate system).

Consider a particular fluid-mechanics experiment performed in a laboratory, and consider a second experiment, that is similar to the first, but differs in some respect. For example: the second experiment could be performed at a different time; the apparatus could be placed in a different location; it could be orientated differently; it could be placed on a moving platform;

a different fluid could be used; or a second apparatus that is geometrically similar to the first, but of a different scale, could be constructed. For each of these differences we can ask whether the velocity fields in the two experiments are similar. That is, are the velocity fields the same when they are appropriately scaled and referred to appropriate coordinate systems? These questions can be answered by studying the *transformation properties* (also called *invariance properties* or *symmetries*) of the Navier–Stokes equations.

These are important considerations in the modelling of turbulent flows. A model will be qualitatively incorrect unless its transformation properties are consistent with those of the Navier–Stokes equations.

Figure 2.2(a) is a sketch of the apparatus considered in the first (reference) experiment. The size of the apparatus is characterized by the length scale $\mathcal{L}$, and the initial and boundary conditions on the velocity are characterized by the velocity scale $\mathcal{U}$. The coordinate system (denoted by E, with orthonormal basis vectors e_i) has its origin and axes fixed relative to the apparatus, which is at rest in an inertial frame.

The length scale $\mathcal{L}$ and the velocity scale $\mathcal{U}$ are used to define the non-dimensional independent variables

$$\hat{x} = x/\mathcal{L}, \quad \hat{t} = t\mathcal{U}/\mathcal{L}, \tag{2.76}$$

and dependent variables

$$\hat{U}(\hat{x},\hat{t}) = U(x,t)/\mathcal{U}, \quad \hat{p}(\hat{x},\hat{t}) = p(x,t)/(\rho\mathcal{U}^2). \tag{2.77}$$

On applying these simple scaling transformations to the continuity equation Eq. (2.19), the Navier–Stokes equations Eq. (2.35), and the Poisson equation Eq. (2.42), we obtain

$$\frac{\partial \hat{U}_i}{\partial \hat{x}_i} = 0, \tag{2.78}$$

$$\frac{\partial \hat{U}_j}{\partial \hat{t}} + \hat{U}_i \frac{\partial \hat{U}_j}{\partial \hat{x}_i} = \frac{1}{\text{Re}} \frac{\partial^2 \hat{U}_j}{\partial \hat{x}_i \partial \hat{x}_i} - \frac{\partial \hat{p}}{\partial \hat{x}_j}, \tag{2.79}$$

$$\frac{\partial^2 \hat{p}}{\partial \hat{x}_i \partial \hat{x}_i} = -\frac{\partial \hat{U}_i}{\partial \hat{x}_j} \frac{\partial \hat{U}_j}{\partial \hat{x}_i}, \tag{2.80}$$

where the Reynolds number is

$$\text{Re} \equiv \mathcal{U}\mathcal{L}/\nu. \tag{2.81}$$

Evidently, the Reynolds number is the only parameter appearing in these equations.

Reynolds-number similarity

The experiment shown in Fig. 2.2(b) has a different length scale $\mathcal{L}_b$, velocity scale $\mathcal{U}_b$, and fluid properties, ν_b and ρ_b. If the scaled variables are defined in an analogous way ($\hat{x} = x/\mathcal{L}_b$, $\hat{U} = U/\mathcal{U}_b$, etc.) then the boundary conditions (expressed in terms of $\hat{U}(\hat{x},\hat{t})$) in two experiments are the same, and the transformed Navier–Stokes equations are the same as Eqs. (2.78)–(2.80), except that Re is replaced by

$$\text{Re}_b \equiv \mathcal{U}_b\,\mathcal{L}_b/\nu_b. \tag{2.82}$$

Thus, if the Reynolds numbers are the same ($\text{Re} = \text{Re}_b$), then the scaled velocity fields $\hat{U}(\hat{x},\hat{t})$ are also the same, because they are governed by identical equations with identical initial and boundary conditions. This is the property of *Reynolds-number similarity*.

The scaled Euler equations are the same as Eq. (2.79), but with the omission of the term in Re. The scaled velocity fields $\hat{U}(\hat{x},\hat{t})$ given by the Euler equations are therefore the same, irrespective of $\mathcal{L}_b$, $\mathcal{U}_b$, and ρ_b: they exhibit *scale similarity* and the Euler equations are said to be *invariant with respect to scale transformations*.

Time and space invariance

The simplest invariance properties of the Navier–Stokes equations are their invariances with respect to shifts in time and space. As depicted in Fig. 2.2(c) we consider the second experiment performed a time T later than the reference experiment, with the apparatus translated by an amount X. The velocity field in the second experiment is referred to the $\bar{E}$ coordinate system shown in the Fig. 2.2(c), which has orthonormal basis vectors $\bar{e}_i$. With the scaled independent variables defined by

$$\hat{x} = \bar{x}/\mathcal{L} = (x - X)/\mathcal{L}, \tag{2.83}$$

$$\hat{t} = (t - T)\mathcal{U}/\mathcal{L}, \tag{2.84}$$

it is trivial to show that the transformed Navier–Stokes equations are identical to Eqs. (2.78)–(2.80).

Rotational and reflectional invariance

Figure 2.2(d) shows the apparatus with a different orientation than that in the reference experiment; the appropriate $\bar{E}$ coordinate system is obtained by a rotation of the reference (E) coordinate axes. Figure 2.2(e) shows a different apparatus, constructed to be the mirror image of the reference apparatus. In this case, the appropriate $\bar{E}$ coordinate system is obtained by a reflection of a coordinate axis.

These coordinate transformations – rotations and reflections of the axes – are precisely those considered in Cartesian tensors (see Appendix A). With $a_{ij} \equiv e_i \cdot \bar{e}_j$ being the direction cosines, the scaled variables are

$$\hat{x}_i = \bar{x}_i/\mathcal{L} = a_{ji}x_j/\mathcal{L}, \tag{2.85}$$

$$\hat{U}_i = a_{ji}U_j. \tag{2.86}$$

It follows immediately from the fact that the Navier–Stokes equations can be written in Cartesian tensor notation that the transformed equations are identical to those in the reference system (Eqs. (2.78)–(2.80)). Thus the Navier–Stokes equations are invariant with respect to rotations and reflections of the coordinate axes.

In these considerations it is important to distinguish between two kinds of 'rotations.' Here we are considering the $\bar{E}$ coordinate system obtained by a fixed rotation of the E coordinate axes. By 'fixed' we mean that the direction cosines a_{ij} do not depend on time. In contrast, we consider below rotating frames, so that the direction cosines are time dependent.

The invariance with respect to reflections has a physical significance and a mathematical consequence which are discussed at greater length in Appendix A. The physical significance is that the Navier–Stokes equations contain no bias toward right-handed or left-handed motions. Of course such bias can occur in a flow – most dramatically in a tornado – but it arises from the initial or boundary conditions, or from frame rotation, not from the equations of motion (expressed in an inertial frame).

Any equation written in Cartesian tensor notation ensures invariance under rotations and reflections of coordinate axes. In contrast, an equation written in vector notation and involving pseudovectors (e.g., vorticity), or written in suffix notation using the alternating symbol ε_{ijk}, does not ensure these invariance properties.

Time reversal

Analogous to the reflection of a coordinate axis (e.g., $\bar{x}_2 = -x_2$), we can consider the reversal of time by defining

$$\hat{t} = -t\mathcal{U}/\mathcal{L}, \tag{2.87}$$

$$\hat{U}(\hat{x}, \hat{t}) = -U(x, t)/\mathcal{U}. \tag{2.88}$$

It is readily shown that the corresponding transformed Navier–Stokes equations are the same as Eqs. (2.78)–(2.80), except that the sign of the viscous term (proportional to Re^{-1}) is altered. Thus, the Navier–Stokes equations are not invariant under a time reversal; but the Euler equations are.

Galilean invariance

The remaining topics in this section are concerned with moving frames. We consider first, as depicted in Fig. 2.2(f), the apparatus moving at a *fixed* velocity V, so that both coordinate systems (E and $\bar{E}$) are in inertial frames. The transformations between the coordinate systems are

$$\bar{x} = x - Vt, \quad \bar{t} = t, \tag{2.89}$$

$$\bar{U}(\bar{x}, \bar{t}) = U(x, t) - V. \tag{2.90}$$

A quantity that is the same in different inertial frames is said to be *Galilean invariant*. From Eqs. (2.89) and (2.90) we obtain

$$\frac{\partial \bar{U}_i}{\partial \bar{x}_j} = \frac{\partial U_i}{\partial x_j}, \tag{2.91}$$

$$\frac{\partial \bar{U}_i}{\partial \bar{t}} = \frac{\partial U_i}{\partial t} + V_j \frac{\partial U_i}{\partial x_j}, \tag{2.92}$$

$$\frac{D\bar{U}_i}{D\bar{t}} \equiv \frac{\partial \bar{U}_i}{\partial \bar{t}} + \bar{U}_j \frac{\partial \bar{U}_i}{\partial \bar{x}_j} = \frac{DU_i}{Dt}, \tag{2.93}$$

showing that the velocity gradients and the fluid acceleration are Galilean invariant, whereas the velocity and its partial time derivative are not. Other quantities that are Galilean invariant include scalars such as $\phi(x, t)$ and pressure $p(x, t)$, and quantities related to velocity gradients, e.g., S_{ij}, Ω_{ij}, and the vorticity ω.

It is simply shown that the transformed Navier–Stokes equations (written for $\hat{U} = \bar{U}/\mathcal{U}$ in terms of $\hat{x} \equiv \bar{x}/\mathcal{L}$, etc.) are identical to Eqs. (2.78)–(2.80), and hence are Galilean invariant. Just like all phenomena described by classical mechanics, the behavior of fluid flows is the same in all inertial frames.

EXERCISE

2.14   Which of the following are Galilean invariant:

(a) a streamline (which by definition is a curve that is everywhere parallel to the velocity vector),

(b) a vortex line (which by definition is a curve that is everywhere parallel to the vorticity vector),

(c) the *helicity*, which is defined as $U \cdot \omega$
(d) the *enstrophy*, which is defined as $\omega \cdot \omega$
(e) material lines, surfaces, and volumes and,
(f) for a scalar field; $\partial\phi/\partial t$, $\partial\phi/\partial x_i$, and $D\phi/Dt$?

Extended Galilean invariance

A peculiar property of the Navier–Stokes equations is that they are invariant under rectilinear accelerations of the frame. We consider, as depicted in Fig. 2.2(g), the second experiment being performed on a platform moving at a variable velocity $V(t)$, but with no rotation of the frame, so that the coordinate directions (e.g., e_1 and $\bar{e}_1$) remain parallel. With the transformed variables $\bar{x}$, $\bar{t}$, and $\bar{U}$ defined by Eqs. (2.89) and (2.90), the transformed Navier–Stokes equations are

$$\frac{\partial \bar{U}_j}{\partial \bar{t}} + \bar{U}_i \frac{\partial \bar{U}_j}{\partial \bar{x}_i} = \nu \frac{\partial^2 \bar{U}_j}{\partial \bar{x}_i \partial \bar{x}_i} - \frac{1}{\rho} \frac{\partial p}{\partial \bar{x}_j} - A_j, \qquad (2.94)$$

where the additional term on the right-hand side is the acceleration of the frame, $A = dV/dt$. The last two terms can be written

$$\frac{1}{\rho} \frac{\partial p}{\partial \bar{x}_j} + A_j = \frac{1}{\rho} \frac{\partial}{\partial \bar{x}_j}(p + \rho \bar{x}_i A_i), \qquad (2.95)$$

showing that the frame acceleration can be absorbed in a modified pressure. Consequently the Navier–Stokes equations for the transformed variables

$$\hat{U} \equiv \bar{U}/\mathcal{U}, \quad \hat{p} \equiv (p + \rho \bar{x} \cdot A)/(\rho \mathcal{U}^2), \qquad (2.96)$$

are identical to Eqs. (2.76)–(2.80). Thus the scaled velocity $\hat{U}$ and modified pressure $\hat{p}$ fields in the experiment in the frame with arbitrary rectilinear acceleration are identical to those in the inertial flame. This is *extended Galilean invariance* (which applies only to constant-density flows).

Frame rotation

Finally, we consider the second experiment being performed in a non-inertial rotating frame, Fig. 2.2(h). In the $\bar{E}$ coordinate system, the time-dependent basis vectors $\bar{e}_i(t)$ evolve by

$$\frac{d}{dt}\bar{e}_i = \tilde{\Omega}_{ij}\bar{e}_j, \qquad (2.97)$$

where $\tilde{\Omega}_{ij}(t) = -\tilde{\Omega}_{ji}(t)$ is the rate of rotation of the frame. Note that, in this case, the direction cosines $a_{ij}(t) \equiv e_i \cdot \bar{e}_j(t)$ are time-dependent.

The Navier–Stokes equations transformed to the non-inertial frame are the same as Eq. (2.94), but with the frame acceleration $-A_j$ replaced by the fictitious force

$$F_j = -\bar{x}_i \tilde{\Omega}_{ik} \tilde{\Omega}_{kj} - 2\bar{U}_i \tilde{\Omega}_{ij} - \bar{x}_i \frac{\mathrm{d}\tilde{\Omega}_{ij}}{\mathrm{d}\bar{t}} \tag{2.98}$$

(see Exercise 2.15). The three contributions to F represent the centrifugal force, the Coriolis force, and the angular acceleration force. The centrifugal force can be absorbed into a modified pressure, but the remaining two forces cannot. As is well known in meteorology and turbomachinery, Coriolis forces can have significant effects on flows in rotating frames.

A quantity that is the same in rotating and non-rotating frames is said to possess *material-frame indifference*. Evidently, the Navier–Stokes equations do not have this property.

The effect of frame rotation is also evident in the vorticity equation. In the non-inertial $\bar{E}$ coordinate system, the equation for the evolution of the vorticity

$$\bar{\omega}_i \equiv \varepsilon_{ijk} \frac{\partial \bar{U}_k}{\partial \bar{x}_j}, \tag{2.99}$$

obtained from the Navier–Stokes equations (i.e., Eq. (2.94) with F_j in place of $-A_j$), is

$$\frac{\partial \bar{\omega}_i}{\partial \bar{t}} + \bar{U}_j \frac{\partial \bar{\omega}_i}{\partial \bar{x}_j} = \nu \frac{\partial^2 \bar{\omega}_i}{\partial \bar{x}_j \partial \bar{x}_j} + \bar{\omega}_j \frac{\partial \bar{U}_i}{\partial \bar{x}_j} - 2\varepsilon_{ijk} \frac{\partial \bar{U}_\ell}{\partial \bar{x}_j} \tilde{\Omega}_{\ell k} - \varepsilon_{ijk} \frac{\mathrm{d}\tilde{\Omega}_{jk}}{\mathrm{d}t}. \tag{2.100}$$

Evidently, because of the last two terms – which correspond to Coriolis and angular acceleration forces – the vorticity equation in a rotating frame is different than that in an inertial frame (Eq. (2.60)).

EXERCISE

2.15   Let $X(t)$ be the position of a moving point relative to the origin of the E coordinate system in an inertial frame. Let $Y(t) = \bar{e}_i(t) Y_i(t)$ be the position of the same point relative to the non-inertial frame $\bar{E}$. The origin of the $\bar{E}$ frame moves with velocity $V(t)$, and its basis vectors $\bar{e}_i$ evolve according to Eq. (2.97). If the origins are coincident at time $t = 0$, then

$$X(t) = Y(t) + \int_0^t V(t')\,\mathrm{d}t'. \tag{2.101}$$

Show that the velocity and acceleration (relative to the inertial frames) are

$$\dot{X} = V + \bar{e}_j(\dot{Y}_j + Y_i\tilde{\Omega}_{ij}), \tag{2.102}$$

$$\ddot{X} = \dot{V} + \bar{e}_j(\ddot{Y}_j + Y_i\tilde{\Omega}_{ik}\tilde{\Omega}_{kj} + 2\dot{Y}_i\tilde{\Omega}_{ij} + Y_i\dot{\tilde{\Omega}}_{ij}), \tag{2.103}$$

where an overdot indicates differentiation with respect to time.

Two-dimensional flows

Another peculiar property of the Navier–Stokes equations is that, for two-dimensional flow (in the x_1–x_2 plane, say), they are invariant with respect to steady rotations of the frame in the plane of the flow (i.e., rotations about the x_3 axis), see Speziale (1981). For two-dimensional flows, it is sometimes useful to re-express the Navier–Stokes equations in terms of streamfunction and vorticity. The streamfunction $\psi(x_1, x_2, t)$ is such that the velocities are given by

$$U_1 = \frac{\partial \psi}{\partial x_2}, \qquad U_2 = -\frac{\partial \psi}{\partial x_1}, \tag{2.104}$$

and the only non-zero component of the vorticity is

$$\omega_3 = \frac{\partial U_2}{\partial x_1} - \frac{\partial U_1}{\partial x_2}. \tag{2.105}$$

For steady rotations of the frame, the final term in Eq. (2.100) is zero, and, for the two-dimensional flows considered, explicit evaluation of the penultimate term reveals that it too is zero (see Exercise 2.17). Thus, for this special case, the vorticity is unaffected by frame rotation, and it follows that the Navier–Stokes equations exhibit material-frame indifference (in this restricted sense).

EXERCISES _____

2.16 For two-dimensional flow, with $U_3 = 0$ and U_1 and U_2 given by Eqs. (2.104), show that the divergence of velocity is zero for all streamfunctions. Show that the streamfunction and vorticity are related by the Poisson equation

$$\left(\frac{\partial^2}{\partial x_1^2} + \frac{\partial^2}{\partial x_2^2} \right) \psi = -\omega_3. \tag{2.106}$$

2.17 With reference to the penultimate term in Eq. (2.100), consider the quantity

$$\Omega_i^* \equiv \varepsilon_{ijk} \frac{\partial \bar{U}_\ell}{\partial \bar{x}_j} \tilde{\Omega}_{\ell k}, \tag{2.107}$$

for a two-dimensional flow (in the x_1–x_2 plane), and for frame rotations in the same plane. Which components of $\partial \bar{U}_\ell / \partial \bar{x}_j$ and $\tilde{\Omega}_{\ell k}$ are zero? Show that Ω_1^* and Ω_2^* are zero. Obtain the result

$$\Omega_3^* = \tilde{\Omega}_{12}\left(\frac{\partial \bar{U}_1}{\partial \bar{x}_1} + \frac{\partial \bar{U}_2}{\partial \bar{x}_2}\right), \qquad (2.108)$$

and hence argue that $\mathbf{\Omega}^*$ is zero for the class of flows considered.

3

The statistical description of turbulent flows

3.1 The random nature of turbulence

In a turbulent flow, the velocity field $U(x,t)$ is *random*. What does this statement mean? Why is it so?

As a first step we need to understand the word 'random.' Consider a fluid-flow experiment that can be repeated many times under a specified set of conditions, C, and consider an event A, such as $A \equiv \{U < 10 \text{ m s}^{-1}\}$, where U is a specified component of velocity at a specified position and time (measured from the initiation of the experiment). If the event A inevitably occurs, then A is *certain* or *sure*. If the event A cannot occur, then it is *impossible*. The third possibility is that A may occur or it may but need not occur. In this case the event A is *random*. Then, in the example $A \equiv \{U < 10 \text{ m s}^{-1}\}$, U is a *random variable*.

A mistake that is sometimes made is to attribute incorrectly additional significance to the designation 'random,' and then to dispute the fact that turbulence is a random phenomenon. That the event A is random means only that it is neither certain nor impossible. That U is a random variable means only that it does not have a unique value – the same every time the experiment is repeated under the same set of conditions, C. Figure 3.1 illustrates the values $U^{(n)}(n = 1, 2, \ldots, 40)$ taken by the random variable U on 40 repetitions of the experiment.

The next issue to resolve is the consistency between the random nature of turbulent flows, and the deterministic nature of classical mechanics embodied in the Navier–Stokes equations. If the equations of motion are deterministic, why are the solutions random? The answer lies in the combination of two observations.

(i) In any turbulent flow there are, unavoidably, perturbations in initial conditions, boundary conditions, and material properties.

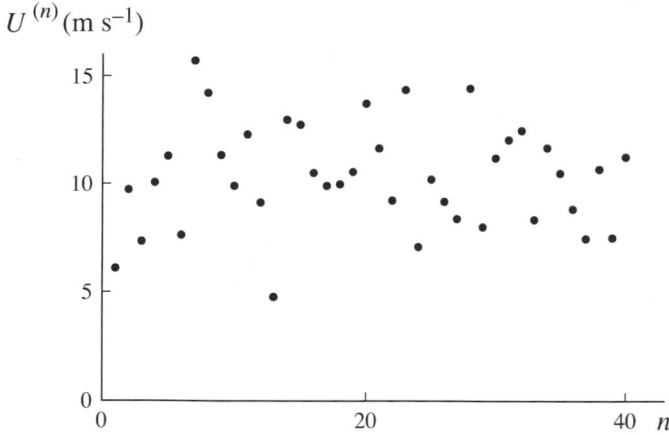

Fig. 3.1. A sketch of the value $U^{(n)}$ of the random velocity variable U on the nth repetition of a turbulent-flow experiment.

(ii) Turbulent flow fields display an acute sensitivity to such perturbations.

At the outset of our discussion on randomness, we considered 'a fluid-flow experiment that can be repeated many times *under a specified set of conditions C.*' An example is the flow of pure water at $20\,°C$ through a smooth straight pipe. It should be appreciated that the conditions, C, thus defined are incomplete: in practice there are, inevitably, perturbations from these nominal conditions. There can be perturbations in boundary conditions, for example, through vibration of the apparatus, or from the detailed finish of nominally smooth surfaces. There can be perturbations in fluid properties caused by small inhomogeneities in temperature or by the presence of impurities, and there can be perturbations in the initial state of the flow. With care and effort these perturbations can be reduced, but they cannot be eliminated. Consequently, the nominal conditions C are incomplete, and hence do not uniquely determine the evolution of the turbulent flow.

The presence of perturbations does not by itself explain the random nature of turbulent flows – for, indeed, such perturbations are also present in laminar flows. However, at the high Reynolds numbers of turbulent flows, the evolution of the flow field is extremely sensitive to small changes in initial conditions, boundary conditions, and material properties. This sensitivity is well understood in the study of dynamical systems, and has been popularized in books on chaos (e.g., Gleick (1988) and Moon (1992)). It is now demonstrated using the Lorenz equations.

Lorenz (1963) studied a time-dependent system, characterized by three state variables, $x(t), y(t)$, and $z(t)$. These variables evolve according to the

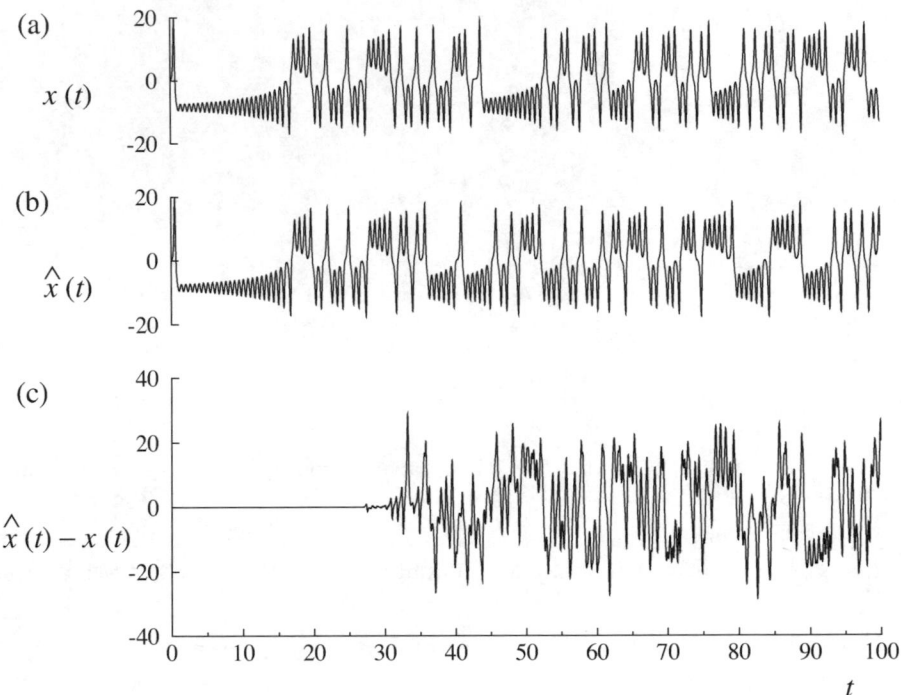

Fig. 3.2. Time histories from the Lorenz equations (Eqs. (3.1)): (a) $x(t)$ from the initial condition Eq. (3.2); (b) $\hat{x}(t)$ from the slightly different initial condition Eq. (3.3); and (c) the difference $\hat{x}(t) - x(t)$.

ordinary differential equations

$$\dot{x} = \sigma(y - x),$$
$$\dot{y} = \rho x - y - xz,$$
$$\dot{z} = -\beta z + xy, \tag{3.1}$$

where the coefficients are $\sigma = 10$, $\beta = \frac{8}{3}$, and $\rho = 28$. For the initial condition

$$[x(0), y(0), z(0)] = [0.1, 0.1, 0.1], \tag{3.2}$$

Fig. 3.2(a) shows the time history $x(t)$ obtained from the numerical integration of Eqs. (3.1). The result obtained – denoted by $\hat{x}(t)$ – with the slightly different initial condition

$$[x(0), y(0), z(0)] = [0.100\,001, 0.1, 0.1], \tag{3.3}$$

is shown in Fig. 3.2(b). It may be observed that (as expected) $x(t)$ and $\hat{x}(t)$ are initially indistinguishable, but by $t = 35$ they are quite different. This observation is made clearer in Fig. 3.2(c), which shows the difference $\hat{x}(t) - x(t)$.

A consequence of this extreme sensitivity to initial conditions is that – beyond some point – the state of the system cannot be predicted. In this example, if the initial state is known only to within 10^{-6}, then Fig. 3.2 clearly shows that no useful prediction can be made beyond $t = 35$.

This example serves to demonstrate that a simple set of deterministic equations – much simpler than the Navier–Stokes equations – can exhibit acute sensitivity to initial conditions, and hence unpredictability.

The qualitative behavior of the Lorenz system depends on the coefficients. In particular, for the fixed values $\sigma = 10$ and $\beta = \frac{8}{3}$, the behavior depends on ρ. If ρ is less than a critical value $\rho^* \approx 24.74$, then the system goes to a stable fixed point, i.e., the state variables $[x(t), y(t), z(t)]$ tend asymptotically to fixed values. However, for $\rho > \rho^*$ (e.g., $\rho = 28$ as in Fig. 3.2) chaotic behavior ensues. Again, there is a similarity to the Navier–Stokes equations, which (with steady boundary conditions) have steady solutions at sufficiently low Reynolds number, but chaotic, turbulent solutions at high Re. Further discussions of the Lorenz equations, dynamical systems and equations, dynamical systems, and chaos are contained in the books of Guckenheimer and Holmes (1983), Moon (1992), and Gleick (1988).

3.2 Characterization of random variables

For a laminar flow, we can use theory (i.e., the Navier–Stokes equations) to calculate U (a particular component of the velocity at a specified position and time), and we can perform an experiment to measure U. From a century of experience, we have a high degree of confidence that the calculated and measured values of U will agree (to within small numerical and experimental errors).

The Navier–Stokes equations apply equally to turbulent flows, but here the aim of theory must be different. Since U is a random variable, its value is inherently unpredictable: a theory that predicts a particular value for U is almost certain to be wrong. A theory can, however, aim at determining the *probability* of events such as $A \equiv \{U < 10 \text{ m s}^{-1}\}$.

In this section we develop the concepts and tools used to characterize a random variable such as U. In particular U is completely characterized by its *probability density function* (PDF). The random velocity field $U(x, t)$ in a turbulent flow is a much more complicated mathematical object than the single random variable U. In subsequent sections we introduce some quantities used to characterize sets of random variables (e.g., U_1, U_2, and U_3), random functions of time (e.g., $U(t)$), and random functions of position (e.g., $U(x)$).

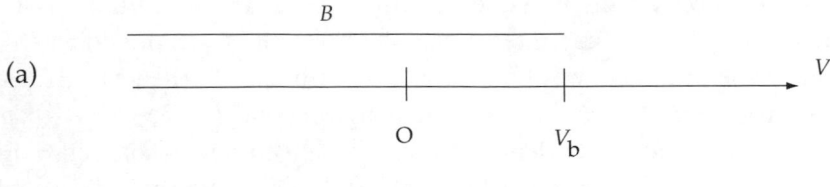

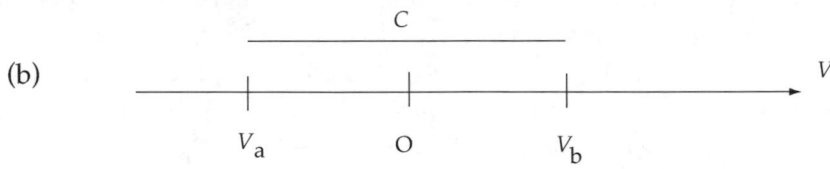

Fig. 3.3. Sketches of the sample space of U showing the regions corresponding to the events (a) $B \equiv \{U < V_b\}$, and (b) $C \equiv \{V_a \leq U < V_b\}$.

Sample space

In order to be able to discuss more general events than $A \equiv \{U < 10 \text{ m s}^{-1}\}$, we introduce an independent velocity variable V, which is referred to as the *sample-space* variable corresponding to U. As illustrated in Fig. 3.3, different events such as

$$B \equiv \{U < V_b\}, \tag{3.4}$$

$$C \equiv \{V_a \leq U < V_b\}, \quad \text{for} \quad V_a < V_b, \tag{3.5}$$

correspond to different regions of the sample space.

Probability

The *probability* of the event B, for example, is written

$$p = P(B) = P\{U < V_b\}. \tag{3.6}$$

For the moment, the reader's intuitive understanding of probability is sufficient: p is a real number $(0 \leq p \leq 1)$ signifying the likelihood of the occurrence of the event. For an impossible event p is zero; for a sure event p is unity. (Probability is discussed further in Section 3.8.)

The cumulative distribution function

The probability of any event can be determined from the *cumulative distribution function* (CDF), which is defined by

$$F(V) \equiv P\{U < V\}. \tag{3.7}$$

For example, we have

$$P(B) = P\{U < V_\mathrm{b}\} = F(V_\mathrm{b}), \tag{3.8}$$

$$P(C) = P\{V_\mathrm{a} \leq U < V_\mathrm{b}\} = P\{U < V_\mathrm{b}\} - P\{U < V_\mathrm{a}\}$$
$$= F(V_\mathrm{b}) - F(V_\mathrm{a}). \tag{3.9}$$

The three basic properties of the CDF are

$$F(-\infty) = 0, \tag{3.10}$$

since $\{U < -\infty\}$ is impossible;

$$F(\infty) = 1, \tag{3.11}$$

since $\{U < \infty\}$ is certain; and,

$$F(V_\mathrm{b}) \geq F(V_\mathrm{a}), \quad \text{for} \ \ V_\mathrm{b} > V_\mathrm{a}, \tag{3.12}$$

since the probability of every event is non-negative, i.e.

$$F(V_\mathrm{b}) - F(V_\mathrm{a}) = P\{V_\mathrm{a} \leq U < V_\mathrm{b}\} \geq 0. \tag{3.13}$$

The third property (Eq. (3.12)) expresses the fact that the CDF is a non-decreasing function.

The probability density function

The *probability density function* (PDF) is defined to be the derivative of the CDF:

$$f(V) \equiv \frac{\mathrm{d}F(V)}{\mathrm{d}V}. \tag{3.14}$$

It follows simply from the properties of the CDF that the PDF is non-negative

$$f(V) \geq 0, \tag{3.15}$$

it satisfies the normalization condition

$$\int_{-\infty}^{\infty} f(V)\,\mathrm{d}V = 1, \tag{3.16}$$

and $f(-\infty) = f(\infty) = 0$. Further, from Eq. (3.13) it follows that the probability of the random variable being in a particular interval equals the integral

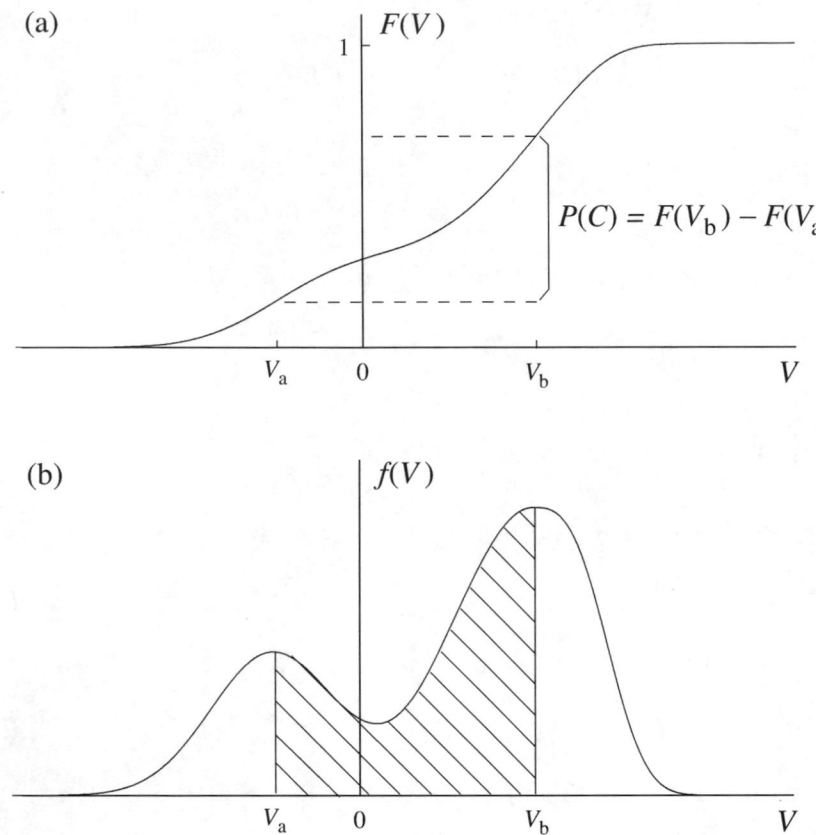

Fig. 3.4. Sketches of (a) the CDF of the random variable U showing the probability of the event $C \equiv \{V_a \leq U < V_b\}$, and (b) the corresponding PDF. The shaded area in (b) is the probability of C.

of the PDF over that interval:

$$P\{V_a \leq U < V_b\} = F(V_b) - F(V_a)$$
$$= \int_{V_a}^{V_b} f(V)\, dV. \qquad (3.17)$$

Figure 3.4 provides a graphical interpretation of this equation.

For an infinitesimal interval, Eq. (3.17) becomes

$$P\{V \leq U < V + dV\} = F(V + dV) - F(V)$$
$$= f(V)\, dV. \qquad (3.18)$$

Thus the PDF $f(V)$ is the probability *per unit distance* in the sample space – hence the term 'probability *density* function.' The PDF $f(V)$ has the dimensions of the inverse of U, whereas the CDF and the product $f(V)\, dV$

are non-dimensional. Under a change of variables, the transformation rule for densities (such as PDF's) is different than that for functions: see Exercise 3.9 on page 49.

It is emphasized that the PDF $f(V)$ (or equally the CDF) fully characterizes the random variable U. Two or more random variables that have the same PDF are said to be *identically distributed*, or equivalently *statistically identical*.

Means and moments

The *mean* (or *expectation*) of the random variable U is defined by

$$\langle U \rangle \equiv \int_{-\infty}^{\infty} V f(V) \, dV. \tag{3.19}$$

It is the probability-weighted average of all possible values of U. More generally, if $Q(U)$ is a function of U, the mean of $Q(U)$ is

$$\langle Q(U) \rangle \equiv \int_{-\infty}^{\infty} Q(V) f(V) \, dV. \tag{3.20}$$

Even when the condition is not stated explicitly, it should be understood (here and below) that the mean $\langle Q(U) \rangle$ exists only if the integral in Eq. (3.20) converges absolutely.

The rules for taking means are quite simple. If $Q(U)$ and $R(U)$ are functions of U, and if a and b are constants, then

$$\langle [aQ(U) + bR(U)] \rangle = a\langle Q(U) \rangle + b\langle R(U) \rangle, \tag{3.21}$$

as may readily be verified from Eq. (3.20). Thus the angled brackets $\langle \ \rangle$ behave as a linear operator. While U, $Q(U)$, and $R(U)$ are all random variables, $\langle U \rangle$, $\langle Q(U) \rangle$, and $\langle R(U) \rangle$ are not. Hence the mean of the mean is the mean: $\langle \langle U \rangle \rangle = \langle U \rangle$.

The *fluctuation* in U is defined by

$$u \equiv U - \langle U \rangle, \tag{3.22}$$

and the *variance* is defined to be the mean-square fluctuation:

$$\mathrm{var}(U) \equiv \langle u^2 \rangle = \int_{-\infty}^{\infty} (V - \langle U \rangle)^2 f(V) \, dV. \tag{3.23}$$

The square-root of the variance is the *standard deviation*

$$\mathrm{sdev}(U) = \sqrt{\mathrm{var}\ (U)} = \langle u^2 \rangle^{1/2}, \tag{3.24}$$

and is also denoted by u' and σ_u, and is also referred to as the r.m.s. (*root mean square*) of U.

The n*th central moment* is defined to be

$$\mu_n \equiv \langle u^n \rangle = \int_{-\infty}^{\infty} (V - \langle U \rangle)^n f(V) \, dV. \tag{3.25}$$

Evidently we have $\mu_0 = 1$, $\mu_1 = 0$, and $\mu_2 = \sigma_u^2$.

(In contrast, the n*th* moment about the origin – or the n*th* raw moment – is defined to be $\langle U^n \rangle$.)

EXERCISES

3.1   With Q and R being random variables, and a and b being constants, use Eq. (3.20) to verify the relations

$$\langle a \rangle = a, \quad \langle aQ \rangle = a\langle Q \rangle, \tag{3.26}$$

$$\langle Q + R \rangle = \langle Q \rangle + \langle R \rangle, \quad \langle \langle Q \rangle \rangle = \langle Q \rangle, \tag{3.27}$$

$$\langle \langle Q \rangle \langle R \rangle \rangle = \langle Q \rangle \langle R \rangle, \quad \langle \langle Q \rangle R \rangle = \langle Q \rangle \langle R \rangle, \tag{3.28}$$

$$\langle q \rangle = 0, \quad \langle q \langle R \rangle \rangle = 0, \tag{3.29}$$

where $q \equiv Q - \langle Q \rangle$.

3.2   Let Q be defined by

$$Q = a + bU, \tag{3.30}$$

where U is a random variable, and a and b are constants. Show that

$$\langle Q \rangle = a + b\langle U \rangle, \tag{3.31}$$

$$\text{var}(Q) = b^2 \, \text{var}(U), \tag{3.32}$$

$$\text{sdev}(Q) = b \, \text{sdev}(U). \tag{3.33}$$

Show also that

$$\text{var}(U) = \langle U^2 \rangle - \langle U \rangle^2. \tag{3.34}$$

Standardization

It is often convenient to work in terms of standardized random variables, which, by definition, have zero mean and unit variance. The standardized random variable $\hat{U}$ corresponding to U is

$$\hat{U} \equiv (U - \langle U \rangle)/\sigma_u, \tag{3.35}$$

and its PDF – the standardized PDF of U – is

$$\hat{f}(\hat{V}) = \sigma_u f(\langle U \rangle + \sigma_u \hat{V}). \tag{3.36}$$

The moments of $\hat{U}$ – the standardized moments of U – are

$$\hat{\mu}_n = \frac{\langle u^n \rangle}{\sigma_u^n} = \frac{\mu_n}{\sigma_u^n} = \int_{-\infty}^{\infty} \hat{V}^n \hat{f}(\hat{V}) \, d\hat{V}. \tag{3.37}$$

Evidently we have $\hat{\mu}_0 = 1$, $\hat{\mu}_1 = 0$ and $\hat{\mu}_2 = 1$. The third standardized moment $\hat{\mu}_3$ is called the *skewness*, and the fourth $\hat{\mu}_4$ is the *flatness* or *kurtosis*.

EXERCISE _____

3.3   Show that the standardized moments of U and Q (defined by Eq. (3.30)) are identical.

The characteristic function

The *characteristic function* of the random variable U is defined by

$$\Psi(s) \equiv \langle e^{iUs} \rangle = \int_{-\infty}^{\infty} f(V) e^{iVs} \, dV. \tag{3.38}$$

It may be recognized that the integral in Eq. (3.38) is an inverse Fourier transform: $\Psi(s)$ and $f(V)$ form a Fourier-transform pair, and consequently they contain the same information.

The characteristic function is a mathematical device that facilitates some derivations and proofs. Its properties are described in Appendix I. Characteristic functions are used extensively in Chapter 12, but not before. Consequently a study of Appendix I can be deferred.

3.3 Examples of probability distributions

To consolidate the notions developed, and to illustrate some qualitatively different behaviors, we now give some specific examples of probability distributions. These distributions are encountered in later chapters.

The uniform distribution

If U is uniformly distributed in the interval $a \leq V < b$, then the PDF of U is

$$f(V) = \begin{cases} \dfrac{1}{b-a}, & \text{for } a \leq V < b, \\ 0, & \text{for } V < a \text{ and } V \geq b. \end{cases} \tag{3.39}$$

This PDF and the corresponding CDF are shown in Fig. 3.5.

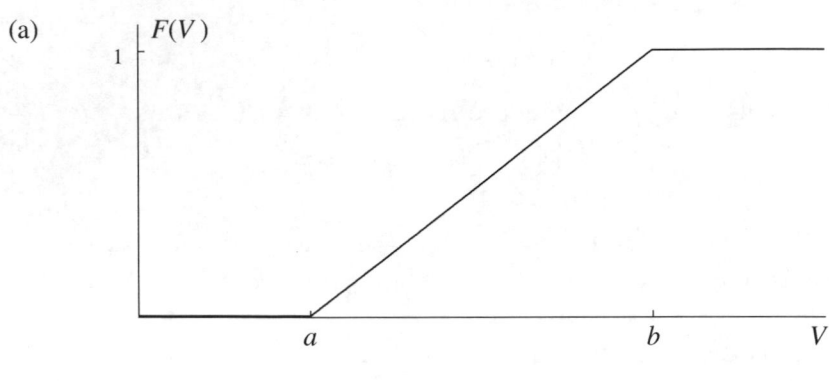

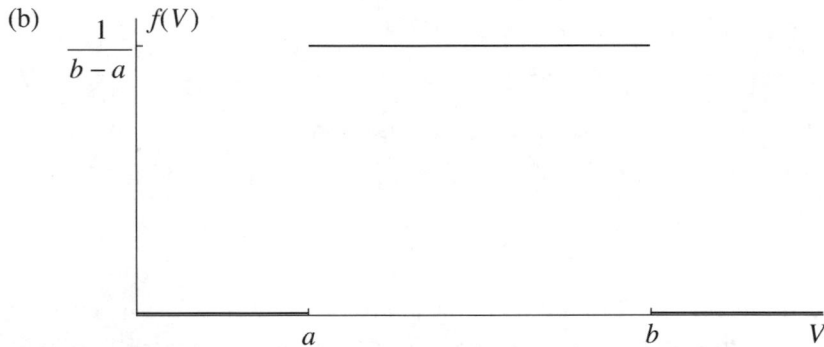

Fig. 3.5. The CDF (a) and the PDF (b) of a uniform random variable (Eq. (3.39)).

EXERCISE

3.4   For the uniform distribution Eq. (3.39) show that

(a) $\langle U \rangle = \frac{1}{2}(a+b)$,

(b) $\mathrm{var}(U) = \frac{1}{12}(b-a)^2$,

(c) $\hat{\mu}_3 = 0$, and

(d) $\hat{\mu}_4 = \frac{9}{5}$.

The exponential distribution

If U is exponentially distributed with parameter λ, then its PDF (see Fig. 3.6) is

$$f(V) = \begin{cases} \dfrac{1}{\lambda}\exp(-V/\lambda), & \text{for } V \geq 0, \\ 0, & \text{for } V < 0. \end{cases} \qquad (3.40)$$

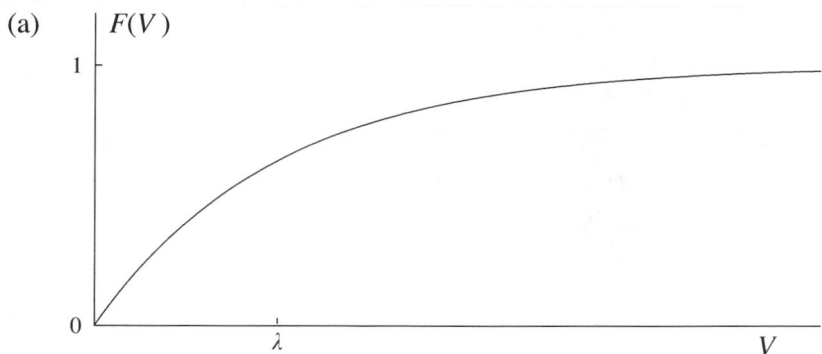

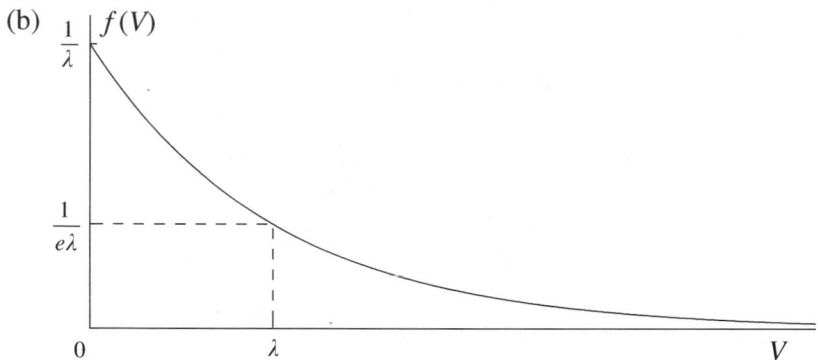

Fig. 3.6. The CDF (a) and PDF (b) of an exponentially distributed random variable (Eq. (3.40)).

3.5   For the exponential distribution Eq. (3.40) show that

(a) the normalization condition is satisfied,
(b) $\langle U \rangle = \lambda$,
(c) $\langle U^n \rangle = n\lambda \langle U^{n-1} \rangle = n!\lambda^n$, for $n \geq 1$,
(d) $F(V) = 1 - \exp(-V/\lambda)$, for $V > 0$,
    $= 0$, for $V \leq 0$, and
(e) $\text{Prob}\{U \geq a\lambda\} = e^{-a}$, for $a \geq 0$.

The normal distribution

Of fundamental importance in probability theory is the normal or *Gaussian* distribution. If U is normally distributed with mean μ and standard deviation σ, then the PDF of U is

$$f(V) = \mathcal{N}(V; \mu, \sigma^2) \equiv \frac{1}{\sigma\sqrt{2\pi}} \exp[-\tfrac{1}{2}(V - \mu)^2/\sigma^2]. \qquad (3.41)$$

Here $\mathcal{N}(V; \mu, \sigma^2)$ – or sometimes $\mathcal{N}(\mu, \sigma^2)$ – denotes the normal distribution with mean μ and variance σ^2. We can also write

$$U \overset{\text{D}}{=} \mathcal{N}(\mu, \sigma^2), \tag{3.42}$$

to indicate that U is *equal in distribution* to a normal random variable, i.e., the PDF of U is given by Eq. (3.41).

If U is normally distributed according to Eq. (3.41) then

$$\hat{U} \equiv (U - \mu)/\sigma \tag{3.43}$$

is a *standardized Gaussian random variable* with PDF

$$\hat{f}(V) = \mathcal{N}(V; 0, 1) = \frac{1}{\sqrt{2\pi}} e^{-V^2/2}. \tag{3.44}$$

This PDF and the corresponding CDF

$$\hat{F}(V) = \int_{-\infty}^{V} \frac{1}{\sqrt{2\pi}} e^{-x^2/2} \, \mathrm{d}x = \tfrac{1}{2}\left[1 + \operatorname{erf}(V/\sqrt{2})\right] \tag{3.45}$$

are shown in Fig. 3.7.

EXERCISE

3.6   By considering the quantity

$$\int_{-\infty}^{\infty} \frac{\mathrm{d}}{\mathrm{d}V}\left(\frac{V^n}{\sqrt{2\pi}} e^{-V^2/2}\right) \mathrm{d}V, \tag{3.46}$$

obtain a recurrence relation for the standardized moments $\hat{\mu}_n$ of the Gaussian distribution. Show that the odd moments ($\hat{\mu}_3$, $\hat{\mu}_5$, ...) are zero, that the kurtosis is

$$\hat{\mu}_4 = 3, \tag{3.47}$$

and that the *superskewness* is

$$\hat{\mu}_6 = 15. \tag{3.48}$$

The log-normal distribution

We again take U to be normally distributed with mean μ and variance σ^2. Then the positive random variable

$$Y = e^U \tag{3.49}$$

is, by definition, log-normally distributed.

The CDF $F_Y(y)$ and PDF $f_Y(y)$ of Y can be deduced from those of U,

(a)

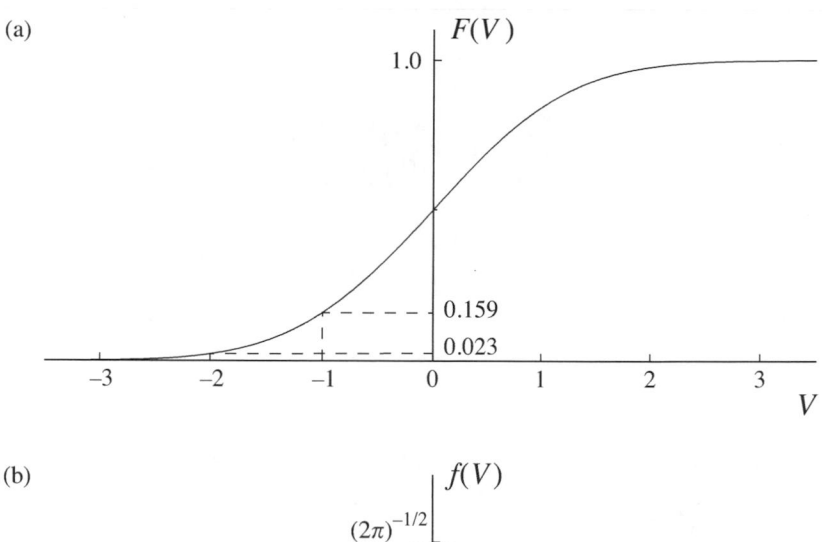

(b)

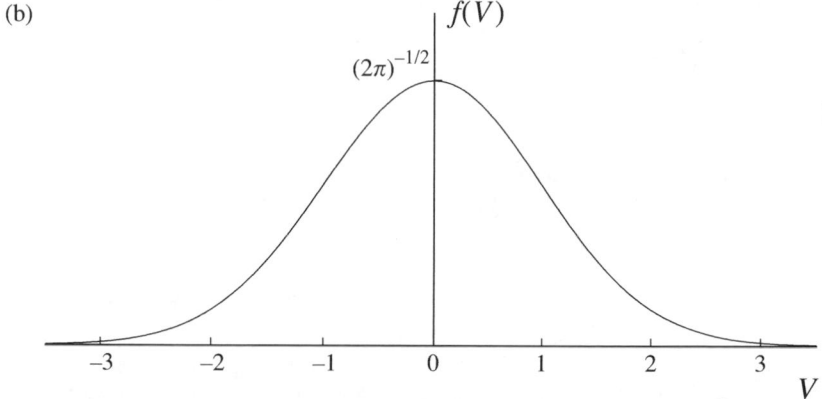

Fig. 3.7. The CDF (a) and PDF (b) of a standardized Gaussian random variable.

namely $F(V)$ and $f(V)$ given by Eq. (3.41). Since Y is positive, the sample space can be taken to be the positive real line, i.e., $y \geq 0$. Starting from the definition of the CDF, we obtain

$$F_Y(y) = P\{Y < y\} = P\{e^U < y\} = P\{U < \ln y\}$$
$$= F(\ln y). \tag{3.50}$$

The PDF is then obtained by differentiating with respect to y:

$$f_Y(y) = \frac{\mathrm{d}}{\mathrm{d}y} F_Y(y) = \frac{1}{y} f(\ln y)$$

$$= \frac{1}{y\sigma\sqrt{2\pi}} \exp\left[-\tfrac{1}{2}(\ln y - \mu)^2/\sigma^2\right]. \tag{3.51}$$

Figure 3.8 shows the PDF $f_Y(y)$ and the CDF $F_Y(y)$ for $\langle Y \rangle = 1$ and various values of the variance. It may be seen that different values of σ^2

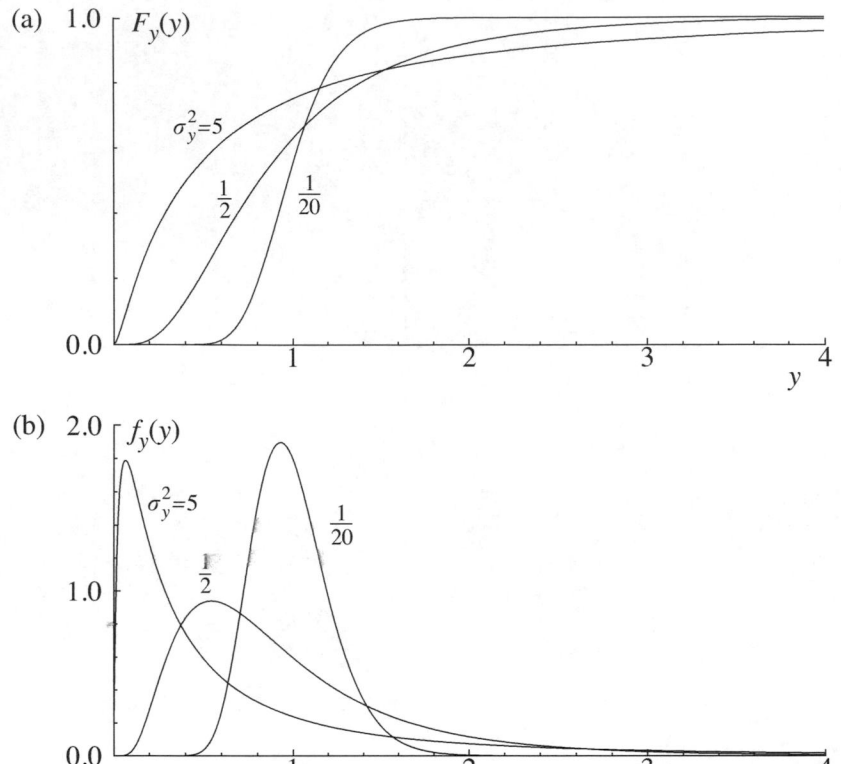

Fig. 3.8. The CDF (a) and PDF (b) of the log-normal random variable Y with $\langle Y \rangle = 1$ and $\mathrm{var}(Y) = \frac{1}{20}, \frac{1}{2}$, and 5.

produce different shapes of PDF. In particular, a large value of σ^2 leads to a PDF with a long tail, which is most clearly seen in the CDF's slow approach to unity. As shown in Exercise 3.7, the normalized variance $\mathrm{var}(Y/\langle Y \rangle)$ increases as e^{σ^2}.

Equations (3.50) and (3.51) illustrate the transformation rules for PDFs and CDFs. These are further developed in Exercise 3.9.

EXERCISES

3.7   Show that the raw moments of Y (defined by Eq. (3.49)) are

$$\langle Y^n \rangle = \exp(n\mu + \tfrac{1}{2}n^2\sigma^2). \qquad (3.52)$$

(Hint: evaluate $\int_{-\infty}^{\infty} e^{nV} f(V)\,\mathrm{d}V$.)

Show that the specification

$$\mu = -\tfrac{1}{2}\sigma^2 \qquad (3.53)$$

results in $\langle Y \rangle$ being unity, and that the variance of Y is

$$\mathrm{var}(Y) = \langle Y \rangle^2 (e^{\sigma^2} - 1). \tag{3.54}$$

3.8   The random variable Z is defined by

$$Z \equiv aY^b, \tag{3.55}$$

where Y is a log-normal random variable, and a and b are positive constants. Show that Z is also log-normal with

$$\mathrm{var}(\ln Z) = b^2 \, \mathrm{var}(\ln Y). \tag{3.56}$$

3.9   The random variable U has the CDF $F(V)$ and PDF $f(V)$. The random variable Y is defined by

$$Y = Q(U), \tag{3.57}$$

where $Q(V)$ is a monotonically increasing function. Following the steps in Eqs. (3.50) and (3.51), show that the CDF $F_Y(y)$ and PDF $f_Y(y)$ for Y are given by

$$F_Y(y) = F(V), \tag{3.58}$$

$$f_Y(y) = f(V) \Big/ \frac{\mathrm{d}Q(V)}{\mathrm{d}V}, \tag{3.59}$$

where

$$y \equiv Q(V). \tag{3.60}$$

Show that the corresponding results for $Q(V)$ being a monotonically decreasing function are

$$F_Y(y) = 1 - F(V), \tag{3.61}$$

$$f_Y(y) = -f(V) \Big/ \frac{\mathrm{d}Q(V)}{\mathrm{d}V}. \tag{3.62}$$

Show that Eqs. (3.59) and (3.62) can be written in the common form

$$f_Y(y)\,\mathrm{d}y = f(V)\,\mathrm{d}V, \tag{3.63}$$

where $\mathrm{d}V$ and

$$\mathrm{d}y \equiv \left| \frac{\mathrm{d}Q(V)}{\mathrm{d}V} \right| \mathrm{d}V \tag{3.64}$$

are corresponding infinitesimal intervals.

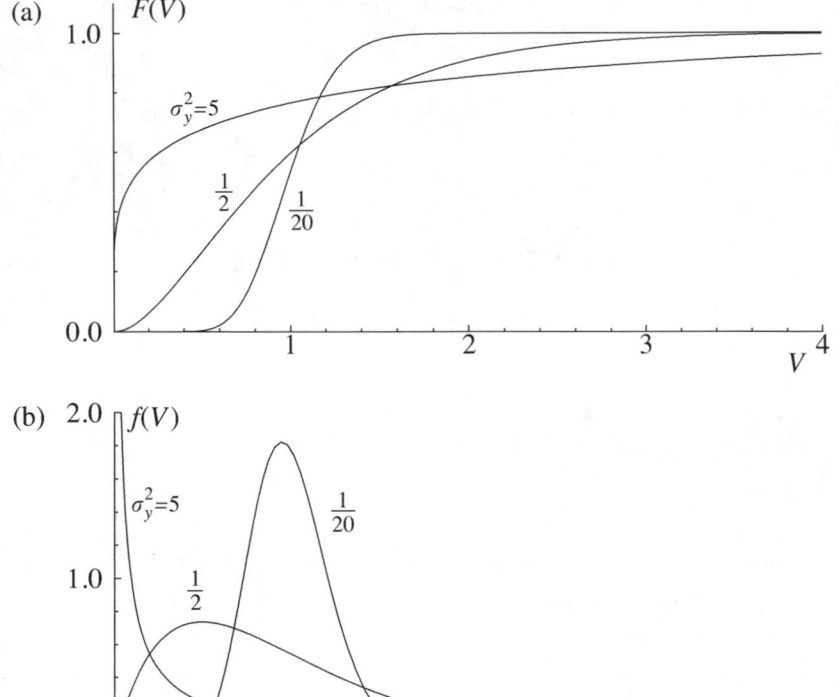

Fig. 3.9. The CDF (a) and PDF (b) for the gamma distribution with mean $\mu = 1$ and variance $\sigma^2 = \frac{1}{20}, \frac{1}{2}$, and 5.

The gamma distribution

The positive random variable U, with mean μ and variance σ^2, has a gamma distribution if its PDF is

$$f(V) = \frac{1}{\Gamma(\alpha)} \left(\frac{\alpha}{\mu}\right)^\alpha V^{\alpha-1} \exp\left(-\frac{\alpha V}{\mu}\right), \tag{3.65}$$

where α is defined by

$$\alpha \equiv \left(\frac{\mu}{\sigma}\right)^2, \tag{3.66}$$

and $\Gamma(\alpha)$ is the gamma function

$$\Gamma(\alpha) \equiv \int_0^\infty x^{\alpha-1} e^{-x}\, \mathrm{d}x. \tag{3.67}$$

For $\alpha = 1$, this becomes the exponential distribution and the value of the PDF at the origin is $f(0) = 1/\mu$. For larger values of α (smaller normalized variance) the PDF is zero at the origin, whereas for small values of α it is infinite – as is evident in Fig. 3.9.

EXERCISE _____

3.10   Use the substitution $x = \alpha V / \mu$ to show that the normalized raw moments of the gamma distribution are

$$\int_0^\infty \left(\frac{V}{\mu}\right)^n f(V)\,\mathrm{d}V = \frac{1}{\alpha^n \Gamma(\alpha)} \int_0^\infty x^{n+\alpha-1} e^{-x}\,\mathrm{d}x$$

$$= \frac{\Gamma(n+\alpha)}{\alpha^n \Gamma(\alpha)} = \frac{(n+\alpha-1)!}{\alpha^n(\alpha-1)!}, \qquad (3.68)$$

where the last expression applies for integer n and α.
Verify the consistency of this result for $n = 0, 1$, and 2.

Delta-function distributions

Suppose that U is a random variable that takes the value a with probability p, and the value b $(b > a)$ with probability $1 - p$. It is straightforward to deduce the CDF of U:

$$F(V) = P\{U < V\} = \begin{cases} 0, & \text{for } V \le a, \\ p, & \text{for } a < V \le b, \\ 1, & \text{for } V > b, \end{cases} \qquad (3.69)$$

see Fig. 3.10. This can be written in terms of Heaviside functions as

$$F(V) = pH(V - a) + (1 - p)H(V - b). \qquad (3.70)$$

The corresponding PDF (obtained by differentiating Eq. (3.70)) is

$$f(V) = p\delta(V - a) + (1 - p)\delta(V - b), \qquad (3.71)$$

see Fig. 3.10. (The properties of Dirac delta functions and Heaviside functions are reviewed in Appendix C.)

A random variable that can take only a finite number of values is a *discrete random variable* (as opposed to a continuous random variable). Although the tools presented in this section are aimed at describing continuous random variables, evidently (with the aid of Heaviside and Dirac delta functions) discrete random variables can also be treated. Furthermore, if U is a *sure variable*, with probability one of having the value a, its CDF and PDF are consistently given by

$$F(V) = H(V - a), \qquad (3.72)$$

$$f(V) = \delta(V - a). \qquad (3.73)$$

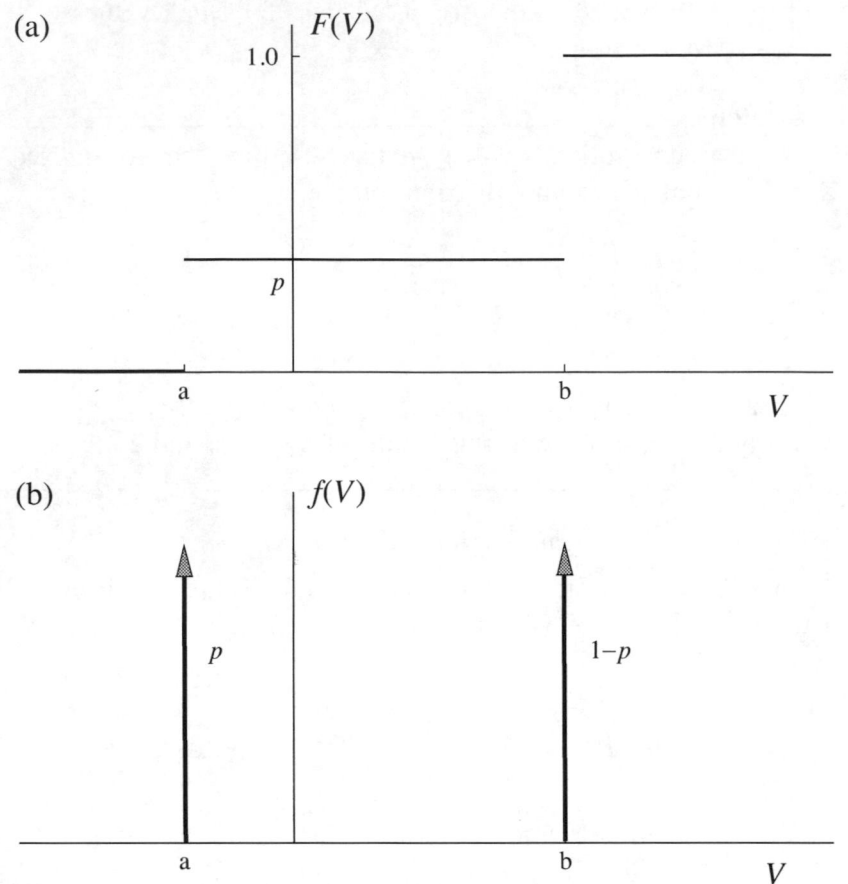

Fig. 3.10. The CDF (a) and the PDF (b) of the discrete random variable U, Eq. (3.69).

3.11   Let U be the outcome of the toss of a fair die, i.e., $U = 1, 2, 3, 4, 5$, or 6 with equal probability. Show that the CDF and PDF of U are

$$F(V) = \frac{1}{6} \sum_{n=1}^{6} H(V - n),\qquad(3.74)$$

$$f(V) = \frac{1}{6} \sum_{n=1}^{6} \delta(V - n).\qquad(3.75)$$

Sketch these distributions.

3.12   Let $f_\phi(\psi)$ be the PDF of a scalar ϕ that satisfies the boundedness condition $\phi_{\min} \le \phi \le \phi_{\max}$. For a given value of the mean $\langle \phi \rangle$, the maximum possible value of the variance $\langle \phi'^2 \rangle$ occurs when $f_\phi(\psi)$ adopts the double-delta-function distribution

$$f_\phi(\psi) = p\delta(\phi_{max} - \psi) + (1 - p)\delta(\phi_{min} - \psi). \qquad (3.76)$$

For this distribution show that

$$p = \frac{\langle \phi \rangle - \phi_{min}}{\phi_{max} - \phi_{min}}, \qquad (3.77)$$

$$\langle \phi'^2 \rangle = (\phi_{max} - \langle \phi \rangle)(\langle \phi \rangle - \phi_{min}). \qquad (3.78)$$

Note: for $\phi_{min} = 0$, $\phi_{max} = 1$, these results are $p = \langle \phi \rangle$ and $\langle \phi'^2 \rangle = \langle \phi \rangle(1 - \langle \phi \rangle)$.

The Cauchy distribution

The mean, variance, and other moments are defined as integrals of the PDF (Eq. (3.20)). We have implicitly assumed that all such integrals converge; and, indeed, with few exceptions, this is true for PDFs encountered in turbulence research. It is useful to have a simple counter-example: this is provided by the Cauchy distribution.

The PDF of the Cauchy distribution centered at c and with half-width w is

$$f(V) = \frac{w/\pi}{(V - c)^2 + w^2}. \qquad (3.79)$$

For large V, f varies as V^{-2}, and hence the integral of $Vf(V)$ diverges as $\ln V$. Hence, although the distribution is symmetric about its center $V = c$, nevertheless the mean (defined by Eq. (3.19)) does not exist. The variance is infinite.

Figure 3.11 shows the Cauchy density (Eq. (3.79)) and the corresponding CDF

$$F(V) = \frac{1}{2} + \frac{1}{\pi} \arctan\left(\frac{V - c}{w}\right), \qquad (3.80)$$

for $c = 0$, $w = 1$.

EXERCISE

3.13   The PDF sketched in Fig. 3.12 has mean zero and unit variance (i.e., it is standardized). Show that the variables defined in the sketch are given by

$$a^2 = \tfrac{6}{11}(1 + 2\sqrt{3}), \quad b = \sqrt{3}\,a, \quad h = \frac{1}{a + \tfrac{1}{2}b}. \qquad (3.81)$$

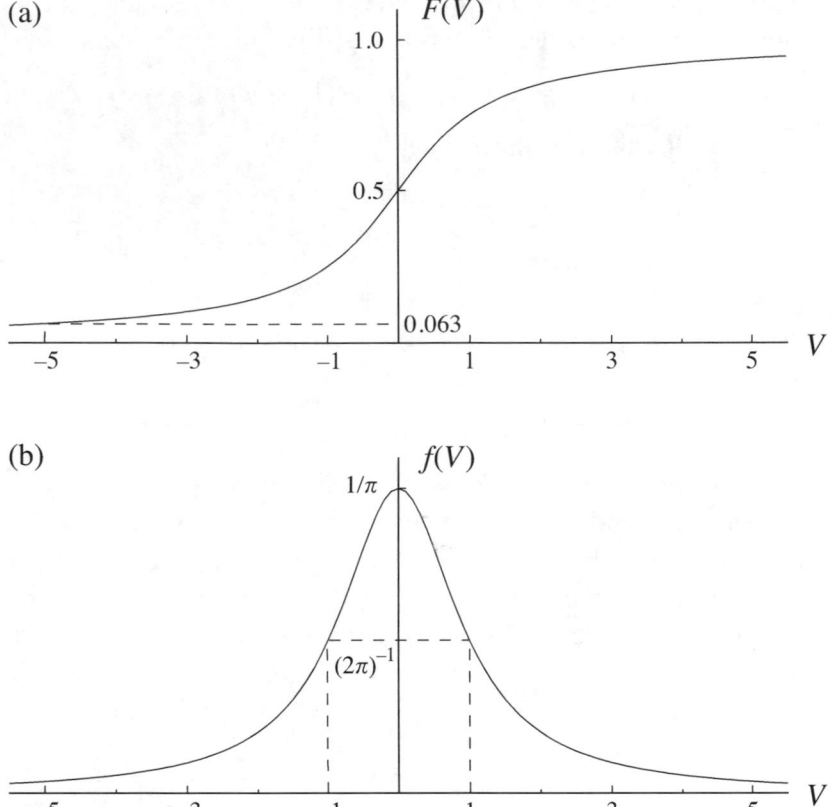

Fig. 3.11. The CDF (a) and PDF (b) for the Cauchy distribution (Eqs. (3.79) and (3.80)) with $c = 0$, $w = 1$.

3.4 Joint random variables

In this section the results obtained for the single random variable U are extended to two or more random variables. We take as an example the components of velocity (U_1, U_2, U_3) at a particular position and time in a turbulent flow.

The sample-space variables corresponding to the random variables $U = \{U_1, U_2, U_3\}$ are denoted by $V = \{V_1, V_2, V_3\}$. For the two components U_1 and U_2, Fig. 3.13 shows a *scatter plot* consisting in the $N = 100$ points $(V_1, V_2) = (U_1^{(n)}, U_2^{(n)}), n = 1, 2, \ldots, N$, where $(U_1^{(n)}, U_2^{(n)})$ are the values of (U_1, U_2) on the nth repetition of the experiment. The CDF of the joint random variables (U_1, U_2) is defined by

$$F_{12}(V_1, V_2) \equiv P\{U_1 < V_1, U_2 < V_2\}. \tag{3.82}$$

It is the probability of the sample point $(V_1, V_2) = (U_1, U_2)$ lying within the

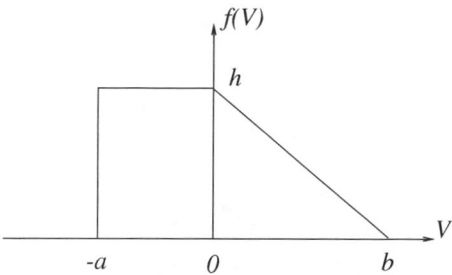

Fig. 3.12. A sketch of the standardized PDF in Exercise 3.13.

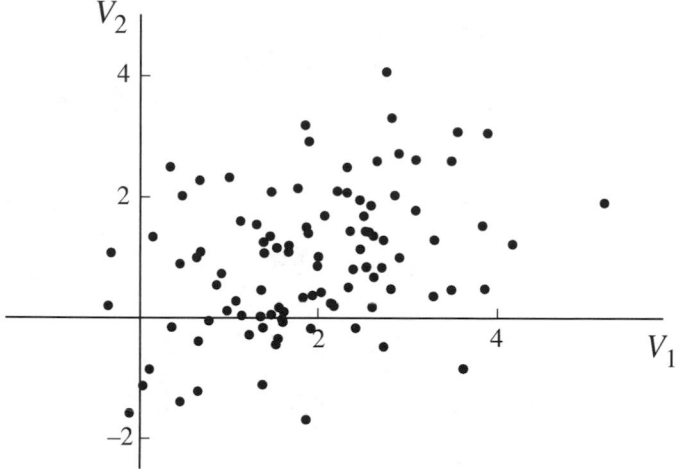

Fig. 3.13. A scatter plot in the V_1–V_2 sample space of 100 samples of the joint random variables (U_1, U_2). (In this example U_1 and U_2 are jointly normal with $\langle U_1 \rangle = 2$, $\langle U_2 \rangle = 1$, $\langle u_1^2 \rangle = 1$, $\langle u_2^2 \rangle = \frac{5}{16}$, and $\rho_{12} = 1/\sqrt{5}$.)

shaded area of Fig. 3.14. Clearly, $F_{12}(V_1, V_2)$ is a non-decreasing function of each of its arguments:

$$F_{12}(V_1 + \delta V_1, V_2 + \delta V_2) \geq F_{12}(V_1, V_2), \quad \text{for all } \delta V_1 \geq 0 \text{ and } \delta V_2 \geq 0. \quad (3.83)$$

Other properties of the CDF are

$$F_{12}(-\infty, V_2) = P\{U_1 < -\infty, \ U_2 < V_2\} = 0, \quad (3.84)$$

since $\{U_1 < -\infty\}$ is impossible; and

$$F_{12}(\infty, V_2) = P\{U_1 < \infty, \ U_2 < V_2\}$$
$$= P\{U_2 < V_2\} = F_2(V_2), \quad (3.85)$$

since $\{U_1 < \infty\}$ is certain. The CDF $F_2(V_2)$ of the single random variable

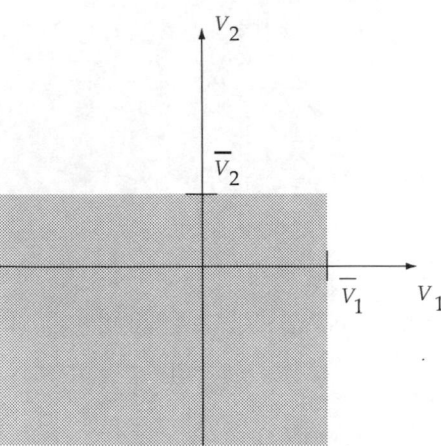

Fig. 3.14. The V_1–V_2 sample space showing the region corresponding to the event
$\{U_1 < \overline{V}_1, U_2 < \overline{V}_2\}$.

U_2 (defined in Eq. (3.85)) is called the *marginal CDF*. Similarly, the marginal
CDF of U_1 is $F_1(V_1) = F_{12}(V_1, \infty)$.

The joint PDF (JPDF) of U_1 and U_2 is defined by

$$f_{12}(V_1, V_2) \equiv \frac{\partial^2}{\partial V_1 \, \partial V_2} F_{12}(V_1, V_2). \qquad (3.86)$$

Its fundamental property, illustrated in Fig. 3.15, is

$$P\{V_{1a} \leq U_1 < V_{1b}, \; V_{2a} \leq U_2 \leq V_{2b}\} = \int_{V_{1a}}^{V_{1b}} \int_{V_{2a}}^{V_{2b}} f_{12}(V_1, V_2) \, \mathrm{d}V_2 \, \mathrm{d}V_1. \quad (3.87)$$

Other properties, that can readily be deduced, are

$$f_{12}(V_1, V_2) \geq 0, \qquad (3.88)$$

$$\int_{-\infty}^{\infty} f_{12}(V_1, V_2) \, \mathrm{d}V_1 = f_2(V_2), \qquad (3.89)$$

$$\int_{-\infty}^{\infty} \int_{-\infty}^{\infty} f_{12}(V_1, V_2) \, \mathrm{d}V_1 \, \mathrm{d}V_2 = 1, \qquad (3.90)$$

where $f_2(V_2)$ is the *marginal* PDF of U_2.

If $Q(U_1, U_2)$ is a function of the random variables, its mean is defined by

$$\langle Q(U_1, U_2) \rangle \equiv \int_{-\infty}^{\infty} \int_{-\infty}^{\infty} Q(V_1, V_2) f_{12}(V_1, V_2) \, \mathrm{d}V_1 \, \mathrm{d}V_2. \qquad (3.91)$$

The means $\langle U_1 \rangle$ and $\langle U_2 \rangle$, and the variances $\langle u_1^2 \rangle$ and $\langle u_2^2 \rangle$, can be determined
from this equation, or equally, from the marginal PDFs $f_1(V_1)$ and $f_2(V_2)$
(see Exercise 3.15). Here u_1 and u_2 are the fluctuations, e.g., $u_1 \equiv U_1 - \langle U_1 \rangle$.

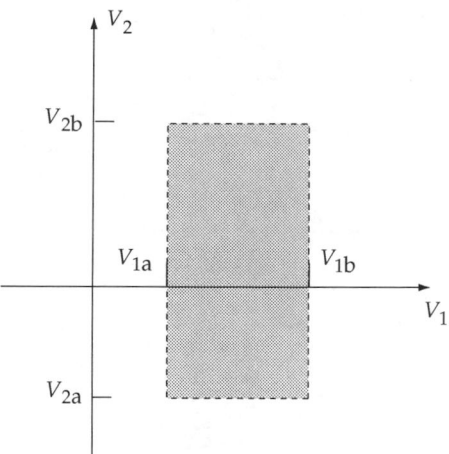

Fig. 3.15. The V_1–V_2 sample space showing the region corresponding to the event $\{V_{1a} \le U_1 < V_{1b},\ V_{2a} \le U_2 < V_{2b}\}$, see Eq. (3.87).

The *covariance* of U_1 and U_2 is the mixed second moment

$$\operatorname{cov}(U_1, U_2) = \langle u_1 u_2 \rangle = \int_{-\infty}^{\infty} \int_{-\infty}^{\infty} (V_1 - \langle U_1 \rangle)(V_2 - \langle U_2 \rangle) f_{12}(V_1, V_2)\, \mathrm{d}V_1\, \mathrm{d}V_2,$$
(3.92)

and the *correlation coefficient* is

$$\rho_{12} \equiv \langle u_1 u_2 \rangle / [\langle u_1^2 \rangle \langle u_2^2 \rangle]^{1/2}.$$
(3.93)

As illustrated by the scatter plot in Fig. 3.13, a positive correlation coefficient arises when positive excursions from the mean for one random variable (e.g., $u_1 > 0$) are preferentially associated with positive excursions for the other (i.e., $u_2 > 0$). Conversely, if positive excursions for u_1 are preferentially associated with negative excursions of u_2, as in Fig. 3.16, then the correlation coefficient is negative. In general, we have the *Cauchy–Schwarz inequality*

$$-1 \le \rho_{12} \le 1,$$
(3.94)

see Exercise 3.16.

If the correlation coefficient ρ_{12} is zero (which implies that the covariance $\langle u_1 u_2 \rangle$ is zero) then the random variables U_1 and U_2 are *uncorrelated*. In contrast, if ρ_{12} is unity, U_1 and U_2 are *perfectly correlated*; and, if ρ_{12} equals -1, they are *perfectly negatively correlated*. Examples of these correlations are given in Exercise 3.17.

For the scatter plot shown in Fig. 3.16, it is clear that the samples with $U_1 \approx V_{1a}$ and those with $U_1 \approx V_{1b}$ are likely to have significantly different values of U_2. This is confirmed in Fig. 3.17, which shows $f_{12}(V_1, V_2)$ for

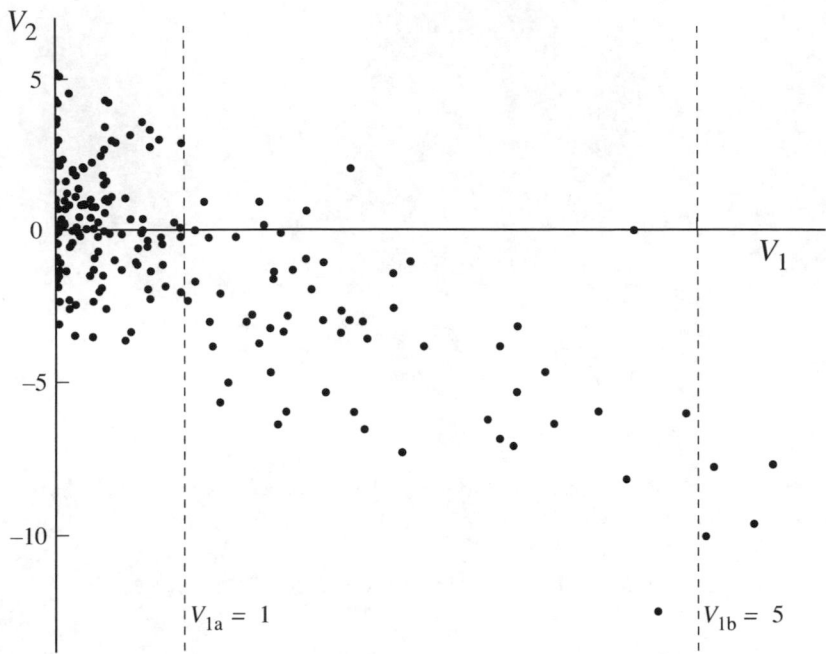

Fig. 3.16. A scatter plot of negatively correlated random variables ($\langle U_1 \rangle = 1$, $\langle U_2 \rangle = -1$, $\langle u_1^2 \rangle = 2$, $\langle u_2^2 \rangle = 12$, and $\rho_{12} = -\sqrt{2/3}$).

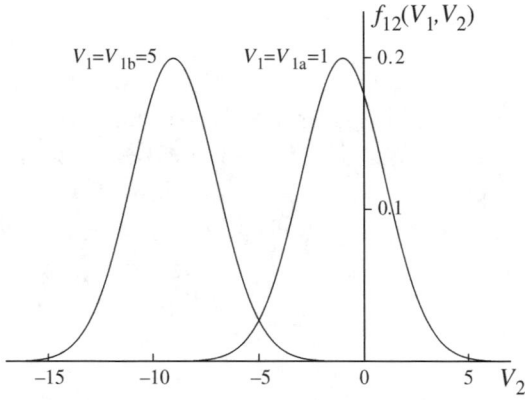

Fig. 3.17. The joint PDF of the distribution shown in Fig. 3.16, plotted against V_2 for $V_1 = V_{1a} = 1$ and $V_1 = V_{1b} = 5$.

$V_1 = V_{1a}$ and $V_1 = V_{1b}$. For fixed V_{1a}, $f_{12}(V_{1a}, V_2)$ indicates how U_2 is distributed for samples (U_1, U_2) with $U_1 = V_{1a}$. These ideas are made precise by defining *conditional* PDFs: the PDF of U_2 conditional on $U_1 = V_1$ is

$$f_{2|1}(V_2|V_1) \equiv f_{12}(V_1, V_2)/f_1(V_1). \tag{3.95}$$

This is simply the joint PDF f_{12}, scaled so that it satisfies the normalization condition

$$\int_{-\infty}^{\infty} f_{2|1}(V_2|V_1)\,dV_2 = 1. \tag{3.96}$$

For given V_1, if $f_1(V_1)$ is zero, then $f_{2|1}(V_2|V_1)$ is undefined. Otherwise it is readily verified that $f_{2|1}(V_2|V_1)$ satisfies all the conditions of a PDF (i.e., it is non-negative, and satisfies the normalization condition, Eq. (3.96)). (A word on notation: '$|V_1$' is an abbreviation for '$|U_1 = V_1$,' and is read 'conditional on $U_1 = V_1$,' or 'given $U_1 = V_1$', or 'given V_1.')

For a function $Q(U_1, U_2)$, the *conditional mean* (conditional on V_1) $\langle Q|V_1 \rangle$ is defined by

$$\langle Q(U_1, U_2)|U_1 = V_1 \rangle \equiv \int_{-\infty}^{\infty} Q(V_1, V_2)f_{2|1}(V_2|V_1)\,dV_2. \tag{3.97}$$

The concept of *independence* is of paramount importance. If U_1 and U_2 are independent, then knowledge of the value of either one of them provides no information about the other. Consequently, 'conditioning' has no effect, and the conditional and marginal PDFs are the same:

$$f_{2|1}(V_2|V_1) = f_2(V_2), \quad \text{for } U_1 \text{ and } U_2 \text{ independent.} \tag{3.98}$$

Hence (from Eq. (3.95)) the joint PDF is the product of the marginals:

$$f_{12}(V_1, V_2) = f_1(V_1)f_2(V_2), \qu\text{for } U_1 \text{ and } U_2 \text{ independent.} \tag{3.99}$$

Independent random variables are uncorrelated; but, in general, the converse is not true.

EXERCISES _____

3.14 Show that the properties of the joint PDF Eqs. (3.87)–(3.90) follow from the definitions of the CDF (Eq. (3.82)) and joint PDF (Eq. (3.86)).

3.15 Show that, for a function $R(U_1)$ of U_1 alone, the definition of the mean $\langle R(U_1) \rangle$ in terms of the joint PDF f_{12} (Eq. (3.91)) is consistent with its definition in terms of the marginal PDF f_1 (Eq. (3.20)).

3.16 By considering the quantity $(u_1/u_1' \pm u_2/u_2')^2$, establish the Cauchy–Schwarz inequality

$$-1 \le \rho_{12} \le 1, \tag{3.100}$$

where u_1' and u_2' are the standard deviations of U_1 and U_2, and ρ_{12} is the correlation coefficient.

3.17   Let U_1 and U_3 be uncorrelated random variables, and let U_2 be defined by

$$U_2 = a + bU_1 + cU_3, \qquad (3.101)$$

where a, b, and c are constants. Show that the correlation coefficient ρ_{12} is

$$\rho_{12} = \frac{b}{(b^2 + c^2 \langle u_3^2 \rangle / \langle u_1^2 \rangle)^{1/2}}. \qquad (3.102)$$

Hence show that U_1 and U_2 are

(a) uncorrelated ($\rho_{12} = 0$) if b is zero and c is non-zero,
(b) perfectly correlated ($\rho_{12} = 1$) if c is zero and b is positive, and
(c) perfectly negatively correlated ($\rho_{12} = -1$) if c is zero and b is negative.

3.18   For the sum of two random variables, obtain the result

$$\mathrm{var}(U_1 + U_2) = \mathrm{var}(U_1) + \mathrm{var}(U_2) + 2\,\mathrm{cov}(U_1, U_2). \qquad (3.103)$$

For the sum of N *independent* random variables obtain the result

$$\mathrm{var}\left(\sum_{i=1}^{N} U_i\right) = \sum_{i=1}^{N} \mathrm{var}(U_i). \qquad (3.104)$$

3.19   Let U_1 be a standardized Gaussian random variable, and let U_2 be defined by $U_2 = |U_1|$. Sketch the possible values of (U_1, U_2) in the V_1–V_2 sample space. Show that U_1 and U_2 are uncorrelated. Argue that the conditional PDF of U_2 is

$$f_{2|1}(V_2|V_1) = \delta(V_2 - |V_1|), \qquad (3.105)$$

and hence that U_2 and U_1 are not independent.

3.20   For any function $R(U_1)$, starting from Eq. (3.97), verify the result

$$\langle R(U_1)|V_1 \rangle = R(V_1). \qquad (3.106)$$

3.21   Show that the unconditional mean can be obtained from the conditional mean by

$$\langle Q(U_1, U_2) \rangle = \int_{-\infty}^{\infty} \langle Q|V_1 \rangle f_1(V_1)\,\mathrm{d}V_1. \qquad (3.107)$$

3.5 Normal and joint-normal distributions

In this section we introduce the *central-limit theorem* which (among other things) shows that the normal or Gaussian distribution (Eq. (3.41)) plays a central role in probability theory. Then the joint-normal distribution and its special properties are described. Many of the results given are most easily obtained via characteristic functions (Appendix I).

We begin by examining *ensemble averages*. Let U denote a component of velocity at a particular position and time in a repeatable turbulent-flow experiment, and let $U^{(n)}$ denote U on the nth repetition. Each repetition is performed under the same nominal conditions, and there is no dependence between different repetitions. Hence, the random variables $\{U^{(1)}, U^{(2)}, U^{(3)}, \ldots\}$ are independent and have the same distribution (i.e., that of U): they are said to be *independent and identically distributed* (i.i.d.).

The *ensemble average* (over N repetitions) is defined by

$$\langle U \rangle_N \equiv \frac{1}{N} \sum_{n=1}^{N} U^{(n)}. \tag{3.108}$$

The ensemble average is itself a random variable, and it is simple to show that its mean and variance are

$$\langle \langle U \rangle_N \rangle = \langle U \rangle, \tag{3.109}$$

$$\mathrm{var}(\langle U \rangle_N) = \frac{1}{N}\,\mathrm{var}(U) = \frac{\sigma_u^2}{N}. \tag{3.110}$$

Consequently (see Exercise 3.22) $\hat{U}$ defined by

$$\hat{U} = [\langle U \rangle_N - \langle U \rangle]N^{1/2}/\sigma_u \tag{3.111}$$

is a standardized random variable (i.e., $\langle \hat{U} \rangle = 0$, $\langle \hat{U}^2 \rangle = 1$).

The *central-limit theorem* states that, as N tends to infinity, the PDF of $\hat{U}$, $\hat{f}(V)$, tends to the standardized normal distribution

$$\hat{f}(V) = \frac{1}{\sqrt{2\pi}} \exp\left(-\tfrac{1}{2}V^2\right), \tag{3.112}$$

(see Fig. 3.7 on page 47 and Exercise I.3 on page 709). This result depends on $\{U^{(1)}, U^{(2)}, \ldots, U^{(N)}\}$ being i.i.d. but the only restriction it places on the underlying random variable, U, is that it have finite variance.

We turn now to the *joint-normal distribution*, which is important both in probability theory and in turbulent flows. For example, in experiments on homogeneous turbulence the velocity components and a conserved passive scalar $\{U_1, U_2, U_3, \phi\}$ are found to be joint-normally distributed (see Fig. 5.46

on page 175). The definition and properties of the joint-normal distribution are now given for a general set of D random variables $U = \{U_1, U_2, \ldots, U_D\}$. For $D = 2$ or 3, U can be thought of as components of velocity in a turbulent flow.

It is convenient to use matrix notation. The mean and fluctuation of the *random vector* U are denoted by

$$\mu = \langle U \rangle, \tag{3.113}$$

$$u = U - \langle U \rangle. \tag{3.114}$$

The (symmetric $D \times D$) covariance matrix is then

$$C = \langle uu^{\mathrm{T}} \rangle, \tag{3.115}$$

If $U = \{U_1, U_2, U_3\}$ is the velocity, then the covariance matrix is a second-order tensor with components $C_{ij} = \langle u_i u_j \rangle$.

If $U = \{U_1, U_2, \ldots, U_D\}$ is joint-normally distributed, then (by definition) its joint PDF is

$$f(V) = [(2\pi)^D \det(C)]^{-1/2} \exp\left[-\tfrac{1}{2}(V - \mu)^{\mathrm{T}} C^{-1}(V - \mu)\right]. \tag{3.116}$$

Note that the V-dependence of the joint PDF is contained in the quadratic form

$$g(V) \equiv (V - \mu)^{\mathrm{T}} C^{-1}(V - \mu). \tag{3.117}$$

For $D = 2$, a constant value of g – corresponding to a constant probability density – is an ellipse in the V_1–V_2 plane. For $D = 3$, a constant-probability-surface is an ellipsoid in V-space.

We now examine the pair $\{U_1, U_2\}$ of joint-normal random variables (i.e., $D = 2$) in more detail. Figure 3.18 shows a scatter plot and constant-probability-density lines for a particular choice of μ and C.

In terms of the variances $\langle u_1^2 \rangle$ and $\langle u_2^2 \rangle$, and the correlation coefficient ρ_{12}, the joint normal PDF (Eq. (3.116)) is

$$f_{12}(V_1, V_2) = \left[4\pi^2 \langle u_1^2 \rangle \langle u_2^2 \rangle (1 - \rho_{12}^2)\right]^{-1/2} \exp\left[\frac{-1}{2(1 - \rho_{12}^2)}\right.$$
$$\times \left.\left(\frac{(V_1 - \langle U_1 \rangle)^2}{\langle u_1^2 \rangle} - \frac{2\rho_{12}(V_1 - \langle U_1 \rangle)(V_2 - \langle U_2 \rangle)}{(\langle u_1^2 \rangle \langle u_2^2 \rangle)^{1/2}} + \frac{(V_2 - \langle U_2 \rangle)^2}{\langle u_2^2 \rangle}\right)\right]. \tag{3.118}$$

From this equation, the following properties can be deduced.

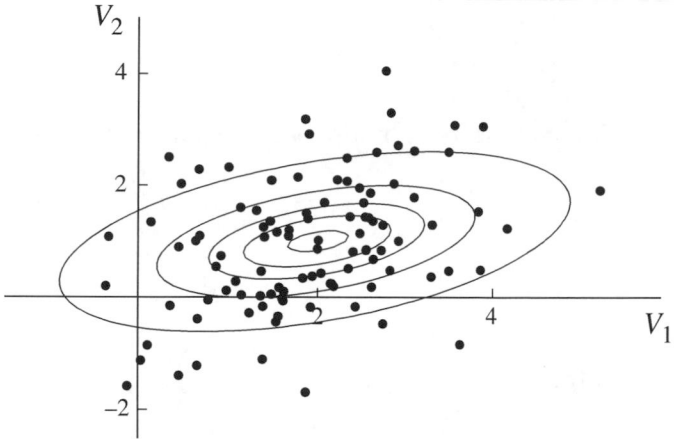

Fig. 3.18. A scatter plot and constant-probability density lines in the V_1–V_2 plane for joint-normal random variables (U_1, U_2) with $\langle U_1 \rangle = 2$, $\langle U_2 \rangle = 1$, $\langle u_1^2 \rangle = 1$, $\langle u_2^2 \rangle = \frac{5}{16}$, and $\rho_{12} = 1/\sqrt{5}$.

(i) The marginal PDFs of U_1 and U_2 ($f_1(V_1)$ and $f_2(V_2)$) are Gaussian.

(ii) If U_1 and U_2 are uncorrelated (i.e., $\rho_{12} = 0$), then they are also independent (since then $f_{12}(V_1, V_2) = f_1(V_1)f_2(V_2)$). This is a special property of the joint-normal distribution: in general, lack of correlation does not imply independence.

(iii) The conditional mean of U_1 is

$$\langle U_1 | U_2 = V_2 \rangle = \langle U_1 \rangle + \frac{\langle u_1 u_2 \rangle}{\langle u_2^2 \rangle} (V_2 - \langle U_2 \rangle). \qquad (3.119)$$

(iv) The conditional variance of U_1 is

$$\langle (U_1 - \langle U_1 | V_2 \rangle)^2 | V_2 \rangle = \langle u_1^2 \rangle (1 - \rho_{12}^2). \qquad (3.120)$$

(v) The conditional PDF $f_{1|2}(V_1 | V_2)$ is Gaussian.

Returning to the general case of $U = \{U_1, U_2, \dots, U_D\}$ being joint normal, additional insight is gained by considering linear transformations of U. An essential result (see Appendix I) is that, if U is joint normal, then a random vector $\hat{U}$ formed by a general linear transformation of U is also joint normal.

Because the covariance matrix $\mathbf{C}$ is symmetric, it can be diagonalized by a unitary transformation, defined by a unitary matrix $\mathbf{A}$. (The properties of a unitary matrix are

$$\mathbf{A}^{\mathrm{T}}\mathbf{A} = \mathbf{A}\mathbf{A}^{\mathrm{T}} = \mathbf{I}, \qquad (3.121)$$

where $\mathbf{I}$ is the $D \times D$ identity matrix.) That is, there is a unitary matrix $\mathbf{A}$

such that

$$\mathbf{A}^\mathrm{T}\mathbf{C}\mathbf{A} = \mathbf{\Lambda}, \tag{3.122}$$

where $\mathbf{\Lambda}$ is the diagonal matrix containing the eigenvalues of $\mathbf{C}$

$$\mathbf{\Lambda} = \begin{bmatrix} \lambda_1 & 0 & \cdots & 0 \\ 0 & \lambda_2 & \cdots & 0 \\ \vdots & \vdots & \ddots & \vdots \\ 0 & 0 & \cdots & \lambda_D \end{bmatrix}. \tag{3.123}$$

Consequently the transformed random vector

$$\hat{\boldsymbol{u}} \equiv \mathbf{A}^\mathrm{T}\boldsymbol{u} \tag{3.124}$$

has a diagonal covariance matrix $\mathbf{\Lambda}$:

$$\hat{\mathbf{C}} = \langle \hat{\boldsymbol{u}}\hat{\boldsymbol{u}}^\mathrm{T} \rangle = \langle \mathbf{A}^\mathrm{T}\boldsymbol{u}\boldsymbol{u}^\mathrm{T}\mathbf{A} \rangle = \mathbf{A}^\mathrm{T}\mathbf{C}\mathbf{A} = \mathbf{\Lambda}. \tag{3.125}$$

There are several observations to be made and results to be deduced from this transformation.

(i) If $\boldsymbol{U}$ is the velocity vector, then $\hat{\boldsymbol{u}}$ is the fluctuating velocity in a particular coordinate system – namely the principal axes of $\langle u_i u_j \rangle$.

(ii) The eigenvalues of $\mathbf{C}$, λ_i, are

$$\lambda_i = \langle \hat{u}_{(i)}\hat{u}_{(i)} \rangle \geq 0, \tag{3.126}$$

(where bracketed suffixes are excluded from the summation convention). Thus, since each eigenvalue is non-negative, $\mathbf{C}$ is symmetric positive semi-definite.

(iii) That the covariance matrix $\hat{\mathbf{C}}$ is diagonal indicates that the transformed random variables $\{\hat{u}_1, \hat{u}_2, \dots, \hat{u}_D\}$ are uncorrelated.

These three observations apply irrespective of whether $\boldsymbol{U}$ is joint normal. In addition we have the following

(iv) If $\boldsymbol{U}$ is joint normal, then $\{\hat{u}_1, \hat{u}_2, \dots, \hat{u}_D\}$ are *independent* Gaussian random variables.

EXERCISES

3.22   From the definition of the ensemble average (Eq. (3.108)) show that

$$\langle \langle U \rangle_N^2 \rangle = \langle U \rangle^2 + \frac{1}{N}\,\mathrm{var}(U), \tag{3.127}$$

and hence verify Eq. (3.110). Hint:

$$\langle U \rangle_N^2 = \frac{1}{N^2} \sum_{n=1}^{N} \sum_{m=1}^{N} U^{(n)} U^{(m)}. \tag{3.128}$$

3.23   Obtain an explicit expression for the kurtosis of $\langle U \rangle_N$ in terms of N and the kurtosis of U. Comment on the result in light of the central-limit theorem.

3.24   Show that, for large N, the ensemble mean (Eq. (3.108)) can be written

$$\langle U \rangle_N = \langle U \rangle + N^{-1/2} u' \xi,$$

where $u' = \mathrm{sdev}(U)$ and ξ is a standardized Gaussian random variable.

3.25   Let U be a joint-normal random vector with mean $\boldsymbol{\mu}$ and positive-definite covariance matrix $\mathbf{C} = \mathbf{A}\boldsymbol{\Lambda}\mathbf{A}^{\mathrm{T}}$, where $\mathbf{A}$ is unitary and $\boldsymbol{\Lambda}$ is diagonal. Show that the random variable

$$\hat{\boldsymbol{u}} \equiv \mathbf{C}^{-1/2}(U - \boldsymbol{\mu})$$

is a *standardized joint normal*, i.e., it has mean zero, identity covariance, and joint PDF

$$\hat{f}(\hat{V}) = \left(\frac{1}{2\pi} \right)^{D/2} \exp \left(-\tfrac{1}{2} \hat{V}^{\mathrm{T}} \hat{V} \right). \tag{3.129}$$

3.26   A Gaussian random-number generator produces a sequence of independent standardized Gaussian random numbers: $\xi^{(1)}$, $\xi^{(2)}$, $\xi^{(3)}$,.... How can these be used to generate a joint-normal random vector U with specified mean $\boldsymbol{\mu}$ and covariance matrix $\mathbf{C}$?

     (Hint: this can be achieved in a number of ways, the best of which involves the *Cholesky decomposition*, i.e., a symmetric semi-definite matrix can be decomposed as $\mathbf{C} = \mathbf{L}\mathbf{L}^{\mathrm{T}}$, where $\mathbf{L}$ is lower triangular.)

3.6 Random processes

As an example of a random variable, we considered (in Section 3.2) a component of velocity U in a repeatable turbulent-flow experiment, at a particular location and time (relative to the initiation of the experiment). The random variable U is completely characterized by its PDF, $f(V)$. Consider now the same velocity, but as a function of time, i.e., $U(t)$. Such a time-dependent random variable is called a *random process*. Figure 3.19 illustrates

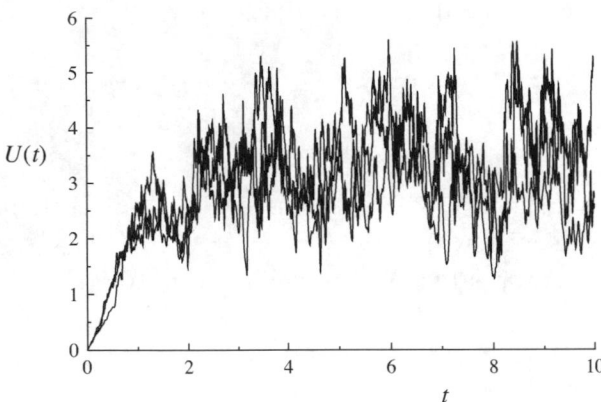

$U(t)$

Fig. 3.19. Sample paths of $U(t)$ from three repetitions of a turbulent-flow experiment.

sample paths (i.e., values of $U(t)$) obtained in different repetitions of the experiment.

How can a random process be characterized? At each point in time, the random variable $U(t)$ is characterized by its *one-time CDF*

$$F(V,t) \equiv P\{U(t) < V\},\tag{3.130}$$

or, equivalently, by the one-time PDF

$$f(V;t) \equiv \frac{\partial F(V,t)}{\partial V}.\tag{3.131}$$

However, these quantities contain no joint information about $U(t)$ at two or more times. To illustrate this limitation, Fig. 3.20 shows sample paths of five different random processes, each with the same one-time PDF. Clearly, radically different behavior (qualitatively and quantitatively) is possible, but is not represented by the one-time PDF. The N-time joint CDF of the process $U(t)$ is defined by

$$F_N(V_1,t_1;V_2,t_2;\ldots;V_N,t_N) \equiv P\{U(t_1) < V_1, U(t_2) < V_2, \ldots, U(t_N) < V_N\},$$
$$\tag{3.132}$$

where $\{t_1,t_2,\ldots,t_N\}$ are specified time points, and $f_N(V_1,t_1;V_2,t_2;\ldots;V_N,t_N)$ is the corresponding N-time joint PDF. To completely characterize the random process, it is necessary to know this joint PDF for *all* instants of time, which is, in general, an impossible task.

Considerable simplification occurs if the process is *statistically stationary*, as are many (but certainly not all) turbulent flows. A process is statistically stationary if all multi-time statistics are invariant under a shift in time, i.e.,

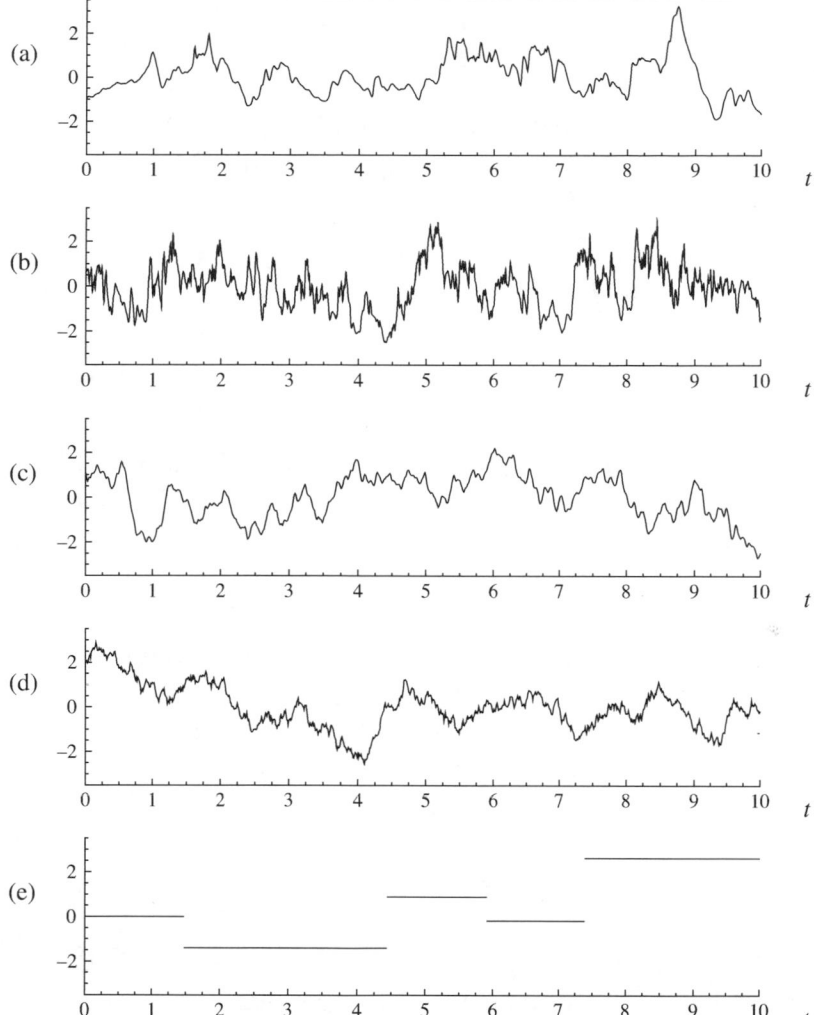

Fig. 3.20. Sample paths of five statistically stationary random processes. The one-time PDF of each is a standardized Gaussian. (a) A measured turbulent velocity. (b) A measured turbulent velocity of a higher frequency than that of (a). (c) A Gaussian process with the same spectrum as that of (a). (d) An Ornstein–Uhlenbeck process (see Chapter 12) with the same integral timescale as that of (a). (e) A jump process with the same spectrum as that of (d).

for all positive time intervals T, and all choices of $\{t_1, t_2, \ldots, t_N\}$, we have

$$f(V_1, t_1 + T; V_2, t_2 + T; \ldots; V_N, t_N + T) = f(V_1, t_1; V_2, t_2, \ldots, V_N, t_N).$$
$$(3.133)$$

After a laminar flow has been initiated, it can pass through an initial transient period and then reach a steady state, in which the flow variables

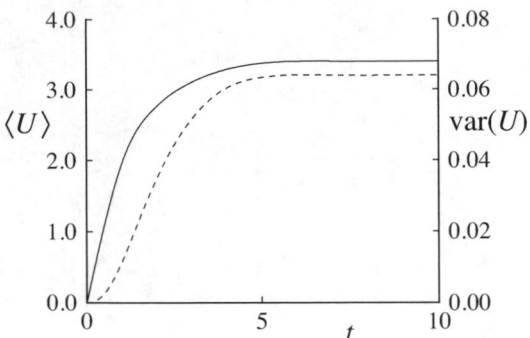

Fig. 3.21. The mean $\langle U(t)\rangle$ (solid line) and variance $\langle u(t)^2\rangle$ of the process shown in Fig. 3.19.

are independent of time. A turbulent flow, after an initial transient period, can reach a statistically stationary state in which, even though the flow variables (e.g., $U(t)$) vary with time, the statistics are independent of time. This is the case for the process shown in Fig. 3.19. The mean $\langle U(t)\rangle$ and variance $\langle u(t)^2\rangle$ of this process are shown in Fig. 3.21. Evidently, after $t \approx 5$, the statistics become independent of time, even though the process itself $U(t)$ continues to vary significantly.

For a statistically stationary process, the simplest multi-time statistic that can be considered is the *autocovariance*

$$R(s) \equiv \langle u(t)u(t+s)\rangle, \tag{3.134}$$

or, in normalized form, the *autocorrelation function*

$$\rho(s) \equiv \langle u(t)u(t+s)\rangle / \langle u(t)^2\rangle, \tag{3.135}$$

where $u(t) \equiv U(t) - \langle U\rangle$ is the fluctuation. (Note that, in view of the assumed statistical stationarity, the mean $\langle U\rangle$, the variance $\langle u^2\rangle$, $R(s)$, and $\rho(s)$ do not depend upon t.) The autocorrelation function is the correlation coefficient between the process at times t and $t + s$. Consequently it has the properties

$$\rho(0) = 1, \tag{3.136}$$

$$|\rho(s)| \le 1. \tag{3.137}$$

Further, putting $t' = t + s$, we obtain

$$\begin{aligned}\rho(s) &= \langle u(t'-s)u(t')\rangle / \langle u^2\rangle \\ &= \rho(-s), \end{aligned} \tag{3.138}$$

i.e., $\rho(s)$ is an even function.

If $U(t)$ is periodic with period T (i.e., $U(t + T) = U(t)$), then so also is

$\rho(s)$ (i.e., $\rho(s + T) = \rho(s)$). However, for processes arising in turbulent flows, we expect the correlation to diminish as the lag time s increases. Usually $\rho(s)$ decreases sufficiently rapidly that the integral

$$\bar{\tau} \equiv \int_0^\infty \rho(s)\,\mathrm{d}s \tag{3.139}$$

converges: then $\bar{\tau}$ is the *integral timescale* of the process.

Figure 3.22 shows the autocorrelation functions for the five processes given in Fig. 3.20. Notice in particular that the high-frequency process (b) has a narrower autocorrelation function (and hence a smaller $\bar{\tau}$) than does the low-frequency process (a). By construction, process (c) has the same autocorrelation as that of (a). Processes (d) and (e) both have the autocorrelation function $\rho(s) = \exp(-|s|/\bar{\tau})$, with the same integral timescale as that of process (a). Hence, apart from (b), all the processes have the same integral timescale.

The autocovariance $R(s) \equiv \langle u(t)u(t + s) \rangle = \langle u(t)^2 \rangle \rho(s)$ and (twice) the *frequency spectrum* $E(\omega)$ form a Fourier-transform pair:

$$E(\omega) \equiv \frac{1}{\pi} \int_{-\infty}^\infty R(s)e^{-i\omega s}\,\mathrm{d}s$$

$$= \frac{2}{\pi} \int_0^\infty R(s)\cos(\omega s)\,\mathrm{d}s, \tag{3.140}$$

and

$$R(s) = \frac{1}{2} \int_{-\infty}^\infty E(\omega)e^{i\omega s}\,\mathrm{d}\omega$$

$$= \int_0^\infty E(\omega)\cos(\omega s)\,\mathrm{d}\omega. \tag{3.141}$$

(The definitions and properties of Fourier transforms are given in Appendix D.) Clearly $R(s)$ and $E(\omega)$ contain the same information, just in different forms. Because $R(s)$ is real and even, so also is $E(\omega)$.

As discussed more fully in Appendix E, the velocity fluctuation $u(t)$ has a spectral representation as the weighted sum of *Fourier modes* of different frequencies ω, i.e., $e^{i\omega t} = \cos(\omega t) + i\sin(\omega t)$. The fundamental property of the frequency spectrum is that (for $\omega_a < \omega_b$) the integral

$$\int_{\omega_a}^{\omega_b} E(\omega)\,\mathrm{d}\omega \tag{3.142}$$

is the contribution to the variance $\langle u(t)^2 \rangle$ of all modes in the frequency range

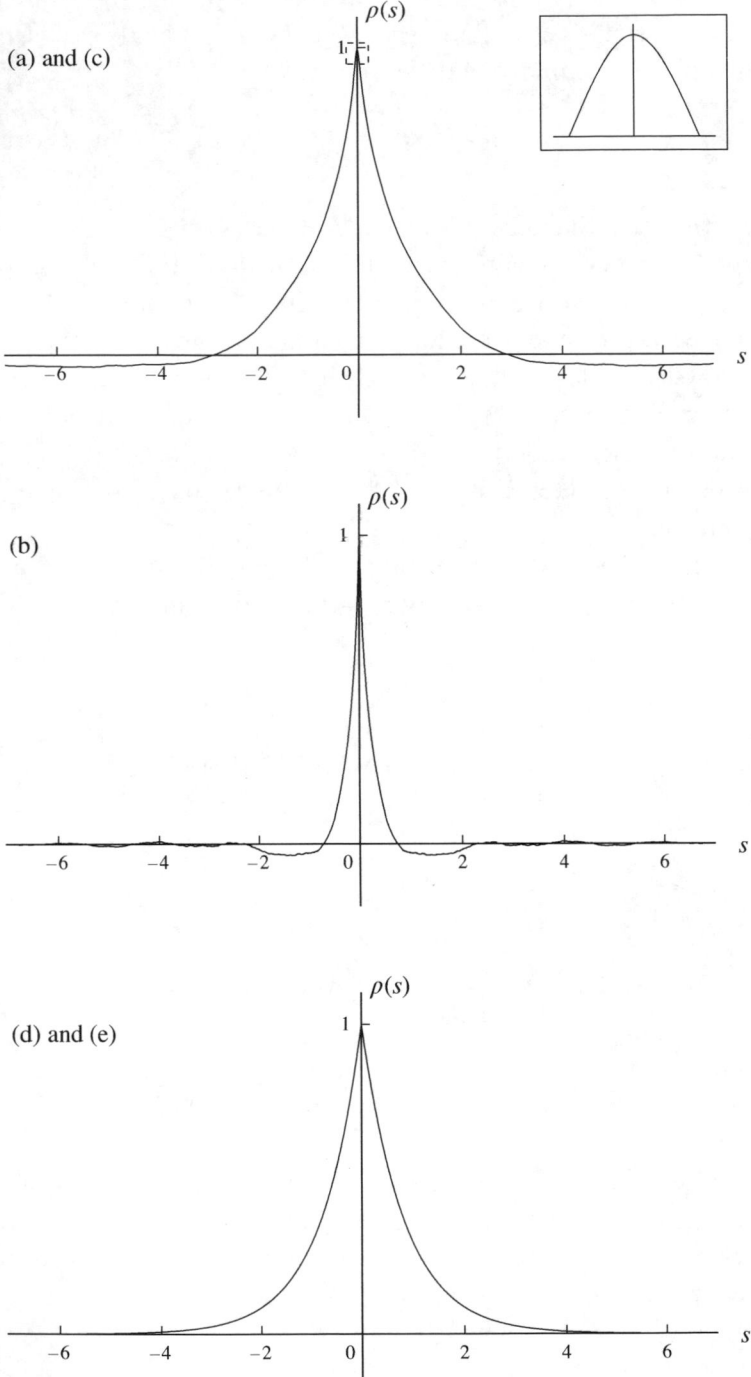

Fig. 3.22. Autocorrelation functions of the processes shown in Fig. 3.20. As the inset shows, for processes (a) and (c) the autocorrelation function is smooth at the origin.

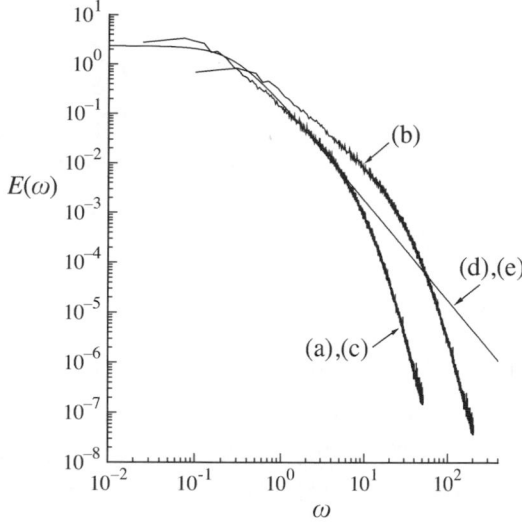

Fig. 3.23. Spectra of processes shown in Fig. 3.20.

$\omega_a \leq \omega < \omega_b$. In particular the variance is

$$R(0) = \langle u(t)^2 \rangle = \int_0^\infty E(\omega)\,d\omega, \qquad (3.143)$$

as is evident from Eq. (3.141) with $s = 0$.

A further simple connection between the spectrum and the autocorrelation is that the integral timescale is given by

$$\bar{\tau} = \frac{\pi E(0)}{2\langle u^2 \rangle}, \qquad (3.144)$$

as is readily verified by setting $\omega = 0$ in Eq. (3.140). A more complete explanation of the spectral representation and interpretation of the frequency spectrum is given in Appendix E.

Figure 3.23 shows the spectra of the stationary random processes given in Fig. 3.20. The high-frequency process (b), having a smaller integral timescale than that of process (a), has a correspondingly smaller value of the spectrum at the origin (Eq. 3.144) – but its spectrum extends to higher frequencies.

In practice, the autocorrelation function or the spectrum is usually the only quantity used to characterize the multi-time properties of a random process. However, it should be appreciated that the one-time PDF and the autocorrelation function provide only a partial characterization of the process. This point is amply demonstrated by processes (d) and (e) in Fig. 3.20. The two processes are qualitatively quite different and yet they have the same one-time PDF (Gaussian) and the same autocorrelation function ($\rho(s) = e^{-|s|/\bar{\tau}}$). To repeat, in general, *the one-time PDF and the autocorrelation function do not completely characterize a random process.*

A *Gaussian process* is an important but very special case. If a process is Gaussian then, by definition, the general N-time PDF (Eq. (3.133)) is joint normal. Now the joint-normal distribution is fully characterized by its means $\langle U(t_n) \rangle$, and its covariances $\langle u(t_n)u(t_m) \rangle$. For a statistically stationary process, we have

$$\langle u(t_n)u(t_m) \rangle = R(t_n - t_m) = \langle u(t)^2 \rangle \, \rho(t_n - t_m). \tag{3.145}$$

Hence a statistically stationary Gaussian process is completely characterized by its mean $\langle U(t) \rangle$, its variance $\langle u(t)^2 \rangle$, and the autocorrelation function $\rho(s)$ (or equivalently the spectrum $E(\omega)$).

In Fig. 3.20, process (c) is defined to be the Gaussian process with the same spectrum as that of the turbulent velocity, process (a). Some differences between processes (a) and (c) may be discernible; and these differences can be clearly revealed by, for example, examining the sample paths of $\ddot{U}(t) \equiv \mathrm{d}^2 U(t)/\mathrm{d}t^2$, see Fig. 3.24. For the Gaussian process (c) it follows that $\ddot{U}(t)$ is also Gaussian and so the kurtosis of $\ddot{U}(t)$ is 3. However, for the turbulent velocity, process (a), $\ddot{U}(t)$ is far from Gaussian, and has a kurtosis of 11.

Random processes arising from turbulence (e.g., process (a)) are differentiable, i.e., for each sample path the following limit exists:

$$\frac{\mathrm{d}U(t)}{\mathrm{d}t} = \lim_{\Delta t \downarrow 0} \left(\frac{U(t + \Delta t) - U(t)}{\Delta t} \right). \tag{3.146}$$

In this case, taking the mean and taking the limit commute, so that

$$\left\langle \frac{\mathrm{d}U(t)}{\mathrm{d}t} \right\rangle = \left\langle \lim_{\Delta t \downarrow 0} \left(\frac{U(t + \Delta t) - U(t)}{\Delta t} \right) \right\rangle$$

$$= \lim_{\Delta t \downarrow 0} \left(\frac{\langle U(t + \Delta t) \rangle - \langle U(t) \rangle}{\Delta t} \right)$$

$$= \frac{\mathrm{d}\langle U(t) \rangle}{\mathrm{d}t}. \tag{3.147}$$

Synthetic processes (such as processes (d) and (e)) need not be differentiable (i.e., the limit Eq. (3.146) does not exist). It may be observed that the spectra of these processes decay as $E(\omega) \sim \omega^{-2}$ at high frequencies, and that (correspondingly) their autocorrelation function $\rho(s) = e^{-|s|/\bar{\tau}}$ is not differentiable at the origin.

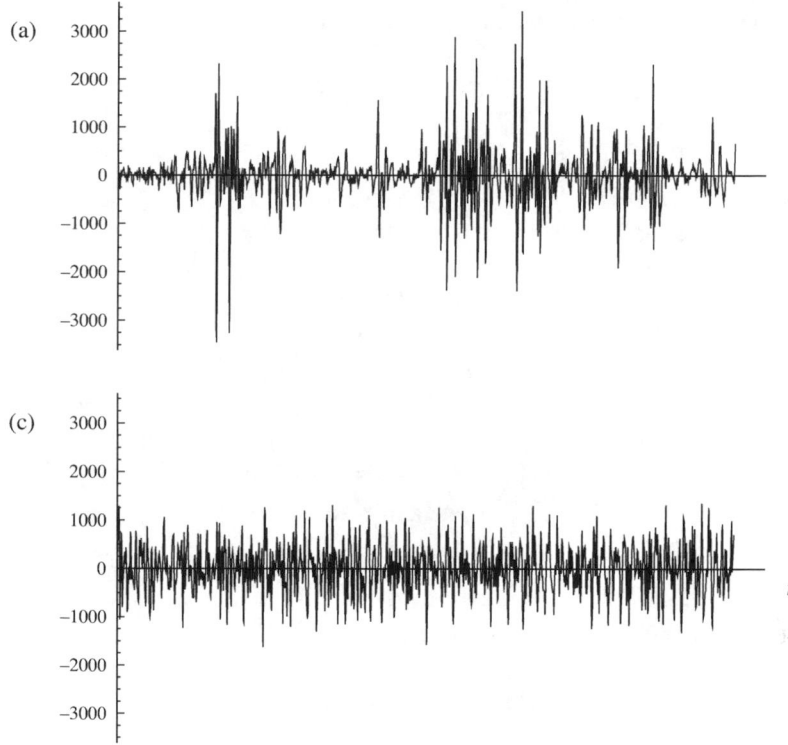

Fig. 3.24. Sample paths of $\ddot{U}(t)$ for processes (a) and (c) shown in Fig. 3.20.

Process (d) is an *Ornstein–Uhlenbeck process*, which is the canonical example of a *diffusion process*. Such processes are used in PDF methods and are described in Chapter 12 and Appendix J.

EXERCISES

In the following exercises, $u(t)$ is a zero-mean, statistically stationary, differentiable random process with autocovariance $R(s)$, autocorrelation function $\rho(s)$, and spectrum $E(\omega)$.

3.27   Show that $u(t)$ and $\dot{u}(t)$ are uncorrelated, and that $u(t)$ and $\ddot{u}(t)$ are negatively correlated.

3.28   Show that

$$\left\langle u^2 \frac{d^3 u}{dt^3} \right\rangle = -2\langle u\dot{u}\ddot{u}\rangle = 2\langle (\dot{u})^3 \rangle + 2\langle u\dot{u}\ddot{u}\rangle = \langle (\dot{u})^3 \rangle. \tag{3.148}$$

3.29   Show that at the origin $(s = 0)$ $dR(s)/ds$ is zero, and $d^2 R(s)/ds^2$ is negative.

3.30   Show that the autocovariance $B(s)$ of the process $\dot{u}(t)$ is

$$B(s) = -\frac{d^2 R(s)}{ds^2}.$$ (3.149)

3.31   Show that the integral timescale of $\dot{u}(t)$ is zero.

3.32   Show that the spectrum of $\dot{u}(t)$ is $\omega^2 E(\omega)$.

3.33   If $u(t)$ is a Gaussian process, show that

$$\langle \dot{u}(t)|u(t) = v \rangle = 0,$$ (3.150)

$$\langle \ddot{u}(t)|u(t) = v \rangle = -v\langle \dot{u}(t)^2 \rangle / \langle u^2 \rangle.$$ (3.151)

3.7 Random fields

In a turbulent flow, the velocity $U(x, t)$ is a time-dependent random vector field. It can be described – i.e., partially characterized – by extensions of the tools presented in the previous sections.

One-point statistics

The one-point, one-time joint CDF of velocity is

$$F(V, x, t) = P\{U_i(x, t) < V_i, \ i = 1, 2, 3\},$$ (3.152)

and then the joint PDF is

$$f(V; x, t) = \frac{\partial^3 F(V, x, t)}{\partial V_1 \partial V_2 \partial V_3}.$$ (3.153)

At each point and time this PDF fully characterizes the random velocity vector, but it contains no joint information at two or more times or positions. In terms of this PDF, the mean velocity field is

$$\langle U(x, t) \rangle = \int\!\!\!\int\!\!\!\int_{-\infty}^{\infty} V f(V; x, t) \, dV_1 \, dV_2 \, dV_3$$ (3.154)

$$= \int V f(V; x, t) \, dV.$$ (3.155)

The second line of this equation introduces an abbreviated notation: $\int (\ ) \, dV$ is written for

$$\int\!\!\!\int\!\!\!\int_{-\infty}^{\infty} (\ ) \, dV_1 \, dV_2 \, dV_3.$$

The fluctuating velocity field is defined by

$$u(x,t) \equiv U(x,t) - \langle U(x,t) \rangle. \tag{3.156}$$

The (one-point, one-time) covariance of the velocity is $\langle u_i(x,t)u_j(x,t) \rangle$. For reasons given in the next chapter, these covariances are called *Reynolds stresses*, and are written $\langle u_i u_j \rangle$, with the dependences on x and t being understood.

A word on notation: the semi-colon in $f(V;x,t)$ indicates that f is a *density* with respect to the sample-space variables that appear to the left of the semi-colon (i.e., V_1, V_2, and V_3), whereas f is a *function* with respect to the remaining variables (i.e., x_1, x_2, x_3, and t). This distinction is useful because densities and functions have different transformation properties (see Exercise 3.9 on page 49).

Turbulent velocity fields are differentiable, and (as discussed in Section 3.6) differentiation and taking the mean commute:

$$\left\langle \frac{\partial U_i}{\partial t} \right\rangle = \frac{\partial \langle U_i \rangle}{\partial t}, \tag{3.157}$$

$$\left\langle \frac{\partial U_i}{\partial x_j} \right\rangle = \frac{\partial \langle U_i \rangle}{\partial x_j}. \tag{3.158}$$

N-point statistics

The N-point, N-time joint PDF can be defined as a simple extension of Eq. (3.132). Let $\{(x^{(n)}, t^{(n)}), n = 1, 2, \ldots, N\}$ be a specified set of positions and times. Then we define

$$f_N(V^{(1)}, x^{(1)}, t^{(1)}, V^{(2)}, x^{(2)}, t^{(2)}; \ldots ; V^{(N)}, x^{(N)}, t^{(N)}) \tag{3.159}$$

to be the joint PDF of $U(x,t)$ at these N space–time points. To determine this N-point PDF for *all* space–time points is obviously impossible, and hence in practice a random velocity field cannot be fully characterized.

Turbulent velocity fields are found *not* to be Gaussian: a Gaussian field is fully characterized by the mean $\langle U(x,t) \rangle$ and the autocovariance $\langle u_i(x^{(1)}, t^{(1)})u_j(x^{(2)}, t^{(2)}) \rangle$.

Statistical stationarity and homogeneity

The random field $U(x,t)$ is *statistically stationary* if all statistics are invariant under a shift in time. In terms of the N-point PDF, this means that f_N is unchanged if $(x^{(n)}, t^{(n)})$ is replaced by $(x^{(n)}, t^{(n)} + T)$ for all N points, where T is the time shift.

Similarly, the field is *statistically homogeneous* if all statistics are invariants

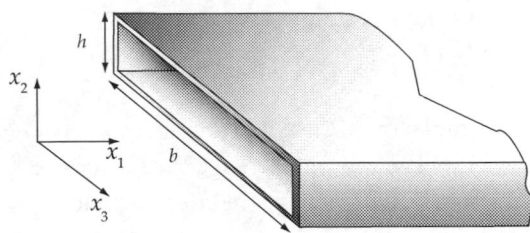

Fig. 3.25. A sketch of a turbulent-channel-flow apparatus.

under a shift in position. Then f_N is unchanged if $(x^{(n)}, t^{(n)})$ is replaced by $(x^{(n)} + X, t^{(n)})$, for all N points, where X is the shift in position. If the velocity field $U(x, t)$ is statistically homogeneous, it follows that the mean velocity $\langle U \rangle$ is uniform; and, with an appropriate choice of frame, $\langle U \rangle$ can be taken to be zero. The definition of *homogeneous turbulence* is less restrictive: specifically, in homogeneous turbulence the *fluctuating* velocity field $u(x, t)$ is statistically homogeneous. It is consistent with this definition for the mean velocity gradients $\partial \langle U_i \rangle / \partial x_j$ to be non-zero, but uniform (see Section 5.4.5). A good approximation to homogeneous turbulence can be achieved in wind-tunnel experiments; and homogeneous turbulence is the simplest class of flows to study using direct numerical simulation.

In a similar way, turbulent flows can be statistically two-dimensional or one-dimensional. For example, Fig. 3.25 is a sketch of a channel flow apparatus. For a large aspect ratio ($b/h \gg 1$), and remote from the end walls ($|x_3|/b \ll 1$), the statistics of the flow vary little in the spanwise (x_3) direction. To within an approximation, then, the velocity field $U(x, t)$ is statistically two-dimensional – statistics being independent of x_3. Sufficiently far down the channel ($x_1/h \gg 1$) the flow becomes (statistically) fully developed. Then the velocity field is statistically one-dimensional, with statistics being independent both of x_1 and of x_3. Similarly, the turbulent flow in a pipe is *statistically axisymmetric* in that (in polar-cylindrical coordinates) all statistics are independent of the circumferential coordinate.

It should be emphasized that, even if a flow is statistically homogeneous or one-dimensional, nevertheless all three components of $U(x, t)$ vary in all three coordinate directions and time. It is only the statistics that are independent of some coordinate directions.

Isotropic turbulence

A statistically homogeneous field $U(x, t)$ is, by definition, statistically invariant under translations (i.e., shifts in the origin of the coordinate system). If the field is also statistically invariant under rotations and reflections of the coordinate system, then it is (statistically) *isotropic*. The concept of isotropy is extremely important in turbulence: hundreds of wind-tunnel experiments have been performed on (approximately) isotropic turbulence, and much of turbulence theory centers on it. In terms of the N-point PDF (Eq. (3.159)), in isotropic turbulence f_N is unchanged if $U(x^{(n)}, t^{(n)})$ is replaced by $\overline{U}(\overline{x}^{(n)}, t^{(n)})$, where $\overline{x}$ and $\overline{U}$ denote the position and velocity in any coordinate system obtained by rotation and reflections of the coordinate axes.

Two-point correlation

The simplest statistic containing some information on the spatial structure of the random field is the two-point, one-time autocovariance

$$R_{ij}(r, x, t) \equiv \langle u_i(x, t) u_j(x + r, t) \rangle, \tag{3.160}$$

which is often referred to as the *two-point correlation*. From this it is possible to define various integral lengthscales, for example

$$L_{11}(x, t) \equiv \frac{1}{R_{11}(0, x, t)} \int_0^\infty R_{11}(e_1 r, x, t) \, dr, \tag{3.161}$$

where e_1 is the unit vector in the x_1-coordinate direction.

Wavenumber spectra

For homogeneous turbulence the two-point correlation $R_{ij}(r, t)$ is independent of x, and the information it contains can be re-expressed in terms of the *wavenumber spectrum*. The spatial Fourier mode

$$e^{i\kappa \cdot x} = \cos(\kappa \cdot x) + i \sin(\kappa \cdot x), \tag{3.162}$$

is a function that varies sinusoidally (with wavelength $\ell = 2\pi/|\kappa|$) in the direction of the wavenumber vector κ, and that is constant in planes normal to κ. The velocity spectrum tensor $\Phi_{ij}(\kappa, t)$ is the Fourier transform of the two-point correlation

$$\Phi_{ij}(\kappa, t) = \frac{1}{(2\pi)^3} \int\!\!\!\int\!\!\!\int_{-\infty}^{\infty} e^{-i\kappa \cdot r} R_{ij}(r, t) \, dr, \tag{3.163}$$

and the inverse transform is

$$R_{ij}(r, t) = \int\!\!\!\int\!\!\!\int_{-\infty}^{\infty} e^{i\kappa \cdot r} \Phi_{ij}(\kappa, t) \, d\kappa, \tag{3.164}$$

where $d\mathbf{r}$ and $d\boldsymbol{\kappa}$ are written for $dr_1\, dr_2\, dr_3$ and $d\kappa_1\, d\kappa_2\, d\kappa_3$, respectively. Setting $\mathbf{r} = 0$ in this equation yields

$$R_{ij}(0,t) = \langle u_i u_j \rangle = \int\!\!\!\int\!\!\!\int_{-\infty}^{\infty} \Phi_{ij}(\boldsymbol{\kappa},t)\, d\boldsymbol{\kappa}. \qquad (3.165)$$

and so $\Phi_{ij}(\boldsymbol{\kappa},t)$ represents the contribution to the covariance $\langle u_i u_j \rangle$ of velocity modes with wavenumber $\boldsymbol{\kappa}$.

The two-point correlation and the spectrum contain two different kinds of directional information. The dependences of $R_{ij}(\mathbf{r},t)$ on $\mathbf{r}$, and of $\Phi_{ij}(\boldsymbol{\kappa},t)$ on $\boldsymbol{\kappa}$, give information about the directional dependence of correlation; while the components of R_{ij} and Φ_{ij} give information about the directions of the velocities.

A useful quantity, especially for qualitative discussions, is the *energy spectrum function*:

$$E(\kappa,t) \equiv \int\!\!\!\int\!\!\!\int_{-\infty}^{\infty} \tfrac{1}{2}\Phi_{ii}(\boldsymbol{\kappa},t)\delta(|\boldsymbol{\kappa}| - \kappa)\, d\boldsymbol{\kappa}, \qquad (3.166)$$

which may be viewed as $\Phi_{ij}(\boldsymbol{\kappa},t)$ stripped of all directional information. Integration of Eq. (3.166) over all scalar wavenumbers, κ, yields

$$\int_0^{\infty} E(\kappa,t)\, d\kappa = \tfrac{1}{2}R_{ii}(0,t) = \tfrac{1}{2}\langle u_i u_i \rangle. \qquad (3.167)$$

Thus, $E(\kappa,t)\, d\kappa$ represents the contribution to the turbulent kinetic energy $\tfrac{1}{2}\langle u_i u_i \rangle$ from all modes with $|\boldsymbol{\kappa}|$ in the range $\kappa \le |\boldsymbol{\kappa}| < \kappa + d\kappa$. Velocity spectra in turbulence are examined in some detail in Section 6.5.

EXERCISES

3.34   From the substitution $\mathbf{x}' = \mathbf{x} + \mathbf{r}$ and the definition of the two-point correlation (Eq. (3.160)), show that

$$R_{ij}(\mathbf{r},\mathbf{x},t) = R_{ji}(-\mathbf{r},\mathbf{x}',t), \qquad (3.168)$$

and hence, for a statistically homogeneous field,

$$R_{ij}(\mathbf{r},t) = R_{ji}(-\mathbf{r},t). \qquad (3.169)$$

3.35   If $\mathbf{u}(\mathbf{x},t)$ is divergence-free (i.e., $\nabla \cdot \mathbf{u} = 0$), show that the two-point correlation (Eq. (3.160)) satisfies

$$\frac{\partial}{\partial r_j} R_{ij}(\mathbf{x}, \mathbf{r}, t) = 0. \tag{3.170}$$

Show that, if, in addition, $\mathbf{u}(\mathbf{x}, t)$ is statistically homogeneous, then

$$\frac{\partial}{\partial r_j} R_{ij}(\mathbf{r}, t) = \frac{\partial}{\partial r_i} R_{ij}(\mathbf{r}, t) = 0. \tag{3.171}$$

3.8 Probability and averaging

Having developed the tools to describe random variables, random processes, and random fields, we now return to the starting point in order to clarify the notion of *probability*, on which everything has been built. Physical quantities such as density and velocity are defined operationally (e.g., in Section 2.1), so that (at least in principle) their values can be determined by measurement. Operational definitions of probability – for example, in terms of time averages or ensemble averages – although they are often used, are unsatisfactory. Instead, in modern treatments, probability theory is *axiomatic*. The purpose of this section is to describe this axiomatic approach, and to explain the connection to measurable quantities (such as time averages). For the sake of simplicity, we start the discussion in the context of a coin-tossing experiment.

Consider a coin that can be tossed any number of times, with the two possible outcomes 'heads' and 'tails.' We *define* the variable p to be the probability of 'heads.' (It is assumed that each toss is statistically independent and indistinguishable from every other toss.)

Suppose that an experiment in which the coin is tossed $N = 1,000,000$ times is patiently performed. The fraction of tosses resulting in heads is a random variable denoted by p_N. In this particular experiment, suppose that the measured value of p_N is 0.5024.

The coin-tossing experiment is an example of Bernoulli trials, for which there is a complete theory. For example, suppose that we hypothesize that the coin is 'fair,' i.e., $p = \frac{1}{2}$. Then a simple statistical calculation shows that (for $N = 1,000,000$) with 99% probability p_N lies in the range

$$0.4987 < p_N < 0.5013.$$

Since the measured value $p_N = 0.5024$ lies outside this range, we can have high confidence that the hypothesis $p = \frac{1}{2}$ is false. Instead, a further statistical calculation, based on the observed value of p_N, shows that, with 99% confidence, p lies in the range

$$0.5011 < p < 0.5037.$$

To summarize:

(i) p is *defined* to be the probability of 'heads,'
(ii) p_N is the *measured* frequency of 'heads,'
(iii) given a hypothesis about p, a *range* for p_N can be predicted, and
(iv) given the measured value of p_N, a *confidence interval* for p can be determined.

The two most important points to appreciate are that p cannot be measured – it can only be estimated with some confidence level; and that, although p_N tends to p as N tends to infinity, this is not taken as the definition of p.

In considering the velocity $U(t)$ as a turbulent flow, we *define* $f(V; t)$ to be its PDF, and then define the mean by

$$\langle U(t) \rangle \equiv \int_{-\infty}^{\infty} V f(V; t) \, dV. \tag{3.172}$$

In turbulent-flow experiments and simulations, several kinds of *averaging* are used to define other means that can be related to $\langle U(t) \rangle$. For statistically stationary flows the time average (over a time interval T) is defined by

$$\langle U(t) \rangle_T \equiv \frac{1}{T} \int_{t}^{t+T} U(t') \, dt'. \tag{3.173}$$

For flows that can be repeated or replicated N times, the ensemble average is defined by

$$\langle U(t) \rangle_N \equiv \frac{1}{N} \sum_{n=1}^{N} U^{(n)}(t), \tag{3.174}$$

where $U^{(n)}(t)$ is the measurement on the nth realization. In simulations of homogeneous turbulence in a cubic domain of side $\mathcal{L}$, the spatial average of $U(\boldsymbol{x}, t)$ is defined by

$$\langle U(t) \rangle_{\mathcal{L}} \equiv \frac{1}{\mathcal{L}^3} \int_0^{\mathcal{L}} \int_0^{\mathcal{L}} \int_0^{\mathcal{L}} U(\boldsymbol{x}, t) \, dx_1 \, dx_2 \, dx_3. \tag{3.175}$$

Similar spatial averages can be defined for statistically one- and two-dimensional flows.

These averages $\langle U \rangle_T$, $\langle U \rangle_N$, and $\langle U \rangle_{\mathcal{L}}$ are (like p_N) random variables. They can be used to *estimate* $\langle U \rangle$, but not to measure it with certainty. Most importantly, $\langle U \rangle$ is well defined for all flows, even those that are not stationary

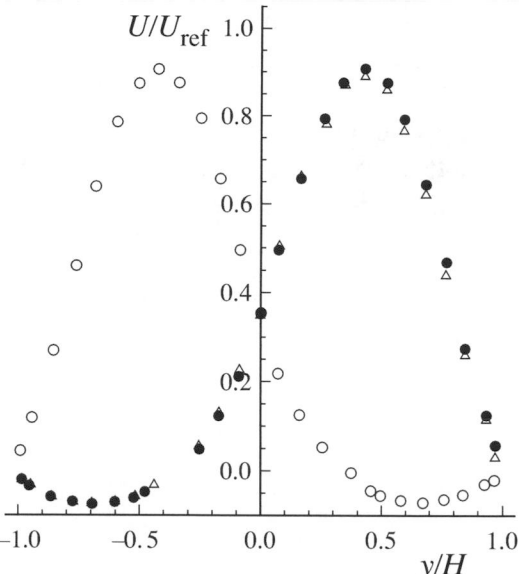

Fig. 3.26. Velocity profiles measured by Durst *et al.* (1974) in the steady laminar flow downstream of a symmetric expansion in a rectangular duct. The geometry and boundary conditions are symmetric about the plane $y = 0$. Symbols: $\bigcirc$, stable state 1; $\triangle$, stable state 2; $\bullet$, reflection of profile 1 about the y axis.

or homogeneous, or that cannot be repeated or replicated. For statistically stationary flows (barring exceptional circumstances) $\langle U \rangle_T$ tends to $\langle U \rangle$ as T tends to infinity, but this is not taken as the definition of the mean.

EXERCISES

3.36   In a turbulent-flow experiment the ensemble mean $\langle U \rangle_N$ obtained from $N = 1,000$ measurements is 11.24 m s^{-1}, and the standard deviation of U is estimated to be 2.5 m s^{-1}. Determine the 95% confidence interval for $\langle U \rangle$.

3.37   For a statistically stationary flow show that

$$\mathrm{var}(\langle U(t) \rangle_T) = \frac{\mathrm{var}(U)}{T^2} \int_0^T \int_0^T \rho(t - s)\, \mathrm{d}s\, \mathrm{d}t,$$

where $\rho(s)$ is the autocorrelation function of $U(t)$. Assuming that the integral timescale $\bar{\tau}$ exists and is positive (Eq. (3.139)), obtain the long-time result

$$\mathrm{var}(\langle U(t) \rangle_T) \sim \frac{2\bar{\tau}}{T} \mathrm{var}(U). \qquad (3.176)$$

3.38   Figure 3.26 shows velocity profiles measured in the steady laminar flow downstream of a symmetric expansion in a rectangular duct. Although the geometry and boundary conditions are symmetric about the plane $y = 0$, the flow is not symmetric. Each time the flow is started from rest, after an initial transient, the flow reaches one of two stable steady states. For this flow, discuss the relationship among the expectation $\langle U \rangle$, the time average $\langle U \rangle_T$, and the ensemble average $\langle U \rangle_N$.

4

Mean-flow equations

4.1 Reynolds equations

In the previous chapter, various statistical quantities were introduced to describe turbulent velocity fields – means, PDFs, two-point correlations, etc. It is possible to derive equations for the evolution of all of these quantities, starting from the Navier–Stokes equations that govern the underlying turbulent velocity field $U(x,t)$. The most basic of these equations (first derived by Reynolds (1894)) are those that govern the mean velocity field $\langle U(x,t) \rangle$.

The decomposition of the velocity $U(x,t)$ into its mean $\langle U(x,t) \rangle$ and the fluctuation

$$u(x,t) \equiv U(x,t) - \langle U(x,t) \rangle \tag{4.1}$$

is referred to as the *Reynolds decomposition*, i.e.,

$$U(x,t) = \langle U(x,t) \rangle + u(x,t). \tag{4.2}$$

It follows from the continuity equation (Eq. (2.19))

$$\nabla \cdot U = \nabla \cdot (\langle U \rangle + u) = 0 \tag{4.3}$$

that both $\langle U(x,t) \rangle$ and $u(x,t)$ are solenoidal. For the mean of this equation is simply

$$\nabla \cdot \langle U \rangle = 0, \tag{4.4}$$

and then by subtraction we obtain

$$\nabla \cdot u = 0. \tag{4.5}$$

(Note that taking the mean and differentiation commute so that $\langle \nabla \cdot U \rangle = \nabla \cdot \langle U \rangle$ and also $\langle \nabla \cdot u \rangle = \nabla \cdot \langle u \rangle = 0$.)

Taking the mean of the momentum equation (Eq. (2.35)) is less simple because of the nonlinear convective term. The first step is to write the substantial derivative in conservative form,

$$\frac{DU_j}{Dt} = \frac{\partial U_j}{\partial t} + \frac{\partial}{\partial x_i}(U_i U_j),$$ (4.6)

so that the mean is

$$\left\langle \frac{DU_j}{Dt} \right\rangle = \frac{\partial \langle U_j \rangle}{\partial t} + \frac{\partial}{\partial x_i}\langle U_i U_j \rangle.$$ (4.7)

Then, substituting the Reynolds decomposition for U_i and U_j, the nonlinear term becomes

$$\langle U_i U_j \rangle = \langle ((\langle U_i \rangle + u_i)(\langle U_j \rangle + u_j) \rangle$$

$$= \langle \langle U_i \rangle \langle U_j \rangle + u_i \langle U_j \rangle + u_j \langle U_i \rangle + u_i u_j \rangle$$

$$= \langle U_i \rangle \langle U_j \rangle + \langle u_i u_j \rangle.$$ (4.8)

For reasons soon to be given, the velocity covariances $\langle u_i u_j \rangle$ are called *Reynolds stresses*. Thus, from the previous two equations, we obtain

$$\left\langle \frac{DU_j}{Dt} \right\rangle = \frac{\partial \langle U_j \rangle}{\partial t} + \frac{\partial}{\partial x_i}\left(\langle U_i \rangle \langle U_j \rangle + \langle u_i u_j \rangle \right)$$

$$= \frac{\partial \langle U_j \rangle}{\partial t} + \langle U_i \rangle \frac{\partial \langle U_j \rangle}{\partial x_i} + \frac{\partial}{\partial x_i}\langle u_i u_j \rangle,$$ (4.9)

the second step following from $\partial \langle U_i \rangle / \partial x_i = 0$ (Eq. (4.4)).

The final result can be usefully re-expressed by defining the *mean substantial derivative*

$$\frac{\bar{D}}{\bar{D}t} \equiv \frac{\partial}{\partial t} + \langle U \rangle \cdot \nabla.$$ (4.10)

For any property $Q(x,t)$, $\bar{D}Q/\bar{D}t$ represents its rate of change following a point moving with the local mean velocity $\langle U(x,t) \rangle$. In terms of this derivative, Eq. (4.9) is

$$\left\langle \frac{DU_j}{Dt} \right\rangle = \frac{\bar{D}}{\bar{D}t}\langle U_j \rangle + \frac{\partial}{\partial x_i}\langle u_i u_j \rangle.$$ (4.11)

Evidently the mean of the substantial derivative $\langle DU_j/Dt \rangle$ does not equal the mean substantial derivative of the mean $\bar{D}\langle U_j \rangle/\bar{D}t$.

It is now a simple matter to take the mean of the momentum equation (Eq. (2.35)) since the other terms are linear in U and p. The result is the

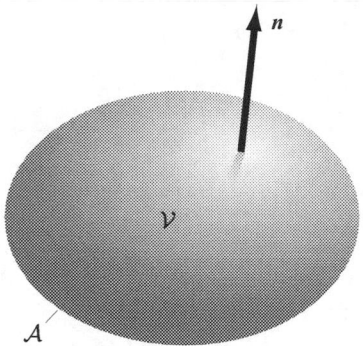

Fig. 4.1. A sketch of a control volume $\mathcal{V}$, with bounding control surface $\mathcal{A}$, showing the outward pointing unit normal $\boldsymbol{n}$.

mean-momentum or *Reynolds equations*

$$\frac{\bar{D}\langle U_j\rangle}{\bar{D}t} = v\,\nabla^2\langle U_j\rangle - \frac{\partial\langle u_i u_j\rangle}{\partial x_i} - \frac{1}{\rho}\frac{\partial\langle p\rangle}{\partial x_j}. \tag{4.12}$$

In appearance, the Reynolds equations (Eq. (4.12)) and the Navier–Stokes equations (Eq. (2.35)) are the same, except for the term in the Reynolds stresses – a crucial difference.

Like $p(\boldsymbol{x}, t)$, the mean pressure field $\langle p(\boldsymbol{x}, t)\rangle$ satisfies a Poisson equation. This may be obtained either by taking the mean of $\nabla^2 p$ (Eq. (2.42)), or by taking the divergence of the Reynolds equations:

$$-\frac{1}{\rho}\nabla^2\langle p\rangle = \left\langle \frac{\partial U_i}{\partial x_j}\frac{\partial U_j}{\partial x_i}\right\rangle$$

$$= \frac{\partial\langle U_i\rangle}{\partial x_j}\frac{\partial\langle U_j\rangle}{\partial x_i} + \frac{\partial^2\langle u_i u_j\rangle}{\partial x_i\,\partial x_j}. \tag{4.13}$$

EXERCISES _____

4.1   Obtain from the Reynolds equations (Eq. (4.12)) an equation for the rate of change of mean momentum in a fixed control volume $\mathcal{V}$ (see Fig. 4.1). Where possible express terms as integrals over the bounding control surface $\mathcal{A}$.

4.2   For a random field $\phi(\boldsymbol{x}, t)$, obtain the results

$$\frac{D\phi}{Dt} = \frac{\bar{D}\phi}{\bar{D}t} + \nabla\cdot(\boldsymbol{u}\,\phi), \tag{4.14}$$

$$\left\langle \frac{D\phi}{Dt}\right\rangle = \frac{\bar{D}\langle\phi\rangle}{\bar{D}t} + \nabla\cdot\langle\boldsymbol{u}\,\phi\rangle. \tag{4.15}$$

4.3   The mean rate of strain $\bar{S}_{ij}$ and mean rate of rotation $\bar{\Omega}_{ij}$ are defined by

$$\bar{S}_{ij} \equiv \frac{1}{2}\left(\frac{\partial\langle U_i\rangle}{\partial x_j} + \frac{\partial\langle U_j\rangle}{\partial x_i}\right), \tag{4.16}$$

$$\bar{\Omega}_{ij} \equiv \frac{1}{2}\left(\frac{\partial\langle U_i\rangle}{\partial x_j} - \frac{\partial\langle U_j\rangle}{\partial x_i}\right). \tag{4.17}$$

Obtain the results

$$\bar{S}_{ij} = \langle S_{ij}\rangle, \quad \bar{\Omega}_{ij} = \langle\Omega_{ij}\rangle, \tag{4.18}$$

$$\frac{\partial\langle U_i\rangle}{\partial x_j}\frac{\partial\langle U_j\rangle}{\partial x_i} = \bar{S}_{ij}\bar{S}_{ij} - \bar{\Omega}_{ij}\bar{\Omega}_{ij}, \tag{4.19}$$

$$\frac{\partial\bar{S}_{ij}}{\partial x_i} = \frac{1}{2}\frac{\partial^2\langle U_j\rangle}{\partial x_i\,\partial x_i}. \tag{4.20}$$

4.2 Reynolds stresses

Evidently the Reynolds stresses $\langle u_i u_j\rangle$ play a crucial role in the equations for the mean velocity field $\langle U\rangle$. If $\langle u_i u_j\rangle$ were zero, then the equations for $U(x,t)$ and $\langle U(x,t)\rangle$ would be identical. The very different behavior of $U(x,t)$ and $\langle U(x,t)\rangle$ (see, e.g., Fig. 1.4 on page 6) are therefore attributable to the effect of the Reynolds stresses. Some of their properties are now described.

Interpretation as stresses

The Reynolds equations can be rewritten

$$\rho\frac{\bar{D}\langle U_j\rangle}{\bar{D}t} = \frac{\partial}{\partial x_i}\left[\mu\left(\frac{\partial\langle U_i\rangle}{\partial x_j} + \frac{\partial\langle U_j\rangle}{\partial x_i}\right) - \langle p\rangle\delta_{ij} - \rho\langle u_i u_j\rangle\right]. \tag{4.21}$$

This is the general form of a momentum conservation equation (cf. Eq. (2.31)), with the term in square brackets representing the sum of three stresses: the viscous stress, the isotropic stress $-\langle p\rangle\delta_{ij}$ from the mean pressure field, and the apparent stress arising from the fluctuating velocity field, $-\rho\langle u_i u_j\rangle$. Even though this apparent stress is $-\rho\langle u_i u_j\rangle$, it is convenient and conventional to refer to $\langle u_i u_j\rangle$ as the Reynolds stress.

The viscous stress (i.e., force per unit area) ultimately stems from momentum transfer at the molecular level. So also the Reynolds stress stems from momentum transfer by the fluctuating velocity field. Referring to Fig. 4.1, the rate of gain of momentum within a fixed control volume $\mathcal{V}$ due to flow through the bounding surface $\mathcal{A}$ is

$$\dot{M} = \iint_{\mathcal{A}} \rho U(-U \cdot n) \, dA. \tag{4.22}$$

(The momentum per unit volume is ρU, and the volume flow rate per unit area into $\mathcal{V}$ through $\mathcal{A}$ is $-U \cdot n$.) The mean of the j component of this equation is

$$
\begin{aligned}
\langle \dot{M}_j \rangle &= \iint_{\mathcal{A}} -\rho(\langle U_i \rangle \langle U_j \rangle + \langle u_i u_j \rangle) n_i \, dA \\
&= \iiint_{\mathcal{V}} -\rho \frac{\partial}{\partial x_i} (\langle U_i \rangle \langle U_j \rangle + \langle u_i u_j \rangle) \, dV, \tag{4.23}
\end{aligned}
$$

the last step following from the divergence theorem. Thus, for the control volume $\mathcal{V}$, the Reynolds stress as it appears in the Reynolds equations (i.e., $-\rho \, \partial \langle u_i u_j \rangle / \partial x_i$) arises from the mean momentum flux due to the fluctuating velocity on the boundary $\mathcal{A}$, $-\rho \langle u_i u_j \rangle n_i$.

The closure problem

For a general statistically three-dimensional flow, there are four independent equations governing the mean velocity field; namely three components of the Reynolds equations (Eq. (4.12)) together with either the mean continuity equation (Eq. (4.4)) or the Poisson equation for $\langle p \rangle$ (Eq. (4.13)). However, these four equations contain more than four unknowns. In addition to $\langle U \rangle$ and $\langle p \rangle$ (four quantities), there are also the Reynolds stresses.

This is a manifestation of the *closure problem*. In general, the evolution equations (obtained from the Navier–Stokes equations) for a set of statistics contain additional statistics to those in the set considered. Consequently, in the absence of separate information to determine the additional statistics, the set of equations cannot be solved. Such a set of equations – with more unknowns than equations – is said to be *unclosed*. The Reynolds equations are unclosed: they cannot be solved unless the Reynolds stresses are somehow determined.

Tensor properties

The Reynolds stresses are the components of a second-order tensor,[1] which is obviously symmetric, i.e., $\langle u_i u_j \rangle = \langle u_j u_i \rangle$. The diagonal components ($\langle u_1^2 \rangle = \langle u_1 u_1 \rangle$, $\langle u_2^2 \rangle$, and $\langle u_3^2 \rangle$) are *normal stresses*, while the off-diagonal components (e.g., $\langle u_1 u_2 \rangle$) are *shear stresses*.

---

[1] The properties of second-order tensors are reviewed in Appendix B.

The *turbulent kinetic energy* $k(x, t)$ is defined to be half the trace of the Reynolds stress tensor:

$$k \equiv \tfrac{1}{2}\langle \boldsymbol{u} \cdot \boldsymbol{u} \rangle = \tfrac{1}{2}\langle u_i u_i \rangle. \tag{4.24}$$

It is the mean kinetic energy per unit mass in the fluctuating velocity field.

In the principal axes of the Reynolds stress tensor, the shear stresses are zero, and the normal stresses are the eigenvalues, which are non-negative (i.e., $\langle u_1^2 \rangle \geq 0$). Thus the Reynolds stress tensor is symmetric positive semi-definite. In general, all eigenvalues are strictly positive; but, in special or extreme circumstances, one or more of the eigenvalues can be zero.

Anisotropy

The distinction between shear stresses and normal stresses is dependent on the choice of coordinate system. An intrinsic distinction can be made between isotropic and anisotropic stresses. The isotropic stress is $\tfrac{2}{3}k\delta_{ij}$, and then the deviatoric anisotropic part is

$$a_{ij} \equiv \langle u_i u_j \rangle - \tfrac{2}{3}k\delta_{ij}. \tag{4.25}$$

The normalized anisotropy tensor – used extensively below – is defined by

$$b_{ij} = \frac{a_{ij}}{2k} = \frac{\langle u_i u_j \rangle}{\langle u_\ell u_\ell \rangle} - \tfrac{1}{3}\delta_{ij}. \tag{4.26}$$

In terms of these anisotropy tensors, the Reynolds stress tensor is

$$\begin{aligned}
\langle u_i u_j \rangle &= \tfrac{2}{3}k\delta_{ij} + a_{ij} \\
&= 2k(\tfrac{1}{3}\delta_{ij} + b_{ij}).
\end{aligned} \tag{4.27}$$

It is only the anisotropic component a_{ij} that is effective in transporting momentum. For we have

$$\rho \frac{\partial \langle u_i u_j \rangle}{\partial x_i} + \frac{\partial \langle p \rangle}{\partial x_j} = \rho \frac{\partial a_{ij}}{\partial x_i} + \frac{\partial}{\partial x_j}(\langle p \rangle + \tfrac{2}{3}\rho k), \tag{4.28}$$

showing that the isotropic component ($\tfrac{2}{3}k$) can be absorbed in a modified mean pressure.

Irrotational motion

An essential feature of turbulent flows is that they are rotational. Consider instead an irrotational random velocity field – such as (to within an approximation) the flow of water waves. The vorticity is zero, and so in turn the mean vorticity, the fluctuating vorticity, and $\partial u_i/\partial x_j - \partial u_j/\partial x_i$ are also zero. Hence we have

$$\left\langle u_i \left(\frac{\partial u_i}{\partial x_j} - \frac{\partial u_j}{\partial x_i} \right) \right\rangle = \frac{\partial}{\partial x_j} (\tfrac{1}{2} \langle u_i u_i \rangle) - \frac{\partial}{\partial x_i} \langle u_i u_j \rangle = 0, \qquad (4.29)$$

from which follows the *Corrsin–Kistler equation* (Corrsin and Kistler 1954)

$$\frac{\partial}{\partial x_i} \langle u_i u_j \rangle = \frac{\partial k}{\partial x_j} \qquad (4.30)$$

for irrotational flow. In this case the Reynolds stress $\langle u_i u_j \rangle$ has the same effect as the isotropic stress $k \delta_{ij}$, which can be absorbed in a modified pressure. In other words, the Reynolds stresses arising from an irrotational field $u(x, t)$ have absolutely no effect on the mean velocity field.

Symmetries

For some flows, symmetries in the flow geometry determine properties of the Reynolds stresses.

Consider a statistically two-dimensional flow in which statistics are independent of x_3, and which is statistically invariant under reflections of the x_3 coordinate axis. For the PDF of velocity $f(V; x, t)$, these two conditions imply that

$$\frac{\partial f}{\partial x_3} = 0, \qquad (4.31)$$

$$f(V_1, V_2, V_3; x_1, x_2, x_3, t) = f(V_1, V_2, -V_3; x_1, x_2, -x_3, t). \qquad (4.32)$$

At $x_3 = 0$, this last equation yields $\langle U_3 \rangle = -\langle U_3 \rangle$, i.e., $\langle U_3 \rangle = 0$; it similarly yields $\langle u_1 u_3 \rangle = 0$ and $\langle u_2 u_3 \rangle = 0$. The first equation (Eq. (4.31)) indicates that these relations hold for all x. Thus, for such a statistically two-dimensional flow, $\langle U_3 \rangle$ is zero and the Reynolds-stress tensor is

$$\begin{bmatrix} \langle u_1^2 \rangle & \langle u_1 u_2 \rangle & 0 \\ \langle u_1 u_2 \rangle & \langle u_2^2 \rangle & 0 \\ 0 & 0 & \langle u_3^2 \rangle \end{bmatrix}. \qquad (4.33)$$

In addition to being statistically two-dimensional, the turbulent channel flow sketched in Fig. 3.25 on page 76 is statistically symmetric about the plane $x_2 = 0$. This symmetry implies that

$$f(V_1, V_2, V_3; x_1, x_2, x_3, t) = f(V_1, -V_2, V_3; x_1, -x_2, x_3, t), \qquad (4.34)$$

from which it follows that $\langle U_2 \rangle$ and $\langle u_1 u_2 \rangle$ are odd functions of x_2, whereas $\langle U_1 \rangle$ and the normal stresses are even functions.

4.4   Each of the following equations is incorrect. Why?

(a)
$$\langle u_i u_j \rangle = \begin{bmatrix} 0.5 & 0.1 & 0 \\ 0.1 & 0.3 & 0.1 \\ 0 & 0.1 & -0.1 \end{bmatrix}$$

(b)
$$\langle u_i u_j \rangle = \begin{bmatrix} 0.21 & -0.05 & 0.01 \\ -0.06 & 0.5 & 0 \\ 0.01 & 0 & 1.0 \end{bmatrix}$$

(c)
$$\langle u_i u_j \rangle = \begin{bmatrix} 1 & 1.5 & 0.2 \\ 1.5 & 1 & 0 \\ 0.2 & 0 & 1 \end{bmatrix}$$

(d)
$$a_{ij} = \begin{bmatrix} 1.8 & 0.2 & 0 \\ 0.2 & -1.6 & 0.1 \\ 0 & 0.1 & -0.3 \end{bmatrix}$$

(e)
$$b_{ij} = \begin{bmatrix} -0.4 & 0 & 0.1 \\ 0 & 0.2 & 0 \\ 0.1 & 0 & 0.2 \end{bmatrix}.$$

4.5   In an experiment on homogeneous turbulent shear flow (in which
$\partial \langle U_1 \rangle / \partial x_2$ is the only non-zero mean velocity gradient) the Reynolds
stresses (normalized by k) are measured to be

$$\frac{\langle u_i u_j \rangle}{k} = \begin{bmatrix} 1.08 & -0.32 & 0 \\ -0.32 & 0.40 & 0 \\ 0 & 0 & 0.52 \end{bmatrix}. \tag{4.35}$$

(a) Determine the corresponding anisotropy tensors a_{ij}/k and b_{ij}.
(b) What is the correlation coefficient between u_1 and u_2?
(c) A matrix of the form of Eq. (4.35) can be transformed into
principal axes by a unitary matrix of the form

$$\mathbf{A} = \begin{bmatrix} \cos\theta & \sin\theta & 0 \\ -\sin\theta & \cos\theta & 0 \\ 0 & 0 & 1 \end{bmatrix}.$$

That is, for particular values of the angle θ, $A_{ki}\langle u_k u_j \rangle A_{j\ell}$ is
diagonal. Determine the angle $\theta = \theta_R$ ($0 \le \theta_R \le \pi/2$) that
transforms $\langle u_i u_j \rangle$ to principal axes.

(d) Determine the angle $\theta = \theta_S$ ($0 \leq \theta_S \leq \pi/2$) that transforms the mean rate of strain $\bar{S}_{ij}$ to principal axes.

(e) Determine the eigenvalues of $\langle u_i u_j \rangle / k$.

4.3 The mean scalar equation

Just as the most basic description of the turbulent velocity field $U(x,t)$ is provided by the mean velocity $\langle U(x,t) \rangle$, so also the most basic description of a conserved passive scalar field $\phi(x,t)$ is provided by its mean $\langle \phi(x,t) \rangle$. The conservation equation for $\langle \phi(x,t) \rangle$ is obtained by the same procedure as that used to obtain the Reynolds equations.

The fluctuating scalar field is defined by

$$\phi'(x,t) = \phi(x,t) - \langle \phi(x,t) \rangle, \tag{4.36}$$

so that the Reynolds decomposition of the scalar field is

$$\phi(x,t) = \langle \phi(x,t) \rangle + \phi'(x,t). \tag{4.37}$$

The conservation equation for $\phi(x,t)$ (Eq. (2.54)) can be written

$$\frac{\partial \phi}{\partial t} + \nabla \cdot (U\phi) = \Gamma \nabla^2 \phi. \tag{4.38}$$

The only nonlinear term is that involving the convective flux $U\phi$, the mean of which is

$$\begin{aligned}
\langle U\phi \rangle &= \langle ((\langle U \rangle + u)(\langle \phi \rangle + \phi')) \rangle \\
&= \langle U \rangle \langle \phi \rangle + \langle u\phi' \rangle.
\end{aligned} \tag{4.39}$$

The velocity–scalar covariance $\langle u\phi' \rangle$ is a vector, which is called the *scalar flux*: it represents the flux (flow rate per unit area) of the scalar due to the fluctuating velocity field (see Exercise 4.6). Thus, taking the mean of Eq. (4.38), we obtain

$$\frac{\partial \langle \phi \rangle}{\partial t} + \nabla \cdot (\langle U \rangle \langle \phi \rangle + \langle u\phi' \rangle) = \Gamma \nabla^2 \langle \phi \rangle, \tag{4.40}$$

or, in terms of the mean substantial derivative (Eq. (4.10)),

$$\frac{\bar{D} \langle \phi \rangle}{\bar{D}t} = \nabla \cdot (\Gamma \nabla \langle \phi \rangle - \langle u\phi' \rangle). \tag{4.41}$$

Evidently, in this mean-scalar equation, the scalar fluxes play an analogous role to that of the Reynolds stresses in the Reynolds equations. In particular they give rise to a closure problem: even if $\langle U \rangle$ is known, Eq. (4.41) cannot be solved for $\langle \phi \rangle$, without a prescription for $\langle u\phi' \rangle$.

EXERCISES

4.6   Let $\Phi(t)$ be the integral of a conserved passive scalar field $\phi(x, t)$ over a fixed control volume $\mathcal{V}$ (see Fig. 4.1). Obtain an equation for $d\langle \Phi(t) \rangle / dt$. Where possible express each term as an integral over the bounding control surface $\mathcal{A}$, and describe the significance of the term.

4.7   Consider a statistically two-dimensional flow in which the statistics of the velocity and scalar fields are independent of x_3 and are invariant under a reflection of the x_3 coordinate axis. Write down the symmetry conditions satisfied by the one-point joint PDF of $U(x, t)$ and $\phi(x, t)$. Show that $\partial \langle \phi \rangle / \partial x_3$ and $\langle u_3 \phi' \rangle$ are zero. If, further, $x_2 = 0$ is a plane of statistical symmetry, show that $\langle \phi \rangle$ is an even function of x_2 and that $\langle u_2 \phi' \rangle$ is an odd function.

4.4 Gradient-diffusion and turbulent-viscosity hypotheses

In the historical development of a scientific field of inquiry, it is usual for there to be a succession of models proposed to describe the phenomena being studied. Often – such as in the study of turbulent flows – the early models are simple, but are subsequently found to be lacking both in physical content and in predictive accuracy. Later models may be superior in physical content and predictive accuracy, but lack simplicity. In spite of their flaws, it is valuable to have an appreciation for the early, simple models. One reason is that the behavior implied by the models may be determined by simple reasoning or simple analysis – as opposed to the numerical solutions usually required for more complex models. Second, the simple models can provide a reference against which the phenomena being studied – and also more complex models – can be compared.

It is in this spirit that we introduce the *gradient-diffusion hypothesis*, the *turbulent-viscosity hypothesis*, and related ideas. These are valuable concepts, whose limitations should always be borne in mind.

The scalar flux $\langle u\phi' \rangle$ vector gives both the direction and the magnitude of the turbulent transport of the conserved scalar ϕ. According to the gradient-diffusion hypothesis, this transport is down the mean scalar gradient – that is, in the direction of $-\nabla \langle \phi \rangle$. Thus, according to the hypothesis, there is a positive scalar $\Gamma_T(x, t)$ – the *turbulent diffusivity* – such that

$$\langle u\phi' \rangle = -\Gamma_T \nabla \langle \phi \rangle. \tag{4.42}$$

With the *effective diffusivity* defined as the sum of the molecular and turbulent diffusivities

$$\Gamma_{\text{eff}}(x, t) = \Gamma + \Gamma_T(x, t), \tag{4.43}$$

the mean scalar conservation equation (Eq. (4.41)) incorporating the gradient-diffusion hypothesis (Eq. (4.42)) is

$$\frac{\bar{D}\langle \phi \rangle}{\bar{D}t} = \nabla \cdot (\Gamma_{\text{eff}} \nabla \langle \phi \rangle). \tag{4.44}$$

It may be seen, then, that this equation is the same as the conservation equation for ϕ (Eq. (2.54)) but with $\langle U \rangle$, $\langle \phi \rangle$, and Γ_{eff} in place of U, ϕ, and Γ.

Mathematically, the gradient-diffusion hypothesis (Eq. (4.42)) is analogous to Fourier's law of heat conduction and Fick's law of molecular diffusion. Similarly, the *turbulent-viscosity hypothesis* – introduced by Boussinesq in 1877 – is mathematically analogous to the stress–rate-of-strain relation for a Newtonian fluid (Eq. (2.32)). According to the hypothesis, the deviatoric Reynolds stress $(-\rho \langle u_i u_j \rangle + \frac{2}{3}\rho k \delta_{ij})$ is proportional to the mean rate of strain,

$$-\rho \langle u_i u_j \rangle + \tfrac{2}{3}\rho k \delta_{ij} = \rho \nu_T \left(\frac{\partial \langle U_i \rangle}{\partial x_j} + \frac{\partial \langle U_j \rangle}{\partial x_i} \right)$$
$$= 2\rho \nu_T \bar{S}_{ij}, \tag{4.45}$$

where the positive scalar coefficient ν_T is the *turbulent viscosity* (also called the *eddy viscosity*).

The mean-momentum equation incorporating the turbulent-viscosity hypothesis (i.e., Eq. (4.45) substituted into Eq. (4.12)) is

$$\frac{\bar{D}}{\bar{D}t}\langle U_j \rangle = \frac{\partial}{\partial x_i}\left[\nu_{\text{eff}}\left(\frac{\partial \langle U_i \rangle}{\partial x_j} + \frac{\partial \langle U_j \rangle}{\partial x_i} \right) \right] - \frac{1}{\rho}\frac{\partial}{\partial x_j}(\langle p \rangle + \tfrac{2}{3}\rho k), \tag{4.46}$$

where

$$\nu_{\text{eff}}(x, t) = \nu + \nu_T(x, t), \tag{4.47}$$

is the *effective viscosity*. This is the same as the Navier–Stokes equations with $\langle U \rangle$ and ν_{eff} in place of U and ν, with $\langle p \rangle + \frac{2}{3}\rho k$ the modified mean pressure.

The gradient-diffusion and turbulent-viscosity hypotheses have been introduced without justification or criticism so far. A thorough appraisal of the hypotheses is postponed to Chapter 10. At this stage, the following observations suffice.

(i) The gradient-diffusion hypothesis implies that the scalar flux vector is aligned with the mean scalar gradient vector. Even in simple turbulent flows this is found not be the case. For example, in an experiment on homogeneous turbulent shear flow (Tavoularis and Corrsin 1981) the angle between $\nabla\langle\phi\rangle$ and $-\langle u\phi'\rangle$ was measured to be $65°$.

(ii) Similarly, the turbulent-viscosity hypothesis implies that the anisotropy tensor a_{ij} is aligned with the mean rate-of-strain tensor, i.e.,

$$a_{ij} \equiv \langle u_i u_j\rangle - \tfrac{2}{3}k\delta_{ij}$$
$$= -v_\mathrm{T}\left(\frac{\partial\langle U_i\rangle}{\partial x_j} + \frac{\partial\langle U_j\rangle}{\partial x_i}\right)$$
$$= -2v_\mathrm{T}\bar{S}_{ij}. \tag{4.48}$$

Being symmetric and deviatoric, both a_{ij} and the mean rate of strain have five independent components. According to the turbulent-viscosity hypothesis, these five components are related to each other through the scalar coefficient v_T. Again, even in simple shear flow, it is found that this alignment does not occur (see Exercise 4.5).

(iii) An important class of flows consists of those that can be described by the two-dimensional turbulent-boundary-layer equations (which are presented in Chapter 5). In these flows, the mean velocity is predominantly in the x_1-coordinate direction, while variations in mean quantities are predominantly in the x_2-coordinate direction. Only one component $\langle u_2\phi'\rangle$ of the scalar flux, and one Reynolds stress $\langle u_1u_2\rangle$, appear in the boundary-layer equations. Consequently, the gradient-diffusion hypothesis reduces to

$$\langle u_2\phi'\rangle = -\Gamma_\mathrm{T}\frac{\partial\langle\phi\rangle}{\partial x_2}, \tag{4.49}$$

and the turbulent-viscosity hypothesis to

$$\langle u_1u_2\rangle = -v_\mathrm{T}\frac{\partial\langle U_1\rangle}{\partial x_2}. \tag{4.50}$$

Both of these equations relate a single covariance to a single gradient. Providing that the covariance and the gradient have opposite signs – which is almost always the case – then, rather than being hypotheses or assumptions, these equations can be taken as definitions of Γ_T and v_T.

(iv) Specification of $v_\mathrm{T}(x,t)$ and $\Gamma_\mathrm{T}(x,t)$ solves the closure problem. That is, if v_T and Γ_T can somehow be specified, then the mean flow equations for $\langle U\rangle$ (Eq. (4.46)) and for $\langle\phi\rangle$ (Eq. (4.44)) can be solved.

(v) At high Reynolds number, and remote from walls, v_T and Γ_T are found to scale with the velocity scale $\mathcal{U}$ and the length scale $\mathcal{L}$ of the flow, independent from the molecular properties of the fluid v and Γ. Consequently, the ratios v_T/v and Γ_T/Γ both increase linearly with Reynolds number, and so (in the given circumstances) the molecular transport is negligible.

(vi) The *turbulent Prandtl number* σ_T is defined by

$$\sigma_T = v_T/\Gamma_T. \tag{4.51}$$

In most simple turbulent flows σ_T is found to be of order unity.

EXERCISES _____

4.8 Show that, according to the gradient-diffusion hypothesis, in a statistically stationary flow the maximum and minimum values of $\langle \phi \rangle$ occur at the boundaries.
(Hint: modify the boundedness argument used in Section 2.6 to apply it to Eq. (4.44).)

4.9 Show that, in order for the turbulent-viscosity hypothesis (Eq. (4.45)) to yield non-negative normal stresses, it is necessary and sufficient for the turbulent viscosity to satisfy

$$v_T \leq \frac{k}{3S_\lambda}, \tag{4.52}$$

where S_λ is the largest eigenvalue of the mean rate-of-strain tensor.

5

Free shear flows

The most commonly studied turbulent free shear flows are jets, wakes, and mixing layers. As the name 'free' implies, these flows are remote from walls, and the turbulent flow arises because of mean-velocity differences.

We begin by examining the round jet. By combining experimental observations (Section 5.1) with the Reynolds equations (Section 5.2), a good deal can be learned, not only about the round jet, but also about the behavior of turbulent flows in general. In Section 5.3, we study the turbulent kinetic energy in the round jet, and the important processes of production and dissipation of energy. Other self-similar free shear flows are briefly described in Section 5.4; and further observations about the behavior of free shear flows are made in Section 5.5.

5.1 The round jet: experimental observations

5.1.1 A description of the flow

We have already encountered the round jet in Chapter 1, for example, Figs. 1.1–1.4. The ideal experimental configuration and the coordinate system employed are shown in Fig. 5.1. A Newtonian fluid steadily flows through a nozzle of diameter d, which produces (approximately) a flat-topped velocity profile, with velocity U_J. The jet from the nozzle flows into an ambient of the same fluid, which is at rest at infinity. The flow is statistically stationary and axisymmetric. Hence statistics depend on the axial and radial coordinates (x and r), but are independent of time and of the circumferential coordinate, θ. The velocity components in the x, r, and θ coordinate directions are denoted by U, V, and W.

In the ideal experiment, the flow is completely defined by U_J, d, and v, and hence the only non-dimensional parameter is the Reynolds number, defined

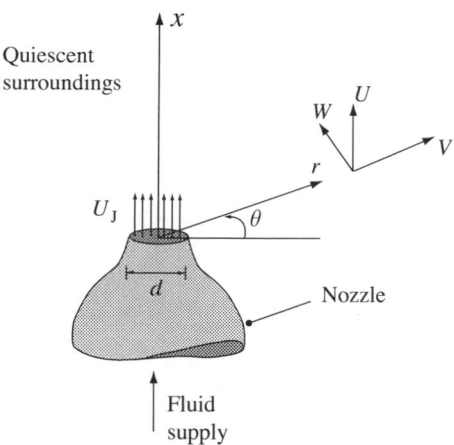

Fig. 5.1. A sketch of a round-jet experiment, showing the polar-cylindrical coordinate system employed.

by $Re = U_J d / v$. (In practice, the details of the nozzle and the surroundings have some effect, as discussed by Schneider (1985) and Hussein *et al.* (1994).)

5.1.2 *The mean velocity field*

As might be expected from the visual appearance of the flow (Figs. 1.1 and 1.2 on pages 4 and 5), the mean velocity is predominantly in the axial direction. Measured radial profiles of the mean axial velocity are shown in Fig. 5.2. (Note that $r = 0$ is the axis, about which the profile of $\langle U \rangle$ is symmetric.) Not shown in Fig. 5.2 is the initial development region ($0 \leq x/d \leq 25$, say), in which the profile changes from being (approximately) square to the rounded shape seen in Fig. 5.2. The mean circumferential velocity is zero (i.e., $\langle W \rangle = 0$), while – as shown in Exercise 5.5 – the mean radial velocity $\langle V \rangle$ is smaller than $\langle U \rangle$ by an order of magnitude.

The axial velocity

In terms of the mean axial velocity field $\langle U(x, r, \theta) \rangle$ (which is independent of θ), the centerline velocity is

$$U_0(x) \equiv \langle U(x, 0, 0) \rangle, \tag{5.1}$$

and the jet's half-width $r_{1/2}(x)$ is defined such that

$$\langle U(x, r_{1/2}(x), 0) \rangle = \tfrac{1}{2} U_0(x). \tag{5.2}$$

Two clear observations from Fig. 5.2 are that, with increasing axial distance, the jet decays (i.e., $U_0(x)$ decreases), and that it spreads (i.e., $r_{1/2}(x)$ increases).

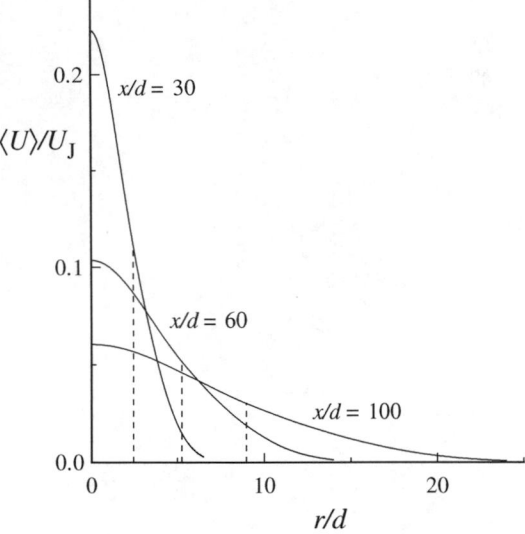

Fig. 5.2. Radial profiles of mean axial velocity in a turbulent round jet, Re = 95,500. The dashed lines indicate the half-width, $r_{1/2}(x)$, of the profiles. (Adapted from the data of Hussein *et al.* (1994).)

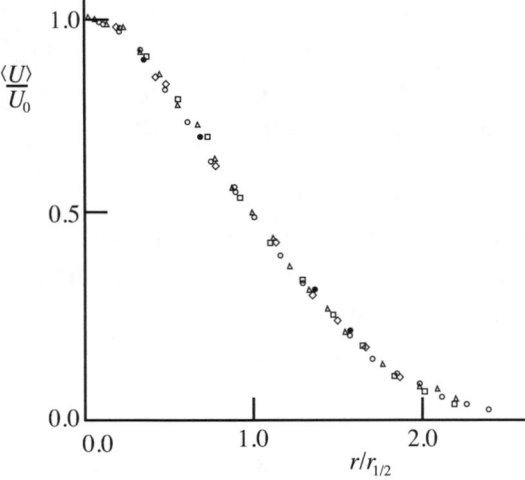

Fig. 5.3. Mean axial velocity against radial distance in a turbulent round jet, Re $\approx 10^5$; measurements of Wygnanski and Fiedler (1969). Symbols: $\circ$, $x/d = 40$; $\triangle$, $x/d = 50$; $\square$, $x/d = 60$; $\diamond$, $x/d = 75$; $\bullet$, $x/d = 97.5$.

As the jet decays and spreads, the mean velocity profiles change, as shown in Fig. 5.2, but the *shape* of the profiles does not change. Beyond the developing region ($x/d > 30$, say), the profiles of $\langle U \rangle / U_0(x)$, plotted against $r/r_{1/2}(x)$ collapse onto a single curve. Figure 5.3 shows the experimental data of Wygnanski and Fiedler (1969) plotted this way for x/d between 40 and 100. The important conclusion is that the mean velocity profile becomes *self-similar*.

Self-similarity

Self-similarity is an important concept that arises in several different contexts in the study of turbulent flows. To explore the general ideas, consider a quantity $Q(x, y)$ that depends on two independent variables (i.e., x and y). As functions of x, characteristic scales $Q_0(x)$ and $\delta(x)$ are defined for the dependent variable Q and the independent variable y, respectively. Then scaled variables are defined by

$$\xi \equiv \frac{y}{\delta(x)}, \qquad (5.3)$$

$$\tilde{Q}(\xi, x) \equiv \frac{Q(x, y)}{Q_0(x)}. \qquad (5.4)$$

If the scaled dependent variable is independent of x, i.e., there is a function $\hat{Q}(\xi)$ such that

$$\tilde{Q}(\xi, x) = \hat{Q}(\xi), \qquad (5.5)$$

then $Q(x, y)$ is self-similar. In this case, $Q(x, y)$ can be expressed in terms of functions of single independent variables – $Q_0(x)$, $\delta(x)$, and $\hat{Q}(\xi)$.

Several comments and qualifications are in order:

(i) the scales $Q_0(x)$ and $\delta(x)$ must be chosen appropriately – they usually have power-law dependences on x;

(ii) in some circumstances, more general transformations are required, e.g.,

$$\tilde{Q}(\xi, x) \equiv [Q(x, y) - Q_\infty(x)]/Q_0(x);$$

(iii) self-similar behavior may be observed (to within a good approximation) over a range of x (but not for all x); and

(iv) if a self-similar quantity $Q(x, y)$ is governed by a partial differential equation, then $Q_0(x), \delta(x)$, and $\hat{Q}(\xi)$ are governed by ordinary differential equations.

Axial variation of scales

Returning to the round jet, to complete the picture we need to determine the axial variation of $U_0(x)$ and $r_{1/2}(x)$. Figure 5.4 shows the inverse of $U_0(x)$, specifically $U_J/U_0(x)$, plotted against x/d. Evidently, over the x/d range considered, the experimental data lie on a straight line. The intercept of this

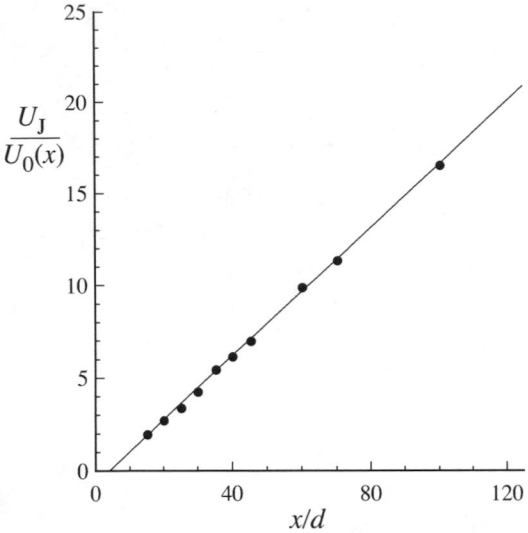

Fig. 5.4. The variation with axial distance of the mean velocity along the centerline in a turbulent round jet, Re = 95,500: symbols, experimental data of Hussein *et al.* (1994); and line, Eq. (5.6) with $x_0/d = 4$ and $B = 5.8$.

line with the abscissa defines the *virtual origin*, denoted by x_0; so that the straight line in Fig. 5.4 corresponds to

$$\frac{U_0(x)}{U_J} = \frac{B}{(x - x_0)/d}, \tag{5.6}$$

where B is an empirical constant. (Obviously the straight-line behavior and Eq. (5.6) do not hold in the developing region close to the nozzle.)

It is found that the jet spreads linearly: the spreading rate

$$S \equiv \frac{dr_{1/2}(x)}{dx} \tag{5.7}$$

is a constant. Or, put another way, the empirical law for $r_{1/2}(x)$ is

$$r_{1/2}(x) = S(x - x_0), \tag{5.8}$$

for x in the self-similar region. We shall see in Section 5.2 that momentum conservation implies that the product $r_{1/2}(x)U_0(x)$ is independent of x; and so the variations $r_{1/2} \sim x$ and $U_0 \sim x^{-1}$ go hand in hand. These variations also show that the local Reynolds number, defined by

$$\mathrm{Re}_0(x) \equiv r_{1/2}(x)U_0(x)/\nu, \tag{5.9}$$

is independent of x.

Table 5.1. *The spreading rate S (Eq. (5.7)) and velocity-decay constant B (Eq. (5.6)) for turbulent round jets (from Panchapakesan and Lumley (1993a))*

	Panchapakesan and Lumley (1993a)	Hussein *et al.* (1994), hot-wire data	Hussein *et al.* (1994), laser-Doppler data
Re	11,000	95,500	95,500
S	0.096	0.102	0.094
B	6.06	5.9	5.8

Reynolds number

In the ideal round-jet experiment, the only non-dimensional parameter is the jet's Reynolds number, Re. We should ask, therefore; how the self-similar profile shape, the velocity-decay constant B, and the spreading rate S vary with Re. The answer is simple and profound: there is no dependence on Re. Table 5.1 shows that, for jets with Re differing by a factor of almost ten, the small differences in the measured values of B and S are within experimental uncertainties. Also, from visual observations, the spreading rate for jets with Re larger by a factor of a thousand is the same (see Fig. 1.1 on page 4 and Mungal and Hollingsworth (1989)). It is evident from Fig. 1.2 on page 5 that the Reynolds number does affect the flow: the small-scale structures are smaller at larger Reynolds number. However, to repeat, the mean velocity profile and the spreading rate are independent of Re.

Summary

In the self-similar region ($x/d > 30$) of high-Reynolds-number turbulent jets (Re $> 10^4$), the centerline velocity $U_0(x)$ and the half width $r_{1/2}(x)$ vary according to Eqs. (5.6) and (5.8). The empirical constants in these laws are independent of Re: for definiteness we take their values to be $B = 5.8$ and $S = 0.094$ (see Table 5.1). The cross-stream similarity variable can be taken to be either

$$\xi \equiv r/r_{1/2}, \tag{5.10}$$

or

$$\eta \equiv r/(x - x_0), \tag{5.11}$$

the two being related by $\eta = S\xi$. The self-similar mean velocity profile is defined by

$$f(\eta) = \bar{f}(\xi) = \langle U(x, r, 0)\rangle / U_0(x), \tag{5.12}$$

and is shown in Fig. 5.5.

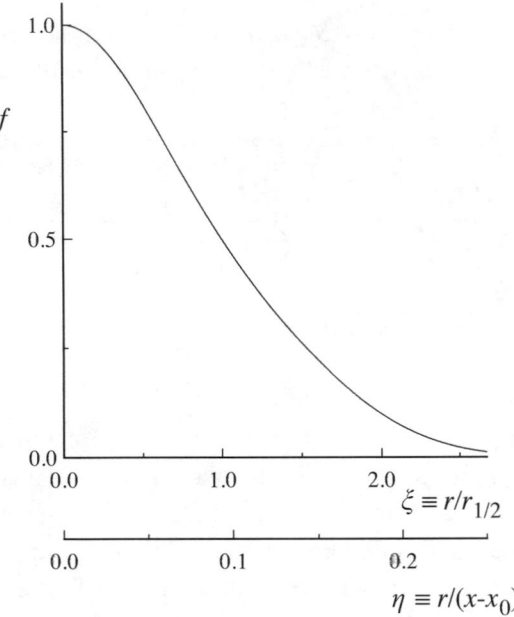

Fig. 5.5. The self-similar profile of the mean axial velocity in the self-similar round jet: curve fit to the LDA data of Hussein *et al.* (1994).

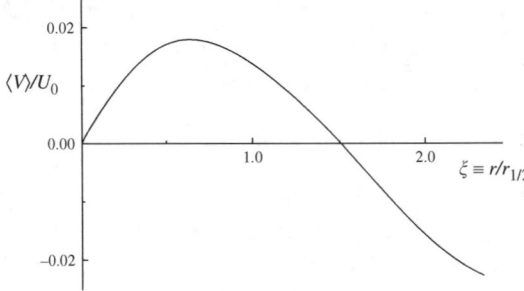

Fig. 5.6. The mean lateral velocity in the self-similar round jet. From the LDA data of Hussein *et al.* (1994).

The lateral velocity

In the self-similar region of the round jet, the mean lateral velocity $\langle V \rangle$ can be determined from $\langle U \rangle$ via the continuity equation (see Exercises 5.4 and 5.5). Figure 5.6 shows the self-similar profile of $\langle V \rangle / U_0$ obtained in this way. It should be observed that $\langle V \rangle$ is very small – less than U_0 by a factor of 40. Notice also that $\langle V \rangle$ is negative at the edge of the jet, indicating that ambient fluid is flowing into the jet and being *entrained*.

EXERCISES

5.1 From the empirical laws for $U_0(x)$ and $r_{1/2}(x)$ (taking $x_0 = 0$), show that

$$\frac{dU_0}{dx} = -\frac{U_0}{x}, \qquad (5.13)$$

and hence

$$\frac{r_{1/2}}{U_0} \frac{dU_0}{dx} = -S. \qquad (5.14)$$

5.2 From the self-similar velocity profile $f(\eta)$ in the turbulent round jet (Eqs. (5.11) and (5.12)) show that

$$\frac{r_{1/2}}{U_0} \frac{\partial \langle U \rangle}{\partial x} = -S(\eta f)', \qquad (5.15)$$

$$\frac{r_{1/2}}{U_0} \frac{\partial \langle U \rangle}{\partial r} = Sf' \qquad (5.16)$$

(where a prime denotes differentiation with respect to η).

5.3 An approximation to the self-similar velocity profile is

$$f(\eta) = (1 + a\eta^2)^{-2}. \qquad (5.17)$$

Show that (from the definition of $r_{1/2}$) the constant a is given by

$$a = (\sqrt{2} - 1)/S^2 \approx 47. \qquad (5.18)$$

Show that, according to this approximation,

$$\frac{r_{1/2}}{U_0} \frac{\partial \langle U \rangle}{\partial r} = -4a\eta S/(1 + a\eta^2)^3, \qquad (5.19)$$

and that

$$\frac{r_{1/2}}{U_0} \left(\frac{\partial \langle U \rangle}{\partial r} \right)_{r=r_{1/2}} = -2 + \sqrt{2} \approx -0.59. \qquad (5.20)$$

(Note that $(\partial \langle U \rangle / \partial r)_{r=r_{1/2}}$ is about six times $(\partial \langle U \rangle / \partial x)_{r=0}$ at the same axial location.)

5.4 For the turbulent round jet, in polar-cylindrical coordinates, the mean continuity equation is

$$\frac{\partial \langle U \rangle}{\partial x} + \frac{1}{r} \frac{\partial}{\partial r}(r \langle V \rangle) = 0. \qquad (5.21)$$

Show that, if $\langle U \rangle$ is self-similar with

$$\langle U \rangle / U_0 = f(\eta), \qquad (5.22)$$

then $\langle V \rangle$ is also self-similar with

$$\langle V \rangle / U_0 = h(\eta), \tag{5.23}$$

where $f(\eta)$ and $h(\eta)$ are related by

$$\eta(f\eta)' = (h\eta)'. \tag{5.24}$$

5.5   If the self-similar axial velocity profile $f(\eta)$ is given by Eq. (5.17), show from Eq. (5.24) that the lateral velocity profile $h(\eta)$ is

$$h(\eta) = \tfrac{1}{2}(\eta - a\eta^3)/(1 + a\eta^2)^2. \tag{5.25}$$

Show that, according to this equation, the lateral velocity at the half-width is

$$\langle V \rangle_{r=r_{1/2}} = U_0 h(S) = (\tfrac{1}{2} - \tfrac{1}{4}\sqrt{2})S U_0 \approx 0.014 U_0. \tag{5.26}$$

Show that, for large $r/r_{1/2}$, Eq. (5.25) implies that

$$\langle V \rangle \sim -\frac{U_0 S}{2(\sqrt{2} - 1)}\frac{1}{(r/r_{1/2})}$$
$$\approx -0.1 U_0 / (r/r_{1/2}). \tag{5.27}$$

5.6   Let $f(\hat{U}, \hat{V}, \hat{W}; x, r, \theta)$ denote the joint PDF of U, V, and W in the turbulent round jet: $\hat{U}$, $\hat{V}$, and $\hat{W}$ are the sample-space variables. The flow is statistically axisymmetric:

$$\frac{\partial f}{\partial \theta} = 0, \tag{5.28}$$

and it is invariant under a reflection of the circumferential coordinate direction:

$$f(\hat{U}, \hat{V}, \hat{W}; x, r, \theta) = f(\hat{U}, \hat{V}, -\hat{W}; x, r, -\theta). \tag{5.29}$$

Show that, for $\theta = 0$, Eq. (5.29) implies that $\langle W \rangle$, $\langle UW \rangle$, and $\langle VW \rangle$ are zero, and that (in view of Eq. (5.28)) these quantities are zero everywhere.

Draw a sketch of the r–θ plane and show that (for given x, r, and θ) the directions corresponding to $V(x, r, \theta), -V(x, r, \theta + \pi)$, $W(x, r, \theta - \pi/2)$, and $-W(x, r, \theta + \pi/2)$ are the same. Hence argue that (for statistically axisymmetric flows)

$$f(\hat{U}, \pm\hat{V}, \pm\hat{W}; x, 0, \theta) = f(\hat{U}, \pm\hat{W}, \pm\hat{V}; x, 0, 0). \tag{5.30}$$

Show that on the axis $\langle V^2 \rangle$ and $\langle W^2 \rangle$ are equal, and that $\langle V \rangle$ and $\langle UV \rangle$ are zero.

5.1.3 Reynolds stresses

The fluctuating velocity components in the x, r, and θ coordinate directions are denoted by u, v, and w. In the turbulent round jet, the Reynolds-stress tensor is

$$\begin{bmatrix} \langle u^2 \rangle & \langle uv \rangle & 0 \\ \langle uv \rangle & \langle v^2 \rangle & 0 \\ 0 & 0 & \langle w^2 \rangle \end{bmatrix}. \tag{5.31}$$

That is, because of the circumferential symmetry, $\langle uw \rangle$ and $\langle vw \rangle$ are zero (see Exercise 5.6). The geometry of the flow also dictates that the normal stresses are even functions of r, while the shear stress $\langle uv \rangle$ is an odd function. As the axis $r = 0$ is approached, the radial V and circumferential W components of velocity become indistinguishable. Hence $\langle v^2 \rangle$ and $\langle w^2 \rangle$ are equal on the axis.

Consider the r.m.s. axial velocity on the centerline

$$u_0'(x) \equiv \langle u^2 \rangle_{r=0}^{1/2}. \tag{5.32}$$

How does $u_0'(x)$ vary with x? Or, in terms of non-dimensional quantities, how does $u_0'(x)/U_0(x)$ vary with x/d and Re? Again the answer is simple, but very revealing. After the development region, $u_0'(x)/U_0(x)$ tends asymptotically to a constant value of approximately 0.25 (see, e.g., Panchapakesan and Lumley (1993a)). Thus, like $U_0(x)$, $u_0'(x)$ decays as x^{-1}. There is some variation in u_0'/U_0 from experiment to experiment, but no systematic dependence on Re has been documented.

As might be expected from the above observations, it is found that the Reynolds stresses become self-similar. That is, the profiles of $\langle u_i u_j \rangle / U_0(x)^2$ plotted against $r/r_{1/2}$ or $\eta \equiv r/(x - x_0)$ collapse for all x beyond the development region. Figure 5.7 shows the self-similar profiles measured by Hussein *et al.* (1994). Some important observations from these data are the following.

(i) On the centerline, the r.m.s. velocity is about 25% of the mean.
(ii) Toward the edge of the jet, although the Reynolds stresses decay (with increasing $r/r_{1/2}$), the ratio of the r.m.s. to the local mean increases without bound (see Fig. 5.8).
(iii) The Reynolds stresses exhibit significant anisotropy, which is revealed both by the shear stress and by the differences in the normal stresses.
(iv) The relative magnitude of the shear stress can be quantified by the ratio $\langle uv \rangle / k$, and by the u–v correlation coefficient, ρ_{uv} (see Fig. 5.9). The two curves have the same shape with a flat central portion, with $\langle uv \rangle / k \approx 0.27$ and $\rho_{uv} \approx 0.4$.

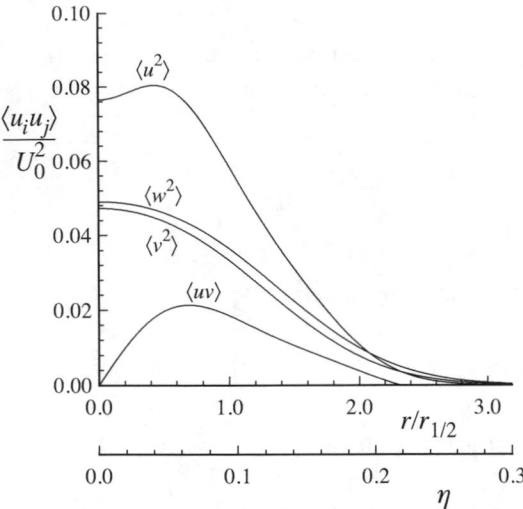

Fig. 5.7. Profiles of Reynolds stresses in the self-similar round jet: curve fit to the LDA data of Hussein *et al.* (1994).

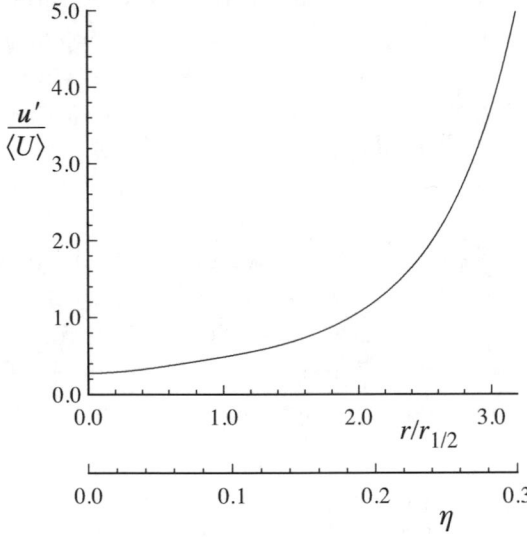

Fig. 5.8. The profile of the local turbulence intensity – $\langle u^2 \rangle^{1/2}/\langle U \rangle$ – in the self-similar round jet. From the curve fit to the experimental data of Hussein *et al.* (1994).

(v) The shear stress is positive where $\partial \langle U \rangle / \partial r$ is negative, and goes to zero where $\partial \langle U \rangle / \partial r$ goes to zero. Hence, for this flow, there is a positive turbulent viscosity ν_{T} such that

$$\langle uv \rangle = -\nu_{\mathrm{T}} \frac{\partial \langle U \rangle}{\partial r}. \tag{5.33}$$

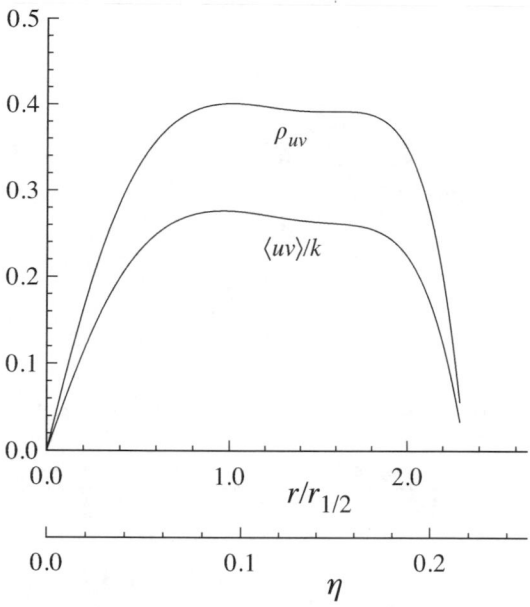

Fig. 5.9. Profiles of $\langle uv \rangle / k$ and the u–v correlation coefficient ρ_{uv} in the self-similar round jet. From the curve fit to the experimental data of Hussein *et al.* (1994).

(vi) Since the profiles of $\langle uv \rangle$ and $\partial \langle U \rangle / \partial r$ are self-similar, evidently the profile of turbulent viscosity defined by Eq. (5.33) is also self-similar. Specifically

$$\nu_T(x, r) = U_0(x)\, r_{1/2}(x)\, \hat{\nu}_T(\eta), \qquad (5.34)$$

where $\hat{\nu}_T$ is the normalized profile – which is shown in Fig. 5.10. It may be observed that $\hat{\nu}_T$ is fairly uniform over the bulk of the jet – within 15% of 0.028 for $0.1 < r/r_{1/2} < 1.5$ – but that it decreases to zero toward the edge.

(vii) The turbulent viscosity has dimensions of velocity times length. Consequently a local lengthscale, $l(x, r)$, can be defined by

$$\nu_T = u'l, \qquad (5.35)$$

where $u'(x, r)$ is the local r.m.s. axial velocity $\langle u^2 \rangle^{1/2}$. Clearly l is self-similar. The profile of $l/r_{1/2}$ (Fig. 5.11) is quite flat, being within 15% of the value 0.12 over most of the jet ($0.1 < r/r_{1/2} < 2.1$).

The lengthscale, l, defined in Eq. (5.35) is a derived quantity, rather than being directly measurable and having a clear physical significance. On the other hand, the integral lengthscales (obtained from the two-point

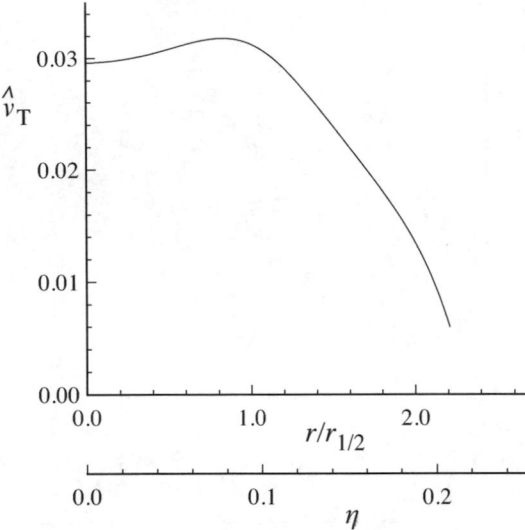

Fig. 5.10. The normalized turbulent diffusivity $\hat{v}_T$ (Eq. (5.34)) in the self-similar round jet. From the curve fit to the experimental data of Hussein *et al.* (1994).

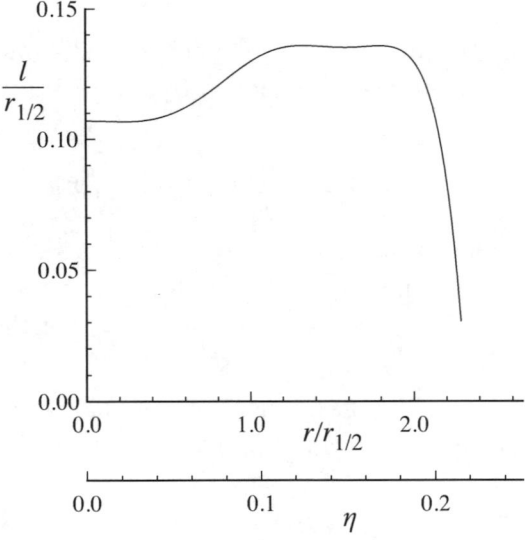

Fig. 5.11. The profile of the lengthscale defined by Eq. (5.35) in the self-similar round jet. From the curve fit to the experimental data of Hussein *et al.* (1994).

velocity correlations) are measurable and characterize the distance over which the fluctuating velocity field is correlated. Wygnanski and Fiedler (1969) measured two-point correlations of the axial velocity, and found them to be self-similar for $x/d > 30$. The longitudinal and lateral correlations are

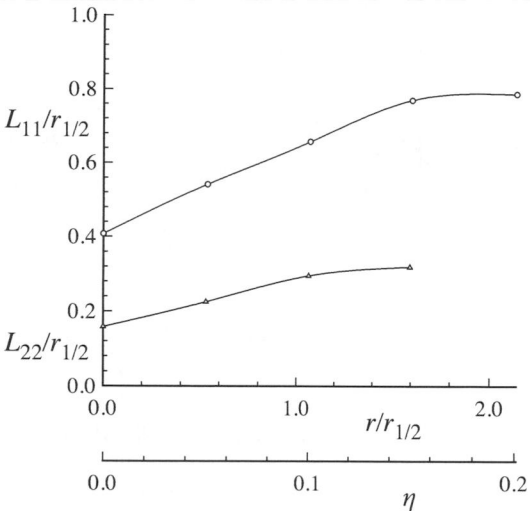

Fig. 5.12. Self-similar profiles of the integral lengthscales in the turbulent round jet. From Wygnanski and Fiedler (1969).

defined by

$$\bar{R}_1(x,r,s) \equiv \frac{\langle u(x+\frac{1}{2}s,r,\theta)u(x-\frac{1}{2}s,r,\theta)\rangle}{[\langle u(x+\frac{1}{2}s,r,\theta)^2\rangle\langle u(x-\frac{1}{2}s,r,\theta)^2\rangle]^{1/2}}, \tag{5.36}$$

$$\bar{R}_2(x,r,s) \equiv \frac{\langle u(x,r+\frac{1}{2}s,\theta)u(x,r-\frac{1}{2}s,\theta)\rangle}{[\langle u(x,r+\frac{1}{2}s,\theta)^2\rangle\langle u(x,r-\frac{1}{2}s,\theta)^2\rangle]^{1/2}}, \tag{5.37}$$

and then the corresponding integral lengthscales are

$$L_{11}(x,r) \equiv \int_0^\infty \bar{R}_1(x,r,s)\,\mathrm{d}s, \tag{5.38}$$

$$L_{22}(x,r) \equiv \int_0^\infty \bar{R}_2(x,r,s)\,\mathrm{d}s. \tag{5.39}$$

Figure 5.12 shows the measured self-similar profiles of these integral lengthscales. It may be seen that L_{11} and L_{22} are typically $0.7r_{1/2}$ and $0.3r_{1/2}$, respectively – considerably larger than $l \approx 0.1r_{1/2}$. It should also be appreciated that there can be significant correlation for separation distances greater than L_{11}, as Fig. 5.13 shows. Some further characteristics of the self-similar round jet are described in Section 5.4.

EXERCISES

5.7   At $r/r_{1/2} = 1$, the principal axes of the mean rate-of-strain tensor are at approximately 45° to the x axis. Show (from the measurements in

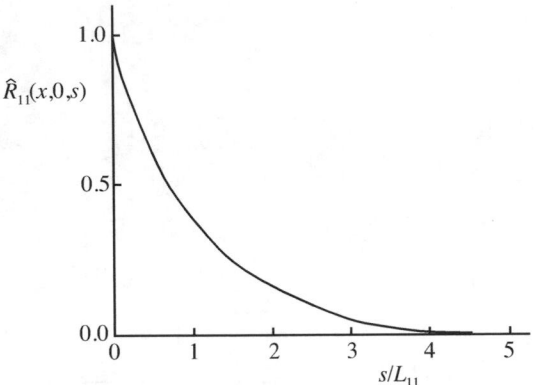

Fig. 5.13. The longitudinal autocorrelation of the axial velocity in the self-similar round jet. From Wygnanski and Fiedler (1969).

Fig. 5.7) that the principal axes of the Reynolds-stress tensor are at an angle of less than 30° to the x axis.

5.8   Compare L_{11} with the visible widths of the jets in Figs. 1.1 and 1.2.

5.9   In polar-cylindrical coordinates ($x, r,$ and θ), the continuity equation is

$$\frac{\partial U}{\partial x} + \frac{1}{r}\frac{\partial}{\partial r}(rV) + \frac{1}{r}\frac{\partial W}{\partial \theta} = 0, \tag{5.40}$$

(where $U, V,$ and W are the velocities in the three coordinate directions); and the Navier–Stokes equations are

$$\frac{\partial U}{\partial t} + U\frac{\partial U}{\partial x} + V\frac{\partial U}{\partial r} + \frac{W}{r}\frac{\partial U}{\partial \theta} = -\frac{1}{\rho}\frac{\partial p}{\partial x} + \nu\nabla^2 U, \tag{5.41}$$

$$\frac{\partial V}{\partial t} + U\frac{\partial V}{\partial x} + V\frac{\partial V}{\partial r} + \frac{W}{r}\frac{\partial V}{\partial \theta} - \frac{W^2}{r} = -\frac{1}{\rho}\frac{\partial p}{\partial r}$$
$$+ \nu\left(\nabla^2 V - \frac{V}{r^2} - \frac{2}{r^2}\frac{\partial W}{\partial \theta}\right), \tag{5.42}$$

$$\frac{\partial W}{\partial t} + U\frac{\partial W}{\partial x} + V\frac{\partial W}{\partial r} + \frac{W}{r}\frac{\partial W}{\partial \theta} + \frac{VW}{r} = -\frac{1}{r\rho}\frac{\partial p}{\partial \theta}$$
$$+ \nu\left(\nabla^2 W + \frac{2}{r^2}\frac{\partial V}{\partial \theta} - \frac{W}{r^2}\right), \tag{5.43}$$

where

$$\nabla^2 f = \frac{\partial^2 f}{\partial x^2} + \frac{1}{r}\frac{\partial}{\partial r}\left(r\frac{\partial f}{\partial r}\right) + \frac{1}{r^2}\frac{\partial^2 f}{\partial \theta^2}, \tag{5.44}$$

(see Batchelor (1967)). In *non-swirling* statistically axisymmetric flows,

$\langle W \rangle$, $\langle uw \rangle$, and $\langle vw \rangle$ are zero. Show that the Reynolds equations for such flows are:

$$\frac{\partial \langle U \rangle}{\partial x} + \frac{1}{r} \frac{\partial}{\partial r}(r \langle V \rangle) = 0, \qquad (5.45)$$

$$\frac{\bar{D} \langle U \rangle}{\bar{D}t} = -\frac{1}{\rho} \frac{\partial \langle p \rangle}{\partial x} - \frac{\partial}{\partial x} \langle u^2 \rangle - \frac{1}{r} \frac{\partial}{\partial r}(r \langle uv \rangle) + v \nabla^2 \langle U \rangle, \qquad (5.46)$$

$$\frac{\bar{D} \langle V \rangle}{\bar{D}t} = -\frac{1}{\rho} \frac{\partial \langle p \rangle}{\partial r} - \frac{\partial}{\partial x} \langle uv \rangle - \frac{1}{r} \frac{\partial}{\partial r}(r \langle v^2 \rangle) + \frac{\langle w^2 \rangle}{r}$$
$$+ v \left(\nabla^2 \langle V \rangle - \frac{\langle V \rangle}{r^2} \right), \qquad (5.47)$$

where

$$\frac{\bar{D}}{\bar{D}t} = \frac{\partial}{\partial t} + \langle U \rangle \frac{\partial}{\partial x} + \langle V \rangle \frac{\partial}{\partial r}. \qquad (5.48)$$

5.2 The round jet: mean momentum

5.2.1 Boundary-layer equations

In the turbulent round jet, there is a dominant mean-flow direction (x), the mean lateral velocity is relatively small ($|\langle V \rangle| \approx 0.03 |\langle U \rangle|$), the flow spreads gradually ($dr_{1/2}/dx \approx 0.1$), and so (for means) axial gradients are small compared with lateral gradients. These features – which are shared by all free shear flows – allow boundary-layer equations to be used in place of the full Reynolds equations. Of course, the turbulent boundary-layer equations also apply to turbulent boundary layers, and to some other wall-bounded flows as well. These flows are discussed in Chapter 7.

We begin by considering statistically two-dimensional, stationary flows in which x is the dominant direction of flow, gradients of means are predominately in the y direction, and statistics do not vary in the z direction. The velocity components are U, V, and W, with $\langle W \rangle$ zero. Examples of such flows are sketched in Fig. 5.14. We consider cases – such as in Fig. 5.14 – in which the upper boundary ($y \rightarrow \infty$) is quiescent or a non-turbulent free stream. For each flow, as functions of x we can define $\delta(x)$ to be the characteristic flow width, $U_c(x)$ to be the characteristic convective velocity, and $U_s(x)$ to be the characteristic velocity difference.

For these flows, the mean continuity and momentum equations are

$$\frac{\partial \langle U \rangle}{\partial x} + \frac{\partial \langle V \rangle}{\partial y} = 0, \qquad (5.49)$$

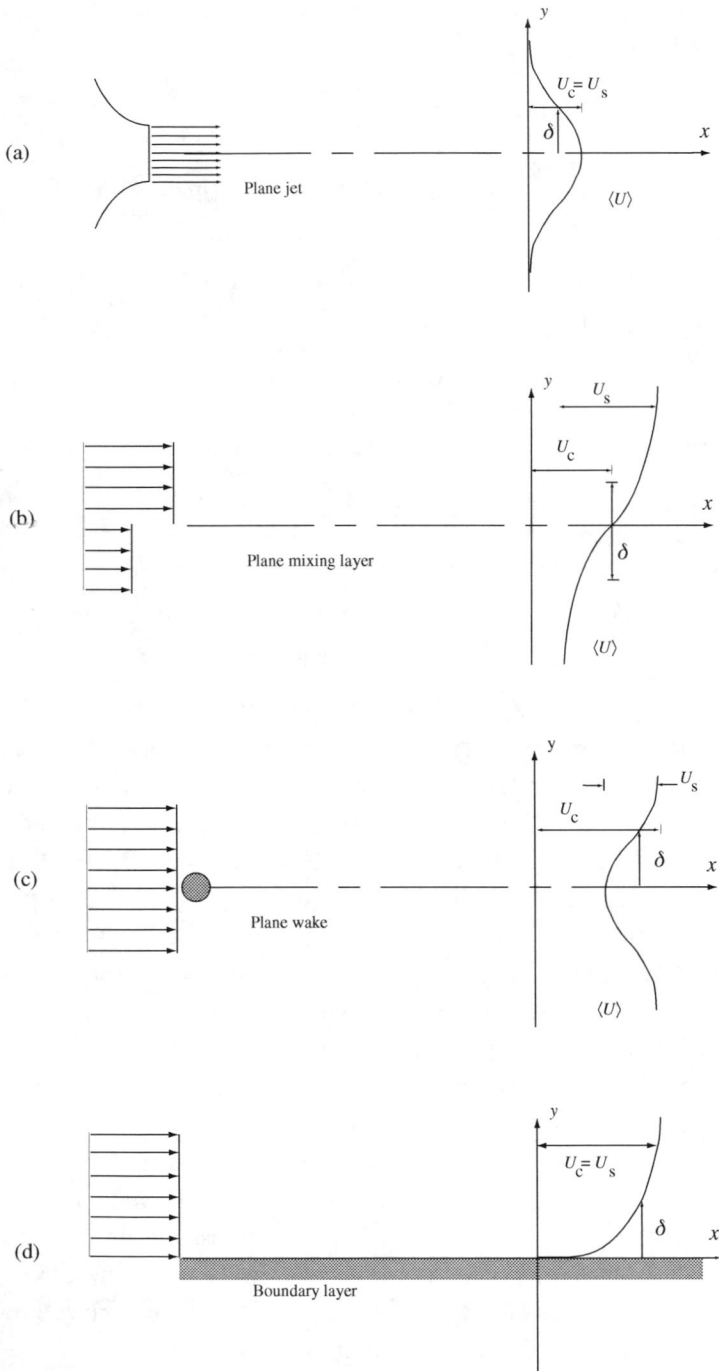

Fig. 5.14. Sketches of plane two-dimensional shear flows showing the characteristic flow width $\delta(x)$, the characteristic convective velocity U_c, and the characteristic velocity difference U_s.

$$\langle U \rangle \frac{\partial \langle U \rangle}{\partial x} + \langle V \rangle \frac{\partial \langle U \rangle}{\partial y} = -\frac{1}{\rho} \frac{\partial \langle p \rangle}{\partial x} + \left\{ v \frac{\partial^2 \langle U \rangle}{\partial x^2} \right\}$$
$$+ v \frac{\partial^2 \langle U \rangle}{\partial y^2} - \frac{\partial \langle u^2 \rangle}{\partial x} - \frac{\partial \langle uv \rangle}{\partial y}, \qquad (5.50)$$

$$\left\{ \langle U \rangle \frac{\partial \langle V \rangle}{\partial x} \right\} + \left\{ \langle V \rangle \frac{\partial \langle V \rangle}{\partial y} \right\} = -\frac{1}{\rho} \frac{\partial \langle p \rangle}{\partial y} + \left\{ v \frac{\partial^2 \langle V \rangle}{\partial x^2} \right\} \qquad (5.51)$$
$$+ \left\{ v \frac{\partial^2 \langle V \rangle}{\partial y^2} \right\} - \frac{\partial \langle uv \rangle}{\partial x} - \frac{\partial \langle v^2 \rangle}{\partial y}.$$

These equations also apply to laminar flow, in which case the Reynolds stresses are zero. The terms in braces ($\{ \; \}$) are neglected in the boundary-layer approximation.

The turbulent boundary-layer equations are obtained simply by neglecting the terms in braces – for the same reasons that they are neglected in the laminar case – and by neglecting the axial derivatives of the Reynolds stresses, on the grounds that they are small compared with the lateral gradients. The lateral momentum equation (Eq. (5.51)) then becomes

$$\frac{1}{\rho} \frac{\partial \langle p \rangle}{\partial y} + \frac{\partial \langle v^2 \rangle}{\partial y} = 0. \qquad (5.52)$$

In the free stream ($y \to \infty$) the pressure is denoted by $p_0(x)$, and $\langle v^2 \rangle$ is zero. Consequently the equation can be integrated to yield

$$\langle p \rangle / \rho = p_0 / \rho - \langle v^2 \rangle, \qquad (5.53)$$

and then the axial pressure gradient is

$$\frac{1}{\rho} \frac{\partial \langle p \rangle}{\partial x} = \frac{1}{\rho} \frac{dp_0}{dx} - \frac{\partial \langle v^2 \rangle}{\partial x}. \qquad (5.54)$$

For flows with quiescent or uniform free streams, the pressure gradient dp_0/dx is zero. In general, dp_0/dx is given in terms of the free-stream velocity by Bernoulli's equation.

In the axial-momentum equation, Eq. (5.50), neglecting the term in braces and substituting Eq. (5.54), we obtain

$$\langle U \rangle \frac{\partial \langle U \rangle}{\partial x} + \langle V \rangle \frac{\partial \langle U \rangle}{\partial y} = v \frac{\partial^2 \langle U \rangle}{\partial y^2} - \frac{1}{\rho} \frac{dp_0}{dx} - \frac{\partial \langle uv \rangle}{\partial y} - \frac{\partial}{\partial x}(\langle u^2 \rangle - \langle v^2 \rangle). \quad (5.55)$$

The first and last terms on the right-hand side require further discussion.

In turbulent free shear flows, $v \, \partial^2 \langle U \rangle / \partial y^2$ is of order $v U_s / \delta^2$, so that, compared with the dominant terms in Eq. (5.55), it is of order Re^{-1}, and hence is negligible. On the other hand, close to the wall in a turbulent boundary layer, the velocity derivatives are very large, and do not scale with U_s and δ. In this case the viscous term $v\,\partial^2\langle U\rangle/\partial y^2$ is of leading order in Eq. (5.55).

In the laminar boundary-layer equations, the axial diffusion term $v\,\partial^2 U/\partial x^2$ is of relative order Re^{-1}, and therefore negligibly small. The comparable axial-stress-gradient term in turbulent boundary-layer flows is the final term in Eq. (5.55). It is consistent to neglect this term; but it should be appreciated that this is not an insignificant approximation. As Exercise 5.11 illustrates, in free shear flows the neglected term can be on the order of 10% of the dominant terms in the equation.

In summary: for statistically two-dimensional, stationary flows that are bounded by quiescent fluid or a uniform stream, the turbulent boundary-layer equations consist of the continuity equation (Eq. (5.49)) and the axial momentum equation

$$\langle U\rangle \frac{\partial\langle U\rangle}{\partial x} + \langle V\rangle \frac{\partial\langle U\rangle}{\partial y} = v\frac{\partial^2\langle U\rangle}{\partial y^2} - \frac{\partial}{\partial y}\langle uv\rangle. \tag{5.56}$$

Except near walls, the viscous term is negligible. The mean pressure distribution is given by Eq. (5.53).

For statistically axisymmetric, stationary non-swirling flows – such as the round jet or the wake behind a sphere – the corresponding turbulent boundary-layer equations are

$$\frac{\partial\langle U\rangle}{\partial x} + \frac{1}{r}\frac{\partial(r\langle V\rangle)}{\partial r} = 0, \tag{5.57}$$

$$\langle U\rangle \frac{\partial\langle U\rangle}{\partial x} + \langle V\rangle \frac{\partial\langle U\rangle}{\partial r} = \frac{v}{r}\frac{\partial}{\partial r}\left(r\frac{\partial\langle U\rangle}{\partial r}\right) - \frac{1}{r}\frac{\partial}{\partial r}(r\langle uv\rangle). \tag{5.58}$$

The mean pressure distribution is

$$\langle p\rangle/\rho = p_0/\rho - \langle v^2\rangle + \int_r^\infty \frac{\langle v^2\rangle - \langle w^2\rangle}{r'}\,dr', \tag{5.59}$$

and the axial-stress-gradient term, neglected on the right-hand side of Eq. (5.58), is

$$-\frac{\partial}{\partial x}\left(\langle u^2\rangle - \langle v^2\rangle + \int_r^\infty \frac{\langle v^2\rangle - \langle w^2\rangle}{r'}\,dr'\right). \tag{5.60}$$

EXERCISES

5.10   Starting from the Reynolds equations in polar-cylindrical coordinates (Eqs. (5.45)–(5.47)), verify that the corresponding boundary-layer equations are Eqs. (5.57)–(5.60)).

5.11   For the centerline of the self-similar round jet, obtain the following estimates for terms in – or omitted from – the boundary-layer equations:

$$\frac{r_{1/2}}{U_0^2}\left(\langle U\rangle\frac{\partial\langle U\rangle}{\partial x}\right)_{r=0} = -S \approx -0.094,$$

$$\frac{r_{1/2}}{U_0^2}\left(\frac{1}{r}\frac{\partial(r\langle uv\rangle)}{\partial r}\right)_{r=0} = \frac{r_{1/2}}{U_0^2}\left(2\frac{\partial\langle uv\rangle}{\partial r}\right)_{r=0} \approx 0.1,$$

$$\frac{r_{1/2}}{U_0^2}\left(\frac{\partial\langle u^2\rangle}{\partial x}\right)_{r=0} = -2S\frac{\langle u^2\rangle_{r=0}}{U_0^2} \approx -0.014. \tag{5.61}$$

5.2.2 Flow rates of mass, momentum, and energy

We return to the turbulent round jet to make some fundamental observations that stem from conservation of momentum. Neglecting the viscous term and multiplying by r, the momentum equation (Eq. (5.58)) becomes

$$\frac{\partial}{\partial x}(r\langle U\rangle^2) + \frac{\partial}{\partial r}(r\langle U\rangle\langle V\rangle + r\langle uv\rangle) = 0. \tag{5.62}$$

(The continuity equation Eq. (5.57) is used to write the convective term in conservative form.) Integrating with respect to r we obtain

$$\frac{\mathrm{d}}{\mathrm{d}x}\int_0^\infty r\langle U\rangle^2\,\mathrm{d}r = -\left[r\langle U\rangle\langle V\rangle + r\langle uv\rangle\right]_0^\infty$$

$$= 0, \tag{5.63}$$

since, for large r, $\langle UV\rangle$ tends to zero more rapidly than does r^{-1}. The momentum flow rate of the mean flow is

$$\dot{M}(x) \equiv \int_0^\infty 2\pi r\rho\langle U\rangle^2\,\mathrm{d}r. \tag{5.64}$$

We see then from Eq. (5.63) that *the momentum flow rate is conserved*: $\dot{M}(x)$ is independent of x. The same conclusion holds for all jets (issuing into quiescent surroundings or uniform streams) and wakes (in uniform streams).

The mean velocity profile in the self-similar round jet can be written

$$\langle U(x,r,0)\rangle = U_0(x)\bar{f}(\xi), \tag{5.65}$$

where

$$\xi \equiv r/r_{1/2}(x), \tag{5.66}$$

and $\bar{f}(\xi)$ is the similarity profile (as a function of ξ rather than η, Eq. (5.12)). The momentum flow rate (Eq. (5.64)) can then be rewritten

$$\dot{M} = 2\pi\rho \left(r_{1/2}U_0\right)^2 \int_0^\infty \xi\bar{f}(\xi)^2 \, d\xi. \qquad (5.67)$$

The integral is a non-dimensional constant, determined by the shape of the profile, but independent of x. Since $\dot{M}$ is independent of x, the product $r_{1/2}(x)U_0(x)$ must therefore also be independent of x. Given the experimental observation that the jet spreads linearly ($dr_{1/2}/dx = S = $ constant), it is inevitable then that the mean velocity $U_0(x)$ decays as x^{-1}.

For the self-similar round jet, the flow rates of mass $\dot{m}(x)$ and kinetic energy $\dot{E}(x)$ associated with the mean velocity field are

$$\dot{m}(x) \equiv \int_0^\infty 2\pi r\rho\langle U\rangle \, dr$$

$$= 2\pi\rho r_{1/2} \left(r_{1/2}U_0\right) \int_0^\infty \xi\bar{f}(\xi) \, d\xi, \qquad (5.68)$$

$$\dot{E}(x) \equiv \int_0^\infty 2\pi r\rho\tfrac{1}{2}\langle U\rangle^3 \, dr$$

$$= \frac{\pi\rho}{r_{1/2}} \left(r_{1/2}U_0\right)^3 \int_0^\infty \xi\bar{f}(\xi)^3 \, d\xi. \qquad (5.69)$$

Since the integrals of $\bar{f}$ and the product $r_{1/2}U_0$ are independent of x, it may be seen that the mass flow rate is linearly proportional to $r_{1/2}$ – and therefore to x – and the energy flow rate is inversely proportional to $r_{1/2}$ (and x).

5.2.3 Self-similarity

For the turbulent round jet, an empirical observation is that the profiles of $\langle U\rangle/U_0(x)$ and $\langle u_iu_j\rangle/U_0(x)^2$ as functions $\xi \equiv r/r_{1/2}(x)$ become self-similar (i.e., independent of x). The self-similar profile of $\langle U\rangle$ is $\bar{f}(\xi)$, Eq. (5.65), and we define that of $\langle uv\rangle$ to be

$$\bar{g}(\xi) \equiv \langle uv\rangle/U_0(x)^2. \qquad (5.70)$$

We now show, from the boundary-layer equations, that this self-similar behavior implies that the jet spreads linearly ($dr_{1/2}/dx = S = $ constant), and consequently that $U_0(x)$ decays as x^{-1} – as is of course observed.

Assuming the flow to be self-similar, and neglecting the viscous term, the boundary-layer momentum equation (Eq. (5.58)) can be written

$$[\xi \bar{f}^2] \left\{ \frac{r_{1/2}}{U_0} \frac{\mathrm{d}U_0}{\mathrm{d}x} \right\} - \left[\bar{f}' \int_0^\xi \xi \bar{f} \,\mathrm{d}\xi \right] \left\{ \frac{r_{1/2}}{U_0} \frac{\mathrm{d}U_0}{\mathrm{d}x} + 2 \frac{\mathrm{d}r_{1/2}}{\mathrm{d}x} \right\} = -[(\xi \bar{g})'], \quad (5.71)$$

where a prime denotes differentiation with respect to ξ. (The steps involved in deriving this equation are given in Exercise 5.12.) The terms in square brackets ([ ]) depend only on ξ, while those in braces ({ }) depend only on x. Since the right-hand side depends only on ξ, there can be no x dependence on the left-hand side. Hence the terms in braces are independent of x, i.e.,

$$\frac{r_{1/2}}{U_0} \frac{\mathrm{d}U_0}{\mathrm{d}x} = C, \quad (5.72)$$

$$\frac{r_{1/2}}{U_0} \frac{\mathrm{d}U_0}{\mathrm{d}x} + 2 \frac{\mathrm{d}r_{1/2}}{\mathrm{d}x} = C + 2S, \quad (5.73)$$

where C and S are constants. (This argument depends upon the fact that the ξ-dependent terms are not identically zero.)

By eliminating C from the above two equations we obtain

$$\frac{\mathrm{d}r_{1/2}}{\mathrm{d}x} = S, \quad (5.74)$$

showing that the linear spreading rate of the jet is an inevitable consequence of self-similarity. Equation (5.72) implies that $U_0(x)$ varies as a power of x, but it does not identify the power. However, given that $r_{1/2}$ varies linearly with x, we have already observed that conservation of momentum requires that $U_0(x)$ vary as x^{-1}. From this it follows that the constant C is

$$C \equiv \frac{r_{1/2}}{U_0} \frac{\mathrm{d}U_0}{\mathrm{d}x} = -S. \quad (5.75)$$

EXERCISE _____

5.12　Starting from the equations $\langle U \rangle = U_0(x)\bar{f}(\xi)$ and $\xi = r/r_{1/2}(x)$, show that the derivatives of $\langle U \rangle$ in a self-similar round jet are

$$\frac{r_{1/2}}{U_0} \frac{\partial \langle U \rangle}{\partial x} = \bar{f} \left(\frac{r_{1/2}}{U_0} \frac{\mathrm{d}U_0}{\mathrm{d}x} \right) - \xi \bar{f}' \left(\frac{\mathrm{d}r_{1/2}}{\mathrm{d}x} \right), \quad (5.76)$$

$$\frac{r_{1/2}}{U_0} \frac{\partial \langle U \rangle}{\partial r} = \bar{f}', \quad (5.77)$$

a prime denoting differentiation with respect to ξ. From the continuity equation (Eq. (5.57)), show that the mean lateral velocity is

$$\frac{\langle V \rangle}{U_0} = -\frac{1}{r U_0} \int_0^r \hat{r} \frac{\partial \langle U \rangle}{\partial x} \, \mathrm{d}\hat{r}$$

$$= -\frac{1}{\xi} \int_0^\xi \hat{\xi} \bar{f} \left(\frac{r_{1/2}}{U_0} \frac{\mathrm{d} U_0}{\mathrm{d} x} \right) - \hat{\xi}^2 \bar{f}' \left(\frac{\mathrm{d} r_{1/2}}{\mathrm{d} x} \right) \mathrm{d}\hat{\xi}$$

$$= \xi \bar{f} \left(\frac{\mathrm{d} r_{1/2}}{\mathrm{d} x} \right) - \left(\frac{r_{1/2}}{U_0} \frac{\mathrm{d} U_0}{\mathrm{d} x} + 2 \frac{\mathrm{d} r_{1/2}}{\mathrm{d} x} \right) \frac{1}{\xi} \int_0^\xi \hat{\xi} \bar{f} \, \mathrm{d}\hat{\xi}, \quad (5.78)$$

where $\hat{r}$ and $\hat{\xi}$ are integration variables. By substituting these relations into the boundary-layer momentum equation (Eq. (5.58), neglecting the viscous term), verify Eq. (5.71).

5.2.4 Uniform turbulent viscosity

The turbulent boundary-layer equations exhibit the closure problem: there are two equations (continuity and axial momentum), involving three dependent variables $\langle U \rangle, \langle V \rangle$, and $\langle uv \rangle$. This closure problem is overcome if the turbulent viscosity $\nu_T(x, r)$ can be specified, for then the shear stress is determined by

$$\langle uv \rangle = -\nu_T \frac{\partial \langle U \rangle}{\partial r}. \quad (5.79)$$

For the self-similar round jet we have observed, first, that ν_T scales with $r_{1/2}$ and U_0, i.e.,

$$\nu_T(x, r) = r_{1/2}(x) U_0(x) \hat{\nu}_T(\eta), \quad (5.80)$$

and, second, that over the bulk of the jet $\hat{\nu}_T(\eta)$ is within 15% of the value 0.028. It is reasonable, therefore, to investigate the solution to the boundary-layer equations with $\hat{\nu}_T(\eta)$ taken to be constant, independent of η. In fact, since the product $r_{1/2}(x) U_0(x)$ is independent of x, this corresponds to taking ν_T to be uniform – independent of both x and r – and so the boundary-layer momentum equation becomes

$$\langle U \rangle \frac{\partial \langle U \rangle}{\partial x} + \langle V \rangle \frac{\partial \langle U \rangle}{\partial r} = \frac{\nu_T}{r} \frac{\partial}{\partial r} \left(r \frac{\partial \langle U \rangle}{\partial r} \right). \quad (5.81)$$

(In view of the high Reynolds number assumed, the viscous term has been neglected, although it can be retained simply by replacing ν_T by ν_{eff}.) This is precisely the laminar boundary-layer equation with $\langle U \rangle, \langle V \rangle$, and ν_T in place of U, V, and ν.

The solution to Eq. (5.81) (together with the continuity equation Eq. (5.57))

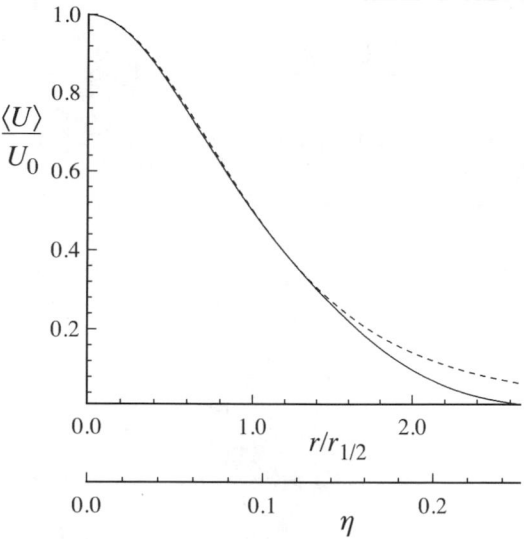

Fig. 5.15. The mean velocity profile in the self-similar round jet: solid line, curve fit to the experimental data of Hussein *et al.* (1994); dashed line, uniform turbulent viscosity solution (Eq. 5.82).

was first obtained by Schlichting (1933). Here we give the solution, discuss some of its consequences, and then give the derivation.

In terms of the similarity profile $f(\eta) = \langle U \rangle / U_0$ with $\eta = r/(x - x_0)$, the solution is

$$f(\eta) = \frac{1}{(1 + a\eta^2)^2}, \tag{5.82}$$

where the coefficient a is given in terms of the spreading rate S by

$$a = \left(\sqrt{2} - 1\right)/S^2, \tag{5.83}$$

(see Exercise 5.3).

This profile (with $S = 0.094$) is compared with the measurements of $\langle U \rangle / U_0$ in Fig. 5.15. There is good agreement between the profiles except at the edge of the jet, where the empirically determined turbulent viscosity $\hat{v}_T(\eta)$ decays to zero (see Fig. 5.10). The spreading rate is determined by the specified normalized viscosity according to

$$S = 8\left(\sqrt{2} - 1\right)\hat{v}_T. \tag{5.84}$$

The spreading rate $S = 0.094$ is obtained with $\hat{v}_T \approx 0.028$, which (not surprisingly) is the average value obtained from the measurements (Fig. 5.10).

The value of $\hat{v}_T$ is sometimes expressed in terms of the *turbulent Reynolds number*

$$R_T \equiv \frac{U_0(x)r_{1/2}(x)}{\nu_T} = \frac{1}{\hat{\nu}_T} \approx 35. \tag{5.85}$$

Thus, in the uniform-turbulent-viscosity approximation, the mean velocity field in the turbulent round jet is the same as the velocity field in a laminar jet of Reynolds number 35.

The solution for uniform turbulent viscosity

The Stokes stream function $\psi(x,r)$ is introduced with

$$\langle U \rangle = \frac{1}{r}\frac{\partial \psi}{\partial r}, \tag{5.86}$$

$$\langle V \rangle = -\frac{1}{r}\frac{\partial \psi}{\partial x}, \tag{5.87}$$

so that the continuity equation (Eq. (5.57)) is automatically satisfied. With x measured from the virtual origin (so that $\eta = r/x$), Eq. (5.86) leads to

$$\psi = \int_0^r r\langle U \rangle \, dr$$
$$= x^2 U_0(x) \int_0^\eta \eta f(\eta) \, d\eta. \tag{5.88}$$

Since $x^2 U_0(x)$ varies linearly with x, it is evident that there is a self-similar scaled stream function $F(\eta)$ such that

$$\psi = \nu_T x F(\eta). \tag{5.89}$$

(The constant ν_T is included so that $F(\eta)$ is non-dimensional.)

From the above equations we obtain

$$\langle U \rangle = \frac{\nu_T}{x}\frac{F'}{\eta}, \tag{5.90}$$

$$\langle V \rangle = \frac{\nu_T}{x}\left(F' - \frac{F}{\eta}\right), \tag{5.91}$$

where $F' = dF/d\eta$. To satisfy the condition that $\langle V \rangle$ is zero on the axis, F must satisfy

$$F(0) = F'(0) = 0. \tag{5.92}$$

All the terms in the boundary-layer equation (Eq. (5.81)) can be expressed in terms of F and its derivatives. After simplification the result is

$$\frac{FF'}{\eta^2} - \frac{F'^2}{\eta} - \frac{FF''}{\eta} = \left(F'' - \frac{F'}{\eta}\right)'. \tag{5.93}$$

The left-hand side is $(-FF'/\eta)'$, so that the equation can be integrated to yield

$$FF' = F' - \eta F''. \tag{5.94}$$

In view of Eq. (5.92), the constant of integration is zero. The equation can be rewritten

$$(\tfrac{1}{2}F^2)' = 2F' - (\eta F')', \tag{5.95}$$

and then integrated a second time, the constant of integration again being zero:

$$\tfrac{1}{2}F^2 = 2F - \eta F', \tag{5.96}$$

or

$$\frac{1}{2F - \tfrac{1}{2}F^2} \frac{\mathrm{d}F}{\mathrm{d}\eta} = \frac{1}{\eta}. \tag{5.97}$$

Integrating a third time, with a constant of integration c, we obtain

$$\tfrac{1}{2}\ln\left(\frac{F}{4 - F}\right) = \ln\eta + c. \tag{5.98}$$

Setting $a = e^{2c}$, the solution is

$$F(\eta) = \frac{4a\eta^2}{1 + a\eta^2}. \tag{5.99}$$

By differentiating this solution, we find the mean velocity profile (Eq. (5.90)) to be

$$\langle U \rangle = \frac{8a\nu_{\mathrm{T}}}{x} \frac{1}{(1 + a\eta^2)^2}. \tag{5.100}$$

Hence the centerline velocity is

$$U_0(x) = \frac{8a\nu_{\mathrm{T}}}{x}, \tag{5.101}$$

and the self-similar profile is

$$f(\eta) = \frac{1}{(1 + a\eta^2)^2}. \tag{5.102}$$

The constant a and the turbulent viscosity ν_{T} can be related to the spreading rate $S = r_{1/2}/x$. Noting that $r = r_{1/2}$ corresponds to $\eta = S$, from the definition of $r_{1/2}$ we require $f(S) = \tfrac{1}{2}$. This leads to

$$a = \left(\sqrt{2} - 1\right)/S^2. \tag{5.103}$$

Then, from Eq. (5.101), we obtain

$$\hat{v}_T = \frac{S}{8\left(\sqrt{2}-1\right)}. \tag{5.104}$$

5.3 The round jet: kinetic energy

The decomposition of the kinetic energy

The kinetic energy of the fluid (per unit mass) is

$$E(x,t) \equiv \tfrac{1}{2}U(x,t) \cdot U(x,t). \tag{5.105}$$

The mean of E can be decomposed into two parts:

$$\langle E(x,t)\rangle = \bar{E}(x,t) + k(x,t), \tag{5.106}$$

where $\bar{E}(x,t)$ is the kinetic energy of the mean flow

$$\bar{E} \equiv \tfrac{1}{2}\langle U\rangle \cdot \langle U\rangle, \tag{5.107}$$

and $k(x,t)$ is the turbulent kinetic energy

$$k \equiv \tfrac{1}{2}\langle u \cdot u\rangle = \tfrac{1}{2}\langle u_i u_i\rangle. \tag{5.108}$$

(This decomposition may be verified by substituting the Reynolds decomposition $U = \langle U\rangle + u$ into Eq. (5.105) and taking the mean.)

The turbulent kinetic energy k determines the isotropic part of the Reynolds stress tensor (i.e., $\tfrac{2}{3}k\delta_{ij}$); but we also find that the anisotropic part scales with k. For example, over much of the turbulent round jet we observe $\langle uv\rangle \approx 0.27k$ (Fig. 5.9), and a mathematical bound on the shear stress is $|uv| \le k$ (Exercise 5.13). Consequently k is a quantity of considerable importance. In this section we consider the processes in turbulent flows that generate and dissipate turbulent kinetic energy. This leads also to a consideration of $E, \langle E\rangle$, and $\bar{E}$.

EXERCISES

5.13 From the Cauchy–Schwarz inequality and the definition of k, show that

$$|\langle uv\rangle| \le k. \tag{5.109}$$

5.14 Show that (for incompressible flow)

$$U \cdot \frac{DU}{Dt} = \frac{DE}{Dt} = \frac{\partial E}{\partial t} + \nabla \cdot (UE). \tag{5.110}$$

From the momentum equation

$$\rho \frac{DU_j}{Dt} = \frac{\partial \tau_{ij}}{\partial x_i}, \tag{5.111}$$

where τ_{ij} is the stress tensor, show that

$$\rho \frac{DE}{Dt} - \frac{\partial}{\partial x_i}(U_j \tau_{ij}) = -S_{ij}\tau_{ij}, \tag{5.112}$$

where $S_{ij} \equiv \frac{1}{2}(\partial U_i/\partial x_j + \partial U_j/\partial x_i)$ is the rate-of-strain tensor. For a Newtonian fluid, τ_{ij} is given by

$$\tau_{ij} = -p\delta_{ij} + 2\rho v S_{ij}. \tag{5.113}$$

Show then that the kinetic-energy equation is

$$\frac{DE}{Dt} + \nabla \cdot \boldsymbol{T} = -2v S_{ij}S_{ij}, \tag{5.114}$$

where

$$T_i \equiv U_i p/\rho - 2v U_j S_{ij}. \tag{5.115}$$

The instantaneous kinetic energy

The equation for the evolution of E, obtained from the Navier–Stokes equations, is

$$\frac{DE}{Dt} + \nabla \cdot \boldsymbol{T} = -2v S_{ij}S_{ij}, \tag{5.116}$$

where $S_{ij} \equiv \frac{1}{2}(\partial U_i/\partial x_j + \partial U_j/\partial x_i)$ is the rate-of-strain tensor, and

$$T_i \equiv U_i p/\rho - 2v U_j S_{ij}, \tag{5.117}$$

is the flux of energy, see Exercise 5.14. The integral of the equation for E over a fixed control volume is

$$\frac{d}{dt} \iiint_{\mathcal{V}} E \, d\mathcal{V} + \iint_{A} (\boldsymbol{U}E + \boldsymbol{T}) \cdot \boldsymbol{n} \, d\mathcal{A} = -\iiint_{\mathcal{V}} 2v S_{ij}S_{ij} \, d\mathcal{V}. \tag{5.118}$$

The surface integral accounts for inflow, outflow, and work done on the control surface: it represents a transfer of E from one region to another. It is important to observe that there is no 'source' of energy within the flow. The quantity $S_{ij}S_{ij}$ – being the sum of squares of components – is positive (or zero if all components of S_{ij} are zero). Consequently the right-hand side is a 'sink' of energy: it represents viscous dissipation – the conversion of mechanical energy into internal energy (heat).

5.15   By expanding $\nabla^2(\frac{1}{2}U_iU_i)$, show that

$$U \cdot \nabla^2 U = \nabla^2 E - \frac{\partial U_i}{\partial x_j}\frac{\partial U_i}{\partial x_j}. \qquad (5.119)$$

Using this result, obtain from the Navier–Stokes equation (Eq. (2.35)) an alternative form of the kinetic-energy equation:

$$\frac{DE}{Dt} + \nabla \cdot \widetilde{T} = -v\frac{\partial U_i}{\partial x_j}\frac{\partial U_i}{\partial x_j}, \qquad (5.120)$$

where

$$\widetilde{T} = Up/\rho - v\,\nabla E. \qquad (5.121)$$

5.16   With S_{ij} being the rate-of-strain tensor and $\Omega_{ij} = \frac{1}{2}(\partial U_i/\partial x_j - \partial U_j/\partial x_i)$ being the rate-of-rotation tensor, show that

$$2S_{ij}S_{ij} = \frac{\partial U_i}{\partial x_j}\frac{\partial U_i}{\partial x_j} + \frac{\partial U_i}{\partial x_j}\frac{\partial U_j}{\partial x_i}, \qquad (5.122)$$

$$2\Omega_{ij}\Omega_{ij} = \frac{\partial U_i}{\partial x_j}\frac{\partial U_i}{\partial x_j} - \frac{\partial U_i}{\partial x_j}\frac{\partial U_j}{\partial x_i}. \qquad (5.123)$$

5.17   Show that Eqs. (5.114) and (5.120) are identical.

5.18   Show that

$$\left\langle \frac{DE}{Dt} \right\rangle = \frac{\bar{D}\langle E \rangle}{\bar{D}t} + \nabla \cdot \langle uE \rangle, \qquad (5.124)$$

and hence (from Eq. (5.114)) show that

$$\frac{\bar{D}\langle E \rangle}{\bar{D}t} + \nabla \cdot (\langle uE \rangle + \langle T \rangle) = -\bar{\varepsilon} - \varepsilon, \qquad (5.125)$$

where $\bar{\varepsilon}$ and ε are defined by Eqs. (5.127) and (5.128).

The mean kinetic energy

The equation for the mean kinetic energy $\langle E \rangle$ is simply obtained by taking the mean of Eq. (5.116):

$$\frac{\bar{D}\langle E \rangle}{\bar{D}t} + \nabla \cdot (\langle uE \rangle + \langle T \rangle) = -\bar{\varepsilon} - \varepsilon, \qquad (5.126)$$

see Exercise 5.18. The two terms on the right-hand side are

$$\bar{\varepsilon} \equiv 2v\bar{S}_{ij}\bar{S}_{ij}, \qquad (5.127)$$

$$\varepsilon \equiv 2v\langle s_{ij}s_{ij} \rangle, \qquad (5.128)$$

where $\bar{S}_{ij}$ and s_{ij} are the mean and fluctuating rates of strain:

$$\bar{S}_{ij} = \langle S_{ij} \rangle = \frac{1}{2} \left(\frac{\partial \langle U_i \rangle}{\partial x_j} + \frac{\partial \langle U_j \rangle}{\partial x_i} \right), \tag{5.129}$$

$$s_{ij} = S_{ij} - \langle S_{ij} \rangle = \frac{1}{2} \left(\frac{\partial u_i}{\partial x_j} + \frac{\partial u_j}{\partial x_i} \right). \tag{5.130}$$

The first contribution, $\bar{\varepsilon}$, is the dissipation due to the mean flow: in general it is of order Re^{-1} compared with the other terms, and therefore negligible. As we shall see, the second contribution, ε, is of central importance.

Mean-flow and turbulent kinetic energy

The equations for $\bar{E} \equiv \frac{1}{2} \langle U \rangle \cdot \langle U \rangle$ and $k \equiv \frac{1}{2} \langle u \cdot u \rangle$ can be written

$$\frac{\bar{\mathrm{D}} \bar{E}}{\bar{\mathrm{D}} t} + \nabla \cdot \bar{T} = -\mathcal{P} - \bar{\varepsilon}, \tag{5.131}$$

$$\frac{\bar{\mathrm{D}} k}{\bar{\mathrm{D}} t} + \nabla \cdot T' = \mathcal{P} - \varepsilon, \tag{5.132}$$

see Exercises 5.19 and 5.20 (where $\bar{T}$ and T' are defined by Eqs. (5.136) and 5.140). The quantity

$$\mathcal{P} \equiv -\langle u_i u_j \rangle \frac{\partial \langle U_i \rangle}{\partial x_j}, \tag{5.133}$$

is generally positive, and hence is a 'source' in the k equation: it is called the production of turbulent kinetic energy – or simply *production*.

EXERCISES

5.19 Starting from the Reynolds equation (Eq. (4.12)) show that the mean-kinetic-energy equation (for $\bar{E} \equiv \frac{1}{2} \langle U \rangle \cdot \langle U \rangle$) is

$$\frac{\bar{\mathrm{D}} \bar{E}}{\bar{\mathrm{D}} t} + \nabla \cdot \bar{T} = -\mathcal{P} - \bar{\varepsilon}, \tag{5.134}$$

where

$$\mathcal{P} \equiv -\langle u_i u_j \rangle \frac{\partial \langle U_i \rangle}{\partial x_j}, \tag{5.135}$$

$$\bar{T}_i \equiv \langle U_j \rangle \langle u_i u_j \rangle + \langle U_i \rangle \langle p \rangle / \rho - 2\nu \langle U_j \rangle \bar{S}_{ij}. \tag{5.136}$$

5.20   By subtracting the Reynolds equations (Eq. (4.12)) from the Navier–Stokes equation (Eq. (2.35)), show that the fluctuating velocity $u(x, t)$ evolves by

$$\frac{\partial u_j}{\partial t} + \frac{\partial}{\partial x_i}(U_i U_j - \langle U_i U_j \rangle) = v\,\nabla^2 u_j - \frac{1}{\rho}\frac{\partial p'}{\partial x_j}, \tag{5.137}$$

or

$$\frac{\bar{\mathrm{D}} u_j}{\bar{\mathrm{D}} t} = -u_i\frac{\partial \langle U_j \rangle}{\partial x_i} + \frac{\partial}{\partial x_i}\langle u_i u_j \rangle + v\,\nabla^2 u_j - \frac{1}{\rho}\frac{\partial p'}{\partial x_j}, \tag{5.138}$$

where p' is the fluctuating pressure field ($p' = p - \langle p \rangle$). Hence show that the turbulent kinetic energy evolves by

$$\frac{\bar{\mathrm{D}} k}{\bar{\mathrm{D}} t} + \nabla \cdot \boldsymbol{T}' = \mathcal{P} - \varepsilon, \tag{5.139}$$

where

$$T_i' \equiv \tfrac{1}{2}\langle u_i u_j u_j \rangle + \langle u_i p' \rangle/\rho - 2v\langle u_j s_{ij} \rangle. \tag{5.140}$$

Production

The equations for $\bar{E}$ and k clearly show the important role played by production. The action of the mean velocity gradients working against the Reynolds stresses removes kinetic energy from the mean flow ($-\mathcal{P}$ in Eq. (5.131) for $\bar{E}$) and transfers it to the fluctuating velocity field ($\mathcal{P}$ in Eq. (5.133) for k).

Some observations concerning production are the following.

(i)   Only the symmetric part of the velocity-gradient tensor affects production, i.e.,

$$\mathcal{P} = -\langle u_i u_j \rangle \bar{S}_{ij}. \tag{5.141}$$

(ii) Only the anisotropic part of the Reynolds-stress tensor affects production, i.e.,

$$\mathcal{P} = -a_{ij}\bar{S}_{ij}, \tag{5.142}$$

where $a_{ij} = \langle u_i u_j \rangle - \tfrac{2}{3}k\delta_{ij}$.

(iii) According to the turbulent-viscosity hypothesis (i.e., $a_{ij} = -2v_\mathrm{T}\bar{S}_{ij}$, Eq. (4.48)) the production is

$$\mathcal{P} = 2v_\mathrm{T}\bar{S}_{ij}\bar{S}_{ij} \geq 0. \tag{5.143}$$

It may be observed that this expression for $\mathcal{P}$ is the same as the dissipation by the mean flow $\bar{\varepsilon}$ (Eq. (5.127)), but with v_T replacing v.

(iv) In the boundary-layer approximation – in which all mean velocity gradients are neglected except for $\partial \langle U \rangle / \partial y$ (or $\partial \langle U \rangle / \partial r$) – the production is

$$\mathcal{P} = -\langle uv \rangle \frac{\partial \langle U \rangle}{\partial y}, \tag{5.144}$$

(or $\mathcal{P} = -\langle uv \rangle \, \partial \langle U \rangle / \partial r$).

(v) According to the turbulent-viscosity hypothesis, in the boundary-layer approximation the production is

$$\mathcal{P} = \nu_{\mathrm{T}} \left(\frac{\partial \langle U \rangle}{\partial y} \right)^2. \tag{5.145}$$

EXERCISE

5.21 Show that the production $\mathcal{P}$ is bounded by

$$|\mathcal{P}| \le 2k\mathcal{S}_\lambda,$$

where $\mathcal{S}_\lambda$ is the largest absolute value of the eigenvalues of the mean rate-of-strain tensor.

Dissipation

In the k equation, the sink ε is the *dissipation of turbulent kinetic energy*, or simply *dissipation*. The fluctuating velocity gradients ($\partial u_i / \partial x_j$) working against the fluctuating deviatoric stresses ($2\nu s_{ij}$) transform kinetic energy into internal energy. (As illustrated in Exercise 5.22, the resulting rise in temperature is almost always negligibly small.) It may be seen from its definition, $\varepsilon \equiv 2\nu \langle s_{ij} s_{ij} \rangle$, that dissipation is non-negative.

To understand the most important characteristic of ε, we return to the self-similar round jet. We have seen that the profiles of $\langle U \rangle / U_0$ and $\langle u_i u_j \rangle / U_0^2$ (as functions of $\xi = r/r_{1/2}$) are self-similar and independent of Re (for sufficiently large x/d and Re). Consequently k/U_0^2 and

$$\widehat{\mathcal{P}} \equiv \mathcal{P} / (U_0^3 / r_{1/2}) \approx -\frac{\langle uv \rangle}{U_0^2} \frac{r_{1/2}}{U_0} \frac{\partial \langle U \rangle}{\partial r} \tag{5.146}$$

are also self-similar and independent of Re. In the balance equation for k, since both $\bar{\mathrm{D}} k / \bar{\mathrm{D}} t$ and $\mathcal{P}$ scale as $U_0^3 / r_{1/2}$, it is almost inevitable that ε has the same scaling: that is

$$\hat{\varepsilon} \equiv \varepsilon / (U_0^3 / r_{1/2}) \tag{5.147}$$

is self-similar, independent of Re. This is confirmed by measurements (e.g., Panchapakesan and Lumley (1993a) and Hussein *et al.* (1994)).

At first sight, this behavior of ε presents a puzzle. Suppose that two high-Reynolds-number round-jet experiments are performed – denoted by a and b – with the same nozzle diameter d and jet velocity U_J, but with fluids of different viscosities ν_a and ν_b. At a given x (in the self-similar region) the velocities $U_0(x)$ and half-widths $r_{1/2}(x)$ in the two experiments are the same. Consequently, at given (x, r), the dissipation in the two experiments is also the same, i.e.,

$$\varepsilon_a(x, r) = \varepsilon_b(x, r) = \hat{\varepsilon}(r/r_{1/2}(x)) \frac{U_0^3(x)}{r_{1/2}(x)}. \qquad (5.148)$$

However, by its definition $\varepsilon \equiv 2\nu \langle s_{ij} s_{ij} \rangle$, ε is directly proportional to ν, which is different in the two experiments! How then can ε_a and ε_b be equal? The seed of the answer can be found in Fig. 1.2 on page 5. It may be seen that the jet with the higher Reynolds number has a finer scale of small structure, and, plausibly, therefore, steeper gradients and higher values of $s_{ij} = \frac{1}{2}(\partial u_i/\partial x_j + \partial u_j/\partial x_i)$.

EXERCISE _____

5.22   Consider a self-similar round jet of an ideal gas. Because of dissipative heating, the centerline temperature $T_0(x)$ is very slightly higher than the temperature of the ambient, T_∞. Use a simple energy balance to obtain the rough estimate

$$\frac{T_0 - T_\infty}{T_\infty} \approx \frac{U_J U_0 - U_0^2}{C_p T_\infty},$$

and hence obtain the estimate for the maximum temperature excess

$$\frac{T_{0,\mathrm{max}} - T_\infty}{T_\infty} \approx \frac{\mathrm{Ma}^2}{4(\gamma - 1)},$$

where Ma is the Mach number based on U_J. (C_p is the constant-pressure specific heat, and γ is the ratio of specific heats.)

_____

Kolmogorov scales

In Chapter 6 we shall see that the characteristic scales of the smallest turbulent motions are the *Kolmogorov scales*. These are the length (η), time (τ_η) and velocity (u_η) scales formed from ε and ν:

$$\eta \equiv \left(\frac{\nu^3}{\varepsilon} \right)^{1/4}, \qquad (5.149)$$

$$\tau_\eta \equiv \left(\frac{\nu}{\varepsilon}\right)^{1/2}, \tag{5.150}$$

$$u_\eta \equiv (\nu\varepsilon)^{1/4}. \tag{5.151}$$

From these definitions and Eq. (5.147) it follows that, compared with the mean-flow scales $r_{1/2}$ and U_0, the Kolmogorov scales vary with the Reynolds number $\mathrm{Re}_0 \equiv U_0 r_{1/2}/\nu$ according to

$$\eta/r_{1/2} = \mathrm{Re}_0^{-3/4} \hat{\varepsilon}^{-1/4}, \tag{5.152}$$

$$\tau_\eta/(r_{1/2}/U_0) = \mathrm{Re}_0^{-1/2} \hat{\varepsilon}^{-1/2}, \tag{5.153}$$

$$u_\eta/U_0 = \mathrm{Re}_0^{-1/4} \hat{\varepsilon}^{1/4}. \tag{5.154}$$

(Recall that $\hat{\varepsilon}$ is non-dimensional and independent of the Reynolds number.) Thus, consistent with visual observations, we see that (relative to the mean flow scales) the smallest motions decrease in size and timescale as the Reynolds number increases.

Two revealing identities stemming from the definitions of the Kolmogorov scales are

$$\frac{\eta u_\eta}{\nu} = 1, \tag{5.155}$$

$$\nu \left(\frac{u_\eta}{\eta}\right)^2 = \frac{\nu}{\tau_\eta^2} = \varepsilon. \tag{5.156}$$

The first shows that – however large Re_0 is – the Reynolds number based on the Kolmogorov scales is unity, indicating that motions on these scales are strongly affected by viscosity. The second shows that the velocity gradients scale in such a way (as $u_\eta/\eta = 1/\tau_\eta$) that ε is independent of ν.

The solution to the puzzle mentioned above is, therefore, that the mean-square strain rate $\langle s_{ij}s_{ij}\rangle$ scales as τ_η^{-2} – inversely proportional to ν – so that $\varepsilon_a = \nu_a\langle s_{ij}s_{ij}\rangle_a$ and $\varepsilon_b = \nu_b\langle s_{ij}s_{ij}\rangle_b$ are equal. The remaining question – why do the small-scale turbulent motions scale this way? – is addressed in Chapter 6.

The budget of the turbulent kinetic energy

For the self-similar round jet, Fig. 5.16 shows the turbulent kinetic-energy budget. The quantities plotted are the four terms in the k equation (Eq. (5.132)) normalized by $U_0^3/r_{1/2}$. The contributions are production, $\mathcal{P}$; dissipation, $-\varepsilon$; mean-flow convection, $-\bar{D}k/\bar{D}t$, and turbulent transport, $-\nabla \cdot \mathbf{T}'$. (Production and mean-flow convection can be measured reliably,

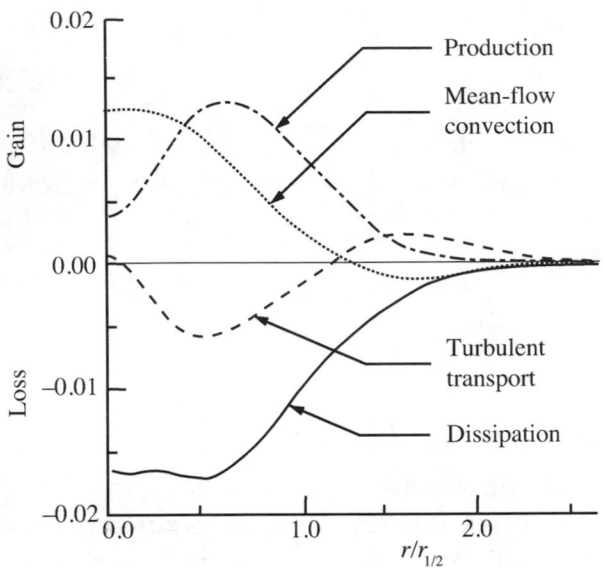

Fig. 5.16. The turbulent-kinetic-energy budget in the self-similar round jet. Quantities are normalized by U_0 and $r_{1/2}$. (From Panchapakesan and Lumley (1993a).)

and there is agreement (to within 20%, say) between different investigations. However, the other two terms are subject to considerable uncertainty, with measurements in different experiments varying by a factor of two or more.) Throughout the jet, dissipation is a dominant term. The production peaks at $r/r_{1/2} \approx 0.6$, where the ratio $\mathcal{P}/\varepsilon$ is about 0.8. On the centerline, $-\langle uv \rangle \, \partial \langle U \rangle / \partial r$ is zero (and varies as r^2), so that the production there is due to the term $-(\langle u^2 \rangle - \langle v^2 \rangle) \partial \langle U \rangle / \partial x$ (which is neglected in the boundary-layer approximation). At the edge of the jet $\mathcal{P}/\varepsilon$ goes to zero, and it is the turbulent transport that balances ε.

Comparison of scales

It is informative to evaluate and compare different rates and timescales associated with the mean flow and k. This is done in Table 5.2 and Fig. 5.17. The timescales τ and $\tau_{\mathcal{P}}$ provide measures of the lifetime of the turbulence in the jet. It takes a time τ to dissipate an amount of energy k at the constant rate ε; and similarly a time $\tau_{\mathcal{P}}$ to produce k at the rate $\mathcal{P}$. These timescales are large and approximately equal: they are comparable to the flight time from the virtual origin τ_J of a particle moving on the centerline at speed $U_0(x)$; and they are about three times the timescale of the imposed shear S^{-1}. *Turbulence is long-lived.*

Figure 5.18 shows a comparison of lengthscales. While the integral scales

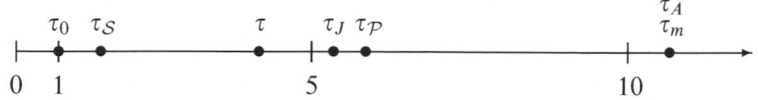

Fig. 5.17. Timescales in the self-similar round jet in units of τ_0. See Table 5.2 for definitions.

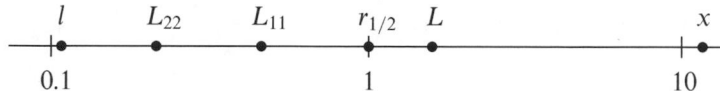

Fig. 5.18. Lengthscales in the self-similar round jet in units of $r_{1/2}$. L_{11} and L_{12} are the longitudinal and lateral integral scales; $L \equiv k^{3/2}/\varepsilon$; $l = \nu_T/u'$; evaluated at $r/r_{1/2} \approx 0.7$. (Note the logarithmic scale.)

Table 5.2. *Timescales, rates, and ratios in the self-similar round jet: the first four entries are evaluated from $U_0(x)$, $r_{1/2}(x)$ and the spreading rate S; the remaining entries are estimated from experimental data at $r/r_{1/2} \approx 0.7$, where $\langle uv \rangle$ and $|\partial\langle U\rangle/\partial r|$ peak*

Definition	Description	Timescale	Value in self-similar round jet, normalized by τ_0
$\tau_0 = r_{1/2}/U_0$	Reference timescale used for normalization	τ_0	1
$\tau_J = \frac{1}{2}x/U_0$	Mean flight time from virtual origin	τ_J	5.3
$\Omega_m = \dfrac{U_0}{\dot{m}}\dfrac{\mathrm{d}\dot{m}}{\mathrm{d}x}$	Entrainment rate	$\tau_m = \Omega_m^{-1}$	10.6
$\Omega_A = \left\| \dfrac{\mathrm{d}U_0}{\mathrm{d}x} \right\|$	Axial strain rate	$\tau_A = \Omega_A^{-1}$	10.6
$S = (2\bar{S}_{ij}\bar{S}_{ij})^{1/2}$ $\approx \left\| \dfrac{\partial\langle U\rangle}{\partial r} \right\|$	Strain rate	$\tau_S = S^{-1}$	1.7
$\omega = \varepsilon/k$	Turbulence decay rate	$\tau = \omega^{-1} = k/\varepsilon$	4.5
$\Omega_P = \mathcal{P}/k$	Turbulence-production rate	$\tau_P = \Omega_P^{-1}$	5.7
$\mathcal{P}/\varepsilon$	Ratio of production to dissipation		0.8
$S/\omega = Sk/\varepsilon$ $= \tau/\tau_S$	Ratio of strain rate to decay rate		2.6

L_{11} and L_{22} have a direct physical significance, $l \equiv v_{\mathrm{T}}/u'$ and

$$L \equiv k^{3/2}/\varepsilon \tag{5.157}$$

do not.

5.23   Referring to Table 5.2, verify that

$$\tau_{\mathrm{J}} = \tfrac{1}{2}x/U_0, \tag{5.158}$$

$$\tau_0\Omega_m = S,$$

$$\tau_0\Omega_{\mathrm{A}} = S.$$

5.24   For the normalized velocity profile $f(\eta) = 1/(1+a\eta^2)^2$, show that the
maximum of $|\partial\langle U\rangle/\partial r|$ is

$$\frac{r_{1/2}}{U_0}\left|\frac{\partial\langle U\rangle}{\partial r}\right|_{\mathrm{max}} = \frac{25}{54}\sqrt{\left(\sqrt{50}-5\right)} \approx 0.67,$$

and occurs at

$$r/r_{1/2} = \left(\sqrt{50}-5\right)^{-1/2} \approx 0.69.$$

Pseudo-dissipation

The *pseudo-dissipation* $\tilde{\varepsilon}$ is defined by

$$\tilde{\varepsilon} \equiv v\left\langle\frac{\partial u_i}{\partial x_j}\frac{\partial u_i}{\partial x_j}\right\rangle, \tag{5.159}$$

and is related to the true dissipation ε by

$$\tilde{\varepsilon} = \varepsilon - v\frac{\partial^2\langle u_i u_j\rangle}{\partial x_i\,\partial x_j}, \tag{5.160}$$

(see Exercise 5.25). In virtually all circumstances, the final term in Eq. (5.160)
is small (at most a few percent of ε) and consequently the distinction between
ε and $\tilde{\varepsilon}$ is seldom important. Indeed, many authors refer to $\tilde{\varepsilon}$ as 'dissipation.'
As Exercise 5.27 illustrates, some equations have a simpler form when $\tilde{\varepsilon}$ is
used in place of ε.

EXERCISES

5.25 Obtain the following relationship between the dissipation ε and the pseudo-dissipation $\tilde{\varepsilon}$:

$$\varepsilon \equiv 2\nu \langle s_{ij}s_{ij} \rangle = \nu \left\langle \frac{\partial u_i}{\partial x_j} \frac{\partial u_i}{\partial x_j} + \frac{\partial u_i}{\partial x_j} \frac{\partial u_j}{\partial x_i} \right\rangle$$

$$= \tilde{\varepsilon} + \nu \frac{\partial^2 \langle u_i u_j \rangle}{\partial x_i \partial x_j}. \tag{5.161}$$

5.26 The dissipation and viscous-diffusion terms in the turbulent-kinetic-energy equation arise from the expression $\nu \langle u_i \nabla^2 u_i \rangle$. Show that this can be re-expressed in the alternative forms

$$\nu \langle u_i \nabla^2 u_i \rangle = \nu \left\langle u_i \frac{\partial}{\partial x_j} \left(\frac{\partial u_i}{\partial x_j} \right) \right\rangle$$

$$= \nu \nabla^2 k - \tilde{\varepsilon}, \tag{5.162}$$

and

$$\nu \langle u_i \nabla^2 u_i \rangle = \nu \left\langle u_i \frac{\partial}{\partial x_j} \left(\frac{\partial u_i}{\partial x_j} + \frac{\partial u_j}{\partial x_i} \right) \right\rangle$$

$$= 2\nu \frac{\partial}{\partial x_j} \langle u_i s_{ij} \rangle - \varepsilon. \tag{5.163}$$

5.27 Show that the turbulent-kinetic-energy equation (Eq. (5.139)) can alternatively be written

$$\frac{\bar{D}k}{\bar{D}t} + \frac{\partial}{\partial x_i} \left[\tfrac{1}{2} \langle u_i u_j u_j \rangle + \langle u_i p' \rangle / \rho \right] = \nu \nabla^2 k + \mathcal{P} - \tilde{\varepsilon}. \tag{5.164}$$

5.28 In homogeneous isotropic turbulence, the fourth-order tensor

$$\left\langle \frac{\partial u_i}{\partial x_j} \frac{\partial u_k}{\partial x_\ell} \right\rangle$$

is isotropic, and hence can be written

$$\left\langle \frac{\partial u_i}{\partial x_j} \frac{\partial u_k}{\partial x_\ell} \right\rangle = \alpha \delta_{ij} \delta_{k\ell} + \beta \delta_{ik} \delta_{j\ell} + \gamma \delta_{i\ell} \delta_{jk}, \tag{5.165}$$

where $\alpha, \beta,$ and γ are scalars. In view of the continuity equation $\partial u_i / \partial x_i = 0$, show that a relation between the scalars is

$$3\alpha + \beta + \gamma = 0. \tag{5.166}$$

By considering $(\partial/\partial x_j)\langle u_i\, \partial u_j/\partial x_\ell\rangle$ (which is zero on account of homogeneity) show that

$$\left\langle \frac{\partial u_i}{\partial x_j} \frac{\partial u_j}{\partial x_\ell}\right\rangle$$

is zero, and hence

$$\alpha + \beta + 3\gamma = 0. \tag{5.167}$$

Show that Eq. (5.165) then becomes

$$\left\langle \frac{\partial u_i}{\partial x_j} \frac{\partial u_k}{\partial x_\ell}\right\rangle = \beta\left(\delta_{ik}\delta_{j\ell} - \tfrac{1}{4}\delta_{ij}\delta_{k\ell} - \tfrac{1}{4}\delta_{i\ell}\delta_{jk}\right). \tag{5.168}$$

Show that

$$\left\langle \left(\frac{\partial u_1}{\partial x_1}\right)^2\right\rangle = \tfrac{1}{2}\beta, \quad \left\langle \left(\frac{\partial u_1}{\partial x_2}\right)^2\right\rangle = 2\left\langle \left(\frac{\partial u_1}{\partial x_1}\right)^2\right\rangle, \tag{5.169}$$

$$\left\langle \frac{\partial u_1}{\partial x_1} \frac{\partial u_2}{\partial x_2}\right\rangle = \left\langle \frac{\partial u_1}{\partial x_2} \frac{\partial u_2}{\partial x_1}\right\rangle = -\tfrac{1}{2}\left\langle \left(\frac{\partial u_1}{\partial x_1}\right)^2\right\rangle, \tag{5.170}$$

$$\varepsilon = v\left\langle \frac{\partial u_i}{\partial x_j} \frac{\partial u_i}{\partial x_j}\right\rangle = \tfrac{15}{2}v\beta = 15v\left\langle \left(\frac{\partial u_1}{\partial x_1}\right)^2\right\rangle. \tag{5.171}$$

5.4 Other self-similar flows

We have looked in some detail at the mean velocity and Reynolds stresses in the self-similar round jet. We now examine briefly the other classical free shear flows – the plane jet, the mixing layer, the plane and axisymmetric wakes – and also homogeneous shear flow (i.e., homogeneous turbulence subjected to a constant and uniform mean shear). Then, in Section 5.5, some other features of these flows are described.

5.4.1 *The plane jet*

The ideal plane jet (which is sketched in Fig. 5.14 on page 112) is statistically two-dimensional. The dominant direction of mean flow is x, the cross-stream coordinate is y, and statistics are independent of the spanwise coordinate, z. There is statistical symmetry about the plane $y = 0$. In laboratory experiments, there is a rectangular nozzle with slot height d (in the y direction), and width w (in the z direction). The aspect ratio w/d must be large (typically

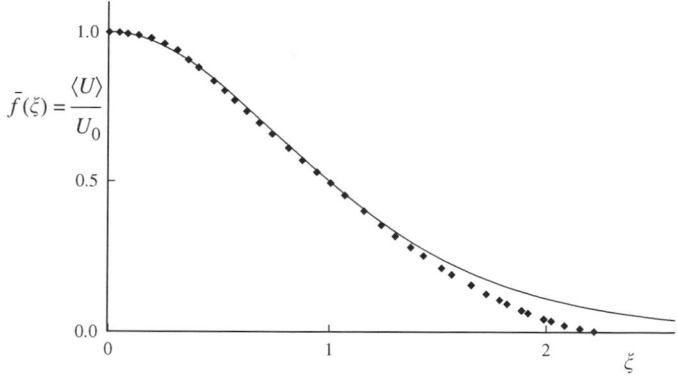

Fig. 5.19. The mean velocity profile in the self-similar plane jet. Symbols, experimental data of Heskestad (1965); line, uniform turbulent-viscosity solution, Eq. (5.187) (with permission of ASME).

50) so that, for $z = 0$ (to a good approximation), the flow is statistically two-dimensional and free of end effects, at least for x/w not too large.

Just as in the round jet, the centerline velocity, $U_0(x)$, and half-width, $y_{1/2}(x)$, are defined by

$$U_0(x) \equiv \langle U(x,0,0) \rangle, \tag{5.172}$$

$$\tfrac{1}{2} U_0(x) \equiv \langle U(x, y_{1/2}(x), 0) \rangle. \tag{5.173}$$

In experiments (e.g., Heskestad (1965), Bradbury (1965), and Gutmark and Wygnanski (1976)) it is found that the mean-velocity and Reynolds-stress profiles become self-similar (beyond about $x/d = 40$) when they are scaled with $U_0(x)$ and $y_{1/2}(x)$. These profiles are shown in Figs. 5.19 and 5.20. It may be seen that the profile shapes and the levels of the Reynolds stresses are comparable to those observed in the round jet.

The variation of $y_{1/2}(x)$ is found to be linear, i.e.,

$$\frac{\mathrm{d} y_{1/2}}{\mathrm{d}x} = S, \tag{5.174}$$

where the rate of spreading S is a constant, $S \approx 0.10$. However, in contrast to the round jet, it is found that $U_0(x)$ varies as $x^{-1/2}$. As is now shown, these variations are consequences of self-similarity.

In conservative form, the boundary-layer equation (neglecting the viscous term) is

$$\frac{\partial}{\partial x}\langle U \rangle^2 + \frac{\partial}{\partial y}(\langle U \rangle \langle V \rangle) = -\frac{\partial}{\partial y}\langle uv \rangle. \tag{5.175}$$

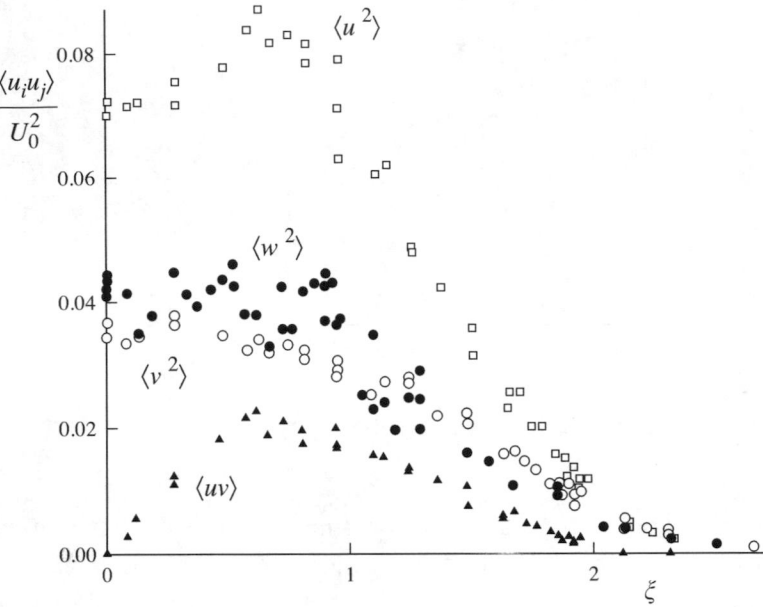

Fig. 5.20. Reynolds-stress profiles in the self-similar plane jet. From the measurements of Heskestad (1965) (with permission of ASME).

Integrating with respect to y, we obtain

$$\frac{\mathrm{d}}{\mathrm{d}x} \int_{-\infty}^{\infty} \langle U \rangle^2 \, \mathrm{d}y = 0, \qquad (5.176)$$

since $\langle U \rangle$ and $\langle uv \rangle$ are zero for $y \to \pm\infty$. Hence the momentum flow rate (per unit span)

$$\dot{M} \equiv \int_{-\infty}^{\infty} \rho \langle U \rangle^2 \, \mathrm{d}y, \qquad (5.177)$$

is conserved (independent of x).

In the self-similar region, the mean axial-velocity profile is

$$\langle U \rangle = U_0(x)\bar{f}(\xi), \qquad (5.178)$$

where

$$\xi \equiv y/y_{1/2}(x), \qquad (5.179)$$

and so the momentum flow rate is

$$\dot{M} = \rho U_0(x)^2 y_{1/2}(x) \int_{-\infty}^{\infty} \bar{f}(\xi)^2 \, \mathrm{d}\xi. \qquad (5.180)$$

Evidently the product $U_0(x)^2 y_{1/2}(x)$ is independent of x – consistent with the observations – from which we deduce that

$$\frac{y_{1/2}}{U_0} \frac{\mathrm{d}U_0}{\mathrm{d}x} = -\frac{1}{2} \frac{\mathrm{d}y_{1/2}}{\mathrm{d}x}. \tag{5.181}$$

On substituting the self-similar profiles for $\langle U \rangle$, Eq. (5.178), and $\langle uv \rangle$,

$$\langle uv \rangle = U_0^2 \bar{g}(\xi), \tag{5.182}$$

into the boundary-layer equation, we obtain

$$\frac{1}{2} \frac{\mathrm{d}y_{1/2}}{\mathrm{d}x} \left(\bar{f}^2 + \bar{f}' \int_0^\xi \bar{f} \, \mathrm{d}\xi \right) = \bar{g}'. \tag{5.183}$$

(This involves the same manipulations as those for the round jet, see Exercise 5.12.) In Eq. (5.183), since the right-hand side and the term in parentheses are independent of x, it follows that $\mathrm{d}y_{1/2}/\mathrm{d}x$ must also be independent of x. Thus, self-similarity requires that the rate of spreading $S \equiv \mathrm{d}y_{1/2}/\mathrm{d}x$ be constant; and then conservation of momentum requires that U_0 vary as $x^{-1/2}$ (Eq. (5.180)).

The self-similarity of $\langle U \rangle$ and $\langle uv \rangle$ imply the self-similarity of the turbulent viscosity, i.e.,

$$\nu_{\mathrm{T}}(x, y) = U_0(x) y_{1/2}(x) \hat{\nu}_{\mathrm{T}}(\xi). \tag{5.184}$$

It may be seen that ν_{T} for the self-similar plane jet increases as $x^{1/2}$, as does the local Reynolds number

$$\mathrm{Re}_0(x) = U_0(x) y_{1/2}(x)/\nu. \tag{5.185}$$

On the other hand, the turbulent Reynolds number

$$\mathrm{R}_{\mathrm{T}} = U_0(x) y_{1/2}(x)/\nu_{\mathrm{T}}(x, y_{1/2}), \tag{5.186}$$

is independent of x.

If the turbulent viscosity is taken to be uniform across the flow (i.e., $\hat{\nu}_{\mathrm{T}} = $ constant), then the self-similar form of the boundary-layer equation (Eq. (5.183)) can be solved (see below) to yield

$$\bar{f}(\xi) = \mathrm{sech}^2(\alpha \xi), \tag{5.187}$$

where $\alpha = \frac{1}{2} \ln(1 + \sqrt{2})^2$. This result is compared with experimental data in Fig. 5.19. Just like with the round jet, the agreement between the profiles is excellent, except at the edge, where the diminishing turbulent viscosity causes the experimental profiles to tend to zero more rapidly than Eq. (5.187). The scaled turbulent viscosity $\hat{v}_\mathrm{T}$ that yields the observed rate of spreading $S \approx 0.1$ corresponds to

$$R_\mathrm{T} = \frac{1}{\hat{v}_\mathrm{T}} \approx 31,$$

(see Eq. (5.200)). This can be compared with $R_\mathrm{T} \approx 35$ for the round jet.

The solution for uniform turbulent viscosity

We now obtain the solution to the self-similar boundary-layer equation (Eq. (5.183)), with the shear stress given by the turbulent-viscosity hypothesis and with $\hat{v}_\mathrm{T}$ taken to be uniform, i.e., independent of ξ. The equation is

$$\tfrac{1}{2}S \left(\bar{f}^2 + \bar{f}' \int_0^\xi \bar{f} \, \mathrm{d}\xi \right) = -\hat{v}_\mathrm{T} \bar{f}''. \tag{5.188}$$

It is convenient to substitute

$$F(\xi) \equiv \int_0^\xi \bar{f}(s) \, \mathrm{d}s. \tag{5.189}$$

Since $\bar{f}(\xi)$ is an even function, $F(\xi)$ is odd. In particular $F(0) = F''(0) = 0$. With this substitution we obtain

$$\tfrac{1}{2}S[(F')^2 + F''F] = -\hat{v}_\mathrm{T} F'''. \tag{5.190}$$

Noting that the term in square brackets is

$$(F')^2 + F''F = (FF')' = \tfrac{1}{2}(F^2)'', \tag{5.191}$$

we can integrate twice to obtain

$$\tfrac{1}{4}SF^2 = -\hat{v}_\mathrm{T} F' + a + b\xi. \tag{5.192}$$

Since F^2 and F' are even functions, the integration constant b is zero, while the boundary condition $F'(0) = 1$ determines $a = \hat{v}_\mathrm{T}$. Defining

$$\alpha = \sqrt{\frac{S}{4\hat{v}_\mathrm{T}}}, \tag{5.193}$$

Eq. (5.192) then becomes

$$F' = 1 - (\alpha F)^2, \tag{5.194}$$

which can be integrated to yield

$$F = \frac{1}{\alpha} \tanh(\alpha\xi), \tag{5.195}$$

and hence

$$\bar{f} = F' = \operatorname{sech}^2(\alpha\xi). \tag{5.196}$$

From the definition of $y_{1/2}$ we have

$$\bar{f}(1) = \tfrac{1}{2} = \operatorname{sech}^2(\alpha), \tag{5.197}$$

from which we obtain

$$\alpha = \tfrac{1}{2}\ln\left(1 + \sqrt{2}\right)^2 \approx 0.88. \tag{5.198}$$

This, together with Eq. (5.193), relates the spreading rate S to the scaled turbulent viscosity $\hat{v}_{\mathrm{T}}$ by

$$S = [\ln(1 + \sqrt{2})^2]^2 \hat{v}_{\mathrm{T}}, \tag{5.199}$$

or

$$R_{\mathrm{T}} = \frac{1}{\hat{v}_{\mathrm{T}}} = \frac{[\ln(1 + \sqrt{2})^2]^2}{S} \approx 31, \tag{5.200}$$

taking the experimental value $S \approx 0.1$.

5.4.2 The plane mixing layer

As sketched in Fig. 5.14 on page 112, the mixing layer is the turbulent flow that forms between two uniform, nearly parallel streams of different velocities, U_{h} and U_{l} ($U_{\mathrm{h}} > U_{\mathrm{l}} \geq 0$). Such a mixing layer forms at the edge (and in the initial region) of a plane jet ($U_{\mathrm{h}} = U_{\mathrm{J}}$) flowing into quiescent surroundings ($U_{\mathrm{l}} = 0$). Alternatively the flow can be created in a wind tunnel, with a splitter plate separating the two streams for $x < 0$, and then the mixing layer develops for $x > 0$.

Just as in the plane jet, the dominant direction of flow is x; the cross-stream coordinate is y; and statistics are independent of the spanwise coordinate, z. In contrast to the round and plane jets, for the mixing layer there are two imposed velocities, U_{h} and U_{l}. Consequently the flow depends on the non-dimensional parameter $U_{\mathrm{l}}/U_{\mathrm{h}}$, and two characteristic velocities can be defined: the characteristic convection velocity

$$U_{\mathrm{c}} \equiv \tfrac{1}{2}(U_{\mathrm{h}} + U_{\mathrm{l}}), \tag{5.201}$$

and the characteristic velocity difference

$$U_{\mathrm{s}} \equiv U_{\mathrm{h}} - U_{\mathrm{l}}. \tag{5.202}$$

All the velocities mentioned ($U_{\mathrm{h}}, U_{\mathrm{l}}, U_{\mathrm{c}}$ and U_{s}) are constant – independent of x.

The characteristic width of the flow $\delta(x)$ can be defined in a number of

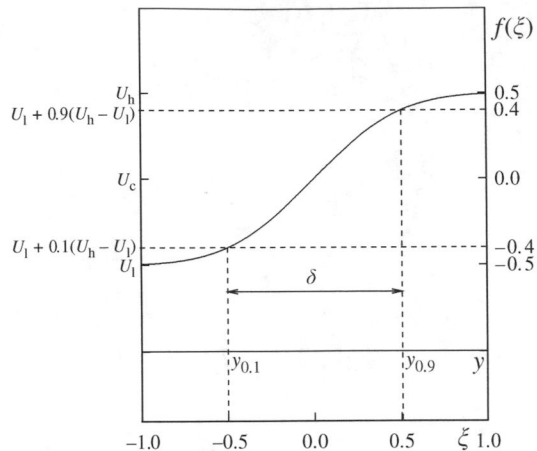

Fig. 5.21. A sketch of the mean velocity $\langle U \rangle$ against y, and of the scaled mean velocity profile $f(\xi)$, showing the definitions of $y_{0.1}$, $y_{0.9}$, and δ.

ways based on the mean velocity profile $\langle U(x, y, z) \rangle$ – which is independent of z. For $0 < \alpha < 1$ we define the cross-stream location $y_\alpha(x)$ such that

$$\langle U(x, y_\alpha(x), 0) \rangle = U_1 + \alpha(U_h - U_1), \tag{5.203}$$

and then take $\delta(x)$ to be

$$\delta(x) = y_{0.9}(x) - y_{0.1}(x), \tag{5.204}$$

see Fig. 5.21. In addition, a reference lateral position $\bar{y}(x)$ is defined by[1]

$$\bar{y}(x) = \tfrac{1}{2}[y_{0.9}(x) + y_{0.1}(x)]. \tag{5.205}$$

The scaled cross-stream coordinate ξ is then defined by

$$\xi = [y - \bar{y}(x)]/\delta(x), \tag{5.206}$$

and the scaled velocity by

$$f(\xi) = (\langle U \rangle - U_c)/U_s. \tag{5.207}$$

From these definitions we have $f(\pm\infty) = \pm\tfrac{1}{2}$, and $f(\pm\tfrac{1}{2}) = \pm0.4$ – see Fig. 5.21.

For the case $U_1/U_h = 0$, there are many experiments that confirm that the mixing layer is self-similar (e.g., Wygnanski and Fiedler (1970) and Champagne *et al.* (1976)). Figure 5.22 shows that the mean velocity profiles measured at different axial locations collapse when they are scaled according

---

[1] Note that a different definition of $\bar{y}(x)$ is used in Exercises 5.29–5.32 where the equations governing self-similar mixing layers are developed.

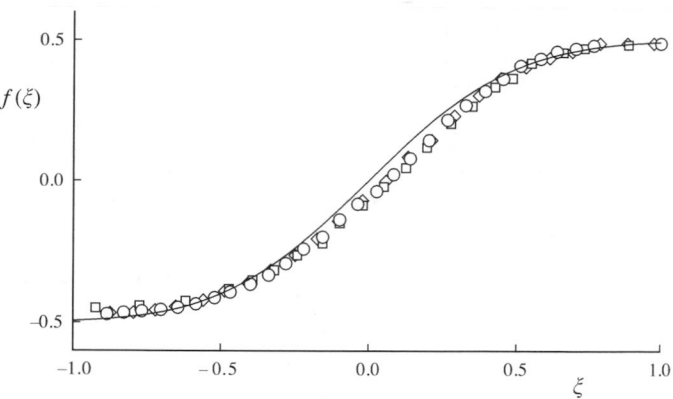

Fig. 5.22. Scaled velocity profiles in a plane mixing layer. Symbols, experimental data of Champagne *et al.* (1976) ($\diamond, x = 39.5$ cm; $\square, x = 49.5$ cm; $\circ, x = 59.5$ cm); line, error-function profile (Eq. (5.224)) shown for reference.

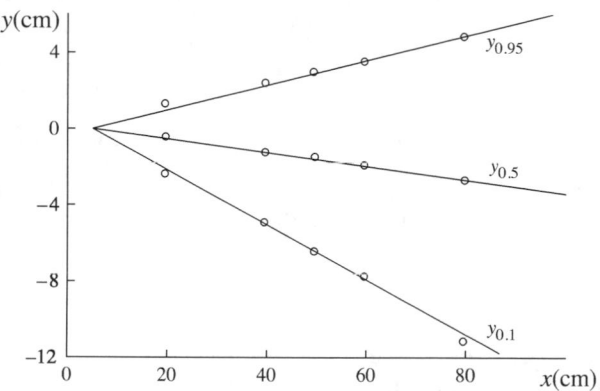

Fig. 5.23. Axial variations of $y_{0.1}, y_{0.5}$, and $y_{0.95}$ in the plane mixing layer, showing the linear spreading. Experimental data of Champagne *et al.* (1976).

to Eq. (5.207); and Figure 5.23 clearly shows that the mixing layer spreads linearly. The experiments also show that the Reynolds stresses are self-similar. It should be noted that the flow is not symmetric about $y = 0$ – nor even about $\xi = 0$ – and it spreads preferentially into the low-speed stream.

Just as with the round and plane jets, the linear spreading of the mixing layer is an inevitable consequence of self-similarity. As shown in Exercise 5.29, with a different definition of $\bar{y}(x)$, and with $g(\xi) \equiv \langle uv \rangle / U_s^2$ being the scaled shear stress, the boundary-layer equation for the self-similar mixing layer is

$$\left(\frac{U_c}{U_s} \frac{d\delta}{dx} \right) \left(\xi + \frac{U_s}{U_c} \int_0^\xi f(\hat{\xi}) \, d\hat{\xi} \right) f' = g'. \tag{5.208}$$

Since nothing in this equation depends upon x except for δ, the spreading rate $d\delta/dx$ and the parameter

$$S \equiv \frac{U_c}{U_s}\frac{d\delta}{dx} \qquad (5.209)$$

must be constant – independent of x – as is observed.

To an observer travelling in the x direction at the speed U_c, the fractional growth rate of the mixing layer is $U_c\, d\ln\delta/dx$. If this rate is normalized by the local timescale δ/U_s, the resulting non-dimensional parameter is

$$\frac{\delta}{U_s}U_c\frac{d\ln\delta}{dx} = \frac{U_c}{U_s}\frac{d\delta}{dx} = S. \qquad (5.210)$$

As may therefore be expected, it is found that S is approximately independent of the velocity ratio, so that $d\delta/dx$ varies (approximately) as U_s/U_c. There is considerable variation in the measured value of S from one experiment to another, which is attributable – at least in part – to the state of the flow as it leaves the splitter plate $(x = 0)$. The range of reported values is from $S \approx 0.06$ to $S \approx 0.11$ (Dimotakis 1991). In the experiment of Champagne et $al.$ (1976), the value is $S \approx 0.097$.

An interesting limit to consider is $U_s/U_c \to 0$, corresponding to $U_l/U_h \to 1$. In this limit, the term in U_s/U_c in the boundary layer equation Eq. (5.208) vanishes. The equation then corresponds to

$$U_c\frac{\partial\langle U\rangle}{\partial x} = -\frac{\partial\langle uv\rangle}{\partial y}, \qquad (5.211)$$

with the remaining convective terms, $(\langle U\rangle - U_c)\partial\langle U\rangle/\partial x$ and $\langle V\rangle\,\partial\langle U\rangle/\partial y$, being negligible compared with $U_c\,\partial\langle U\rangle/\partial x$. An observer travelling in the x direction at speed U_c sees the two streams ($y \to \infty$ and $y \to -\infty$) moving to the right and left, with velocities $\frac{1}{2}U_s$ and $-\frac{1}{2}U_s$, respectively. Gradients of means in the x direction are vanishingly small (of order U_s/U_c) compared with gradients in the y direction. The thickness of the layer grows in time at the rate $S U_s$. Thus, in the moving frame, as U_s/U_c tends to zero, the flow becomes statistically one-dimensional and time-dependent. It is called the *temporal mixing layer* (as opposed to the *spatial mixing layer* in laboratory coordinates). The temporal mixing layer is statistically symmetric about $y = 0$.

A direct numerical simulation (DNS) of the temporal mixing layer is described by Rogers and Moser (1994). The Navier–Stokes equations are solved by a spectral method with $512 \times 210 \times 192$ modes (in the $x, y,$ and z directions). After an initial transient, the mixing layer becomes self-similar, and the width δ increases linearly with time. The observed spreading parameter $S \approx 0.062$ is toward the low end of the experimentally observed range (0.06–0.11). In experiments it is difficult to approach the limit $U_l/U_h \to$

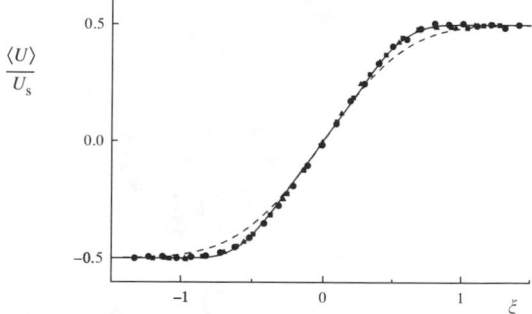

Fig. 5.24. The scaled mean velocity profile in self-similar plane mixing layers. Symbols, experiment of Bell and Mehta (1990) ($U_l/U_h = 0.6$); solid line, DNS data for the temporal mixing layer (Rogers and Moser 1994); dashed line, error-function profile with width chosen to match data in the center of the layer.

1 corresponding to the temporal mixing layer. However, the experiment of Bell and Mehta (1990) with $U_l/U_h = 0.6$ produces results similar to those for the temporal mixing layer of Rogers and Moser (1994). In the experiment the spreading parameter is $S \approx 0.069$.

Figure 5.24 shows that the scaled mean velocity profile for the temporal mixing layer and that for the spatial mixing layer with $U_l/U_h = 0.6$ are indistinguishable. Also shown in Fig. 5.24 is an error-function profile, which is the constant-turbulent-viscosity solution for the temporal mixing layer (see Exercise 5.33). Just as with jets, compared with the measurements, the mean velocity profile given by the constant-turbulent-viscosity solution tends more slowly to the free-stream velocity. Figure 5.25 shows the scaled Reynolds-stress profiles, which are little different for the two mixing layers.

For mixing layers, since U_s is fixed and δ varies linearly with x, the Reynolds number $\mathrm{Re}_0(x) \equiv U_s\delta/\nu$ and the turbulent viscosity also increase linearly with x. The flow rate of turbulent kinetic energy $\dot{K}(x) = \int_{-\infty}^{\infty}\langle U\rangle k\,\mathrm{d}y$ scales as $U_c U_s^2 \delta$ and hence also increases linearly with x. This is in contrast to jets and wakes, in which $\dot{K}$ decreases with x. Because $\dot{K}$ increases with x in the mixing layer, averaged across the flow, production $\mathcal{P}$ must exceed dissipation ε. In the center of the layer Rogers and Moser (1994) observe $\mathcal{P}/\varepsilon \approx 1.4$.

In the following exercises, the similarity equations for spatial and temporal mixing layers are developed. The temporal mixing layer is symmetric (i.e., $\bar{y}$ is zero and $f(\xi)$ is an odd function), and the free streams are parallel (i.e., $\langle V\rangle_{y=\infty} = \langle V\rangle_{y=-\infty} = 0$). The spatial mixing layer is not symmetric, it spreads preferentially into the low-speed stream ($\mathrm{d}\bar{y}/\mathrm{d}x$ is negative), and it entrains fluid (i.e., with $\langle V\rangle_{y=\infty}$ being zero, $\langle V\rangle_{y=-\infty}$ is positive). Consequently the two free streams are not exactly parallel. (In experiments, the free streams can be

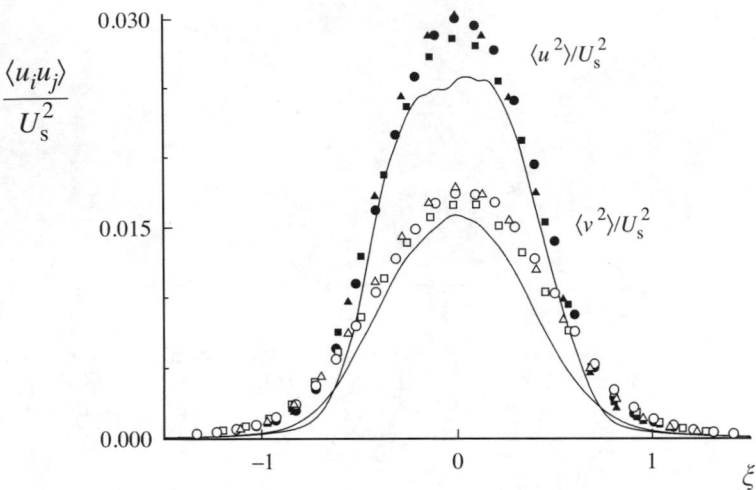

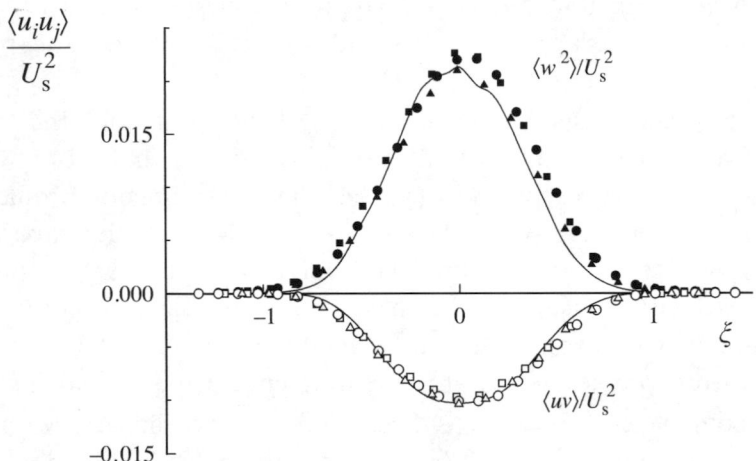

Fig. 5.25. Scaled Reynolds-stress profiles in self-similar plane mixing layers. Symbols, experiment of Bell and Mehta (1990) ($U_l/U_h = 0.6$); solid line, DNS data for the temporal mixing layer (Rogers and Moser 1994).

maintained at approximately uniform velocities by adjusting the inclination of the wind tunnel's walls.)

EXERCISES _____

5.29   From the definition of ξ and the scaled velocity $f(\xi)$ (Eqs. (5.206) and (5.207)) show that

$$\frac{\partial \langle U \rangle}{\partial x} = -\frac{U_s}{\delta} f'\left(\xi \frac{\mathrm{d}\delta}{\mathrm{d}x} + \frac{\mathrm{d}\bar{y}}{\mathrm{d}x} \right) \qquad (5.212)$$

in a self-similar mixing layer, irrespective of the definition of $\bar{y}(x)$. Hence, from the mean continuity equation, show that

$$
\begin{aligned}
\left(\langle V \rangle_{y=\infty} - \langle V \rangle_{y=-\infty} \right) / U_s \\
= \frac{d\bar{y}}{dx} + \frac{d\delta}{dx} \left(-\int_{-\infty}^{0} (\tfrac{1}{2} + f) \, d\xi + \int_{0}^{\infty} (\tfrac{1}{2} - f) \, d\xi \right) \\
= \frac{d\delta}{dx} \left\{ -\int_{-\infty}^{y=0} (\tfrac{1}{2} + f) \, d\xi + \int_{y=0}^{\infty} (\tfrac{1}{2} - f) \, d\xi \right\}.
\end{aligned}
\tag{5.213}
$$

Let the coordinate system be chosen so that the x axis is parallel to the velocity in the high-speed free stream, and consequently $\langle V \rangle_{y=\infty}$ is zero. Show then that the lateral mean velocity is

$$
\frac{\langle V \rangle}{U_s} = \frac{d\delta}{dx} \left(\xi(f - \tfrac{1}{2}) + \int_{\xi}^{\infty} (f - \tfrac{1}{2}) \, d\xi \right) + \frac{d\bar{y}}{dx} (f - \tfrac{1}{2}).
\tag{5.214}
$$

Hence show that the boundary-layer equation can be written

$$
\left[\left(\frac{U_c}{U_s} + \frac{1}{2} \right) \left(\xi \frac{d\delta}{dx} + \frac{d\bar{y}}{dx} \right) - \frac{d\delta}{dx} \int_{\xi}^{\infty} (f - \tfrac{1}{2}) \, d\xi \right] f' = g',
\tag{5.215}
$$

where

$$
g(\xi) \equiv \langle uv \rangle / U_s^2.
\tag{5.216}
$$

The preceding results apply to any specification of $\bar{y}(x)$. We now make the particular specification that $\bar{y}(x)$ *is the location of the peak shear stress* $|g|$. By considering Eq. (5.215) at $\xi = 0$, show that

$$
\begin{aligned}
\frac{d\bar{y}}{dx} &= -\frac{d\delta}{dx} \int_{0}^{\infty} (\tfrac{1}{2} - f) \, d\xi \Big/ \left(\frac{U_c}{U_s} + \frac{1}{2} \right) \\
&= -\frac{U_s}{U_h} \frac{d\delta}{dx} \int_{0}^{\infty} (\tfrac{1}{2} - f) \, d\xi.
\end{aligned}
\tag{5.217}
$$

Show that, with this specification of $\bar{y}(x)$, the boundary-layer equation becomes

$$
\frac{d\delta}{dx} \left(\frac{U_c}{U_s} \xi + \int_{0}^{\xi} f \, d\xi \right) f' = g'.
\tag{5.218}
$$

5.30 Show that the entrainment velocity is

$$
\langle V \rangle_{y=-\infty} = U_s \frac{d\delta}{dx} \left(\int_{-\infty}^{0} (\tfrac{1}{2} + f) \, d\xi - \frac{U_1}{U_h} \int_{0}^{\infty} (\tfrac{1}{2} - f) \, d\xi \right).
\tag{5.219}
$$

Discuss the sign of $\langle V \rangle_{y=-\infty}$ for the temporal and spatial mixing layers.

5.31   By integrating Eq. (5.218) from $\xi = -\infty$ to $\xi = \infty$, obtain the relation

$$\frac{U_c}{U_s}\left(\int_0^\infty \left(\tfrac{1}{2}-f\right)\mathrm{d}\xi - \int_{-\infty}^0 \left(\tfrac{1}{2}+f\right)\mathrm{d}\xi\right)$$
$$+ \int_0^\infty f\left(\tfrac{1}{2}-f\right)\mathrm{d}\xi + \int_{-\infty}^0 -f\left(\tfrac{1}{2}+f\right)\mathrm{d}\xi = 0. \qquad (5.220)$$

Hence show that the spatial mixing layer cannot be symmetric.

5.32   If $f(\xi)$ is *approximated* by an error function, then

$$\int_{-\infty}^0 \left(\tfrac{1}{2}+f\right)\mathrm{d}\xi = \int_0^\infty \left(\tfrac{1}{2}-f\right)\mathrm{d}\xi = I_0 \approx 0.24. \qquad (5.221)$$

With this approximation, obtain the result

$$-\frac{\mathrm{d}\bar{y}}{\mathrm{d}x} = \frac{\langle V\rangle_{y=-\infty}}{U_s} = \frac{U_s}{U_h}\frac{\mathrm{d}\delta}{\mathrm{d}x} I_0. \qquad (5.222)$$

For a mixing layer with $U_h/U_l = 2$, taking the spreading rate S to be 0.09, evaluate the right-hand side of Eq. (5.222) and show that, in the low-speed stream, the angle between the streamlines and the x axis is about $\tfrac{1}{2}°$.

5.33   With the turbulent-viscosity hypothesis, $g = -\hat{v}_T f'$, and assuming $\hat{v}_T$ to be uniform, show that the momentum equation (Eq. 5.208) for the temporal mixing layer reduces to

$$S\xi f' = -\hat{v}_T f''. \qquad (5.223)$$

Show that the solution to this equation (satisfying the appropriate boundary conditions) is

$$f(\xi) = \int_0^\xi \frac{1}{\sigma\sqrt{2\pi}}\exp\left(-\tfrac{1}{2}\zeta^2/\sigma^2\right)\mathrm{d}\zeta$$
$$= \tfrac{1}{2}\,\mathrm{erf}\left(\frac{\xi}{\sigma\sqrt{2}}\right), \qquad (5.224)$$

where

$$\sigma^2 = \hat{v}_T/S, \qquad (5.225)$$

and that the condition $f\left(\pm\tfrac{1}{2}\right) = \pm 0.4$ is satisfied by

$$\sigma = \left[2\sqrt{2}\,\mathrm{erf}^{-1}\left(\tfrac{4}{5}\right)\right]^{-1} \approx 0.3902. \qquad (5.226)$$

5.34   For the self-similar temporal mixing layer, starting from the momentum equation (Eq. (5.218)), show that the normalized shear stress at the center of the layer is given by

$$-g(0) = S \int_0^\infty \xi f' \, d\xi = S \int_0^\infty \left(\tfrac{1}{2} - f \right) d\xi. \tag{5.227}$$

If $f(\xi)$ is approximated by the error function profile (Eq. (5.224)), show that

$$-g(0) = \frac{S\sigma}{\sqrt{2\pi}} \approx 0.156S. \tag{5.228}$$

How well do the measured values of $g(0)$ and S agree with this relation?

5.4.3 *The plane wake*

As sketched in Fig. 5.14(c) (on page 112), a plane wake is formed when a uniform stream (of velocity U_c in the x direction) flows over a cylinder (that is aligned with the z axis). The flow is statistically stationary, two-dimensional, and symmetric about the plane $y = 0$.

The characteristic convective velocity is the free-stream velocity U_c, while the characteristic velocity difference is

$$U_s(x) \equiv U_c - \langle U(x,0,0) \rangle. \tag{5.229}$$

The half-width, $y_{1/2}(x)$, is defined such that

$$\langle U(x, \pm y_{1/2}, 0) \rangle = U_c - \tfrac{1}{2} U_s(x). \tag{5.230}$$

As expected, with increasing downstream distance, the wake spreads ($y_{1/2}$ increases) and decays (U_s/U_c decreases toward zero).

Just as with the mixing layer, there are two different velocity scales, U_s and U_c. In the mixing layer, these have a constant ratio, independent of x. In the wake, the ratio evolves as U_s/U_c decays. Because of this, the flow cannot be exactly self-similar; but it does become asymptotically self-similar in the far wake as U_s/U_c tends to zero. In experiments, self-similar behavior is observed when this velocity ratio is less than about $\frac{1}{10}$.

With $\xi \equiv y/y_{1/2}(x)$ being the scaled cross-stream variable, the self-similar *velocity defect* $f(\xi)$ is defined by

$$f(\xi) = [U_c - \langle U(x,y,0) \rangle]/U_s(x), \tag{5.231}$$

so that the mean velocity is

$$\langle U \rangle = U_c - U_s(x)f(\xi). \tag{5.232}$$

From these definitions we have $f(0) = 1$ and $f(\pm 1) = \frac{1}{2}$.

The momentum-deficit flow rate (per unit span), $\dot{M}(x)$, is defined by

$$\dot{M}(x) = \int_{-\infty}^{\infty} \rho\langle U \rangle (U_c - \langle U \rangle)\,dy, \tag{5.233}$$

and, in the self-similar region, this is

$$\dot{M}(x) = \rho U_c U_s(x)y_{1/2}(x) \int_{-\infty}^{\infty} \left(1 - \frac{U_s}{U_c}f(\xi)\right)f(\xi)\,d\xi. \tag{5.234}$$

Application of the momentum theorem to the cylinder and wake (see Batchelor (1967)) shows that the momentum deficit flow rate $\dot{M}(x)$ is conserved (independent of x) and equals the drag (per unit span) on the cylinder (see also Exercise 5.35). Consequently, it is evident from Eq. (5.234) that, in the far wake ($U_s/U_c \to 0$), the product $U_s(x)y_{1/2}(x)$ is independent of x.

In the far wake, the mean convection term $\bar{D}\langle U \rangle/\bar{D}t$ reduces to $U_c\,\partial\langle U \rangle/\partial x$, so the boundary-layer equation becomes

$$U_c \frac{\partial\langle U \rangle}{\partial x} = -\frac{\partial\langle uv \rangle}{\partial y}. \tag{5.235}$$

In terms of the similarity variables $f(\xi)$ and $g(\xi) \equiv \langle uv \rangle/U_s^2$, this equation is

$$S(\xi f)' = -g', \tag{5.236}$$

where the spreading parameter is

$$S \equiv \frac{U_c}{U_s}\frac{dy_{1/2}}{dx}, \tag{5.237}$$

(see Exercise 5.36). Equation (5.236) shows that self-similarity dictates that S be constant. This, together with the constancy of $U_s(x)y_{1/2}(x)$, implies that $U_s(x)$ and $y_{1/2}(x)$ vary as $x^{-1/2}$ and $x^{1/2}$, respectively. Note that Eq. (5.236) can be integrated to yield a simple relationship among the shear stress, the mean velocity, and the spreading rate:

$$g = -S\xi f. \tag{5.238}$$

The turbulent viscosity ν_T scales with $U_s(x)y_{1/2}(x)$, and is therefore independent of x. With the assumption of a constant turbulent viscosity, i.e.,

$$\nu_T = \hat{\nu}_T U_s(x)y_{1/2}(x), \tag{5.239}$$

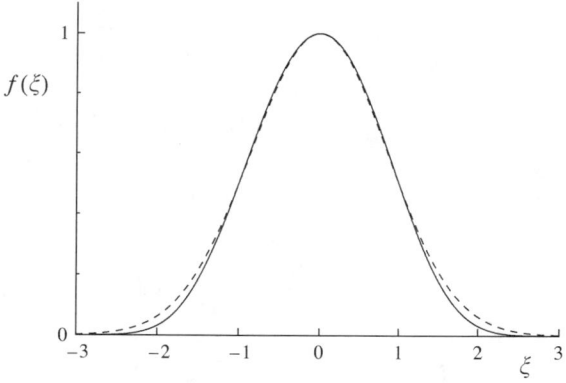

Fig. 5.26. The normalized velocity defect profile in the self-similar plane wake. Solid line, from experimental data of Wygnanski *et al.* (1986); dashed line, constant-turbulent-viscosity solution, Eq. (5.240).

the solution to Eq. (5.238) is

$$f(\xi) = \exp(-\alpha \xi^2), \tag{5.240}$$

where $\alpha = \ln 2 \approx 0.693$ (see Exercise 5.37). The turbulent Reynolds number is

$$R_T \equiv \frac{U_s(x) y_{1/2}(x)}{\nu_T} = \frac{1}{\hat{\nu}_T} = \frac{2 \ln 2}{S}. \tag{5.241}$$

Revealing experiments were reported by Wygnanski, Champagne, and Marasli (1986). In addition to a circular cylinder, these investigators also used a symmetric airfoil and a thin rectangular plate to generate plane wakes. In each case the flow is found (convincingly) to be self-similar, with the mean velocity profile shown in Fig. 5.26. It may be seen that the constant-turbulent-viscosity solution agrees well with the data, except at the edges – as is the case with jets.

However, it appears that the wakes from the different generators do not tend to precisely the same self-similar state. The spreading-rate parameters are $S_{\text{plate}} = 0.073$, $S_{\text{cylinder}} = 0.083$, and $S_{\text{airfoil}} = 0.103$. Consistent with Eq. (5.238), the shear-stress profiles exhibit the same level of difference. For the plate the peak axial r.m.s. velocity is $\langle u^2 \rangle_{\text{max}}^{1/2} / U_s = 0.32$, whereas for the airfoil it is 0.41. As discussed by George (1989), the different states observed are completely consistent with self-similarity. However, they indicate that, as the turbulent fluid is convected downstream, it retains information about how the wake was generated, rather than tending to a universal state.

5.35   Starting from the boundary-layer equations, show that

$$\frac{\partial}{\partial x}[\langle U \rangle (U_c - \langle U \rangle)] + \frac{\partial}{\partial y}[\langle V \rangle (U_c - \langle U \rangle)] = \frac{\partial}{\partial y}\langle uv \rangle \qquad (5.242)$$

for the plane wake; and hence show that the momentum deficit flow rate is conserved (Eq. (5.233)).

5.36   For the self-similar plane wake, since $y_{1/2}(x)U_s(x)$ is independent of x, show that

$$\frac{y_{1/2}U_c}{U_s^2}\frac{dU_s}{dx} = -S, \qquad (5.243)$$

where the spreading parameter S is defined by Eq. (5.237). From Eq. (5.232) show that

$$\frac{\partial \langle U \rangle}{\partial x} = -f\frac{dU_s}{dx} + \frac{U_s}{y_{1/2}}\frac{dy_{1/2}}{dx}\xi f', \qquad (5.244)$$

and hence

$$\frac{U_c y_{1/2}}{U_s^2}\frac{\partial \langle U \rangle}{\partial x} = S(\xi f)'. \qquad (5.245)$$

With the scaled shear stress being

$$g(\xi) = \langle uv \rangle / U_s^2, \qquad (5.246)$$

show that the (approximate) boundary-layer equation

$$U_c\frac{\partial \langle U \rangle}{\partial x} = -\frac{\partial \langle uv \rangle}{\partial y} \qquad (5.247)$$

can be written

$$S(\xi f)' = -g'. \qquad (5.248)$$

5.37   Show that the turbulent viscosity hypothesis (with uniform ν_T) amounts to

$$g = \hat{\nu}_T f'. \qquad (5.249)$$

Substituting Eq. (5.249) into Eq. (5.248), obtain the solution

$$f(\xi) = \exp(-\alpha\xi^2), \qquad (5.250)$$

where

$$\alpha = \frac{S}{2\hat{\nu}_T}. \qquad (5.251)$$

Show that $\alpha = \ln 2$, and hence that

$$\frac{1}{\hat{v}_T} = \frac{2 \ln 2}{S}. \tag{5.252}$$

5.4.4 The axisymmetric wake

The analysis of the axisymmetric wake parallels closely that of the plane wake. However, the experimental data reveal striking differences.

An axisymmetric wake forms behind a round object – a sphere, spheroid, or disk, for example – held in a uniform stream, flowing with velocity U_c in the x direction. The flow is statistically axisymmetric, with statistics depending on x and r, but being independent of θ. The centerline velocity deficit $U_s(x)$ and flow half-width $r_{1/2}(x)$ are defined in the obvious manner.

Just as with the plane wake, self-similarity is possible only as U_s/U_c tends to zero, and then the spreading parameter $S = (U_c/U_s)\,dr_{1/2}/dx$ is constant. For this flow, however, the momentum deficit flow rate – which equals the drag on the body – is proportional to $\rho U_c U_s r_{1/2}^2$. As a consequence U_s varies as $x^{-2/3}$ and $r_{1/2}$ as $x^{1/3}$, so that the Reynolds number decreases as $x^{-1/3}$. The assumption that the turbulent viscosity is uniform across the flow leads to the same mean velocity-deficit profile as that for the plane wake (Eqs. (5.239)–(5.241)).

Uberoi and Freymuth (1970) reported measurements made in the wake of a sphere (of diameter d), with Reynolds number $\mathrm{Re}_d \equiv U_c d/v = 8,600$. After a development distance ($x/d < 50$), self-similarity in the mean velocity and Reynolds stresses is observed over the range of x/d examined ($50 < x/d < 150$). The measured mean velocity-deficit profile is compared with the constant-turbulent-viscosity solution in Fig. 5.27, and the profiles of r.m.s. velocities are shown in Fig. 5.28. It should be observed that the peak value of $\langle u^2 \rangle^{1/2}/U_s$ is about 0.9, much higher than those in the other flows we have examined. Correspondingly, the spreading parameter is $S \approx 0.51$ – at least five times larger than that observed in plane wakes.

The balance of the turbulent kinetic energy (Fig. 5.29) is also substantially different than those of other flows. The dominant term is convection from upstream (i.e., $-\langle U \rangle \, \partial k/\partial x$), with dissipation ε and lateral transport each being about half as large. In contrast, at its peak, the production $\mathcal{P}$ is just 20% of ε, and 15% of convection. The dominance of convection, and the relatively small amount of production, suggest that the turbulence is strongly influenced by conditions upstream.

This hypothesis is strengthened by the observation that the measured

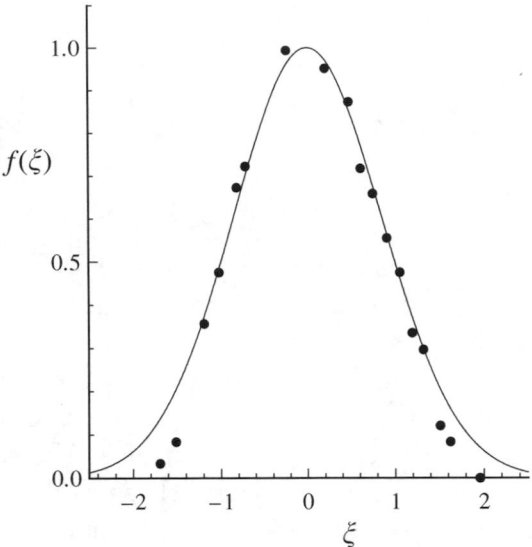

Fig. 5.27. Mean velocity-deficit profiles in a self-similar axisymmetric wake. Symbols, experimental data of Uberoi and Freymuth (1970); line, constant-turbulent-viscosity solution $f(\xi) = \exp(-\xi^2 \ln 2)$.

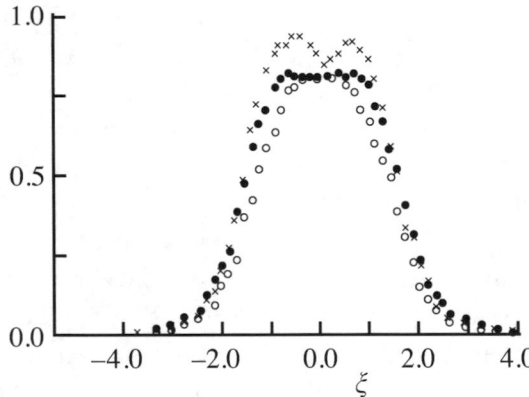

Fig. 5.28. R.m.s. velocity profiles in a self-similar axisymmetric wake. Experimental data of Uberoi and Freymuth (1970): $\times, \langle u^2 \rangle^{1/2}/U_s; \bullet, \langle v^2 \rangle^{1/2}/U_s; \circ, \langle w^2 \rangle^{1/2}/U_s$.

spreading parameter and turbulence level depend very significantly on the geometry of the body that generates the wake (see Table 5.3). On going from streamlined bodies to bluff bodies, S increases by a factor of ten, and the relative turbulence intensity by a factor of three. These observations are discussed further in Section 5.5.4.

Of the free shear flows examined in this chapter, only in the axisymmetric wake does the Reynolds number decrease with x (as $x^{-1/3}$). Consequently,

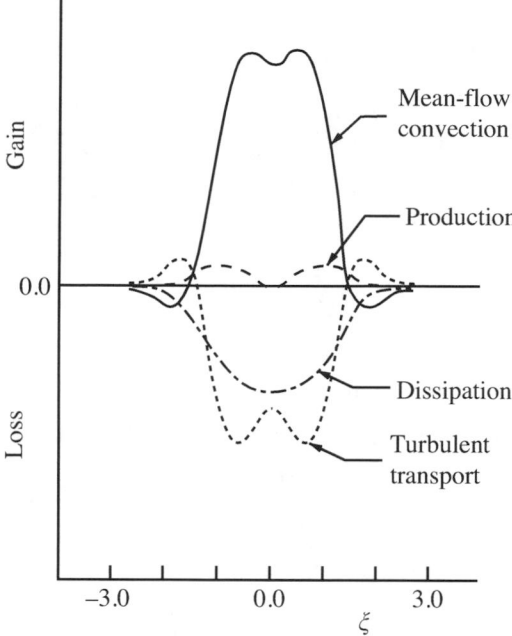

Fig. 5.29. The turbulent kinetic energy budget in a self-similar axisymmetric wake. Experimental data of Uberoi and Freymuth (1970).

Table 5.3. *The spreading parameter and turbulence intensity for axisymmetric wakes behind various bodies*

Body	Spreading parameter S	Turbulence intensity on centerline $\langle u^2 \rangle_0^{1/2}/U_s$	Investigation
49% blockage-screen	0.064	0.3	Cannon and Champagne (1991)
6:1 spheroid	0.11	0.3	Chevray (1968)
84% blockage-screen	0.34	0.75	Cannon and Champagne (1991)
Sphere	0.51	0.84	Uberoi and Freymuth (1970)
Disk	0.71	1.1	Cannon and Champagne (1991)
Disk	0.8	0.94	Carmody (1964)

only over a limited range of x can self-similarity (independent of Re) be expected; for, at sufficiently large x, the flow can be assumed to relaminarizes. The laminar wake admits the same self-similar velocity profile, but with U_s and $r_{1/2}$ varying as x^{-1} and $x^{1/2}$, respectively. Some experimental data (e.g., Cannon and Champagne (1991)) suggest that modest departures from self-similarity (based on high-Reynolds-number scaling) occur.

EXERCISES

5.38 Starting from the approximate boundary-layer equation for the far axisymmetric wake

$$U_c \frac{\partial \langle U \rangle}{\partial x} = -\frac{1}{r} \frac{\partial}{\partial r}(r \langle uv \rangle), \tag{5.253}$$

show that the momentum-deficit flow rate

$$\dot{M} \equiv \int_0^\infty 2\pi r \rho U_c (U_c - \langle U \rangle)\, dr \tag{5.254}$$

is conserved.

For the self-similar wake, re-express $\dot{M}$ in terms of U_s and $f(\xi) = (U_c - \langle U \rangle)/U_s$, where $\xi = r/r_{1/2}$. Hence show that

$$\frac{r_{1/2} U_c}{U_s^2} \frac{dU_s}{dx} = -2S, \tag{5.255}$$

where the spreading parameter is

$$S \equiv \frac{U_c}{U_s} \frac{dr_{1/2}}{dx}. \tag{5.256}$$

5.39 Starting from Eq. (5.253), show that

$$-S(2\xi f + \xi^2 f') = (\xi g)' \tag{5.257}$$

for the self-similar axisymmetric wake, where $g(\xi) \equiv \langle uv \rangle / U_s^2$. Hence show that

$$g = -Sf\xi. \tag{5.258}$$

Show that, if the uniform-turbulent-viscosity hypothesis is invoked (i.e., $g = \hat{v}_T f'$), the solution to Eq. (5.258) is

$$f(\xi) = \exp(-\xi^2 \ln 2), \tag{5.259}$$

and that

$$S = 2 \ln 2\, \hat{v}_T. \tag{5.260}$$

5.4.5 Homogeneous shear flow

As visual observation suggests, the free shear flows we have examined are inherently statistically inhomogeneous. In the center of the temporal mixing layer the mean shear rate $\partial \langle U \rangle / \partial y$ appears to be fairly uniform, but the Reynolds stresses exhibit appreciable spatial variation. The turbulent

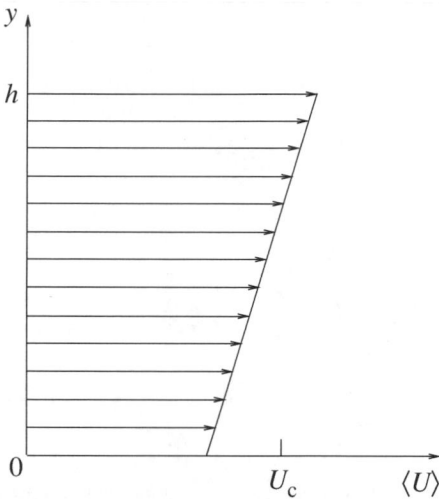

Fig. 5.30. A sketch of the mean velocity profile in homogeneous shear flow.

kinetic energy does not change with time, yet production exceeds dissipation by 40%. This excess of energy produced is transported outward, and the transport processes involved depend essentially on statistical inhomogeneity. In contrast to these flows, it is informative to study homogeneous turbulence, from which this type of transport process is absent. By the definition of homogeneous turbulence, the fluctuating components of velocity $u(x,t)$ and pressure $p'(x,t)$ are statistically homogeneous. It follows that imposed mean velocity gradients $\partial \langle U_i \rangle / \partial x_j$ must be uniform, although they can vary with time (see Exercises 5.40 and 5.41). Here we examine homogeneous shear flow in which the single imposed mean velocity gradient $S = \partial \langle U \rangle / \partial y$ is constant.

Homogeneous shear flow can be reasonably well approximated in wind-tunnel experiments. By controlling the flow resistance upstream, a turbulent flow with the mean velocity profile sketched in Fig. 5.30 can be produced. (The mean flow is entirely in the x direction, i.e., $\langle V \rangle = \langle W \rangle = 0$, and $\langle U \rangle$ varies only in the y direction.) At the beginning of the flow ($x/h = 0$), the Reynolds stresses are uniform normal to the direction of flow, and this uniformity persists downstream. Figure 5.31 shows the axial variation of the Reynolds stresses measured by Tavoularis and Corrsin (1981). In spite of this axial variation, in a frame moving with the mean velocity U_c, the turbulence is approximately homogeneous. Direct numerical simulations of homogeneous shear flow have also been performed (e.g., Rogallo (1981) and Rogers and Moin (1987)), with results in broad agreement with the experiments.

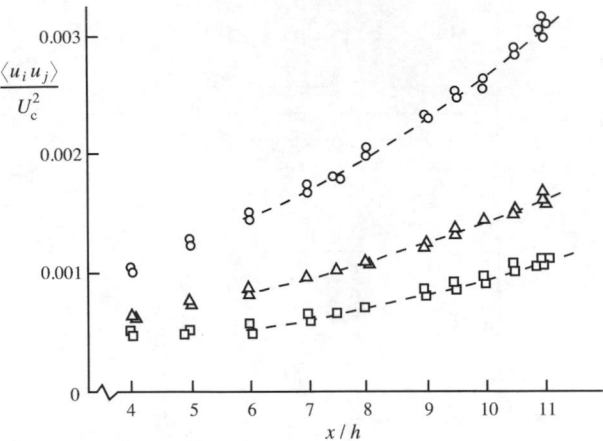

Fig. 5.31. Reynolds stresses against axial distance in the homogeneous-shear-flow experiment of Tavoularis and Corrsin (1981): $\bigcirc$, $\langle u^2 \rangle$; $\square$, $\langle v^2 \rangle$; $\triangle$, $\langle w^2 \rangle$.

The important conclusion from these studies is that, after a development time, homogeneous turbulent shear flow becomes self-similar. That is, when statistics are normalized by the imposed shear rate S and the kinetic energy $k(t)$, they become independent of time. Table 5.4 compares some of these statistics at two locations in the experiments of Tavoularis and Corrsin (1981) and from the DNS study of Rogers and Moin (1987). Between $x/h = 7.5$ and $x/h = 11.0$, the kinetic energy increases by 65%, yet the normalized Reynolds stresses barely change at all. The turbulence timescale $\tau = k/\varepsilon$ does not change appreciably, but is in a fixed proportion to the imposed mean-flow timescale S^{-1}. Between the two measurement locations the longitudinal integral lengthscale L_{11} increases by 30%, but remains constant when scaled by S and k.

The equation for the evolution of the turbulent kinetic energy is, simply,

$$\frac{\mathrm{d}k}{\mathrm{d}t} = \mathcal{P} - \varepsilon, \tag{5.261}$$

see Exercise 5.40. This can be rewritten

$$\frac{\tau}{k}\frac{\mathrm{d}k}{\mathrm{d}t} = \frac{\mathcal{P}}{\varepsilon} - 1, \tag{5.262}$$

which – since τ and $\mathcal{P}/\varepsilon$ are constant – has the solution

$$k(t) = k(0)\exp\left[\frac{t}{\tau}\left(\frac{\mathcal{P}}{\varepsilon} - 1\right)\right]. \tag{5.263}$$

Thus, since $\mathcal{P}/\varepsilon \approx 1.7$ is greater than unity, the kinetic energy grows exponentially in time. Consequently, both ε and $L \equiv k^{3/2}/\varepsilon = k^{1/2}/\tau$ also grow

Table 5.4. *Statistics in homogeneous turbulent shear flow from the experiments of Tavoularis and Corrsin (1981) and the DNS of Rogers and Moin (1987)*

	Tavoularis and Corrsin		Rogers and Moin
	$x/h = 7.5$	$x/h = 11.0$	$St = 8.0$
$\langle u^2 \rangle/k$	1.04	1.07	1.06
$\langle v^2 \rangle/k$	0.37	0.37	0.32
$\langle w^2 \rangle/k$	0.58	0.56	0.62
$-\langle uv \rangle/k$	0.28	0.28	0.33
$-\rho_{uv}$	0.45	0.45	0.57
Sk/ε	6.5	6.1	4.3
$\mathcal{P}/\varepsilon$	1.8	1.7	1.4
$L_{11}S/k^{1/2}$	4.0	4.0	3.7
$L_{11}/(k^{3/2}/\varepsilon)$	0.62	0.66	0.86

exponentially. Additional experiments and DNS on homogeneous shear flow have been performed by Tavoularis and Karnik (1989), de Souza *et al.* (1995), and Lee *et al.* (1990).

EXERCISES

5.40   By subtracting the Reynolds equations (Eq. (4.12)) from the Navier–Stokes equations (Eq. (2.35)), show that the fluctuating velocity $u(x,t)$ evolves by

$$\frac{Du_j}{Dt} = -u_i \frac{\partial \langle U_j \rangle}{\partial x_i} + \frac{\partial}{\partial x_i} \langle u_i u_j \rangle + v \, \nabla^2 u_j - \frac{1}{\rho} \frac{\partial p'}{\partial x_j}. \tag{5.264}$$

Show that, for homogeneous turbulence,

$$\left\langle u_j \frac{Du_j}{Dt} \right\rangle = \frac{dk}{dt}, \tag{5.265}$$

$$\left\langle u_j \frac{\partial p'}{\partial x_j} \right\rangle = 0, \tag{5.266}$$

$$v \langle u_j \, \nabla^2 u_j \rangle = -\varepsilon, \tag{5.267}$$

and hence that the kinetic energy evolves by

$$\frac{dk}{dt} = \mathcal{P} - \varepsilon, \tag{5.268}$$

where

$$\mathcal{P} \equiv -\langle u_i u_j \rangle \frac{\partial \langle U_j \rangle}{\partial x_i}. \tag{5.269}$$

(Note that this implies that a necessary condition for homogeneous turbulence is that the mean velocity gradients be uniform.)

5.41   By differentiating with respect to x_k the Reynolds equations (Eq. (4.12)) written for $\langle U_j \rangle$, show that, in homogeneous turbulence, the velocity gradients evolve by

$$\frac{d}{dt} \frac{\partial \langle U_j \rangle}{\partial x_k} + \frac{\partial \langle U_i \rangle}{\partial x_k} \frac{\partial \langle U_j \rangle}{\partial x_i} = -\frac{1}{\rho} \frac{\partial^2 \langle p \rangle}{\partial x_j \partial x_k}. \tag{5.270}$$

Hence show that the mean rate of strain

$$\bar{S}_{jk} \equiv \frac{1}{2} \left(\frac{\partial \langle U_j \rangle}{\partial x_k} + \frac{\partial \langle U_k \rangle}{\partial x_j} \right)$$

and rotation

$$\bar{\Omega}_{jk} \equiv \frac{1}{2} \left(\frac{\partial \langle U_j \rangle}{\partial x_k} - \frac{\partial \langle U_k \rangle}{\partial x_j} \right)$$

evolve by

$$\frac{d\bar{S}_{jk}}{dt} + \bar{S}_{ik}\bar{S}_{ji} + \bar{\Omega}_{ik}\bar{\Omega}_{ji} = -\frac{1}{\rho} \frac{\partial^2 \langle p \rangle}{\partial x_j \, \partial x_k}, \tag{5.271}$$

$$\frac{d\bar{\Omega}_{jk}}{dt} + \bar{S}_{ik}\bar{\Omega}_{ji} + \bar{\Omega}_{ik}\bar{S}_{ji} = 0. \tag{5.272}$$

(Note that the mean pressure field is of the form $\langle p(\boldsymbol{x}, t) \rangle = A(t) + B_i(t)x_i + C_{ij}(t)x_ix_j$, and that $C_{ij}(t)$ can be chosen to produce any desired evolution $d\bar{S}_{jk}/dt$. On the other hand, the evolution of the mean rotation rate is entirely determined by $\bar{S}_{ij}$ and $\bar{\Omega}_{ij}$.)

5.4.6 Grid turbulence

In the absence of mean velocity gradients, homogeneous turbulence decays because there is no production ($\mathcal{P} = 0$). A good approximation to decaying homogeneous turbulence can be achieved in wind-tunnel experiments by passing a uniform stream (of velocity U_0 in the x direction) through a grid similar to that shown in Fig. 5.32 (which is characterized by the mesh spacing M). In the laboratory frame, the flow is statistically stationary and (in the center of the flow) statistics vary only in the x direction. In the frame moving with the mean velocity U_0, the turbulence is (to an adequate approximation) homogeneous, and it evolves with time ($t = x/U_0$).

Figure 5.33 shows measurements of $\langle u^2 \rangle$ and $\langle v^2 \rangle$ from the grid-turbulence experiments of Comte-Bellot and Corrsin (1966). The symmetries in the

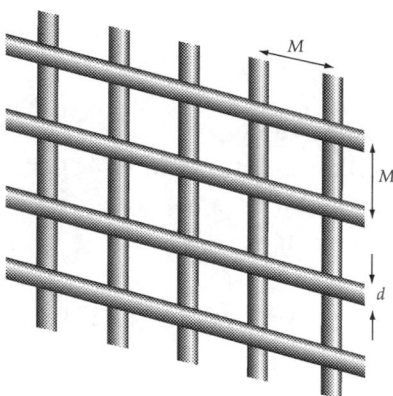

Fig. 5.32. A sketch of a turbulence-generating grid composed of bars of diameter d, with mesh spacing M.

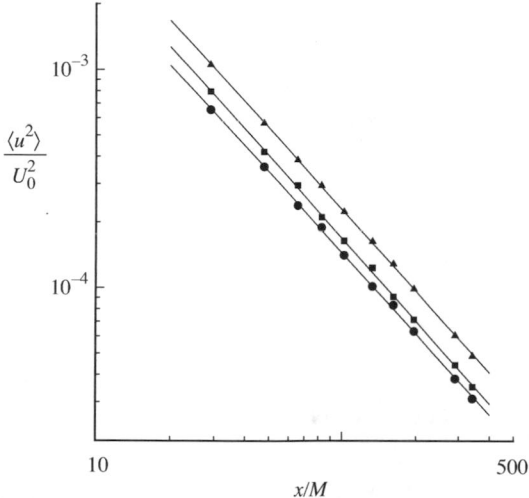

Fig. 5.33. The decay of Reynolds stresses in grid turbulence: squares, $\langle u^2 \rangle / U_0^2$; circles $\langle v^2 \rangle / U_0^2$; triangles k / U_0^2; lines, proportional to $(x/M)^{-1.3}$. (From Comte-Bellot and Corrsin (1966).)

(ideal) experiment dictate that $\langle v^2 \rangle$ and $\langle w^2 \rangle$ are equal, and that all of the shear stresses are zero. It may be seen that the r.m.s. axial velocity $\langle u^2 \rangle^{1/2}$ is 10% greater than the lateral r.m.s. $\langle v^2 \rangle^{1/2}$. (Comte-Bellot and Corrsin (1966) demonstrated a modification to the experiment that yields equal normal stresses, thus providing a better approximation to the ideal of homogeneous isotropic turbulence.)

It is evident from Fig. 5.33 that the normal stresses and k decay as power laws, which, in the laboratory frame, can be written

$$\frac{k}{U_0^2} = A\left(\frac{x - x_0}{M}\right)^{-n},$$
(5.273)

where x_0 is the virtual origin. Values of the decay exponent n between 1.15 and 1.45 are reported in the literature; but Mohamed and LaRue (1990) suggest that nearly all of the data are consistent with $n = 1.3$ (and $x_0 = 0$). (The value of A varies widely depending on the geometry of the grid and the Reynolds number.)

In the moving frame, the power law (Eq. (5.273)) can be written

$$k(t) = k_0\left(\frac{t}{t_0}\right)^{-n},$$
(5.274)

where t_0 is an arbitrary reference time, and k_0 is the value of k at that time. On differentiating we obtain

$$\frac{dk}{dt} = -\left(\frac{nk_0}{t_0}\right)\left(\frac{t}{t_0}\right)^{-(n+1)}.$$
(5.275)

Now, for decaying homogeneous turbulence, the exact equation for the evolution of k (Eq. (5.261)) reduces to

$$\frac{dk}{dt} = -\varepsilon.$$
(5.276)

Hence, a comparison of the last two equations shows that ε also decays as a power law:

$$\varepsilon(t) = \varepsilon_0\left(\frac{t}{t_0}\right)^{-(n+1)},$$
(5.277)

with $\varepsilon_0 = nk_0/t_0$. The decay exponents for other quantities are given in Exercise 5.42.

As the turbulence decays, the Reynolds number decreases so that eventually effects of viscosity dominate. This leads to the *final period of decay*, discussed in Section 6.3, in which the decay exponent is $n = \frac{5}{2}$ (see Exercise 6.10 on page 205).

In a sense, grid turbulence (as an approximation to homogeneous isotropic turbulence) is the most fundamental turbulent flow, and consequently it has been studied extensively both experimentally and theoretically. However, in another sense it is pathological: in contrast to turbulent shear flows, there is no turbulence-production mechanism (downstream of the grid).

'Active' grids, consisting of an array of moving flaps, that produce significantly higher turbulence levels, and thus higher Reynolds numbers, have been developed (e.g., Makita (1991), and Mydlarski and Warhaft (1996)).

5.42   For grid turbulence, given the decay laws for k and ε (Eqs. (5.274) and (5.277), and taking $n = 1.3$), verify the following behaviors:

$$\tau \equiv \frac{k}{\varepsilon} \sim t,$$

$$L \equiv \frac{k^{3/2}}{\varepsilon} \sim \left(\frac{t}{t_0}\right)^{(1-n/2)} = \left(\frac{t}{t_0}\right)^{0.35},$$

$$\frac{k^{1/2}L}{\nu} \sim \left(\frac{t}{t_0}\right)^{(1-n)} = \left(\frac{t}{t_0}\right)^{-0.3}.$$

(Note that the Reynolds number $k^{1/2}L/\nu$ decreases. The increase of L and τ should not be misunderstood. It is not that the turbulent motions become larger and slower. Rather, the smaller, faster motions decay more rapidly, leaving behind the larger, slower motions.)

5.5 Further observations

Our examination of free shear flows has focused mainly on the mean velocity field and the Reynolds stresses. In this section we first extend these considerations to a conserved passive scalar ϕ, by considering its mean $\langle\phi\rangle$, its variance $\langle\phi'^2\rangle$, and the scalar flux $\langle u\phi'\rangle$. Then we examine quantities and phenomena that are not described by first and second moments.

5.5.1 A conserved scalar

The behavior of a conserved passive scalar $\phi(x,t)$ has been studied experimentally in all of the free shear flows discussed in this chapter. In jets, the jet fluid can have a small temperature excess, or a slightly different chemical composition than the ambient. It is convenient to normalize the resulting scalar field so that ϕ is unity in the jet and zero in the ambient. Such a normalized scalar is sometimes referred to (especially in the combustion literature) as the *mixture fraction*. Similarly, in mixing layers, ϕ can be normalized to be zero and one in the two streams.

For plane flows, the boundary-layer equation for the mean $\langle\phi\rangle$ is

$$\langle U\rangle \frac{\partial\langle\phi\rangle}{\partial x} + \langle V\rangle \frac{\partial\langle\phi\rangle}{\partial y} = \Gamma \frac{\partial^2\langle\phi\rangle}{\partial y^2} - \frac{\partial\langle v\phi'\rangle}{\partial y}. \qquad (5.278)$$

This equation is very similar to that for the axial velocity $\langle U\rangle$ (Eq. (5.56)), and similar conserved quantities arise from it. Taking the plane jet as

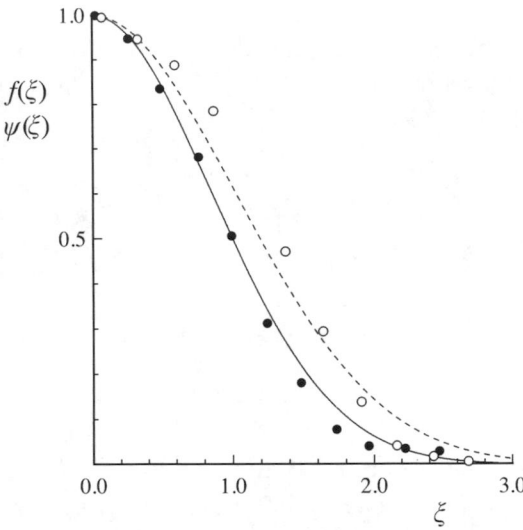

Fig. 5.34. The normalized mean velocity deficit $f(\xi)$ and scalar, $\varphi(\xi) = \langle\phi\rangle/\langle\phi\rangle_{y=0}$ in the self-similar plane wake. Symbols (solid f, open φ) experimental data of Fabris (1979); solid line, $f(\xi) = \exp(-\xi^2 \ln 2)$; dashed line, $\varphi(\xi) = \exp(-\xi^2 \sigma_T \ln 2)$ with $\sigma_T = 0.7$.

an example, we have seen that the momentum flow rate $\int_{-\infty}^{\infty} \rho\langle U\rangle^2 \, \mathrm{d}y$ is conserved, from which we deduced that $\langle U\rangle$ scales as $x^{-1/2}$. Similarly, it follows from Eq. (5.278) that the scalar flow rate $\int_{-\infty}^{\infty} \rho\langle U\rangle\langle\phi\rangle \, \mathrm{d}y$ is also conserved, and hence $\langle\phi\rangle$ must scale in the same way as $\langle U\rangle$, i.e., as $x^{-1/2}$. The same conclusion applies to all the self-similar free shear flows: $\langle\phi\rangle$ scales with x in the same way as $\langle U\rangle$ does.

The lateral profiles of $\langle\phi\rangle$ are again similar to those of $\langle U\rangle$, but in all cases they are found to be somewhat wider. For example, Fig. 5.34 shows the profiles of $\langle U\rangle$ and $\langle\phi\rangle$ measured by Fabris (1979) in the self-similar plane wake. The assumption of a constant turbulent viscosity ν_T leads to the normalized velocity profile $f(\xi) = \exp(-\xi^2 \ln 2)$ shown in Fig. 5.34 (see Eq. (5.240)). Similarly, the assumption of a constant turbulent diffusivity

$$\Gamma_T = \nu_T/\sigma_T, \qquad (5.279)$$

leads to the normalized scalar profile

$$\langle\phi\rangle/\langle\phi\rangle_{y=0} = \exp(-\xi^2 \sigma_T \ln 2) \qquad (5.280)$$

(see Exercise 5.43). This profile, with the turbulent Prandtl number set to $\sigma_T = 0.7$, agrees quite well with the data. (As shown in Exercise 5.43, the ratio of the widths of the scalar and velocity profiles is proportional to $\sigma_T^{-1/2}$).

It should be appreciated that, in self-similar free shear flows, the gradient-

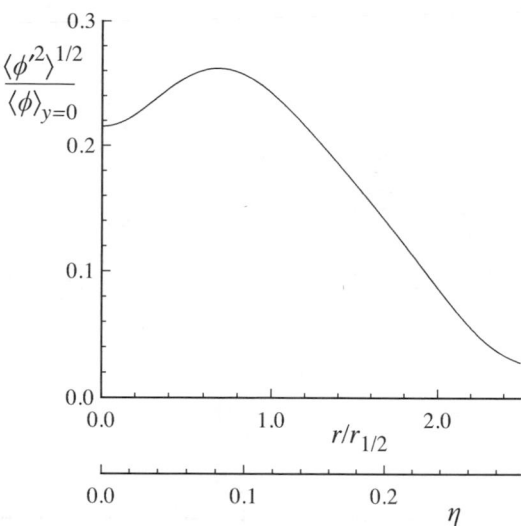

Fig. 5.35. Normalized r.m.s. scalar fluctuations in a round jet. From the experimental data of Panchapakesan and Lumley (1993b).

diffusion and turbulent-viscosity hypotheses are successful only in a limited sense – namely in approximating the lateral fluxes $\langle v\phi' \rangle$ and $\langle uv \rangle$. In the plane wake, $|\partial\langle\phi\rangle/\partial x|$ is vanishingly small compared with $|\partial\langle\phi\rangle/\partial y|$ as U_s/U_c tends to zero. Yet the experimental data (e.g., Fabris (1979)) show $|\langle u\phi' \rangle|$ to be comparable to $|\langle v\phi' \rangle|$. The lack of alignment between $\nabla\langle\phi\rangle$ and $-\langle u\phi' \rangle$ is also clearly demonstrated in homogeneous shear flow. Tavoularis and Corrsin (1981) imposed a mean scalar gradient in the y direction, and yet found $|\langle u\phi' \rangle|$ to be more than twice $|\langle v\phi' \rangle|$: the angle between $\nabla\langle\phi\rangle$ and $-\langle u\phi' \rangle$ is 65°.

As the turbulent kinetic energy $k = \frac{1}{2}\langle \boldsymbol{u} \cdot \boldsymbol{u} \rangle$ characterizes the energy in the fluctuating velocity field, the scalar variance $\langle\phi'^2\rangle$ – or the r.m.s. $\langle\phi'^2\rangle^{1/2}$ – characterizes the level of scalar fluctuations. Figure 5.35 shows the r.m.s. measured by Panchapakesan and Lumley (1993b) in a helium jet in air.[2] The shape of the profile and the level of the fluctuations (up to 25%) are comparable to those of the r.m.s axial velocity $\langle u^2\rangle^{1/2}$ (see Fig. 5.7).

The equation for the evolution of the scalar variance can be written

$$\frac{\bar{\mathrm{D}}\langle\phi'^2\rangle}{\bar{\mathrm{D}}t} + \nabla \cdot \mathcal{T}_\phi = \mathcal{P}_\phi - \varepsilon_\phi \tag{5.281}$$

(see Exercise 5.44), where the *scalar-variance production* is

$$\mathcal{P}_\phi = -2\langle \boldsymbol{u}\phi' \rangle \cdot \nabla\langle\phi\rangle, \tag{5.282}$$

---

[2] Over the range of the measurements $x/d = 50$–120, the mean density varies by as much as 15%, which has some effect on the flow; i.e., ϕ is not entirely passive.

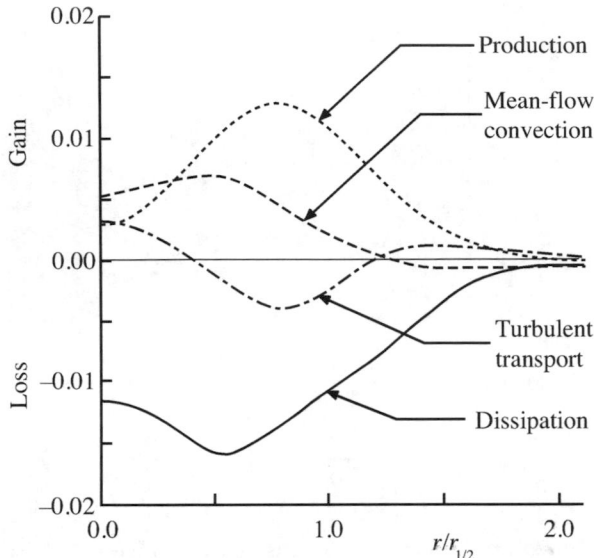

Fig. 5.36. The scalar-variance budget in a round jet: terms in Eq. (5.281) normalized by $\langle\phi\rangle_{y=0}$, U_s and $r_{1/2}$. (From the experimental data of Panchapakesan and Lumley (1993b).)

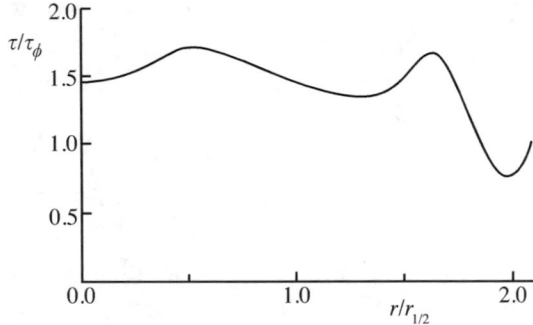

Fig. 5.37. The scalar-to-velocity timescale ratio for a round jet. (From the experimental data of Panchapakesan and Lumley (1993b).)

the *scalar dissipation* is

$$\varepsilon_\phi = 2\Gamma\langle\nabla\phi' \cdot \nabla\phi'\rangle, \tag{5.283}$$

and the flux is

$$\mathcal{T}_\phi = \langle u\phi'^2\rangle - \Gamma\nabla\langle\phi'^2\rangle. \tag{5.284}$$

(Sometimes the equation is written for $\frac{1}{2}\langle\phi'^2\rangle$, and the factors of 2 are omitted from the definitions of $\mathcal{P}_\phi$ and ε_ϕ.)

For the round jet, the terms in Eq. (5.281) are shown in Fig. 5.36. A

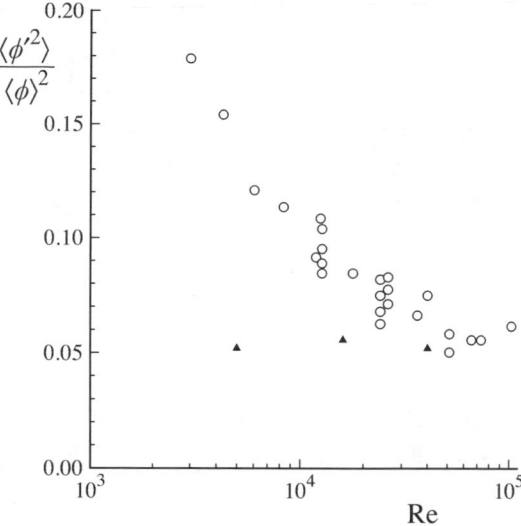

Fig. 5.38. The normalized scalar variances on the axes of self-similar round jets at various Reynolds numbers. Triangles, air jets (experiments of Dowling and Dimotakis (1990)); circles, water jets (experiments of Miller (1991)). (From Miller (1991).)

comparison with Fig. 5.16 shows that the balance of the terms is very similar to the kinetic-energy balance. Where $\mathcal{P}_\phi$ peaks, the ratio $\mathcal{P}_\phi/\varepsilon_\phi$ is 0.85.

The quantity $\tau \equiv k/\varepsilon$ defines a characteristic timescale of the velocity fluctuations, and similarly $\tau_\phi \equiv \langle \phi'^2 \rangle / \varepsilon_\phi$ defines the analogous scalar timescale. The profile of the timescale ratio τ/τ_ϕ in the round jet is shown in Fig. 5.37. Over most of the profile τ/τ_ϕ is within 15% of 1.5. In many other shear flows this timescale ratio is found to be in the range 1.5–2.5 (Béguier *et al.* 1978).

What is the influence of the molecular diffusivity Γ? This question can be rephrased in terms of the Reynolds number $\mathcal{U}\mathcal{L}/\nu$, and the Prandtl or Schmidt number $\sigma \equiv \nu/\Gamma$. Values of σ from 0.3 (for helium in air) to 1,500 (for dyes in water) are encountered in experiments. Figure 5.38 provides valuable information to address the question. It shows the scalar variance normalized by the mean scalar on the centerline of self-similar round jets at various Reynolds numbers. For the air jet ($\sigma \approx 1$), there is no influence of Re over the decade studied. For the water jet ($\sigma \sim 10^3$), the scalar variance decreases with increasing Re, and appears to tend asymptotically to the same value as that in the air jet. These data are therefore consistent with the view that, at sufficiently high Reynolds number (here Re $>$ 30,000), the means and variances of the velocities and scalar are unaffected by Re and σ.

As is evident from its definition (Eq. (5.283)), scalar dissipation is inherently a molecular process. However, as Fig. 1.2 on page 5 illustrates, in a given flow (characterized by $\mathcal{U}, \mathcal{L}, \nu$, and σ), as the Reynolds number increases, so also do the normalized scalar gradients $\mathcal{L} \nabla \phi$. As a consequence, instead of scaling with $\Gamma/\mathcal{L}^2$ as is vaguely suggested by its definition, ε_ϕ scales as $\mathcal{U}/\mathcal{L}$, independent of Γ (at sufficiently high Reynolds number). The processes causing this behavior are described in the next chapter.

With ψ being a specified value of the scalar, the points x satisfying the equation

$$\phi(x, t) = \psi, \tag{5.285}$$

define an *isoscalar surface*. At high Reynolds numbers, such surfaces exhibit a fractal nature over an intermediate range of scales, with a fractal dimension of about 2.36 (see, e.g., Sreenivasan (1991)). Careful examination reveals, however, that the surface geometry departs significantly from a perfect fractal (Frederiksen *et al.* 1997).

EXERCISES

5.43   For the self-similar plane wake, neglecting the molecular diffusivity Γ, the mean scalar equation (Eq. (5.278)) is

$$U_c \frac{\partial \langle \phi \rangle}{\partial x} = -\frac{\partial}{\partial y} \langle v \phi' \rangle. \tag{5.286}$$

Let $\phi_0(x)$ denote the value of $\langle \phi \rangle$ on the plane of symmetry; and, with $\xi \equiv y/y_{1/2}$, let the self-similar scalar profile be

$$\varphi(\xi) = \langle \phi \rangle / \phi_0. \tag{5.287}$$

By integrating Eq. (5.286) over all y, show that $U_c \phi_0(x) y_{1/2}(x)$ is conserved, and hence that

$$-\frac{U_c y_{1/2}}{U_s \phi_0} \frac{\mathrm{d}\phi_0}{\mathrm{d}x} = S \equiv \frac{U_c}{U_s} \frac{\mathrm{d}y_{1/2}}{\mathrm{d}x}. \tag{5.288}$$

Show that

$$\frac{U_c y_{1/2}}{U_s \phi_0} \frac{\partial \langle \phi \rangle}{\partial x} = -S(\xi \varphi)'. \tag{5.289}$$

The constant-turbulent-diffusivity hypothesis can be written

$$-\langle v \phi' \rangle = \Gamma_T \frac{\partial \langle \phi \rangle}{\partial y} = \frac{\nu_T}{\sigma_T} \frac{\partial \langle \phi \rangle}{\partial y} = \frac{\hat{\nu}_T}{\sigma_T} U_s(x) y_{1/2}(x) \frac{\partial \langle \phi \rangle}{\partial y}, \tag{5.290}$$

where σ_T is the turbulent Prandtl number. Show that, with this assumption, the scalar equation (Eq. (5.286)) becomes

$$-S(\xi \varphi)' = \frac{\hat{\nu}_T}{\sigma_T} \varphi''. \tag{5.291}$$

Show that the solution to this equation is

$$\varphi(\xi) = \exp(-\beta \xi^2), \tag{5.292}$$

where

$$\beta = \frac{S \sigma_T}{2 \hat{v}_T}. \tag{5.293}$$

Use the results of Exercise 5.37 to show that this can be rewritten

$$\varphi(\xi) = \exp(-\xi^2 \sigma_T \ln 2)$$
$$= f\left(\xi \sqrt{\sigma_T}\right). \tag{5.294}$$

Let $\xi_{1/2}$ be the half-width of the normalized scalar profile (i.e., $\varphi(\xi_{1/2}) = \frac{1}{2}$). Show that $\xi_{1/2}$ and σ_T are related by

$$\sigma_T = 1/\xi_{1/2}^2. \tag{5.295}$$

5.44 From the Reynolds decompositions $U = \langle U \rangle + u$ and $\phi = \langle \phi \rangle + \phi'$, show that

$$U\phi - \langle U\phi \rangle = U\phi' + u\langle \phi \rangle - \langle u\phi' \rangle. \tag{5.296}$$

Hence, from the conservation equation for $\phi(x,t)$ (Eq. (4.38)), show that the scalar fluctuation evolves by

$$\frac{D\phi'}{Dt} = -u \cdot \nabla \langle \phi \rangle + \nabla \cdot \langle u\phi' \rangle + \Gamma \nabla^2 \phi'. \tag{5.297}$$

Show that

$$2\phi' \nabla^2 \phi' = \nabla^2(\phi'^2) - 2 \nabla \phi' \cdot \nabla \phi'. \tag{5.298}$$

By multiplying Eq. (5.297) by $2\phi'$ and taking the mean, show that the scalar variance evolves according to Eq. (5.281).

5.5.2 Intermittency

Visual observations of free shear flows at an instant of time (e.g., Figs. 1.1 and 1.2 on pages 4 and 5) suggest that there is a sharp (but highly irregular) interface between the turbulent flow and the ambient fluid. Many experiments, starting with those of Corrsin (1943), have confirmed this picture: there is a highly contorted moving surface – called the *viscous superlayer* – that separates regions of turbulent and non-turbulent flow. Regions of turbulent flow are characterized by large vorticity: the r.m.s. vorticity ω', like the strain rate, scales with the Kolmogorov scales, i.e., $\omega' \sim 1/\tau_\eta \sim (U_s/\delta)\mathrm{Re}^{1/2}$. In contrast, the non-turbulent flow is essentially irrotational. At a fixed location

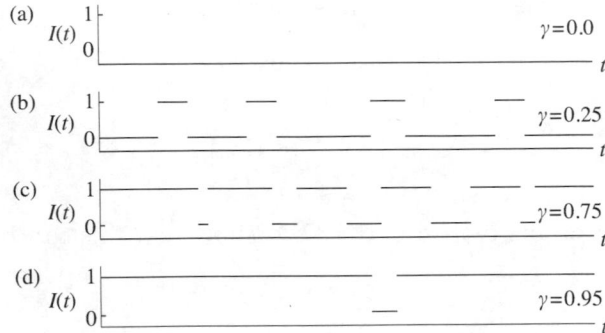

Fig. 5.39. A sketch of the intermittency function versus time in a free shear flow (a) in the irrotational non-turbulent surroundings, (b) in the outer part of the intermittent region, (c) in the inner part of the intermittent region, and (d) close to the center of the flow.

toward the edge of a free shear flow, the motion is sometimes turbulent and sometimes non-turbulent – the flow there is *intermittent*.

To avoid misconceptions, it is emphasized that there are no discontinuities across the viscous superlayer. In the mathematical sense, the vorticity and all other fields vary smoothly across the layer: but the layer is very thin compared with the flow width δ.

The starting point for the quantitative description of intermittency is the indicator function – or *intermittency function* – $I(x,t)$. This is defined to be $I = 1$ in turbulent flow, and $I = 0$ in non-turbulent flow. Operationally it can be obtained in terms of the Heaviside function as

$$I(x,t) = H(|\omega(x,t)| - \omega_{\text{thresh}}), \qquad (5.299)$$

where ω_{thresh} is a small positive threshold. Figure 5.39 shows a sketch of time series of $I(t)$ at various locations in a free shear flow.

The *intermittency factor* $\gamma(x,t)$ is the probability that the flow at (x,t) is turbulent:

$$\gamma(x,t) = \langle I(x,t) \rangle = \text{Prob}\{|\omega(x,t)| > \omega_{\text{thresh}}\}. \qquad (5.300)$$

For all the free shear flows, experiments show that the profiles of γ become self-similar. As an example, Fig. 5.40 shows the profile of γ measured in the plane wake. In other flows, the profiles are similar in shape, but in jets and mixing layers the intermittent region ($0.1 < \gamma < 0.9$, say) occupies a smaller fraction of the flow's width.

The intermittency function can be used to obtain *conditional statistics*. We take the scalar $\phi(x,t)$ as an example. With ψ being the sample-space variable, the PDF of $\phi(x,t)$ is denoted by $f_\phi(\psi;x,t)$. This can be decomposed into a

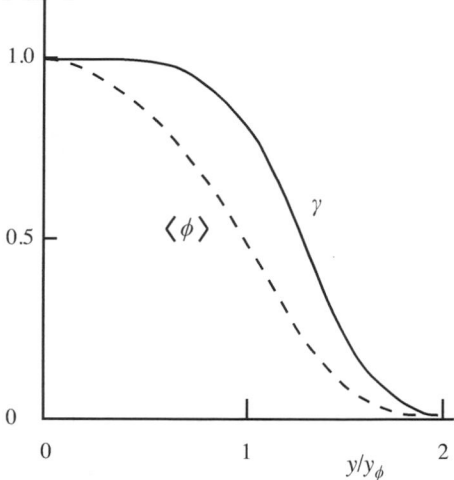

Fig. 5.40. The profile of the intermittency factor in the self-similar plane wake. The mean scalar profile is shown for comparison: y_ϕ is the half-width. (From the experimental data of LaRue and Libby (1974, 1976).)

contribution from turbulent flow, $\gamma(x,t)f_T(\psi;x,t)$, and a contribution from non-turbulent flow;

$$f_\phi = \gamma f_T + (1 - \gamma)f_N. \tag{5.301}$$

Thus f_T and f_N are the PDFs of ϕ conditional upon turbulent flow ($I = 1$), and upon non-turbulent flow ($I = 0$), respectively.

In the non-turbulent region, just as the vorticity is essentially zero, so also the scalar is essentially equal to the free-stream value, ϕ_∞. In fact, the condition $|\phi - \phi_\infty| > \phi_{\text{thresh}}$ is often used as an alternative indicator of turbulent flow. Thus, to within an approximation, the non-turbulent PDF is

$$f_N(\psi;x,t) = \delta(\psi - \phi_\infty). \tag{5.302}$$

The *turbulent* mean and variance of ϕ are defined by

$$\langle \phi(x,t) \rangle_T = \int_{-\infty}^{\infty} \psi f_T(\psi;x,t)\,\mathrm{d}\psi, \tag{5.303}$$

$$(\phi'_T)^2 = \int_{-\infty}^{\infty} (\psi - \langle \phi \rangle_T)^2 f_T\,\mathrm{d}\psi. \tag{5.304}$$

The non-turbulent moments $\langle \phi \rangle_N$ and $(\phi'_N)^2$ are defined similarly, and if f_N is given by Eq. (5.302) they are $\langle \phi \rangle_N = \phi_\infty$ and $\phi'_N = 0$. It follows from Eq. (5.301) that the unconditional mean and variance are given in terms of

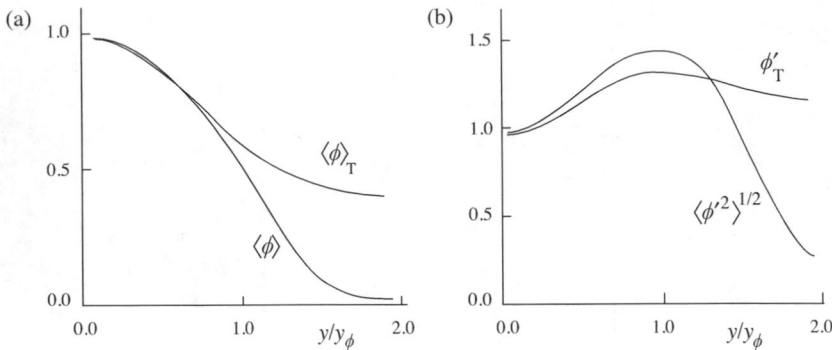

Fig. 5.41. A comparison of normalized unconditional and turbulent mean (a) and r.m.s. (b) scalar profiles in the self-similar plane wake. From the experimental data of LaRue and Libby (1974).

the conditional moments by

$$\langle \phi \rangle = \gamma \langle \phi \rangle_{\mathrm{T}} + (1 - \gamma)\langle \phi \rangle_{\mathrm{N}}, \qquad (5.305)$$

$$\langle \phi'^2 \rangle = \gamma(\phi'_{\mathrm{T}})^2 + (1 - \gamma)(\phi'_{\mathrm{N}})^2 + \gamma(1 - \gamma)(\langle \phi \rangle_{\mathrm{T}} - \langle \phi \rangle_{\mathrm{N}})^2. \qquad (5.306)$$

Notice that the unconditional variance contains a contribution from the difference between the conditional means.

For the turbulent wake, ϕ_∞, $\langle \phi \rangle_{\mathrm{N}}$ and ϕ'_{N} are all zero. The turbulent mean $\langle \phi \rangle_{\mathrm{T}}$ and r.m.s. ϕ'_{T} are compared with their unconditional counterparts in Fig. 5.41. As may be seen, both in terms of the mean and in terms of the r.m.s., the state of the turbulent flow is much more uniform across the flow than is suggested by the unconditional profiles.

Having described some of the features of the intermittent region, we now return to some of the fundamental questions concerning intermittency. Why is the turbulent/non-turbulent interface so sharp? What is the nature and behavior of the viscous superlayer? What are the characteristics of the fluctuations in velocity in the non-turbulent region?

A feature of all turbulent shear flows is that they *entrain* fluid. In the round jet, for example, the mass flow rate increases linearly with the axial distance x. Since the intermittency factor is self-similar, it follows that the mass flow rate of turbulent fluid also increases linearly with x. Consequently, at $x/d = 100$, 80% of the fluid in turbulent motion has been entrained between $x/d = 20$ and $x/d = 100$. The entrained fluid originates from the ambient, where ω and ϕ are zero. Only by molecular diffusion can ω and ϕ depart from zero[3] and, at the high Reynolds numbers of turbulent flows,

---

[3] Where ω is zero, so also is the vortex-stretching term $\omega \cdot \nabla U$.

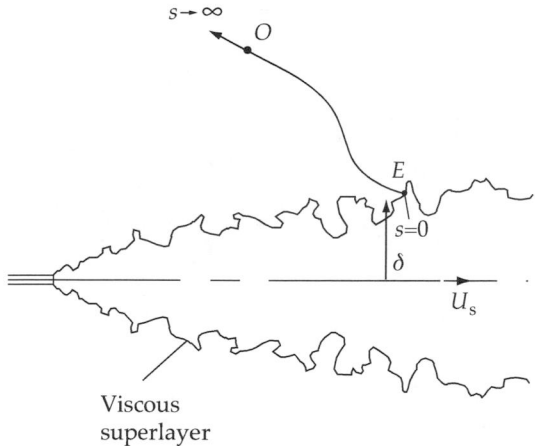

Fig. 5.42. A sketch of a turbulent round jet showing the viscous superlayer, and the path of a fluid particle from a point in the quiescent ambient, O, to the superlayer, E.

molecular diffusion can have a significant effect only if gradients are very steep. Hence, in the viscous superlayer, $|\omega|$ and ϕ rise steeply from zero, and the thickness of the layer (being inversely proportional to $|\nabla \phi|$) is small.

Some characteristics of the viscous superlayer are known from experiments (e.g., LaRue and Libby (1976)), but some are not. The large-scale turbulent motions convect the superlayer so that the PDF of its lateral position is quite accurately given by a Gaussian distribution. Correspondingly, the profile of γ is close to an error function (see Fig. 5.40). The turbulent motions also deform the superlayer so that it is randomly corrugated, and over an intermediate range of length scales it is approximately fractal, with a fractal dimension of 2.36 (Sreenivasan *et al.* 1989). Since the mass flow rate of turbulent fluid increases with downstream distance, it follows that the superlayer propagates relative to the fluid into the irrotational ambient: the product of the superlayer's area and this propagation speed gives the volumetric entrainment rate, which scales as $U_s \delta^2$ (independent of Re).

A consistent picture emerges if (as first suggested by Corrsin and Kistler (1954)) we suppose that the propagation speed u_e of the superlayer relative to the fluid scales with the Kolmogorov velocity $u_\eta \sim U_s \text{Re}^{-1/4}$ (Eq. (5.154)). Then, so that the overall entrainment rate is independent of Re, the superlayer's area scales as $\delta^2 \text{Re}^{1/4}$. The following crude model then suggests that the superlayer thickness scales with the Kolmogorov scale $\eta \sim \delta \text{Re}^{-3/4}$ (Eq. (5.152)).

Figure 5.42 is a sketch of a turbulent round jet showing the path of a fluid particle from a point O in the ambient to a point E in the viscous

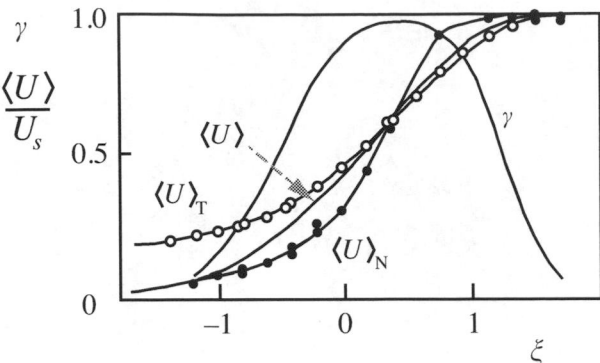

Fig. 5.43. Profiles of the intermittency factor γ, the unconditional mean axial velocity $\langle U \rangle$ and the turbulent $\langle U \rangle_\mathrm{T}$ and non-turbulent $\langle U \rangle_\mathrm{N}$ conditional mean velocities in a self-similar mixing layer. From the experimental data of Wygnanski and Fiedler (1970).

superlayer. Let s denote the arclength along the fluid particle's path measured from E in the direction of O. The scalar equation $\mathrm{D}\phi/\mathrm{D}t = \Gamma \nabla^2 \phi$ is crudely approximated along this path by the ordinary differential equation

$$-u_\mathrm{e} \frac{\mathrm{d}\phi}{\mathrm{d}s} = \Gamma \frac{\mathrm{d}^2 \phi}{\mathrm{d}s^2}. \tag{5.307}$$

The appropriate boundary conditions are $\phi(\infty) = 0$ (in the ambient) and $\phi(0) = \phi_0$ of order unity, in the superlayer. The solution to Eq. (5.307) is

$$\phi(s) = \phi_0 \exp(-s/\Lambda), \tag{5.308}$$

where the lengthscale Λ – proportional to the estimated thickness of the viscous superlayer – is $\Lambda = \Gamma/u_\mathrm{e}$. Relative to the local flow width δ, this lengthscale is

$$\frac{\Lambda}{\delta} = \frac{\nu}{U_\mathrm{s}\delta} \left(\frac{\Gamma}{\nu} \right) \left(\frac{U_\mathrm{s}}{u_\mathrm{e}} \right) = c\mathrm{Re}^{-3/4}, \tag{5.309}$$

where c is a constant, and the Reynolds number is $U_\mathrm{s}\delta/\nu$. (Recall that $u_\mathrm{e}/U_\mathrm{s}$ scales as $\mathrm{Re}^{-1/4}$.) Thus, according to this model, the superlayer thickness scales with the Kolmogorov scales, as $\delta\mathrm{Re}^{-3/4}$. The experimental evidence (e.g., LaRue and Libby (1976)) is consistent with this picture, but insufficient to be deemed convincing confirmation.

We have focused on the vorticity ω and the scalar ϕ because, in the non-turbulent region, their values can be changed only by molecular effects ($\nu \nabla^2 \omega$ and $\Gamma \nabla^2 \phi$), and these are negligible. The velocity, on the other hand, is affected also by pressure gradients. The turbulent motions in the core of

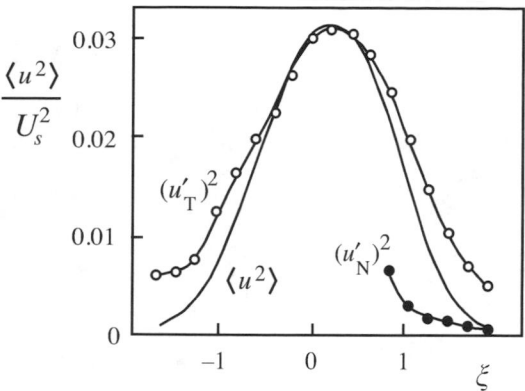

Fig. 5.44. Profiles of unconditional $\langle u^2 \rangle$, turbulent $(u'_T)^2$, and non-turbulent $(u'_N)^2$ variances of axial velocity in a self-similar mixing layer. From the experimental data of Wygnanski and Fiedler (1970).

the flow induce pressure fluctuations which lead to velocity fluctuations in the non-turbulent region.

Figure 5.43 shows conditional and unconditional mean velocities measured in a self-similar mixing layer. It may be seen that the non-turbulent velocity $\langle U \rangle_N$ is appreciably different than the free-stream velocity. For the same flow, Fig. 5.44 shows the conditional and unconditional variances of the axial velocity. As may be seen, the r.m.s. of the non-turbulent velocity u'_N is not insignificant. A remarkably successful theory due to Phillips (1955) indicates that the non-turbulent normal stresses (e.g., $(u'_N)^2$) decrease with lateral distance (y) as y^{-4}.

EXERCISE

5.45 Consider the non-turbulent irrotational flow outside the the temporal mixing layer. (The flow is statistically homogeneous in the $x = x_1$ and $z = x_3$ directions.) Show from the Corrsin–Kistler equation (Section 4.2, Eq. (4.30)) that

$$\langle v^2 \rangle = \langle u^2 \rangle + \langle w^2 \rangle. \tag{5.310}$$

(This relation is found experimentally to be accurate for other free shear flows, e.g., Wygnanski and Fiedler (1970).)

5.5.3 PDFs and higher moments

An examination of one-point PDFs measured in shear flows leads to a simple conclusion: in homogeneous shear flow with a uniform mean scalar gradient

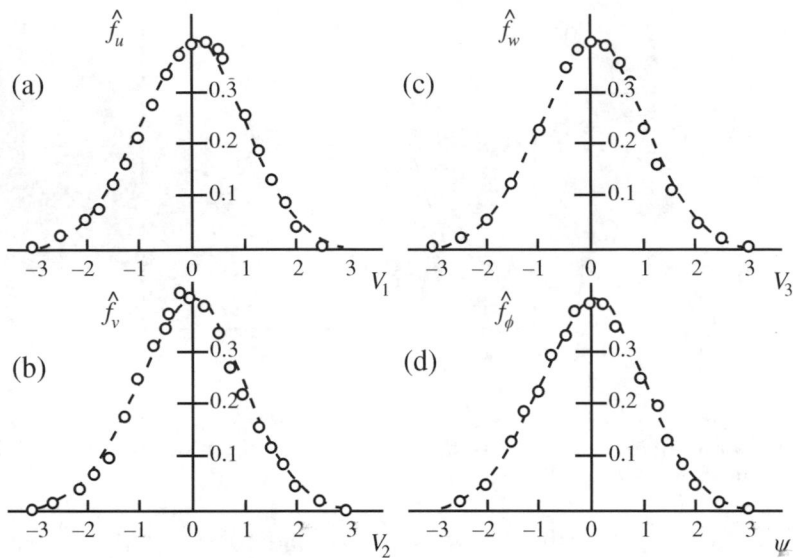

Fig. 5.45. Standardized PDFs of (a) u, (b) v, (c) w, and (d) ϕ in homogeneous shear flow. Dashed lines are standardized Gaussians. (From Tavoularis and Corrsin (1981).)

the joint PDF of velocity and the scalar is joint normal; whereas in free shear flows the PDFs are not Gaussian. Figure 5.45 shows the standardized marginal PDFs of u, v, w and ϕ measured by Tavoularis and Corrsin (1981) in homogeneous shear flow. No appreciable departure from the standardized Gaussian is evident. For the same flow, Fig. 5.46 shows velocity–velocity and velocity–scalar joint PDFs. Again, these are accurately described by joint-normal distributions. This is an important and valuable observation.

In free shear flows the picture is quite different. Figure 5.47 shows the scalar PDFs $f_\phi(\psi;\xi)$ in the temporal mixing layer. The scalar values in the two streams are $\phi = 0$ and $\phi = 1$; and, because of the boundedness property (see Section 2.6), $\phi(x,t)$ everywhere lies between zero and unity. Consequently $f_\phi(\psi;\xi)$ is zero for $\psi < 0$ and $\psi > 1$.

As may be seen from Fig. 5.47, in the center of the layer there is a broad, roughly bell-shaped, distribution that spans the entire range of values. As the measurement location moves toward the high-speed stream, the PDF moves to higher values of ψ, and develops a spike of increasing magnitude at the upper bound $\psi = 1$.

Qualitatively similar PDF shapes are found in jets (e.g., Dahm and Dimotakis (1990)) and wakes (e.g., LaRue and Libby (1974)). The measurements of Dowling and Dimotakis (1990) in the round jet are particularly valuable in clearly showing the self-similarity of scalar PDFs.

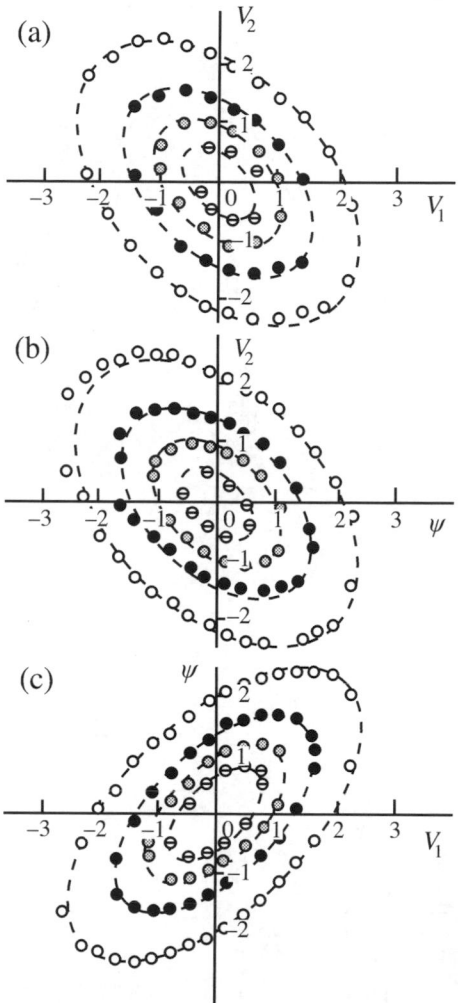

Fig. 5.46. Contour plots of joint PDFs of standardized variables measured in homogeneous shear flow: (a) u and v, (b) ϕ and v, (c) u and ϕ. Contour values are $0.15, 0.10, 0.05$, and 0.01. Dashed lines are corresponding contours for joint-normal distributions with the same correlation coefficients. (From Tavoularis and Corrsin (1981).)

Toward the edge of free shear flows, the changing shape of the PDF strongly influences the higher moments. Figure 5.48 shows the skewness

$$S_\phi \equiv \langle \phi'^3 \rangle / \langle \phi'^2 \rangle^{3/2},$$

and the kurtosis

$$K_\phi \equiv \langle \phi'^4 \rangle / \langle \phi'^2 \rangle^2,$$

of the scalar measured in a plane wake. In the center of the flow, these are

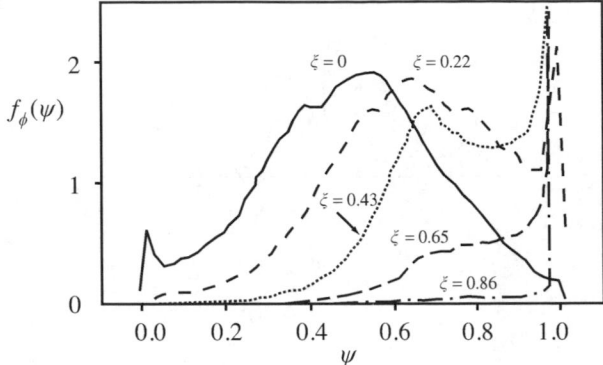

Fig. 5.47. PDFs of a conserved passive scalar in the self-similar temporal mixing layer at various lateral positions. From direct numerical simulations of Rogers and Moser (1994).

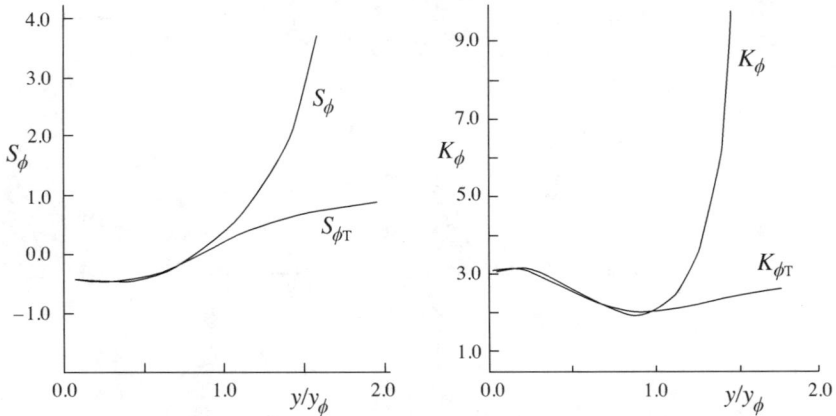

Fig. 5.48. Profiles of unconditional (S_ϕ and K_ϕ) and conditional turbulent ($S_{\phi T}$ and $K_{\phi T}$) skewness and kurtosis of a conserved passive scalar in the self-similar plane wake. From the experimental data of LaRue and Libby (1974).

not too far from the Gaussian values ($S_\phi = 0, K_\phi = 3$), but they increase to much larger values at the edge of the flow. Also shown in Fig. 5.48 are the values $S_{\phi T}$ and $K_{\phi T}$ obtained by conditionally averaging in the turbulent region. These remain closer to the Gaussian values throughout the flow.

We now turn to the PDF f_u of the axial velocity $U(x, t)$ in free shear flows. Figure 5.49 shows $f_u(V)$ at various cross-stream locations in a temporal mixing layer. In the center of the layer the familiar bell-shaped curve is observed; but at the edges, the PDF is considerably skewed toward the velocities within the layer. Unlike the scalar ϕ, the velocities are not subject to boundedness conditions. Also it is evident from Fig. 5.49 that, toward the

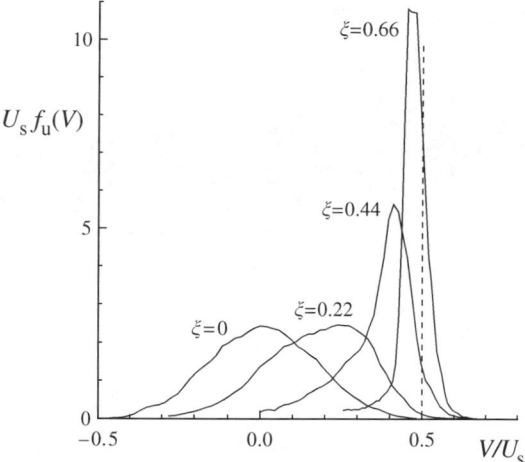

Fig. 5.49. PDFs of axial velocity in a temporal mixing layer. The distance from the center of the layer is $\xi = y/\delta$. The dashed line corresponds to the freestream velocity U_h. From the DNS data of Rogers and Moser (1994).

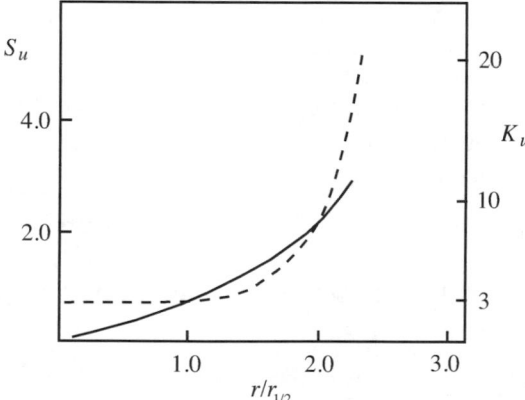

Fig. 5.50. Profiles of skewness (solid line) and kurtosis (dashed line) in the self-similar round jet. From the experimental data of Wygnanski and Fiedler (1969).

edge of the layer, velocities higher than the free-stream velocity occur with significant probability.

As is the case for the scalar ϕ, the skewness S_u and the kurtosis K_u of the axial velocity deviate significantly from Gaussian values (0 and 3) toward the edges of the flow. For example, Fig. 5.50 shows the values measured by Wygnanski and Fiedler (1969) in the self-similar round jet.

In summary: in homogeneous turbulence, the velocity–scalar joint PDF is joint normal. In the center of free shear flows, the PDFs are bell-shaped, but not exactly Gaussian, and the departure from Gaussianity becomes

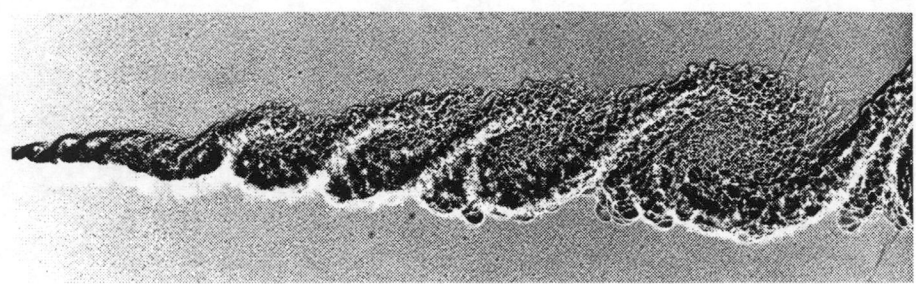

Fig. 5.51. A visualization of the flow of a plane mixing layer. A spark shadow graph of a mixing layer between helium (upper) $U_h = 10.1$ m s^{-1} and nitrogen (lower) $U_l = 3.8$ m s^{-1} at a pressure of 8 atm. (From Brown and Roshko (1974).)

pronounced in the intermittent region toward the edge of the flow. In the intermittent region the PDF of the scalar has a spike at the bound, whereas the PDF of velocity is unbounded.

5.5.4 Large-scale turbulent motion

Important though they are, the one-point statistics we have examined (e.g., $\langle U \rangle$, $\langle u_i u_j \rangle$, γ, etc.) provide only a very limited description of turbulent flows. Measurements of the two-point velocity correlations (e.g., Fig. 5.13) show that the velocity is appreciably correlated over distances comparable to the flow width. However, even these two-point statistics give little information about the structure of the instantaneous flow fields that give rise to these long-range correlations.

Some information on the large-scale turbulent motions is provided by flow visualization. Examples for a mixing layer and axisymmetric wakes are shown in Figs. 5.51 and 5.52. The predominant features visible in the mixing layer are the large 'rollers.' These are found to be regions of concentrated spanwise vorticity that are coherent over substantial spanwise distances. As the rollers are convected downstream they grow in size and spacing, and hence decrease in number. A roller can merge with an adjacent roller in a 'pairing' process, or it can be torn apart and its vorticity absorbed by adjacent rollers.

Large-scale motions have been studied in other free shear flows. Because they are of the size of the flow's width, it is inevitable that they are strongly influenced by the flow's geometry and boundary conditions, and consequently are different in different flows. To some extent, stability theory has been successful in explaining the structure of the large-scale motions, hence

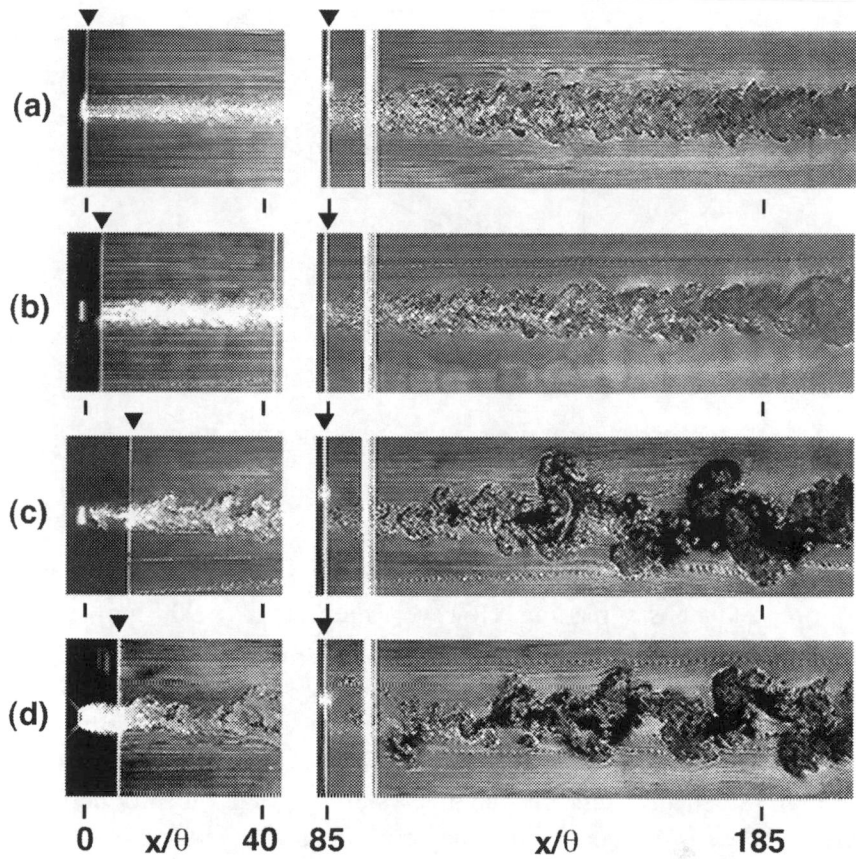

Fig. 5.52. A visualization of statistically axisymmetric wakes: (a) 50%-blockage screen, (b) 60%, (c) 85%, (d) 100% (i.e., a disk). The momentum thickness of the wake is θ; smoke wires are located at $x/\theta = 10$ and $x/\theta = 85$. (From Cannon *et al.* (1993) with permission of Springer-Verlag.)

supporting the view that these motions arise from the instabilities inherent in the various flows.

There are several phenomena that are plausibly explained in terms of the large-scale motions. As a clear example, Oster and Wygnanski (1982) performed experiments on turbulent mixing layers both with and without small-amplitude forcing. The forcing consists in oscillating a flap at the trailing edge of the splitter plate. The amplitude of the flap's motion (typically 1.5 mm) is quite small, so that in the initial region of the mixing layer ($x < 100$ mm) the effects on the mean-velocity and Reynolds-stress profiles are imperceptible. However, as Fig. 5.53 shows, further downstream there is a large effect on the growth of the layer. After some distance ($x \approx 300$ mm)

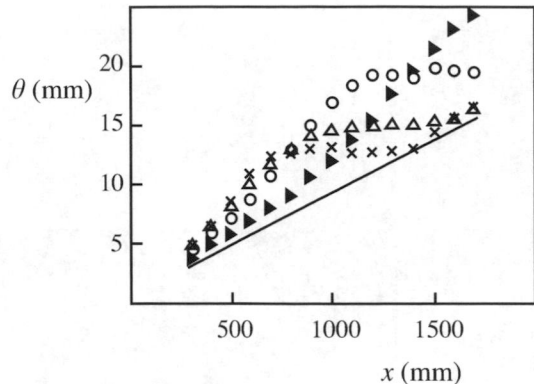

θ (mm)

x (mm)

Fig. 5.53. The thickness of the mixing layer, θ, against the axial distance, x, for various forcing frequencies, f: —, 0 Hz; $\times$, 60 Hz; $\triangle$, 50 Hz; $\circ$, 40 Hz; $\blacktriangle$, 30 Hz. (From Oster and Wygnanski (1982).)

the layer forced at 60 Hz begins to spread at almost twice the rate at which the unforced layer is spreading. However, then (at $x \approx 900$ mm) the layer ceases to grow, and in fact contracts slightly, before resuming its growth (at $x \approx 1,500$ mm). At lower forcing frequencies the same phenomenon occurs, but at larger distances downstream, where the characteristic timescale of the layer δ/U_s is larger. In the region where the layer ceases to grow, it is found that the shear stress $\langle uv \rangle$ changes sign, so that the production $\mathcal{P}$ is negative: energy is extracted from the turbulence and returned to the mean flow.

It is clear that the turbulent-viscosity hypothesis cannot explain these observations, especially the negative production $\mathcal{P}$. However, a plausible picture connecting the forcing, the inherent instabilities in the flow, and the large-scale turbulent motions emerges. Forcing at frequency f can be expected to excite modes of lengthscale ℓ proportional to U_s/f. Close to the splitter plate the layer thickness δ is small compared with ℓ. As δ increases toward ℓ, the larger-scale motions are excited, and pairings or amalgamations of rollers are promoted, thus increasing the scale of the motions and the width of the layer. A resonance is achieved, whereby the large-scale motions lock on to the scale of the forcing. Subsequent pairing or amalgamation leading to larger scales is inhibited, and hence the layer ceases to grow. In this region, the structure of the large-scale motions is found to resemble closely the patterns predicted by stability theory.

For unforced mixing layers, the spreading parameter S varies considerably from experiment to experiment, from $S \approx 0.06$ to $S \approx 0.11$ (Dimotakis 1991). Oster and Wygnanski (1982) suggested that this is due to the sensitivity of the mixing layer to the spectrum of small, uncontrollable disturbances of the flow, that depend upon the particular apparatus.

In the near field of wakes of bluff bodies (e.g., cylinders and spheres) there are large-scale motions with preferred frequencies, the clearest example being vortex shedding from a cylinder. These motions depend, of course, on the geometry of the body – they are different for screens, spheres, and disks. It is evident from Fig. 5.52 that differences in the large-scale motions persist into the far wake. Again, these motions have been linked to basic instabilities, and deemed responsible for the large differences among observed spreading rates (Cannon and Champagne (1991), see Table 5.3).

At a minimum there are two lessons to be learned from studies of large-scale structures. The first is that turbulence and turbulent flows can exhibit a much richer range of behavior than is admitted by the turbulent-viscosity hypothesis. The second is that turbulent processes are non-local in space and time: turbulence has a long memory, and its behavior at a point can be strongly influenced by the flow remote from that point.

6

The scales of turbulent motion

In examining free shear flows, we have observed that the turbulent motions range in size from the width of the flow δ to much smaller scales, which become progressively smaller (relative to δ) as the Reynolds number increases. We have also seen the importance of the turbulent kinetic energy and of the anisotropy in the Reynolds stresses. In this chapter we consider how the energy and anisotropy are distributed among the various scales of motion; and we examine the different physical processes occurring on these scales.

Two repeating themes in the chapter are the *energy cascade* and the *Kolmogorov hypotheses*. In brief, the idea of the energy cascade (introduced by Richardson (1922)) is that kinetic energy enters the turbulence (through the production mechanism) at the largest scales of motion. This energy is then transferred (by inviscid processes) to smaller and smaller scales until, at the smallest scales, the energy is dissipated by viscous action. Kolmogorov (1941b) added to and quantified this picture. In particular he identified the smallest scales of turbulence to be those that now bear his name.

In the first section, the energy cascade and Kolmogorov hypotheses are described in more detail. Then various statistics that discriminate among the various scales of motion are examined; namely, structure functions (Section 6.2), two-point correlations (Section 6.3), and spectra (Section 6.5). As a prelude to the discussion of spectra, in Section 6.4 the turbulent velocity field is expressed as the sum of Fourier modes, and the evolution of these modes according to the Navier–Stokes equations is deduced. The remaining sections give the spectral view of the energy cascade (Section 6.6), and discuss limitations of the Kolmogorov hypotheses (Section 6.7).

6.1 The energy cascade and Kolmogorov hypotheses

We consider a fully turbulent flow at high Reynolds number with characteristic velocity $\mathcal{U}$ and lengthscale $\mathcal{L}$. It is emphasized that the Reynolds number $Re = \mathcal{U}\mathcal{L}/\nu$ is large; and in fact the concepts are easiest to grasp if a *very* high Reynolds number is considered.

6.1.1 The energy cascade

The first concept in Richardson's view of the energy cascade is that the turbulence can be considered to be composed of *eddies* of different sizes. Eddies of size ℓ have a characteristic velocity $u(\ell)$ and timescale $\tau(\ell) \equiv \ell/u(\ell)$. An 'eddy' eludes precise definition, but it is conceived to be a turbulent motion, localized within a region of size ℓ, that is at least moderately coherent over this region. The region occupied by a large eddy can also contain smaller eddies.

The eddies in the largest size range are characterized by the lengthscale ℓ_0 which is comparable to the flow scale $\mathcal{L}$, and their characteristic velocity $u_0 \equiv u(\ell_0)$ is on the order of the r.m.s. turbulence intensity $u' \equiv (\frac{2}{3}k)^{1/2}$ which is comparable to $\mathcal{U}$. The Reynolds number of these eddies $Re_0 \equiv u_0\ell_0/\nu$ is therefore large (i.e., comparable to Re), so the direct effects of viscosity are negligibly small.

Richardson's notion is that the large eddies are unstable and break up, transferring their energy to somewhat smaller eddies. These smaller eddies undergo a similar break-up process, and transfer their energy to yet smaller eddies. This *energy cascade* – in which energy is transferred to successively smaller and smaller eddies – continues until the Reynolds number $Re(\ell) \equiv u(\ell)\ell/\nu$ is sufficiently small that the eddy motion is stable, and molecular viscosity is effective in dissipating the kinetic energy. Richardson (1922) succinctly summarized the matter thus:

> Big whorls have little whorls,
> Which feed on their velocity;
> And little whorls have lesser whorls,
> And so on to viscosity
> (in the molecular sense).

One reason that this picture is of importance is that it places dissipation at the end of a sequence of processes. The *rate* of dissipation ε is determined, therefore, by the first process in the sequence, which is the transfer of energy from the largest eddies. These eddies have energy of order u_0^2 and timescale $\tau_0 = \ell_0/u_0$, so the rate of transfer of energy can be supposed to scale as $u_0^2/\tau_0 = u_0^3/\ell_0$. Consequently, consistent with the experimental observations in free shear flows, this picture of the cascade indicates that ε scales as u_0^3/ℓ_0, independent of ν (at the high Reynolds numbers being considered).

6.1.2 The Kolmogorov hypotheses

Several fundamental questions remain unanswered. What is the size of the smallest eddies that are responsible for dissipating the energy? As ℓ decreases, do the characteristic velocity and timescales $u(\ell)$ and $\tau(\ell)$ increase, decrease, or remain the same? (The assumed decrease of the Reynolds number $u(\ell)\ell/\nu$ with ℓ is not sufficient to determine these trends.)

These questions and more are answered by the theory advanced by Kolmogorov (1941b)[1] which is stated in the form of three hypotheses. A consequence of the theory – which Kolmogorov used to motivate the hypotheses – is that both the velocity and timescales $u(\ell)$ and $\tau(\ell)$ decrease as ℓ decreases.

The first hypothesis concerns the isotropy of the small-scale motions. In general, the large eddies are anisotropic and are affected by the boundary conditions of the flow. Kolmogorov argued that the directional biases of the large scales are lost in the chaotic scale-reduction process, by which energy is transferred to successively smaller and smaller eddies. Hence (approximately stated):

> **Kolmogorov's hypothesis of local isotropy.** At sufficiently high Reynolds number, the small-scale turbulent motions ($\ell \ll \ell_0$) are statistically isotropic.

(The term 'local isotropy' means isotropy only at small scales, and is defined more precisely in Section 6.1.4.) It is useful to introduce a lengthscale ℓ_{EI} (with $\ell_{EI} \approx \frac{1}{6}\ell_0$, say) as the demarcation between the anisotropic large eddies ($\ell > \ell_{EI}$) and the isotropic small eddies ($\ell < \ell_{EI}$). (Justification for this specification of ℓ_{EI}, and of other scales introduced below, is provided in Section 6.5.)

Just as the directional information of the large scales is lost as the energy passes down the cascade, Kolmogorov argued that all information about the geometry of the large eddies – determined by the mean flow field and boundary conditions – is also lost. As a consequence, the statistics of the small-scale motions are in a sense universal – similar in every high-Reynolds-number turbulent flow.

On what parameters does this statistically universal state depend? In the energy cascade (for $\ell < \ell_{EI}$) the two dominant processes are the transfer of energy to successively smaller scales, and viscous dissipation. A plausible hypothesis, then, is that the important parameters are the rate at which the small scales receive energy from the large scales (which we denote by $\mathcal{T}_{EI}$), and the kinematic viscosity ν. As we shall see, the dissipation rate ε

---

[1] An English translation of this paper is reproduced as Kolmogorov (1991) in a special issue of the *Proceedings of the Royal Society* published to mark the fiftieth anniversary of the original publication. The other papers in this issue, which relate to the Kolmogorov hypotheses, are also of interest.

is determined by the energy transfer rate $\mathcal{T}_{EI}$, so that these two rates are nearly equal, i.e., $\varepsilon \approx \mathcal{T}_{EI}$. Consequently, the hypothesis that the statistically universal state of the small scales is determined by v and the rate of energy transfer from the large scales $\mathcal{T}_{EI}$ can be stated as:

> **Kolmogorov's first similarity hypothesis.** In every turbulent flow at sufficiently high Reynolds number, the statistics of the small-scale motions ($\ell < \ell_{EI}$) have a universal form that is uniquely determined by v and ε.

The size range $\ell < \ell_{EI}$ is referred to as the *universal equilibrium range*. In this range, the timescales $\ell/u(\ell)$ are small compared with ℓ_0/u_0, so that the small eddies can adapt quickly to maintain a dynamic equilibrium with the energy-transfer rate $\mathcal{T}_{EI}$ imposed by the large eddies.

Given the two parameters ε and v, there are (to within multiplicative constants) unique length, velocity, and time scales that can be formed. These are the Kolmogorov scales:

$$\eta \equiv (v^3/\varepsilon)^{1/4}, \tag{6.1}$$

$$u_\eta \equiv (\varepsilon v)^{1/4}, \tag{6.2}$$

$$\tau_\eta \equiv (v/\varepsilon)^{1/2}. \tag{6.3}$$

Two identities stemming from these definitions clearly indicate that the Kolmogorov scales characterize the very smallest, dissipative eddies. First, the Reynolds number based on the Kolmogorov scales is unity, i.e., $\eta u_\eta/v = 1$, which is consistent with the notion that the cascade proceeds to smaller and smaller scales until the Reynolds number $u(\ell)\ell/v$ is small enough for dissipation to be effective. Second, the dissipation rate is given by

$$\varepsilon = v(u_\eta/\eta)^2 = v/\tau_\eta^2, \tag{6.4}$$

showing that $(u_\eta/\eta) = 1/\tau_\eta$ provides a consistent characterization of the velocity gradients of the dissipative eddies.

Having identified the Kolmogorov scales, we can now state a consequence of the hypotheses that demonstrates their potency, and clarifies the meaning of the phrases 'similarity hypothesis' and 'universal form.' Consider a point x_0 in a high-Reynolds-number turbulent flow at a time t_0. In terms of the Kolmogorov scales at (x_0, t_0), non-dimensional coordinates are defined by

$$y \equiv (x - x_0)/\eta, \tag{6.5}$$

and the non-dimensional velocity-difference field is defined by

$$w(y) \equiv [U(x, t_0) - U(x_0, t_0)]/u_\eta. \tag{6.6}$$

It is not possible to form a non-dimensional parameter from ε and ν; so (on dimensional grounds) the 'universal form' of the statistics of the non-dimensional field $w(y)$ cannot depend on ε and ν. Consequently, according to the Kolmogorov hypotheses stated above, when the non-dimensional velocity field $w(y)$ is examined on not too large a scale (specifically $|y| < \ell_{EI}/\eta$), it is statistically isotropic and statistically identical at all points (x_0, t_0) in all high-Reynolds-number turbulent flows. *On the small scales, all high-Reynolds-number turbulent velocity fields are statistically similar; that is, they are statistically identical when they are scaled by the Kolmogorov scales (Eqs. (6.5) and (6.6)).*

The ratios of the smallest to largest scales are readily determined from the definitions of the Kolmogorov scales and from the scaling $\varepsilon \sim u_0^3/\ell_0$. The results are

$$\eta/\ell_0 \sim \mathrm{Re}^{-3/4}, \tag{6.7}$$

$$u_\eta/u_0 \sim \mathrm{Re}^{-1/4}, \tag{6.8}$$

$$\tau_\eta/\tau_0 \sim \mathrm{Re}^{-1/2}. \tag{6.9}$$

Evidently, at high Reynolds number, the velocity scales and timescales of the smallest eddies (u_η and τ_η) are – as previously supposed – small compared with those of the largest eddies (u_0 and τ_0).

Inevitably, as is evident from flow visualization (e.g., Fig. 1.2 on page 5), the ratio η/ℓ_0 decreases with increasing Re. As a consequence, at sufficiently high Reynolds number, there is a range of scales ℓ that are very small compared with ℓ_0, and yet very large compared with η, i.e., $\ell_0 \gg \ell \gg \eta$. Since eddies in this range are much bigger than the dissipative eddies, it may be supposed that their Reynolds number $\ell u(\ell)/\nu$ is large, and consequently that their motion is little affected by viscosity. Hence, following from this and from the first similarity hypothesis, we have (approximately stated):

> **Kolmogorov's second similarity hypothesis**. In every turbulent flow at sufficiently high Reynolds number, the statistics of the motions of scale ℓ in the range $\ell_0 \gg \ell \gg \eta$ have a universal form that is uniquely determined by ε, independent of ν.

It is convenient to introduce a lengthscale ℓ_{DI} (with $\ell_{DI} = 60\eta$, say), so that the range in the above hypothesis can be written $\ell_{EI} > \ell > \ell_{DI}$. This lengthscale ℓ_{DI} splits the universal equilibrium range ($\ell < \ell_{EI}$) into two subranges: the *inertial subrange* ($\ell_{EI} > \ell > \ell_{DI}$) and the *dissipation range* ($\ell < \ell_{DI}$). As the names imply, according to the second similarity hypothesis, motions in the inertial subrange are determined by inertial effects – viscous

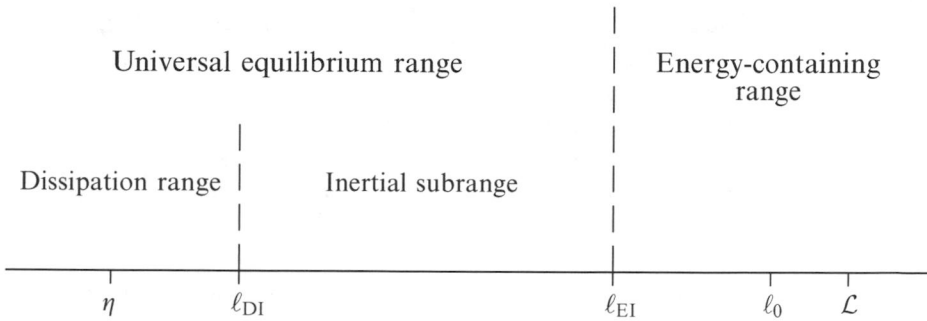

Fig. 6.1. Eddy sizes ℓ (on a logarithmic scale) at very high Reynolds number, showing the various lengthscales and ranges.

effects being negligible – whereas only motions in the dissipation range experience significant viscous effects, and so are responsible for essentially all of the dissipation. The various lengthscales and ranges are sketched in Fig. 6.1. (We shall see that the bulk of the energy is contained in the larger eddies in the size range $\ell_{EI} = \frac{1}{6}\ell_0 < \ell < 6\ell_0$, which is therefore called the *energy-containing range*. The suffixes EI and DI indicate that ℓ_{EI} is the demarcation line between energy (E) and inertial (I) ranges, as ℓ_{DI} is that between the dissipation (D) and inertial (I) subranges.)

Lengthscales, velocity scales, and timescales cannot be formed from ε alone. However, given an eddy size ℓ (in the inertial subrange), characteristic velocity scales and timescales for the eddy are those formed from ε and ℓ:

$$u(\ell) = (\varepsilon\ell)^{1/3} = u_\eta(\ell/\eta)^{1/3} \sim u_0(\ell/\ell_0)^{1/3}, \tag{6.10}$$

$$\tau(\ell) = (\ell^2/\varepsilon)^{1/3} = \tau_\eta(\ell/\eta)^{2/3} \sim \tau_0(\ell/\ell_0)^{2/3}. \tag{6.11}$$

A consequence, then, of the second similarity hypothesis is that (in the inertial subrange) the velocity scales and timescales $u(\ell)$ and $\tau(\ell)$ decrease as ℓ decreases.

In the conception of the energy cascade, a quantity of central importance – denoted by $\mathcal{T}(\ell)$ – is the rate at which energy is transferred from eddies larger than ℓ to those smaller than ℓ. If this transfer process is accomplished primarily by eddies of size comparable to ℓ, then $\mathcal{T}(\ell)$ can be expected to be of order $u(\ell)^2/\tau(\ell)$. The identity

$$u(\ell)^2/\tau(\ell) = \varepsilon, \tag{6.12}$$

stemming from Eqs. (6.10) and (6.11), is particularly revealing, therefore, since it suggests that $\mathcal{T}(\ell)$ is independent of ℓ (for ℓ in the inertial subrange). As we shall see, this is the case, and furthermore $\mathcal{T}(\ell)$ is equal to ε. Hence

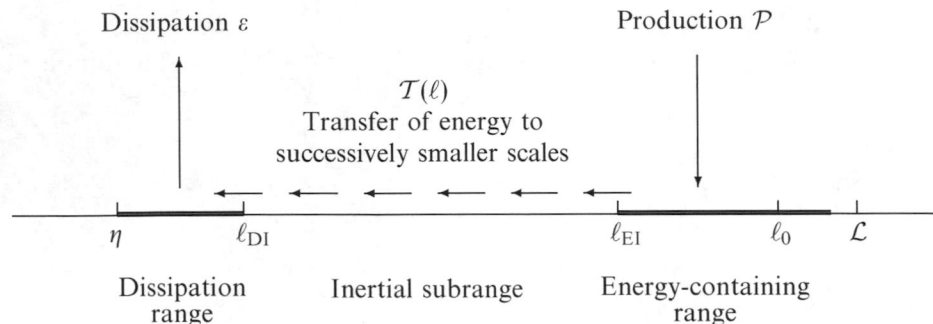

Fig. 6.2. A schematic diagram of the energy cascade at very high Reynolds number.

we have

$$\mathcal{T}_{\mathrm{EI}} \equiv \mathcal{T}(\ell_{\mathrm{EI}}) = \mathcal{T}(\ell) = \mathcal{T}_{\mathrm{DI}} \equiv \mathcal{T}(\ell_{\mathrm{DI}}) = \varepsilon, \qquad (6.13)$$

(for $\ell_{\mathrm{EI}} > \ell > \ell_{\mathrm{DI}}$). That is, the rate of energy transfer from the large scales, $\mathcal{T}_{\mathrm{EI}}$, determines the constant rate of energy transfer through the inertial subrange, $\mathcal{T}(\ell)$; hence the rate at which energy leaves the inertial subrange and enters the dissipation range $\mathcal{T}_{\mathrm{DI}}$; and hence the dissipation rate ε. This picture is sketched in Fig. 6.2.

6.1.3 The energy spectrum

It remains to be determined how the turbulent kinetic energy is distributed among the eddies of different sizes. This is most easily done for homogeneous turbulence by considering the energy spectrum function $E(\kappa)$ introduced in Chapter 3 (Eq. (3.166)).

Recall from Section 3.7 that motions of lengthscale ℓ correspond to wavenumber $\kappa = 2\pi/\ell$, and that the energy in the wavenumber range $(\kappa_{\mathrm{a}}, \kappa_{\mathrm{b}})$ is

$$k_{(\kappa_{\mathrm{a}},\kappa_{\mathrm{b}})} = \int_{\kappa_{\mathrm{a}}}^{\kappa_{\mathrm{b}}} E(\kappa)\,\mathrm{d}\kappa. \qquad (6.14)$$

In Section 6.5, $E(\kappa)$ is considered in some detail, and one result of interest here is that the contribution to the dissipation rate ε from motions in the range $(\kappa_{\mathrm{a}}, \kappa_{\mathrm{b}})$ is

$$\varepsilon_{(\kappa_{\mathrm{a}},\kappa_{\mathrm{b}})} = \int_{\kappa_{\mathrm{a}}}^{\kappa_{\mathrm{b}}} 2\nu\kappa^2 E(\kappa)\,\mathrm{d}\kappa. \qquad (6.15)$$

It follows from Kolmogorov's first similarity hypothesis that, in the universal equilibrium range ($\kappa > \kappa_{\mathrm{EI}} \equiv 2\pi/\ell_{\mathrm{EI}}$) the spectrum is a universal function of ε and v. From the second hypothesis it follows that, in the inertial range $(\kappa_{EI} < \kappa < \kappa_{DI} \equiv 2\pi/\ell_{DI})$, the spectrum is

$$E(\kappa) = C\varepsilon^{2/3}\kappa^{-5/3}, \tag{6.16}$$

where C is a universal constant. (These assertions are justified in Section 6.5.)

To understand some basic features of the Kolmogorov $-\frac{5}{3}$ spectrum, we consider the general power-law spectrum

$$E(\kappa) = A\kappa^{-p}, \tag{6.17}$$

where A and p are constants. The energy contained in wavenumbers greater than κ is

$$k_{(\kappa,\infty)} \equiv \int_{\kappa}^{\infty} E(\kappa')\,d\kappa' = \frac{A}{p-1}\kappa^{-(p-1)}, \tag{6.18}$$

for $p > 1$, while the integral diverges for $p \le 1$. Similarly the dissipation in wavenumbers less than κ is

$$\varepsilon_{(0,\kappa)} \equiv \int_{0}^{\kappa} 2v\kappa'^2 E(\kappa')\,d\kappa' = \frac{2vA}{3-p}\kappa^{3-p}, \tag{6.19}$$

for $p < 3$, while the integral diverges for $p \ge 3$. Thus, $p = \frac{5}{3}$, corresponding to the Kolmogorov spectrum, is around the middle of the range $(1,3)$ for which the integrals $k_{(\kappa,\infty)}$ and $\varepsilon_{(0,\kappa)}$ converge. The amount of energy in the high wavenumbers decreases as $k_{(\kappa,\infty)} \sim \kappa^{-2/3}$ as κ increases, whereas the dissipation in the low wavenumbers decreases as $\varepsilon_{(0,\kappa)} \sim \kappa^{4/3}$ as κ decreases toward zero.

Although the Kolmogorov $-\frac{5}{3}$ spectrum applies only to the inertial range, the observations made are consistent with the notion that the bulk of the energy is in the large scales ($\ell > \ell_{EI}$ or $\kappa < 2\pi/\ell_{EI}$), and that the bulk of the dissipation is in the small scales ($\ell < \ell_{DI}$ or $\kappa > 2\pi/\ell_{DI}$).

6.1.4 Restatement of the Kolmogorov hypotheses

In order to deduce precise consequences from them, it is worthwhile to provide here more precise statements of the Kolmogorov (1941) hypotheses. Kolmogorov presented these in terms of an N-point distribution in the four-dimensional x–t space. Here, however, we consider the N-point distribution in physical space (x) at a fixed time t – which is sufficiently general for most purposes.

Consider a simple domain $\mathcal{G}$ within the turbulent flow, and let $x^{(0)}$, $x^{(1)},\ldots,x^{(N)}$ be a specified set of points within $\mathcal{G}$. New coordinates and velocity differences are defined by

$$y \equiv x - x^{(0)}, \tag{6.20}$$

$$v(y) \equiv U(x,t) - U(x^{(0)},t), \tag{6.21}$$

and the joint PDF of v at the N points $y^{(1)}, y^{(2)}, \ldots, y^{(N)}$ is denoted by f_N.

> **The definition of local homogeneity.** The turbulence is locally homogeneous in the domain $\mathcal{G}$, if for every fixed N and $y^{(n)}(n = 1, 2, \ldots, N)$, the N-point PDF f_N is independent of $x^{(0)}$ and $U(x^{(0)}, t)$.
>
> **The definition of local isotropy.** The turbulence is locally isotropic in the domain $\mathcal{G}$ if it is locally homogeneous and if in addition the PDF f_N is invariant with respect to rotations and reflections of the coordinate axes.
>
> **The hypothesis of local isotropy.** In any turbulent flow with a sufficiently large Reynolds number (Re $= \mathcal{U}\mathcal{L}/\nu$), the turbulence is, to a good approximation, locally isotropic if the domain $\mathcal{G}$ is sufficiently small (i.e., $|y^{(n)}| \ll \mathcal{L}$, for all n) and is not near the boundary of the flow or its other singularities.
>
> **The first similarity hypothesis.** For locally isotropic turbulence, the N-point PDF f_N is uniquely determined by the viscosity ν and the dissipation rate ε.
>
> **The second similarity hypothesis.** If the moduli of the vectors $y^{(m)}$ and of their differences $y^{(m)} - y^{(n)}$ ($m \neq n$) are large compared with the Kolmogorov scale η, then the N-point PDF f_N is uniquely determined by ε and does not depend on ν.

It is important to observe that the hypotheses apply specifically to velocity differences. The use of the N-point PDF f_N allows the hypotheses to be applied to any turbulent flow, whereas statements in terms of wavenumber spectra apply only to flows that are statistically homogeneous (in at least one direction).

For inhomogeneous flows, local isotropy is possible only 'to a good approximation' (as stated in the hypothesis). For example, taking $y^{(1)} = e\ell$ and $y^{(2)} = -e\ell$ (where ℓ is a specified length and e a specified unit vector), we have

$$\langle v(y^{(1)}) - v(y^{(2)}) \rangle = \langle U(y^{(1)}) \rangle - \langle U(y^{(2)}) \rangle$$

$$\approx 2\frac{\ell}{\mathcal{L}} e \cdot \mathcal{L}\nabla\langle U \rangle. \tag{6.22}$$

Evidently this simple statistic is not exactly isotropic, but instead has a

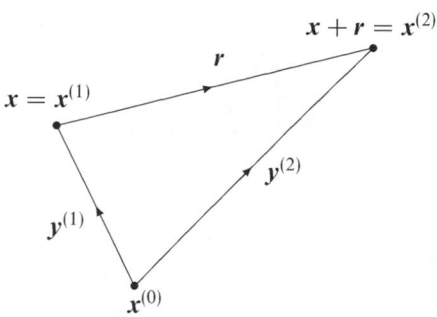

Fig. 6.3. A sketch showing the points x and $x+r$ in terms of $x^{(n)}$ and $y^{(n)}$. All points are within the domain $\mathcal{G}$.

small anisotropic component – of order $\ell/\mathcal{L}$ – arising from large-scale inhomogeneities.

6.2 Structure functions

To illustrate the correct application of the Kolmogorov hypotheses, we consider – as did Kolmogorov (1941b) – the second-order velocity structure functions. The predictions of the hypotheses are deduced, and then compared with experimental data.

By definition, the second-order velocity structure function is the covariance of the difference in velocity between two points $x+r$ and x:

$$D_{ij}(r, x, t) \equiv \langle [U_i(x+r, t) - U_i(x, t)][U_j(x+r, t) - U_j(x, t)] \rangle. \qquad (6.23)$$

It is rather obvious that the Kolmogorov hypotheses are applicable to this statistic, but this can be verified by re-expressing it in terms of the position and velocity differences y and v defined by Eqs. (6.20) and (6.21):

$$D_{ij}(y^{(2)} - y^{(1)}, x^{(0)} + y^{(1)}, t) = \langle [v_i(y^{(2)}) - v_i(y^{(1)})][v_j(y^{(2)}) - v_j(y^{(1)})] \rangle, \qquad (6.24)$$

see Fig. 6.3. We assume that all other conditions (e.g., sufficiently large Reynolds number) are satisfied.

The first implication of the hypothesis of local isotropy is that (for $r \equiv |r| \ll \mathcal{L}$) D_{ij} is independent of $x^{(0)}$. Equation (6.24) shows, then, that D_{ij} does not depend on its second argument, i.e., $D_{ij}(r, x, t)$ is independent of x. It is then evident from its definition (Eq. (6.23)) that $D_{ij}(r, t)$ depends on $r = y^{(2)} - y^{(1)}$, but not on $y^{(1)}$ and $y^{(2)}$ separately. Thus $D_{ij}(r, t)$ is an isotropic function of r.

To within scalar multiples, the only second-order tensors that can be

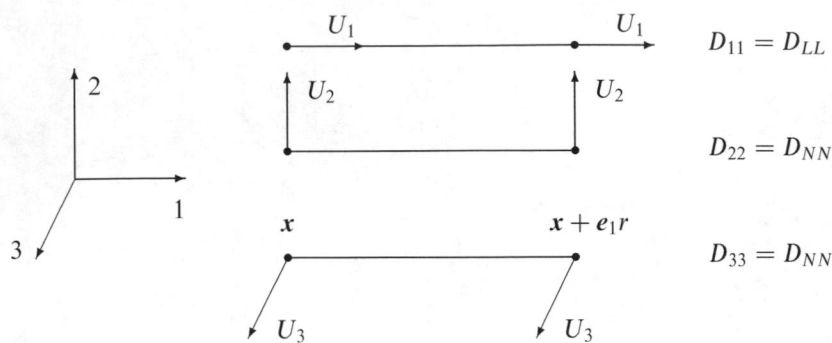

Fig. 6.4. A sketch of the velocity components involved in the longitudinal and transverse structure functions for $r = e_1 r$.

formed from the vector r are δ_{ij} and $r_i r_j$. Consequently D_{ij} can be written

$$D_{ij}(r,t) = D_{NN}(r,t)\delta_{ij} + [D_{LL}(r,t) - D_{NN}(r,t)]\frac{r_i r_j}{r^2}, \qquad (6.25)$$

where the scalar functions D_{LL} and D_{NN} are called, respectively, the longitudinal and transverse structure functions. If the coordinate system is chosen so that r is in the x_1 direction (i.e., $r = e_1 r$) then we obtain

$$\begin{aligned} D_{11} &= D_{LL}, \quad D_{22} = D_{33} = D_{NN}, \\ D_{ij} &= 0, \quad \text{for } i \neq j, \end{aligned} \qquad (6.26)$$

which, together with Fig. 6.4, shows the significance of D_{LL} and D_{NN}.

In homogeneous turbulence with $\langle U \rangle = 0$, a consequence of the continuity equation is

$$\frac{\partial}{\partial r_i} D_{ij}(r,t) = 0, \qquad (6.27)$$

see Exercise 6.1. In the present context, this equation also applies (to a good approximation) because of local homogeneity. It then follows from Eqs. (6.25) and (6.27) (see Exercise 6.2) that D_{NN} is uniquely determined by D_{LL} according to

$$D_{NN}(r,t) = D_{LL}(r,t) + \frac{1}{2} r \frac{\partial}{\partial r} D_{LL}(r,t). \qquad (6.28)$$

Thus, in locally isotropic turbulence, $D_{ij}(r,t)$ is determined by the single scalar function $D_{LL}(r,t)$.

According to the first similarity hypothesis, given r ($|r| \ll \mathcal{L}$), D_{ij} is uniquely determined by ε and ν. The quantity $(\varepsilon r)^{2/3}$ has dimensions of velocity squared, and so can be used to make D_{ij} non-dimensional. There is only one independent non-dimensional group that can be formed from r, ε

and v, which can conveniently be taken to be $r\varepsilon^{1/4}/v^{3/4} = r/\eta$. Thus, according to the first similarity hypothesis, there is a universal, non-dimensional function $\widehat{D}_{LL}(r/\eta)$ such that (subject to the conditions of the hypothesis)

$$D_{LL}(r,t) = (\varepsilon r)^{2/3}\widehat{D}_{LL}(r/\eta). \tag{6.29}$$

According to the second similarity hypothesis, for large r/η ($\mathcal{L} \gg r \gg \eta$), D_{LL} is independent of v. In this case there is no non-dimensional group that can be formed from ε and r, so D_{LL} is given by

$$D_{LL}(r,t) = C_2(\varepsilon r)^{2/3}, \tag{6.30}$$

where C_2 is a universal constant . (This implies that, for large r/η, $\widehat{D}_{LL}$ tends asymptotically to the constant value C_2.) The transverse structure function is, from Eq. (6.28),

$$D_{NN}(r,t) = \tfrac{4}{3}D_{LL}(r,t) = \tfrac{4}{3}C_2(\varepsilon r)^{2/3}, \tag{6.31}$$

and hence, from Eq. (6.25), D_{ij} is given by

$$D_{ij}(\boldsymbol{r},t) = C_2(\varepsilon r)^{2/3}\left(\frac{4}{3}\delta_{ij} - \frac{1}{3}\frac{r_i r_j}{r^2}\right). \tag{6.32}$$

Thus, in the inertial subrange ($\mathcal{L} \gg r \gg \eta$), the Kolmogorov hypotheses are sufficient to determine the second-order structure function in terms of ε, r and the universal constant C_2.

The predictions of the Kolmogorov hypotheses embodied in Eq. (6.32) have been tested by Saddoughi and Veeravalli (1994) in a turbulent boundary layer – claimed to be the highest-Reynolds-number boundary layer ever attained in a laboratory. The principal direction of flow is $x = x_1$, and the distance normal to the wall is $y = x_2$. Measurements are reported at $y = 400$ mm, at a location where the boundary layer thickness is $\delta = 1{,}090$ mm, the Reynolds number (based on δ) is 3.6×10^6, and the Kolmogorov scale is measured to be $\eta = 0.09$ mm. Taking $\mathcal{L} = \delta$ as the characteristic flow length, the ratio $\mathcal{L}/\eta \approx 12{,}000$ shows that there is a very large separation of scales.

With $\boldsymbol{r} = \boldsymbol{e}_1 r$, and $\mathcal{L} \gg r \gg \eta$, the prediction from Eq. (6.32) can be written

$$D_{11}/(\varepsilon r)^{2/3} = C_2, \tag{6.33}$$

$$D_{22}/(\varepsilon r)^{2/3} = D_{33}/(\varepsilon r)^{2/3} = \tfrac{4}{3}C_2, \tag{6.34}$$

$$D_{ij} = 0, \quad \text{for } i \neq j. \tag{6.35}$$

Figure 6.5 shows the measured structure functions divided by $(\varepsilon r)^{2/3}$, so that

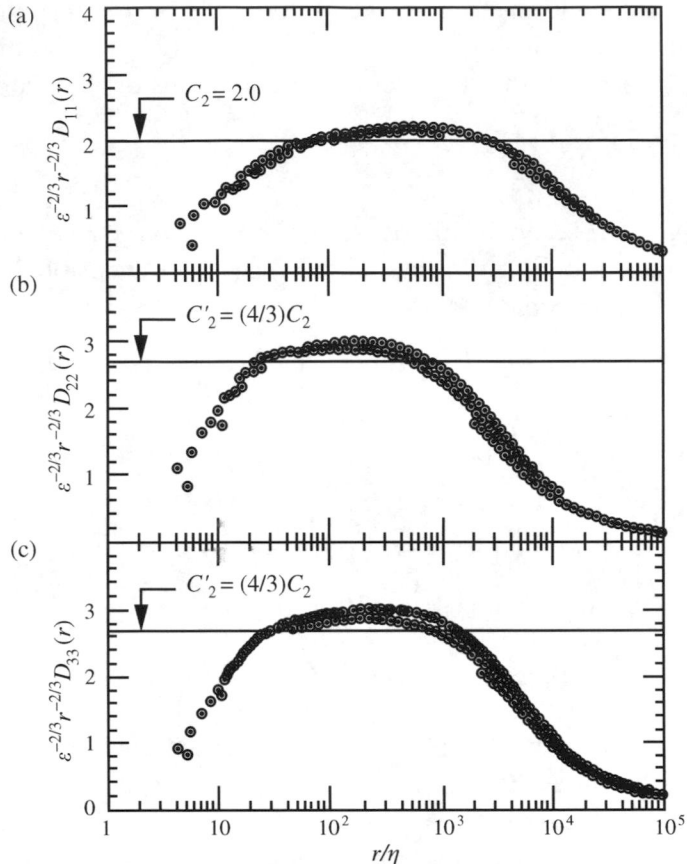

Fig. 6.5. Second-order velocity structure functions measured in a high-Reynolds-number turbulent boundary layer. The horizontal lines show the predictions of the Kolmogorov hypotheses in the inertial subrange, Eqs. (6.33) and (6.34). (From Saddoughi and Veeravalli (1994).)

the above predictions can readily be examined. Taking the value $C_2 = 2.0$ suggested by these and other data, we draw the following conclusions.

(i) For $7,000 \, \eta \approx \frac{1}{2}\mathcal{L} > r > 20\eta$, $D_{11}/(\varepsilon r)^{2/3}$ is within $\pm 15\%$ of C_2.

(ii) There is no perceptible difference between D_{22} and D_{33}.

(iii) For $1,200 \, \eta \approx \frac{1}{10}\mathcal{L} > r > 12\eta$, $D_{22}/(\varepsilon r)^{2/3}$ is within $\pm 15\%$ of $\frac{4}{3}C_2$.

Over the ranges of r given above, D_{11} and D_{22} change by factors of 50 and 20, respectively, and so $\pm 15\%$ variations can be considered small in comparison.

Clearly, these experimental observations provide substantial support for the Kolmogorov hypotheses. However, other flows need to be considered in order to test the universality of C_2, and there are many other statistics that can be examined. Further comparisons with experimental data are given in Section 6.5.

EXERCISES

6.1　For homogeneous turbulence with $\langle U \rangle = 0$, show that the structure function $D_{ij}(r)$ (Eq. (6.23)) and the two-point correlation $R_{ij}(r)$ (Eq. (3.160)) are related by

$$D_{ij}(r) = 2R_{ij}(0) - R_{ij}(r) - R_{ji}(r)$$
$$= 2R_{ij}(0) - R_{ij}(r) - R_{ij}(-r). \tag{6.36}$$

Show that a consequence of the continuity equation is

$$\frac{\partial D_{ij}}{\partial r_i} = \frac{\partial D_{ij}}{\partial r_j} = 0. \tag{6.37}$$

(Hint: refer to Exercises 3.34 and 3.35 on page 78.)

6.2　Differentiate Eq. (6.25) to obtain

$$\frac{\partial D_{ij}}{\partial r_i} = \frac{r_j}{r^2}\left(r\frac{\partial D_{LL}}{\partial r} + 2(D_{LL} - D_{NN})\right). \tag{6.38}$$

Hence verify Eq. (6.28).

6.3　Show that, for small r ($r \ll \eta$), in isotropic turbulence the second-order velocity structure functions are

$$D_{LL}(r) = r^2\left\langle\left(\frac{\partial u_1}{\partial x_1}\right)^2\right\rangle = \frac{r^2\varepsilon}{15\nu}, \tag{6.39}$$

$$D_{NN}(r) = r^2\left\langle\left(\frac{\partial u_2}{\partial x_1}\right)^2\right\rangle = \frac{2r^2\varepsilon}{15\nu} = 2D_{LL}(r). \tag{6.40}$$

6.3 Two-point correlation

The Kolmogorov hypotheses, and deductions drawn from them, have no direct connection to the Navier–Stokes equations (although, as in the previous section, the continuity equation is usually invoked). Although, in the description of the energy cascade, the transfer of energy to successively smaller scales has been identified as a phenomenon of prime importance, the precise mechanism by which this transfer takes place has not been identified or quantified. It is natural, therefore, to try to extract from the Navier–Stokes equations useful information about the energy cascade. The earliest attempts (outlined in this section) are those of Taylor (1935a) and of von Kármán and Howarth (1938), which are based on the two-point correlation. The next two sections give the view from wavenumber space in terms of the energy spectrum – the Fourier transform of the two-point correlation.

Autocorrelation functions

Consider homogeneous isotropic turbulence, with zero mean velocity, r.m.s. velocity $u'(t)$, and dissipation rate $\varepsilon(t)$. Because of homogeneity, the two-point correlation

$$R_{ij}(\boldsymbol{r}, t) \equiv \langle u_i(\boldsymbol{x} + \boldsymbol{r}, t) u_j(\boldsymbol{x}, t) \rangle, \tag{6.41}$$

is independent of $\boldsymbol{x}$. At the origin it is

$$R_{ij}(0, t) = \langle u_i u_j \rangle = u'^2 \delta_{ij}. \tag{6.42}$$

There is neither production nor transport, so the equation for the evolution of the turbulent kinetic energy $k(t) = \frac{3}{2} u'(t)^2$ (Eq. (5.132)) reduces to

$$\frac{\mathrm{d}k}{\mathrm{d}t} = -\varepsilon. \tag{6.43}$$

Just as with the structure function D_{ij}, a consequence of isotropy is that R_{ij} can be expressed in terms of two scalar functions $f(r, t)$ and $g(r, t)$:

$$R_{ij}(\boldsymbol{r}, t) = u'^2 \left(g(r, t) \delta_{ij} + [f(r, t) - g(r, t)] \frac{r_i r_j}{r^2} \right), \tag{6.44}$$

(cf. Eq. (6.25)). With $\boldsymbol{r} = \boldsymbol{e}_1 r$, this equation becomes

$$R_{11}/u'^2 = f(r, t) = \langle u_1(\boldsymbol{x} + \boldsymbol{e}_1 r, t) u_1(\boldsymbol{x}, t) \rangle / \langle u_1^2 \rangle,$$
$$R_{22}/u'^2 = g(r, t) = \langle u_2(\boldsymbol{x} + \boldsymbol{e}_1 r, t) u_2(\boldsymbol{x}, t) \rangle / \langle u_2^2 \rangle, \tag{6.45}$$
$$R_{33} = R_{22}, \qquad R_{ij} = 0, \quad \text{for } i \neq j,$$

thus identifying f and g as the longitudinal and transverse autocorrelation functions, respectively. (Note that f and g are non-dimensional with $f(0, t) = g(0, t) = 1$.) Again in parallel with the properties of D_{ij}, the continuity equation implies that $\partial R_{ij}/\partial r_j = 0$ (see Exercise 3.35), which, in combination with Eq. (6.44), leads to

$$g(r, t) = f(r, t) + \frac{1}{2} r \frac{\partial}{\partial r} f(r, t). \tag{6.46}$$

Thus, in isotropic turbulence the two-point correlation $R_{ij}(\boldsymbol{r}, t)$ is completely determined by the longitudinal autocorrelation function $f(r, t)$. Figure 6.6 shows the measurements of $f(r, t)$ in nearly isotropic grid-generated turbulence obtained by Comte-Bellot and Corrsin (1971).

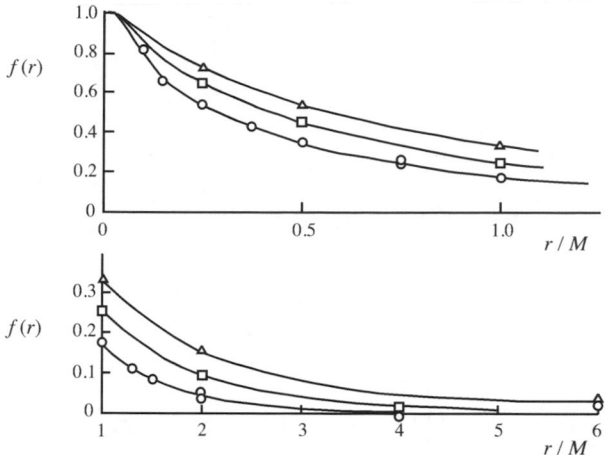

Fig. 6.6. Measurements of the longitudinal velocity autocorrelation functions $f(r,t)$ in grid turbulence: $x_1/M = 42, \circ; 98, \square; 172, \triangle$. (From Comte-Bellot and Corrsin (1971).)

There are two distinct longitudinal lengthscales, $L_{11}(t)$ and $\lambda_f(t)$, that can be defined from f; and then there are corresponding transverse lengthscales $L_{22}(t)$ and $\lambda_g(t)$ defined from g.

Integral lengthscales

The first of the lengthscales obtained from $f(r,t)$ is the *longitudinal integral scale*

$$L_{11}(t) \equiv \int_0^\infty f(r,t)\,dr, \tag{6.47}$$

which we have already encountered (e.g., in Section 5.1, Fig. 5.13 on page 110). The integral scale $L_{11}(t)$ is simply the area under the curve of $f(r,t)$, so inspection of Fig. 6.6 immediately reveals that L_{11} grows with time (in grid turbulence). As previously observed, L_{11} is characteristic of the larger eddies. In isotropic turbulence, the transverse integral scale

$$L_{22}(t) \equiv \int_0^\infty g(r,t)\,dr \tag{6.48}$$

is just half of $L_{11}(t)$ (see Exercise 6.4).

EXERCISES

6.4   Show that Eq. (6.46) can be rewritten

$$g(r,t) = \tfrac{1}{2}\left(f(r,t) + \frac{\partial}{\partial r}[rf(r,t)] \right), \tag{6.49}$$

and hence that, in isotropic turbulence, the transverse integral scale

$$L_{22}(t) \equiv \int_0^\infty g(r,t)\,dr \qquad (6.50)$$

is half of the longitudinal scale, i.e.,

$$L_{22}(t) = \tfrac{1}{2}L_{11}(t). \qquad (6.51)$$

6.5   Show from Eq. (6.46) that

$$\int_0^\infty rg(r,t)\,dr = 0, \qquad (6.52)$$

(assuming that $f(r,t)$ decays more rapidly than r^{-2} for large r).

Taylor microscales

The second lengthscale obtained from $f(r,t)$ is the *longitudinal Taylor microscale* $\lambda_f(t)$. Since $f(r,t)$ is an even function of r and no greater than unity, the first derivative at the origin $f'(0,t) = (\partial f/\partial r)_{r=0}$ is zero, while the second derivative $f''(0,t) = (\partial^2 f/\partial r^2)_{r=0}$ is non-positive. As we shall see, in turbulence $f''(0)$ is strictly negative, so $\lambda_f(t)$ defined by

$$\lambda_f(t) = \left[-\tfrac{1}{2}f''(0,t)\right]^{-1/2} \qquad (6.53)$$

is real, positive, and has dimensions of length.

A geometric construction makes this abstruse definition clear. Let $p(r)$ be the parabola osculating $f(r)$ at $r = 0$ (i.e., the parabola with $p(0) = f(0)$, $p'(0) = f'(0)$, and $p''(0) = f''(0)$). Evidently $p(r)$ is

$$\begin{aligned} p(r) &= 1 + \tfrac{1}{2}f''(0)r^2 \\ &= 1 - r^2/\lambda_f^2. \end{aligned} \qquad (6.54)$$

Thus, as sketched in Fig. 6.7, the osculating parabola intersects the axis at $r = \lambda_f$.

As the following manipulation shows, $f''(0,t)$ (and hence $\lambda_f(t)$) is related to velocity derivatives:

$$\begin{aligned} -u'^2 f''(0,t) &= -u'^2 \lim_{r \to 0} \frac{\partial^2}{\partial r^2} f(r,t) \\ &= -\lim_{r \to 0} \frac{\partial^2}{\partial r^2} \langle u_1(x + e_1 r, t) u_1(x,t) \rangle \\ &= -\lim_{r \to 0} \left\langle \left(\frac{\partial^2 u_1}{\partial x_1^2}\right)_{x + e_1 r} u_1(x,t) \right\rangle \end{aligned}$$

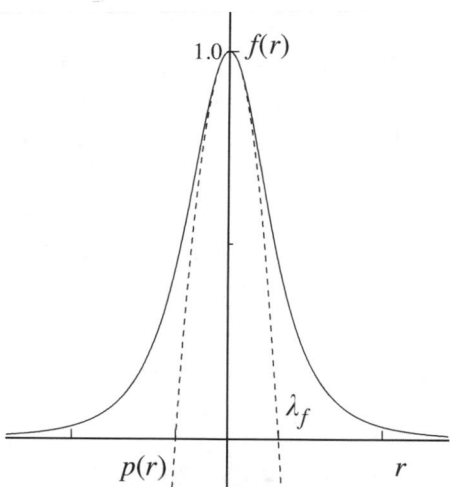

Fig. 6.7. A sketch of the longitudinal velocity autocorrelation function showing the definition of the Taylor microscale λ_f.

$$= -\left\langle \left(\frac{\partial^2 u_1}{\partial x_1^2} \right) u_1 \right\rangle$$

$$= -\left\langle \frac{\partial}{\partial x_1} \left(u_1 \frac{\partial u_1}{\partial x_1} \right) - \left(\frac{\partial u_1}{\partial x_1} \right)^2 \right\rangle$$

$$= \left\langle \left(\frac{\partial u_1}{\partial x_1} \right)^2 \right\rangle. \tag{6.55}$$

Thus we obtain

$$\left\langle \left(\frac{\partial u_1}{\partial x_1} \right)^2 \right\rangle = \frac{2u'^2}{\lambda_f^2}. \tag{6.56}$$

The transverse Taylor microscale $\lambda_g(t)$, defined by

$$\lambda_g(t) = \left[-\tfrac{1}{2} g''(0, t) \right]^{-1/2}, \tag{6.57}$$

is, in isotropic turbulence, equal to $\lambda_f(t)/\sqrt{2}$ (see Exercise 6.6). It then follows from these two equations and the relation $\varepsilon = 15\nu \langle (\partial u_1/\partial x_1)^2 \rangle$ (Eq. (5.171)) that the dissipation is given by

$$\varepsilon = 15\nu u'^2 / \lambda_g^2. \tag{6.58}$$

In a classic paper marking the start of the study of isotropic turbulence, Taylor (1935a) defined λ_g and obtained the above equation for ε. He then stated that 'λ_g may roughly be regarded as a measure of the diameter of the smallest eddies which are responsible for the dissipation of energy.' This deduction from Eq. (6.58) is incorrect, because it incorrectly supposes that u' is the characteristic velocity of the dissipative eddies. Instead, the characteristic length and velocity scales of the smallest eddies are the Kolmogorov scales η and u_η.

To determine the relationship between the Taylor and Kolmogorov scales, we define $L \equiv k^{3/2}/\varepsilon$ to be the lengthscale characterizing the large eddies, and the turbulence Reynolds number to be

$$\mathrm{Re}_L \equiv \frac{k^{1/2}L}{v} = \frac{k^2}{\varepsilon v}. \tag{6.59}$$

Then the microscales are given by

$$\lambda_g/L = \sqrt{10}\,\mathrm{Re}_L^{-1/2}, \tag{6.60}$$

$$\eta/L = \mathrm{Re}_L^{-3/4}, \tag{6.61}$$

$$\lambda_g = \sqrt{10}\,\eta^{2/3}L^{1/3}. \tag{6.62}$$

Thus, at high Reynolds number, λ_g is intermediate in size between η and L.

The Taylor scale does not have a clear physical interpretation. It is, however, a well-defined quantity that is often used. In particular, the Taylor-scale Reynolds number

$$R_\lambda \equiv u'\lambda_g/v, \tag{6.63}$$

is traditionally used to characterize grid turbulence. Observe, from Eq. (6.60), that R_λ varies as the square-root of the integral-scale Reynolds number

$$R_\lambda = \left(\tfrac{20}{3}\mathrm{Re}_L\right)^{1/2}. \tag{6.64}$$

In addition, it may be observed that the ratio

$$\lambda_g/u' = (15v/\varepsilon)^{1/2} = \sqrt{15}\,\tau_\eta \tag{6.65}$$

correctly characterizes the timescale of the small eddies.

EXERCISE

6.6   Show from Eq. (6.46) that

$$g''(r,t) = 2f''(r,t) + \tfrac{1}{2}rf'''(r,t), \tag{6.66}$$

and hence that the transverse Taylor microscale

$$\lambda_g(t) \equiv \left[-\tfrac{1}{2}g''(0,t)\right]^{-1/2} \tag{6.67}$$

is related to the longitudinal scale $\lambda_f(t)$ by

$$\lambda_g(t) = \lambda_f(t)/\sqrt{2}. \tag{6.68}$$

Show that

$$\left\langle \left(\frac{\partial u_1}{\partial x_2}\right)^2\right\rangle = \frac{2u'^2}{\lambda_g^2}. \tag{6.69}$$

The Kármán–Howarth equation

von Kármán and Howarth (1938) obtained from the Navier–Stokes equation an evolution equation for $f(r,t)$. We outline here the principal steps, the result, and some implications: a detailed derivation can be found in the original work or in standard references (e.g., Hinze (1975), Monin and Yaglom (1975)).

The time derivative of $R_{ij}(r, x, t)$ can be expressed as

$$\frac{\partial}{\partial t} R_{ij}(r, t) = \frac{\partial}{\partial t}\langle u_i(x + r, t)u_j(x, t)\rangle$$

$$= \left\langle u_j(x, t)\frac{\partial}{\partial t}u_i(x + r, t)\right\rangle$$

$$+ \left\langle u_i(x + r, t)\frac{\partial}{\partial t}u_j(x, t)\right\rangle, \tag{6.70}$$

and then the Navier–Stokes equations, i.e.,

$$\frac{\partial u_j}{\partial t} = -\frac{\partial(u_i u_j)}{\partial x_i} - \frac{1}{\rho}\frac{\partial p}{\partial x_j} + v\frac{\partial^2 u_j}{\partial x_i \partial x_i}, \tag{6.71}$$

can be used to eliminate the time derivatives on the right-hand side of Eq. (6.70). Three types of terms arise, corresponding to the convection, pressure-gradient, and viscous terms in Eq. (6.71). For isotropic turbulence the pressure-gradient term in the equation for $R_{ij}(r, t)$ is zero.

The convective term involves two-point triple velocity correlations, such as

$$\bar{S}_{ijk}(r, t) \equiv \langle u_i(x, t)u_j(x, t)u_k(x + r, t)\rangle. \tag{6.72}$$

Just as R_{ij} is uniquely determined by f (Eq. (6.44)), in isotropic turbulence $\bar{S}_{ijk}$ is uniquely determined by the longitudinal correlation

$$\bar{k}(r, t) = \bar{S}_{111}(e_1 r, t)/u'^3$$

$$= \langle u_1(x, t)^2 u_1(x + e_1 r, t)\rangle/u'^3. \tag{6.73}$$

It can be shown that $\bar{k}(r, t)$ is an odd function of r, and that the continuity equation implies that $\bar{k}'(0, t) = 0$, so that its series expansion is

$$\bar{k}(r, t) = \bar{k}'''r^3/3! + \bar{k}^v r^5/5!\dots \tag{6.74}$$

($\bar{k}^v$ is the fifth derivative of $k(r, t)$ at $r = 0$).

By this procedure, an exact equation for $f(r,t)$ is obtained from the Navier–Stokes equations: it is the *Kármán–Howarth equation*

$$\frac{\partial}{\partial t}(u'^2 f) - \frac{u'^3}{r^4}\frac{\partial}{\partial r}(r^4 \bar{k}) = \frac{2\nu u'^2}{r^4}\frac{\partial}{\partial r}\left(r^4 \frac{\partial f}{\partial r}\right). \tag{6.75}$$

The principal observations to be made are the following.

(i) There is a closure problem. This single equation involves two unknown functions, $f(r,t)$ and $\bar{k}(r,t)$.

(ii) The terms in $\bar{k}$ and ν represent inertial and viscous processes, respectively.

(iii) At $r = 0$, the term in $\bar{k}$ vanishes (on account of Eq. (6.74)); while, from the fact that f is even in r, we obtain

$$\left[\frac{1}{r^4}\frac{\partial}{\partial r}\left(r^4 \frac{\partial f}{\partial r}\right)\right]_{r=0} = 5f''(0,t) = -\frac{5}{\lambda_g(t)^2}. \tag{6.76}$$

Hence, for $r = 0$, the Kármán–Howarth equation reduces to ($\frac{2}{3}$ times) the kinetic-energy equation:

$$\frac{\mathrm{d}}{\mathrm{d}t}u'(t)^2 = -10\nu\frac{u'^2}{\lambda_g^2} = -\tfrac{2}{3}\varepsilon. \tag{6.77}$$

(iv) In the Richardson–Kolmogorov view of the energy cascade at high Reynolds number, the transfer of energy from larger to smaller scales is an inertial process (at least for $r \gg \eta$). Consequently, this transfer of energy to smaller scales is accomplished by the term in $\bar{k}$ in the Kármán–Howarth equation.

(v) If $u(x,t)$ were a Gaussian field then $\bar{k}(r,t)$ – like all third moments – would be zero. Hence the energy cascade depends on non-Gaussian aspects of the velocity field.

EXERCISES

6.7   By following a procedure similar to Eq. (6.55), show that

$$u'^3\bar{k}'''(0,t) = \left\langle\left(\frac{\partial u_1}{\partial x_1}\right)^3\right\rangle$$

$$= S\left(\frac{\varepsilon}{15\nu}\right)^{3/2}, \tag{6.78}$$

where S is the skewness of the velocity derivative (Eq. (6.85)). (Hint: see also Eq. (3.148).)

6.8 Show that, in isotropic turbulence, the longitudinal structure function and autocorrelation function are related by

$$u'(t)^2 f(r,t) = u'(t)^2 - \tfrac{1}{2} D_{LL}(r,t). \tag{6.79}$$

With the third-order structure function being defined by

$$D_{LLL}(r,t) = \langle [u_1(\boldsymbol{x}+\boldsymbol{e}_1 r,t) - u_1(\boldsymbol{x},t)]^3 \rangle, \tag{6.80}$$

show that

$$u'(t)^3 \bar{k}(r,t) = \tfrac{1}{6} D_{LLL}(r,t). \tag{6.81}$$

6.9 Show that the Kármán–Howarth equation (Eq. (6.75)) re-expressed in terms of structure functions is

$$\frac{\partial}{\partial t} D_{LL} + \frac{1}{3r^4} \frac{\partial}{\partial r}(r^4 D_{LLL}) = \frac{2v}{r^4} \frac{\partial}{\partial r}\left(r^4 \frac{\partial D_{LL}}{\partial r}\right) - \tfrac{4}{3}\varepsilon. \tag{6.82}$$

Integrate this equation to obtain

$$\frac{3}{r^4} \int_0^r s^4 \frac{\partial}{\partial t} D_{LL}(s,t)\,\mathrm{d}s = 6v\,\frac{\partial D_{LL}}{\partial r} - D_{LLL} - \tfrac{4}{5}\varepsilon r. \tag{6.83}$$

Further observations

The Kármán–Howarth equation has been studied extensively, and many more results have been obtained than can be mentioned here. Some of the better known and most informative results are now given; more comprehensive accounts are provided by Batchelor (1953), Monin and Yaglom (1975), and Hinze (1975).

The skewness of the velocity derivative

The quantity $\bar{k}'''(0,t)$, which determines $\bar{k}(r,t)$ to leading order (Eq. (6.74)), can be re-expressed as

$$u'^3 \bar{k}'''(0,t) = \left\langle \left(\frac{\partial u_1}{\partial x_1}\right)^3 \right\rangle$$

$$= S\left(\frac{\varepsilon}{15v}\right)^{3/2}$$

$$= -\frac{2}{35}\left\langle \omega_i \omega_j \frac{\partial u_i}{\partial x_j} \right\rangle, \tag{6.84}$$

where

$$S \equiv \left\langle \left(\frac{\partial u_1}{\partial x_1}\right)^3 \right\rangle \Big/ \left\langle \left(\frac{\partial u_1}{\partial x_1}\right)^2 \right\rangle^{3/2}, \tag{6.85}$$

is the *velocity-derivative skewness* (which is found to be negative), see Exercise 6.7. So, there is a connection among this skewness, vortex stretching, and the transfer of energy between different scales.

The Kolmogorov $\frac{4}{5}$ law

The Kármán–Howarth equation can be re-expressed in terms of the structure functions $D_{LL}(r,t)$ and $D_{LLL}(r,t)$,

$$D_{LLL}(r,t) = \langle [u_1(x + e_1 r, t) - u_1(x,t)]^3 \rangle. \tag{6.86}$$

The result (see Exercise 6.9) is the *Kolmogorov equation* (Kolmogorov 1941a)

$$\frac{3}{r^4} \int_0^r s^4 \frac{\partial}{\partial t} D_{LL}(s,t)\, ds = 6v \frac{\partial D_{LL}}{\partial r} - D_{LLL} - \tfrac{4}{5}\varepsilon r, \tag{6.87}$$

from which several useful results can be deduced. Kolmogorov argued that in locally isotropic turbulence, the unsteady term on the left-hand side is zero, and that the viscous term is negligible in the inertial subrange. This leads to the Kolmogorov $\frac{4}{5}$ law:

$$D_{LLL}(r,t) = -\tfrac{4}{5}\varepsilon r. \tag{6.88}$$

Kolmogorov further argued that the structure-function skewness

$$S' \equiv D_{LLL}(r,t)/D_{LL}(r,t)^{3/2}, \tag{6.89}$$

is constant, leading to

$$D_{LL}(r,t) = \left(\frac{-4}{5S'} \right)^{2/3} (\varepsilon r)^{2/3}, \tag{6.90}$$

which is the same as the prediction from the Kolmogorov hypotheses (Eq. (6.30)). For $D_{LL}(r,t)$ in the inertial subrange, this development shows, therefore, the consistency between the Kolmogorov hypotheses and the Navier–Stokes equations. It also relates the constant C_2 to the skewness S'.

The Loitsyanskii integral

On multiplying the Kármán–Howarth equation (Eq. (6.75)) by r^4 and integrating between zero and R we obtain

$$\frac{d}{dt} \int_0^R u'^2 r^4 f(r,t)\, dr - u'^3 R^4 \bar{k}(R,t) = 2v u'^2 R^4 f'(R,t). \tag{6.91}$$

Loitsyanskii (1939) considered the limit $R \to \infty$, and assumed that f and $\bar{k}$ decrease sufficiently rapidly with r that the *Loitsyanskii integral*

$$B_2 \equiv \int_0^\infty u'^2 r^4 f(r,t)\,\mathrm{d}r \qquad (6.92)$$

converges, so that the terms in $\bar{k}(R,t)$ and $f'(R,t)$ vanish. With these assumptions, Eq. (6.91) indicates that B_2 does not change with time, and so B_2 became known as the *Loitsyanskii invariant*. However, the assumptions made are incorrect. Depending on how the isotropic turbulence is created, the Loitsyanskii integral can be finite or it can diverge (Saffman 1967). When it is finite, it is found that the term in $\bar{k}$ in Eq. (6.91) does not vanish as R tends to infinity, and in fact B_2 increases with time (see, e.g., Chasnov (1993)).

The final period of decay

As isotropic turbulence decays the Reynolds number decreases so that inertial effects diminish relative to viscous processes. Eventually, when the Reynolds number is sufficiently small, inertial effects become negligible.

For this *final period of decay*, Batchelor and Townsend (1948) showed that the Kármán–Howarth equation – with neglect of the inertial term – admits the self-similar solution

$$f(r,t) = \exp[-r^2/(8\nu t)], \qquad (6.93)$$

which is in excellent agreement with experimental data. It is emphasized that this solution applies to very low Reynolds number – much lower than is generally of interest.

EXERCISES

6.10 Verify that Eq. (6.93) satisfies the Kármán–Howarth equation for the final period in which the inertial term involving $\bar{k}$ is negligible. Show that the turbulent kinetic energy decays as $k \sim t^{-5/2}$ in the final period.

6.11 For homogeneous isotropic turbulence, consider the sixth-order tensor

$$\mathcal{H}_{ijkpqr} \equiv \left\langle \frac{\partial u_i}{\partial x_p} \frac{\partial u_j}{\partial x_q} \frac{\partial u_k}{\partial x_r} \right\rangle. \qquad (6.94)$$

This is an isotropic tensor and hence can be written as the sum of scalar coefficients multiplying Kronecker delta products, e.g., $a_1 \delta_{ip} \delta_{jq} \delta_{kr}$. There are 15 distinct Kronecker delta products (corresponding to different orderings of the suffixes). Argue, however, that in view of symmetries (e.g., $\mathcal{H}_{ijkpqr} = \mathcal{H}_{jikqpr}$), a general representation is

$$\mathcal{H}_{ijkpqr} = a_1 \delta_{ip} \delta_{jq} \delta_{kr} + a_2(\delta_{ip}\delta_{jk}\delta_{qr} + \delta_{jq}\delta_{ik}\delta_{pr} + \delta_{kr}\delta_{ij}\delta_{pq})$$
$$+ a_3(\delta_{ip}\delta_{jr}\delta_{qk} + \delta_{jq}\delta_{ir}\delta_{pk} + \delta_{kr}\delta_{iq}\delta_{pj}) + a_4(\delta_{iq}\delta_{pk}\delta_{jr} + \delta_{ir}\delta_{pj}\delta_{qk})$$
$$+ a_5(\delta_{ij}\delta_{pk}\delta_{qr} + \delta_{ij}\delta_{qk}\delta_{pr} + \delta_{ik}\delta_{pj}\delta_{qr}$$
$$+ \delta_{ik}\delta_{rj}\delta_{pq} + \delta_{jk}\delta_{qi}\delta_{pr} + \delta_{jk}\delta_{ri}\delta_{pq}). \tag{6.95}$$

Show that the continuity equation implies that $\mathcal{H}_{ijkiqr} = 0$, which leads to the relations

$$3a_1 + 2a_2 + 2a_3 = 0,$$
$$3a_2 + 4a_5 = 0, \tag{6.96}$$
$$3a_3 + 2a_4 + 2a_5 = 0.$$

By considering the quantity

$$\frac{\partial}{\partial x_i} \left\langle u_k \frac{\partial u_i}{\partial x_j} \frac{\partial u_j}{\partial x_k} \right\rangle$$

(which is zero in homogeneous turbulence) show that $\mathcal{H}_{ijkjki}$ is zero, and that this leads to the relation

$$a_1 + 3a_2 + 9a_3 + 10a_4 + 12a_5 = 0. \tag{6.97}$$

Show that the four relations Eqs. (6.96)–(6.97) determine all of the coefficients in terms of a_1 as

$$a_2 = -\tfrac{4}{3}a_1, \quad a_3 = -\tfrac{1}{6}a_1, \quad a_4 = -\tfrac{3}{4}a_1, \quad a_5 = a_1. \tag{6.98}$$

Hence show that

$$\left\langle \left(\frac{\partial u_1}{\partial x_1}\right)^3 \right\rangle = \mathcal{H}_{111111} = a_1 = S\left(\frac{\varepsilon}{15\nu}\right)^{3/2}, \tag{6.99}$$

where S is the velocity-derivative skewness (see Eqs. (6.84) and (6.85)). (Thus, in isotropic turbulence, the 729 components of $\mathcal{H}_{ijkpqr}$ are completely determined by the velocity-derivative skewness S, and the Kolmogorov timescale $\tau_\eta = (\nu/\varepsilon)^{1/2}$.)

Use Eqs. (6.95) and (6.99) to obtain the results

$$\left\langle \omega_i \omega_j \frac{\partial u_i}{\partial x_j} \right\rangle = -\tfrac{35}{2}a_1, \tag{6.100}$$

$$\left\langle \frac{\partial u_i}{\partial x_k} \frac{\partial u_i}{\partial x_q} \frac{\partial u_k}{\partial x_q} \right\rangle = \mathcal{H}_{iikkqq} = \tfrac{35}{2}a_1. \tag{6.101}$$

6.4 Fourier modes

For isotropic turbulence, the Kármán–Howarth equation (Eq. (6.75)), which stems from the Navier–Stokes equations, fully describes the dynamics of the two-point velocity correlation. It does not, however, provide a very clear picture of the processes involved in the energy cascade. Some further insights can be gained by examining the Navier–Stokes equations in wavenumber space. In this section we examine the behavior of discrete Fourier modes dictated by the Navier–Stokes equations for homogeneous turbulence in which the mean velocity is zero.

The first subsection provides the mathematical background for the representation of the velocity field as the three-dimensional Fourier series

$$u(x, t) = \sum_{\kappa} e^{i\kappa \cdot x} \hat{u}(\kappa, t). \tag{6.102}$$

(This extends the material of Appendix E to three-dimensional vector fields.) Then the equation for the evolution of the Fourier modes $\hat{u}(\kappa, t)$ is deduced from the Navier–Stokes equations. Finally, the balance equation for the kinetic energy at wavenumber κ,

$$\hat{E}(\kappa, t) = \langle \hat{u}^*(\kappa, t) \cdot \hat{u}(\kappa, t) \rangle, \tag{6.103}$$

is derived and discussed.

In addition to the insights that it provides on the energy cascade, there are other motivations for studying the Fourier representation of the Navier–Stokes equations. As is discussed in Chapter 9, direct numerical simulations (DNS) of homogeneous turbulence are usually performed in wavenumber space (i.e., by solving for $\hat{u}(\kappa, t)$); and *rapid-distortion theory* (RDT) – which in Chapter 11 is applied to homogeneous turbulence subjected to very large mean velocity gradients – is also set in wavenumber space.

6.4.1 Fourier-series representation

An implication of the Fourier series Eq. (6.102) is that the turbulent velocity field is periodic. Accordingly we consider the cube $0 \le x_i \le \mathcal{L}$ in physical space, where the length of the side $\mathcal{L}$ is large compared with the turbulent integral scale L_{11}. Then the velocity field is supposed to be periodic, i.e.,

$$u(x + N\mathcal{L}, t) = u(x, t), \tag{6.104}$$

for all integer vectors N. The effects of this artificially imposed periodicity vanish as $\mathcal{L}/L_{11}$ tends to infinity.

In the x_1 direction, Fourier modes are of the form

$$\cos(2\pi n_1 x_1 / \mathcal{L}) = \cos(\kappa_0 n_1 x_1), \tag{6.105}$$

and $\sin(\kappa_0 n_1 x_1)$, for integer n_1, where κ_0 is the lowest wavenumber:

$$\kappa_0 \equiv 2\pi / \mathcal{L}. \tag{6.106}$$

Or, in complex form, the Fourier modes are

$$e^{i\kappa_0 n_1 x_1} = \cos(\kappa_0 n_1 x_1) + i\sin(\kappa_0 n_1 x_1), \tag{6.107}$$

for positive and negative integers n_1. Similarly, in the other two directions the Fourier modes are $e^{i\kappa_0 n_2 x_2}$ and $e^{i\kappa_0 n_3 x_3}$; and the general three-dimensional mode is just the product of the one-dimensional modes. By defining the *wavenumber vector*

$$\boldsymbol{\kappa} = \kappa_0 \boldsymbol{n} = \kappa_0(\boldsymbol{e}_1 n_1 + \boldsymbol{e}_2 n_2 + \boldsymbol{e}_3 n_3), \tag{6.108}$$

we can write the general Fourier mode succinctly as

$$e^{i\boldsymbol{\kappa} \cdot \boldsymbol{x}} = e^{i\kappa_0 n_1 x_1} e^{i\kappa_0 n_2 x_2} e^{i\kappa_0 n_3 x_3}. \tag{6.109}$$

The Fourier mode given by Eq. (6.109) can be interpreted in terms of the magnitude $\kappa \equiv |\boldsymbol{\kappa}|$ and direction $\boldsymbol{e} \equiv \boldsymbol{\kappa}/\kappa$ of the wavenumber vector. Let s be a coordinate in physical space in the direction of $\boldsymbol{e}$, i.e., $s = \boldsymbol{e} \cdot \boldsymbol{x}$. Then we observe that

$$\boldsymbol{\kappa} \cdot \boldsymbol{x} = \kappa \boldsymbol{e} \cdot \boldsymbol{x} = \kappa s. \tag{6.110}$$

Thus $e^{i\boldsymbol{\kappa} \cdot \boldsymbol{x}}$ is constant in the plane normal to $\boldsymbol{\kappa}$ (constant s), while in the direction of $\boldsymbol{\kappa}$ it varies as a one-dimensional Fourier mode of wavenumber κ. As an illustration, Fig. 6.8 shows the Fourier mode with $(n_1, n_2, n_3) = (4, 2, 0)$.

The Fourier modes are orthonormal. To state this property simply in equations, we introduce two definitions. First, given two wavenumber vectors $\boldsymbol{\kappa}$ and $\boldsymbol{\kappa}'$, we define

$$\delta_{\boldsymbol{\kappa},\boldsymbol{\kappa}'} = \begin{cases} 1, & \text{if } \boldsymbol{\kappa} = \boldsymbol{\kappa}', \\ 0, & \text{if } \boldsymbol{\kappa} \neq \boldsymbol{\kappa}'. \end{cases} \tag{6.111}$$

Second, we denote by $\langle \ \rangle_{\mathcal{L}}$ the volume average over the cube $0 \leq x_i \leq \mathcal{L}$. The orthonormality property is then

$$\langle e^{i\boldsymbol{\kappa} \cdot \boldsymbol{x}} e^{-i\boldsymbol{\kappa}' \cdot \boldsymbol{x}} \rangle_{\mathcal{L}} = \delta_{\boldsymbol{\kappa},\boldsymbol{\kappa}'} \tag{6.112}$$

(see Exercise 6.12).

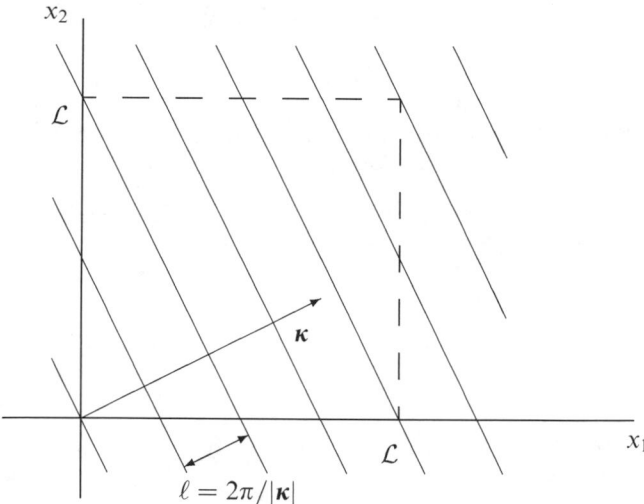

Fig. 6.8. A sketch of the Fourier mode corresponding to $\kappa = \kappa_0(4, 2, 0)$. The oblique lines show the crests, where $\Re(e^{i\kappa \cdot x}) = \cos \kappa \cdot x$ is unity.

For a periodic function $g(x)$ (e.g., a component of velocity at a given time), its Fourier series is

$$g(x) = \sum_{\kappa} e^{i\kappa \cdot x} \hat{g}(\kappa), \qquad (6.113)$$

where the sum is over the infinite number of discrete wavenumbers $\kappa = \kappa_0 n$, and $\hat{g}(\kappa)$ is the complex Fourier coefficient at wavenumber κ. Since $g(x)$ is real, $\hat{g}(\kappa)$ satisfies conjugate symmetry,

$$\hat{g}(\kappa) = \hat{g}^*(-\kappa), \qquad (6.114)$$

where an asterisk denotes the complex conjugate.

Given $g(x)$, the Fourier coefficients can be determined from the orthogonality condition (Eq. (6.111)):

$$\langle g(x) e^{-i\kappa' \cdot x} \rangle_{\mathcal{L}} = \left\langle \sum_{\kappa} \hat{g}(\kappa) e^{i\kappa \cdot x} e^{-i\kappa' \cdot x} \right\rangle_{\mathcal{L}}$$

$$= \sum_{\kappa} \hat{g}(\kappa) \delta_{\kappa,\kappa'} = \hat{g}(\kappa'). \qquad (6.115)$$

It is convenient to define the operator $\mathcal{F}_{\kappa}\{\ \}$ by

$$\mathcal{F}_{\kappa}\{g(x)\} = \langle g(x) e^{-i\kappa \cdot x} \rangle_{\mathcal{L}}$$

$$= \frac{1}{\mathcal{L}^3} \int_0^{\mathcal{L}} \int_0^{\mathcal{L}} \int_0^{\mathcal{L}} g(x) e^{-i\kappa \cdot x} \, dx_1 \, dx_2 \, dx_3, \qquad (6.116)$$

so that the previous equation can be written, simply, as

$$\mathcal{F}_\kappa\{g(x)\} = \hat{g}(\kappa). \tag{6.117}$$

Thus the operator $\mathcal{F}_\kappa\{\ \}$ determines the coefficient of the Fourier mode of wavenumber κ.

One of the principal reasons for invoking the Fourier representation is the form taken by derivatives. In Eq. (6.116), if $g(x)$ is replaced by $\partial g/\partial x_j$, we obtain

$$
\begin{aligned}
\mathcal{F}_\kappa\left\{\frac{\partial g(x)}{\partial x_j}\right\} &= \left\langle \frac{\partial g}{\partial x_j} e^{-i\kappa \cdot x} \right\rangle_\mathcal{L} \\
&= \left\langle -g(x)\frac{\partial}{\partial x_j} e^{-i\kappa \cdot x} \right\rangle_\mathcal{L} \\
&= \left\langle i\kappa_j g(x) e^{-i\kappa \cdot x} \right\rangle_\mathcal{L} \\
&= i\kappa_j \hat{g}(\kappa).
\end{aligned}
\tag{6.118}
$$

Differentiation with respect to x_j in physical space corresponds to multiplication by $i\kappa_j$ in wavenumber space.

The Fourier series of the turbulent velocity field is

$$u(x,t) = \sum_\kappa e^{i\kappa \cdot x} \hat{u}(\kappa, t), \tag{6.119}$$

where the Fourier coefficients of velocity are

$$\hat{u}_j(\kappa, t) = \mathcal{F}_\kappa\{u_j(x,t)\}. \tag{6.120}$$

The Fourier modes $e^{i\kappa \cdot x}$ are non-random and fixed in time. Hence the time-dependent, random nature of the turbulent velocity field $u(x,t)$ implies that the Fourier coefficients $\hat{u}(\kappa, t)$ are time-dependent and random. Since the mean $\langle u(x,t)\rangle$ is zero, it follows from Eq. (6.120) that the means $\langle \hat{u}(\kappa, t)\rangle$ are also zero. Note that, for each κ, $\hat{u}(\kappa, t)$ is a complex vector that satisfies conjugate symmetry, i.e.,

$$\hat{u}(\kappa, t) = \hat{u}^*(-\kappa, t). \tag{6.121}$$

EXERCISES

6.12   Show that, for integer n,

$$\int_0^\mathcal{L} e^{2\pi i n x/\mathcal{L}}\, \mathrm{d}x = \begin{cases} 0, & \text{for } n \neq 0, \\ \mathcal{L}, & \text{for } n = 0, \end{cases} \tag{6.122}$$

and hence establish the orthonormality property Eq. (6.112).

6.13   Show from Eq. (6.118) that

$$\mathcal{F}_{\kappa}\{\nabla^2 g(x)\} = -\kappa^2 \hat{g}(\kappa). \tag{6.123}$$

6.14   Given that the volume-average velocity $\langle u(x,t)\rangle_{\mathcal{L}}$ is zero, show that the coefficient of the zeroth Fourier mode is zero:

$$\hat{u}(0,t) = 0. \tag{6.124}$$

6.15   Show that the Fourier coefficient $\hat{\omega}(\kappa)$ of the vorticity $\omega = \nabla \times u$ is

$$\hat{\omega}(\kappa) = \mathcal{F}_{\kappa}\{\omega(x)\} = i\kappa \times \hat{u}(\kappa). \tag{6.125}$$

Show that $\kappa, \hat{u}(\kappa)$, and $\hat{\omega}(\kappa)$ are mutually orthogonal.

6.16   In general, an incompressible velocity field can be written as the sum of an irrotational component $\nabla\phi$ and a rotational component $\nabla \times B$, where $\phi(x)$ is the velocity potential, and $B(x)$ is the vector potential, which is divergence free ($\nabla \cdot B = 0$):

$$u = \nabla\phi + \nabla \times B. \tag{6.126}$$

For a periodic velocity field with $\langle u \rangle_{\mathcal{L}} = 0$, show that ϕ is uniform, and obtain a relationship between $\hat{B}(\kappa) \equiv \mathcal{F}_{\kappa}\{B(x)\}$ and $\hat{\omega}(\kappa) \equiv \mathcal{F}_{\kappa}\{\omega(x)\}$.

6.4.2 The evolution of Fourier modes

In wavenumber space, the divergence of velocity is

$$\mathcal{F}_{\kappa}\left\{\frac{\partial u_j}{\partial x_j}\right\} = i\kappa_j \hat{u}_j = i\kappa \cdot \hat{u}, \tag{6.127}$$

so that the continuity equation $\nabla \cdot u = 0$ indicates that $\hat{u}$ is normal to κ:

$$\kappa \cdot \hat{u} = 0. \tag{6.128}$$

There is reason to examine in more detail the orientation of a vector (such as $\hat{u}$) relative to the wavenumber κ. Any vector $\hat{G}$ can be decomposed into a component $\hat{G}^{\parallel}$ that is parallel to κ, and a component $\hat{G}^{\perp}$ that is normal to κ, i.e., $\hat{G} = \hat{G}^{\parallel} + \hat{G}^{\perp}$, see Fig. 6.9. With $e = \kappa/\kappa$ being the unit vector in the direction of κ, we have

$$\hat{G}^{\parallel} = e\,(e \cdot \hat{G}) = \kappa\,(\kappa \cdot \hat{G})/\kappa^2, \tag{6.129}$$

or

$$\hat{G}_j^{\parallel} = \frac{\kappa_j \kappa_k}{\kappa^2}\hat{G}_k. \tag{6.130}$$

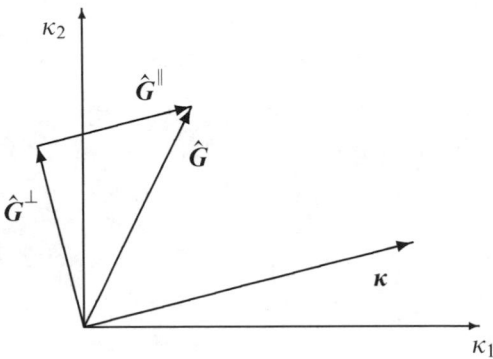

Fig. 6.9. A sketch (in two-dimensional wavenumber space) showing the decomposition of any vector $\hat{\boldsymbol{G}}$ into a component $\hat{\boldsymbol{G}}^{\|}$ parallel to $\boldsymbol{\kappa}$, and a component $\hat{\boldsymbol{G}}^{\perp}$ perpendicular to $\boldsymbol{\kappa}$.

Hence, from $\hat{\boldsymbol{G}}^{\perp} = \hat{\boldsymbol{G}} - \hat{\boldsymbol{G}}^{\|}$, we obtain

$$\hat{\boldsymbol{G}}^{\perp} = \hat{\boldsymbol{G}} - \boldsymbol{\kappa}(\boldsymbol{\kappa} \cdot \hat{\boldsymbol{G}})/\kappa^2, \tag{6.131}$$

or

$$\hat{G}_j^{\perp} = P_{jk}\hat{G}_k, \tag{6.132}$$

where the *projection tensor* $P_{jk}(\boldsymbol{\kappa})$ is

$$P_{jk} \equiv \delta_{jk} - \frac{\kappa_j \kappa_k}{\kappa^2}. \tag{6.133}$$

The principal observation is that the projection tensor $P_{jk}(\boldsymbol{\kappa})$ determines $\hat{G}_j^{\perp} = P_{jk}\hat{G}_k$ to be the projection of $\hat{\boldsymbol{G}}$ onto the plane normal to $\boldsymbol{\kappa}$. This projection tensor is used below in writing the Navier–Stokes equations in wavenumber space.

The equation for the evolution of the velocity vector in wavenumber space $\hat{\boldsymbol{u}}(\boldsymbol{\kappa}, t)$ is obtained by applying the operator $\mathcal{F}_{\boldsymbol{\kappa}}\{\ \}$ (Eq. (6.116)) term by term to the Navier–Stokes equations:

$$\frac{\partial u_j}{\partial t} + \frac{\partial(u_j u_k)}{\partial x_k} = \nu \frac{\partial^2 u_j}{\partial x_k \, \partial x_k} - \frac{1}{\rho}\frac{\partial p}{\partial x_j}. \tag{6.134}$$

The time derivative is simply

$$\mathcal{F}_{\boldsymbol{\kappa}}\left\{\frac{\partial u_j}{\partial t}\right\} = \frac{\mathrm{d}\hat{u}_j}{\mathrm{d}t}, \tag{6.135}$$

while, for the viscous term, application of Eq. (6.118) twice produces

$$\mathcal{F}_{\boldsymbol{\kappa}}\left\{\nu \frac{\partial^2 u_j}{\partial x_k \, \partial x_k}\right\} = -\nu \kappa^2 \hat{u}_j. \tag{6.136}$$

With $\hat{p}(\boldsymbol{\kappa}, t) \equiv \mathcal{F}_\kappa\{p(\boldsymbol{x}, t)/\rho\}$ being the Fourier coefficient of the dynamic pressure (p/ρ), the pressure-gradient term is

$$\mathcal{F}_\kappa\left\{-\frac{1}{\rho}\frac{\partial p}{\partial x_j}\right\} = -i\kappa_j \hat{p}. \tag{6.137}$$

For the moment, the nonlinear convection term is written

$$\mathcal{F}_\kappa\left\{\frac{\partial}{\partial x_k}(u_j u_k)\right\} = \hat{G}_j(\boldsymbol{\kappa}, t), \tag{6.138}$$

thus defining the Fourier coefficients $\hat{\boldsymbol{G}}$. On combining these results we obtain

$$\frac{d\hat{u}_j}{dt} + \nu\kappa^2 \hat{u}_j = -i\kappa_j \hat{p} - \hat{G}_j. \tag{6.139}$$

When Eq. (6.139) is multiplied by κ_j, the left-hand side vanishes (on account of the continuity equation $\kappa_j \hat{u}_j = 0$, Eq. (6.128)), leaving

$$\kappa^2 \hat{p} = i\kappa_j \hat{G}_j. \tag{6.140}$$

This is, in wavenumber space, the Poisson equation for pressure obtained from the Navier–Stokes equations, i.e.,

$$\mathcal{F}_\kappa\{-\nabla^2 p\} = \mathcal{F}_\kappa\left\{\frac{\partial}{\partial x_j}\left(\frac{\partial}{\partial x_k}(u_j u_k)\right)\right\}. \tag{6.141}$$

By solving Eq. (6.140) for $\hat{p}$, we obtain for the pressure term in Eq. (6.139)

$$-i\kappa_j \hat{p} = \frac{\kappa_j \kappa_k}{\kappa^2}\hat{G}_k = \hat{G}_j^{\parallel}. \tag{6.142}$$

That is, the pressure term exactly balances $-\hat{\boldsymbol{G}}^{\parallel}$, the component of $-\hat{\boldsymbol{G}}$ in the direction of $\boldsymbol{\kappa}$. What remains, then, on the right-hand side of Eq. (6.139) is $-\hat{\boldsymbol{G}}^{\perp}$, the component of $-\hat{\boldsymbol{G}}$ perpendicular to $\boldsymbol{\kappa}$:

$$\frac{d\hat{u}_j}{dt} + \nu\kappa^2 \hat{u}_j = -\left(\delta_{jk} - \frac{\kappa_j \kappa_k}{\kappa^2}\right)\hat{G}_k$$
$$= -P_{jk}\hat{G}_k = -\hat{G}_j^{\perp}. \tag{6.143}$$

The viscous term in this equation has a simple effect. Consider, for example, the final period of decay of isotropic turbulence in which the Reynolds number is so low that convection is negligible relative to the effects of viscosity. Then, from a specified initial condition $\hat{\boldsymbol{u}}(\boldsymbol{\kappa}, 0)$, Eq. (6.143) with neglect of the term $\hat{\boldsymbol{G}}^{\perp}$ has the solution

$$\hat{\boldsymbol{u}}(\boldsymbol{\kappa}, t) = \hat{\boldsymbol{u}}(\boldsymbol{\kappa}, 0)e^{-\nu\kappa^2 t}. \tag{6.144}$$

Thus, in the final period of decay, each Fourier coefficient evolves independently of all other modes, decaying exponentially with time at the rate $\nu\kappa^2$. High-wavenumber modes decay more rapidly than do low-wavenumber modes.

Expressed in terms of $\hat{u}(\kappa)$ (the dependence on t being implicit), the nonlinear convective term is

$$\hat{G}_j(\kappa,t) \equiv \mathcal{F}_\kappa\left\{\frac{\partial}{\partial x_k}(u_j u_k)\right\} = i\kappa_k \mathcal{F}_\kappa\{u_j u_k\}$$

$$= i\kappa_k \mathcal{F}_\kappa\left\{\left(\sum_{\kappa'} \hat{u}_j(\kappa')e^{i\kappa'\cdot x}\right)\left(\sum_{\kappa''} \hat{u}_k(\kappa'')e^{i\kappa''\cdot x}\right)\right\}$$

$$= i\kappa_k \sum_{\kappa'}\sum_{\kappa''} \hat{u}_j(\kappa')\hat{u}_k(\kappa'')\langle e^{i(\kappa'+\kappa'')\cdot x}e^{-i\kappa\cdot x}\rangle_L$$

$$= i\kappa_k \sum_{\kappa'}\sum_{\kappa''} \hat{u}_j(\kappa')\hat{u}_k(\kappa'')\delta_{\kappa,\kappa'+\kappa''}$$

$$= i\kappa_k \sum_{\kappa'} \hat{u}_j(\kappa')\hat{u}_k(\kappa-\kappa'). \tag{6.145}$$

(The six steps in this development invoke Eqs. (6.138), (6.118), (6.119), (6.116), (6.112) and (6.111), respectively.) On substituting this result into Eq. (6.143) we obtain, in its final form, the evolution equation for $\hat{u}(\kappa,t)$:

$$\left(\frac{d}{dt} + \nu\kappa^2\right)\hat{u}_j(\kappa,t) = -i\kappa_\ell P_{jk}(\kappa)\sum_{\kappa'} \hat{u}_k(\kappa',t)\hat{u}_\ell(\kappa-\kappa',t). \tag{6.146}$$

The left-hand side involves $\hat{u}$ only at κ. In contrast, the right-hand side involves $\hat{u}$ at κ' and κ'', such that $\kappa' + \kappa'' = \kappa$; and, in fact, the contributions from $\kappa' = \kappa$ and $\kappa'' = \kappa$ are zero. Thus in wavenumber space the convection term is nonlinear and non-local, involving the interaction of *wavenumber triads*, κ, κ', and κ'', such that $\kappa' + \kappa'' = \kappa$.

EXERCISE

6.17   Let κ^a, κ^b, and κ^c be three wavenumber vectors such that

$$\kappa^a + \kappa^b + \kappa^c = 0, \tag{6.147}$$

and define $a(t) = \hat{u}(\kappa^a, t)$, $b(t) = \hat{u}(\kappa^b, t)$, and $c(t) = \hat{u}(\kappa^c, t)$. Consider a periodic velocity field, which at time $t = 0$ has non-zero Fourier coefficient only at the six wavenumbers $\pm\kappa_a, \pm\kappa_b$, and $\pm\kappa_c$. Consider the initial evolution of the velocity field governed by the Euler equations. Use Eq. (6.146) with $\nu = 0$ to show that, at $t = 0$;

(a)

$$\frac{\mathrm{d}a_j}{\mathrm{d}t} = -i\kappa_\ell^{\mathrm{a}} P_{jk}(\boldsymbol{\kappa}^{\mathrm{a}})(b_k^* c_\ell^* + c_k^* b_\ell^*), \qquad (6.148)$$

(b)

$$\frac{\mathrm{d}}{\mathrm{d}t}\left(\tfrac{1}{2}\boldsymbol{a}\cdot\boldsymbol{a}^*\right) = -\Im\{\boldsymbol{a}\cdot\boldsymbol{b}\ \ \boldsymbol{\kappa}^{\mathrm{a}}\cdot\boldsymbol{c} + \boldsymbol{a}\cdot\boldsymbol{c}\ \ \boldsymbol{\kappa}^{\mathrm{a}}\cdot\boldsymbol{b}\}, \qquad (6.149)$$

(c)

$$\frac{\mathrm{d}}{\mathrm{d}t}(\boldsymbol{a}\cdot\boldsymbol{a}^* + \boldsymbol{b}\cdot\boldsymbol{b}^* + \boldsymbol{c}\cdot\boldsymbol{c}^*) = 0, \qquad (6.150)$$

(d) there are 24 modes with non-zero rates of change. Sketch their locations in wavenumber space.

6.4.3 *The kinetic energy of Fourier modes*

For the periodic case being considered, Eq. (6.146) is the Navier–Stokes equations in wavenumber space. It is a deterministic set of ordinary differential equations for the Fourier coefficients $\hat{u}(\boldsymbol{\kappa}, t)$. In order to describe the turbulence statistically, we now consider various means.

Since (by assumption) the mean velocity $\langle U(x,t)\rangle$ is zero everywhere, the Fourier coefficients of $\langle U(x,t)\rangle$ – i.e., $\langle \hat{u}(\boldsymbol{\kappa},t)\rangle$ – are also zero. The next simplest statistic to consider is the covariance of two Fourier coefficients, i.e.,

$$\langle \hat{u}_i(\boldsymbol{\kappa}',t)\hat{u}_j(\boldsymbol{\kappa},t)\rangle. \qquad (6.151)$$

It is shown in Exercise 6.18 that these coefficients are uncorrelated, unless the wavenumbers sum to zero, i.e., $\boldsymbol{\kappa}' + \boldsymbol{\kappa} = 0$, or equivalently, $\boldsymbol{\kappa}' = -\boldsymbol{\kappa}$. Thus, all the covariance information is contained in

$$\hat{R}_{ij}(\boldsymbol{\kappa},t) \equiv \langle \hat{u}_i^*(\boldsymbol{\kappa},t)\hat{u}_j(\boldsymbol{\kappa},t)\rangle$$
$$= \langle \hat{u}_i(-\boldsymbol{\kappa},t)\hat{u}_j(\boldsymbol{\kappa},t)\rangle. \qquad (6.152)$$

It is readily shown that $\hat{R}_{ij}(\boldsymbol{\kappa},t)$ are the Fourier coefficients of the two-point velocity correlation

$$\hat{R}_{ij}(\boldsymbol{\kappa},t) = \mathcal{F}_{\boldsymbol{\kappa}}\{R_{ij}(x,t)\} \qquad (6.153)$$

(see Exercise 6.18) so that $R_{ij}(x,t)$ has the Fourier series

$$R_{ij}(r,t) \equiv \langle u_i(x,t)u_j(x+r,t)\rangle$$
$$= \sum_{\boldsymbol{\kappa}} \hat{R}_{ij}(\boldsymbol{\kappa},t)e^{i\boldsymbol{\kappa}\cdot r}, \qquad (6.154)$$

(see Exercise 6.19).

The *velocity-spectrum tensor* is defined by

$$\Phi_{ij}(\bar{\kappa}, t) \equiv \sum_{\kappa} \delta(\bar{\kappa} - \kappa) \hat{R}_{ij}(\kappa, t), \tag{6.155}$$

where $\bar{\kappa}$ is a continuous wavenumber variable. Evidently the spectrum tensor is the Fourier transform of the two-point correlation, since Eqs. (6.154) and (6.155) yield

$$R_{ij}(r, t) = \iiint\limits_{-\infty}^{\infty} \Phi_{ij}(\bar{\kappa}, t) e^{i\bar{\kappa}\cdot r} \, d\bar{\kappa}. \tag{6.156}$$

Setting $r = 0$ in Eqs. (6.154) and (6.156), we then obtain

$$R_{ij}(0, t) = \langle u_i u_j \rangle = \sum_{\kappa} \hat{R}_{ij}(\kappa, t) = \iiint\limits_{-\infty}^{\infty} \Phi_{ij}(\bar{\kappa}, t) \, d\bar{\kappa}. \tag{6.157}$$

Thus $\hat{R}_{ij}(\kappa, t)$ is identified as the contribution to the Reynolds stress from the Fourier mode with wavenumber κ; while

$$\iiint\limits_{\mathcal{K}} \Phi_{ij}(\bar{\kappa}, t) \, d\bar{\kappa}$$

is the contribution to $\langle u_i u_j \rangle$ from modes in a specified region $\mathcal{K}$ of wavenumber space. Other properties of $\hat{R}_{ij}$ and Φ_{ij} are given in Exercise 6.20.

Of particular interest is the kinetic energy of the Fourier mode, defined as

$$\hat{E}(\kappa, t) = \tfrac{1}{2} \langle \hat{u}_i^*(\kappa, t) \hat{u}_i(\kappa, t) \rangle = \tfrac{1}{2} \hat{R}_{ii}(\kappa, t). \tag{6.158}$$

The turbulent kinetic energy is

$$k(t) = \tfrac{1}{2} \langle u_i u_i \rangle = \sum_{\kappa} \tfrac{1}{2} \hat{R}_{ii}(\kappa, t) = \sum_{\kappa} \hat{E}(\kappa, t) = \iiint\limits_{-\infty}^{\infty} \tfrac{1}{2} \Phi_{ii}(\bar{\kappa}, t) \, d\bar{\kappa}, \tag{6.159}$$

which identifies $\hat{E}(\kappa, t)$ as the contribution to k from wavenumber κ.

The dissipation rate $\varepsilon(t)$ is also related to $\hat{E}(\kappa, t)$, by

$$\varepsilon(t) = -\nu \langle u_j \nabla^2 u_j \rangle = -\nu \lim_{r \to 0} \frac{\partial^2}{\partial r_k \, \partial r_k} R_{jj}(r, t)$$

$$= -\nu \lim_{r \to 0} \sum_{\kappa} e^{i\kappa \cdot r} (-\kappa_k \kappa_k) \hat{R}_{jj}(\kappa, t)$$

$$= \sum_{\kappa} 2\nu \kappa^2 \hat{E}(\kappa, t) = \iiint\limits_{-\infty}^{\infty} 2\nu \bar{\kappa}^2 \tfrac{1}{2} \Phi_{ii}(\bar{\kappa}, t) \, d\bar{\kappa}, \tag{6.160}$$

(see also Exercise 6.23). Thus $\hat{E}(\kappa, t)$ and $2\nu\kappa^2 \hat{E}(\kappa, t)$ are the contributions to the kinetic energy and dissipation rate, respectively, from the Fourier mode $\boldsymbol{\kappa}$.

The evolution equation for $\hat{E}(\boldsymbol{\kappa}, t)$ can be deduced from that for $\hat{u}(\boldsymbol{\kappa}, t)$ (Eq. (6.146)):

$$\frac{\mathrm{d}}{\mathrm{d}t}\hat{E}(\boldsymbol{\kappa}, t) = \widehat{T}(\boldsymbol{\kappa}, t) - 2\nu\kappa^2 \hat{E}(\boldsymbol{\kappa}, t), \qquad (6.161)$$

where

$$\widehat{T}(\boldsymbol{\kappa}) = \kappa_\ell P_{jk}(\boldsymbol{\kappa}) \Re \left\{ i \sum_{\boldsymbol{\kappa}'} \langle \hat{u}_j(\boldsymbol{\kappa}) \hat{u}_k^*(\boldsymbol{\kappa}') \hat{u}_\ell^*(\boldsymbol{\kappa} - \boldsymbol{\kappa}') \rangle \right\}, \qquad (6.162)$$

and $\Re\{\ \}$ denotes the real part. When it is summed over all $\boldsymbol{\kappa}$, the left-hand side of Eq. (6.161) is $\mathrm{d}k/\mathrm{d}t$, while the last term on the right-hand side sums to $-\varepsilon$. For isotropic turbulence $\mathrm{d}k/\mathrm{d}t$ equals $-\varepsilon$, and so (as can be confirmed directly) the sum of $\widehat{T}$ is zero:

$$\sum_{\boldsymbol{\kappa}} \widehat{T}(\boldsymbol{\kappa}, t) = 0. \qquad (6.163)$$

Thus the term $\widehat{T}(\boldsymbol{\kappa}, t)$ represents a *transfer* of energy between modes.

There is a direct correspondence between Eq. (6.161) for $\hat{E}(\boldsymbol{\kappa}, t)$ and the Kármán–Howarth equation for $f(r, t)$, Eq. (6.75). They contain essentially the same information, but expressed differently. An advantage of the formulation in terms of Fourier modes is that it provides a clear quantification of the energy at different scales of motion, and that an explicit expression for the energy-transfer rate $\widehat{T}(\boldsymbol{\kappa}, t)$, which plays a central role in the energy cascade, is obtained. Indeed, using direct numerical simulations of isotropic turbulence, it is possible to measure $\widehat{T}(\boldsymbol{\kappa}, t)$ and related quantities (see e.g., Domaradzki (1992)).

EXERCISES _____

6.18   Show that the covariance of two Fourier coefficients of velocity can be expressed as

$$\langle \hat{u}_i(\boldsymbol{\kappa}', t) \hat{u}_j(\boldsymbol{\kappa}, t) \rangle = \langle \mathcal{F}_{\boldsymbol{\kappa}'}\{u_i(\boldsymbol{x}', t)\} \mathcal{F}_{\boldsymbol{\kappa}}\{u_j(\boldsymbol{x}, t)\} \rangle$$
$$= \langle \langle u_i(\boldsymbol{x}', t) e^{-i\boldsymbol{\kappa}' \cdot \boldsymbol{x}'} \rangle_\mathcal{L} \langle u_j(\boldsymbol{x}, t) e^{-i\boldsymbol{\kappa} \cdot \boldsymbol{x}} \rangle_\mathcal{L} \rangle$$
$$= \frac{1}{\mathcal{L}^6} \int_0^\mathcal{L} \cdots \int_0^\mathcal{L} \langle u_i(\boldsymbol{x}', t) u_j(\boldsymbol{x}, t) \rangle e^{-i(\boldsymbol{\kappa}' \cdot \boldsymbol{x}' + \boldsymbol{\kappa} \cdot \boldsymbol{x})} \, \mathrm{d}\boldsymbol{x} \, \mathrm{d}\boldsymbol{x}'.$$

$$(6.164)$$

With the substitution $\boldsymbol{x} = \boldsymbol{x}' + \boldsymbol{r}$, and from the fact that in homogeneous turbulence the two-point correlation $R_{ij}(\boldsymbol{r}, t)$ is independent of position, show that the last result can be re-expressed as

$$\langle \hat{u}_i(\boldsymbol{\kappa}', t)\hat{u}_j(\boldsymbol{\kappa}, t)\rangle = \langle R_{ij}(\boldsymbol{r}, t)e^{-i\boldsymbol{\kappa}\cdot\boldsymbol{r}}\rangle_\mathcal{L}\langle e^{-i\boldsymbol{x}'(\boldsymbol{\kappa}'+\boldsymbol{\kappa})}\rangle_\mathcal{L}$$

$$= \mathcal{F}_\kappa\{R_{ij}(\boldsymbol{r}, t)\}\delta_{\boldsymbol{\kappa}, -\boldsymbol{\kappa}'}. \tag{6.165}$$

(Hint: see Eq. (E.22).) Hence, by setting $\boldsymbol{\kappa}' = -\boldsymbol{\kappa}$, verify Eq. (6.153).

6.19   Through the substitutions

$$u_i(\boldsymbol{x}) = \sum_{\boldsymbol{\kappa}'} e^{i\boldsymbol{\kappa}'\cdot\boldsymbol{x}}\,\hat{u}_i(\boldsymbol{\kappa}') = \sum_{\boldsymbol{\kappa}'} e^{-i\boldsymbol{\kappa}'\cdot\boldsymbol{x}}\,\hat{u}_i^*(\boldsymbol{\kappa}'), \tag{6.166}$$

$$u_j(\boldsymbol{x}+\boldsymbol{r}) = \sum_{\boldsymbol{\kappa}} e^{i\boldsymbol{\kappa}\cdot(\boldsymbol{x}+\boldsymbol{r})}\hat{u}_j(\boldsymbol{\kappa}), \tag{6.167}$$

show that, in homogeneous turbulence,

$$R_{ij}(\boldsymbol{r}) = \langle R_{ij}(\boldsymbol{r})\rangle_\mathcal{L} = \sum_{\boldsymbol{\kappa}} e^{i\boldsymbol{\kappa}\cdot\boldsymbol{r}}\langle \hat{u}_i^*(\boldsymbol{\kappa})\hat{u}_j(\boldsymbol{\kappa})\rangle, \tag{6.168}$$

hence establishing Eq. (6.154).

6.20   From the definition of $\hat{R}_{ij}(\boldsymbol{\kappa})$ (Eq. (6.152)) show that

$$\hat{R}_{ij}(\boldsymbol{\kappa}) \geq 0, \quad \text{for } i = j, \tag{6.169}$$

$$\hat{R}_{ii}(\boldsymbol{\kappa}) \geq 0. \tag{6.170}$$

From conjugate symmetry show that

$$\hat{R}_{ij}(\boldsymbol{\kappa}) = \hat{R}_{ji}(-\boldsymbol{\kappa}) = \hat{R}_{ji}^*(\boldsymbol{\kappa}). \tag{6.171}$$

From the incompressibility condition $\boldsymbol{\kappa}\cdot\hat{\boldsymbol{u}}(\boldsymbol{\kappa}) = 0$ show that

$$\kappa_i\hat{R}_{ij}(\boldsymbol{\kappa}) = \kappa_j\hat{R}_{ij}(\boldsymbol{\kappa}) = 0. \tag{6.172}$$

Note that all of these properties also apply to the velocity-spectrum tensor $\Phi_{ij}(\boldsymbol{\kappa})$.

6.21   Let $\boldsymbol{Y}$ be any constant vector, and define $\hat{g}(\boldsymbol{\kappa}) = \boldsymbol{Y}\cdot\hat{\boldsymbol{u}}(\boldsymbol{\kappa})$. Obtain the result

$$Y_iY_j\hat{R}_{ij}(\boldsymbol{\kappa}) = \langle \hat{g}^*(\boldsymbol{\kappa})\hat{g}(\boldsymbol{\kappa})\rangle \geq 0, \tag{6.173}$$

to show that both $\hat{R}_{ij}(\boldsymbol{\kappa})$ and $\Phi_{ij}(\bar{\boldsymbol{\kappa}})$ are positive semi-definite, i.e.,

$$Y_iY_j\hat{R}_{ij}(\boldsymbol{\kappa}) \geq 0, \qquad Y_iY_j\Phi_{ij}(\bar{\boldsymbol{\kappa}}) \geq 0, \tag{6.174}$$

for all $\boldsymbol{Y}$. (This is a stronger result than Eq. (6.169).)

6.22 Show that $\hat{E}(\boldsymbol{\kappa})$ (Eq. (6.158)) is real, non-negative, with

$$\hat{E}(\boldsymbol{\kappa}) = \hat{E}(-\boldsymbol{\kappa}). \tag{6.175}$$

6.23 Starting from the spectral representation for $\boldsymbol{u}(\boldsymbol{x})$ (Eq. (6.119)), show that the spectral representation of $\partial u_i / \partial x_k$ is

$$\frac{\partial u_i}{\partial x_k} = \sum_{\boldsymbol{\kappa}} i \kappa_k \hat{u}_i(\boldsymbol{\kappa}) e^{i\boldsymbol{\kappa}\cdot\boldsymbol{x}}. \tag{6.176}$$

Hence show the relations

$$\left\langle \frac{\partial u_i}{\partial x_k} \frac{\partial u_j}{\partial x_\ell} \right\rangle = \sum_{\boldsymbol{\kappa}} \kappa_k \kappa_\ell \hat{R}_{ij}(\boldsymbol{\kappa})$$

$$= \int\!\!\!\int\!\!\!\int_{-\infty}^{\infty} \bar{\kappa}_k \bar{\kappa}_\ell \Phi_{ij}(\bar{\boldsymbol{\kappa}})\, \mathrm{d}\bar{\boldsymbol{\kappa}}, \tag{6.177}$$

$$\varepsilon = \sum_{\boldsymbol{\kappa}} 2\nu\kappa^2 \hat{E}(\boldsymbol{\kappa})$$

$$= \int\!\!\!\int\!\!\!\int_{-\infty}^{\infty} 2\nu\bar{\kappa}^2 \tfrac{1}{2}\Phi_{ii}(\bar{\boldsymbol{\kappa}})\, \mathrm{d}\bar{\boldsymbol{\kappa}}. \tag{6.178}$$

6.5 Velocity spectra

In the previous section, the velocity-spectrum tensor $\Phi_{ij}(\boldsymbol{\kappa}, t)$ is defined (for homogeneous turbulence) as the Fourier transform of the two-point velocity correlation $R_{ij}(\boldsymbol{r})$. (We now use $\boldsymbol{\kappa}$ for the continuous wavenumber variable, in place of $\bar{\boldsymbol{\kappa}}$ used above.) In Section 6.5.1 the properties of $\Phi_{ij}(\boldsymbol{\kappa}, t)$ are reviewed and related quantities are introduced; primarily, the energy-spectrum function $E(\kappa, t)$ and the one-dimensional spectra $E_{ij}(\kappa_1, t)$. The Kolmogorov hypotheses have implications for the forms of these spectra at high wavenumber (i.e., in the universal equilibrium range). These implications are presented in Section 6.5.2, and experimentally measured spectra are presented as further tests of the hypotheses. Section 6.6 describes the energy cascade in wavenumber space in terms of the energy-spectrum function $E(\kappa, t)$.

6.5.1 Definitions and properties

The velocity-spectrum tensor

In homogeneous turbulence, the two-point velocity correlation and the velocity-spectrum tensor form a Fourier-transform pair:

$$\Phi_{ij}(\boldsymbol{\kappa}) = \frac{1}{(2\pi)^3} \iiint\limits_{-\infty}^{\infty} R_{ij}(\boldsymbol{r}) e^{-i\boldsymbol{\kappa}\cdot\boldsymbol{r}} \, d\boldsymbol{r}, \tag{6.179}$$

$$R_{ij}(\boldsymbol{r}) = \iiint\limits_{-\infty}^{\infty} \Phi_{ij}(\boldsymbol{\kappa}) e^{i\boldsymbol{\kappa}\cdot\boldsymbol{r}} \, d\boldsymbol{\kappa}. \tag{6.180}$$

Here $\boldsymbol{\kappa} = \{\kappa_1, \kappa_2, \kappa_3\}$ is the (continuous) wavenumber vector; and, to abbreviate the notation, the dependences of R_{ij} and Φ_{ij} on time are not shown explicitly. The velocity-spectrum tensor $\Phi_{ij}(\boldsymbol{\kappa})$ is a complex quantity that has the properties

$$\Phi_{ij}(\boldsymbol{\kappa}) = \Phi_{ji}^*(\boldsymbol{\kappa}) = \Phi_{ji}(-\boldsymbol{\kappa}), \tag{6.181}$$

$$\kappa_i \Phi_{ij}(\boldsymbol{\kappa}) = \kappa_j \Phi_{ij}(\boldsymbol{\kappa}) = 0. \tag{6.182}$$

Equation (6.181) stems from the symmetry properties of $R_{ij}(\boldsymbol{r})$ and from the fact that $R_{ij}(\boldsymbol{r})$ is real; while Eq. (6.182) is a result of incompressibility (see Exercise 6.20). In addition $\Phi_{ij}(\boldsymbol{\kappa})$ is positive semi-definite, i.e.,

$$\Phi_{ij}(\boldsymbol{\kappa}) Y_i Y_j \geq 0, \tag{6.183}$$

for all vectors $\boldsymbol{Y}$ (see Exercise 6.21). It then follows that the diagonal components of $\Phi_{ij}(\boldsymbol{\kappa})$ (i.e., $i = j$) are real and non-negative, and therefore so also is the trace:

$$\Phi_{ii}(\boldsymbol{\kappa}) = \Phi_{ii}^*(\boldsymbol{\kappa}) \geq 0. \tag{6.184}$$

The velocity-spectrum tensor $\Phi_{ij}(\boldsymbol{\kappa})$ is a useful quantity to consider because (as shown in Section 6.4.3) it represents the *Reynolds-stress density* in wavenumber space: that is, $\Phi_{ij}(\boldsymbol{\kappa})$ is the contribution (per unit volume in wavenumber space) from the Fourier mode $e^{i\boldsymbol{\kappa}\cdot\boldsymbol{x}}$ to the Reynolds stress $\langle u_i u_j \rangle$. In particular, on setting $\boldsymbol{r} = 0$ in Eq. (6.180) we obtain

$$R_{ij}(0) = \langle u_i u_j \rangle = \iiint\limits_{-\infty}^{\infty} \Phi_{ij}(\boldsymbol{\kappa}) \, d\boldsymbol{\kappa}. \tag{6.185}$$

(Note that Φ_{ij} has dimensions of (velocity)^2/(wavenumber)^3, or equivalently (velocity)$^2 \times$ (length)^3.)

The information contained in $\Phi_{ij}(\kappa)$ can be considered in three parts. First, the subscripts (i and j) give the directions of the velocity in physical space. So, for example, $\Phi_{22}(\kappa)$ pertains entirely to the field $u_2(x)$. Second, the wavenumber direction $\kappa/|\kappa|$ gives the direction in physical space of the Fourier mode. Third, the wavenumber's magnitude determines the lengthscale of the mode, i.e., $\ell = 2\pi/|\kappa|$ (see Fig. 6.8).

Velocity-derivative information is also contained in $\Phi_{ij}(\kappa)$, in particular,

$$\left\langle \frac{\partial u_i}{\partial x_k} \frac{\partial u_j}{\partial x_\ell} \right\rangle = \int\!\!\!\int\!\!\!\int_{-\infty}^{\infty} \kappa_k \kappa_\ell \Phi_{ij}(\kappa)\,\mathrm{d}\kappa, \tag{6.186}$$

so that the dissipation rate is

$$\varepsilon = \int\!\!\!\int\!\!\!\int_{-\infty}^{\infty} 2\nu\kappa^2 \tfrac{1}{2}\Phi_{ii}(\kappa)\,\mathrm{d}\kappa, \tag{6.187}$$

(see Exercise 6.23). The relationship between $\Phi_{ij}(\kappa)$ and the integral length-scales is given below (Eqs. (6.210) and (6.213)).

The energy-spectrum function

Being a second-order tensor function of a vector, $\Phi_{ij}(\kappa)$ contains a great deal of information. A simpler though less complete description is provided by the energy-spectrum function $E(\kappa)$, which is a scalar function of a scalar.

The energy-spectrum function is obtained from $\Phi_{ij}(\kappa)$ by removing all directional information. The information about the direction of the velocities is removed by considering (half) the trace, i.e., $\tfrac{1}{2}\Phi_{ii}(\kappa)$. The information about the direction of the Fourier modes is removed by integrating over all wavenumbers κ of magnitude $|\kappa| = \kappa$. To express this mathematically, we denote by $S(\kappa)$ the sphere in wavenumber space, centered at the origin, with radius κ; and integration over the surface of this sphere is denoted by $\oint(\ )\,\mathrm{d}S(\kappa)$. Thus the energy spectrum function is defined as

$$E(\kappa) = \oint \tfrac{1}{2}\Phi_{ii}(\kappa)\,\mathrm{d}S(\kappa). \tag{6.188}$$

Alternatively, on account of the sifting property of the Dirac delta function (see Eq. (C.11)), an equivalent expression is

$$E(\kappa) = \int\!\!\!\int\!\!\!\int_{-\infty}^{\infty} \tfrac{1}{2}\Phi_{ii}(\kappa)\delta(|\kappa| - \kappa)\,\mathrm{d}\kappa, \tag{6.189}$$

where κ is here an independent variable (i.e., independent of κ).

The properties of $E(\kappa)$ follow straightforwardly from those of $\Phi_{ij}(\boldsymbol{\kappa})$: $E(\kappa)$ is real, non-negative, and, for negative κ, it is undefined according to Eq. (6.188), or zero according to Eq. (6.189). Integration of $E(\kappa)$ over all κ is the same as integration of $\frac{1}{2}\Phi_{ii}(\boldsymbol{\kappa})$ over all $\boldsymbol{\kappa}$. Thus, from Eqs. (6.185) and (6.187), we obtain for the turbulent kinetic energy

$$k = \int_0^\infty E(\kappa)\,d\kappa, \tag{6.190}$$

and for the dissipation

$$\varepsilon = \int_0^\infty 2\nu\kappa^2 E(\kappa)\,d\kappa. \tag{6.191}$$

Evidently, $E(\kappa)\,d\kappa$ is the contribution to k from all wavenumbers $\boldsymbol{\kappa}$ in the infinitesimal shell $\kappa \le |\boldsymbol{\kappa}| < \kappa + d\kappa$ in wavenumber space.

In general, $\Phi_{ij}(\boldsymbol{\kappa})$ contains much more information than does $E(\kappa)$; but, in isotropic turbulence, $\Phi_{ij}(\boldsymbol{\kappa})$ is completely determined by $E(\kappa)$. If the turbulence is isotropic, the directional information in $\Phi_{ij}(\boldsymbol{\kappa})$ can depend only on $\boldsymbol{\kappa}$, and, to within scalar multiples, the only second-order tensors that can be formed from $\boldsymbol{\kappa}$ are δ_{ij} and $\kappa_i\kappa_j$. Consequently, in isotropic turbulence, $\Phi_{ij}(\boldsymbol{\kappa})$ is given by

$$\Phi_{ij}(\boldsymbol{\kappa}) = A(\kappa)\delta_{ij} + B(\kappa)\kappa_i\kappa_j, \tag{6.192}$$

where $A(\kappa)$ and $B(\kappa)$ are scalar functions of κ. These scalar functions are readily determined (see Exercise 6.25), to yield the result that, in isotropic turbulence, the velocity-spectrum tensor is

$$\begin{aligned}\Phi_{ij}(\boldsymbol{\kappa}) &= \frac{E(\kappa)}{4\pi\kappa^2}\left(\delta_{ij} - \frac{\kappa_i\kappa_j}{\kappa^2}\right) \\ &= \frac{E(\kappa)}{4\pi\kappa^2}P_{ij}(\boldsymbol{\kappa}),\end{aligned} \tag{6.193}$$

where $P_{ij}(\boldsymbol{\kappa})$ is the projection tensor (Eq. (6.133)).

If it is assumed that $\Phi_{ij}(\boldsymbol{\kappa})$ is analytic at the origin, then $E(\kappa)$ varies as κ^4 for small κ (see Exercise 6.26). However, it is possible for $\Phi_{ij}(\boldsymbol{\kappa})$ to be non-analytic, with $E(\kappa)$ varying as κ^2 (Saffman 1967). In direct numerical simulations both κ^2 and κ^4 behaviors can be obtained (Chasnov 1995). There have been suggestions (e.g., Reynolds (1987)) that grid turbulence produces κ^2 behavior, but the evidence is not conclusive.

EXERCISES _____

6.24 Show that

$$\oint d\mathcal{S}(\boldsymbol{\kappa}) = 4\pi\kappa^2, \tag{6.194}$$

$$\oint \kappa_i\kappa_j d\mathcal{S}(\boldsymbol{\kappa}) = \tfrac{4}{3}\pi\kappa^4\delta_{ij}. \tag{6.195}$$

(Hint: argue that the integral in Eq. (6.195) must be isotropic, i.e., a scalar multiple of δ_{ij}.)

6.25 Show that application of the incompressibility condition $\kappa_i \Phi_{ij}(\boldsymbol{\kappa}) = 0$ to Eq. (6.192) yields

$$B(\kappa) = -A(\kappa)/\kappa^2. \tag{6.196}$$

Using the results of Exercise (6.24), show that the energy-spectrum function corresponding to Eq. (6.192) is

$$E(\kappa) = 6\pi\kappa^2 A(\kappa) + 2\pi\kappa^4 B(\kappa). \tag{6.197}$$

Hence deduce Eq. (6.193).

6.26 If $\Phi_{ij}(\boldsymbol{\kappa})$ is analytic at the origin, it has an expansion of the form

$$\Phi_{ij}(\boldsymbol{\kappa}) = \Phi_{ij}^{(0)} + \Phi_{ijk}^{(1)}\kappa_k + \Phi_{ijk\ell}^{(2)}\kappa_k\kappa_\ell + \ldots, \tag{6.198}$$

where $\boldsymbol{\Phi}^{(n)}$ are constant tensors. Show that incompressibility dictates $\Phi_{ij}^{(0)} = 0$, and that the positive semi-definiteness of $\Phi_{ij}(\boldsymbol{\kappa})$ (see Eq. (6.183)) dictates $\Phi_{ijk}^{(1)} = 0$. Then show that (to leading order for small κ) the energy-spectrum function is

$$E(\kappa) = \tfrac{4}{3}\pi\kappa^4\Phi_{ijkk}^{(2)} + \ldots. \tag{6.199}$$

Taylor's hypothesis

In direct numerical simulations of turbulence it is possible to extract the velocity-spectrum tensor $\Phi_{ij}(\boldsymbol{\kappa})$ and the energy-spectrum function $E(\kappa)$. To determine these quantities experimentally requires the measurement of the two-point velocity correlation $R_{ij}(\boldsymbol{r})$ for all $\boldsymbol{r}$ – which clearly is not feasible. However, with a single probe (e.g., a hot-wire anemometer) it is possible, to an approximation, to measure $R_{ij}(\boldsymbol{r})$ along a line.

One technique is to use a 'flying hot wire.' The probe is moved rapidly through the turbulence at a constant speed V, along a straight line that we take to be parallel to the x_1 axis (i.e., the probe moves in the direction of the $\boldsymbol{e}_1$ basis vector). If the probe is located at $\boldsymbol{x}_0$ at time $t = 0$, then at time t it is at

$$\boldsymbol{X}(t) \equiv \boldsymbol{x}_0 + \boldsymbol{e}_1 V t, \tag{6.200}$$

and the velocity measured by the probe is

$$\boldsymbol{U}^{(m)}(t) = \boldsymbol{U}(\boldsymbol{X}(t), t) - \boldsymbol{e}_1 V. \tag{6.201}$$

The temporal autocovariance obtained from $U^{(m)}(t)$ is

$$
\begin{aligned}
R_{ij}^{(m)}(s) &\equiv \left\langle \left[U_i^{(m)}(t) - \langle U_i^{(m)}(t) \rangle \right] \left[U_j^{(m)}(t+s) - \langle U_j^{(m)}(t+s) \rangle \right] \right\rangle \\
&= \langle u_i(X(t),t) u_j(X(t+s),t+s) \rangle \\
&= \langle u_i(X(t),t) u_j(X(t) + e_1 r_1, t + r_1/V) \rangle,
\end{aligned}
\tag{6.202}
$$

where $r_1 = Vs$ is the distance moved by the probe in time s. If the turbulence is statistically homogeneous in the x_1 direction, then, in the hypothetical limit of the probe's speed V tending to infinity, we obtain

$$
\begin{aligned}
R_{ij}^{(m)}(s) &= \langle u_i(x_0 + e_1 V t, 0) u_j(x_0 + e_1 V t + e_1 r_1, 0) \rangle \\
&= \langle u_i(x_0, 0) u_j(x_0 + e_1 r_1, 0) \rangle \\
&= R_{ij}(e_1 r_1, x_0, 0).
\end{aligned}
\tag{6.203}
$$

Thus, in these circumstances, the temporal autocovariance measured by the probe yields the spatial autocovariance at $(x_0, 0)$. For the practical case of finite V, clearly Eq. (6.203) is an approximation – one that improves as V increases.

A simpler and much more common technique is to use a single stationary probe. This method is applicable to statistically stationary flows in which (at the measurement location) the turbulence intensity u' is small compared with the mean velocity $\langle U \rangle$, which we take to be in the x_1 direction. In a frame moving with the mean velocity, the probe is moving with velocity $e_1 V = -\langle U \rangle = -e_1 \langle U_1 \rangle$. Hence the flying-hot-wire analysis applies (with $r_1 = -\langle U_1 \rangle s$).

The approximation of spatial correlations by temporal correlations – e.g., Eq. (6.203) – is known as *Taylor's hypothesis* (Taylor 1938) or the *frozen-turbulence approximation*. Its accuracy depends both upon the properties of the flow and on the statistic being measured. In grid turbulence with $u'/\langle U_1 \rangle \ll 1$, it is quite accurate, and higher-order corrections can be made (Lumley 1965). In free shear flows, on the other hand, many experiments (e.g., Tong and Warhaft (1995)) have shown Taylor's hypothesis to fail.

One-dimensional spectra

Nearly all the existing experimental data on turbulence spectra come from stationary hot-wire measurements. The quantities deduced from the measurements (using Taylor's hypothesis) are of the form

$$
R_{11}(e_1 r_1, t) = \langle u_1^2 \rangle f(r_1, t),
\tag{6.204}
$$

$$
R_{22}(e_1 r_1, t) = \langle u_2^2 \rangle g(r_1, t),
\tag{6.205}
$$

where f and g are the longitudinal and transverse autocorrelation functions (Eq. (6.45)). Here we define and deduce the properties of the one-dimensional spectra $E_{ij}(\kappa_1, t)$ obtained from $R_{ij}(e_1 r_1, t)$, and show their relationship to $\Phi_{ij}(\boldsymbol{\kappa}, t)$ and $E(\kappa, t)$ in isotropic turbulence. In the next section, experimental data on $E_{ij}(\kappa_1, t)$ are used to assess the Kolmogorov hypotheses.

The one-dimensional spectra $E_{ij}(\kappa_1)$ are defined to be *twice* the one-dimensional Fourier transform of $R_{ij}(e_1 r_1)$:

$$E_{ij}(\kappa_1) \equiv \frac{1}{\pi} \int_{-\infty}^{\infty} R_{ij}(e_1 r_1) e^{-i\kappa_1 r_1} \, dr_1. \tag{6.206}$$

(Henceforth the dependence on t is not shown explicitly.) For the diagonal components – taking $i = j = 2$ as an example – $R_{22}(e_1 r_1)$ is real, and an even function of r_1. Consequently $E_{22}(\kappa_1)$ is also real and even, so that Eq. (6.206) can be rewritten

$$E_{22}(\kappa_1) = \frac{2}{\pi} \int_0^{\infty} R_{22}(e_1 r_1) \cos(\kappa_1 r_1) \, dr_1, \tag{6.207}$$

with the inversion formula

$$R_{22}(e_1 r_1) = \int_0^{\infty} E_{22}(\kappa_1) \cos(\kappa_1 r_1) \, d\kappa_1, \tag{6.208}$$

(cf. Eqs. (D.8) and (D.9)). The factor of two in the definition of $E_{ij}(\kappa_1)$ is added so that (setting $r_1 = 0$ in Eq. (6.208)) we obtain

$$R_{22}(0) = \langle u_2^2 \rangle = \int_0^{\infty} E_{22}(\kappa_1) \, d\kappa_1. \tag{6.209}$$

The one-dimensional spectrum is related to the velocity-spectrum tensor by

$$E_{22}(\kappa_1) = 2 \int\!\!\int_{-\infty}^{\infty} \Phi_{22}(\boldsymbol{\kappa}) \, d\kappa_2 \, d\kappa_3, \tag{6.210}$$

(see Exercise 6.27). It should be appreciated that $E_{22}(\kappa_1)$ has contributions from *all* wavenumbers $\boldsymbol{\kappa}$ in the plane $e_1 \cdot \boldsymbol{\kappa} = \kappa_1$, so that the wavenumber magnitude $|\boldsymbol{\kappa}|$ of the Fourier modes contributing to $E_{22}(\kappa_1)$ can be appreciably larger than κ_1.

The one-dimensional spectrum $E_{11}(\kappa_1)$ is related to the longitudinal auto-correlation function by

$$E_{11}(\kappa_1) = \frac{2}{\pi} \langle u_1^2 \rangle \int_0^{\infty} f(r_1) \cos(\kappa_1 r_1) \, dr_1, \tag{6.211}$$

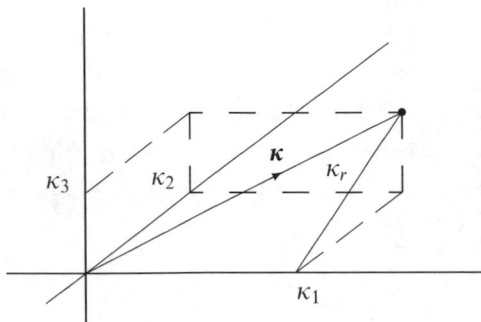

Fig. 6.10. A sketch of wavenumber space showing the definition of the radial coordinate κ_r.

and to the longitudinal structure function by

$$D_{11}(e_1 r_1) = 2 \int_0^\infty E_{11}(\kappa_1)[1 - \cos(\kappa_1 r_1)] \, d\kappa_1. \tag{6.212}$$

With $\kappa_1 = 0$, Eq. (6.211) yields for the longitudinal integral scale

$$L_{11} = \int_0^\infty f(r_1) \, dr_1 = \frac{\pi E_{11}(0)}{2\langle u_1^2 \rangle}. \tag{6.213}$$

There are similar results for the transverse correlations.

In isotropic turbulence the one-dimensional spectra are determined by the energy-spectrum function $E(\kappa)$. Writing Eq. (6.210) for $E_{11}(\kappa_1)$ and substituting Eq. (6.193) for $\Phi_{ij}(\kappa)$, we obtain

$$E_{11}(\kappa_1) = \int\!\!\!\int\limits_{-\infty}^{\infty} \frac{E(\kappa)}{2\pi\kappa^2} \left(1 - \frac{\kappa_1^2}{\kappa^2}\right) d\kappa_2 \, d\kappa_3. \tag{6.214}$$

The integration is over the plane of fixed κ_1, and the integrand is radially symmetric about the κ_1 axis. Hence, introducing the radial coordinate κ_r (see Fig. 6.10)

$$\kappa_r^2 = \kappa_2^2 + \kappa_3^2 = \kappa^2 - \kappa_1^2, \tag{6.215}$$

and noting that $2\pi\kappa_r \, d\kappa_r = 2\pi\kappa \, d\kappa$ (for fixed κ_1), Eq. (6.214) can be rewritten

$$E_{11}(\kappa_1) = \int_{\kappa_1}^\infty \frac{E(\kappa)}{\kappa} \left(1 - \frac{\kappa_1^2}{\kappa^2}\right) d\kappa. \tag{6.216}$$

This formula underscores the previous observation that $E_{11}(\kappa_1)$ contains contributions from wavenumbers κ greater than κ_1 – a phenomenon called *aliasing*. Indeed, it is readily shown (see Exercise 6.28) that $E_{11}(\kappa_1)$ is a monotonically decreasing function of κ_1, so that E_{11} is maximum at zero wavenumber, irrespective of the shape of $E(\kappa)$.

The above formula can be inverted (see Exercise 6.28) to obtain $E(\kappa)$ in terms of $E_{11}(\kappa_1)$ (for isotropic turbulence):

$$E(\kappa) = \tfrac{1}{2}\kappa^3 \frac{d}{d\kappa}\left(\frac{1}{\kappa}\frac{dE_{11}(\kappa)}{d\kappa}\right). \qquad (6.217)$$

Just as the transverse velocity autocorrelation function $g(r)$ is, in isotropic turbulence, determined by its longitudinal counterpart (Eq. (6.46)), so also $E_{22}(\kappa_1)$ is determined by $E_{11}(\kappa_1)$. The relationship can be obtained by taking the cosine Fourier transform of Eq. (6.46) (see also Eq. (6.211)):

$$E_{22}(\kappa_1) = \tfrac{1}{2}\left(E_{11}(\kappa_1) - \kappa_1 \frac{dE_{11}(\kappa_1)}{d\kappa_1}\right). \qquad (6.218)$$

EXERCISES

6.27   From Eq. (6.180) show that

$$R_{22}(e_1 r_1) = \int_{-\infty}^{\infty}\left(\int\!\!\!\int_{-\infty}^{\infty}\Phi_{22}(\boldsymbol{\kappa})\,d\kappa_2\,d\kappa_3\right)e^{i\kappa_1 r_1}\,d\kappa_1, \qquad (6.219)$$

and from Eq. (6.208) show that

$$R_{22}(e_1 r_1) = \int_{-\infty}^{\infty}\tfrac{1}{2}E_{22}(\kappa_1)e^{i\kappa_1 r_1}\,d\kappa_1. \qquad (6.220)$$

Hence verify Eq. (6.210).

6.28   Differentiate Eq. (6.216) to obtain

$$\frac{dE_{11}(\kappa_1)}{d\kappa_1} = -2\kappa_1\int_{\kappa_1}^{\infty}E(\kappa)\kappa^{-3}\,d\kappa, \qquad (6.221)$$

$$\frac{d^2E_{11}(\kappa_1)}{d\kappa_1^2} = \frac{2E(\kappa_1)}{\kappa_1^2} - 2\int_{\kappa_1}^{\infty}E(\kappa)\kappa^{-3}\,d\kappa. \qquad (6.222)$$

Hence verify Eq. (6.217). Use Eq. (6.221) to show that $E_{11}(\kappa_1)$ is a monotonically decreasing function of κ_1, and hence is maximum at $\kappa_1 = 0$.

6.29   Show that in isotropic turbulence,

$$E(\kappa) = -\kappa\frac{d}{d\kappa}\tfrac{1}{2}E_{ii}(\kappa) \qquad (6.223)$$

$$= -\kappa\frac{d}{d\kappa}\left(\tfrac{1}{2}E_{11}(\kappa) + E_{22}(\kappa)\right). \qquad (6.224)$$

6.30   Show that, in isotropic turbulence, the longitudinal integral scale is

$$L_{11} = \frac{\pi}{2\langle u_1^2 \rangle} \int_0^\infty \frac{E(\kappa)}{\kappa} \, d\kappa.$$ (6.225)

6.31   Show that, in isotropic turbulence, $E_{22}(\kappa_1)$ and $E(\kappa)$ are related by

$$E_{22}(\kappa_1) = \frac{1}{2} \int_{\kappa_1}^\infty \frac{E(\kappa)}{\kappa} \left(1 + \frac{\kappa_1^2}{\kappa^2} \right) d\kappa,$$ (6.226)

$$E(\kappa) = -\kappa \left\{ \frac{dE_{22}(\kappa)}{d\kappa} + \int_\kappa^\infty \frac{1}{\kappa_1} \frac{dE_{22}(\kappa_1)}{d\kappa_1} \, d\kappa_1 \right\}.$$ (6.227)

Power-law spectra

In examining the Kolmogorov hypotheses (in Section 6.5.2), we are interested in *power-law spectra* of the form

$$E_{11}(\kappa_1) = C_1 A \kappa_1^{-p},$$ (6.228)

(over some range of κ_1), where C_1 is a constant and A is a normalization factor (e.g., $A = \langle u_1^2 \rangle L_{11}^{1-p}$). Such spectra are examined in detail in Appendix G. If $E_{11}(\kappa_1)$ is given by Eq. (6.228) then it follows from Eq. (6.217) that $E(\kappa)$ is

$$E(\kappa) = C A \kappa^{-p},$$ (6.229)

with

$$C = \tfrac{1}{2} p (2 + p) C_1$$ (6.230)

and from Eq. (6.218) that $E_{22}(\kappa_1)$ is

$$E_{22}(\kappa) = C_1' A \kappa^{-p},$$ (6.231)

with

$$C_1' = \tfrac{1}{2}(1 + p) C_1.$$ (6.232)

Thus the power-law exponent p is the same for the three spectra, and the constants C, C_1, and C_1' are related.

Comparison of spectra

In Section 6.5.3, we introduce a model spectrum (Eq. (6.246)) which provides a reasonably accurate representation of measured turbulence spectra. Figure 6.11 shows the various spectra – $E(\kappa)$, $E_{11}(\kappa_1)$, and $E_{22}(\kappa_1)$ – given by this model for isotropic turbulence at Reynolds number $R_\lambda = 500$. The following observations are made.

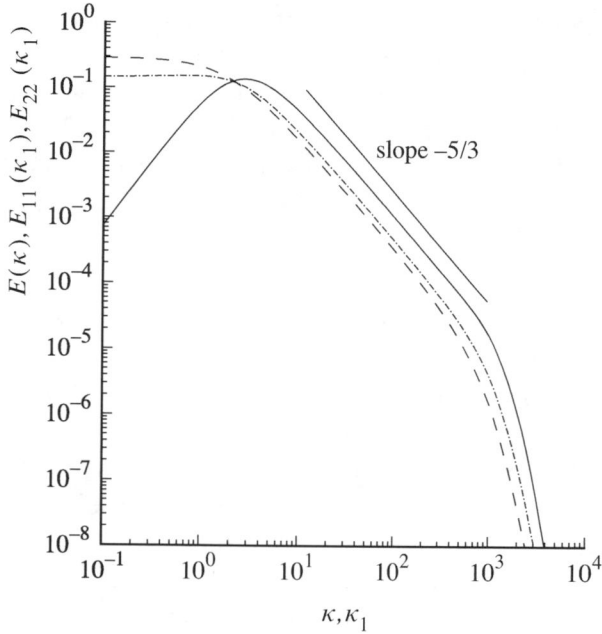

Fig. 6.11. Comparison of spectra in isotropic turbulence at $R_\lambda = 500$: solid line, $E(\kappa)$; dashed line, $E_{11}(\kappa_1)$; dot-dashed line, $E_{22}(\kappa_1)$. From the model spectrum, Eq. (6.246). (Arbitrary units.)

(i) In the center of the wavenumber range, all the spectra exhibit power-law behavior with $p = \frac{5}{3}$. In this range, consistent with Eqs. (6.230) and (6.232), the values of E_{11}, E_{22} and E are in the ratios $1 : \frac{4}{3} : \frac{55}{18}$.

(ii) At high wavenumber, the spectra decay more rapidly than a power of κ, consistent with the underlying velocity field being infinitely differentiable.

(iii) At low wavenumber, $E(\kappa)$ tends to zero as κ^2. In contrast, the one-dimensional spectra are maximum at zero wavenumber. This again illustrates the fact that the one-dimensional spectra contain contributions from wavenumbers κ greater than κ_1 (see Eq. (6.216)).

(iv) At low wavenumber, the one-dimensional spectra E_{11} and E_{22} are in the ratio 2:1 – consistent with the ratios of the integral length scales L_{11} and L_{22} (see Eqs. (6.51) and (6.213)).

6.5.2 *Kolmogorov spectra*

According to the Kolmogorov hypotheses, in any turbulent flow at sufficiently high Reynolds number, the high-wavenumber portion of the velocity

spectra adopts particular universal forms. This conclusion, and the forms of
the *Kolmogorov spectra*, can be obtained via two different routes. The implications of the Kolmogorov hypotheses for the second-order velocity structure
functions are given in Section 6.2 (e.g., Eqs. (6.29) and (6.30)). The first route
is to obtain the Kolmogorov spectra as the appropriate Fourier transforms
of the structure functions. However, we follow the second route, which is
simpler though less rigorous: this is to apply the Kolmogorov hypotheses
directly to the spectra.

Recall that (for any turbulent flow at sufficiently high Reynolds number)
the Kolmogorov hypotheses apply to the velocity field on small lengthscales, specifically in the universal equilibrium range defined by $\ell < \ell_{EI}$. In
wavenumber space the corresponding range is $\kappa > \kappa_{EI} \equiv 2\pi/\ell_{EI}$.

According to the hypothesis of local isotropy, velocity statistics pertaining
to the universal equilibrium range are isotropic. Consequently, for $\kappa > \kappa_{EI}$,
the velocity-spectrum tensor $\Phi_{ij}(\boldsymbol{\kappa})$ is given in terms of the energy-spectrum
function $E(\kappa)$ by Eq. (6.193); and the isotropic relations among $E_{11}(\kappa_1)$,
$E_{22}(\kappa_1)$, and $E(\kappa)$ apply (see Eqs. (6.214)–(6.218)).

According to the first similarity hypothesis, velocity statistics pertaining
to the universal equilibrium range have a universal form that is uniquely
determined by ε and v. Consequently, for $\kappa > \kappa_{EI}$, $E(\kappa)$ is a universal
function of κ, ε, and v. Using ε and v to non-dimensionalize κ and $E(\kappa)$,
simple dimensional analysis shows that this universal relation can be written

$$E(\kappa) = (\varepsilon v^5)^{1/4} \varphi(\kappa\eta)$$
$$= u_\eta^2 \eta \varphi(\kappa\eta), \tag{6.233}$$

where $\varphi(\kappa\eta)$ is a universal non-dimensional function – the *Kolmogorov
spectrum function*. Alternatively, if ε and κ are used to non-dimensionalize
$E(\kappa)$, the relation is

$$E(\kappa) = \varepsilon^{2/3} \kappa^{-5/3} \Psi(\kappa\eta), \tag{6.234}$$

where $\Psi(\kappa\eta)$ is the *compensated Kolmogorov spectrum function*. These universal functions are related by

$$\Psi(\kappa\eta) = (\kappa\eta)^{5/3} \varphi(\kappa\eta), \tag{6.235}$$

and Eqs. (6.233) and (6.234) apply for $\kappa > \kappa_{EI}$, which corresponds to

$$\kappa\eta > \frac{2\pi\eta}{\ell_{EI}}. \tag{6.236}$$

The second similarity hypothesis applies to scales in the inertial subrange,

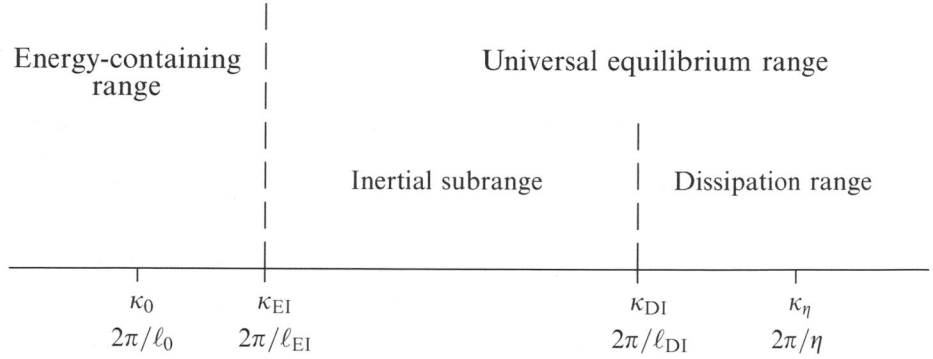

Fig. 6.12. Wavenumbers (on a logarithmic scale) at very high Reynolds number showing the various ranges.

i.e., $\eta \ll \ell \ll \ell_0$, or more precisely $\ell_{DI} < \ell < \ell_{EI}$. The corresponding range in wavenumber space is $\kappa_{EI} < \kappa < \kappa_{DI}$, see Fig. 6.12; or, in terms of $\kappa\eta$,

$$1 \gg \kappa\eta \gg \eta/\ell_0, \tag{6.237}$$

or

$$\kappa_{DI}\eta = \frac{2\pi\eta}{\ell_{DI}} > \kappa\eta > \frac{2\pi\eta}{\ell_{EI}} = \kappa_{EI}\eta. \tag{6.238}$$

In the inertial subrange, according to the second similarity hypothesis, $E(\kappa)$ has a universal form uniquely determined by ε, independent of v. In Eq. (6.234) for $E(\kappa)$, v enters solely through η. Hence the hypothesis implies that, as its argument $\kappa\eta$ tends to zero (i.e., $\kappa\eta \ll 1$, cf. Eq. (6.237)), the function Ψ becomes independent of its argument, i.e., it tends to a constant, C. Hence the second similarity hypothesis predicts that, in the inertial subrange, the energy-spectrum function is

$$E(\kappa) = C\varepsilon^{2/3}\kappa^{-5/3}, \tag{6.239}$$

(i.e., Eq. (6.234) with $\Psi = C$.) This is the famous Kolmogorov $-\frac{5}{3}$ spectrum, and C is a universal Kolmogorov constant. Experimental data support the value $C = 1.5$ (see e.g., Fig. 6.17 below, and Sreenivasan (1995)).

According to the hypothesis, in the inertial subrange, $\Phi_{ij}(\boldsymbol{\kappa})$ is an isotropic tensor function and $E(\kappa)$ is a power-law spectrum (i.e., Eq. (6.228) with $p = \frac{5}{3}$). Consequently, as shown in Section 6.5.1, the one-dimensional spectra are given by

$$E_{11}(\kappa_1) = C_1\varepsilon^{2/3}\kappa_1^{-5/3}, \tag{6.240}$$

$$E_{22}(\kappa_1) = C_1'\varepsilon^{2/3}\kappa_1^{-5/3}, \tag{6.241}$$

where

$$C_1 = \tfrac{18}{55}C \approx 0.49, \tag{6.242}$$

$$C_1' = \tfrac{4}{3}C_1 = \tfrac{24}{55}C \approx 0.65 \tag{6.243}$$

(see Eqs. (6.228)–(6.232)).

Some properties of power-law spectra are given in Appendix G. There is a direct correspondence between the form of $E_{11}(\kappa_1)$ (Eq. (6.240)) and that of the second-order velocity structure function

$$D_{LL}(r) = C_2(\varepsilon r)^{2/3}, \tag{6.244}$$

(Eq. (6.30)) in the inertial range. The powers p and q ($E(\kappa) \sim \kappa^{-p}$, $D_{LL}(r) \sim r^q$) are related by $p = \tfrac{5}{3} = 1 + q = 1 + \tfrac{2}{3}$; and the constants (to an excellent approximation) by

$$C_2 \approx 4C_1 \approx 2.0 \tag{6.245}$$

(see Eq. (G.25)).

6.5.3 A model spectrum

Before examining experimental data to test further the Kolmogorov hypotheses, we introduce a simple model spectrum that is used for comparison. The model for the energy-spectrum function is

$$E(\kappa) = C\varepsilon^{2/3}\kappa^{-5/3}f_L(\kappa L)f_\eta(\kappa \eta), \tag{6.246}$$

where f_L and f_η are specified non-dimensional functions. The function f_L determines the shape of the energy-containing range, and tends to unity for large κL. Similarly, f_η determines the shape of the dissipation range, and it tends to unity for small $\kappa \eta$. In the inertial subrange both f_L and f_η are essentially unity, so the Kolmogorov $-\tfrac{5}{3}$ spectrum with constant C is recovered.

The specification of f_L is

$$f_L(\kappa L) = \left(\frac{\kappa L}{[(\kappa L)^2 + c_L]^{1/2}} \right)^{5/3 + p_0}, \tag{6.247}$$

where p_0 is taken to be 2, and c_L is a positive constant. Clearly f_L tends to unity for large κL, while the exponent $\tfrac{5}{3} + p_0$ leads to $E(\kappa)$ varying as $\kappa^{p_0} = \kappa^2$ for small κL. (With the alternative choice $p_0 = 4$, Eq. (6.247) is known as the *von Kármán spectrum* (von Kármán 1948), which has $E(\kappa) \sim \kappa^4$ for small κ.)

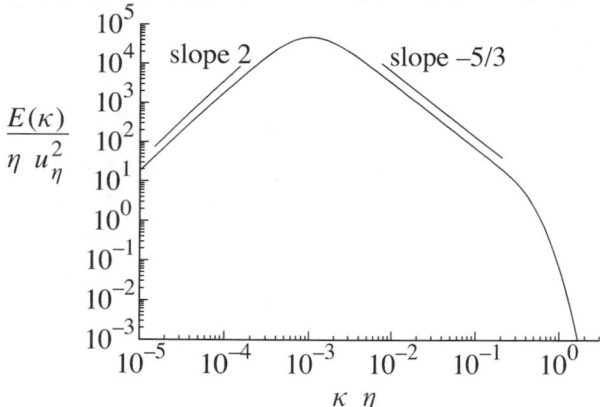

Fig. 6.13. The model spectrum (Eq. (6.246)) for $R_\lambda = 500$ normalized by the Kolmogorov scales.

The specification of f_η is

$$f_\eta(\kappa\eta) = \exp\{-\beta\{[(\kappa\eta)^4 + c_\eta^4]^{1/4} - c_\eta\}\}, \qquad (6.248)$$

where β and c_η are positive constants. Note that, for $c_\eta = 0$, this reduces to

$$f_\eta(\kappa\eta) = \exp(-\beta\kappa\eta). \qquad (6.249)$$

Because the velocity field $u(x)$ is infinitely differentiable, it follows that, for large κ, the energy-spectrum function decays more rapidly than any power of κ (see Appendix G). Hence the exponential decay (as suggested by Kraichnan (1959)). Several experiments support the exponential form with $\beta = 5.2$ (see Saddoughi and Veeravalli (1994)). However, the simple exponential (Eq. (6.249)) departs from unity too rapidly for small $\kappa\eta$, and the value of β is constrained to be $\beta \approx 2.1$ (see Exercise 6.33). These deficiencies are remedied by Eq. (6.248).

For specified values of k, ε, and v, the model spectrum is determined by Eqs. (6.246)–(6.248) with $C = 1.5$ and $\beta = 5.2$. Alternatively, the non-dimensional model spectrum is uniquely determined by a specified value of R_λ. The constants c_L and c_η are determined by the requirements that $E(\kappa)$ and $2v\kappa^2 E(\kappa)$ integrate to k and ε, respectively: at high Reynolds number their values are $c_L \approx 6.78$ and $c_\eta \approx 0.40$ (see Exercise 6.32). For isotropic turbulence, corresponding models for the one-dimensional spectra $E_{11}(\kappa_1)$ and $E_{22}(\kappa_2)$ are obtained from Eqs. (6.216)–(6.218).

Figure 6.13 is a log–log plot of the model spectrum (with Kolmogorov scaling) for $R_\lambda = 500$. The power laws $E(\kappa) \sim \kappa^2$ at low wavenumber and

$E(\kappa) \sim \kappa^{-5/3}$ in the inertial subrange are evident, as is the exponential decay at large κ.

6.32   Show that, at very high Reynolds number, the integral of the model spectrum (Eq. (6.246)) over all κ yields

$$k = C(\varepsilon L)^{2/3} \int_0^\infty (\kappa L)^{-5/3} f_L(\kappa L)\, \mathrm{d}(\kappa L). \qquad (6.250)$$

Show that, with f_L given by Eq. (6.247) (with $p_0 = 2$), the integral in Eq. (6.250) is

$$c_L^{-1/3} \int_0^\infty \frac{x^2}{(x^2 + 1)^{11/6}}\, \mathrm{d}x = c_L^{-1/3} \frac{3\Gamma(\tfrac{1}{3})\Gamma(\tfrac{3}{2})}{5\Gamma(\tfrac{5}{6})} \approx 1.262\, c_L^{-1/3}. \qquad (6.251)$$

Hence show that, for $C = 1.5$, the high-Reynolds-number asymptote of c_L is

$$c_L \approx (1.262\, C)^3 \approx 6.783. \qquad (6.252)$$

6.5.4 Dissipation spectra

In this and the next four subsections, experimental data, the Kolmogorov hypotheses, and the model spectrum are used to examine velocity spectra in turbulent flows. In most of the relevant experiments, Taylor's hypothesis is invoked in order to obtain measurements of the one-dimensional spectra $E_{ij}(\kappa_1)$.

Figure 6.14 is a compilation of measurements of $E_{11}(\kappa_1)$, plotted with Kolmogorov scaling. As is the case with $E(\kappa)$ (Eq. (6.233)), the Kolmogorov hypotheses imply that the scaled spectrum $\varphi_{11} \equiv E_{11}(\kappa_1)/(\varepsilon \nu^5)^{1/4}$ is a universal function of $\kappa_1 \eta$, at sufficiently high Reynolds number, and for $\kappa_1 > \kappa_{\mathrm{EI}}$. The data shown in Fig. 6.14 come from many different flows, with Taylor-scale Reynolds numbers from 23 to 3,180. It may be seen that, for $\kappa_1 \eta > 0.1$, all the data lie on a single curve. The high-Reynolds-number data exhibit power-law behavior for $\kappa_1 \eta < 0.1$, the extent of the power-law region generally increasing with R_λ. Thus the data are consistent with $E_{11}(\kappa_1)/(\varepsilon \nu^5)^{1/4}$ being a universal function of $\kappa_1 \eta$ for $\kappa_1 > \kappa_{\mathrm{EI}}$, with the departures from universal behavior in Fig. 6.14 arising from the energy-containing range $\kappa < \kappa_{\mathrm{EI}}$. The model spectra (also shown in Fig. 6.14 for various R_λ) appear to represent the data quite accurately.

Compensated one-dimensional spectra (i.e., $\kappa_1^{5/3} E_{11}(\kappa_1)$) with Kolmogorov

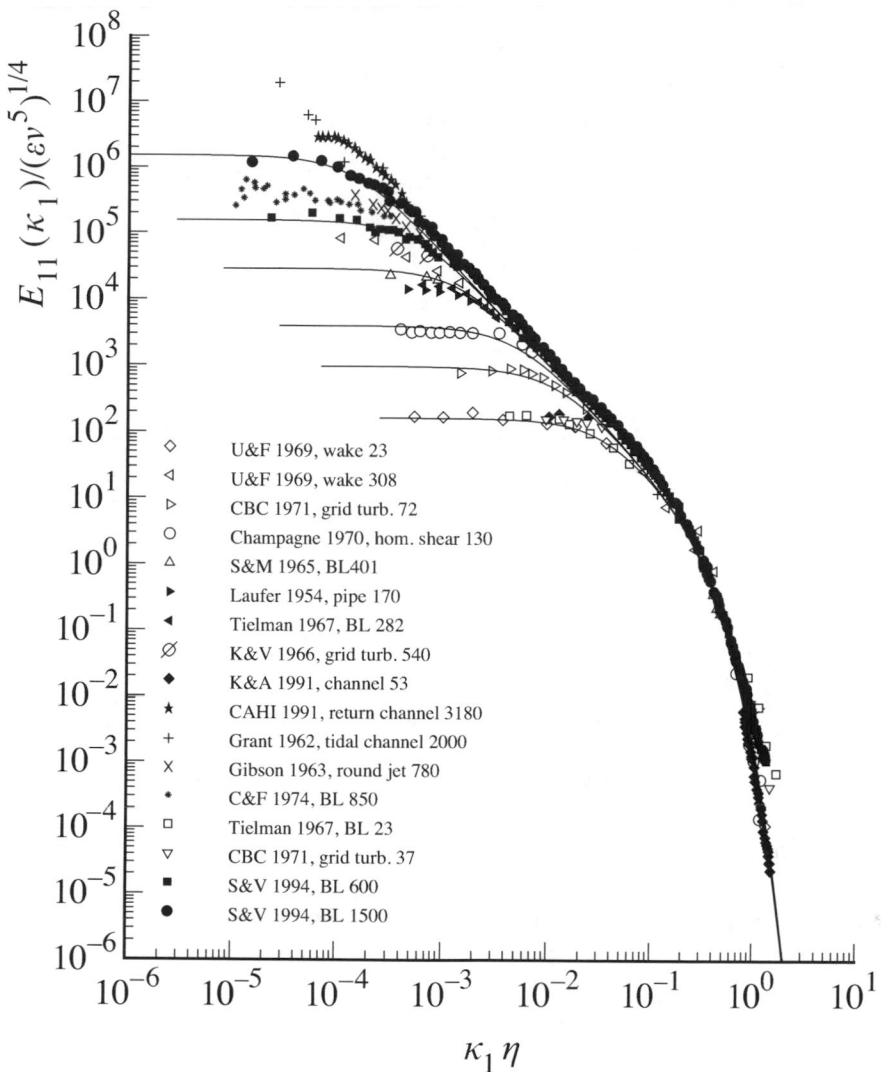

Fig. 6.14. Measurements of one-dimensional longitudinal velocity spectra (symbols), and model spectra (Eq. (6.246)) for $R_\lambda = 30, 70, 130, 300, 600$, and 1,500 (lines). The experimental data are taken from Saddoughi and Veeravalli (1994) where references to the various experiments are given. For each experiment, the final number in the key is the value of R_λ.

scaling are shown in Fig. 6.15 on a linear–log plot, which emphasizes the dissipation range. For $\kappa_1 \eta > 0.1$, there is close agreement between measurements in grid turbulence ($R_\lambda \approx 60$) and in a turbulent boundary layer ($R_\lambda \approx 600$), again supporting the universality of the high-wavenumber spectra. The straight-line behavior evident in this plot for $\kappa_1 \eta > 0.3$ corresponds

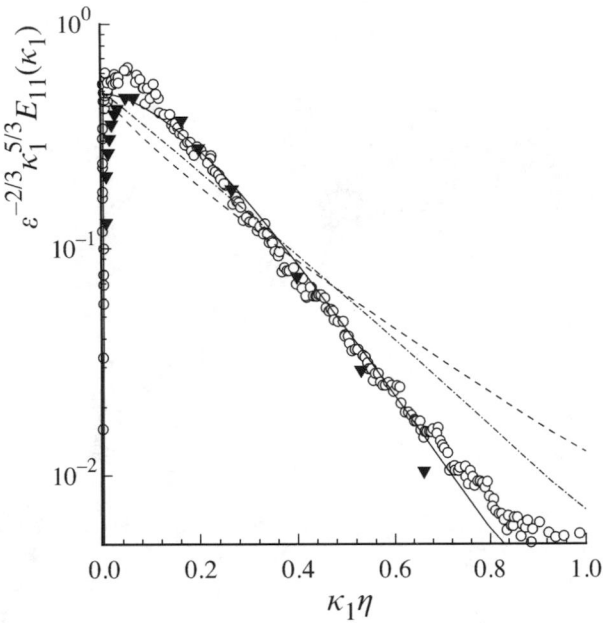

Fig. 6.15. Compensated one-dimensional velocity spectra. Measurements of Comte-Bellot and Corrsin (1971) in grid turbulence at $R_\lambda \approx 60$ (triangles), and of Saddoughi and Veeravalli (1994) in a turbulent boundary layer at $R_\lambda \approx 600$ (circles). Solid line, model spectrum Eq. (6.246) for $R_\lambda = 600$; dashed line, exponential spectrum Eq. (6.253); dot–dashed line, Pao's spectrum, Eq. (6.254).

to exponential decay of the spectrum at the highest wavenumbers. Again, the model spectrum represents the data accurately.

Also shown in Fig. 6.15 are the one-dimensional spectra deduced from two alternative models for $f_\eta(\kappa\eta)$. These are the exponential

$$f_\eta(\kappa\eta) = \exp(-\beta_0\kappa\eta), \qquad (6.253)$$

where β_0 is given by Eq. (6.258), and the *Pao spectrum*

$$f_\eta(\kappa\eta) = \exp[-\tfrac{3}{2}C(\kappa\eta)^{4/3}], \qquad (6.254)$$

(see Pao (1965) and Section 6.6). It is evident from Fig. 6.15 that these alternatives do not represent the data as well as the model spectrum does.

Having established that the model spectrum describes the dissipation range accurately, we now use it to quantify the scales of the dissipative motions. Figure 6.16 shows the dissipative spectrum $D(\kappa) = 2\nu\kappa^2 E(\kappa)$ according to the model for $R_\lambda = 600$, and also the cumulative dissipation

$$\varepsilon_{(0,\kappa)} \equiv \int_0^\kappa D(\kappa')\,\mathrm{d}\kappa'. \qquad (6.255)$$

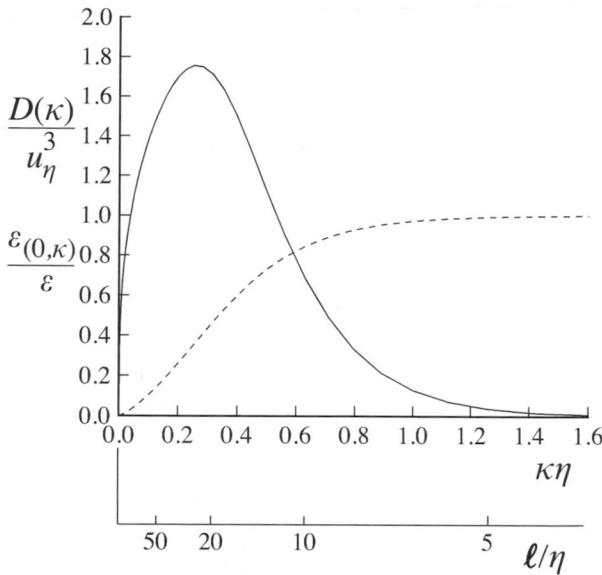

Fig. 6.16. The dissipation spectrum (solid line) and cumulative dissipation (dashed line) corresponding to the model spectrum Eq. (6.246) for $R_\lambda = 600$. $\ell = 2\pi/\kappa$ is the wavelength corresponding to wavenumber κ.

The abscissa shows the wavenumber κ and the corresponding wavelength $\ell = 2\pi/\kappa$, both normalized by the Kolmogorov scale η. Characteristic wavenumbers and wavelengths obtained from these curves are given in Table 6.1. It may be seen that the peak of the dissipation spectrum occurs at $\kappa\eta \approx 0.26$, corresponding to $\ell/\eta \approx 24$, while the centroid (where $\varepsilon_{(0,\kappa)} = \frac{1}{2}\varepsilon$) occurs at $\kappa\eta \approx 0.34$, corresponding to $\ell/\eta \approx 18$. Thus the motions responsible for the bulk of the dissipation ($0.1 < \kappa\eta < 0.75$, or $60 > \ell/\eta > 8$) are considerably larger than the Kolmogorov scale. (There is no inconsistency between this observation and the Kolmogorov hypotheses: the hypotheses imply that the characteristic size of the dissipative motions *scale* with η, not that it be equal to η.) On the basis of these observations we take the demarcation lengthscale between the inertial and dissipative ranges to be $\ell_{\text{DI}} = 60\eta$. (The significance of ℓ_{DI} is illustrated in Figs. 6.2 and 6.12.)

EXERCISE _____

6.33   Show that, at high Reynolds number, the expression for dissipation obtained from integration of the model spectrum (Eq. (6.246)) is

$$\varepsilon = 2C \, \nu \varepsilon^{2/3} \eta^{-4/3} \int_0^\infty (\kappa\eta)^{1/3} f_\eta(\kappa\eta) \, \mathrm{d}(\kappa\eta). \qquad (6.256)$$

Table 6.1. *Characteristic wavenumbers and lengthscales of the dissipation spectrum (based on the model spectrum Eq. (6.246) at* $R_\lambda = 600$*)*

Defining wavenumbers	$\kappa\eta$	ℓ/η
Peak of dissipation spectrum	0.26	24
$\varepsilon_{(0,\kappa)} = 0.1\varepsilon$	0.10	63
$\varepsilon_{(0,\kappa)} = 0.5\varepsilon$	0.34	18
$\varepsilon_{(0,\kappa)} = 0.9\varepsilon$	0.73	8.6

Show that, if f_η is given by the exponential Eq. (6.253), then the integral in Eq. (6.256) is

$$\int_0^\infty x^{1/3} e^{-\beta_0 x} \, \mathrm{d}x = \beta_0^{-4/3} \Gamma(\tfrac{4}{3}). \tag{6.257}$$

Hence show that the high-Reynolds-number asymptote of β_0 is

$$\beta_0 = [2C\,\Gamma(\tfrac{4}{3})]^{3/4} \approx 2.094, \tag{6.258}$$

for $C = 1.5$. Confirm that the Pao spectrum (Eq. (6.254)) satisfies Eq. (6.256).

6.5.5 *The inertial subrange*

The second Kolmogorov hypothesis predicts a $-\tfrac{5}{3}$ spectrum in the inertial subrange. The power-law behavior evident in Fig. 6.14 is best examined by plotting the compensated spectrum $\varepsilon^{-2/3} \kappa_1^{5/3} E_{11}(\kappa_1)$. For then the Kolmogorov hypotheses predict that this quantity adopts the constant value C_1 (see Eq. (6.240)) in the inertial range. The compensated spectra measured in a high-Reynolds-number boundary layer are compared with the Kolmogorov prediction in Fig. 6.17. It may be seen that the data are within 20% of the predicted value over two decades of wavenumbers – over which range $\kappa_1^{5/3}$ increases by more than a factor of 2,000.

The model spectrum is, of course, constructed to yield the Kolmogorov behavior in the inertial subrange, and this behavior is evident in Fig. 6.17. Figure 6.17 also provides some evidence of local isotropy in this distinctly anisotropic turbulent flow: for $\kappa_1 \eta > 2 \times 10^{-3}$, E_{22} and E_{33} are very similar, and (as predicted by local isotropy) the plateau value of their compensated spectra is $\tfrac{4}{3}$ that of E_{11} (Eq. (6.243)). A more direct test (that includes the dissipation range) is to compare the measured value of $E_{22}(\kappa_1)$ with

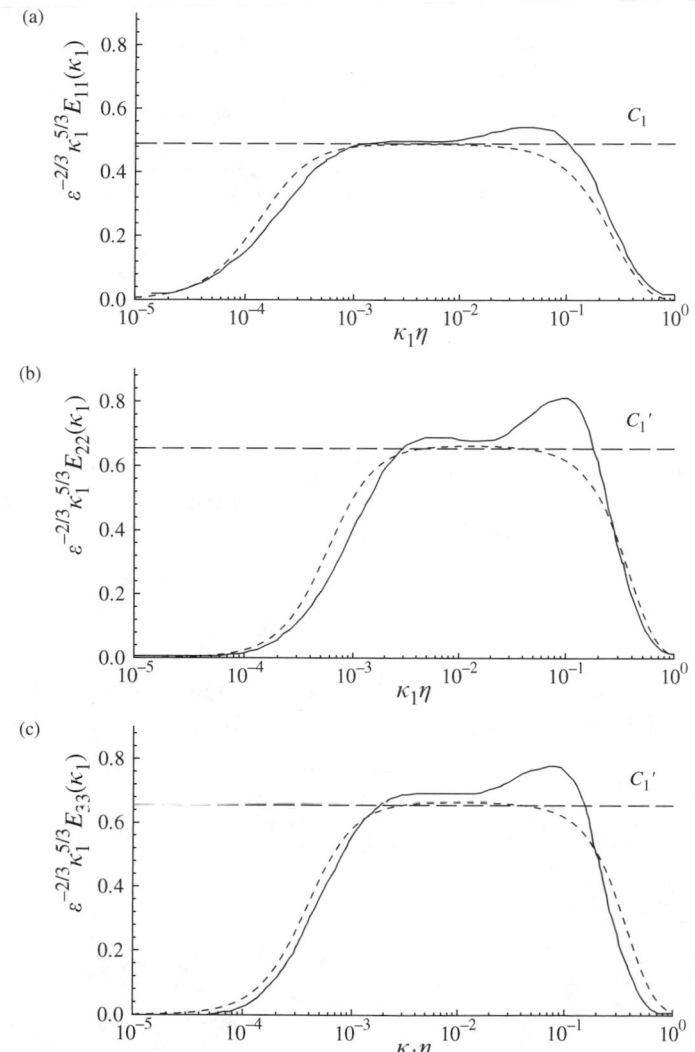

Fig. 6.17. Compensated one-dimensional spectra measured in a turbulent boundary layer at $R_\lambda \approx 1{,}450$. Solid lines, experimental data Saddoughi and Veeravalli (1994); dashed lines, model spectra from Eq. (6.246); long dashed lines, C_1 and C_1' corresponding to Kolmogorov inertial-range spectra. (For E_{11}, E_{22} and E_{33} the model spectra are for $R_\lambda = 1{,}450$, 690, and 910, respectively, corresponding to the measured values of $\langle u_1^2 \rangle$, $\langle u_2^2 \rangle$, and $\langle u_3^2 \rangle$.)

that calculated from $E_{11}(\kappa_1)$ with the assumption of isotropy, Eq. (6.218). Saddoughi and Veeravalli (1994) performed this test and found that the measured and calculated values differ by no more than 10% throughout the equilibrium range.

Table 6.2. *Characteristic wavenumbers and lengthscales of the energy spectrum (based on the model spectrum Eq. (6.246) at* $R_\lambda = 600$*)*

Defining wavenumber	κL_{11}	ℓ/L_{11}
Peak of energy spectrum	1.3	5.0
$k_{(0,\kappa)} = 0.1k$	1.0	6.1
$k_{(0,\kappa)} = 0.5k$	3.9	1.6
$k_{(0,\kappa)} = 0.8k$	15	0.42
$k_{(0,\kappa)} = 0.9k$	38	0.16

6.5.6 The energy-containing range

Two factors make the examination of the energy-containing range more difficult. First, unlike the universal equilibrium range, the energy-containing range depends on the particular flow. Second, the one-dimensional spectra provide little direct information. This is because $E_{11}(\kappa_1)$ contains contributions from all wavenumbers of magnitude greater than κ_1 (i.e., $|\kappa| > \kappa_1$). The energy-spectrum function $E(\kappa)$ is the most informative quantity. For isotropic turbulence this can be obtained experimentally by differentiating one-dimensional spectra (see Eqs. (6.217) and (6.223)) – although this is a poorly conditioned process. With these difficulties in mind, we examine $E(\kappa)$ in grid turbulence (which is reasonably isotropic).

Appropriate scales for normalization are the turbulent kinetic energy k, and the longitudinal integral length scale L_{11}. For, in isotropic turbulence, $E(\kappa)$ has the integral properties

$$\int_0^\infty E(\kappa)\,d\kappa = k, \tag{6.259}$$

$$\int_0^\infty \frac{E(\kappa)}{\kappa}\,d\kappa = \frac{4}{3\pi}kL_{11}. \tag{6.260}$$

With these scalings, Fig. 6.18 shows measurements of $E(\kappa)$ in grid turbulence at $R_\lambda \approx 60$, and also the model spectrum for $R_\lambda = 60$ and $R_\lambda = 1,000$. It may be seen that (with this scaling) the shape of the spectrum does not vary strongly with the Reynolds number, and that the model provides a reasonable representation of the data. Also shown in Fig. 6.18 is the model spectrum Eq. (6.246) with $p_0 = 4$ (which gives $E(\kappa) \sim \kappa^4$ for small κL). Compared with $p_0 = 2$, the shape is little different, and the 10% difference in the peak values is most likely within experimental uncertainties.

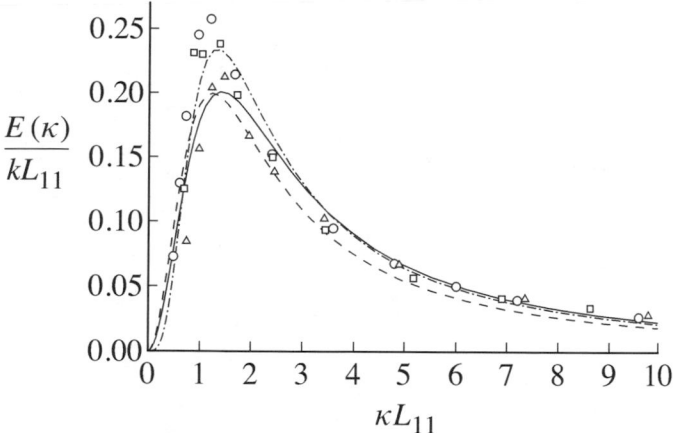

Fig. 6.18. The energy-spectrum function in isotropic turbulence normalized by k and L_{11}. Symbols, grid-turbulence experiments of Comte-Bellot and Corrsin (1971): $\bigcirc, R_\lambda = 71; \square, R_\lambda = 65; \triangle, R_\lambda = 61$. Lines, model spectrum, Eq. (6.246): solid, $p_0 = 2$, $R_\lambda = 60$; dashed, $p_0 = 2$, $R_\lambda = 1,000$; dot–dashed $p_0 = 4$, $R_\lambda = 60$.

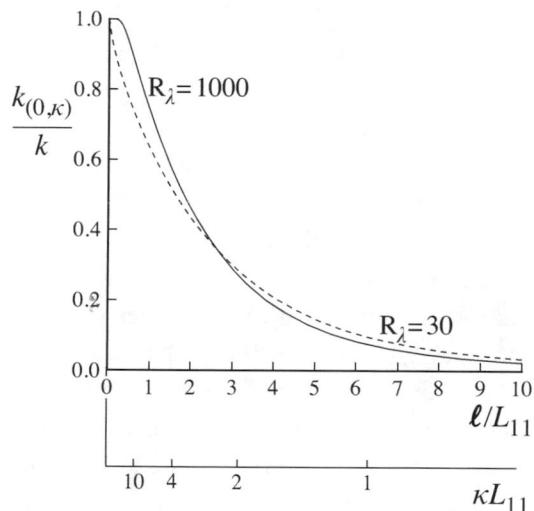

Fig. 6.19. The cumulative turbulent kinetic energy $k_{(0,\kappa)}$ against wavenumber κ and wavelength $\ell = 2\pi/\kappa$ for the model spectrum.

For the model spectrum, Fig. 6.19 shows the cumulative kinetic energy

$$k_{(0,\kappa)} = \int_0^\kappa E(\kappa')\,d\kappa', \qquad (6.261)$$

plotted against $\ell/L_{11} = 2\pi/(\kappa L_{11})$, and some of the numerical characteristics of $k_{(0,\kappa)}$ are given in Table 6.2. The centroid of the spectrum is at $\kappa L_{11} \approx 4$ ($\ell/L_{11} \approx 1\frac{1}{2}$), and 80% of the energy is contained in motions of lengthscale

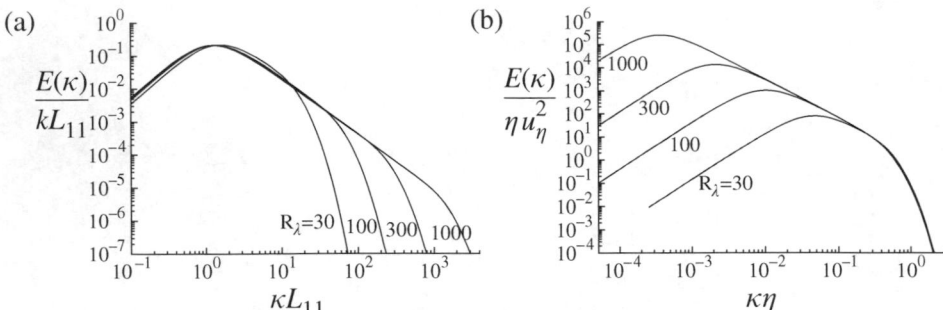

Fig. 6.20. The model spectrum for various Reynolds numbers, scaled by (a) k and L_{11}, and (b) Kolmogorov scales.

$\frac{1}{6}L_{11} < \ell < 6L_{11}$. On this basis we take the lengthscales characterizing the energy-containing motions to be $\ell_0 = L_{11}$ and $\ell_{EI} = \frac{1}{6}L_{11}$.

6.5.7 *Effects of the Reynolds number*

Figure 6.20(a) shows the model spectrum normalized by k and L_{11} for a range of Reynolds numbers. It may be seen that the energy-containing ranges of the spectra ($0.1 < \kappa L_{11} < 10$, say) are very similar, whereas, with increasing R_λ, the extent of the $-\frac{5}{3}$ region increases, and the dissipation range (where the spectrum rolls off) moves to higher values of κL_{11}.

The same spectra, but normalized by the Kolmogorov scales, are shown in Fig. 6.20(b). Now the dissipation ranges ($\kappa \eta > 0.1$, say) are very similar, while the energy-containing range moves to lower values of $\kappa \eta$ as R_λ increases.

Figure 6.21 contrasts high-Reynolds-number ($R_\lambda = 1,000$) and low-Reynolds-number ($R_\lambda = 30$) energy and dissipation spectra. As is the usual practice with log–linear plots such as this, the spectra are multiplied by κ so that the area under the curve $\kappa E(\kappa)$ represents energy. That is, the energy in the wavenumber range (κ_a, κ_b) is

$$k_{(\kappa_a,\kappa_b)} = \int_{\kappa_a}^{\kappa_b} E(\kappa)\,d\kappa = \int_{\kappa_a}^{\kappa_b} \kappa E(\kappa)\,d\ln\kappa. \qquad (6.262)$$

With the Kolmogorov scaling employed, the high-Reynolds-number spectrum contains more energy (i.e., a greater value of k/u_η^2). Consequently the energy spectra are scaled by different numerical factors so that they can be compared on the same plot. The important observation to be made with Fig. 6.21 is that, at low Reynolds number, the energy and dissipation spectra overlap significantly: there is no clear separation of scales.

The overlap between the energy and dissipation spectra can be quantified

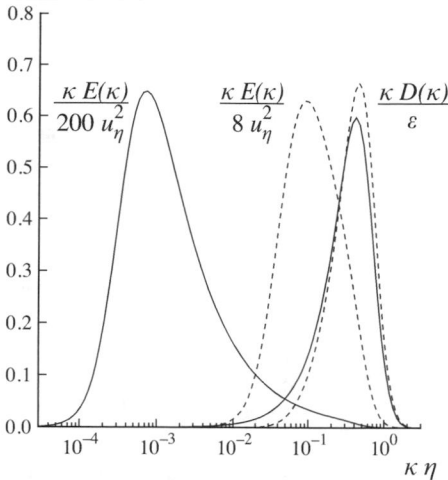

Fig. 6.21. Model energy and dissipation spectra normalized by the Kolmogorov scales at $R_\lambda = 1,000$ (solid lines) and $R_\lambda = 30$ (dashed lines). (Note the scaling of $E(\kappa)$.)

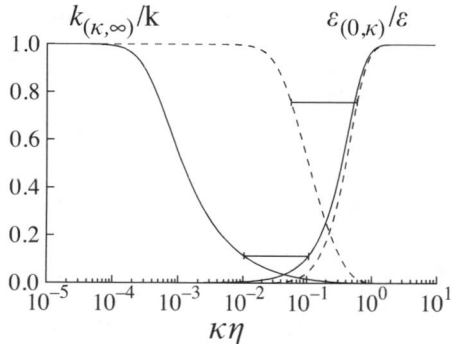

Fig. 6.22. The fraction of the energy at wavenumbers greater than κ $(k_{(\kappa,\infty)}/k)$ and the fraction of the dissipation at wavenumbers less than κ $(\varepsilon_{(0,\kappa)}/\varepsilon)$ for the model spectrum at $R_\lambda = 1,000$ (solid line) and at $R_\lambda = 30$ (dashed line). For the two Reynolds numbers, the horizontal bars identify the 'decade of wavenumbers of most overlap' between the energy and dissipation spectra.

as shown in Fig. 6.22. For $R_\lambda = 30$ and $1,000$, Fig. 6.22 shows the fraction of the energy due to wavenumbers greater than κ (i.e., $k_{(\kappa,\infty)}/k$) and the fraction of dissipation due to wavenumbers less than κ (i.e., $\varepsilon_{(0,\kappa)}/\varepsilon$). If there were a complete separation of scales then, with increasing κ, $k_{(\kappa,\infty)}/k$ would decrease to zero before $\varepsilon_{(0,\kappa)}/\varepsilon$ rose from zero. It may be seen from Fig. 6.22 that there is considerable overlap for $R_\lambda = 30$, whereas at $R_\lambda = 1,000$ there is much less – but it is not negligible.

As shown in Fig. 6.22, for given R_λ, a 'decade of maximum overlap'

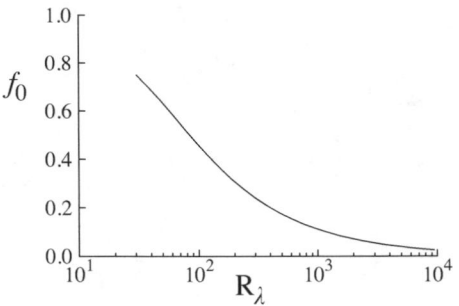

Fig. 6.23. The fraction f_0 of the energy and dissipation contributed by the wavenumber decade of maximum overlap as a function of R_λ for the model spectrum.

$(\kappa_m, 10\kappa_m)$ and an 'overlap fraction' f_0 can be identified and defined by

$$f_0 = k_{(\kappa_m, \infty)}/k = \varepsilon_{(0,10\kappa_m)}/\varepsilon. \qquad (6.263)$$

Thus the decade of wavenumbers $(\kappa_m, 10\kappa_m)$ contributes a fraction a little less than f_0 both to the energy and to the dissipation. For $R_\lambda = 30$ and 1,000 the values of f_0 are 0.75 and 0.11. Figure 6.23 shows f_0 as a function of R_λ for the model spectrum. Evidently very large Reynolds numbers are required in order for there to be a decade of wavenumbers in which both energy and dissipation are negligible.

An important tenet in the picture of the energy cascade is that (at high Reynolds number) the rate of energy dissipation ε scales as u_0^3/ℓ_0, where u_0 and ℓ_0 are characteristic velocity scales and lengthscales of the energy-containing eddies. Taking $u_0 = k^{1/2}$ and $\ell_0 = L_{11}$, this tenet is $\varepsilon \sim k^{3/2}/L_{11}$. Now, from the definition $L \equiv k^{3/2}/\varepsilon$, we have

$$\varepsilon = \frac{k^{3/2}}{L} = \frac{k^{3/2}}{L_{11}} \left(\frac{L_{11}}{L} \right), \qquad (6.264)$$

so that the scaling of ε with $k^{3/2}/L_{11}$ is equivalent to the constancy of L_{11}/L. Figure 6.24 shows the lengthscale ratio L_{11}/L as a function of R_λ. Evidently, according to the model spectrum, at high Reynolds number L_{11}/L tends asymptotically to a value of 0.43. However, this ratio increases significantly as R_λ decreases – for example, it exceeds the asymptotic value by 50% at $R_\lambda = 50$.

Finally, Fig. 6.25 shows the relationship between the different turbulence Reynolds numbers. From the definitions of Re_L, R_λ, and Eq. (6.64) we have

$$Re_L \equiv \frac{k^{1/2}L}{v} = \frac{k^2}{\varepsilon v} = \tfrac{3}{20} R_\lambda^2, \qquad (6.265)$$

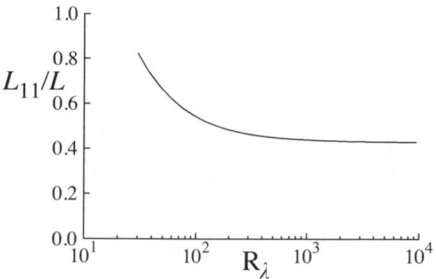

Fig. 6.24. The ratio of the longitudinal integral lengthscale L_{11} to $L = k^{3/2}/\varepsilon$ as a function of the Reynolds number for the model spectrum.

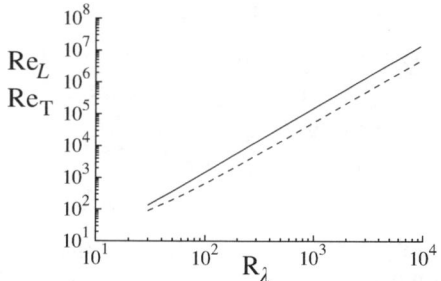

Fig. 6.25. Turbulence Reynolds numbers Re_L (solid line) and Re_T (dashed line) as functions of R_λ for the model spectrum.

Whereas, based on u' and L_{11}, we have

$$Re_T \equiv \frac{u'L_{11}}{v} = \sqrt{\frac{2}{3}}\frac{L_{11}}{L}Re_L \sim \tfrac{1}{20}R_\lambda^2. \tag{6.266}$$

In turbulent flows, the flow Reynolds number $Re = \mathcal{U}\mathcal{L}/v$ is typically an order of magnitude greater than Re_T (e.g., $u'/\mathcal{U} \approx 0.2, L_{11}/\mathcal{L} = 0.5$), leading to the rough estimate $R_\lambda \approx \sqrt{2Re}$.

EXERCISE _____

6.34   For a high Reynolds number, and for wavenumber κ in the inertial subrange, use the Kolmogorov spectrum (Eq. (6.239)) to estimate that the fraction of energy arising from motions of wavenumber greater than κ is

$$1 - \frac{k_{(0,\kappa)}}{k} \approx \tfrac{3}{2}C(\kappa L)^{-2/3} \tag{6.267}$$

$$\approx 1.28(\kappa L_{11})^{-2/3}. \tag{6.268}$$

How does this estimate compare with the values given in Table 6.2?

6.5.8 The shear-stress spectrum

Thus far we have examined the velocity spectra only for isotropic turbulence, or the isotropic portions of the spectra for locally isotropic turbulence. In these cases the shear-stress spectrum $E_{12}(\kappa_1)$ is zero. In simple shear flows with $S \equiv \partial\langle U_1\rangle/\partial x_2 > 0$ being the only significant mean velocity gradient, the mean shear rate S causes the turbulence to be anisotropic. This anisotropy is evident in the Reynolds stresses (e.g., $\langle u_1 u_2\rangle/k \approx -0.3$) and, in view of the relation

$$\langle u_1 u_2\rangle = \int_0^\infty E_{12}(\kappa_1)\,d\kappa_1, \tag{6.269}$$

the spectrum must therefore also be anisotropic over at least part of the wavenumber range. Given the prominent role played by the shear stress both in momentum transport and in production of turbulence energy, it is important to ascertain the contributions to $\langle u_1 u_2\rangle$ from the various scales of motion. The simple, consistent picture that emerges is that (inevitably) the dominant contribution to $\langle u_1 u_2\rangle$ is from wavenumbers in the energy-containing range, and at higher wavenumbers $E_{12}(\kappa_1)$ decays more rapidly than does $E_{11}(\kappa_1)$ (consistent with local isotropy).

If $\tau(\kappa)$ is the characteristic timescale of motions of wavenumber κ, then the influence of the mean shear S is characterized by the non-dimensional parameter $S\tau(\kappa)$. It is reasonable to suppose that, if $S\tau(\kappa)$ is small, then so also is the level of anisotropy created by the mean shear.

In the dissipation range the appropriate timescale is τ_η. Hence, as first suggested by Corrsin (1958), a criterion for the isotropy of the smallest scales is

$$S\tau_\eta \ll 1. \tag{6.270}$$

The parameter $S\tau_\eta$ varies as R_λ^{-1} (see Exercise 6.35), so that Eq. (6.270) amounts to a high-Reynolds-number requirement.

In the inertial subrange, the appropriate timescale is that formed from κ and ε, i.e., $\tau(\kappa) = (\kappa^2\varepsilon)^{-1/3}$. Hence the criterion for isotropy at wavenumber κ is

$$S\tau(\kappa) = S\kappa^{-2/3}\varepsilon^{-1/3} \ll 1. \tag{6.271}$$

With the lengthscale L_S defined by

$$L_S \equiv \varepsilon^{1/2}S^{-3/2}, \tag{6.272}$$

this criterion can be re-expressed as

$$\kappa L_S \gg 1. \tag{6.273}$$

(Exercise 6.35 shows that L_S is typically a sixth of $L \equiv k^{3/2}/\varepsilon$.)

At very high Reynolds number there is a wavenumber range

$$L_S^{-1} \ll \kappa \ll \eta^{-1} \tag{6.274}$$

within the inertial subrange. The small level of anisotropy in this range can be hypothesized to be a small perturbation (caused by S) of the background isotropic state – which is characterized by ε. It follows from this hypothesis that the shear-stress spectrum $E_{12}(\kappa_1)$ is determined by κ_1, ε, and S, and, furthermore, that – as a small perturbation – it varies linearly with S. Dimensional analysis then yields

$$\frac{E_{12}(\kappa_1)}{u_S^2 L_S} = \widehat{E}_{12}(\kappa_1 L_S), \tag{6.275}$$

where $\widehat{E}_{12}$ is a non-dimensional function and the velocity scale u_S is

$$u_S \equiv (\varepsilon/S)^{1/2}. \tag{6.276}$$

(Exercise 6.35 shows that u_S is typically $\frac{1}{2}k^{1/2}$.) The linearity of E_{12} with S then determines $\widehat{E}_{12}$, yielding

$$\frac{E_{12}(\kappa_1)}{u_S^2 L_S} = -C_{12}(\kappa_1 L_S)^{-7/3}, \tag{6.277}$$

or

$$E_{12}(\kappa_1) = -C_{12} S \varepsilon^{1/3} \kappa_1^{-7/3}, \tag{6.278}$$

where C_{12} is a constant. This result is due to Lumley (1967a).

Figure 6.26 shows $E_{12}(\kappa_1)$ (scaled by L_S and u_S) measured at four different locations and Reynolds numbers in turbulent boundary layers. Evidently, for $\kappa_1 L_S > \frac{1}{2}$, the data are in reasonable agreement with Eq. (6.277) with $C_{12} = 0.15$.

It is of course significant that $E_{12}(\kappa_1)$ decays more rapidly than does $E_{11}(\kappa_1)$ (as $\kappa_1^{-7/3}$ compared with $\kappa_1^{-5/3}$) so that the anisotropy decreases with κ_1. This can be seen directly in the spectral coherency $H_{12}(\kappa_1)$, which is the u_1–u_2 correlation coefficient of the Fourier modes. Figure 6.27 shows measurements of $H_{12}(\kappa_1)$ in a turbulent boundary layer. On the basis of this and other data Saddoughi and Veeravalli (1994) propose the criterion

$$\kappa_1 L_S > 3 \tag{6.279}$$

for the locally isotropic region of the spectrum. This is consistent with $\ell_{\mathrm{EI}} = \frac{1}{6}L_{11}$ marking the start of the inertial subrange, since (with some

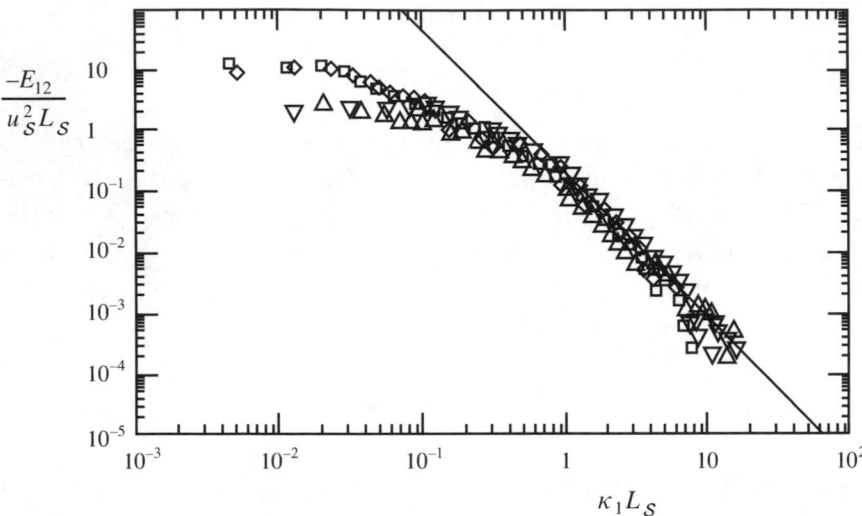

Fig. 6.26. Shear-stress spectra scaled by u_S and L_S: line, Eq. (6.277) with $C_{12} = 0.15$; symbols, experimental data of Saddoughi and Veeravalli (1994) from turbulent boundary layers with $R_\lambda \approx 500$ to 1,450.

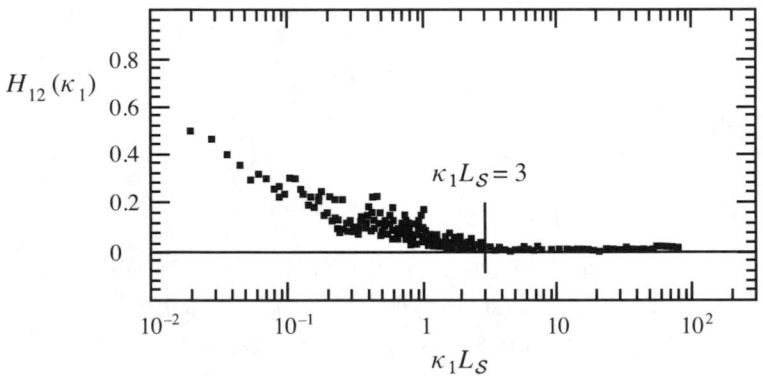

Fig. 6.27. The spectral coherency measured in a turbulent boundary layer at $R_\lambda = 1,400$ (Saddoughi and Veeravalli 1994).

assumptions) the data of Saddoughi and Veeravalli (1994) suggest that $(2\pi/\ell_{EI})L_S \approx 6$.

EXERCISE _____

6.35	For a simple turbulent shear flow with $S = \partial\langle U_1 \rangle/\partial x_2$, $\mathcal{P}/\varepsilon \approx 1$, and $\alpha \equiv -\langle u_1 u_2 \rangle/k \approx 0.3$, obtain the following results:

$$Sk/\varepsilon = \frac{1}{\alpha}\frac{\mathcal{P}}{\varepsilon} \approx 3, \tag{6.280}$$

$$St_\eta = \frac{1}{\alpha} \frac{P}{\varepsilon} \mathrm{Re}_L^{-1/2} \approx 3\mathrm{Re}_L^{-1/2},$$

$$= \sqrt{\frac{20}{3}} \frac{1}{\alpha} \frac{P}{\varepsilon} \mathrm{R}_\lambda^{-1} \approx 9\mathrm{R}_\lambda^{-1}, \qquad (6.281)$$

$$L_S \equiv S^{-3/2} \varepsilon^{1/2} = \left(\frac{P}{\varepsilon}\right)^{-3/2} \alpha^{3/2} L \approx \tfrac{1}{6} L, \qquad (6.282)$$

$$u_S \equiv (\varepsilon/S)^{1/2} = \alpha^{1/2} \left(\frac{P}{\varepsilon}\right)^{-1/2} k^{1/2} \approx \tfrac{1}{2} k^{1/2}. \qquad (6.283)$$

6.6 The spectral view of the energy cascade

In Sections 6.2–6.5 we introduced several statistics used to quantify turbulent motions on various scales, and we examined these statistics through experimental data, the Kolmogorov hypotheses, and a simple model spectrum. We are now in a position to provide a fuller account of the energy cascade than is given in Section 6.1. This section therefore serves to summarize and consolidate the preceding development.

Energy-containing motions

We again consider very-high-Reynolds-number flow, so that there is a clear separation between the energy-containing and dissipative scales of motion (i.e., $L_{11}/\eta \sim \mathrm{Re}^{3/4} \gg 1$). The bulk of the turbulent kinetic energy is contained in motions of lengthscale ℓ, comparable to the integral lengthscale L_{11} ($6L_{11} > \ell > \tfrac{1}{6}L_{11} = \ell_{EI}$, say), whose characteristic velocity is of order $k^{1/2}$. Since their size is comparable to the flow dimensions $\mathcal{L}$, these large-scale motions can be strongly influenced by the geometry of the flow. Furthermore, their timescale $L_{11}/k^{1/2}$ is large compared with the mean-flow timescale (see Table 5.2 on page 131), so that they are significantly affected by the flow's history. In other words, and in contrast to the universal equilibrium range, the energy-containing motions do not have a universal form brought about by a statistical equilibrium.

All of the anisotropy is confined to the energy-containing motions, and consequently so also is all of the production of turbulence. On the other hand, the viscous dissipation is negligible. Instead, during the initial steps in the cascade, energy is removed by inviscid processes and transferred to smaller scales ($\ell < \ell_{EI}$) at a rate $\mathcal{T}_{EI}$, which scales as $k^{3/2}/L_{11}$. This transfer process depends on the non-universal energy-containing motions, and consequently the non-dimensional ratio $\mathcal{T}_{EI}/(k^{3/2}/L_{11})$ is not universal.

The energy-spectrum balance

For homogeneous turbulence (with imposed mean velocity gradients) this picture is quantified by the balance equation for the energy-spectrum function $E(\kappa, t)$. This equation (derived in detail in Hinze (1975) and Monin and Yaglom (1975)) can be written

$$\frac{\partial}{\partial t} E(\kappa, t) = \mathcal{P}_\kappa(\kappa, t) - \frac{\partial}{\partial \kappa} \mathcal{T}_\kappa(\kappa, t) - 2\nu \kappa^2 E(\kappa, t). \qquad (6.284)$$

The three terms on the right-hand side represent production, spectral transfer, and dissipation.

The production spectrum $\mathcal{P}_\kappa$ is given by the product of the mean velocity gradients $\partial \langle U_i \rangle / \partial x_j$ and an anisotropic part of the spectrum tensor. The contribution to the production from the wavenumber range (κ_a, κ_b) is denoted by

$$\mathcal{P}_{(\kappa_a, \kappa_b)} = \int_{\kappa_a}^{\kappa_b} \mathcal{P}_\kappa \, d\kappa, \qquad (6.285)$$

and, to the extent that all of the anisotropy is contained in the energy-containing range, we therefore have

$$\mathcal{P} = \mathcal{P}_{(0,\infty)} \approx \mathcal{P}_{(0,\kappa_{EI})}, \qquad (6.286)$$

$$\mathcal{P}_{(\kappa_{EI},\infty)}/\mathcal{P} \ll 1. \qquad (6.287)$$

In the second term on the right-hand side of Eq. (6.284), $\mathcal{T}_\kappa(\kappa)$ is the *spectral energy transfer rate*: it is the net rate at which energy is transferred from modes of lower wavenumber than κ to those with wavenumbers higher than κ. This is simply related to $\mathcal{T}(\ell)$ – the rate of transfer of energy from eddies larger than ℓ to those smaller than ℓ – by

$$\mathcal{T}(\ell) = \mathcal{T}_\kappa(2\pi/\ell). \qquad (6.288)$$

The rate of gain of energy in the wavenumber range (κ_a, κ_b) due to this spectral transfer is

$$\int_{\kappa_a}^{\kappa_b} -\frac{\partial}{\partial \kappa} \mathcal{T}_\kappa(\kappa) \, d\kappa = \mathcal{T}_\kappa(\kappa_a) - \mathcal{T}_\kappa(\kappa_b). \qquad (6.289)$$

Since $\mathcal{T}_\kappa$ vanishes at zero and infinite wavenumber, this transfer term makes no contribution to the balance of turbulent kinetic energy k.

An exact expression for $\mathcal{T}_\kappa$ can be obtained from the Navier–Stokes equations (see, e.g., Hinze (1975)). There are two contributions: one resulting from interactions of triads of wavenumber modes, similar to Eq. (6.162); the other (examined in detail in Section 11.4) expressing a primarily kinematic

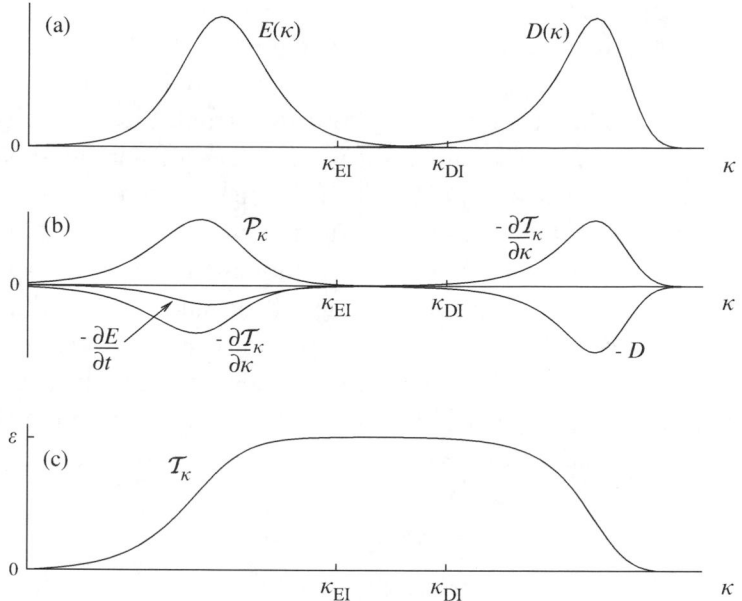

Fig. 6.28. For homogeneous turbulence at very high Reynolds number, sketches of (a) the energy and dissipation spectra, (b) the contributions to the balance equation for $E(\kappa, t)$ (Eq. (6.284)), and (c) the spectral energy-transfer rate.

effect that mean velocity gradients have on the spectrum. The final term in Eq. (6.284) is the dissipation spectrum $D(\kappa, t) = 2\nu\kappa^2 E(\kappa, t)$.

Figure 6.28 is a sketch of the quantities appearing in the balance equation for $E(\kappa, t)$. In the energy-containing range, all the terms are significant except for dissipation. With the approximations $k_{(0,\kappa_{EI})} \approx k$, $\varepsilon_{(0,\kappa_{EI})} \approx 0$ and $\mathcal{P}_{(0,\kappa_{EI})} \approx \mathcal{P}$, when it is integrated over the energy-containing range $(0, \kappa_{EI})$, Eq. (6.284) yields

$$\frac{\mathrm{d}k}{\mathrm{d}t} \approx \mathcal{P} - \mathcal{T}_{EI}, \qquad (6.290)$$

where $\mathcal{T}_{EI} = \mathcal{T}_\kappa(\kappa_{EI})$. In the inertial subrange, spectral transfer is the only significant process so that (when it is integrated from κ_{EI} to κ_{DI}) Eq. (6.284) yields

$$0 \approx \mathcal{T}_{EI} - \mathcal{T}_{DI}, \qquad (6.291)$$

where $\mathcal{T}_{DI} = \mathcal{T}_\kappa(\kappa_{DI})$. Whereas in the dissipation range, spectral transfer balances dissipation so that (when it is integrated from κ_{DI} to infinity) Eq. (6.284) yields

$$0 \approx \mathcal{T}_{DI} - \varepsilon. \qquad (6.292)$$

When they are added together, the last three equations give (without approximation) the turbulent-kinetic-energy equation $dk/dt = \mathcal{P} - \varepsilon$.

The above equations again highlight the essential characteristics of the energy cascade. The rate of energy transfer from the energy-containing range $\mathcal{T}_{EI}$ depends, in a non-universal way, on several factors including the mean velocity gradients and the details of the energy-containing range of the spectrum. However, this transfer rate then establishes an inertial subrange of universal character with $\mathcal{T}_\kappa(\kappa) = \mathcal{T}_{EI}$; and finally the high wavenumber part of the spectrum dissipates the energy at the same rate as that at which it receives it. Thus both $\mathcal{T}_{DI}$ and ε are determined by, and are equal to, $\mathcal{T}_{EI}$. Quite often, when 'dissipation' is being considered – e.g., in characterizing the inertial range spectrum as $E(\kappa) = C\varepsilon^{2/3}\kappa^{-5/3}$ – it is conceptually superior to consider $\mathcal{T}_{EI}$ in place of ε.

The cascade timescale

An analogy of questionable validity is that the flow of energy in the inertial subrange is like the flow of an incompressible fluid through a variable-area duct. The constant flow rate is $\mathcal{T}_{EI}$ (in units of energy per time) while the capacity of the cascade (analogous to the duct's area) is $E(\kappa)$ (in units of energy per wavenumber). So the speed (in units of wavenumber per time) at which the energy travels through the cascade is

$$\dot{\kappa}(\kappa) = \mathcal{T}_{EI}/E(\kappa) = \kappa^{5/3}\varepsilon^{1/3}/C, \qquad (6.293)$$

the latter expression being obtained from the Kolmogorov spectrum and the substitution $\mathcal{T}_{EI} = \varepsilon$. Notice that this speed increases rapidly with increasing wavenumber.

It follows from the solution of the equation $d\kappa/dt = \dot{\kappa}$ that, according to this analogy, the time $t_{(\kappa_a,\kappa_b)}$ that it takes for energy to flow from wavenumber κ_a to the higher wavenumber κ_b is

$$t_{(\kappa_a,\kappa_b)} = \tfrac{3}{2}C\varepsilon^{-1/3}\left(\kappa_a^{-2/3} - \kappa_b^{-2/3}\right)$$
$$= \tau\tfrac{3}{2}C\left[(\kappa_a L)^{-2/3} - (\kappa_b L)^{-2/3}\right]. \qquad (6.294)$$

With the relations $\kappa_{EI} = 2\pi/\ell_{EI}, \ell_{EI} = \tfrac{1}{6}L_{11}$, and $L_{11}/L \approx 0.4$, this formula yields

$$t_{(\kappa_{EI},\infty)} \approx \tfrac{1}{10}\tau, \qquad (6.295)$$

giving the estimate that the lifetime of the energy once it enters the inertial subrange is just a tenth of its total lifetime $\tau = k/\varepsilon$.

Spectral energy-transfer models

In the universal equilibrium range ($\kappa > \kappa_{EI}$), the balance in the spectral energy equation (Eq. (6.284)) is between the energy transfer and the dissipation, see Fig. 6.28(b). Hence (at any time t) Eq. (6.284) reduces to

$$0 = -\frac{d}{d\kappa}\mathcal{T}_\kappa(\kappa) - 2\nu\kappa^2 E(\kappa). \tag{6.296}$$

During the period from 1940 to 1970 many models for the spectral energy transfer rate $\mathcal{T}_\kappa$ were proposed, which allow the form of the spectrum $E(\kappa)$ to be deduced from Eq. (6.296). The proposals of Obukhov (1941), Heisenberg (1948), and many others are reviewed by Panchev (1971). Appropriate to the physics of the cascade, most of these models are non-local in the sense that $\mathcal{T}_\kappa(\kappa)$ is postulated to depend on $E(\kappa')$, for $\kappa' \neq \kappa$. However, to illustrate the approach, we consider the simple local model due to Pao (1965). Similar to Eq. (6.293), the speed of energy transfer $\dot\kappa(\kappa)$ is defined by

$$\dot\kappa(\kappa) \equiv \mathcal{T}_\kappa(\kappa)/E(\kappa). \tag{6.297}$$

The single (though strong) assumption in Pao's model is that $\dot\kappa$ depends solely on ε and κ. Dimensional analysis then determines

$$\mathcal{T}_\kappa(\kappa) = E(\kappa)\dot\kappa(\kappa) = E(\kappa)\alpha^{-1}\varepsilon^{1/3}\kappa^{5/3}, \tag{6.298}$$

where α is a constant. With this expression for $\mathcal{T}_\kappa$, Eq. (6.296) can be integrated (see Exercise 6.36) to yield the Pao spectrum

$$E(\kappa) = C\varepsilon^{2/3}\kappa^{-5/3}\exp\left[-\tfrac{3}{2}C(\kappa\eta)^{4/3}\right], \tag{6.299}$$

cf. Eq. (6.254). This is compared with experimental data in Fig. 6.15.

EXERCISE _____

6.36   Substitute Eq. (6.298) into Eq. (6.296) to obtain

$$\frac{d}{d\kappa}\ln\left[E(\kappa)\kappa^{5/3}\right] = -2\alpha\nu\varepsilon^{-1/3}\kappa^{1/3}, \tag{6.300}$$

and then integrate to obtain

$$\begin{aligned} E(\kappa) &= \beta\kappa^{-5/3}\exp\left(-\tfrac{3}{2}\alpha\nu\varepsilon^{-1/3}\kappa^{4/3}\right), \\ &= \beta\kappa^{-5/3}\exp\left[-\tfrac{3}{2}\alpha(\kappa\eta)^{4/3}\right], \end{aligned} \tag{6.301}$$

where β is a (dimensional) constant of integration. Argue that, for consistency with the Kolmogorov spectrum (for small $\kappa\eta$), β is required to be $\beta = C\varepsilon^{2/3}$. Show that the dissipation given by Eq. (6.301) is

$$\int_0^\infty 2\nu\kappa^2 E(\kappa)\,d\kappa = \varepsilon^{1/3}\beta/\alpha, \tag{6.302}$$

and hence that α is identical to the Kolmogorov constant C. Confirm that, with $\beta = C\varepsilon^{2/3}$ and $\alpha = C$, Eq. (6.301) yields the Pao spectrum, Eq. (6.299).

6.7 Limitations, shortcomings, and refinements

In considerations of turbulent motions of various scales, the notions of the energy cascade, vortex stretching, and the Kolmogorov hypotheses provide an invaluable conceptual framework. However, both conceptually and empirically, there are some shortcomings. Indeed, since around 1960, a major line of research (theoretical, experimental, and computational) has been to examine these shortcomings and to attempt to improve on the Kolmogorov hypotheses. While it is appropriate to provide some discussion of these issues here, it should be appreciated that they have minor impact on the study and modelling of turbulent flows. This is simply because the small scales ($\ell < \ell_{EI}$) contain little energy (and less anisotropy) and so have little direct effect on the flow.

6.7.1 The Reynolds number

A limitation of the Kolmogorov hypotheses is that they apply only to high-Reynolds-number flows, and that a criterion for 'sufficiently high Reynolds number' is not provided. Many laboratory and practical flows have reasonably high Reynolds number (e.g., $\mathrm{Re} \approx 10{,}000, \mathrm{R}_\lambda \approx 150$), and yet even the motions on the dissipative scales are found to be anisotropic (see, e.g., George and Hussein (1991)).

Close scrutiny of the inertial range spectra show that the Kolmogorov $-\frac{5}{3}$ spectrum is approached slowly as the Reynolds number increases. From experiments on grid turbulence at quite high Reynolds number ($\mathrm{R}_\lambda \approx 50$–$500$) Mydlarski and Warhaft (1998) conclude that the inertial-range spectrum is indeed a power law, $E(\kappa) \sim \kappa^{-p}$, but that the exponent p depends on R_λ (see Fig. 6.29). As the curve in Fig. 6.29 illustrates, it is quite plausible that p approaches $\frac{5}{3}$ at very large R_λ, but, at $\mathrm{R}_\lambda \approx 200$ (which is typical of many laboratory flows), p is around 1.5.

It is certainly an oversimplification to suppose that the energy cascade consists of the one-way transfer of energy from eddies of size ℓ to those

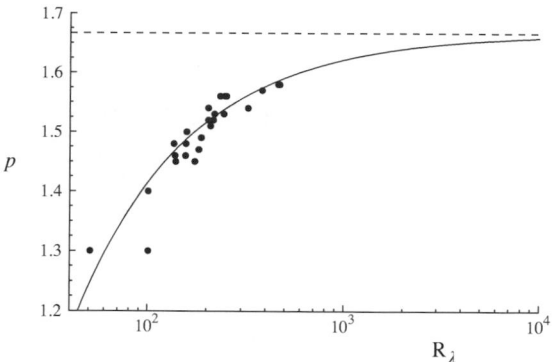

Fig. 6.29. The spectrum power-law exponent p $(E(\kappa) \sim \kappa^{-p})$ as a function of the Reynolds number in grid turbulence: symbols, experimental data of Mydlarski and Warhaft (1998); dashed line, $p = \frac{5}{3}$; solid line, empirical curve $p = \frac{5}{3} - 8R_\lambda^{-3/4}$.

of a somewhat smaller size (e.g., $\frac{1}{2}\ell$), and that this energy transfer depends solely on motions of size ℓ. Spectral energy transfer is almost impossible to measure experimentally (but see Kellogg and Corrsin (1980)), whereas it can be extracted from direct numerical simulations (which are restricted to moderate or low Reynolds numbers). The picture that emerges from the DNS study of Domaradzki and Rogallo (1990) is that there is energy transfer both to smaller and to larger scales, with the net transfer being toward smaller scales. In wavenumber space, the energy transfer is accomplished by *triad interactions*, that is, interactions among three modes with wavenumbers κ^a, κ^b, and κ^c such that $\kappa^a + \kappa^b + \kappa^c = 0$ (see Eq. (6.162)). The DNS results suggest that the transfer is predominantly local (e.g., between modes a and b with $|\kappa^a| \approx |\kappa^b|$), but that it is effected by interactions with a third mode of significantly smaller wavenumber (i.e., $|\kappa^c| \ll |\kappa^a|$). (Further studies have been performed by Domaradzki (1992) and Zhou (1993).)

6.7.2 Higher-order statistics

All the experimental data considered so far in this chapter pertain to second-order velocity statistics (i.e., statistics that are quadratic in velocity). These are the most important quantities since they determine the kinetic energy and the Reynolds stresses.

The simplest examples of higher-order statistics are the normalized velocity-derivative moments

$$M_n = \left\langle \left(\frac{\partial u_1}{\partial x_1}\right)^n \right\rangle \Big/ \left\langle \left(\frac{\partial u_1}{\partial x_1}\right)^2 \right\rangle^{n/2}. \tag{6.303}$$

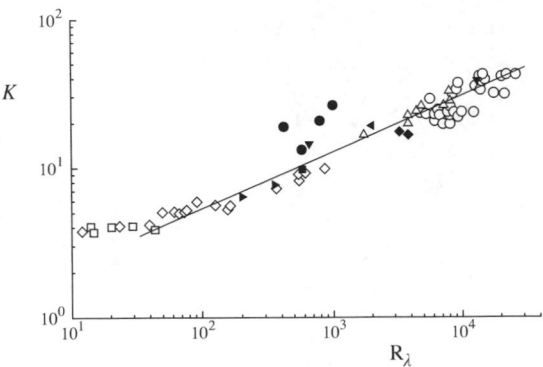

Fig. 6.30. Measurements (symbols) compiled by Van Atta and Antonia (1980) of the velocity-derivative kurtosis as a function of Reynolds number. The solid line is $K \sim R_\lambda^{3/8}$.

For $n = 3$ and $n = 4$ these are the velocity-derivative skewness S and kurtosis K. (Recall that, for a Gaussian random variable, S is zero and K is 3.) According to the Kolmogorov hypotheses, for each n, M_n is a universal constant. However, it is found that S and K are not constant, but increase with Reynolds number. Figure 6.30, for example, shows that measurements of the kurtosis increase from $K \approx 4$ in low-Reynolds-number grid turbulence to $K \approx 40$ at the highest Reynolds numbers measured. In contrast to the Reynolds-number effects discussed previously, here K does not appear to reach an asymptote, but instead the data are consistent with an indefinite increase, possibly as $K \sim R_\lambda^{3/8}$.

The velocity-derivative moments M_n (e.g., the skewness S and the kurtosis K) pertain to the dissipative range. The simplest higher-order statistics pertaining to the inertial subrange are the longitudinal velocity structure functions,

$$D_n(r) \equiv \langle (\Delta_r u)^n \rangle, \tag{6.304}$$

i.e., the moments of the velocity difference defined (at x, t) by

$$\Delta_r u \equiv U_1(x + e_1 r, t) - U_1(x, t). \tag{6.305}$$

Recall that the second- and third-order structure functions $D_2(r)$ and $D_3(r)$ are considered in Sections 6.2 and 6.3, where they are denoted by D_{LL} and $D_{LLL}(r)$.

According to Kolmogorov's second hypothesis, for inertial-range separations ($L \gg r \gg \eta$) $D_n(r)$ depends only on ε and r, and hence dimensional

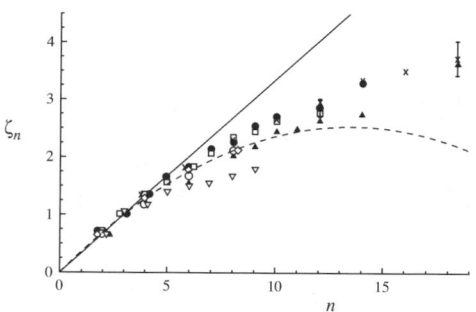

Fig. 6.31. Measurements (symbols) compiled by Anselmet *et al.* (1984) of the longitudinal velocity structure function exponent ζ_n in the inertial subrange, $D_n(r) \sim r^{\zeta_n}$. The solid line is the Kolmogorov (1941) prediction, $\zeta_n = \frac{1}{3}n$; the dashed line is the prediction of the refined similarity hypothesis, Eq. (6.323) with $\mu = 0.25$.

analysis yields

$$D_n(r) = C_n(\varepsilon r)^{n/3}, \tag{6.306}$$

where $C_2, C_3, \ldots$ are constants. Measurements confirm this prediction for $n = 2$ (with $C_2 = 2.0$, see Fig. 6.5 on page 194); and also for $n = 3$, for which the Kolmogorov $\frac{4}{5}$ law (Eq. (6.88)) yields $C_3 = -\frac{4}{5}$. Higher-order structure functions, for n up to 18 have been measured by Anselmet *et al.* (1984). In the inertial subrange, a power-law dependence on r is observed,

$$D_n(r) \sim r^{\zeta_n}, \tag{6.307}$$

but the measured exponents, shown in Fig. 6.31, differ from the Kolmogorov prediction Eq. (6.306), i.e., $\zeta_n = n/3$.

It is instructive to examine the PDFs that underlie these higher-order moments. Figure 6.32 shows the standardized PDF of $\partial u_1/\partial x_1$ measured in the atmospheric boundary layer – a very-high-Reynolds-number flow. This PDF is denoted by $f_Z(z)$, where Z is the standardized derivative

$$Z \equiv \frac{\partial u_1}{\partial x_1} \Big/ \left\langle \left(\frac{\partial u_1}{\partial x_1}\right)^2 \right\rangle^{1/2}. \tag{6.308}$$

Observe that the tails of the distribution (beyond four standard derivations, say) are close to straight lines on this plot, corresponding to *exponential tails*: the dashed lines shown in Fig. 6.32 are the approximations

$$f_Z(z) = 0.2 \exp(-1.1|z|), \quad \text{for } z > 4, \tag{6.309}$$

$$f_Z(z) = 0.2 \exp(-1.0|z|), \quad \text{for } z < -4. \tag{6.310}$$

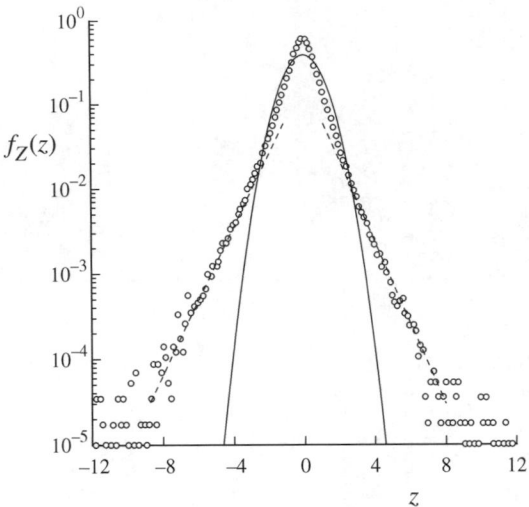

Fig. 6.32. The PDF $f_Z(z)$ of the normalized velocity derivative $Z \equiv (\partial u_1/\partial x_1)/\langle(\partial u_1/\partial x_1)^2\rangle^{1/2}$ measured by Van Atta and Chen (1970) in the atmospheric boundary layer (high Re). The solid line is a Gaussian; the dashed lines correspond to exponential tails (Eqs. (6.309) and (6.310)).

(Note that the slower decay for negative z is consistent with the observed negative skewness S.) This exponential decay is of course much slower than that of the standardized Gaussian, which is also shown in Fig. 6.32.

What is the significance of these tails? First, they correspond to rare events: taking Eq. (6.310) as an approximation for large $|z|$, it follows that there is less than 0.3% probability of $|Z|$ exceeding 5. However, these low-probability tails can make vast contributions to higher moments. Table 6.3 shows the tail contribution ($|Z| > 5$) to the moments

$$M_n^{(5)} \equiv 2\int_5^\infty z^n f_Z(z)\,\mathrm{d}z, \qquad (6.311)$$

for $f_Z(z)$ being given by Eq. (6.310). Observe, for example, that the contribution to the superskewness M_6 is 220, compared with the Gaussian value of 15. Some laboratory measurements of the PDF of $\partial u_1/\partial x_1$ and its moments over a range of Reynolds number are described by Belin *et al.* (1997).

6.7.3 Internal intermittency

The discrepancies between the Kolmogorov predictions and the experimental values of the higher-order moments M_n and $D_n(r)$ are attributed to the phenomenon of *internal intermittency*, and are largely accounted for in the *refined similarity hypotheses* proposed by Obukhov (1962) and Kolmogorov

Table 6.3. *Contributions* $M_n^{(5)}$ *from the exponential tails* ($|Z| > 5$) *of the PDF of Z to the moments* M_n *according to Eqs.* (6.310) *and* (6.311)

Moment n	Tail contribution $M_n^{(5)}$	Gaussian value M_n
0	0.003	1
2	0.1	1
4	4.2	3
6	220	15
8	1.5×10^4	105
10	1.4×10^6	945

(1962). To describe these ideas it is necessary to introduce several quantities related to dissipation.

The *instantaneous dissipation* $\varepsilon_0(x, t)$ is defined by

$$\varepsilon_0 = 2 v s_{ij} s_{ij}, \tag{6.312}$$

and, for a given distance r, the average of ε_0 over a sphere $\mathcal{V}(r)$ of radius r is given by

$$\varepsilon_r(x, t) = \frac{3}{4 \pi r^3} \iiint_{\mathcal{V}(r)} \varepsilon_0(x + r, t) \, dr. \tag{6.313}$$

Unfortunately, for practical purposes, it is impossible to measure ε_0 and ε_r. Instead, one-dimensional *surrogates* are used, namely

$$\hat{\varepsilon}_0 = 15 v \left(\frac{\partial u_1}{\partial x_1} \right)^2, \tag{6.314}$$

$$\hat{\varepsilon}_r(x, t) \equiv \frac{1}{r} \int_0^r \hat{\varepsilon}_0(x + e_1 r, t) \, dr. \tag{6.315}$$

In locally isotropic turbulence, each of these quantities has mean ε. It is generally supposed that the statistics of ε_0 and $\hat{\varepsilon}_0$ are qualitatively similar, but there are certainly substantial quantitative differences.

As early as 1949 (e.g., Batchelor and Townsend (1949)) experiments revealed that the instantaneous dissipation $\hat{\varepsilon}_0$ intermittently attains very large values. The peak value of $\hat{\varepsilon}_0/\varepsilon$ observed increases with Reynolds number: in a laboratory experiment (moderate R_λ) Meneveau and Sreenivasan (1991) observed a peak value of $\hat{\varepsilon}_0/\varepsilon \approx 15$, whereas in the atmosphere's surface layer (high R_λ) the corresponding observation was 50. Kolmogorov (1962) conjectured that mean-square dissipation fluctuations scale as

$$\langle \varepsilon_0^2 \rangle / \varepsilon^2 \sim (L/\eta)^\mu, \tag{6.316}$$

and similarly

$$\langle \varepsilon_r^2 \rangle / \varepsilon^2 \sim (L/r)^{\mu}, \quad \text{for } \eta < r \ll L, \tag{6.317}$$

where μ is a positive constant – *the intermittency exponent*. Experiments confirm Eq. (6.317) for the surrogate $\hat{\varepsilon}_r$, and determine $\mu = 0.25 \pm 0.05$ (Sreenivasan and Kailasnath 1993). It may be seen from the definition of $\hat{\varepsilon}_0$ (Eq. (6.314)) that $\langle \hat{\varepsilon}_0^2 \rangle / \varepsilon^2$ is precisely the velocity-derivative kurtosis K. Thus, taking $\mu = \frac{1}{4}$ and recalling the relation $L/\eta \sim R_{\lambda}^{3/2}$, Eq. (6.316) for the surrogate leads to

$$K \sim R_{\lambda}^{3\mu/2} = R_{\lambda}^{3/8}, \tag{6.318}$$

which is consistent with experimental data (see Fig. 6.30).

6.7.4 Refined similarity hypotheses

Considering the velocity increment $\Delta_r u$ (Eq. (6.305)), the first (original) Kolmogorov hypothesis states that the statistics of $\Delta_r u$ (for $r \ll L$) are universal, determined by the mean dissipation ε and ν. The idea behind the refined similarity hypotheses (Obukhov 1962, Kolmogorov 1962) is that $\Delta_r u$ is influenced not by the mean dissipation ε, but by the local value (averaged over the distance r), namely ε_r. Thus the first *refined* similarity hypothesis is that (for $r \ll L$) the statistics of $\Delta_r u$ *conditional on* ε_r are universal, determined by ε_r and ν. The second refined similarity hypothesis is that, for $\eta \ll r \ll L$, these conditional statistics depend only on ε_r, independent of ν.

Application of the second refined similarity hypothesis to the moments of $\Delta_r u$ yields

$$\langle (\Delta_r u)^n | \varepsilon_r = \epsilon \rangle = C_n (\epsilon r)^{n/3}, \tag{6.319}$$

where C_n are universal constants (cf. Eq. (6.306)), and ϵ is a sample-space variable. The structure function $D_n(r)$, which is the unconditional mean, is then obtained as

$$D_n(r) = \langle (\Delta_r u)^n \rangle = \langle \langle (\Delta_r u)^n | \varepsilon_r \rangle \rangle$$
$$= C_n \langle \varepsilon_r^{n/3} \rangle r^{n/3}. \tag{6.320}$$

For $n = 3$, since $\langle \varepsilon_r \rangle$ equals ε, the original and the refined hypotheses make the same prediction; which, with $C_3 = -\frac{4}{5}$, is the Kolmogorov $\frac{4}{5}$ law (Eq. (6.88)). For $n = 6$, and using Eq. (6.317) for $\langle \varepsilon_r^2 \rangle$, the prediction is

$$D_6(r) \sim \varepsilon^2 L^{\mu} r^{2-\mu}, \tag{6.321}$$

i.e., a power law in r (Eq. (6.307)) with exponent $\zeta_6 = 2 - \mu = 1.75$ for $\mu = 0.25$. This value of ζ_6 is in agreement with the data shown in Fig. 6.31.

For other values of n, $\langle \varepsilon_r^{n/3} \rangle$ can be determined from the PDF of ε_r. Obukhov (1962) and Kolmogorov (1962) conjectured that ε_r is log-normally distributed, i.e., $\ln(\varepsilon_r/\varepsilon)$ has a Gaussian distribution. From this assumption of log-normality, and from the scaling of $\langle \varepsilon_r^2 \rangle$ (Eq. (6.317)), it follows (see Exercise 6.37) that the moments of ε_r scale as

$$\langle \varepsilon_r^m \rangle / \varepsilon^m \sim (L/r)^{m(m-1)\mu/2}. \tag{6.322}$$

Consequently, the structure function (Eq. (6.320)) is predicted to scale as $D_n(r) \sim r^{\zeta_n}$, with

$$\zeta_n = \tfrac{1}{3}n[1 - \tfrac{1}{6}\mu(n-3)]. \tag{6.323}$$

For n not too large ($n \leq 10$, say), this prediction is in reasonable agreement with the data shown in Fig. 6.31. For large n, the discrepancies are attributed to the deficiencies in the log-normal assumption, which has been roundly criticized by Mandelbrot (1974) and others.

For the second-order structure function $D_2(r)$, Eq. (6.323) yields

$$\zeta_2 = \tfrac{2}{3} + \tfrac{1}{9}\mu \approx \tfrac{2}{3} + \tfrac{1}{36}. \tag{6.324}$$

Correspondingly, the inertial-range spectrum is predicted to be a power law $E(\kappa) \sim \kappa^{-p}$ with

$$p = \tfrac{5}{3} + \tfrac{1}{9}\mu \approx \tfrac{5}{3} + \tfrac{1}{36}. \tag{6.325}$$

Hence the predicted modification to the $-\tfrac{5}{3}$ spectrum is very small.

For the velocity-derivative moments, the refined hypotheses yield

$$\left\langle \left(\frac{\partial u_1}{\partial x_1} \right)^n \middle| \varepsilon_r = \epsilon \right\rangle = \bar{C}_n \left(\frac{\epsilon}{\nu} \right)^{n/2}, \tag{6.326}$$

where $\bar{C}_n$ are constants, and hence

$$M_n \equiv \frac{\left\langle \left(\dfrac{\partial u_1}{\partial x_1} \right)^n \right\rangle}{\left\langle \left(\dfrac{\partial u_1}{\partial x_1} \right)^2 \right\rangle^{n/2}} = \frac{\bar{C}_n \langle \varepsilon_r^{n/2} \rangle}{(\bar{C}_2 \varepsilon)^{n/2}}. \tag{6.327}$$

Using the log-normal assumption to evaluate the moments of ε_r (Eq. (6.322)), for the skewness and kurtosis we obtain

$$-S \sim (L/r)^{3\mu/8}, \tag{6.328}$$

$$K \sim (L/r)^{\mu}. \tag{6.329}$$

Hence, irrespective of the value of μ, a prediction is

$$-S \sim K^{3/8}. \tag{6.330}$$

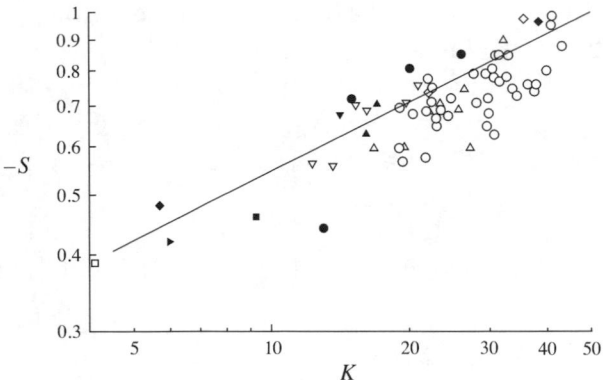

Fig. 6.33. Measurements of the velocity-derivative skewness S and kurtosis K compiled by Van Atta and Antonia (1980). The line is $-S \sim K^{3/8}$.

Figure 6.33 shows that the data are indeed consistent with this prediction, although there is considerable scatter.

The research literature in this area is vast. Useful reviews are provided by Nelkin (1994) and Stolovitzky *et al.* (1995).

EXERCISE _____

6.37   The positive random variable ε_r is log-normally distributed if

$$\phi \equiv \ln(\varepsilon_r/\varepsilon_{\text{ref}}) \tag{6.331}$$

is Gaussian, where ε_{ref} is a positive constant. Let ϕ be Gaussian with variance σ^2.

(a) Show that the moments of ε_r are

$$\langle \varepsilon_r^n \rangle = \varepsilon_{\text{ref}}^n \exp(n\langle \phi \rangle + \tfrac{1}{2}n^2\sigma^2). \tag{6.332}$$

(b) Show that, if ε_{ref} is taken to be $\langle \varepsilon_r \rangle$, then

$$\langle \varepsilon_r^n \rangle/\langle \varepsilon_r \rangle^n = \exp\left[\tfrac{1}{2}\sigma^2 n(n-1)\right]. \tag{6.333}$$

(c) If the mean square of ε_r depends on the parameter r according to

$$\langle \varepsilon_r^2 \rangle/\langle \varepsilon_r \rangle^2 = A(L/r)^\mu, \tag{6.334}$$

where A, L, and μ are positive constants, show that

$$\sigma^2 = \ln A + \mu \ln(L/r), \tag{6.335}$$

$$\langle \varepsilon_r^n \rangle/\langle \varepsilon_r \rangle^n = A^{n(n-1)/2}(L/r)^{\mu n(n-1)/2}. \tag{6.336}$$

6.7.5 Closing remarks

A great deal of research work has focused on the shortcomings of the original Kolmogorov hypotheses, and on the details of the refined similarity hypotheses. It is again emphasized, however, that, in the context of the mean velocity field and Reynolds stresses in turbulent flows, these issues are of minor significance. The turbulence energy and anisotropy are predominantly contained in the large-scale motions. Internal intermittency, on the other hand, concerns rare events that are manifest only in high-order statistics of small-scale quantities. In the context of the energy cascade and turbulent flows, of much more importance than internal intermittency is the question of what determines the rate of energy transfer $\mathcal{T}_{EI}$ from the energy-containing scales. It is a difficult question, because the large scales are not universal. This question is addressed in Chapter 10.

7

Wall flows

In contrast to the free shear flows considered in Chapter 5, most turbulent flows are bounded (at least in part) by one or more solid surfaces. Examples include *internal flows* such as the flow through pipes and ducts; *external flows* such as the flow around aircraft and ships' hulls; and flows in the environment such as the atmospheric boundary layer, and the flow of rivers.

We consider three of the simplest of these flows (sketched in Fig. 7.1), namely: fully developed channel flow; fully developed pipe flow; and the flat-plate boundary layer. In each of these flows the mean velocity vector is (or is nearly) parallel to the wall, and, as we shall see, the near-wall behaviors in each of these cases are very similar. These simple flows are of practical importance and played a prominent role in the historical development of the study of turbulent flows.

Central issues are the forms of the mean velocity profiles, and the *friction laws*, which describe the shear stress exerted by the fluid on the wall. In addition the *mixing length* is introduced in Section 7.1.7; the balance equations for the Reynolds stresses are derived and examined in Section 7.3.5; and the *proper orthogonal decomposition* (POD) is described in Section 7.4.

7.1 Channel flow

7.1.1 A description of the flow

As sketched in Fig. 7.1, we consider the flow through a rectangular duct of height $h = 2\delta$. The duct is long ($L/\delta \gg 1$) and has a large aspect ratio ($b/\delta \gg 1$). The mean flow is predominantly in the axial ($x = x_1$) direction, with the mean velocity varying mainly in the cross-stream ($y = x_2$) direction. The bottom and top walls are at $y = 0$ and $y = 2\delta$, respectively, with the mid-plane being $y = \delta$. The extent of the channel in the spanwise ($z = x_3$) direction is large compared with δ so that (remote from the end walls) the

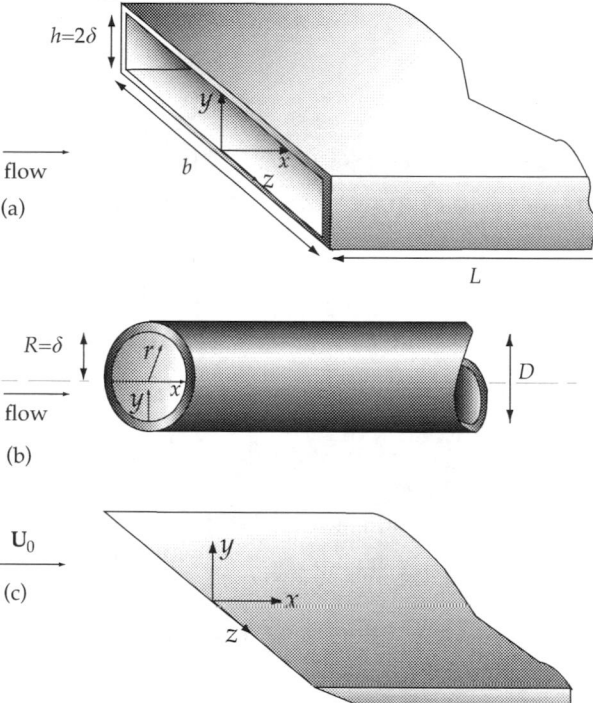

Fig. 7.1. Sketches of (a) channel flow, (b) pipe flow, and (c) a flat-plate boundary layer.

flow is statistically independent of z. The centerline is defined by $y = \delta$, $z = 0$. The velocities in the three coordinate directions are $(U, V, W) = (U_1, U_2, U_3)$ with fluctuations $(u, v, w) = (u_1, u_2, u_3)$. The mean cross-stream velocity $\langle W \rangle$ is zero.

Near the entry of the duct $(x = 0)$ there is a flow-development region. We, however, confine our attention to the *fully developed* region (large x), in which velocity statistics no longer vary with x. Hence the fully developed channel flow being considered is statistically stationary and statistically one-dimensional, with velocity statistics depending only on y. Experiments confirm the natural expectation that the flow is statistically symmetric about the mid-plane $y = \delta$: the statistics of (U, V, W) at y are the same as those of $(U, -V, W)$ at $2\delta - y$ (see Eqs. (4.31)–(4.34)).

The Reynolds numbers used to characterize the flow are

$$\mathrm{Re} \equiv (2\delta)\bar{U}/\nu, \qquad (7.1)$$

$$\mathrm{Re}_0 \equiv U_0\delta/\nu, \qquad (7.2)$$

where $U_0 = \langle U \rangle_{y=\delta}$ is the centerline velocity, and $\bar{U}$ is the bulk velocity

$$\bar{U} \equiv \frac{1}{\delta} \int_0^\delta \langle U \rangle \, dy. \tag{7.3}$$

The flow is laminar for Re < 1,350, and fully turbulent for Re > 1,800, although transitional effects are evident up to Re = 3,000 (see Patel and Head (1969)).

7.1.2 The balance of mean forces

The mean continuity equation reduces to

$$\frac{d\langle V \rangle}{dy} = 0, \tag{7.4}$$

since $\langle W \rangle$ is zero, and $\langle U \rangle$ is independent of x. With the boundary condition $\langle V \rangle_{y=0}$, this dictates that $\langle V \rangle$ is zero for all y, so that the boundary condition at the top wall $\langle V \rangle_{y=2\delta} = 0$ is also satisfied.

The lateral mean-momentum equation reduces to

$$0 = -\frac{d}{dy}\langle v^2 \rangle - \frac{1}{\rho}\frac{\partial \langle p \rangle}{\partial y}, \tag{7.5}$$

which, with the boundary condition $\langle v^2 \rangle_{y=0} = 0$, integrates to

$$\langle v^2 \rangle + \langle p \rangle/\rho = p_w(x)/\rho, \tag{7.6}$$

where $p_w = \langle p(x,0,0) \rangle$ is the mean pressure on the bottom wall. An important deduction from this equation is that the mean axial pressure gradient is uniform across the flow:

$$\frac{\partial \langle p \rangle}{\partial x} = \frac{dp_w}{dx}. \tag{7.7}$$

The axial mean-momentum equation,

$$0 = v\frac{d^2 \langle U \rangle}{dy^2} - \frac{d}{dy}\langle uv \rangle - \frac{1}{\rho}\frac{\partial \langle p \rangle}{\partial x}, \tag{7.8}$$

can be rewritten

$$\frac{d\tau}{dy} = \frac{dp_w}{dx}, \tag{7.9}$$

where the total shear stress $\tau(y)$ is

$$\tau = \rho v\frac{d\langle U \rangle}{dy} - \rho\langle uv \rangle. \tag{7.10}$$

For this flow there is no mean acceleration, so the mean momentum equation

(Eq. (7.9)) amounts to a balance of forces: the axial normal stress gradient is balanced by the cross-stream shear-stress gradient.

Since τ is a function only of y, and p_w is a function only of x, it is evident from Eq. (7.9) that both $d\tau/dy$ and dp_w/dx are constant. The solutions for $\tau(y)$ and dp_w/dx can be written explicitly in terms of the *wall shear stress*

$$\tau_w \equiv \tau(0). \tag{7.11}$$

Because $\tau(y)$ is antisymmetric about the mid-plane, it follows that $\tau(\delta)$ is zero; and at the top wall the stress is $\tau(2\delta) = -\tau_w$. Hence, the solution to Eq. (7.9) is

$$-\frac{dp_w}{dx} = \frac{\tau_w}{\delta}, \tag{7.12}$$

and

$$\tau(y) = \tau_w \left(1 - \frac{y}{\delta} \right). \tag{7.13}$$

The wall shear stress normalized by a reference velocity is called a *skin-friction coefficient*. On the basis of U_0 and $\bar{U}$ we define

$$c_f \equiv \tau_w / \left(\tfrac{1}{2} \rho U_0^2 \right), \tag{7.14}$$

$$C_f \equiv \tau_w / \left(\tfrac{1}{2} \rho \bar{U}^2 \right). \tag{7.15}$$

To summarize: the flow is driven by the drop in pressure between the entrance and the exit of the channel. In the fully developed region there is a constant (negative) mean pressure gradient $\partial \langle p \rangle / \partial x = dp_w/dx$, which is balanced by the shear-stress gradient $d\tau/dy = -\tau_w/\delta$. For a given pressure gradient dp_w/dx and channel half-width δ, the linear shear-stress profile is given by Eqs. (7.12) and (7.13) – independent of the fluid properties (e.g., ρ and v), and independent of the state of fluid motion (i.e., laminar or turbulent). Note that, if the flow is defined by ρ, v, δ, and dp_w/dx, then U_0 and $\bar{U}$ are not known *a priori*. Alternatively, in an experiment $\bar{U}$ can be imposed and then the pressure gradient is unknown. In both cases the skin-friction coefficient is not known *a priori*. Of course, as the following exercise demonstrates, all of these quantities are readily determined for laminar flow.

EXERCISES _____

7.1   For laminar flow, from Eqs. (7.10) and (7.13), show that the mean velocity profile is

$$U(y) = \frac{\tau_w \delta}{2\rho v} \frac{y}{\delta} \left(2 - \frac{y}{\delta} \right). \tag{7.16}$$

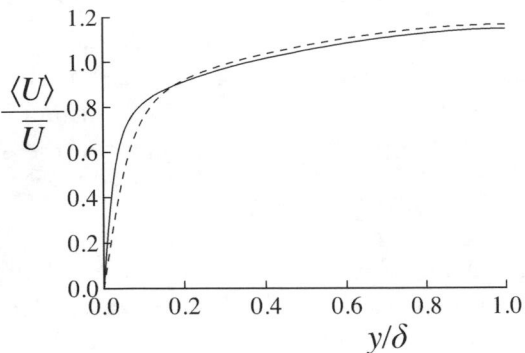

Fig. 7.2. Mean velocity profiles in fully developed turbulent channel flow from the DNS of Kim *et al.* (1987): dashed line, Re = 5,600; solid line, Re = 13,750.

Hence obtain the following results:

$$U_0 = \frac{\tau_{\mathrm{w}}\delta}{2\rho\nu} = \tfrac{3}{2}\bar{U}, \tag{7.17}$$

$$\mathrm{Re} = \tfrac{4}{3}\mathrm{Re}_0, \tag{7.18}$$

$$c_{\mathrm{f}} = \frac{4}{\mathrm{Re}_0} = \frac{16}{3\mathrm{Re}}, \tag{7.19}$$

$$C_{\mathrm{f}} = \frac{9}{\mathrm{Re}_0} = \frac{12}{\mathrm{Re}}. \tag{7.20}$$

7.2   The *friction velocity* is defined by

$$u_\tau \equiv \sqrt{\tau_{\mathrm{w}}/\rho}. \tag{7.21}$$

Show that, in general,

$$c_{\mathrm{f}} = 2(u_\tau/U_0)^2, \tag{7.22}$$

and that for laminar flow

$$\frac{u_\tau}{U_0} = \sqrt{\frac{2}{\mathrm{Re}_0}} = \sqrt{\frac{8}{3\mathrm{Re}}}. \tag{7.23}$$

Evaluate u_τ/U_0 for the upper limit of laminar flow, i.e., Re = 1,350.

7.1.3 The near-wall shear stress

Figure 7.2 shows the mean velocity profiles obtained by Kim *et al.* (1987) from direct numerical simulations of fully developed turbulent channel flow

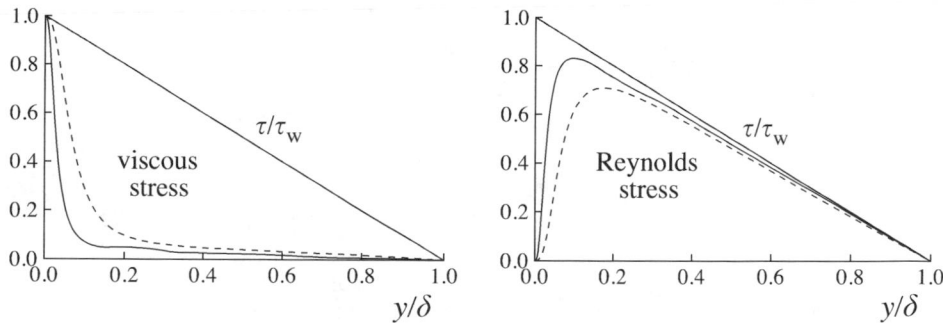

Fig. 7.3. Profiles of the viscous shear stress, and the Reynolds shear stress in turbulent channel flow: DNS data of Kim *et al.* (1987): dashed line, Re = 5,600; solid line, Re = 13,750.

at Re = 5,600 and Re = 13,750.[1] The objective of this and the next subsection is to explain and quantify these profiles.

The total shear stress $\tau(y)$ (Eq. (7.10)) is the sum of the viscous stress $\rho v \, d\langle U \rangle / dy$ and the Reynolds stress $-\rho \langle uv \rangle$. At the wall, the boundary condition $U(x, t) = 0$ dictates that all the Reynolds stresses are zero. Consequently the wall shear stress is due entirely to the viscous contribution, i.e.,

$$\tau_w \equiv \rho v \left(\frac{d\langle U \rangle}{dy} \right)_{y=0}. \tag{7.24}$$

Profiles of the viscous and Reynolds shear stresses are shown in Fig. 7.3.

The important observation that the viscous stress dominates at the wall is in contrast to the situation in free shear flows. There, at high Reynolds number, the viscous stresses are everywhere negligibly small compared with the Reynolds stresses. Also, near the wall, since the viscosity is an influential parameter, the velocity profile depends upon the Reynolds number (as may be observed in Fig. 7.2) – again in contrast to free shear flows.

It is evident that, close to the wall, the viscosity v and the wall shear stress τ_w are important parameters. From these quantities (and ρ) we define *viscous scales* that are the appropriate velocity scales and lengthscales in the near-wall region. These are the *friction velocity*

$$u_\tau \equiv \sqrt{\frac{\tau_w}{\rho}}, \tag{7.25}$$

---

[1] The higher-Reynolds number data are briefly presented by Moser, Kim, and Mansour (1999). A description of DNS of channel flow is given in Chapter 9.

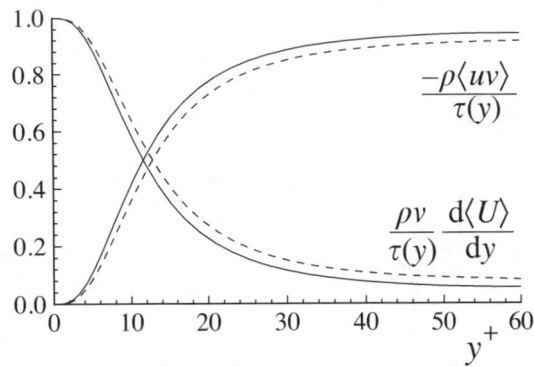

Fig. 7.4. Profiles of the fractional contributions of the viscous and Reynolds stresses to the total stress. DNS data of Kim *et al.* (1987): dashed lines, Re = 5,600; solid lines, Re = 13,750.

and the *viscous lengthscale*

$$\delta_v \equiv v\sqrt{\frac{\rho}{\tau_w}} = \frac{v}{u_\tau}. \qquad (7.26)$$

The Reynolds number based on the viscous scales $u_\tau\delta_v/v$ is identically unity, while the *friction Reynolds number* is defined by

$$\mathrm{Re}_\tau \equiv \frac{u_\tau\delta}{v} = \frac{\delta}{\delta_v}. \qquad (7.27)$$

(In the DNS of Kim *et al.* 1987, the friction Reynolds numbers are $\mathrm{Re}_\tau = 180$ at Re = 5,600, and $\mathrm{Re}_\tau = 395$ at Re = 13,750.)

The distance from the wall measured in viscous lengths – or *wall units* – is denoted by

$$y^+ \equiv \frac{y}{\delta_v} = \frac{u_\tau y}{v}. \qquad (7.28)$$

Notice that y^+ is similar to a local Reynolds number, so its magnitude can be expected to determine the relative importance of viscous and turbulent processes. In support of this supposition, Fig. 7.4 shows the fractional contributions to the total stress from the viscous and Reynolds stresses in the near-wall region of channel flow. When thay are plotted against y^+, the profiles for the two Reynolds numbers almost collapse. The viscous contribution drops from 100% at the wall ($y^+ = 0$) to 50% at $y^+ \approx 12$ and is less than 10% by $y^+ = 50$.

Different regions, or layers, in the near-wall flow are defined on the basis of y^+. In the *viscous wall region* $y^+ < 50$, there is a direct effect of molecular viscosity on the shear stress; whereas, conversely, in the *outer layer* $y^+ > 50$

the direct effect of viscosity is negligible. Within the viscous wall region, in the *viscous sublayer* $y^+ < 5$, the Reynolds shear stress is negligible compared with the viscous stress. As the Reynolds number of the flow increases, the fraction of the channel occupied by the viscous wall region decreases, since δ_v/δ varies as Re_τ^{-1} (Eq. (7.27)).

EXERCISE _____

7.3   An experiment is performed on fully developed turbulent channel flow at $\text{Re} = 10^5$. The fluid is water ($\nu = 1.14 \times 10^{-6}$ m^2 s^{-1}) and the channel half-height is $\delta = 2$ cm. The skin-friction coefficient is found to be $C_f = 4.4 \times 10^{-3}$. Determine: $\bar{U}$, $u_\tau/\bar{U}$, Re_τ, and δ_v/δ. What are the thicknesses of the viscous wall region and of the viscous sublayer, both as fractions of δ and in millimeters?

7.1.4 Mean velocity profiles

Fully developed channel flow is completely specified by ρ, ν, δ, and dp_w/dx; or, equivalently, by ρ, ν, δ, and u_τ, since we have

$$u_\tau = \left(-\frac{\delta}{\rho} \frac{dp_w}{dx} \right)^{1/2}. \tag{7.29}$$

There are just two independent non-dimensional groups that can be formed from $\rho, \nu, \delta, u_\tau$, and y (e.g., y/δ and $\text{Re}_\tau = u_\tau \delta/\nu$) and consequently the mean velocity profile can be written

$$\langle U \rangle = u_\tau F_0 \left(\frac{y}{\delta}, \text{Re}_\tau \right), \tag{7.30}$$

where F_0 is a universal non-dimensional function to be determined.

While this approach to determining the mean velocity profile appears natural, it is, however, preferable to proceed somewhat differently. Instead of $\langle U \rangle$, we consider the velocity gradient $d\langle U \rangle/dy$, which is the dynamically important quantity. The viscous stress and the turbulence production, for example, are both determined by $d\langle U \rangle/dy$. Again on dimensional grounds, $d\langle U \rangle/dy$ depends on just two non-dimensional parameters, so that (without any assumption) we can write

$$\frac{d\langle U \rangle}{dy} = \frac{u_\tau}{y} \Phi \left(\frac{y}{\delta_v}, \frac{y}{\delta} \right), \tag{7.31}$$

where Φ is a universal non-dimensional function. The idea behind the choice of the two parameters is that δ_v is the appropriate lengthscale in the viscous wall region ($y^+ < 50$) while δ is the appropriate scale in the outer layer ($y^+ > 50$). The relation

$$\left(\frac{y}{\delta_v}\right)\Big/\left(\frac{y}{\delta}\right) = \mathrm{Re}_\tau \tag{7.32}$$

shows, as is inevitable, that these two parameters contain the same information as y/δ and Re_τ (Eq. (7.30)).

The law of the wall

Prandtl (1925) postulated that, at high Reynolds number, close to the wall ($y/\delta \ll 1$) there is an *inner layer* in which the mean velocity profile is determined by the viscous scales, independent of δ and U_0. Mathematically, this implies that the function $\Phi(y/\delta_v, y/\delta)$ in Eq. (7.31) tends asymptotically to a function of y/δ_v only, as y/δ tends to zero, so that Eq. (7.31) becomes

$$\frac{\mathrm{d}\langle U\rangle}{\mathrm{d}y} = \frac{u_\tau}{y}\Phi_1\left(\frac{y}{\delta_v}\right), \quad \text{for } \frac{y}{\delta} \ll 1, \tag{7.33}$$

where

$$\Phi_1\left(\frac{y}{\delta_v}\right) = \lim_{y/\delta\to0} \Phi\left(\frac{y}{\delta_v},\frac{y}{\delta}\right). \tag{7.34}$$

With $y^+ \equiv y/\delta_v$ and $u^+(y^+)$ defined by

$$u^+ \equiv \frac{\langle U\rangle}{u_\tau}, \tag{7.35}$$

Eq. (7.33) can alternatively be written

$$\frac{\mathrm{d}u^+}{\mathrm{d}y^+} = \frac{1}{y^+}\Phi_1(y^+). \tag{7.36}$$

The integral of Eq. (7.36) is the *law of the wall*:

$$u^+ = f_w(y^+), \tag{7.37}$$

where

$$f_w(y^+) = \int_0^{y^+} \frac{1}{y'}\Phi_1(y')\,\mathrm{d}y'. \tag{7.38}$$

The important point is not Eq. (7.38), but the fact that (according to Prandtl's hypothesis) u^+ depends solely on y^+ for $y/\delta \ll 1$.

For Reynolds numbers not too close to transition, there is abundant experimental verification that the function f_w is universal, not only for channel flow, but also for pipe flow and boundary layers. As is now shown, the form of the function $f_w(y^+)$ can be determined for small and large values of y^+.

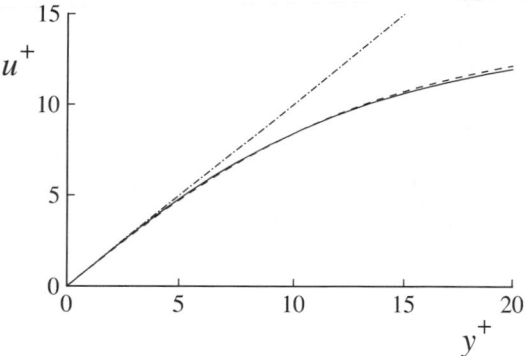

Fig. 7.5. Near-wall profiles of mean velocity from the DNS data of Kim *et al.* (1987): dashed line, Re = 5,600; solid line, Re = 13,750; dot–dashed line, $u^+ = y^+$.

The viscous sublayer

The no-slip condition $\langle U \rangle_{y=0} = 0$ corresponds to $f_w(0) = 0$, while the viscous stress law at the wall (Eq. (7.24)) yields for the derivative

$$f'_w(0) = 1. \tag{7.39}$$

(This is simply a result of the normalization by the viscous scales.) Hence, the Taylor-series expansion for $f_w(y^+)$ for small y^+ is

$$f_w(y^+) = y^+ + \mathcal{O}(y^{+2}). \tag{7.40}$$

(In fact, closer examination reveals that, after the linear term, the next non-zero term is of order y^{+4}, see Exercise 7.9.)

Figure 7.5 shows the profiles of u^+ in the near-wall region obtained from direct numerical simulations. The departures from the linear relation $u^+ = y^+$ are negligible in the viscous sublayer ($y^+ < 5$), but are significant (greater than 25%) for $y^+ > 12$.

The log law

The inner layer is usually defined as $y/\delta < 0.1$. At high Reynolds number, the outer part of the inner layer corresponds to large y^+, i.e., $y^+ \approx 0.1\delta/\delta_v = 0.1\mathrm{Re}_\tau \gg 1$. As has already been discussed, for large y^+ it can be supposed that viscosity has little effect. Hence, in Eq. (7.33), the dependence of $\Phi_I(y/\delta_v)$ on v (through δ_v) vanishes, so that Φ_I adopts a constant value denoted by κ^{-1}:

$$\Phi_I(y^+) = \frac{1}{\kappa}, \qquad \text{for } \frac{y}{\delta} \ll 1 \text{ and } y^+ \gg 1. \tag{7.41}$$

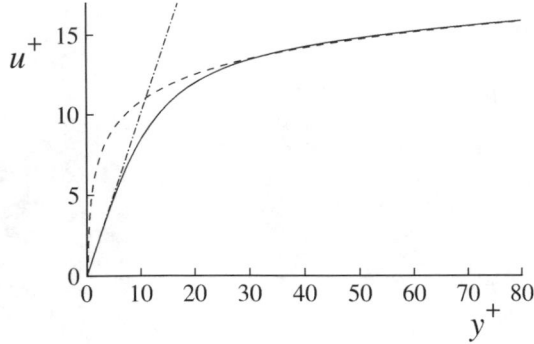

Fig. 7.6. Near-wall profiles of mean velocity: solid line, DNS data of Kim *et al.* (1987): Re = 13,750; dot–dashed line, $u^+ = y^+$; dashed line, the log law, Eqs. (7.43)–(7.44).

Thus, in this region, the mean velocity gradient is

$$\frac{du^+}{dy^+} = \frac{1}{\kappa y^+},$$ (7.42)

which integrates to

$$u^+ = \frac{1}{\kappa} \ln y^+ + B,$$ (7.43)

where B is a constant. This is the logarithmic law of the wall due to von Kármán (1930) – or simply, the *log law* – and κ is the von Kármán constant. In the literature, there is some variation in the values ascribed to the log-law constants, but generally they are within 5% of

$$\kappa = 0.41, \qquad B = 5.2.$$ (7.44)

Figure 7.6 shows a comparison between the log law and the DNS data in the inner part of the channel ($y/\delta < 0.25$). Clearly there is excellent agreement for $y^+ > 30$.

The log law is more clearly revealed in a semi-log plot. Figure 7.7 shows measured profiles of $u^+(y^+)$ for turbulent channel flow at Reynolds numbers between $Re_0 \approx 3,000$ and $Re_0 \approx 40,000$. It may be seen that the data collapse to a single curve – in confirmation of the law of the wall – and that for $y^+ > 30$ the data conform to the log law, except near the channel's mid-plane (the last few data points for each Reynolds number).

The region between the viscous sublayer ($y^+ < 5$) and the *log-law region* ($y^+ > 30$) is called the *buffer layer*. It is the transition region between the viscosity-dominated and the turbulence-dominated parts of the flow. The various regions and layers that are used to describe near-wall flows are summarized in Table 7.1 and Fig. 7.8.

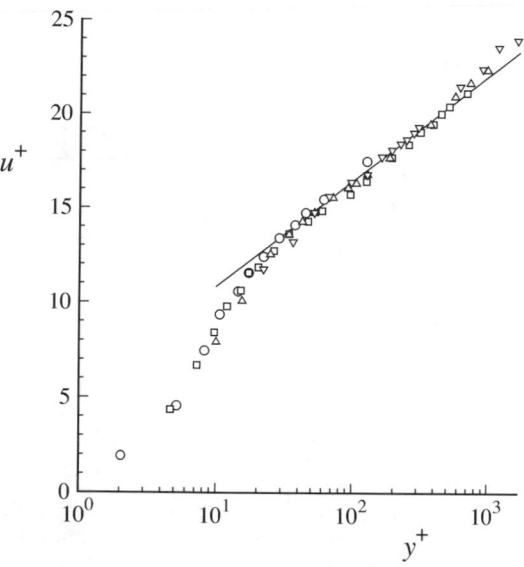

Fig. 7.7. Mean velocity profiles in fully developed turbulent channel flow measured by Wei and Willmarth (1989): $\bigcirc$, $\mathrm{Re}_0 = 2{,}970$; $\square$, $\mathrm{Re}_0 = 14{,}914$; $\triangle$, $\mathrm{Re}_0 = 22{,}776$; $\triangledown$, $\mathrm{Re}_0 = 39{,}582$; line, the log law, Eqs. (7.43)–(7.44).

Table 7.1. *Wall regions and layers and their defining properties*

Region	Location	Defining property
Inner layer	$y/\delta < 0.1$	$\langle U \rangle$ determined by u_τ and y^+, independent of U_0 and δ
Viscous wall region	$y^+ < 50$	The viscous contribution to the shear stress is significant
Viscous sublayer	$y^+ < 5$	The Reynolds shear stress is negligible compared with the viscous stress
Outer layer	$y^+ > 50$	Direct effects of viscosity on $\langle U \rangle$ are negligible
Overlap region	$y^+ > 50$, $y/\delta < 0.1$	Region of overlap between inner and outer layers (at large Reynolds numbers)
Log-law region	$y^+ > 30$, $y/\delta < 0.3$	The log-law holds
Buffer layer	$5 < y^+ < 30$	The region between the viscous sublayer and the log-law region

The velocity-defect law

In the outer layer ($y^+ > 50$), the assumption that $\Phi(y/\delta_v, y/\delta)$ is independent of v implies that, for large y/δ_v, Φ tends asymptotically to a function of y/δ

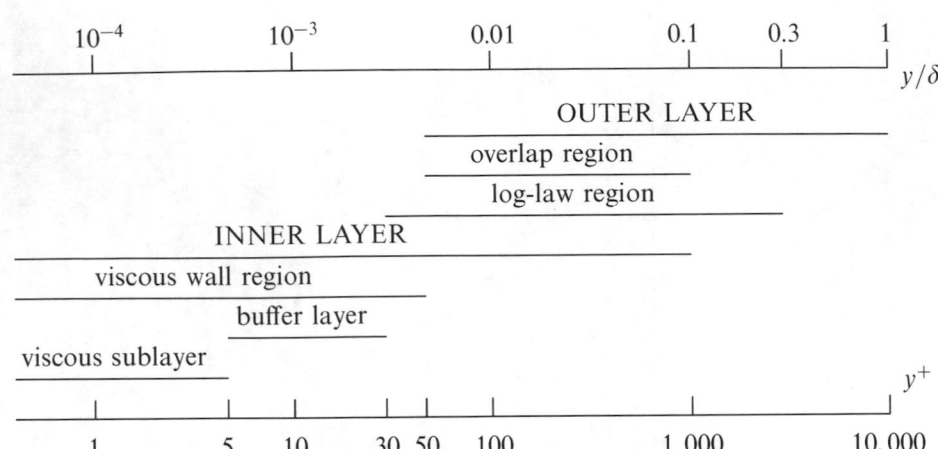

Fig. 7.8. A sketch showing the various wall regions and layers defined in terms of $y^+ = y/\delta_v$ and y/δ, for turbulent channel flow at high Reynolds number ($\mathrm{Re}_\tau = 10^4$).

only, i.e.,

$$\lim_{y/\delta_v \to \infty} \Phi\left(\frac{y}{\delta_v}, \frac{y}{\delta}\right) = \Phi_0\left(\frac{y}{\delta}\right). \tag{7.45}$$

Substituting Φ_0 for Φ in Eq. (7.31) and integrating between y and δ then yields the *velocity-defect law* due to von Kármán (1930):

$$\frac{U_0 - \langle U\rangle}{u_\tau} = F_D\left(\frac{y}{\delta}\right), \tag{7.46}$$

where

$$F_D\left(\frac{y}{\delta}\right) = \int_{y/\delta}^{1} \frac{1}{y'}\Phi_0(y')\,\mathrm{d}y'. \tag{7.47}$$

By definition, the velocity defect is the difference between the mean velocity $\langle U\rangle$ and the centerline value U_0. The velocity-defect law states that this velocity defect normalized by u_τ depends on y/δ only. Unlike the law-of-the-wall function $f_w(y^+)$, here there is no suggestion that $F_D(y/\delta)$ is universal: it is different in different flows.

At sufficiently high Reynolds number (approximately Re > 20,000) there is an *overlap region* between the inner layer ($y/\delta < 0.1$) and the outer layer ($y/\delta_v > 50$) (see Fig. 7.8). In this region, both Eqs. (7.33) and (7.45) are valid, yielding (from Eq. (7.31))

$$\frac{y}{u_\tau}\frac{\mathrm{d}\langle U\rangle}{\mathrm{d}y} = \Phi_I\left(\frac{y}{\delta_v}\right) = \Phi_0\left(\frac{y}{\delta}\right), \quad \text{for } \delta_v \ll y \ll \delta. \tag{7.48}$$

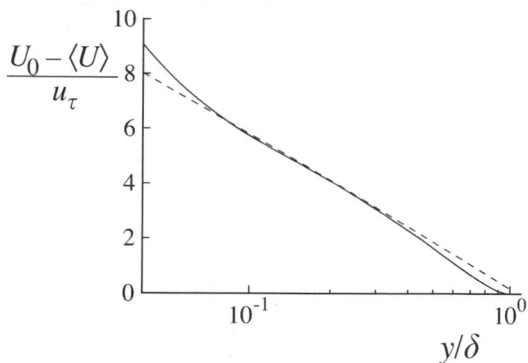

Fig. 7.9. The mean velocity defect in turbulent channel flow. Solid line, DNS of Kim *et al.* (1987), $Re = 13,750$; dashed line, log law, Eqs. (7.43)–(7.44).

This equation can be satisfied in the overlap region only by Φ_1 and Φ_0 being constant, which leads to

$$\frac{y}{u_\tau}\frac{\mathrm{d}\langle U\rangle}{\mathrm{d}y} = \frac{1}{\kappa}, \quad \text{for } \delta_v \ll y \ll \delta. \tag{7.49}$$

This argument, due to Millikan (1938), provides an alternative derivation of the log law. It also establishes the form of the velocity-defect law for small y/δ, i.e.,

$$\frac{U_0 - \langle U\rangle}{u_\tau} = F_\mathrm{D}\left(\frac{y}{\delta}\right) = -\frac{1}{\kappa}\ln\left(\frac{y}{\delta}\right) + B_1, \quad \text{for } \frac{y}{\delta} \ll 1. \tag{7.50}$$

where B_1 is a flow-dependent constant. (The overlap region and the arguments leading to the log law are considered further in Section 7.3.4.)

Figure 7.9 shows the velocity defect in the DNS of turbulent channel flow. It may be seen that the log law is followed quite closely between $y/\delta = 0.08$ ($y^+ \approx 30$) and $y/\delta = 0.3$. Even in the central part of the channel ($0.3 < y/\delta < 1.0$) the deviations from the log law are quite small; but it should be appreciated that the arguments leading to the log law are not applicable in this region.

Let $U_{0,\mathrm{log}}$ denote the value of $\langle U\rangle$ on the centerline obtained by extrapolation of the log law. For $y/\delta = 1$, Eq. (7.50) then yields

$$\frac{U_0 - U_{0,\mathrm{log}}}{u_\tau} = B_1, \tag{7.51}$$

which provides a convenient way of determining B_1. It may be seen from Fig. 7.9 that the difference $U_0 - U_{0,\mathrm{log}}$ is very small – about 1% of U_0 – which makes B_1 difficult to measure. The DNS data yield $B_1 \approx 0.2$, but, from a survey of many measurements, Dean (1978) suggests $B_1 \approx 0.7$. The uncertainty in B_1 is of little consequence: the point is that it is small.

In the outer layer of boundary layers, the deviations from the log law are more substantial. Consequently the velocity defect law is discussed further in that context (Section 7.3).

7.1.5 The friction law and the Reynolds number

Having characterized the mean velocity profile, we are now in a position to determine the Reynolds-number dependence of the skin-friction coefficient and other quantities. The primary task is to establish relationships among the velocities U_0, $\bar{U}$, and u_τ.

A good estimate of the bulk velocity $\bar{U}$ is obtained by using the log law (Eq. (7.50)) to approximate $\langle U \rangle$ over the whole channel. (For consistency at $y = \delta$, this requires taking $B_1 = 0$.) As we have seen, in the center of the channel, the departures from the log law are quite small (Fig. 7.9): near the wall ($y^+ < 30$) the approximation is poor (Fig. 7.6), but this region makes a negligible contribution to the integral of $\langle U \rangle$ (except at very low Reynolds number). The result obtained with this approximation is

$$\frac{U_0 - \bar{U}}{u_\tau} = \frac{1}{\delta} \int_0^\delta \frac{U_0 - \langle U \rangle}{u_\tau} \, dy$$

$$\approx \frac{1}{\delta} \int_0^\delta -\frac{1}{\kappa} \ln\left(\frac{y}{\delta}\right) dy = \frac{1}{\kappa} \approx 2.4. \qquad (7.52)$$

This estimate agrees well with the experimental data which are scattered between 2 and 3 (Dean 1978), and the DNS value of 2.6.

The log law in the inner layer (Eq. (7.43)) can be written

$$\frac{\langle U \rangle}{u_\tau} = \frac{1}{\kappa} \ln\left(\frac{y}{\delta_\nu}\right) + B, \qquad (7.53)$$

whereas in the outer layer it is (Eq. (7.50))

$$\frac{U_0 - \langle U \rangle}{u_\tau} = -\frac{1}{\kappa} \ln\left(\frac{y}{\delta}\right) + B_1. \qquad (7.54)$$

When these two equations are added together the y dependence vanishes to yield

$$\frac{U_0}{u_\tau} = \frac{1}{\kappa} \ln\left(\frac{\delta}{\delta_\nu}\right) + B + B_1$$

$$= \frac{1}{\kappa} \ln\left[\mathrm{Re}_0 \left(\frac{U_0}{u_\tau}\right)^{-1} \right] + B + B_1. \qquad (7.55)$$

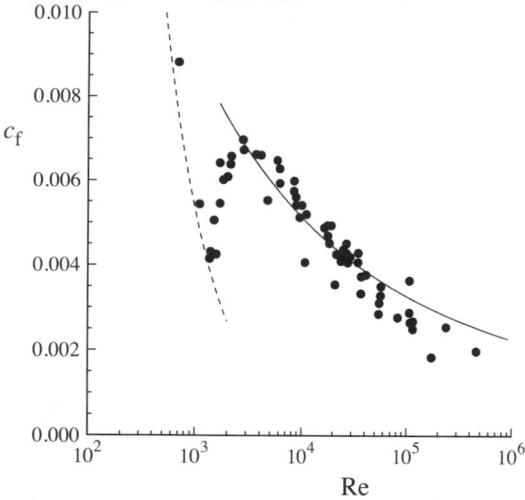

Fig. 7.10. The skin-friction coefficient $c_f \equiv \tau_w/(\tfrac{1}{2}\rho U_0^2)$ against the Reynolds number ($\mathrm{Re} = 2\bar{U}\delta/\nu$) for channel flow: symbols, experimental data compiled by Dean (1978); solid line, from Eq. (7.55); dashed line, laminar friction law, $c_f = 16/(3\mathrm{Re})$.

For given Re_0 this equation can be solved for U_0/u_τ, hence determining the skin-friction coefficient $c_f = \tau_w/(\tfrac{1}{2}\rho U_0^2) = 2(u_\tau/U_0)^2$. With the aid of the approximation Eq. (7.52), $\mathrm{Re} \equiv 2\bar{U}\delta/\nu$ and $C_f \equiv \tau_w/(\tfrac{1}{2}\rho\bar{U}^2)$ can then also be determined.

Figure 7.10 shows the skin friction coefficient c_f obtained from Eq. (7.55) as a function of Re. Also shown is the laminar relation and the experimental data compiled by Dean (1978). For $\mathrm{Re} > 3{,}000$, Eq. (7.55) provides a good representation of the skin-friction coefficient. It is interesting to note that Patel and Head (1969) found that $\mathrm{Re} = 3{,}000$ is the lowest Reynolds number at which a log law with universal constants is observed.

The ratios of the mean flow to viscous scales are shown in Figs. 7.11 and 7.12. The lengthscale ratio $\delta/\delta_\nu = \mathrm{Re}_\tau$ increases almost linearly with Re – a good approximation being $\mathrm{Re}_\tau \approx 0.09\mathrm{Re}^{0.88}$. Consequently, at high Reynolds number the viscous lengthscale can be very small. As an example, for a channel with $\delta = 2$ cm, at $\mathrm{Re} = 10^5$ the viscosity scale is $\delta_\nu \approx 10^{-5}$ m, so the location $y^+ = 100$ is just 1 mm from the wall. Needless to say, there are considerable difficulties in making measurements in the viscous wall region of high-Reynolds-number laboratory flows.

In contrast, the velocity ratios increase very slowly with Re (Fig. 7.12) – a simple approximation being $U_0/u_\tau \approx 5 \log_{10}\mathrm{Re}$. As a consequence, a significant fraction of the increase in mean velocity between the wall and the centerline occurs in the viscous wall region. In the example introduced

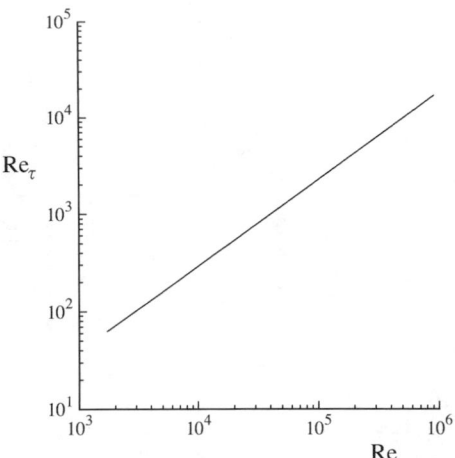

Fig. 7.11. The outer-to-inner lengthscale ratio $\delta/\delta_v = \mathrm{Re}_\tau$ for turbulent channel flow as a function of the Reynolds number (obtained from Eq. (7.55)).

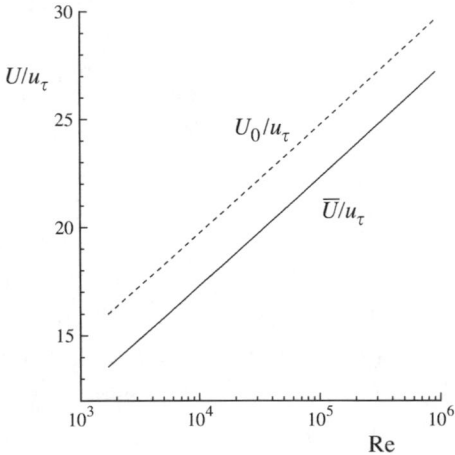

Fig. 7.12. Outer-to-inner velocity-scale ratios for turbulent channel flow as functions of the Reynolds number (obtained from Eq. (7.55)): solid line, $\bar{U}/u_\tau$; dashed line U_0/u_τ.

above ($\delta = 2$ cm, $\mathrm{Re} = 10^5$) it follows that, at $y^+ = 10$ (i.e., $y \approx 0.1$ mm), the mean velocity is over 30% of the centerline value, U_0.

Figure 7.13 shows the Reynolds-number dependence of the y locations that delineate the various regions and layers. According to this plot, a log-law region ($30\delta_v < y < 0.3\delta$) exists for $\mathrm{Re} > 3{,}000$ – in agreement with the experimental observations of Patel and Head (1969). On the other hand, a Reynolds number in excess of 20,000 is required for there to be an overlap region, according to the criterion $50\delta_v < y < 0.1\delta$. As has already been

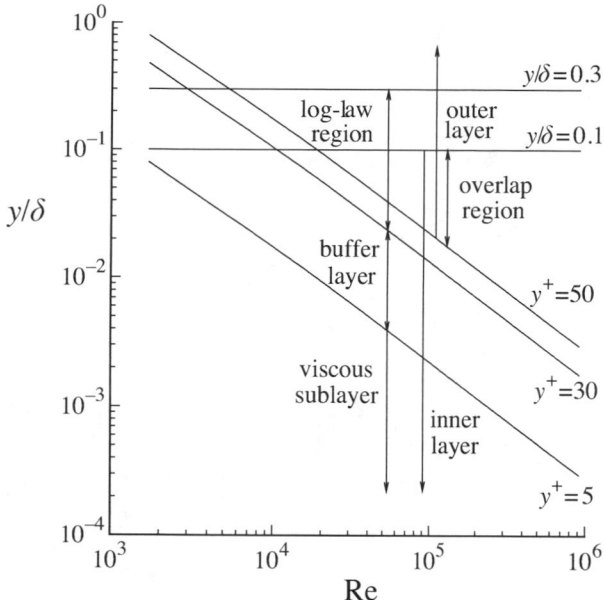

Fig. 7.13. Regions and layers in turbulent channel flow as functions of the Reynolds number.

observed, the log law persists beyond the region suggested by the overlap argument.

7.1.6 Reynolds stresses

Figures 7.14–7.16 show the Reynolds stresses and some related statistics obtained from the DNS of channel flow at Re = 13,750. In order to discuss these statistics, it is useful to divide the flow into three regions: the viscous wall region ($y^+ < 50$); the log-law region ($50\delta_v < y < 0.3\delta$, or $50 < y^+ < 120$ at this Reynolds number); and the *core* ($y > 0.3\delta$).

In the log-law region there is approximate self-similarity. The normalized Reynolds stresses $\langle u_i u_j \rangle / k$ are essentially uniform, as are the production-to-dissipation ratio, $\mathcal{P}/\varepsilon$, and the normalized mean shear rate, Sk/ε (where $S = \partial \langle U \rangle / \partial y$). Their values are given in Table 7.2. It is interesting to observe that the values of $\langle u_i u_j \rangle / k$ are within a few percent of those measured by Tavoularis and Corrsin (1981) in homogeneous shear flow (see Table 5.4 on page 157). Production $\mathcal{P}$ and dissipation ε are almost in balance, the viscous and turbulent transport of k being very small in comparison.

On the centerline, both the mean velocity gradient and the shear stress vanish, so that the production $\mathcal{P}$ is zero. Figure 7.16 shows the gradual

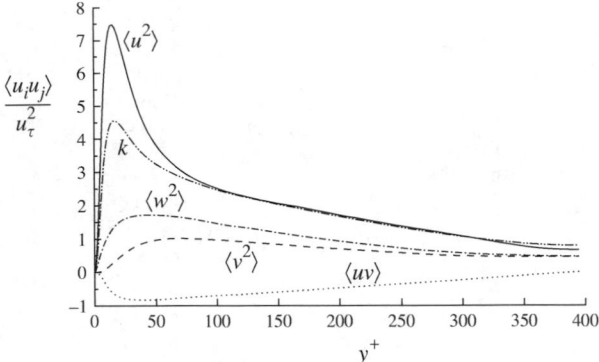

Fig. 7.14. Reynolds stresses and kinetic energy normalized by the friction velocity against y^+ from DNS of channel flow at Re = 13,750 (Kim *et al.* 1987).

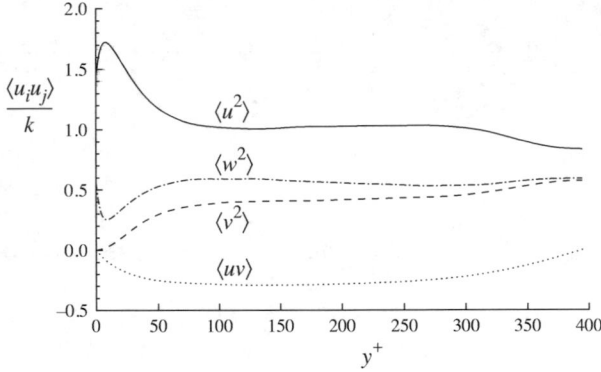

Fig. 7.15. Profiles of Reynolds stresses normalized by the turbulent kinetic energy from DNS of channel flow at Re = 13,750 (Kim *et al.* 1987).

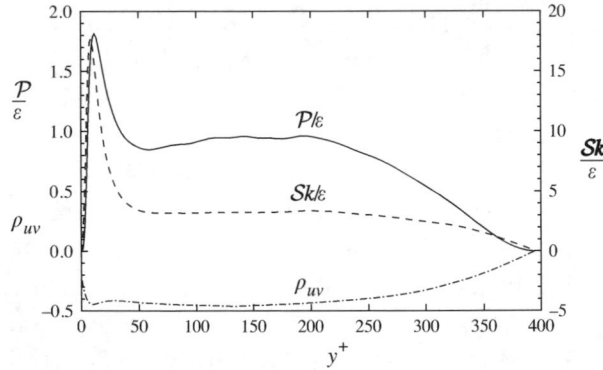

Fig. 7.16. Profiles of the ratio of production to dissipation ($\mathcal{P}/\varepsilon$), normalized mean shear rate ($\mathcal{S}k/\varepsilon$), and shear stress correlation coefficient (ρ_{uv}) from DNS of channel flow at Re = 13,750 (Kim *et al.* 1987).

Table 7.2. *Statistics in turbulent channel flow, obtained from the DNS data of Kim et al. (1987), Re = 13,750*

	Location		
	Peak production $y^+ = 11.8$	Log law $y^+ = 98$	Centerline $y^+ = 395$
$\langle u^2 \rangle / k$	1.70	1.02	0.84
$\langle v^2 \rangle / k$	0.04	0.39	0.57
$\langle w^2 \rangle / k$	0.26	0.59	0.59
$\langle uv \rangle / k$	-0.116	-0.285	0
ρ_{uv}	-0.44	-0.45	0
Sk/ε	15.6	3.2	0
$\mathcal{P}/\varepsilon$	1.81	0.91	0

changes of $\mathcal{P}/\varepsilon$, Sk/ε, and ρ_{uv} from their log-law values to zero on the centerline. Figure 7.15 indicates that the Reynolds stresses are anisotropic on the centerline, but considerably less so than in the log-law region (see also Table 7.2).

The viscous wall region ($y^+ < 50$) contains the most vigorous turbulent activity. The production, dissipation, turbulent kinetic energy and anisotropy all achieve their peak values at y^+ less than 20. We shall examine the behavior in this region in more detail.

The boundary condition $U = 0$ at the wall determines the way in which the Reynolds stresses depart from zero for small y. For fixed x, z, and t, and for small y, the fluctuating velocity components can be written as Taylor series of the forms

$$u = a_1 + b_1 y + c_1 y^2 + \ldots, \tag{7.56}$$

$$v = a_2 + b_2 y + c_2 y^2 + \ldots, \tag{7.57}$$

$$w = a_3 + b_3 y + c_3 y^2 + \ldots. \tag{7.58}$$

The coefficients are zero-mean random variables, and, for fully developed channel flow, they are statistically independent of x, z, and t. For $y = 0$, the no-slip condition yields $u = a_1 = 0$ and $w = a_3 = 0$; and similarly the impermeability condition yields $v = a_2 = 0$. At the wall, since u and w are zero for all x and z, the derivatives $(\partial u / \partial x)_{y=0}$ and $(\partial w / \partial z)_{y=0}$ are also zero. Hence the continuity equation yields

$$\left(\frac{\partial v}{\partial y} \right)_{y=0} = b_2 = 0. \tag{7.59}$$

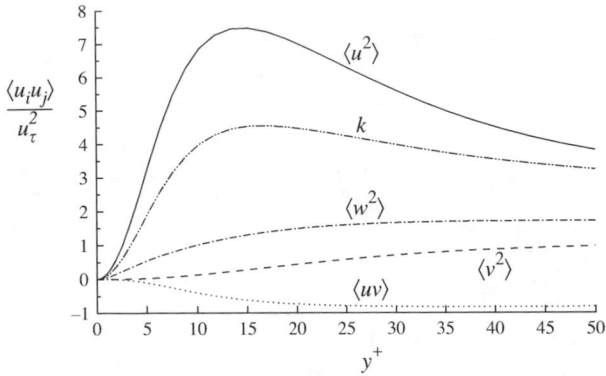

Fig. 7.17. Profiles of Reynolds stresses and kinetic energy normalized by the friction velocity in the viscous wall region of turbulent channel flow: DNS data of Kim *et al.* (1987). Re = 13,750.

The significance of the coefficient b_2 being zero is that, very close to the wall, there is *two-component flow*. That is, to order y, v is zero whereas u and w are non-zero. The resulting motion corresponds to flow in planes parallel to the wall. (This is called two-component flow, rather than two-dimensional flow, because u and w vary in the y direction.)

The Reynolds stresses can be obtained from the expansions (Eqs. (7.56)–(7.58)) simply by taking the means of the products of the series. Taking account of the coefficients that are zero (i.e., a_1, a_2, a_3, and b_2), to leading order in y the Reynolds stresses are

$$\langle u^2 \rangle = \langle b_1^2 \rangle y^2 + \dots, \tag{7.60}$$

$$\langle v^2 \rangle = \langle c_2^2 \rangle y^4 + \dots, \tag{7.61}$$

$$\langle w^2 \rangle = \langle b_3^2 \rangle y^2 + \dots, \tag{7.62}$$

$$\langle uv \rangle = \langle b_1 c_2 \rangle y^3 + \dots. \tag{7.63}$$

Thus, while $\langle u^2 \rangle$, $\langle w^2 \rangle$, and k increase from zero as y^2, $-\langle uv \rangle$ and $\langle v^2 \rangle$ increase more slowly – as y^3 and y^4, respectively. These behaviors can be clearly seen in log–log plots of $\langle u_i u_j \rangle$ against y (not shown), and they are also evident in Fig. 7.17, which shows the profiles of $\langle u_i u_j \rangle$ and k in the viscous wall region.

For fully developed channel flow, the balance equation for turbulent kinetic energy is

$$0 = \mathcal{P} - \tilde{\varepsilon} + v \frac{\mathrm{d}^2 k}{\mathrm{d}y^2} - \frac{\mathrm{d}}{\mathrm{d}y} \langle \tfrac{1}{2} v \boldsymbol{u} \cdot \boldsymbol{u} \rangle - \frac{1}{\rho} \frac{\mathrm{d}}{\mathrm{d}y} \langle v p' \rangle, \tag{7.64}$$

see Exercise 7.4. Figure 7.18 shows the terms in this equation for the viscous

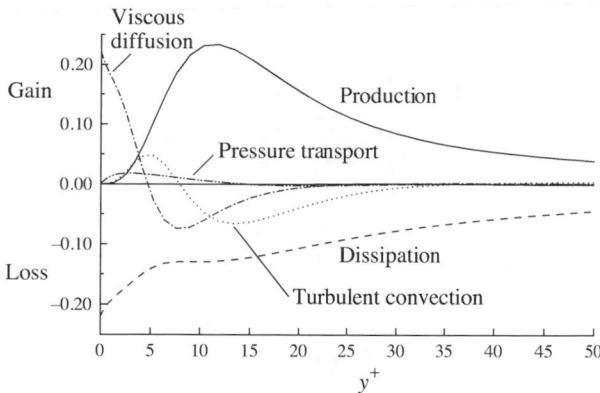

Fig. 7.18. The turbulent-kinetic-energy budget in the viscous wall region of channel flow: terms in Eq. (7.64) normalized by viscous scales. From the DNS data of Kim *et al.* (1987). Re = 13,750.

wall region. In order, the terms are production, pseudo-dissipation, viscous diffusion, turbulent convection, and pressure transport.

Like $-\langle uv \rangle$, the production $\mathcal{P}$ increases from zero as y^3. It reaches its peak value well within the buffer layer, at $y^+ \approx 12$. In fact, it can be shown (Exercise 7.6) that the peak production occurs precisely where the viscous stress and the Reynolds shear stress are equal. Around this peak, production exceeds dissipation ($\mathcal{P}/\varepsilon \approx 1.8$), and the excess energy produced is transported away. Pressure transport is small, while turbulent convection transports energy both toward the wall and into the log-law region. Viscous transport – $\nu\, \mathrm{d}^2 k/\mathrm{d}y^2$ – transports kinetic energy all the way to the wall.

Perhaps surprisingly, the peak dissipation occurs at the wall, where the kinetic energy is zero. Although the fluctuating velocity vanishes at $y = 0$, the fluctuating strain rate s_{ij} and hence the dissipation do not (Exercise 7.7). The dissipation at the wall is balanced by viscous transport,

$$\varepsilon = \tilde{\varepsilon} = \nu\, \frac{\mathrm{d}^2 k}{\mathrm{d}y^2}, \quad \text{for } y = 0, \tag{7.65}$$

the other terms in Eq. (7.64) being zero. (See also Exercises 7.5 and 7.7.)

For fully turbulent flow, the statistics considered here (normalized by the viscous scales) have only a weak dependence on Reynolds number in the inner layer ($y/\delta < 0.1$). Figure 7.19 shows profiles of the r.m.s. of u and v measured at various Reynolds numbers. The peak value of u'/u_τ appears independent of Re; but at $y^+ = 50$ (which is within the inner layer for all but the lowest Reynolds number) the value of u'/u_τ increases by 20% between $\mathrm{Re}_0 = 14{,}914$ and $\mathrm{Re}_0 = 39{,}582$. These and other Reynolds-number effects are discussed by Wei and Willmarth (1989) and Antonia *et al.* (1992).

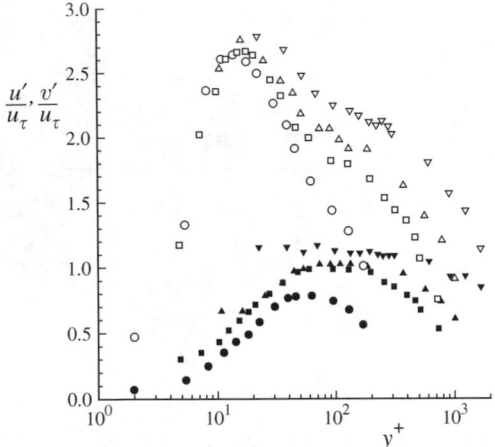

Fig. 7.19. Profiles of r.m.s. velocity measured in channel flow at various Reynolds numbers by Wei and Willmarth (1989). Open symbols: $u'/u_\tau = \langle u^2 \rangle^{1/2}/u_\tau$; $\bigcirc$, $\text{Re}_0 = 2{,}970$; $\square$, $\text{Re}_0 = 14{,}914$; $\triangle$, $\text{Re}_0 = 22{,}776$; $\triangledown$, $\text{Re}_0 = 39{,}582$. Solid symbols: $v'/u_\tau = \langle v^2 \rangle^{1/2}/u_\tau$ at the same Reynolds numbers.

EXERCISES

7.4   Starting from Eqs. (5.139) and (5.164), show that the turbulent kinetic energy equation for fully developed channel flow can be written

$$\frac{\mathrm{d}}{\mathrm{d}y}\left(\tfrac{1}{2}\langle v\boldsymbol{u}\cdot\boldsymbol{u}\rangle + \frac{\langle vp'\rangle}{\rho} - v\frac{\mathrm{d}}{\mathrm{d}y}(k + \langle v^2\rangle) \right) = \mathcal{P} - \varepsilon, \qquad (7.66)$$

or

$$\frac{\mathrm{d}}{\mathrm{d}y}\left(\tfrac{1}{2}\langle v\boldsymbol{u}\cdot\boldsymbol{u}\rangle + \frac{\langle vp'\rangle}{\rho} - v\frac{\mathrm{d}k}{\mathrm{d}y} \right) = \mathcal{P} - \tilde{\varepsilon}. \qquad (7.67)$$

For this flow, determine the relationship between ε and $\tilde{\varepsilon}$ (see Exercise 5.25 on page 133).

7.5   By using the expansions Eqs. (7.56)–(7.58), show that, very close to the wall, the orders of the terms in the kinetic-energy equation are

$$\mathcal{P} = \mathcal{O}(y^3), \qquad \varepsilon = \mathcal{O}(1),$$

$$v\frac{\mathrm{d}^2 k}{\mathrm{d}y^2} = \mathcal{O}(1), \qquad v\frac{\mathrm{d}^2\langle v^2\rangle}{\mathrm{d}y^2} = \mathcal{O}(y^2), \qquad (7.68)$$

$$\frac{\mathrm{d}}{\mathrm{d}y}\langle \tfrac{1}{2}v\boldsymbol{u}\cdot\boldsymbol{u}\rangle = \mathcal{O}(y^3), \qquad \frac{1}{\rho}\frac{\mathrm{d}}{\mathrm{d}y}\langle vp'\rangle = \mathcal{O}(y).$$

7.6   For fully developed turbulent channel flow, show that the Reynolds shear stress can be written

$$-\langle uv \rangle = \frac{\tau_w}{\rho}\left(1 - \frac{y}{\delta}\right) - \nu S, \qquad (7.69)$$

and hence that the production rate is

$$\mathcal{P} = \frac{\tau_w}{\rho}\left(1 - \frac{y}{\delta}\right)S - \nu S^2. \qquad (7.70)$$

From this expression for $\mathcal{P}$, show that the peak production $\check{\mathcal{P}}$ occurs at the location $\check{y}$ where the viscous and Reynolds stresses are equal. Show that this peak value is

$$\frac{\nu\check{\mathcal{P}}}{u_\tau^4} = \tfrac{1}{4}[\tau(\check{y})/\tau_w]^2 < \tfrac{1}{4}. \qquad (7.71)$$

7.7   Show that the fluctuating rate of strain at a stationary solid wall $(y = 0)$ is

$$s_{ij} \equiv \frac{1}{2}\left(\frac{\partial u_i}{\partial x_j} + \frac{\partial u_j}{\partial x_i}\right) = \frac{1}{2}\begin{bmatrix} 0 & \dfrac{\partial u}{\partial y} & 0 \\[2mm] \dfrac{\partial u}{\partial y} & 0 & \dfrac{\partial w}{\partial y} \\[2mm] 0 & \dfrac{\partial w}{\partial y} & 0 \end{bmatrix} = \frac{1}{2}\begin{bmatrix} 0 & b_1 & 0 \\ b_1 & 0 & b_3 \\ 0 & b_3 & 0 \end{bmatrix},$$

$$(7.72)$$

where b_1 and b_3 are the coefficients in Eqs. (7.56) and (7.58). Hence, obtain the following result due to Hanjalić and Launder (1976): for $y = 0$

$$\varepsilon \equiv 2\nu\langle s_{ij}s_{ij} \rangle = \nu(\langle b_1^2 \rangle + \langle b_3^2 \rangle)$$

$$= \nu \frac{\partial^2 k}{\partial y^2} = 2\nu\left(\frac{\partial k^{1/2}}{\partial y}\right)^2. \qquad (7.73)$$

Show that ε and $\tilde{\varepsilon}$ are equal at $y = 0$.

7.8   Let ε_0^+ denote the dissipation at the wall normalized by the viscous scales, i.e.,

$$\varepsilon_0^+ \equiv \varepsilon_{y=0}\frac{\delta_\nu}{u_\tau^3}. \qquad (7.74)$$

Use Fig. 7.18 to estimate the value of ε_0^+. Show that the Kolmogorov scale at the wall is

$$\frac{\eta_{y=0}}{\delta_\nu} = (\varepsilon_0^+)^{-1/4} \approx 1.5. \qquad (7.75)$$

7.9   The expansion for the Reynolds shear stress at the wall (Eq. (7.63))
       can be written

$$\langle uv \rangle = -\sigma u_\tau^2 y^{+3} \dots, \tag{7.76}$$

where the non-dimensional coefficient σ may be assumed to be independent of the Reynolds number. Show from the momentum equation (Eqs. (7.10) and (7.13)) that this implies the following expansion
for $\langle U \rangle$:

$$u^+ = y^+ - \frac{y^{+2}}{2\mathrm{Re}_\tau} - \tfrac{1}{4}\sigma y^{+4} \dots. \tag{7.77}$$

Why does it follow that the expansion for the law of the wall is

$$f_\mathrm{w}(y^+) = y^+ - \tfrac{1}{4}\sigma y^{+4} \dots? \tag{7.78}$$

7.1.7 Lengthscales and the mixing length

Three fundamental properties of the log-law region are the form of the mean
velocity gradient,

$$S = \frac{\mathrm{d}\langle U \rangle}{\mathrm{d}y} = \frac{u_\tau}{\kappa y}, \quad \text{or} \quad \frac{\mathrm{d}u^+}{\mathrm{d}y^+} = \frac{1}{\kappa y^+}; \tag{7.79}$$

the fact that production and dissipation are almost in balance,

$$\mathcal{P}/\varepsilon \approx 1; \tag{7.80}$$

and the near constancy of the normalized Reynolds shear stress,

$$-\langle uv \rangle / k \approx 0.3. \tag{7.81}$$

A fourth property, that follows from these three, is the near constancy of the
turbulence-to-mean-shear timescale ratio

$$\frac{Sk}{\varepsilon} = \left| \frac{k}{\langle uv \rangle} \right| \frac{\mathcal{P}}{\varepsilon} \approx 3. \tag{7.82}$$

From these relations, it is a matter of algebra to deduce that the turbulence
lengthscale $L \equiv k^{3/2}/\varepsilon$ varies as

$$L = \kappa y \frac{|\langle uv \rangle|^{1/2}}{u_\tau} \left(\frac{\mathcal{P}}{\varepsilon} \right) \left| \frac{\langle uv \rangle}{k} \right|^{-3/2}. \tag{7.83}$$

At high Reynolds number, in the overlap region ($50\delta_\nu < y < 0.1\delta$), the
Reynolds stress is essentially constant, so that then L varies linearly with y:

$$L = C_L y, \tag{7.84}$$

with

$$C_L \approx \kappa \left(\frac{\mathcal{P}}{\varepsilon}\right) \left|\frac{\langle uv \rangle}{k}\right|^{-3/2} \approx 2.5. \tag{7.85}$$

Notice that $\mathcal{S}, \mathcal{P}$, and ε vary inversely with y, whereas L and $\tau = k/\varepsilon$ vary linearly with y.

(At the moderate Reynolds numbers accessible in DNS, there is no overlap region, and the shear stress changes appreciably over the log-law region. This, together with imperfections in the approximations Eqs. (7.79)–(7.81), results in Eq. (7.84) providing a poor approximation to L obtained from DNS.)

The turbulent viscosity $\nu_T(y)$ is defined so that the Reynolds shear stress is given by

$$-\langle uv \rangle = \nu_T \frac{\mathrm{d}\langle U \rangle}{\mathrm{d}y}. \tag{7.86}$$

It can be expressed as the product of a velocity scale u^* and a lengthscale ℓ_m:

$$\nu_T = u^* \ell_m. \tag{7.87}$$

One of these scales can be specified at will, and then the other determines ν_T. A propitious (implicit) specification is

$$u^* = |\langle uv \rangle|^{1/2}. \tag{7.88}$$

By substituting Eqs. (7.87) and (7.88) into Eq. (7.86) and taking the absolute value we obtain the explicit relation

$$u^* = \ell_m \left|\frac{\mathrm{d}\langle U \rangle}{\mathrm{d}y}\right|. \tag{7.89}$$

(In the upper half of the channel ($\delta < y < 2\delta$) the velocity gradient $\mathrm{d}\langle U \rangle/\mathrm{d}y$ is negative and the Reynolds stress $\langle uv \rangle$ is positive. The absolute values in Eqs. (7.88) and (7.89) ensure that u^* is non-negative for all y.)

In the overlap region ($50\delta_\nu < y < 0.1\delta$) that occurs at high Reynolds number, the shear stress $-\langle uv \rangle$ differs little from u_τ^2, and the mean velocity gradient is $u_\tau/(\kappa y)$. Consequently, u^* equals u_τ, and then Eq. (7.89) determines ℓ_m to be

$$\ell_m = \kappa y. \tag{7.90}$$

Like $L \equiv k^{3/2}/\varepsilon$, the lengthscale ℓ_m varies linearly with y.

The above relations constitute *Prandtl's mixing-length hypothesis* (Prandtl 1925). In summary, the turbulent viscosity is given by

$$\nu_T = u^* \ell_m = \ell_m^2 \left|\frac{\mathrm{d}\langle U \rangle}{\mathrm{d}y}\right|, \tag{7.91}$$

where ℓ_m is the *mixing length*. In the overlap region, ℓ_m varies linear with y, the constant of proportionality being the Kármán constant, κ.

In order to use the mixing-length hypothesis as a model of turbulence, it is necessary to specify ℓ_m outside the overlap region, i.e., in the viscous wall region and in the core. The discussion of this topic is deferred to Section 7.3.

EXERCISE

7.10　　From Eqs. (7.79)–(7.80), obtain the following estimates for the Kolmogorov scale in the log-law region:

$$\frac{\eta}{\delta_\nu} = (\kappa y^+)^{1/4}, \quad \frac{\eta}{L} = \frac{1}{C_L}\left(\frac{\kappa}{y^{+3}}\right)^{1/4}. \tag{7.92}$$

7.2 Pipe flow

Since Reynolds' experiment in 1883, pipe flow has played an important role in the development of our understanding of turbulent flows. In particular, it is quite simple to measure the drop in pressure over a length of fully developed turbulent pipe flow, and hence to determine the skin-friction coefficient, C_f. In an influential set of experiments performed during the 1930s, Nikuradse[2] measured C_f as a function of the Reynolds number for smooth pipes, and for pipes with varying amounts of wall roughness.

The main purpose of this section is to describe the effects of wall roughness, which are similar in pipe, channel, and boundary-layer flows. First the smooth-wall case is briefly outlined.

7.2.1 The friction law for smooth pipes

We consider the fully developed turbulent flow in a long straight pipe of circular cross section, with internal diameter D – see Fig. 7.1. In polar-cylindrical coordinates (x, r, θ), velocity statistics depend solely on the radial coordinate, r. The mean centerline velocity is denoted by U_0,

$$U_0 \equiv \langle U(x, 0, \theta)\rangle, \tag{7.93}$$

and the bulk velocity is

$$\bar{U} \equiv \frac{1}{\pi R^2}\int_0^R \langle U\rangle 2\pi r\, dr, \tag{7.94}$$

[2] See Schlichting (1979) for a description of Nikuradse's experiments and for references.

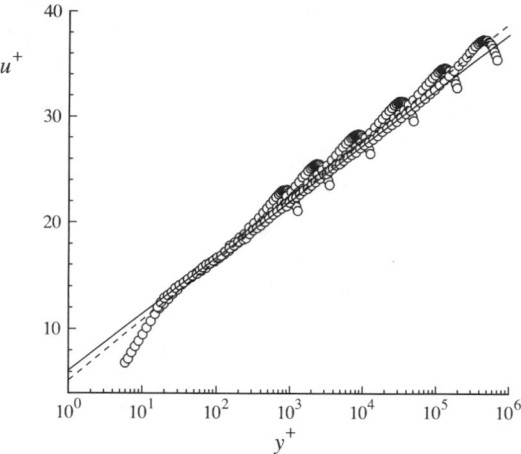

Fig. 7.20. Mean velocity profiles in fully developed turbulent pipe flow. Symbols, experimental data of Zagarola and Smits (1997) at six Reynolds numbers (Re ≈ 32×10^3, 99×10^3, 409×10^3, 1.79×10^6, 7.71×10^6, 29.9×10^6). Solid line, log law with $\kappa = 0.436$ and $B = 6.13$; dashed line, log law with $\kappa = 0.41$, $B = 5.2$.

where $R = \frac{1}{2}D$ is the pipe's radius. We take $\delta \equiv R$ to be the characteristic flow width, and then the conventionally defined Reynolds number is

$$\mathrm{Re} \equiv \frac{\bar{U}D}{\nu} = \frac{2\bar{U}\delta}{\nu}. \qquad (7.95)$$

As for channel flow, we define y to be the distance from the wall, i.e.,

$$y \equiv R - r. \qquad (7.96)$$

There is an abundance of experimental data showing that the mean velocity profile in the inner region ($y/\delta < 0.1$) is in accord with the universal law of the wall $u^+ = f_{\mathrm{w}}(y^+)$. Figure 7.20 shows mean velocity profiles measured by Zagarola and Smits (1997) in fully developed turbulent pipe flow at Reynolds numbers from Re ≈ 30×10^3 to Re ≈ 30×10^6. For comparison, the log law is shown with the standard constants ($\kappa = 0.41$, $B = 5.2$) and with those that best fit the data ($\kappa = 0.436$, $B = 6.13$). It may be seen that, for $y^+ > 30$, the profiles follow the log law for a range of y^+ that increases with Re; and, as expected, the profiles deviate from the log law as the pipe's centerline is approached. The same data restricted to $y/R < 0.1$ are shown in Fig. 7.21. Clearly, for all Reynolds numbers, the measured velocities for $y^+ > 30$ and $y/R < 0.1$ differ little from the log law.

The friction law for pipe flow is traditionally expressed in terms of the

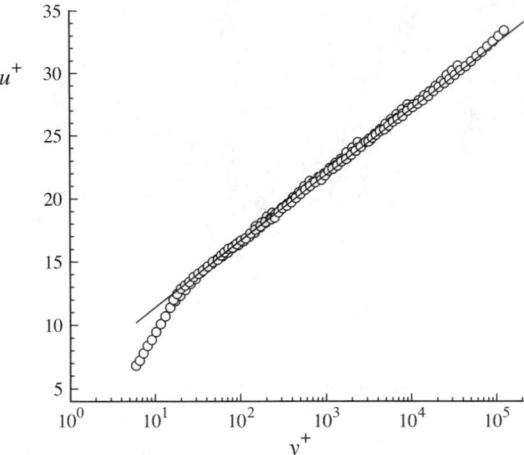

Fig. 7.21. Mean velocity profiles in fully developed turbulent pipe flow. Symbols, experimental data of Zagarola and Smits (1997) for $y/R < 0.1$, for the same values of Re as in Fig. 7.20. Line, log law with $\kappa = 0.436$ and $B = 6.13$.

friction factor f

$$f \equiv \frac{\Delta p\, D}{\frac{1}{2}\rho \bar{U}^2 \mathcal{L}}, \qquad (7.97)$$

where Δp is the drop in pressure over an axial distance $\mathcal{L}$. This is just four times the skin-friction coefficient C_f (see Exercise 7.12). Just as with channel flow, a friction law can be obtained by using the log law to approximate the velocity profile over the whole flow (see Exercises 7.14 and 7.15). With a small adjustment to the constants, the result (Eq. (7.110)) is *Prandtl's friction law for smooth pipes*

$$\frac{1}{\sqrt{f}} = 2.0 \log_{10}(\sqrt{f}\, \text{Re}) - 0.8, \qquad (7.98)$$

which implicitly yields f as a function of Re. As may be seen from Fig. 7.22, this friction law is in excellent agreement with experimental data over the entire range of turbulent Reynolds numbers.

EXERCISES

7.11   Starting from the Reynolds equations in polar-cylindrical coordinates (Eqs. (5.45)–(5.47)), show that, for fully developed turbulent pipe flow, the shear stress

$$\tau(r) \equiv \rho v\, \frac{\mathrm{d}\langle U \rangle}{\mathrm{d}r} - \rho \langle uv \rangle \qquad (7.99)$$

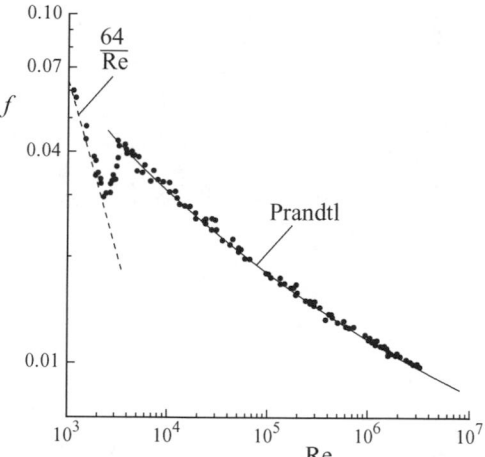

Fig. 7.22. The friction factor f against the Reynolds number for fully developed flow in smooth pipes. Dashed line, Hagen–Poiseuille friction law for laminar flow; solid line, Prandtl friction law for turbulent flow, Eq. (7.98); symbols, measurements compiled by Schlichting (1979). (Reproduced with permission of McGraw-Hill.)

is given by

$$\tau(r) = \tfrac{1}{2}r \frac{\mathrm{d}p_w}{\mathrm{d}x}, \tag{7.100}$$

where $p_w(x)$ is the mean pressure at the wall (cf. Eq. (7.9) for channel flow). Hence obtain the relation

$$-\frac{\mathrm{d}p_w}{\mathrm{d}x} = 2\frac{\tau_w}{R}, \tag{7.101}$$

where the wall shear stress τ_w – a positive quantity – is defined as

$$\tau_w = -\tau(R). \tag{7.102}$$

(Note that, with the coordinate system used here, $\langle uv \rangle$ is positive, and the velocity gradient $\mathrm{d}\langle U \rangle/\mathrm{d}r$ is negative.)

7.12 With the friction factor f being defined by Eq. (7.97), and the skin-friction coefficient C_f by Eq. (7.15), obtain the relations

$$f = 4C_f, \tag{7.103}$$

$$\frac{u_\tau}{\bar{U}} = \sqrt{\frac{f}{8}}, \tag{7.104}$$

where $u_\tau \equiv \sqrt{\tau_w/\rho}$ is the friction velocity.

7.13 For laminar flow, solve Eq. (7.99) to show that the velocity profile is parabolic, and that the centerline velocity U_0 is twice the bulk velocity. Obtain the *Hagen–Poiseuille friction law* for fully developed laminar pipe flow,

$$f = \frac{64}{\mathrm{Re}}. \tag{7.105}$$

7.14   By approximating the mean velocity profile by the logarithmic defect law (Eq. (7.50)) with $B_1 = 0$, i.e.,

$$\frac{U_0 - \langle U \rangle}{u_\tau} = -\frac{1}{\kappa} \ln\left(\frac{y}{R}\right), \tag{7.106}$$

obtain the estimate

$$\frac{U_0 - \bar{U}}{u_\tau} = \frac{3}{2\kappa} \approx 3.66, \tag{7.107}$$

(cf. Eq. (7.52) for channel flow). (According to Schlichting (1979), the value 4.07 is in better agreement with experimental data.)

7.15   With $y \equiv R - r$ being the distance from the wall, the log law Eq. (7.43) is

$$u^+ \equiv \frac{\langle U \rangle}{u_\tau} = \frac{1}{\kappa} \ln\left(\frac{yu_\tau}{\nu}\right) + B. \tag{7.108}$$

By assuming that this holds on the axis, and by using Eqs. (7.104), obtain the friction law

$$\frac{1}{\sqrt{f}} = \frac{1}{2\sqrt{2}\kappa} \ln(\mathrm{Re}\sqrt{f}) - \frac{(3 + 5\ln 2 - 2\kappa B)}{4\sqrt{2}\kappa} \tag{7.109}$$

or

$$\frac{1}{\sqrt{f}} \approx 1.99 \log_{10}(\mathrm{Re}\sqrt{f}) - 0.95, \tag{7.110}$$

for $\kappa = 0.41$ and $B = 5.2$.

7.16   Let $\bar{y}$ denote the y location at which the mean velocity gradient $\mathrm{d}\langle U \rangle/\mathrm{d}y$ is equal to $\bar{U}/\delta$. Assuming that $\bar{y}$ lies in the log-law region, show that

$$\frac{\bar{y}}{\delta} = \frac{1}{\kappa}\sqrt{\frac{f}{8}} \approx 0.86\sqrt{f}, \tag{7.111}$$

$$\bar{y}^+ \equiv \frac{\bar{y}}{\delta_\nu} = \frac{\mathrm{Re}f}{16\kappa} \approx 0.15\mathrm{Re}f. \tag{7.112}$$

Estimate $\bar{y}/\delta$ and $\bar{y}^+$ for $\mathrm{Re} = 10^4$ and 10^6.

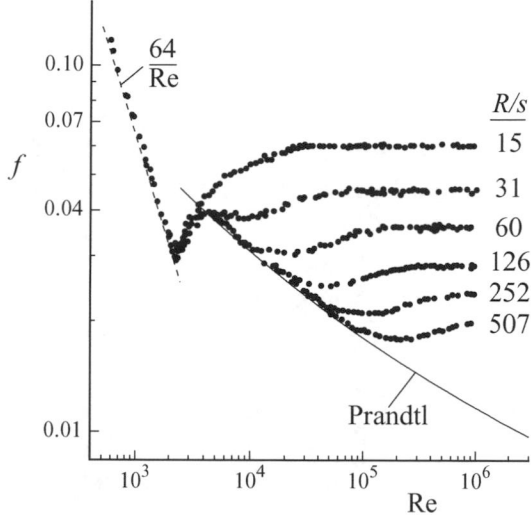

Fig. 7.23. The friction factor f against the Reynolds number for fully developed flow in pipes of various roughnesses. Dashed line, friction law for laminar flow; solid line, Prandtl friction law for turbulent flow in smooth pipes, Eq. (7.98); symbols, measurements of Nikuradse. (Adapted from Schlichting (1979) with permission of McGraw-Hill.)

7.2.2 Wall roughness

Up to this point we have assumed that the walls of channels and pipes are completely smooth. Of course, in practice every surface departs from the ideal to some extent, and to a first approximation this departure is characterized by a lengthscale of protrusions or indentations, s. For a given flow (i.e., given $R, \bar{U}$, and v), the primary questions to address are the following: Is there a value of s, (s^*, say), below which the flow is independent of s, so that the wall is effectively smooth? For $s > s^*$, how does the roughness affect the flow?

Nikuradse performed experiments on pipes with sand glued to the wall as densely as possible, with grain sizes s varying from $s/R = 1/15$ to $s/R = 1/500$. The measurements of the friction factor f are shown in Fig. 7.23. It may be seen that the roughness has little effect in the laminar regime, and apparently little effect on transition. Then the curves for each roughness follow the same line – namely the Prandtl law for smooth pipes – up to some Reynolds number before turning upward, and reaching an asymptote. At the highest Reynolds numbers, the friction factor is independent of Re, with an asymptotic value that increases with s/R.

The observed behavior is explained by the extension of the law of the wall to incorporate roughness. For a given geometry of the surface (so that the roughness is fully characterized by s) the mean velocity gradient can be written

$$\frac{d\langle U \rangle}{dy} = \frac{u_\tau}{y} \bar{\Phi}\left(\frac{y}{\delta_v}, \frac{y}{\delta}, \frac{s}{\delta_v}\right), \tag{7.113}$$

where $\bar{\Phi}$ is a universal non-dimensional function (cf. Eq. (7.31)). Just as before, it is postulated that there is no dependence of $\bar{\Phi}$ on y/δ in the inner layer ($y/\delta < 0.1$).

At high Reynolds number, two extreme cases can be considered. If s/δ_v is very small, there is every reason to suppose that the flow is unaffected by the roughness, and then the standard law of the wall is recovered:

$$\frac{d\langle U \rangle}{dy} = \frac{u_\tau}{y}\Phi_I\left(\frac{y}{\delta_v}\right), \quad \text{for } s \ll \delta_v \text{ and } y \ll \delta, \tag{7.114}$$

where

$$\Phi_I\left(\frac{y}{\delta_v}\right) = \lim_{\substack{y/\delta \to 0 \\ s/\delta_v \to 0}} \bar{\Phi}\left(\frac{y}{\delta_v}, \frac{y}{\delta}, \frac{s}{\delta_v}\right). \tag{7.115}$$

For large y/δ_v, the supposition that the dependence on viscosity vanishes implies that Φ_I tends asymptotically to a constant, $\Phi_I \sim 1/\kappa$, and then Eq. (7.114) integrates to the log law, i.e.,

$$\frac{\langle U \rangle}{u_\tau} = u^+ = \frac{1}{\kappa}\ln\left(\frac{y}{\delta_v}\right) + B, \quad \text{for } s \ll \delta_v \ll y \ll \delta, \tag{7.116}$$

where

$$B \equiv \lim_{y^* \to \infty} \left\{ \int_0^{y^*} \Phi_I(y^+)\frac{dy^+}{y^+} - \frac{1}{\kappa}\ln y^+ \right\}, \tag{7.117}$$

is a universal constant.

In the second extreme case, the roughness scale s is large compared with the viscous scale δ_v. Then the local Reynolds number of the flow over the roughness elements is large ($u_\tau s/\nu = s/\delta_v \gg 1$). The transfer of momentum from the fluid to the wall is accomplished by the drag on the roughness elements, which at high Reynolds number is predominantly by pressure forces, rather than by viscous stresses. It can be supposed, then, that ν and hence δ_v are not relevant parameters, so that Eq. (7.113) can be rewritten

$$\frac{d\langle U \rangle}{dy} = \frac{u_\tau}{y}\Phi_R\left(\frac{y}{s}\right), \quad \text{for } \delta_v \ll s \text{ and } y \ll \delta, \tag{7.118}$$

where Φ_R is a universal non-dimensional function (for a given roughness geometry).

For $y \gg s$ it can be supposed that the turbulence is determined by local

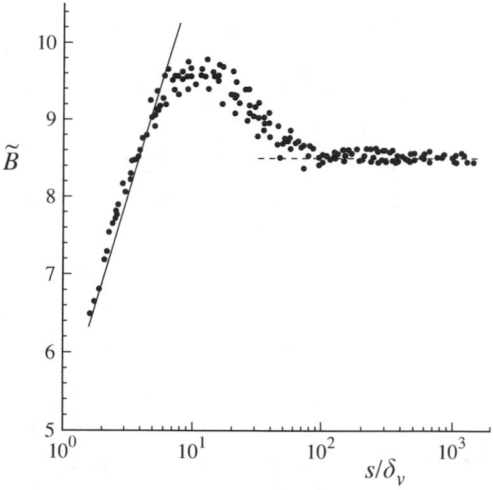

Fig. 7.24. The additive constant in the log law $\tilde{B}$ (Eq. (7.121)) as a function of the roughness scale s normalized by the viscous length δ_v. Dashed line, fully rough $\tilde{B} = 8.5$; solid line, smooth (Eq. (7.122)); symbols, from Nikuradse's data. (Adapted from Schlichting (1979) with permission of McGraw-Hill.)

processes, independent of s – the same processes as those that occur for a smooth wall – which implies that Φ_R tends asymptotically to the constant $1/\kappa$. Then Eq. (7.118) integrates to the log law

$$u^+ = \frac{1}{\kappa} \ln\left(\frac{y}{s}\right) + B_2, \quad \text{for } \delta_v \ll s \ll y \ll \delta, \tag{7.119}$$

where

$$B_2 \equiv \lim_{y^* \to \infty} \left[\int_0^{y^*} \Phi_R\left(\frac{y}{s}\right) \frac{\mathrm{d}y}{y} - \frac{1}{\kappa} \ln\left(\frac{y}{s}\right) \right], \tag{7.120}$$

is a universal constant.

For the general case in which s is comparable to δ_v, similar arguments lead to the conclusion that (for y large compared both with δ_v and with s) there is a log law with constant κ and additive constant $\tilde{B}$ that depends upon s/δ_v, i.e.,

$$u^+ = \frac{1}{\kappa} \ln\left(\frac{y}{s}\right) + \tilde{B}\left(\frac{s}{\delta_v}\right). \tag{7.121}$$

For the smooth wall ($s/\delta_v \ll 1$), Eq. (7.116) corresponds to Eq. (7.121) with

$$\tilde{B}\left(\frac{s}{\delta_v}\right) = B + \frac{1}{\kappa} \ln\left(\frac{s}{\delta_v}\right), \tag{7.122}$$

whereas for a *fully rough* wall ($s/\delta_v \gg 1$) Eq. (7.119) corresponds to

Eq. (7.121) with

$$\tilde{B}\left(\frac{s}{\delta_v}\right) = B_2. \tag{7.123}$$

Experiments indeed confirm the log law (Eq. (7.121)) for rough walls, and the additive constant $\tilde{B}$ has been determined as a function of s/δ_v from Nikuradse's data, see Fig. 7.24. Evidently, for $s/\delta_v > 70$, say, the wall is fully rough with $B_2 = \tilde{B}(\infty) = 8.5$. At the other extreme, the measured values of $\tilde{B}$ agree with Eq. (7.122) up to $s/\delta_v \approx 5$, say, giving this as the limit of *admissible roughness* – the limit s^* below which the wall is effectively smooth.

The log law for the fully rough case leads to an accurate friction law (Eq. (7.124) of Exercise 7.17), giving the friction factor f as a function of the roughness s/R (independent of the Reynolds number).

EXERCISE _____

7.17   Show that the log law for a fully rough wall Eq. (7.119) together with Eq. (7.107) yields the friction law

$$f = 8\left[\frac{1}{\kappa}\ln\left(\frac{R}{s}\right) + B_2 - \frac{3}{2\kappa}\right]^{-2}$$

$$\approx \frac{1}{\left[1.99\log_{10}(R/s) + 1.71\right]^2}. \tag{7.124}$$

Compare this law with the experimental data in Fig. 7.23.
(Schlichting (1979) suggests the slightly modified values of 2.0 and 1.74 in place of 1.99 and 1.71.)

7.3 Boundary layers

The simplest boundary layer to consider is that which is formed when a uniform-velocity non-turbulent stream flows over a smooth flat plate (see Fig. 7.1). Compared with fully developed channel flow with a given mean pressure gradient, the primary differences are:

   (i) the boundary layer develops continuously in the flow direction, with the boundary-layer thickness $\delta(x)$ increasing with x;
   (ii) the wall shear stress $\tau_w(x)$ is not known *a priori*; and
   (iii) the outer part of the flow consists of intermittent turbulent/non-turbulent motion (see Section 5.5.2).

In spite of these differences, the behavior in the inner layer ($y/\delta(x) < 0.1$) is essentially the same as that in channel flow. This is demonstrated, and the behavior in the buffer layer is examined in more detail. In the defect layer ($y/\delta(x) > 0.1$), the departures from the log law are more significant, which warrants a closer examination of the velocity-defect law.

7.3.1 A description of the flow

As sketched in Fig. 7.1, the coordinate system is the same as that used for channel flow. The surface of the plate (i.e., the wall) is at $x_2 = y = 0$ for $x_1 = x \geq 0$, with the leading edge being $x = 0$, $y = 0$. The mean flow is predominantly in the x direction, with the free-stream velocity (outside the boundary layer) being denoted by $U_0(x)$. Statistics vary primarily in the y direction, and are independent of z. Unlike channel flow, however, the boundary layer continually develops, so that statistics depend both upon x and upon y. The velocity components are U, V, and W, with $\langle W \rangle$ being zero.

The free-stream pressure $p_0(x)$ is linked to the velocity $U_0(x)$ by Bernoulli's equation (Eq. (2.67)) – $p_0(x) + \frac{1}{2}\rho U_0(x)^2 = $ constant – so that the pressure gradient is

$$-\frac{\mathrm{d}p_0}{\mathrm{d}x} = \rho U_0 \frac{\mathrm{d}U_0}{\mathrm{d}x}. \tag{7.125}$$

Accelerating flow ($\mathrm{d}U_0/\mathrm{d}x > 0$) corresponds to a negative – or *favorable* – pressure gradient. Conversely, decelerating flow yields a positive, *adverse* pressure gradient, so called because it can lead to separation of the boundary layer from the surface. In aeronautical applications, it is generally desirable for boundary layers to be attached. Most of our attention is focused on the zero-pressure-gradient case, corresponding to $U_0(x)$ being constant.

The *boundary-layer thickness* $\delta(x)$ is generally defined as the value of y at which $\langle U(x, y) \rangle$ equals 99% of the free-stream velocity $U_0(x)$. This is a poorly conditioned quantity, since it depends on the measurement of a small velocity difference. More reliable are integral measures such as the *displacement thickness*

$$\delta^*(x) \equiv \int_0^\infty \left(1 - \frac{\langle U \rangle}{U_0} \right) \mathrm{d}y, \tag{7.126}$$

and the *momentum thickness*

$$\theta(x) \equiv \int_0^\infty \frac{\langle U \rangle}{U_0} \left(1 - \frac{\langle U \rangle}{U_0} \right) \mathrm{d}y. \tag{7.127}$$

Various Reynolds numbers are defined on the basis of these thicknesses and also of x:

$$\text{Re}_x \equiv \frac{U_0 x}{\nu}, \quad \text{Re}_\delta \equiv \frac{U_0 \delta}{\nu}, \quad \text{Re}_{\delta^*} \equiv \frac{U_0 \delta^*}{\nu}, \quad \text{Re}_\theta \equiv \frac{U_0 \theta}{\nu}. \quad (7.128)$$

In a zero-pressure-gradient boundary layer, there is laminar flow from the leading edge ($x = 0$) until the location at which Re_x reaches a critical value $\text{Re}_{\text{crit}} \approx 10^6$ marking the start of transition. (The value of Re_{crit} varies considerably, depending on the nature and level of the disturbances in the free stream, see, e.g., Schlichting (1979).) Transition occurs over some distance (maybe 30% of the distance from the leading edge), after which the boundary layer is fully turbulent. In some experiments a wire or other device is placed across the flow in order to *trip* the laminar boundary layer, i.e., to promote the transition to turbulence. (For more information on the topic of transition the reader is referred to Arnal and Michel (1990), Kachanov (1994) and references therein.)

7.3.2 Mean-momentum equations

Naturally, the boundary-layer equations apply – the flow develops slowly in the x direction, with axial stress gradients being small compared with cross-stream gradients. The lateral mean momentum equation (Eq. (5.52)) integrates to

$$\langle p \rangle + \rho \langle v^2 \rangle = p_0(x). \quad (7.129)$$

Notice that, since $\langle v^2 \rangle$ is zero at the wall, the wall pressure $p_w(x)$ equals the free-stream pressure, $p_0(x)$.

In the boundary-layer approximation, the mean-axial-momentum equation is

$$\langle U \rangle \frac{\partial \langle U \rangle}{\partial x} + \langle V \rangle \frac{\partial \langle U \rangle}{\partial y} = \nu \frac{\partial^2 \langle U \rangle}{\partial y^2} - \frac{\partial \langle uv \rangle}{\partial y} - \frac{1}{\rho} \frac{\mathrm{d} p_0}{\mathrm{d} x}$$

$$= \frac{1}{\rho} \frac{\partial \tau}{\partial y} + U_0 \frac{\mathrm{d} U_0}{\mathrm{d} x}, \quad (7.130)$$

where $\tau(x, y)$ is the total shear stress,

$$\tau = \rho \nu \frac{\partial \langle U \rangle}{\partial y} - \rho \langle uv \rangle, \quad (7.131)$$

(see Eq. (5.55)). At the wall the convective terms are zero, so that the shear stress and pressure gradients balance. If the pressure gradient is zero, then

$$\frac{1}{\rho} \left(\frac{\partial \tau}{\partial y} \right)_{y=0} = \nu \left(\frac{\partial^2 \langle U \rangle}{\partial y^2} \right)_{y=0} = 0 \quad (7.132)$$

(since $\langle uv \rangle$ increases from zero as y^3).

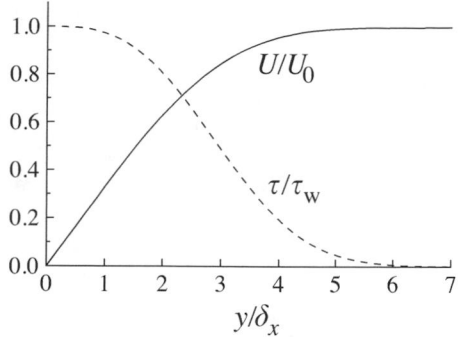

Fig. 7.25. Normalized velocity and shear-stress profiles from the Blasius solution for the zero-pressure-gradient laminar boundary layer on a flat plate: y is normalized by $\delta_x \equiv x/\mathrm{Re}_x^{1/2} = (xv/U_0)^{1/2}$.

The boundary-layer momentum equation (Eq. (7.130)) can be integrated to obtain *von Kármán's integral momentum equation* (see Exercise 7.18). For the zero-pressure-gradient case the result is

$$\tau_w = \frac{d}{dx}(\rho U_0^2 \theta) = \rho U_0^2 \frac{d\theta}{dx}, \qquad (7.133)$$

or, for the skin-friction coefficient,

$$c_f \equiv \frac{\tau_w}{\frac{1}{2}\rho U_0^2} = 2\frac{d\theta}{dx}. \qquad (7.134)$$

Equation (7.133) quantifies the decrease in the momentum-flow rate of the stream – or the increase in the momentum deficit – caused by the wall shear stress.

For the laminar zero-pressure-gradient boundary layer, there is a similarity solution to Eq. (7.130) due to Blasius (1908), described in detail by Schlichting (1979). The scaled velocity $U(x, y)/U_0$ depends solely on the scaled cross-stream coordinate y/δ_x, where the lengthscale δ_x is $\delta_x \equiv (xv/U_0)^{1/2} = x/\mathrm{Re}_x^{1/2}$. This solution is shown in Fig. 7.25. The various thicknesses obtained from the solution are

$$\frac{\delta}{x} \approx 4.9\mathrm{Re}_x^{-1/2}, \quad \frac{\delta^*}{\delta} \approx 0.35, \quad \frac{\theta}{\delta} \approx 0.14. \qquad (7.135)$$

The skin-friction coefficient is

$$c_f \approx 0.664\,\mathrm{Re}_x^{-1/2}. \qquad (7.136)$$

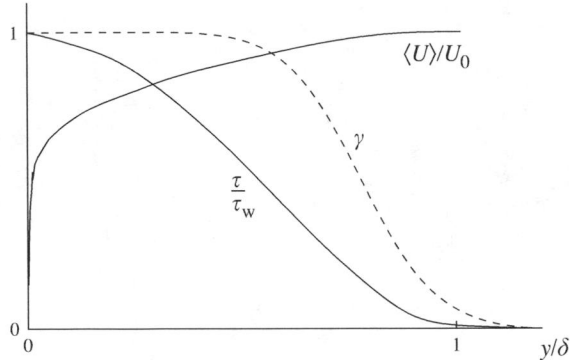

Fig. 7.26. Profiles of the mean velocity, shear stress and intermittency factor in a zero-pressure-gradient turbulent boundary layer, $\mathrm{Re}_\theta = 8{,}000$. (From the experimental data of Klebanoff (1954).)

EXERCISE

7.18   From the mean continuity (Eq. (5.49)) and momentum (Eq. (5.55)) equations, obtain the result

$$\frac{\partial}{\partial x}[\langle U\rangle(U_0 - \langle U\rangle)] + \frac{\partial}{\partial y}[\langle V\rangle(U_0 - \langle U\rangle)] + [U_0 - \langle U\rangle]\frac{\mathrm{d}U_0}{\mathrm{d}x} = -\frac{1}{\rho}\frac{\partial \tau}{\partial y}.$$

(7.137)

Integrate from $y = 0$ to $y = \infty$ to obtain

$$\frac{\mathrm{d}}{\mathrm{d}x}(U_0^2\theta) + \delta^* U_0 \frac{\mathrm{d}U_0}{\mathrm{d}x} = \frac{\tau_\mathrm{w}}{\rho},$$

and hence obtain *von Kármán's integral momentum equation*

$$c_\mathrm{f} \equiv \frac{\tau_\mathrm{w}}{\frac{1}{2}\rho U_0^2} = 2\frac{\mathrm{d}\theta}{\mathrm{d}x} + \frac{(4\theta + 2\delta^*)}{U_0}\frac{\mathrm{d}U_0}{\mathrm{d}x}.$$

(7.138)

(Recall that δ^* and θ are the displacement and momentum thicknesses, Eqs. (7.126) and (7.127).)

7.3.3 Mean velocity profiles

Figure 7.26 shows the profiles of the mean velocity, shear stress and intermittency factor γ (defined in Section 5.5.2) measured by Klebanoff (1954) in a zero-pressure-gradient turbulent boundary layer of momentum-thickness Reynolds number $\mathrm{Re}_\theta = 8{,}000$. Notice that the mean velocity profile rises much more steeply from the wall than does the Blasius profile (Fig. 7.25),

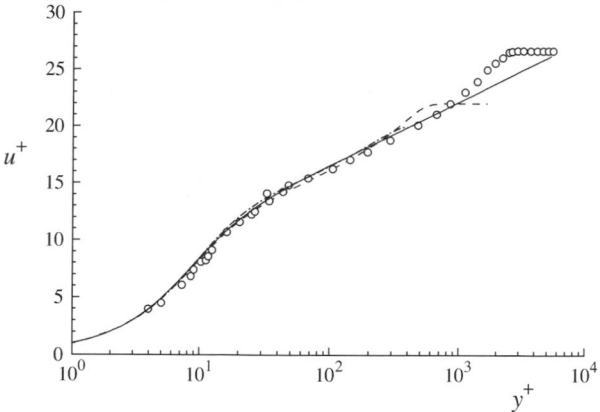

Fig. 7.27. Mean velocity profiles in wall units. Circles, boundary-layer experiments of Klebanoff (1954), $Re_\theta = 8,000$; dashed line, boundary-layer DNS of Spalart (1988), $Re_\theta = 1,410$; dot–dashed line, channel flow DNS of Kim *et al.* (1987), $Re = 13,750$; solid line, van Driest's law of the wall, Eqs. (7.144)–(7.145).

and then it is flatter away from the wall. This 'flatness' of the mean velocity profile is quantified by the shape factor H, defined as the ratio of the displacement and momentum thicknesses:

$$H \equiv \delta^*/\theta. \tag{7.139}$$

For the Blasius profile the flatness factor is $H \approx 2.6$: for the Klebanoff boundary layer it is $H \approx 1.3$. The mean velocity profile appears very similar to that in channel flow (Fig. 7.2); and, just as with channel flow, with increasing Reynolds number, the profiles of $\langle U \rangle / U_0$ plotted against y/δ steepen at the wall and become flatter away from the wall. Correspondingly, the shape factor H decreases with Re.

The shear-stress profile is similar to the corresponding laminar profile (Fig. 7.25), even though the origin of the shear stress is entirely different. The velocity profiles in the various layers are now examined in more detail.

The law of the wall

Velocity profiles, normalized by the viscous scales u_τ and δ_ν, are shown in Fig. 7.27 for three flows: Klebanoff's boundary layer ($Re_\theta = 8,000$); a boundary-layer DNS ($Re_\theta = 1,410$) performed by Spalart (1988); and channel flow ($Re = 13,750$). The agreement between the profiles illustrates the universality of the law of the wall, not only in the log-law region, but also in the buffer layer. In the viscous sublayer ($y^+ < 5$) the law of the wall is $f_w(y^+) \approx y^+$, while the log law holds for y^+ greater than 30 or 50 (and $y/\delta \ll 1$). What form does the law of the wall take in the buffer layer ($5 <$

$y^+ < 50$)? A purely empirical answer – but an inspired one nonetheless – is provided by van Driest (1956), in the context of the mixing-length hypothesis.

For a boundary layer (with $\partial \langle U \rangle / \partial y > 0$), according to the mixing-length hypothesis, the total shear stress is

$$\tau(y)/\rho = v \frac{\partial \langle U \rangle}{\partial y} + v_{\mathrm{T}} \frac{\partial \langle U \rangle}{\partial y}$$

$$= v \frac{\partial \langle U \rangle}{\partial y} + \ell_{\mathrm{m}}^2 \left(\frac{\partial \langle U \rangle}{\partial y} \right)^2 . \tag{7.140}$$

With the definition

$$\ell_{\mathrm{m}}^+ \equiv \ell_{\mathrm{m}}/\delta_v , \tag{7.141}$$

when it is normalized by viscous scales, Eq. (7.140) becomes

$$\frac{\tau}{\tau_{\mathrm{w}}} = \frac{\partial u^+}{\partial y^+} + \left(\ell_{\mathrm{m}}^+ \frac{\partial u^+}{\partial y^+} \right)^2 . \tag{7.142}$$

This is a quadratic equation for $\partial u^+ / \partial y^+$, which has the solution

$$\frac{\partial u^+}{\partial y^+} = \frac{2\tau/\tau_{\mathrm{w}}}{1 + [1 + (4\tau/\tau_{\mathrm{w}})(\ell_{\mathrm{m}}^+)^2]^{1/2}} . \tag{7.143}$$

In the inner layer, the ratio τ/τ_{w} is essentially unity, so that the law of the wall is obtained in terms of the mixing length as the integral of Eq. (7.143):

$$u^+ = f_{\mathrm{w}}(y^+) = \int_0^{y^+} \frac{2 \, dy'}{1 + [1 + 4\ell_{\mathrm{m}}^+(y')^2]^{1/2}} . \tag{7.144}$$

In the log-law region, the appropriate specification of the mixing length is $\ell_{\mathrm{m}} = \kappa y$ or equivalently $\ell_{\mathrm{m}}^+ = \kappa y^+$ (Eq. (7.90)). If the same specification were used in the viscous sublayer, the implied turbulent stress $v_{\mathrm{T}} \, \partial \langle U \rangle / \partial y$ would increase as y^2, whereas $-\langle uv \rangle$ increases more slowly, as y^3. Evidently, then, the specification $\ell_{\mathrm{m}} = \kappa y$ needs to be reduced, or *damped*, near the wall. Accordingly, van Driest (1956) proposed the specification

$$\ell_{\mathrm{m}}^+ = \kappa y^+ [1 - \exp(-y^+/A^+)], \tag{7.145}$$

where A^+ is a constant ascribed the value $A^+ = 26$. The term in square brackets ([ ]) is referred to as the *van Driest damping function*.

The law of the wall given by Eqs. (7.144) and (7.145) is shown in Fig. 7.27. Clearly, it provides an excellent representation of the data.

For large y^+, the damping function tends to unity and the log law is recovered. Notice that (for given κ) there is a one-to-one correspondence between the van Driest constant A^+ and the additive constant B in the log law. With $\kappa = 0.41$, the specification $A^+ = 26$ corresponds to $B = 5.3$.

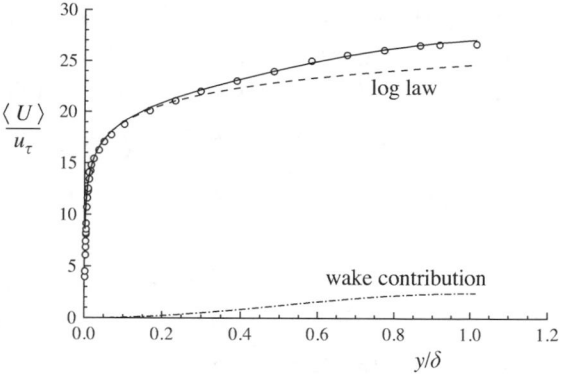

Fig. 7.28. The mean velocity profile in a turbulent boundary layer showing the law of the wake. Symbols, experimental data of Klebanoff (1954); dashed line, log law ($\kappa = 0.41, B = 5.2$); dot–dashed line, wake contribution $\Pi w(y/\delta)/\kappa$ ($\Pi = 0.5$); solid line, sum of log law and wake contribution (Eq. (7.148)).

The universality of the law of the wall has been considered extensively in the literature (see, e.g., Bradshaw and Huang (1995)). In boundary layers and duct flows with various pressure gradients dp_0/dx (and hence shear stress gradients $(\partial \tau/\partial y)_{y=0}$) the log law is observed with κ close to $\kappa = 0.41$. However, A^+ increases significantly when the wall shear-stress gradient $(-\partial \tau/\partial y)_{y=0}$ exceeds $2 \times 10^{-3}\tau_w/\delta_v$ (Huffman and Bradshaw 1972).

EXERCISE

7.19   Show that, according to the mixing-length hypothesis, very close to the wall ($y^+ \ll 1$) the Reynolds shear stress is

$$-\frac{\langle uv \rangle}{u_\tau^2} \approx (\ell_m^+)^2. \qquad (7.146)$$

Show that the van Driest specification (Eq. (7.145)) yields

$$-\frac{\langle uv \rangle}{u_\tau^2} \approx \left(\frac{\kappa}{A^+}\right)^2 y^{+4}. \qquad (7.147)$$

Contrast this result to the correct dependence of $\langle uv \rangle$ on y (for very small y), Eq. (7.63).

The velocity-defect law

In the *defect layer* ($y/\delta > 0.2$, say), the mean velocity deviates from the log law. This may be seen in Fig. 7.27, and more clearly in Fig. 7.28.

From an extensive examination of boundary-layer data, Coles (1956) showed that the mean velocity profile (over the whole boundary layer) is well represented by the sum of two functions. The first function is the law of the wall $f_w(y^+)$, which depends on y/δ_v: the second function, called the *law of the wake*, depends on y/δ. This representation is written

$$\frac{\langle U \rangle}{u_\tau} = f_w\left(\frac{y}{\delta_v}\right) + \frac{\Pi}{\kappa} w\left(\frac{y}{\delta}\right). \tag{7.148}$$

The *wake function* $w(y/\delta)$ is assumed to be universal (the same for all boundary layers) and is defined to satisfy the normalization conditions $w(0) = 0$ and $w(1) = 2$. Coles (1956) tabulated $w(y/\delta)$ (based on experimental data), but a more convenient approximation is

$$w\left(\frac{y}{\delta}\right) = 2\sin^2\left(\frac{\pi}{2}\frac{y}{\delta}\right). \tag{7.149}$$

The non-dimensional quantity Π is called the *wake strength parameter*, and its value is flow dependent.

For Klebanoff's boundary layer, Fig. 7.28 illustrates the representation of $\langle U \rangle/u_\tau$ as the sum of these two 'laws.' The dashed line is the log law (i.e., an excellent approximation to f_w for $y^+ > 50$, $y/\delta < 0.2$), while the dot–dashed line is the law of the wake. Their sum, the solid line, agrees well with the experimental data. As the name implies, the shape of the function $w(y/\delta)$ is similar to the velocity profile in a plane wake, with a symmetry plane at $y = 0$. However, there is no implication of a detailed similarity between these two flows.

Equation (7.148) can also be written in the form of a velocity-defect law. Approximating f_w by the log law, and imposing the condition $\langle U \rangle_{y=\delta} = U_0$, we obtain

$$\frac{U_0 - \langle U \rangle}{u_\tau} = \frac{1}{\kappa}\left\{-\ln\left(\frac{y}{\delta}\right) + \Pi\left[2 - w\left(\frac{y}{\delta}\right)\right]\right\}. \tag{7.150}$$

This law is compared with Klebanoff's data in Fig. 7.29.

With the same approximations, Eq. (7.148) evaluated at $y = \delta$ leads to a friction law:

$$\frac{U_0}{u_\tau} = \frac{1}{\kappa}\ln\left(\frac{\delta u_\tau}{\nu}\right) + B + \frac{2\Pi}{\kappa} \tag{7.151}$$

$$= \frac{1}{\kappa}\ln\left(\mathrm{Re}_\delta \frac{u_\tau}{U_0}\right) + B + \frac{2\Pi}{\kappa}. \tag{7.152}$$

For given Re_δ this equation can be solved for u_τ/U_0, hence determining the skin-friction coefficient $c_f = 2(u_\tau/U_0)^2$. More convenient explicit forms of the friction law can be found in Schlichting (1979). Among these is the

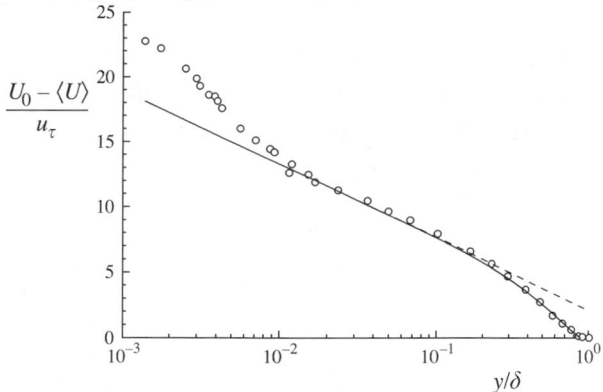

Fig. 7.29. The velocity-defect law. Symbols, experimental data of Klebanoff (1954); dashed line, log law; solid line, sum of log law and wake contribution $\Pi w(y/\delta)/\kappa$.

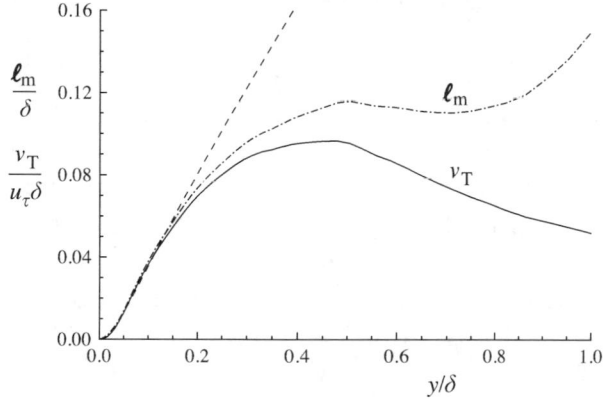

Fig. 7.30. Turbulent viscosity and mixing length deduced from direct numerical simulations of a turbulent boundary layer (Spalart 1988). Solid line, ν_T from DNS; dot–dashed line, ℓ_m from DNS; dashed line ℓ_m and ν_T according to van Driest's specification (Eq. (7.145)).

Schultz–Grunow formula,

$$c_f = 0.370(\log_{10}\mathrm{Re}_x)^{-2.584}. \tag{7.153}$$

In the defect layer, the shear stress $\tau(y)$ is less than τ_w, and the velocity gradient $\partial\langle U\rangle/\partial y$ is greater than the value $u_\tau/(\kappa y)$ given by the log law. Clearly, then, the value of the turbulent viscosity $\nu_T = \tau/(\partial\langle U\rangle/\partial y)$ is less than that given by the log-law formula $\nu_T = u_\tau \kappa y$; and, consequently, the mixing length is less than κy. This is confirmed by Fig. 7.30, which shows $\nu_T(y)$ and $\ell_m(y)$ deduced from the DNS boundary-layer data of Spalart (1988).

In applying the mixing-length model to boundary layers, it is necessary therefore to modify the formula $\ell_m = \kappa y$ in the defect layer. A simple modification, proposed by Escudier (1966), is to set ℓ_m to the minimum of κy and 0.09δ. Other variants of the mixing-length model (e.g., Smith and Cebeci (1967) and Baldwin and Lomax (1978)) achieve a similar effect by different means (see Wilcox (1993)).

7.3.4 The overlap region reconsidered

Given the complexity of near-wall turbulent motions and the processes involved, it is remarkable how the mean velocity profiles in pipe flow, channel flow and boundary layers are well represented by simple formulae – especially the universal log law. However, the empirical success of a theory does not necessarily imply the validity of the assumptions on which it is based: different assumptions may lead to predictions of comparable accuracy.

Over the years, arguments against the log law have been advanced by Barenblatt and Monin (1979), Long and Chen (1981), and George, Castillo, and Knecht (1996), among others. A central issue is the influence of the Reynolds number in the overlap region. The strong assumption made in the argument leading to the log law (see Eq. (7.41)), is that (for $Re \gg 1$, $y^+ \gg 1$, and $y/\delta \ll 1$) $y\partial u^+/\partial y$ is independent of U_0, δ, and v, and hence of the Reynolds number. It is clear from experiments (see, e.g., Fig. 7.19 and Gad-el-Hak and Bandyopadhyay (1994)) that the Reynolds-stress profiles in the overlap region depend on the Reynolds number, and hence the turbulent processes involved are not completely independent of Re.

We now consider weaker alternative assumptions in the context of fully developed turbulent pipe flow at high Reynolds number. The velocity profile in the inner layer is expressed as

$$u^+ = f_1(y^+), \tag{7.154}$$

(cf. Eq. (7.37)), where the function f_1 may depend on the Reynolds number. In the outer layer, the defect law is rewritten

$$\frac{U_0 - \langle U \rangle}{u_o} = F_o(\eta), \tag{7.155}$$

with $\eta \equiv y/\delta$ (cf. Eq. (7.46)), where u_o is a velocity scale for the outer layer, which may be different than u_τ, and the function F_o may depend on the Reynolds number.

In the overlap region ($\delta_v \ll y \ll \delta$), the asymptotic forms of $f_1(y^+)$ (for large y^+) and $F_o(\eta)$ (for small η) must match. There are two forms of velocity

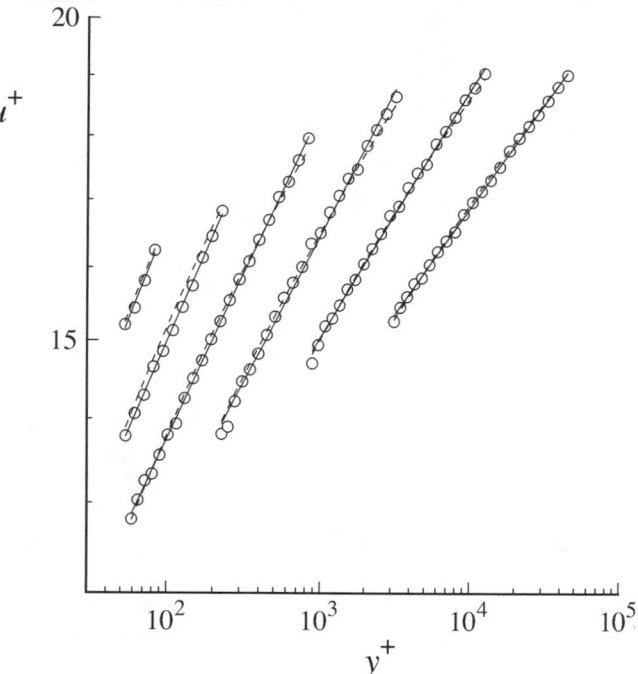

Fig. 7.31. A log–log plot of mean velocity profiles in turbulent pipe flow at six Reynolds number (from left to right: Re $\approx 32 \times 10^3$, 99×10^3, 409×10^3, 1.79×10^6, 7.71×10^6, and 29.9×10^6). The scale for u^+ pertains to the lowest Reynolds number: subsequent profiles are shifted down successively by a factor of 1.1. The range shown is the overlap region, $50\delta_\nu < y < 0.1\,R$. Symbols, experimental data of Zagarola and Smits (1997); dashed lines, log law with $\kappa = 0.436$ and $B = 6.13$; solid lines, power law (Eq. (7.157)) with the power α determined by the best fit to the data.

profile that are consistent with this matching requirement: the *log law*

$$u^+ = \frac{1}{\kappa} \ln y^+ + B, \qquad (7.156)$$

and the *power law*

$$u^+ = C(y^+)^\alpha, \qquad (7.157)$$

(see Exercise 7.20 and Barenblatt (1993)). The assumptions made in this development allow the positive coefficients κ, B, α, and C to depend on the Reynolds number. If, to the contrary, the coefficients are independent of the Reynolds number, then the laws are said to be *universal*.

Figure 7.31 shows measured mean velocity profiles in the overlap region of turbulent pipe flow at various Reynolds numbers. The data are compared with the universal log law (with $\kappa = 0.436$ and $B = 6.13$), and the power law (Eq. (7.157)) with the exponent α determined from the data at each Reynolds

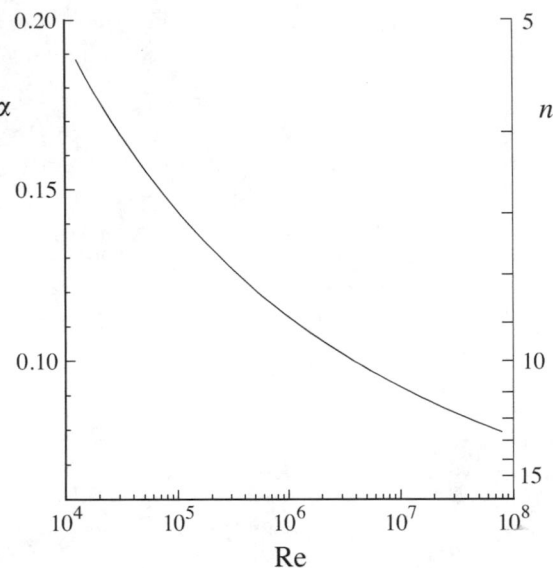

Fig. 7.32. The exponent $\alpha = 1/n$ (Eq. (7.158)) in the power-law relationship $u^+ = C(y^+)^\alpha = C(y^+)^{1/n}$ for pipe flow as a function of the Reynolds number.

number. It is clear that α decreases significantly with Re: Zagarola, Perry, and Smits (1997) showed that the empirically determined values are well approximated by the formula

$$\alpha = \frac{1.085}{\ln \text{Re}} + \frac{6.535}{(\ln \text{Re})^2}, \qquad (7.158)$$

which is shown in Fig. 7.32.

It may be observed from Fig. 7.31 that both the log law and the power law provide quite accurate representations of the measured velocity profiles. Although the data have been carefully scrutinized (e.g., Zagarola *et al.* (1997)), because of the small differences between the log-law and power-law predictions, the conclusions drawn are likely to remain controversial (see, e.g., Barenblatt and Chorin (1998)). Whatever the merits of the underlying assumptions, the log law has the practical advantage of being universal.

Another issue that arises is the appropriate choice of the velocity scale u_o in the outer layer. A universal log law implies that $u_o = u_\tau$; whereas a universal power law (which, as Fig. 7.31 shows, does not exist) would imply that $u_o = U_0$ (see Exercise 7.20). Various suggestions have been made: for example $u_o = U_0 - \bar{U}$ for pipe flow (Zagarola and Smits 1997), and $u_o = U_0$ for boundary layers (George *et al.* 1996).

EXERCISES

7.20   Consider fully developed turbulent pipe flow at high Reynolds number. In the inner and outer layers, the mean velocity profile is given by Eqs. (7.154) and (7.155), respectively.

(a) Show that the matching of $\langle U \rangle$ and $d\langle U \rangle/dy$ given by Eqs. (7.154) and (7.155) yields

$$f_{\mathrm{I}}(y^+) = \frac{U_0}{u_\tau} - \frac{u_0}{u_\tau} F_0(\eta), \tag{7.159}$$

$$y^+ \frac{df_{\mathrm{I}}(y^+)}{dy^+} = -\frac{u_0}{u_\tau} \eta \frac{dF_0(\eta)}{d\eta}. \tag{7.160}$$

In the overlap region ($\delta_\nu \ll y \ll \delta$), different asymptotic forms of f_{I} (for large y^+) and F_0 (for small η) are obtained from the alternative assumptions

$$y^+ \frac{df_{\mathrm{I}}(y^+)}{dy^+} = \frac{1}{\kappa}, \tag{7.161}$$

and

$$y^+ \frac{df_{\mathrm{I}}(y^+)}{dy^+} = \alpha\, C(y^+)^\alpha, \tag{7.162}$$

where κ, C, and α are positive constants (i.e., independent of y, but not necessarily of Re).

(b) Show that general solutions to Eqs. (7.160) and (7.161) are

$$f_{\mathrm{I}}(y^+) = \frac{1}{\kappa} \ln y^+ + B \tag{7.163}$$

(cf. Eq. (7.43)) and

$$F_0(\eta) = \frac{u_\tau}{u_0} \left(-\frac{1}{\kappa} \ln \eta + B_1 \right), \tag{7.164}$$

(cf. Eq. (7.50)) where B and B_1 are constants of integration. Show that Eq. (7.159) then yields the friction law Eq. (7.55).

(c) Show that general solutions to Eqs. (7.160) and (7.162) are

$$f_{\mathrm{I}}(y^+) = C(y^+)^\alpha + b, \tag{7.165}$$

$$F_0(\eta) = b_1 - C_{\mathrm{F}} \eta^\alpha, \tag{7.166}$$

where

$$C_{\mathrm{F}} \equiv C \frac{u_\tau}{u_0} \left(\frac{\delta}{\delta_\nu} \right)^\alpha, \tag{7.167}$$

and b and b_1 are constants of integration. Show that Eq. (7.159) then determines

$$b_1 = \frac{U_0 - bu_\tau}{u_o}. \tag{7.168}$$

(d) Show that Eqs. (7.163) and (7.164) are *universal* log laws (i.e., the coefficients are independent of Re) if, and only if, κ, B, and B_1 are independent of Re, u_o scales with u_τ, and the friction law Eq. (7.55) is satisfied.

(e) Show that Eqs. (7.165) and (7.166) are *universal* power laws if, and only if, C and α are independent of Re, b is zero, u_o scales with U_0, and the friction law

$$\frac{u_\tau}{U_0} = b_2 \left(\frac{U_0 \delta}{\nu} \right)^{-\alpha/(1+\alpha)}, \tag{7.169}$$

is satisfied, for some constant b_2. (Note: experimental data clearly show to the contrary that α depends on Re, see Fig. 7.32.)

(f) Take $\alpha = \frac{1}{7}$ and neglect the variation of $U_0/\bar{U}$ with Re. Show then that Eq. (7.169) yields for the friction factor

$$f = b_3 \mathrm{Re}^{-1/4}, \tag{7.170}$$

where b_3 is a constant.

(Note: with $b_3 = 0.3164$, Eq. (7.170) is the empirical *Blasius resistance formula*, which agrees with the Prandtl formula (Eq. (7.98)) to within 3% for $10^4 < \mathrm{Re} < 10^5$, but is in error by 30% at $\mathrm{Re} = 10^7$. It is stressed that this derivation incorrectly assumes the existence of universal power laws.)

7.21 Consider the power law

$$\frac{\langle U \rangle}{U_0} = \left(\frac{y}{R} \right)^\alpha \tag{7.171}$$

as an approximation to the velocity profile (for $0 \le y \le R$) in fully developed turbulent pipe flow. Show that the area-averaged velocity is

$$\frac{\bar{U}}{U_0} = \frac{2}{(1+\alpha)(2+\alpha)}. \tag{7.172}$$

7.22 As an approximation to the mean velocity profile in a boundary

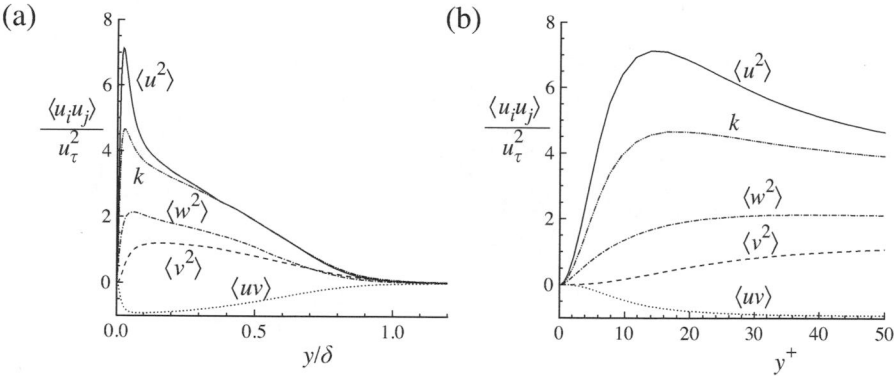

Fig. 7.33. Profiles of Reynolds stresses and kinetic energy normalized by the friction velocity in a turbulent boundary layer at $\mathrm{Re}_\theta = 1,410$: (a) across the boundary layer and (b) in the viscous near-wall region. From the DNS data of Spalart (1988).

layer, consider the power-law profile

$$\frac{\langle U \rangle}{U_0} = \begin{cases} \left(\dfrac{y}{\delta}\right)^{1/n}, & \text{for } \dfrac{y}{\delta} \le 1, \\[2ex] 1, & \text{for } \dfrac{y}{\delta} \ge 1, \end{cases} \tag{7.173}$$

where n is positive. For this profile show that

$$\frac{\delta^*}{\delta} = \frac{1}{n+1}, \tag{7.174}$$

$$\frac{\theta}{\delta} = \frac{n}{(n+1)(n+2)}, \tag{7.175}$$

and hence that the shape factor is

$$H = 1 + \frac{2}{n}. \tag{7.176}$$

7.3.5 Reynolds-stress balances

Both for free shear flows and for channel flow we have examined the balance of the turbulent kinetic energy. Here, using the boundary-layer DNS data of Spalart (1988), we go further to examine the balance of the individual Reynolds stresses. First, the Reynolds-stress profiles and kinetic-energy balance are briefly described.

In the inner layer, the Reynolds-stress profiles (shown in Fig. 7.33) are

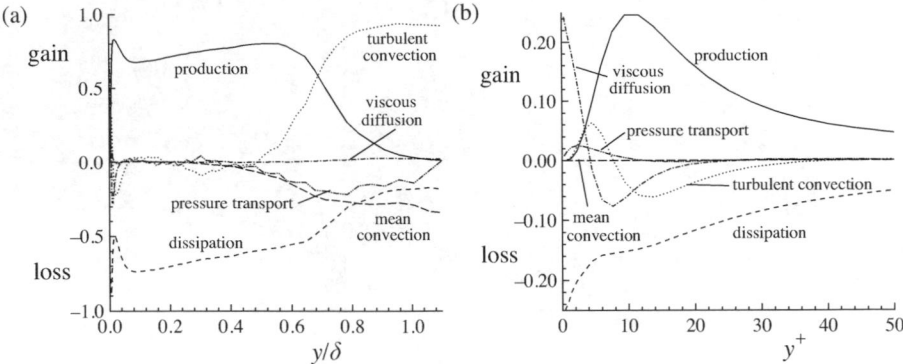

Fig. 7.34. The turbulent-kinetic-energy budget in a turbulent boundary layer at $\mathrm{Re}_\theta =$ 1,410: terms in Eq. (7.177) (a) normalized as a function of y so that the sum of the squares of the terms is unity and (b) normalized by the viscous scales. From the DNS data of Spalart (1988).

little different than those in channel flow (cf. Figs. 7.14 and 7.17). As the edge of the boundary layer is approached, all the Reynolds stresses tend smoothly to zero (corresponding to the non-turbulent free stream). Just as in channel flow, the Reynolds-stress profiles exhibit a weak Reynolds-number dependence (see, e.g., Marušić, Uddin, and Perry (1997)).

The kinetic-energy balance

In the boundary-layer approximation, the equation for the turbulent kinetic energy (Eq. (5.164)) is

$$
0 = - \left(\langle U \rangle \frac{\partial k}{\partial x} + \langle V \rangle \frac{\partial k}{\partial y} \right) + \mathcal{P} - \tilde{\varepsilon} + v \frac{\partial^2 k}{\partial y^2}
$$
$$
- \frac{\partial}{\partial y} \langle \tfrac{1}{2} v \boldsymbol{u} \cdot \boldsymbol{u} \rangle - \frac{1}{\rho} \frac{\partial}{\partial y} \langle v p' \rangle. \tag{7.177}
$$

In order, the terms in this balance equation are mean-flow convection, production, pseudo-dissipation, viscous diffusion, turbulent convection, and pressure transport. This equation is essentially the same as that for channel flow (Eq. (7.64)), but with the addition of the mean-flow-convection term.

The profiles of the various contributions to Eq. (7.177) are shown in Fig. 7.34. In the near-wall region ($y^+ \leq 50$, Fig. 7.34(b)), the profiles are normalized by the viscous scales, and are again very similar to those in channel flow (Fig. 7.18). In this region, mean-flow convection is negligible.

With increasing y/δ, the magnitudes of the terms in Eq. (7.177) decrease. For example, in the log-law region both $\mathcal{P}$ and ε decrease inversely with y. In order to show the relative importance of the terms as functions of y, in

Fig. 7.34(a) the contributions have been normalized locally, so that the sum of their squares is unity. From $y^+ \approx 40$ to $y/\delta \approx 0.4$ the dominant balance is between production and dissipation. Further out in the boundary layer, production becomes small, and the balance is between dissipation and the various transport terms.

The Reynolds-stress equation

The transport equation for the Reynolds stresses deduced from the Navier–Stokes equations (see Exercise 7.23) is

$$0 = -\frac{\bar{D}}{\bar{D}t}\langle u_i u_j \rangle - \frac{\partial}{\partial x_k}\langle u_i u_j u_k \rangle + v\,\nabla^2 \langle u_i u_j \rangle + \mathcal{P}_{ij} + \Pi_{ij} - \varepsilon_{ij}, \qquad (7.178)$$

where $\mathcal{P}_{ij}$ is the *production tensor*

$$\mathcal{P}_{ij} \equiv -\langle u_i u_k \rangle \frac{\partial \langle U_j \rangle}{\partial x_k} - \langle u_j u_k \rangle \frac{\partial \langle U_i \rangle}{\partial x_k}, \qquad (7.179)$$

Π_{ij} is the *velocity-pressure-gradient tensor*

$$\Pi_{ij} \equiv -\frac{1}{\rho}\left\langle u_i \frac{\partial p'}{\partial x_j} + u_j \frac{\partial p'}{\partial x_i} \right\rangle, \qquad (7.180)$$

and ε_{ij} is the *dissipation tensor*

$$\varepsilon_{ij} \equiv 2v \left\langle \frac{\partial u_i}{\partial x_k} \frac{\partial u_j}{\partial x_k} \right\rangle. \qquad (7.181)$$

To relate these symmetric second-order tensors to more familiar quantities, we observe that half the trace of the Reynolds-stress equation is the kinetic energy equation, and, in particular, we have

$$\tfrac{1}{2}\mathcal{P}_{ii} = \mathcal{P}, \quad \tfrac{1}{2}\varepsilon_{ii} = \tilde{\varepsilon}, \qquad (7.182)$$

$$\tfrac{1}{2}\Pi_{ii} = -\frac{\partial}{\partial x_i}\langle u_i\, p'/\rho \rangle. \qquad (7.183)$$

Figures 7.35–7.38 show the profiles of the terms in Eq. (7.178) for each of the non-zero Reynolds stresses. In order, the terms are referred to as mean convection, turbulent convection, viscous diffusion, production, pressure, and dissipation. The normal stresses $\langle u^2 \rangle$, $\langle v^2 \rangle$, and $\langle w^2 \rangle$ are examined first.

Normal-stress balances

In simple shear flows, in which $\partial \langle U \rangle / \partial y$ is the only significant mean velocity gradient, the normal-stress productions are

$$\mathcal{P}_{11} = 2\mathcal{P} = -2\langle uv \rangle \frac{\partial \langle U \rangle}{\partial y}, \qquad (7.184)$$

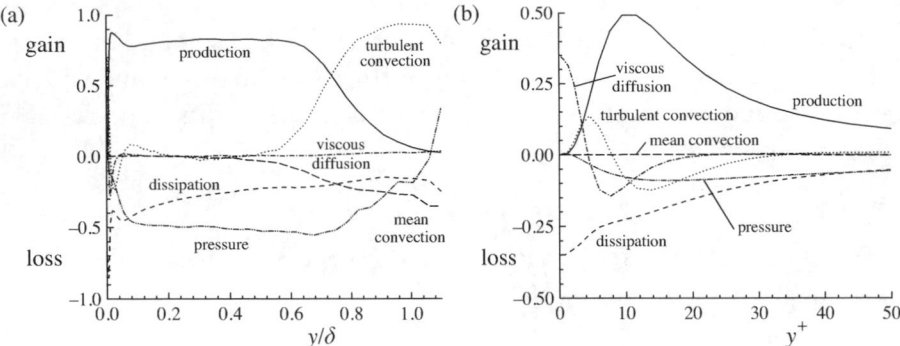

Fig. 7.35. The budget of $\langle u^2 \rangle$ in a turbulent boundary layer: conditions and normalization are the same as those in Fig. 7.34.

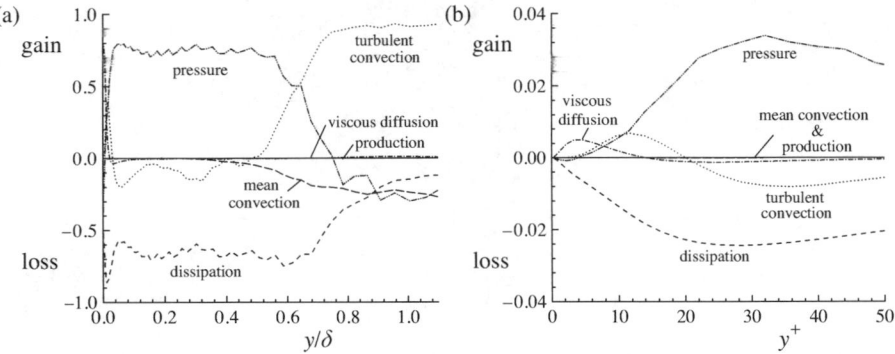

Fig. 7.36. The budget of $\langle v^2 \rangle$ in a turbulent boundary layer: conditions and normalization are the same as those in Fig. 7.34.

$$\mathcal{P}_{22} = \mathcal{P}_{33} = 0, \tag{7.185}$$

i.e., all the kinetic-energy production is in $\langle u^2 \rangle$. As expected, therefore, over most of the boundary layer, $\mathcal{P}_{11}$ is the dominant source of $\langle u^2 \rangle$ (see Fig. 7.35).

In the turbulent kinetic energy balance, p' appears only as a transport term (i.e., $\frac{1}{2}\Pi_{ii} = -\nabla \cdot \langle \boldsymbol{u}\,p'/\rho \rangle$), which is relatively small over most of the boundary layer (see Fig. 7.34). In contrast, in the Reynolds-stress equations the pressure term Π_{ij} plays a central role. Over most of the boundary layer Π_{11} is the dominant sink in the $\langle u^2 \rangle$ balance, while Π_{22} and Π_{33} are the dominant sources in the $\langle v^2 \rangle$ and $\langle w^2 \rangle$ balances. Thus the primary effect of the fluctuating pressure is to *redistribute* the energy among the components – to extract energy from $\langle u^2 \rangle$ and to transfer it to $\langle v^2 \rangle$ and $\langle w^2 \rangle$.

The redistributive effect of the fluctuating pressure is revealed by the

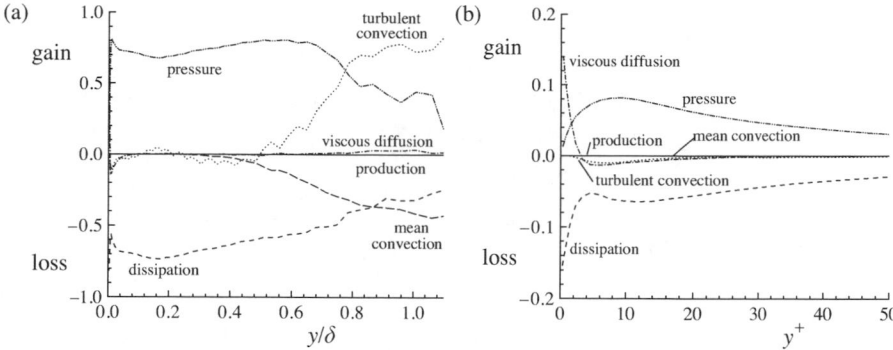

Fig. 7.37. The budget of $\langle w^2 \rangle$ in a turbulent boundary layer: conditions and normalization are the same as those in Fig. 7.34.

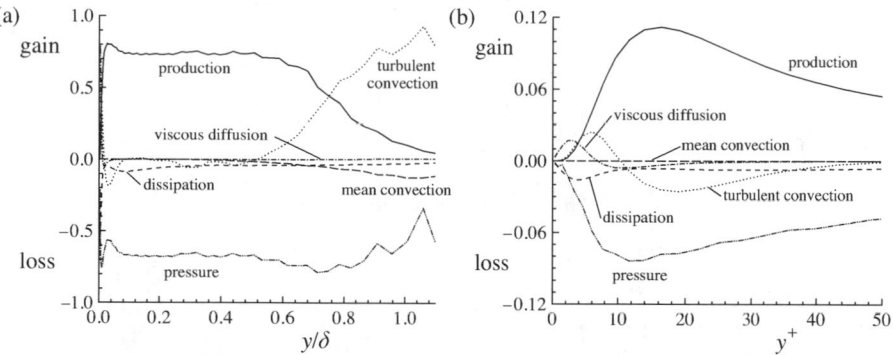

Fig. 7.38. The budget of $-\langle uv \rangle$ in a turbulent boundary layer: conditions and normalization are the same as those in Fig. 7.34.

decomposition

$$\Pi_{ij} = \mathcal{R}_{ij} - \frac{\partial}{\partial x_k} T_{kij}^{(p)}, \tag{7.186}$$

where the *pressure–rate-of-strain tensor* $\mathcal{R}_{ij}$ is

$$\mathcal{R}_{ij} \equiv \left\langle \frac{p'}{\rho} \left(\frac{\partial u_i}{\partial x_j} + \frac{\partial u_j}{\partial x_i} \right) \right\rangle, \tag{7.187}$$

and $T_{kij}^{(p)}$ is the pressure transport (see Exercise 7.24). In view of the continuity equation $\nabla \cdot \boldsymbol{u} = 0$, the pressure–rate-of-strain tensor contracts to zero, and so vanishes from the turbulent-kinetic-energy equation. In the boundary layer, there is a transfer of energy, at the rate $-\mathcal{R}_{11} = \mathcal{R}_{22} + \mathcal{R}_{33}$, from $\langle u^2 \rangle$ to $\langle v^2 \rangle$ and $\langle w^2 \rangle$.

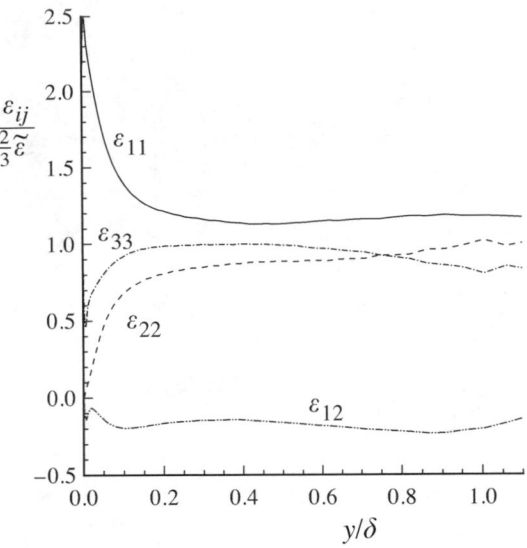

Fig. 7.39. Normalized dissipation components in a turbulent boundary layer at $Re_\theta = 1{,}410$: from the DNS data of Spalart (1988), for which $\delta = 650\delta_\nu$.

The shear-stress balance

Figure 7.38 shows the budget for $-\langle uv \rangle$. Since $\langle uv \rangle$ is negative, a 'gain' in $-\langle uv \rangle$ corresponds to an increase in the magnitude of the shear stress. It may be seen that, over the bulk of the boundary layer ($y^+ \approx 40$ to $y/\delta \approx 0.5$), there is an approximate balance between production (i.e., $-\mathcal{P}_{12} = \langle v^2 \rangle \, \partial \langle U \rangle / \partial y$) and the pressure term $-\Pi_{12}$. In contrast to the normal-stress balances, except near the wall, the dissipation ε_{12} is relatively small.

In locally isotropic turbulence the dissipation tensor is isotropic, i.e.,

$$\varepsilon_{ij} = \tfrac{2}{3} \tilde{\varepsilon} \delta_{ij}. \tag{7.188}$$

Figure 7.39 shows the profiles of ε_{ij} normalized by $\tfrac{2}{3}\tilde{\varepsilon}$. Close to the wall, the anisotropy in ε_{ij} is clearly large; but, for $y/\delta > 0.2$ ($y^+ > 130$), there is approximate isotropy (i.e., the diagonal components of $\varepsilon_{ij}/(\tfrac{2}{3}\tilde{\varepsilon})$ are close to unity, and the off-diagonal component is close to zero). The small but distinct level of anisotropy in ε_{ij} (for $y/\delta > 0.1$) is attributable to the relatively low Reynolds number of the simulations: the boundary-layer experiments of Saddoughi and Veeravalli (1994) clearly show that there is local isotropy at high Reynolds number.

It is clear from the Reynolds-stress budgets that the velocity–pressure-gradient correlation Π_{ij} is an important quantity, as of course are the production $\mathcal{P}_{ij}$ and dissipation ε_{ij}. The Reynolds-stress equations and Π_{ij} are discussed at length in Chapter 11.

7.23 From the Reynolds decomposition $U = \langle U \rangle + u$ and from the definitions $\mathrm{D}/\mathrm{D}t = \partial/\partial t + U \cdot \nabla$ and $\bar{\mathrm{D}}/\bar{\mathrm{D}}t = \partial/\partial t + \langle U \rangle \cdot \nabla$ show that

$$\left\langle u_i \frac{\mathrm{D}u_j}{\mathrm{D}t} + u_j \frac{\mathrm{D}u_i}{\mathrm{D}t} \right\rangle = \frac{\bar{\mathrm{D}}\langle u_i u_j \rangle}{\bar{\mathrm{D}}t} + \frac{\partial}{\partial x_k} \langle u_i u_j u_k \rangle. \qquad (7.189)$$

Hence, from the transport equation for the fluctuating velocity $u(x, t)$ (Eq. (5.138)), obtain the Reynolds-stress equation

$$\frac{\bar{\mathrm{D}}\langle u_i u_j \rangle}{\bar{\mathrm{D}}t} + \frac{\partial}{\partial x_k} \langle u_i u_j u_k \rangle = \mathcal{P}_{ij} + \Pi_{ij} + v \langle u_i \nabla^2 u_j + u_j \nabla^2 u_i \rangle, \qquad (7.190)$$

where $\mathcal{P}_{ij}$ and Π_{ij} are defined by Eqs. (7.179) and (7.180). Show that the viscous term in Eq. (7.190) can be re-expressed as

$$v \langle u_i \nabla^2 u_j + u_j \nabla^2 u_i \rangle = -\varepsilon_{ij} + v \nabla^2 \langle u_i u_j \rangle, \qquad (7.191)$$

where ε_{ij} is defined by Eq. (7.181). Hence verify Eq. (7.178).

7.24 Show that Π_{ij} (Eq. (7.180)) can be decomposed as

$$\Pi_{ij} = \mathcal{R}_{ij} - \frac{\partial T_{kij}^{(p)}}{\partial x_k}, \qquad (7.192)$$

where $\mathcal{R}_{ij}$ is the *pressure–rate-of-strain tensor* defined by Eq. (7.187), and

$$T_{kij}^{(p)} \equiv \frac{1}{\rho} \langle u_i p' \rangle \delta_{jk} + \frac{1}{\rho} \langle u_j p' \rangle \delta_{ik}. \qquad (7.193)$$

Show that $\mathcal{R}_{ij}$ is deviatoric, and that Π_{ij} and $\mathcal{R}_{ij}$ are equal in homogeneous turbulence.

7.25 From the above results, show that the Reynolds-stress equation can be written

$$\frac{\bar{\mathrm{D}}}{\bar{\mathrm{D}}t} \langle u_i u_j \rangle + \frac{\partial}{\partial x_k} T_{kij} = \mathcal{P}_{ij} + \mathcal{R}_{ij} - \varepsilon_{ij}, \qquad (7.194)$$

where the Reynolds-stress flux T_{kij} is

$$T_{kij} = T_{kij}^{(u)} + T_{kij}^{(p)} + T_{kij}^{(v)}, \qquad (7.195)$$

with

$$T_{kij}^{(u)} \equiv \langle u_i u_j u_k \rangle, \qquad T_{kij}^{(v)} \equiv -v \frac{\partial \langle u_i u_j \rangle}{\partial x_k}. \qquad (7.196)$$

7.26   Show that

$$\tfrac{1}{2}\mathcal{P}_{ii} = \mathcal{P}, \qquad \tfrac{1}{2}\varepsilon_{ii} = \tilde{\varepsilon}, \tag{7.197}$$

and hence show that half the trace of the Reynolds-stress equation (Eq. (7.194)) is identical to the kinetic energy equation (Eq. (5.164)).

7.27   For simple shear flow with $\partial\langle U_1\rangle/\partial x_2$ being the only non-zero mean velocity gradient, show that the production tensor $\mathcal{P}$ is

$$\mathcal{P} = \frac{\partial\langle U_1\rangle}{\partial x_2}\begin{bmatrix} -2\langle u_1 u_2\rangle & -\langle u_2^2\rangle & 0 \\ -\langle u_2^2\rangle & 0 & 0 \\ 0 & 0 & 0 \end{bmatrix}. \tag{7.198}$$

7.28   In terms of the Reynolds-stress anisotropy $a_{ij} \equiv \langle u_i u_j\rangle - \tfrac{2}{3}k\delta_{ij}$, show that the production tensor is

$$\mathcal{P}_{ij} = -\frac{2}{3}k\left(\frac{\partial\langle U_i\rangle}{\partial x_j} + \frac{\partial\langle U_j\rangle}{\partial x_i}\right) - a_{ik}\frac{\partial\langle U_j\rangle}{\partial x_k} - a_{jk}\frac{\partial\langle U_i\rangle}{\partial x_k}. \tag{7.199}$$

Give an example of a flow in which the normal-stress production $\mathcal{P}_{11}$ is negative.

7.29   Discuss the effect of the pressure–rate-of-strain tensor on the Reynolds shear stresses

(a) in the principal axes of $\mathcal{R}_{ij}$, and
(b) in the principal axes of $\langle u_i u_j\rangle$.

7.30   From the definition of ε_{ij} (Eq. (7.181)), show that, at a wall,

$$\varepsilon_{ij} = \nu\frac{\partial^2\langle u_i u_j\rangle}{\partial y^2}, \qquad \text{for } y = 0. \tag{7.200}$$

Which components are non-zero at the wall?

7.3.6 *Additional effects*

So far in this section we have considered the zero-pressure-gradient turbulent boundary layer on a flat plate. Now briefly described are the effects of non-zero pressure gradients, and of surface curvature.

Mean pressure gradients

The effects of mean pressure gradients on boundary layers are mentioned, first, because in applications they are present more often than not, and second because the effects can be large. A favorable pressure gradient ($\mathrm{d}p_0/\mathrm{d}x < 0$) corresponds to an accelerating free stream ($\mathrm{d}U_0/\mathrm{d}x > 0$) and occurs, for example, on the forward part of an airfoil. Conversely, an adverse pressure gradient ($\mathrm{d}p_0/\mathrm{d}x > 0$) corresponds to a decelerating free steam ($\mathrm{d}U_0/\mathrm{d}x < 0$) and occurs in a diffuser or on the aft part of an airfoil.

A favorable mean pressure gradient causes the mean velocity profile to steepen, so that the shape factor H decreases and the skin-friction coefficient c_f increases. The width of the intermittent region increases (Fiedler and Head 1966), and non-turbulent fluid can penetrate all the way to the wall. Indeed, at low Reynolds number, a sufficiently strong favorable pressure gradient can cause the boundary layer to relaminarize (Narasimha and Sreenivasan 1979).

As might be expected, a mild adverse pressure gradient has the opposite effect: the mean velocity profile flattens, with the shape factor H increasing and the skin friction coefficient c_f decreasing (Bradshaw 1967, Spalart and Watmuff 1993). However, a strong, prolonged adverse pressure gradient causes the boundary layer to separate, or break away from the surface (Simpson 1989). The separation is accompanied by large-scale unsteadiness of the flow, and reverse flow ($\langle U \rangle < 0$) downstream of the separation.

Surface curvature

Boundary layers on curved surfaces are important in many applications, such as the flow over compressor and turbine blades in turbomachinery. In these applications, the curvature is in the dominant flow direction. On the upper (suction) surface of an airfoil the curvature is convex; whereas there is concave curvature over part of the lower (pressure) surface of a highly cambered airfoil.

In the flow over airfoils, the boundary layer is simultaneously subjected to the effects of curvature and a mean pressure gradient. These effects can be studied separately in laboratory experiments. Muck, Hoffmann, and Bradshaw (1985) performed experiments on a boundary layer, of constant free-stream velocity, that develops on a surface with a plane upstream section followed by a section of convex curvature (of constant radius of curvature R_c). A similar experiment with a concave surface is described by Hoffmann, Muck, and Bradshaw (1985).

In these experiments the ratio of the boundary-layer thickness δ to the radius of curvature R_c is about 0.01. Experiments with convex curvature as large as $\delta/R_\mathrm{c} \approx 0.1$ are reported by Gillis and Johnston (1983).

Rayleigh's criterion (Rayleigh 1916) correctly predicts the stabilizing or destabilizing effect of curvature. For convex curvature, the center of curvature is beneath the surface, and in the boundary layer the angular momentum of the fluid increases with radius (measured from the center of curvature).

According to Rayleigh's criterion, this increasing angular momentum is stabilizing, and indeed the experimental data show that there is a reduction in the Reynolds stresses and skin-friction coefficient compared with the plane boundary layer.

In the Reynolds-stress equations (written in the appropriate curvilinear coordinate system) there are additional production terms (due to curvature) of relative magnitude δ/R_c. However, as pointed out by Bradshaw (1973), the effects of curvature on the turbulence are an order of magnitude larger than can be explained by this direct mechanism.

For flow over a surface of concave curvature, the center of curvature is within the fluid above the surface. In the boundary layer, the angular momentum of the fluid decreases with radius (i.e., it decreases as the surface is approached). This, according to the Rayleigh criterion, is destabilizing. The experimental observations are that longitudinal Taylor–Görtler vortices form, and that the Reynolds stresses and the skin-friction coefficient increase (compared with the plane boundary layer). Again, the magnitude of the effect is much larger than simple scaling arguments suggest.

For boundary layers on mildly curved surfaces, the non-zero mean rates of strain are

$$\bar{S}_{12} = \bar{S}_{21} = \frac{1}{2}\left(\frac{\partial\langle U_1\rangle}{\partial x_2} - \frac{\langle U_1\rangle}{R_c}\right). \tag{7.201}$$

(This is in a local Cartesian coordinate system in which the basis vector e_2 is normal to the surface, e_1 and e_3 are tangential to the surface, and the mean flow is in the e_1 direction.) It may be seen that the curvature creates the *extra rate of strain* $-\frac{1}{2}\langle U_1\rangle/R_c$, which is small compared with $\partial\langle U_1\rangle/\partial x_2$ – of order δ/R_c. It is generally found that turbulent shear flows exhibit a disproportionately large response to such extra rates of strain (Bradshaw 1973).

7.4 Turbulent structures

As is the case with free shear flows, since 1960 a good fraction of the experimental effort on wall-bounded flows has been directed at *turbulent structures* or *quasi-coherent structures*. These structures are identified by flow visualization, by conditional sampling techniques, or by other *eduction* methodologies (described below); but they are difficult to define precisely. The idea is that they are regions of space and time (significantly larger than the smallest flow or turbulence scales) within which the flow field has a characteristic coherent pattern. Different instances of the structure occur at different positions and times, and their flow fields certainly differ in detail: but they possess a common characteristic coherent pattern.

Among the motivations for experimental studies of turbulent structures are the desires

(i) to seek order within apparent chaos,
(ii) to 'explain' patterns seen in flow visualization,
(iii) to 'explain' important 'mechanisms' in the flow in terms of elemental structures, and
(iv) to identify 'important' structures with a view to modifying them in order to achieve engineering goals such as reduction of drag and augmentation of heat transfer.

Without doubt, these studies have yielded valuable results, and have achieved some of their objectives. However, other objectives have not been, and will not be, achieved. The mind imbued with Newtonian mechanics seeks simple deterministic explanations of phenomena. Only in a very limited sense can coherent structures simply 'explain' the behavior of near-wall turbulent flows. There are many structures within the random background, and the deterministic and stochastic interactions among them are far from clear, and unlikely to be simple. The goal of developing a quantitative theory of near-wall turbulence based on the dynamical interaction of a small number of structures has not been attained, and is likely unattainable.

Description of structures in wall flows

Kline and Robinson (1990) and Robinson (1991) provide a useful categorization of quasi-coherent structures in channel flow and boundary layers. The eight categories identified are the following:

1. Low-speed *streaks* in the region $0 < y^+ \leq 10$.
2. *Ejections* of low-speed fluid outward from the wall.
3. *Sweeps* of high-speed fluid toward the wall.
4. *Vortical structures* of several proposed forms.
5. Strong internal *shear layers* in the wall zone ($y^+ \leq 80$).
6. Near-wall *pockets*, observed as areas clear of marked fluid in certain types of flow visualizations.
7. *Backs*: surfaces (of scale δ) across which the streamwise velocity changes abruptly.
8. Large-scale motions in the outer layers (including, for boundary layers, *bulges*, *superlayers*, and *deep valleys* of free-stream fluid).

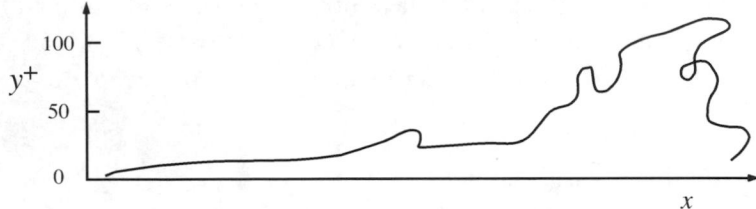

Fig. 7.40. A dye streak in a turbulent boundary layer showing the ejection of low-speed near-wall fluid. (From the experiment of Kline *et al.* (1967).)

Reviews of experimental work and discussion on these structures are provided by Kline and Robinson (1990), Robinson (1991), Sreenivasan (1989), Cantwell (1981), and Gad-el-Hak and Bandyopadhyay (1994), among many others. Here we mention just some of the principal findings.

Numerous flow-visualization experiments have revealed *streaks* in the near-wall region, $y^+ < 40$. In one experiment on a boundary layer in a water channel, Kline *et al.* (1967) used a fine wire placed across the flow (in the z direction) as an electrode to generate tiny hydrogen bubbles. With the wire placed between the wall and $y^+ = 10$, clearly visible in the plane of the wall are long streaks in the streamwise (x) direction, corresponding to an accumulation of hydrogen bubbles. These and subsequent experiments (e.g., Kim, Kline, and Reynolds (1971) and Smith and Metzler (1983)) have determined many of the characteristics of these structures. Near the wall ($y^+ < 7$) the spacing between the streaks is randomly distributed between about $80\delta_v$ and $120\delta_v$, independent of the Reynolds number, and their length (in the x direction) can exceed $1{,}000\delta_v$. The streaks correspond to relatively slow-moving fluid – with streamwise velocity about half of the local mean – while the fluid between the streaks (inevitably) is relatively fast moving.

The streaks have a characteristic behavior, known as *bursting*. With increasing downstream distance, a streak migrates slowly away from the wall; but then, at some point (typically around $y^+ \approx 10$), it turns and moves away from the wall more rapidly – a process referred to as *streak lifting*, or *ejection*. As it is lifted, the streak exhibits a rapid oscillation followed by a *breakdown* into finer-scale motions. Figure 7.40 shows a dye streak from a typical ejection.

With fluid moving away from the wall in ejections, continuity demands a flow toward the wall in some other regions. Corino and Brodkey (1969) identified regions of high-speed fluid (i.e., $u > 0$) moving toward the wall in events called *sweeps*.

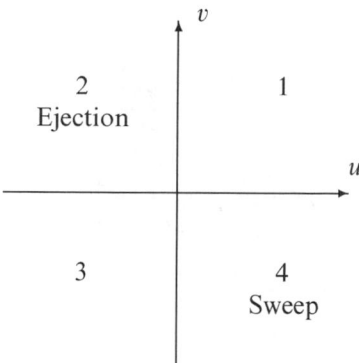

Fig. 7.41. The u–v sample space showing the numbering of the four quadrants, and the quadrants corresponding to ejections and sweeps.

An important issue is the significance of ejections and sweeps in turbulence production. The u–v sample space of the fluctuating velocities is divided by the axes into four quadrants, shown in Fig. 7.41. In quadrants 2 and 4, the product uv is negative, and consequently events in these regions correspond to positive production (recall that $\mathcal{P} = -\langle uv \rangle\, \partial \langle U \rangle / \partial y$). Thus both ejections (quadrant 2) and sweeps (quadrant 4) produce turbulent energy. Measurements of the contributions to $\langle uv \rangle$ from the various quadrants (e.g., Wallace, Eckelmann, and Brodkey (1972) and Willmarth and Lu (1972)) have been cited as evidence for the importance of ejections and sweeps to production. It should be recognized, however, that a quadrant-2 event is not necessarily an ejection; and the simple fact that the u–v correlation coefficient is around -0.5 suggests that quadrant-2 and -4 events are twice as likely as quadrant-1 and -3 events, irrespective of the turbulence structure (see Exercise 7.31).

In the near-wall region ($y^+ < 100$), pairs of counter-rotating streamwise vortices or *rolls* – depicted in Fig. ?? – have been identified as the dominant 'vortical structures' (Bakewell and Lumley 1967, Blackwelder and Haritonidis 1983). Close to the wall, between the rolls, there is a convergence of the flow in the plane of the wall ($\partial W / \partial z < 0$), which accounts for the observed streaks. In the simplified picture of Fig. ??, the fluid moving away from the wall between the rolls has a relatively reduced axial velocity, which leads to the velocity profiles shown in Fig. ??. These profiles contain inflexion points and so are inviscidly unstable, and have been conjectured to be associated with bursting (Holmes, Lumley, and Berkooz (1996)).

Head and Bandyopadhyay (1981) suggested that the dominant vortical structures further out from the wall in the boundary layer are *horseshoe* or *hairpin* vortices as sketched in Fig. 7.43. The cross-stream dimensions scale

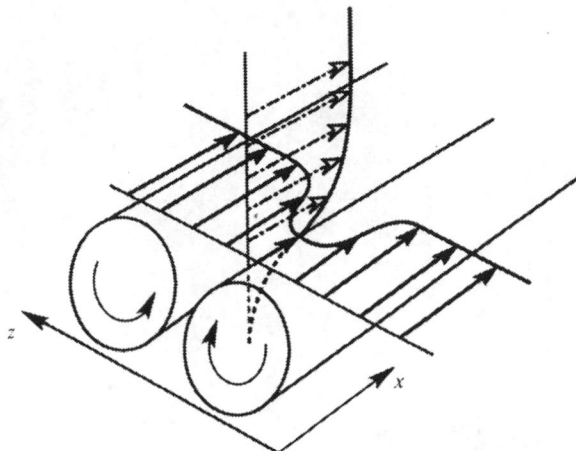

Fig. 7.42. A sketch of counter-rotating rolls in the near-wall region. (From Holmes *et al.* (1996).)

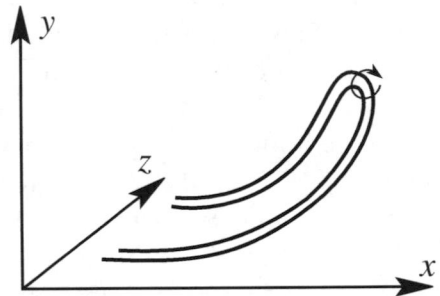

Fig. 7.43. The hairpin vortex suggested by Head and Bandyopadhyay (1981).

with δ_v, but the overall length can be of order δ, so that they are extremely elongated at high Reynolds number. As had previously been predicted by Theodorsen (1952), they are inclined at approximately 45° to the wall.

Head and Bandyopadhyay (1981) suggested that larger structures can be composed of an ensemble of hairpin vortices; and indeed Perry and co-workers have demonstrated that a suitable distribution of such elemental vortical structures can account for measured statistics in boundary layers (Perry and Chong 1982, Perry, Henbest, and Chong 1986, Perry and Marušić 1995).

In the outermost part of boundary layers the flow is intermittent. As described in Section 5.5.2, there is a thin turbulence front – the *viscous super-layer* – separating the turbulent boundary-layer fluid from the irrotational free-stream fluid. The large-scale characteristics of the viscous superlayer can be seen in flow visualizations in which the boundary-layer fluid contains

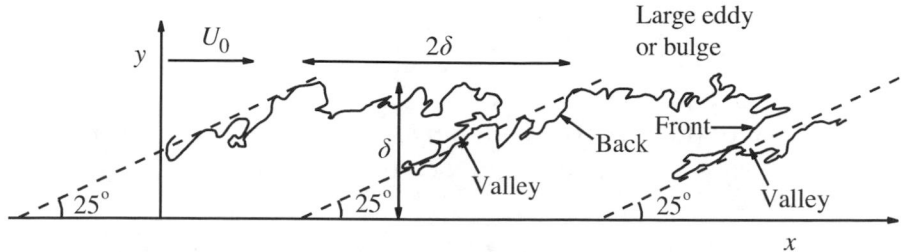

Fig. 7.44. The large-scale features of a turbulent boundary layer at $Re_\theta \approx 4{,}000$. The irregular line – approximating the viscous superlayer – is the boundary between smoke-filled turbulent fluid and clear free-stream fluid. (From the experiment of Falco 1977.)

smoke, while the free-stream fluid is clear. As an example, Fig. 7.44 shows the outline of the smoke-filled region (interpreted as the location of the viscous superlayer) from the experiment of Falco (1977).

The nature and behavior of the features identified in Fig. 7.44 have been quantified in a number of experiments based on conditional sampling of hot-wire measurements (e.g., Corrsin and Kistler (1954), Kovasznay, Kibens, and Blackwelder (1970), Blackwelder and Kovasznay (1972), and Murlis, Tsai, and Bradshaw (1982)). For given x, the y location of the superlayer is (to a good approximation) normally distributed with mean 0.8δ and standard deviation 0.15δ. Correspondingly, the intermittency factor γ has an error-function profile (see Fig. 7.26). There are *valleys* of non-turbulent fluid that penetrate deep into the boundary layer. These valleys separate *large eddies* or *bulges* that are inclined at a characteristic angle of 20–25°. The bulges are typically of length δ to 3δ (in the x direction) and about half as wide (in the z direction). As they are convected downstream – at a speed of $0.9U_0$ to $0.97U_0$ – they slowly evolve, and slowly rotate in the same sense as the mean rotation (clockwise in Fig. 7.44). The large-scale behavior of the bulges is characterized by the outer scales U_0 and δ.

The large eddies and the superlayer also contain finer-scale structures: Falco (1974) called these *typical eddies*, whereas, as previously mentioned, Head and Bandyopadhyay (1981) suggested that they are hairpin vortices inclined at 45°. The measurements of Murlis *et al.* (1982) support these authors' contention that the fine-scale structures have lengths of order δ, but widths of order δ_ν.

EXERCISE

7.31 Suppose that u and v are joint normal with correlation coefficient ρ_{uv}. Show that the probability P_1 of a quadrant-1 event (i.e., $u > 0$

and $v > 0$) is

$$P_1 = \frac{1}{4} + \frac{1}{2\pi} \sin^{-1} \rho_{uv}. \tag{7.202}$$

For joint normal u and v, and for $\rho_{uv} = -0.5$ (typical of a boundary layer), show that the probabilities of quadrant-1 and -2 events are $P_1 = \frac{1}{6}$ and $P_2 = \frac{1}{3}$.

(Hint: consider the joint normal random variables $\check{u} \equiv u$ and $\check{v} \equiv \alpha u + \beta v$, where α and β are chosen so that the $\check{u}$–$\check{v}$ covariance matrix is isotropic.)

Eduction techniques

The quasi-coherent structures described above have been identified mainly through visualizations of low-Reynolds-number turbulent flows (either from experiments or from DNS). Such a subjective approach inevitably leads to controversy over the nature and significance of the structures. As a consequence, over the years, there have been many suggestions for objective eduction techniques, to identify structures and to quantify their importance.

Examples of eduction techniques based on conditional single-point mea-surements are quadrant analysis (Willmarth and Lu 1972) and variable-interval time averaging (VITA) (Blackwelder and Kaplan 1976). The access to flow-field information afforded by DNS allows eduction techniques based on pressure (Robinson 1991), and various invariants of the velocity-gradient tensor (e.g., Blackburn, Mansour, and Cantwell (1996) and Jeong *et al.* (1997)).

Proper orthogonal decomposition (POD)

The POD is an eduction technique based on the two-point velocity correla-tion. It identifies the motions which, on average, contain the most energy. The POD, also known as the Karhunen–Loève decomposition, was intro-duced into the study of turbulent flows by Lumley (1967b), and complete descriptions are given by Berkooz, Holmes, and Lumley (1993) and Holmes *et al.* (1996).

For turbulent flows the POD provides a representation for the fluctuating velocity field $\mathbf{u}(\mathbf{x}, t)$. However, for ease of explanation, it is described here for a random scalar function $u(x)$ in the interval $0 \leq x \leq L$.

An *orthogonal decomposition* of $u(x)$ is

$$u(x) = \sum_{n=1}^{\infty} a_n \varphi_n(x), \tag{7.203}$$

where $\{\varphi_n(x), \ n = 1, 2, \ldots\}$ is a set of real basis functions, satisfying the orthonormality condition

$$\frac{1}{\mathcal{L}} \int_0^{\mathcal{L}} \varphi_n(x)\varphi_{\mathrm{m}}(x)\,\mathrm{d}x = \delta_{nm}, \qquad (7.204)$$

and $\{a_n, n = 1, 2, \ldots\}$ are the basis-function coefficients. These are given by

$$a_{\mathrm{m}} = \frac{1}{\mathcal{L}} \int_0^{\mathcal{L}} u(x)\varphi_{\mathrm{m}}(x)\,\mathrm{d}x, \qquad (7.205)$$

as is readily deduced by multiplying Eq. (7.203) by $\varphi_{\mathrm{m}}(x)$ and integrating. The basis functions $\varphi_n(x)$ are non-random, whereas the coefficients a_n are random. There are infinitely many choices of the basis functions, the most familiar being the Fourier modes, which (to conform with the present notation) can be written

$$\begin{aligned}\varphi_n(x) &= \cos[\pi(n-1)x/\mathcal{L}], && \text{for } n \text{ odd,} \\ &= \sin(\pi n x/\mathcal{L}), && \text{for } n \text{ even.}\end{aligned} \qquad (7.206)$$

The average mean energy of $u(x)$ over the domain is defined by

$$E \equiv \frac{1}{\mathcal{L}} \int_0^{\mathcal{L}} \tfrac{1}{2}\langle u(x)^2 \rangle \,\mathrm{d}x. \qquad (7.207)$$

Substituting Eq. (7.203) for $u(x)$ and invoking Eq. (7.204), we obtain

$$\begin{aligned}E &= \frac{1}{\mathcal{L}} \int_0^{\mathcal{L}} \frac{1}{2} \sum_{n=1}^{\infty} \sum_{m=1}^{\infty} \langle a_n a_m \rangle \varphi_n(x)\varphi_{\mathrm{m}}(x)\,\mathrm{d}x \\ &= \sum_{n=1}^{\infty} \tfrac{1}{2}\langle a_n^2 \rangle.\end{aligned} \qquad (7.208)$$

Thus, $\tfrac{1}{2}\langle a_n^2 \rangle$ is the contribution to the energy from the nth mode $a_n \varphi_n(x)$, and the partial sum

$$E_N \equiv \sum_{n=1}^{N} \tfrac{1}{2}\langle a_n^2 \rangle \qquad (7.209)$$

is the energy contained in the first N modes. This development applies to any orthogonal decomposition; that is, to any choice of $\varphi_n(x)$ consistent with Eq. (7.204).

A defining property of the *proper orthogonal decomposition* is that the basis functions are chosen to maximize the partial sums E_N (for $N = 1, 2, \ldots$): the first N modes of the POD contain more energy than do the first N modes of any other orthogonal decomposition. The nth POD basis function is therefore determined to be the function $\varphi_n(x)$ which satisfies the orthonormality condition (Eq. (7.204) for $m \leq n$), and which maximizes

$$\langle a_n^2 \rangle = \frac{1}{\mathcal{L}^2} \int_0^{\mathcal{L}} \int_0^{\mathcal{L}} \langle u(x)u(y) \rangle \varphi_n(x)\varphi_n(y) \, dx \, dy. \tag{7.210}$$

Evidently, knowledge of the two-point correlation is sufficient to determine $\varphi_n(x)$; and a straightforward application of the calculus of variations (see Holmes *et al.* (1996)) shows that $\varphi_n(x)$ are the eigenfunctions of the integral equation

$$\frac{1}{\mathcal{L}} \int_0^{\mathcal{L}} \langle u(x)u(y) \rangle \varphi_n(y) \, dy = \lambda_{(n)} \varphi_n(x). \tag{7.211}$$

The eigenvalue $\lambda_{(n)}$ is equal to $\langle a_n^2 \rangle$ (see Exercise 7.32); so that the eigenvalues must be ordered in decreasing magnitude ($\lambda_{(1)} \geq \lambda_{(2)} \geq \lambda_{(3)} \ldots$) so that the partial sums E_N are maximal.

The application of POD to turbulent flows requires the extension of these ideas to the three-dimensional velocity field $u(x, t)$ – see Holmes *et al.* (1996). Usually the decomposition is in space only (rather than in time and space), so that the POD is of the form

$$u(x, t) = \sum_{n=1}^{\infty} a_n(t)\varphi_n(x). \tag{7.212}$$

The basis functions $\varphi_n(x)$ are incompressible vector fields that in some sense are 'characteristic eddies' (Moin and Moser 1989). The coefficients $a_n(t)$ are random functions of time.

Bakewell and Lumley (1967) deduced from measured two-point correlations that the first POD basis function in fully developed turbulent pipe flow corresponds to a pair of counter-rotating rolls, similar to those shown in Fig. ??. Moin and Moser (1989) used DNS data for fully developed channel flow to perform and analyze a comprehensive POD. The first eigenfunction is shown to account for 50% of the energy and 75% of the production. Also for channel flow, Sirovich, Ball, and Keefe (1990) examined time series of the coefficients $a_n(t)$ deduced from DNS. Other POD studies are reviewed by Berkooz *et al.* (1993) and Holmes *et al.* (1996).

EXERCISE

7.32   By substituting Eq. (7.203), show that the left-hand side of Eq. (7.211) can be re-expressed as

$$\frac{1}{\mathcal{L}} \int_0^{\mathcal{L}} \langle u(x)u(y) \rangle \varphi_n(y) \, dy = \langle u(x)a_n \rangle = \sum_{m=1}^{\infty} \langle a_m a_n \rangle \varphi_m(x). \tag{7.213}$$

Multiply Eq. (7.211) by $\varphi_\ell(x)$ and integrate to obtain

$$\langle a_\ell a_n \rangle = \lambda_{(n)} \delta_{\ell n}, \qquad (7.214)$$

showing that different POD coefficients are uncorrelated, and that the POD eigenvalues are

$$\lambda_{(n)} = \langle a_n^2 \rangle. \qquad (7.215)$$

Show that the two-point correlation is given in terms of the POD eigenvalues and eigenfunctions as

$$\langle u(x)u(y) \rangle = \sum_{n=1}^{\infty} \lambda_{(n)} \varphi_n(x)\varphi_n(y). \qquad (7.216)$$

Dynamical models

It may be recalled from Section 6.4.2 that the Navier–Stokes equations in wavenumber space consist of an infinite set of ordinary differential equations for the Fourier coefficients (Eq. (6.146)). These equations are nonlinear and involve triadic interactions between the modes. In the same way, for the orthogonal decomposition Eq. (7.212), the Navier–Stokes equations imply an infinite set of nonlinear ordinary differential equations that describe the evolution of the coefficients $a_n(t)$. It is natural to attempt to describe the interactions among the dominant structures in terms of a truncated set of N ordinary differential equations for N coefficients $a_n(t)$ of an orthogonal decomposition.

Aubry et al. (1988) developed a dynamical model for the near-wall region on the basis of the first five POD modes. This model had some success in describing qualitatively some features of near-wall behavior, including the phenomenon of bursting. The further development of this approach is the subject of the book of Holmes, Lumley, and Berkooz (1996).

A somewhat different model, but again based on orthogonal modes, has been developed by Waleffe (1997). This model exhibits a self-sustaining process in which an interaction between streamwise rolls and the mean shear leads to the development of streaks. The streaks then become unstable and break down to recreate rolls.

Conclusions

Experiments and DNS (mainly at low Reynolds numbers) have revealed several quasi-coherent structures in wall-bounded flows (at least eight according to the categorization of Kline and Robinson (1990)). Some objective eduction techniques, notably the proper orthogonal decomposition, have been developed to identify and quantify the characteristics of the structures. The insights gained have been useful, for example in the development of drag-reduction devices (see, e.g., Bushnell and McGinley (1989), Choi (1991), and Schoppa and Hussain (1998)). However, it should be appreciated that a complete quantitative picture of near-wall turbulence behavior is far from being obtained. In particular the dynamics of most of the structures and the interactions between them are largely unknown, and are likely to be complex.

Part two
Modelling and Simulation

8

An introduction to modelling and simulation

8.1 The challenge

In the study of turbulent flows – as in other fields of scientific inquiry – the ultimate objective is to obtain a tractable quantitative theory or model that can be used to calculate quantities of interest and practical relevance. A century of experience has shown the 'turbulence problem' to be notoriously difficult, and there are no prospects of a simple analytic theory. Instead, the hope is to use the ever-increasing power of digital computers to achieve the objective of calculating the relevant properties of turbulent flows. In the subsequent chapters, five of the leading computational approaches to turbulent flows are described and examined.

It is worthwhile at the outset to reflect on the particular properties of turbulent flows that make it difficult to develop an accurate tractable theory or model. The velocity field $U(x,t)$ is three-dimensional, time-dependent, and random. The largest turbulent motions are almost as large as the characteristic width of the flow, and consequently are directly affected by the boundary geometry (and hence are not universal). There is a large range of timescales and lengthscales. Relative to the largest scales, the Kolmogorov timescale decreases as $Re^{-1/2}$, and the Kolmogorov lengthscale as $Re^{-3/4}$. In wall-bounded flows, the most energetic motions (that are responsible for the peak turbulence production) scale with the viscous lengthscale δ_v which is small compared with the outer scale δ, and which decreases (relative to δ) approximately as $Re^{-0.8}$.

Difficulties arise from the nonlinear convective term in the Navier–Stokes equations, and much more so from the pressure-gradient term. When it is expressed in terms of velocity (via the solution to the Poisson equation, Eq. (2.49) on page 19), the pressure-gradient term is both nonlinear and non-local.

335

8.2 An overview of approaches

The methodologies described in the subsequent chapters take the form of sets of partial differential equations, in some cases supplemented by algebraic equations. For a given flow, with the specification of the appropriate initial and boundary conditions, these equations are solved numerically.

In a turbulent-flow *simulation*, equations are solved for a time-dependent velocity field that, to some extent, represents the velocity field $U(x, t)$ for one realization of the turbulent flow. In contrast, in a *turbulence model*, equations are solved for some mean quantities, for example $\langle U \rangle$, $\langle u_i u_j \rangle$, and ε. (The word 'models' is used to refer both to simulations and to turbulence models, when the distinction is not needed.)

The two simulation approaches described are direct numerical simulation (DNS, Chapter 9), and large-eddy simulation (LES, Chapter 13). In DNS, the Navier–Stokes equations are solved to determine $U(x, t)$ for one realization of the flow. Because all lengthscales and timescales have to be resolved, DNS is computationally expensive; and, because the computational cost increases as Re^3, this approach is restricted to flows with low-to-moderate Reynolds number. In LES, equations are solved for a 'filtered' velocity field $\overline{U}(x, t)$, which is representative of the larger-scale turbulent motions. The equations solved include a model for the influence of the smaller-scale motions which are not directly represented.

The approaches described in Chapters 10 and 11 are called Reynolds-averaged Navier–Stokes (RANS), since they involve the solution of the Reynolds equations to determine the mean velocity field $\langle U \rangle$. In the first of these approaches, the Reynolds stresses are obtained from a turbulent-viscosity model. The turbulent viscosity can be obtained from an algebraic relation (such as in the mixing-length model) or it can be obtained from turbulence quantities such as k and ε for which modelled transport equations are solved. In Reynolds-stress models (Chapter 11), modelled transport equations are solved for the Reynolds stresses, thus obviating the need for a turbulent viscosity.

The mean velocity $\langle U \rangle$ and the Reynolds stresses $\langle u_i u_j \rangle$ are the first and second moments of the Eulerian PDF of velocity $f(V; x, t)$. In PDF methods (Chapter 12), a model transport equation is solved for a PDF such as $f(V; x, t)$.

8.3 Criteria for appraising models

The purpose of this section is to provide an overview of the criteria used in appraising models. Historically, many models have been proposed and many are currently in use. It is important to appreciate that there is a broad range of turbulent flows, and also a broad range of questions to be addressed. Consequently it is useful and appropriate to have a broad range of models, that vary in complexity, accuracy, and other attributes.

The principal criteria that can be used to assess different models are the

  (i) level of description,
 (ii) completeness,
(iii) cost and ease of use,
 (iv) range of applicability, and
  (v) accuracy.

As examples to elaborate on these criteria, we consider two models – the mixing-length model and DNS – which are at the extremes of the range of approaches.

Recall that DNS (direct numerical simulation) consists in solving the Navier–Stokes equations to determine the instantaneous velocity field $U(x, t)$ for one realization of the flow. The mixing-length model (applied to statistically stationary two-dimensional boundary-layer flows) consists of the boundary-layer equations for $\langle U(x, y) \rangle$ and $\langle V(x, y) \rangle$, with the Reynolds shear stress and the turbulent viscosity being obtained from the model equations

$$\langle uv \rangle = -v_\mathrm{T} \frac{\partial \langle U \rangle}{\partial y}, \tag{8.1}$$

$$v_\mathrm{T} = \ell_\mathrm{m}^2 \left| \frac{\partial \langle U \rangle}{\partial y} \right|. \tag{8.2}$$

The mixing length $\ell_\mathrm{m}(x, y)$ is specified as a function of position.

The level of description

In DNS, the flow is described by the instantaneous velocity $U(x, t)$, from which all other information can be determined. For example, flow visualizations can be performed to examine turbulent structures, and multi-time, multi-point statistics can be extracted. In the mixing-length model, on the other hand, the description is at the mean-flow level: apart from the specified mixing length, the only quantities represented directly are $\langle U \rangle$ and $\langle V \rangle$. No information is provided about PDFs of velocity, two-point correlations, or turbulence structures, for example. The limited description provided by mean-flow closures (such as the mixing-length model) *is adequate* in many applications. The issue is more that a higher level of description can provide

Table 8.1. *Examples of turbulent flows of various levels of computational difficulty (the difficulty increases downward and to the right)*

Dimensionality (number of directions of statistical inhomogeneity)	Boundary layer (statistically stationary, boundary-layer approximations apply)	Statistically stationary	Not statistically stationary
0			Homogeneous shear flow
1		Fully developed pipe or channel flow; self-similar free shear flows[a]	Temporal mixing layer
2	Flat-plate boundary layer; jet in a co-flow	Flow through a sudden expansion in a two-dimensional duct	Flow over an oscillating cylinder
3	Boundary layer on a wing	Jet in a cross-flow; flow over an aircraft or building	Flow in the cylinder of a reciprocating engine

[a] In similarity variables, turbulence-model equations for two-dimensional self-similar free shear flows have a single independent variable.

a more complete characterization of the turbulence, leading to models of greater accuracy and wider applicability.

Completeness

A model is deemed complete if its constituent equations are free from flow-dependent specifications. One flow is distinguished from another solely by the specification of material properties (i.e., ρ and v) and of initial and boundary conditions. DNS is complete, whereas the mixing-length model is incomplete: the mixing length $\ell_m(x, y)$ has to be specified, and the appropriate specification is flow dependent.

Incomplete models can be useful for flows within a narrow class (e.g., attached boundary layers on airfoils) for which there is a body of semi-empirical knowledge on the appropriate flow-dependent specifications. However, in general, completeness is clearly desirable.

Cost and ease of use

For all but the simplest of flows, numerical methods are required to solve the model equations. The difficulty of performing a turbulence-model calculation depends both on the flow and on the model.

Table 8.1 provides a categorization of turbulent flows according to their geometries. The computational difficulty increases with the statistical dimensionality of the flow: it decreases if the flow is statistically stationary, and decreases further if boundary-layer equations can be used. In some approaches (e.g., DNS) the computational cost is a rapidly increasing function of the Reynolds number of the flow; whereas in others (e.g., the mixing length) the increase in cost with Reynolds number is insignificant or nonexistent.

The task of performing a turbulent-flow calculation for a particular flow can be considered in two parts. First, the computer program to solve the model equations has to be obtained or developed, and set up for the flow at hand (e.g., by specifying appropriate boundary conditions). Second, the computer program is executed to perform the calculation, and the required results are extracted. The cost and difficulty of the first step depend on the available software and algorithms, and on the complexity of the model. The effort required to develop a computer program for a particular class of flows and models can be very significant, and is therefore a substantial impediment to the evaluation and use of new models requiring new programs. It is, however, a 'one-time cost.'

The cost and difficulty of the second part – performing the computation – depend on the scale of computer required (e.g., a workstation or a super-computer), on the amount of human time and skill needed to perform the computation, and on the computer resources consumed. These are 'recurrent costs.'

In terms of computer time consumed, what computational cost is acceptable? The answer can vary by a factor of a million, depending on the context. Peterson *et al.* (1989) suggest that about 200 h of CPU time on the most powerful supercomputers is the upper reasonable limit on 'large-scale research' calculations. (The channel-flow DNS of Kim *et al.* (1987) required 250 h.) Very few calculations of this scale are performed. For 'applications' Peterson *et al.* (1989) suggest that a more reasonable time is 15 min CPU time on the most powerful supercomputers, which corresponds to 25 h of CPU time on a workstation of one hundredth the speed. To perform engineering design studies on a workstation – requiring 'repetitive' turbulent-flow calculations – CPU times of a minute or less per calculation are desirable. The ratio of the sizes of these 'repetitive,' 'application,' and 'large-scale research' computations is $1:1.5 \times 10^3:1.2 \times 10^6$.

The amount of computation measured in *flops* (floating-point operations) that can be performed in a given time is determined by the speed of the computer, measured in *megaflops*, *gigaflops*, or *teraflops*, i.e., $10^6, 10^9$, or 10^{12} flops per second.[1] Figure 8.1 shows the peak speed of the largest

[1] Note that *flops* is number of operations, whereas *megaflops* is a rate, i.e., the number of operations (in millions) per second.

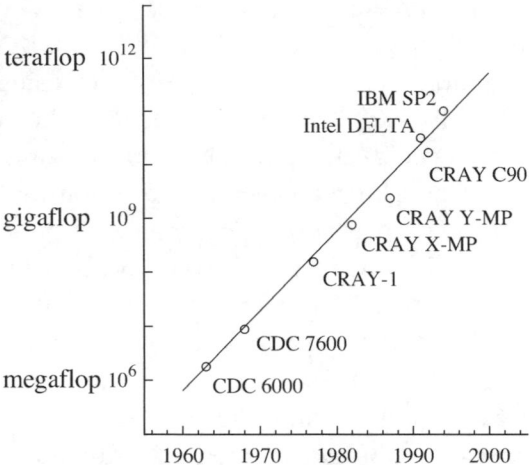

Fig. 8.1. The speed (*flops* per second) of the fastest supercomputers against the year of their introduction. The line shows a growth rate of a factor of 30 per decade. (Adapted from Foster (1995) with permission of Addison Wesley Longman.)

supercomputers over a 30-year period. It may be seen that the speed has increased exponentially, by a factor of 30 per decade. This is a remarkable rate of increase – a factor of a thousand in 20 years, and a million in 40 years. Although there is no sound basis for extrapolation beyond a few years, it is nevertheless generally supposed that this trend will continue (Foster 1995). Consequently, today's 'research' approaches may be feasible for 'applications' in 20 years, and for 'repetitive' calculations in 40 years. On the other hand, this 40 year span between 'large-scale research' and 'repetitive' computations again illustrates the need for a range of models, differing in their computational requirements.

(The absolute speeds shown in Fig. 8.1 need to be viewed with caution. The speed achieved in practice may be less than the peak speed by one or even two orders of magnitude. Typically only a fraction (e.g., one eighth) of the processors of a parallel computer are used, and only 20–50% of the peak speed is achieved on each processor. It should also be appreciated that, while here the focus of the discussion is on CPU time, memory can also be a limiting factor.)

Range of applicability

Not all models are applicable to all flows. For example, there are many models based on velocity spectra or two-point correlations, which are applicable to homogeneous turbulence only. (Such models are not considered here, but are described in the books of Lesieur (1990) and McComb (1990).)

As a second example, particular mixing-length models typically make assumptions about the flow geometry in the specification of the mixing length, so that their applicability is confined to flows of that geometry. Computational requirements place another – though nonetheless real – limitation on the applicability of some models. In particular, for DNS the computational requirements rise so steeply with Reynolds number that the approach is applicable only to flows of low or moderate Reynolds number. This limitation is examined in more detail in Chapter 9.

In this book, attention is focused on the velocity field in constant-density flows. It should be appreciated, however, that, in many flows to which turbulence models are applied, there are additional phenomena, such as heat and mass transfer, chemical reactions, buoyancy, compressibility, and multiphase flow. An important consideration, therefore, is the extent to which the approaches considered here are applicable to – or can be extended to – these more complex flows.

It is emphasized that, in these considerations, we separate *applicability* from *accuracy*. A model is applicable to a flow if the model equations are well posed and can be solved, irrespective of whether the solutions are accurate.

Accuracy

It goes without saying that accuracy is a desirable attribute of any model. In application to a particular flow, the accuracy of a model can be determined by comparing model calculations with experimental measurements. This process of model testing is of fundamental importance and deserves careful consideration. As shown in Fig. 8.2, the process consists in a number of steps, several of which introduce errors.

For a number of reasons, the boundary conditions in the calculations need not correspond exactly to those of the measured flow. A flow may be approximately two-dimensional, but may be assumed to be exactly so in the calculations. Boundary conditions on some properties will not necessarily be known, and so have to be estimated; or, they may be taken from experimental data that contain some measurement error.

The numerical solution of the model equations inevitably contains numerical error. This may be from a number of sources, but it is often dominated by spatial truncation error. In a finite-difference or finite-element method, for example, this error scales as a positive power of the grid spacing, Δx, while the computational cost increases with $1/\Delta x$ – possibly as a positive power. Many published turbulence-model calculations contain significant numerical errors, either because the available computer resources do not allow a

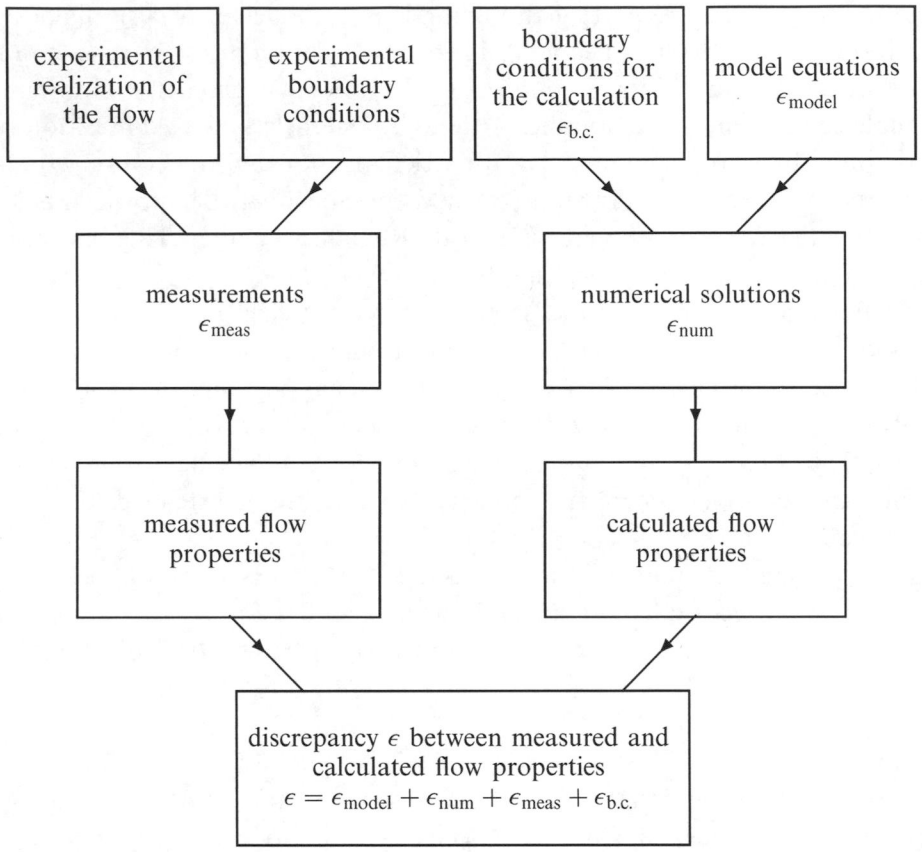

Fig. 8.2. A chart illustrating that the discrepancy ϵ between measured and calculated flow properties stems from inaccuracies of the model, ϵ_{model}; numerical errors, ϵ_{num}; measurement errors ϵ_{meas}; and discrepancies in the boundary conditions, $\epsilon_{b.c.}$. (The equation given for ϵ is merely suggestive: the errors do not add linearly.)

sufficiently fine grid spacing, or, put bluntly, because the calculations are performed with insufficient care or regard for numerical accuracy.

In summary, as depicted in Fig. 8.2, the discrepancy between measured and calculated flow properties arises from

(i) inaccuracies of the model,
(ii) numerical error,
(iii) measurement error, and
(iv) discrepancies in the boundary conditions.

The important conclusion is that a comparison between measured and calculated flow properties determines the accuracy of the model only if the errors arising from (ii)–(iv) are relatively small. In particular, there is a danger of drawing false conclusions about the accuracy of a model from calculations containing large or unquantified numerical errors. These issues are discussed further by Coleman and Stern (1997).

Final remarks

The suitability of a particular model for a particular turbulent-flow problem depends on a weighted combination of the criteria discussed above; and the relative weighting of the importances of the various criteria depends significantly on the problem. Consequently, as mentioned at the outset, now and into the future, there is no one 'best' model, but rather there is a range of models that can usefully be applied to the broad range of turbulent-flow problems.

9

Direct numerical simulation

Direct numerical simulation (DNS) consists in solving the Navier–Stokes equations, resolving all the scales of motion, with initial and boundary conditions appropriate to the flow considered. Each simulation produces a single realization of the flow. The DNS approach was infeasible until the 1970s when computers of sufficient power became available. Even though it is a latecomer among modelling approaches, it is logical to discuss DNS first. Conceptually it is the simplest approach and, when it can be applied, it is unrivalled in accuracy and in the level of description provided. However, it is important to appreciate that the cost is extremely high; and the computer requirements increase so rapidly with Reynolds number that the applicability of the approach is limited to flows of low or moderate Reynolds numbers.

In this chapter, we first describe DNS applied to homogeneous turbulence and examine in some detail the computational requirements. Then we consider DNS for inhomogeneous turbulent flows, for which rather different numerical methods are required.

9.1 Homogeneous turbulence

For homogeneous turbulence, *pseudo-spectral* methods (pioneered by Orszag and Patterson (1972) and Rogallo (1981)) are the preferred numerical approach, because of their superior accuracy. The rudiments of these methods are described in Section 9.1.1, which allows the computational cost of DNS to be estimated (in Section 9.1.2).

9.1.1 Pseudo-spectral methods

In a DNS of homogeneous isotropic turbulence, the solution domain is a cube of side $\mathcal{L}$, and the velocity field $u(x, t)$ is represented as a finite Fourier series

$$u(x,t) = \sum_{\kappa} e^{i\kappa \cdot x} \hat{u}(\kappa, t), \tag{9.1}$$

see Section 6.4.1. In total N^3 wavenumbers are represented, where the even number N determines the size of the simulation, and consequently the Reynolds number that can be attained. Typically N is chosen to be rich in powers of 2 (e.g., $N = 128$ or $N = 192$). In magnitude, the lowest non-zero wavenumber is $\kappa_0 = 2\pi/\mathcal{L}$, and the N^3 wavenumbers represented are

$$\kappa = \kappa_0 n = \kappa_0(e_1 n_1 + e_2 n_2 + e_3 n_3), \tag{9.2}$$

for integer values of n_i between $-\frac{1}{2}N + 1$ and $\frac{1}{2}N$. In each direction, the largest wavenumber represented is

$$\kappa_{\max} = \tfrac{1}{2}N\kappa_0 = \frac{\pi N}{\mathcal{L}}. \tag{9.3}$$

This spectral representation is equivalent to representing $u(x,t)$ in physical space on an N^3 grid of uniform spacing

$$\Delta x = \frac{\mathcal{L}}{N} = \frac{\pi}{\kappa_{\max}}. \tag{9.4}$$

The discrete Fourier transform (DFT, see Appendix F) gives a one-to-one mapping between the Fourier coefficients $\hat{u}(\kappa, t)$ and the velocities $u(x,t)$ at the N^3 grid nodes; and the fast Fourier transform (FFT) can be used to transform between wavenumber space (i.e., $\hat{u}(\kappa, t)$) and physical space (i.e., $u(x,t)$) in on the order of $N^3 \log N$ operations.

A *spectral method* involves advancing the Fourier modes $\hat{u}(\kappa, t)$ in small time steps Δt according to the Navier–Stokes equations in wavenumber space (Eq. (6.146)). Summing the triad interactions in this equation requires on the order of N^6 operations. To avoid this large cost, in *pseudo-spectral* methods the nonlinear terms in the Navier–Stokes equations are evaluated differently: the velocity field is transformed into physical space; the nonlinear terms (i.e., $u_i u_j$) are formed; and then they are transformed to wavenumber space. This procedure requires on the order of $N^3 \log N$ operations. An *aliasing error* (see Appendix F) that must be removed or controlled is also introduced.

The main numerical and computational issues in pseudo-spectral methods are the time-stepping strategy, the control of aliasing errors, and implementation on distributed-memory parallel computers. Each of these issues is elegantly treated in the algorithm developed by Rogallo (1981), which forms the basis of many other DNS codes for homogeneous turbulence.

In addition to isotropic turbulence, DNS has been applied to homogeneous turbulence with a variety of imposed mean velocity gradients (see Exercise 5.41 on page 158). This includes isotropic turbulence subjected to a mean rotation (Bardina, Ferziger, and Rogallo 1985), homogeneous shear flow (Rogers and Moin 1987, Lee *et al.* 1990), and irrotational mean straining (Lee and Reynolds 1985). For these cases, the periodic solution domain and the wavenumber vectors are distorted by the mean deformation, as described in Section 11.4.

9.1.2 The computational cost

The computational cost of a simulation is largely determined by the resolution requirements. The box size $\mathcal{L}$ must be large enough to represent the energy-containing motions; and the grid spacing Δx must be small enough to resolve the dissipative scales. In addition, the time step Δt used to advance the solution is limited by considerations of numerical accuracy. These factors are now examined in greater detail.

For isotropic turbulence with a given spectrum, a reasonable lower limit on $\mathcal{L}$ is eight integral length scales ($\mathcal{L} = 8L_{11}$), which, in terms of the lowest wavenumber κ_0, implies that

$$\kappa_0 L_{11} = \frac{\pi}{4} \approx 0.8. \tag{9.5}$$

From Fig. 6.18 (on page 241) and Table 6.2 (on page 240) it may be observed that the peak of the energy spectrum is at $\kappa L_{11} \approx 1.3$, and that 10% of the energy is at wavenumbers below $\kappa L_{11} \approx 1.0$: with $\kappa_0 L_{11} \approx 0.8$, approximately 5% of the energy at the low-wavenumber end of the spectrum is not resolved.

The Fourier representation of the velocity field implies that one must impose periodic boundary conditions on the solution domain. Figure 9.1 shows how these artificial conditions are manifested in the autocorrelation function $f(r)$. With infinite low-wavenumber resolution ($\mathcal{L}/L_{11} \to \infty$), the autocorrelation function tends monotonically to zero, whereas for $\mathcal{L}/L_{11} = 8$ it is periodic with significant differences apparent for $|r|/L_{11} > 3$. The effect of a finite box size has not been studied systematically; but, at a price, the artificial effects can be reduced by increasing $\mathcal{L}/L_{11}$.

The resolution of the smallest, dissipative motions (which are characterized by the Kolmogorov scale η) requires a sufficiently small grid spacing $\Delta x/\eta$ or, correspondingly, a sufficiently large maximum wavenumber $\kappa_{\max}\eta$. It may be observed from Fig. 6.16 (on page 237) that the dissipation spectrum is extremely small beyond $\kappa\eta = 1.5$, and indeed experience shows that $\kappa_{\max}\eta \geq 1.5$ is the criterion for good resolution of the smallest scales (see,

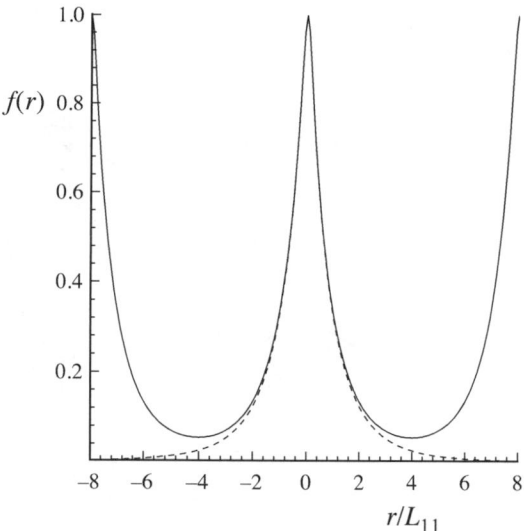

Fig. 9.1. The effect of periodicity on the longitudinal velocity autocorrelation function. Dashed line, $f(r)$ for the model spectrum at $R_\lambda = 40$; solid line, $f(r)$ for the periodic velocity field ($\mathcal{L} = 8L_{11}$), with approximately the same spectrum.

e.g., Yeung and Pope (1989)). The corresponding grid spacing in physical space is

$$\frac{\Delta x}{\eta} = \frac{\pi}{1.5} \approx 2.1. \tag{9.6}$$

This may appear large, but recall that η underestimates the size of the dissipative motions (see Fig. 6.16 and Table 6.1 on page 238).

The two spatial-resolution requirements $\mathcal{L}/L_{11} = 8$ and $\kappa_{max}\eta = 1.5$ determine the necessary number of Fourier modes (or grid nodes) N^3 as a function of the Reynolds number. The above equations yield

$$N = 2\frac{\kappa_{max}}{\kappa_0} = 2\frac{\kappa_{max}\eta}{\kappa_0 L_{11}} \left(\frac{L_{11}}{L}\right) \left(\frac{L}{\eta}\right) = \frac{12}{\pi} \left(\frac{L_{11}}{L}\right) \left(\frac{L}{\eta}\right), \tag{9.7}$$

(where $L \equiv k^{3/2}/\varepsilon$). Using the model spectrum to obtain L_{11}/L, the value of N obtained is shown as a function of the Reynolds number in Fig. 9.2. At high Reynolds number L_{11}/L has the asymptotic value 0.43 (see Fig. 6.24 on page 245), so that Eq. (9.7) becomes

$$N \sim 1.6 \left(\frac{L}{\eta}\right) = 1.6\,\mathrm{Re}_L^{3/4} \approx 0.4\,\mathrm{R}_\lambda^{3/2}. \tag{9.8}$$

Hence the total number of modes increases as

$$N^3 \sim 4.4\,\mathrm{Re}_L^{9/4} \approx 0.06\,\mathrm{R}_\lambda^{9/2}. \tag{9.9}$$

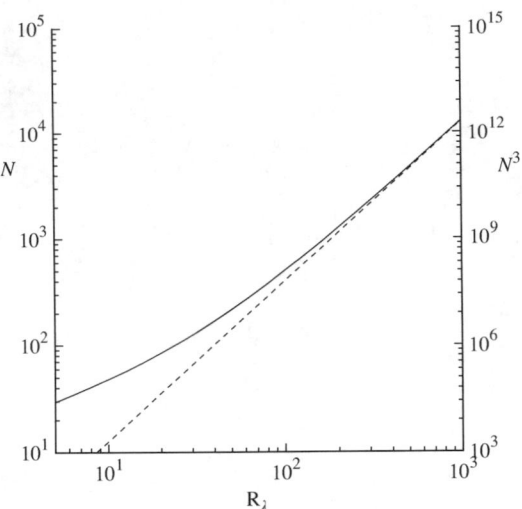

Fig. 9.2. The number of Fourier modes (or grid nodes) N in each direction required for adequate resolution of isotropic turbulence. Solid line, Eq. (9.7); dashed line, asymptote, Eq. (9.8). The right-hand axis shows the total number of modes required, N^3.

(This is in reasonable agreement with the estimate $N^3 \sim 0.1\,\mathrm{R}_\lambda^{9/2}$ made by Reynolds (1990).)

For the advance of the solution in time to be accurate, it is necessary that a fluid particle move only a fraction of the grid spacing Δx in a time step Δt. In practice, the *Courant number* thus imposed is found to be approximately

$$\frac{k^{1/2}\,\Delta t}{\Delta x} = \frac{1}{20}. \tag{9.10}$$

The duration of a simulation is typically on the order of four times the turbulence time scale, $\tau = k/\varepsilon$. Hence the number of time steps required is

$$M = \frac{4\tau}{\Delta t} = 80\frac{L}{\Delta x} = \frac{120}{\pi}\frac{L}{\eta} \approx 9.2\mathrm{R}_\lambda^{3/2}. \tag{9.11}$$

(Exercise 9.1 gives an alternative estimate for Δt and hence for M.)

To an approximation, the number of floating-point operations required to perform a simulation is proportional to the product of the number of modes and the number of steps, N^3M (mode-steps). The preceding results yield

$$N^3M \sim 160\mathrm{Re}_L^3 \approx 0.55\mathrm{R}_\lambda^6, \tag{9.12}$$

showing the very steep rise with the Reynolds number.

To complete the picture, with justification to follow, we suppose that 1,000 floating point operations per mode per time step are needed. Then the time

Table 9.1. *Estimates, for DNS of isotropic turbulence at various Reynolds numbers, of modes required in each direction, N (Eq. (9.7)); total number of modes, N^3; number of time steps, M (Eq. (9.11)); number of mode-steps, N^3M; and the time to perform a simulation at 1 gigaflop (assuming 1,000 operations per mode per step)*

R_λ	Re_L	N	N^3	M	N^3M	CPU	Time
25	94	104	1.1×10^6	1.2×10^3	1.3×10^9	20	min
50	375	214	1.0×10^7	3.3×10^3	3.2×10^{10}	9	h
100	1,500	498	1.2×10^8	9.2×10^3	1.1×10^{12}	13	days
200	6,000	1,260	2.0×10^9	2.6×10^4	5.2×10^{13}	20	months
400	24,000	3,360	3.8×10^{10}	7.4×10^4	2.8×10^{15}	90	years
800	96,000	9,218	7.8×10^{11}	2.1×10^5	1.6×10^{17}	5,000	years

in days, T_G, needed to perform a simulation at a computing rate of 1 gigaflop is

$$T_G = \frac{10^3 N^3 M}{10^9 \times 60 \times 60 \times 24} \sim \left(\frac{Re_L}{800}\right)^3 \approx \left(\frac{R_\lambda}{70}\right)^6. \quad (9.13)$$

Figure 9.3 shows this estimate as a function of R_λ, together with timings of a DNS code that lend support to the assumptions made. A summary of these estimates is given in Table 9.1.

The obvious conclusion from these estimates is that the computational cost increases so steeply with the Reynolds number (as R_λ^6 or Re_L^3) that it is impracticable to go much higher than $R_\lambda \approx 100$ with gigaflop computers. The ordinate (T_G) in Fig. 9.3 has another interpretation: it is the factor of improvement (over a gigaflop computer, and in numerical methods) that is needed to perform a DNS in one day. Thus, to achieve $R_\lambda = 1,000$, a millionfold improvement is needed.

It is revealing to examine the distribution of the computational effort over the various scales of turbulent motion. In the three-dimensional wavenumber space, the modes represented lie within the cube of side $2\kappa_{max}$. In a well-resolved simulation ($\kappa_{max}\eta = 1.5$, say), only modes within the sphere of radius κ_{max} are dynamically significant. As depicted in Fig. 9.4, the dissipation range corresponds to the spherical shell of wavenumbers $|\kappa|$ between κ_{DI} and κ_{max}, where $\kappa_{DI} = 0.1/\eta$ is the wavenumber of the largest dissipative motions (see Section 6.5.4). Within the sphere of radius κ_{DI} lie the energy-containing and (at sufficiently high Reynolds number) the inertial-range motions. In Fig. 9.4 the inner sphere of radius κ_E corresponds to the peak in the energy spectrum for isotropic turbulence at $R_\lambda = 70$: at higher Reynolds number, this sphere

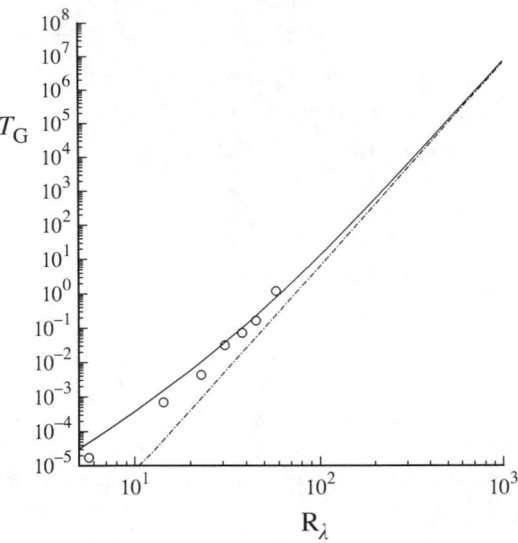

Fig. 9.3. The time in days T_G required to perform DNS of homogeneous isotropic turbulence on a gigaflop computer as a function of the Reynolds number. Solid line, estimate from Eqs. (9.7), (9.11), and (9.13); dashed line, asymptote $(R_\lambda/70)^6$; symbols, based on DNS timings for a 40-node IBM SP2.

is smaller. It is a matter of simple arithmetic (Exercise 9.2) to show that 99.98% of the modes represented have wavenumbers $|\kappa|$ greater than κ_{DI}; less than 0.02% of the modes represent motions in the energy-containing range or in the inertial subrange.

EXERCISES

9.1   The time step Δt dictated by the Courant number (Eq. (9.10)) is a restriction imposed by the numerical methods that are currently employed. The intrinsic restriction on Δt imposed by the turbulence is that $\Delta t/\tau_\eta$ should be small. If the Courant-number restriction (Eq. (9.10)) is replaced by

$$\frac{\Delta t}{\tau_\eta} = 0.1, \tag{9.14}$$

obtain the revised estimates

$$M = 4\sqrt{15}\,R_\lambda, \tag{9.15}$$

$$N^3 M \sim 0.93 R_\lambda^{11/2}, \tag{9.16}$$

$$T_G \sim \left(\frac{R_\lambda}{100}\right)^{11/2}. \tag{9.17}$$

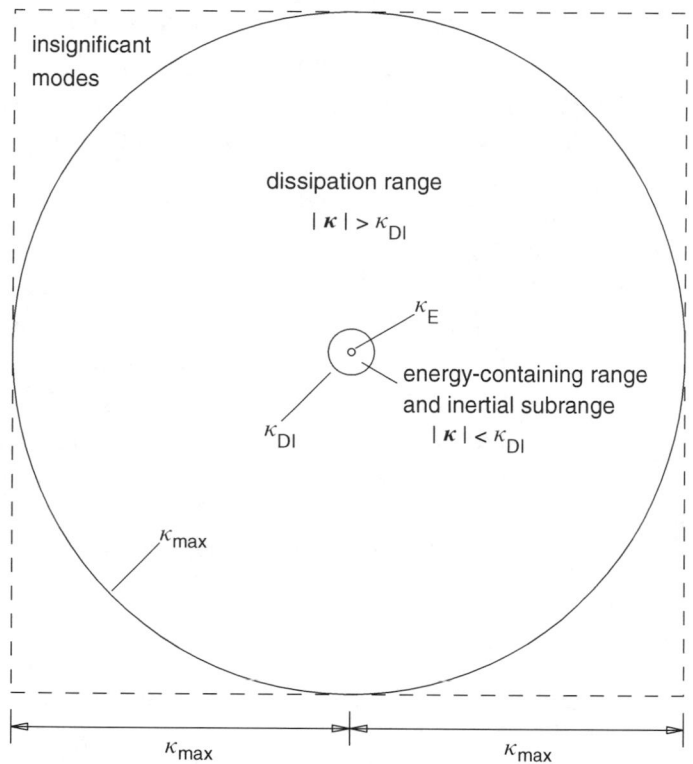

Fig. 9.4. The solution domain in wavenumber space for a pseudo-spectral DNS of isotropic turbulence. The modes represented lie within the cube of side $2\kappa_{max}$ (dashed line). The three spheres shown are: of radius κ_{max}, the maximum wavenumber resolved in all directions ($\kappa_{max}\eta = 1.5$); of radius κ_{DI}, the wavenumber of the largest dissipative motions ($\kappa_{DI}\eta = 0.1$); and of radius κ_E, the wavenumber corresponding to the peak of the energy spectrum at $R_\lambda = 70$ ($\kappa_E L_{11} = 1.3$). Only 0.016% of the modes represented lie within the sphere of radius κ_{DI}, corresponding to motions in the energy-containing range and in the inertial subrange.

(For $R_\lambda = 1,000$, this estimate yields $T_G \approx 0.3 \times 10^6$, compared with the value $T_G = 8.5 \times 10^6$ given by Eq. (9.13).)

9.2   Consider DNS of homogeneous isotropic turbulence in which the small-scale resolution is characterized by $\kappa_{max}\eta = 1.5$, as depicted in Fig. 9.4. The dissipation range is defined (see Section 6.5.4) as $|\kappa| > \kappa_{DI}$, where $\kappa_{DI}\eta = 0.1$. Show that the fraction of the N^3 wavenumber modes represented that are *not* in the dissipation range is $15^{-3}\pi/6 \approx 0.00016$; and correspondingly that 99.98% of the modes are in the dissipation range. Estimate that, at $R_\lambda = 70$, the peak of the energy spectrum occurs at $\kappa_E \approx 0.01\kappa_{max}$. What fraction of the modes is insignificant, i.e., with $|\kappa| > \kappa_{max}$?

9.1.3 Artificial modifications and incomplete resolution

The high cost of DNS has motivated several different approaches to attaining higher Reynolds numbers (with a given number of modes) by incompletely resolving either the low-wavenumber or the high-wavenumber modes.

Low-wavenumber forcing

Statistically stationary homogeneous turbulence can be obtained in DNS by artificially *forcing* the low-wavenumber modes, so as to supply energy to them. The energy-containing motions are, therefore, unnatural and are not governed by the Navier–Stokes equations. However, insofar as the small-scale motions are universal, useful information about them can be extracted. As an example, in a simulation using 256^3 modes and forcing, it is possible to achieve $R_\lambda = 180$ with a $-\frac{5}{3}$ spectrum over a decade of wavenumbers (Overholt and Pope 1996): without forcing, the Reynolds number that can be attained is $R_\lambda = 60$ (Fig. 9.2). Chen *et al.* (1993) reported a forced 512^3 simulation with Reynolds number $R_\lambda = 202$.

Large-eddy simulation

In wavenumber space, the number of discrete wavenumber modes $\kappa = \kappa_0 n$ of magnitude $|\kappa|$ less than κ increases as κ^3. Consequently, in a fully resolved DNS, the vast majority of modes is in the dissipative range (see Fig. 9.4 and Exercise 9.2). This observation provides strong motivation for approaches that reduce the resolution requirements in the dissipative range, especially if the energy-containing scales are of primary interest.

In large-eddy simulation (LES) only the energy-containing motions are resolved, and the effects of the unresolved modes are modelled. This LES approach is the subject of Chapter 13.

Hyperviscosity

A different approach is to replace the viscous term $v\,\nabla^2 u$ by $v_H(-1)^{m+1}\,\nabla^{2m}u$, for integer $m > 1$, where v_H is a hyperviscosity coefficient (see, e.g., Borue and Orszag (1996)). In wavenumber space, this corresponds to changing $v\kappa^2\hat{u}(\kappa)$ to $v_H\kappa^{2m}\hat{u}(\kappa)$, which causes the dissipation range to be narrower, hence requiring fewer modes to resolve.

Sparse-mode methods

In this approach, the energy-containing motions ($|\kappa| < \kappa_{EI}$) are fully resolved, whereas in shells of higher wavenumber ($2^{m-1}\kappa_{EI} \leq |\kappa| < 2^m\kappa_{EI}$, $m = 1, 2, \ldots$) only a fraction 2^{-3m} of the Fourier modes is represented (see, e.g., Meneguzzi *et al.* (1996)).

Caveat

It should be appreciated that all of the approaches mentioned above – forcing, LES, hyperviscosity, and use of sparse modes – amount to substantial departures from the Navier–Stokes equations. They are not, therefore, direct numerical simulations of turbulence, but rather each is a model whose accuracy is not known *a priori*.

9.2 Inhomogeneous flows

Compared with homogeneous turbulence, the principal differences in applying DNS to inhomogeneous flows are that

(i) Fourier representations cannot be used in directions of inhomogeneity,
(ii) physical boundary conditions (as opposed to periodic conditions) are required, and
(iii) near-wall motions, characterized by the viscous length scale δ_v, impose an additional resolution requirement.

Some of these issues are now illustrated for particular flows.

9.2.1 Channel flow

In the DNS of channel flow performed by Kim *et al.* (1987), the solution domain is a rectangular box with dimensions $\mathcal{L}_x \times h \times \mathcal{L}_z$. The flow is statistically homogeneous in the mean-flow (x) and spanwise (z) directions, allowing Fourier representations with N_x and N_z modes, respectively. Correspondingly, in physical space, there are grid nodes with uniform spacings $\Delta x = \mathcal{L}_x/N_x$ and $\Delta z = \mathcal{L}_z/N_z$ in the x and z directions. As is the case for homogeneous turbulence, the periodic boundary conditions implied by the Fourier representation are artificial, but their influence is small provided that the periods $\mathcal{L}_x$ and $\mathcal{L}_z$ are sufficiently large compared with flow scales.

In the cross-stream direction, the solution domain extends from the bottom $y = 0$ to the top wall $y = h$. At both boundaries the no-slip condition $U = 0$ applies.

The essential characteristic of the N_x and N_z Fourier modes in the x and z directions is that they provide a set of orthogonal basis functions. In the nonperiodic y direction, the same is achieved through the use of N_y Chebyshev polynomials $T_n(\xi)$ (which are defined on the interval $-1 \le \xi \le 1$). Thus the velocity field $U(x, y, z, t)$ is represented as the sum of $N_x \times N_y \times N_z$ modes of the form

$$\widehat{U}(n_x, n_y, n_z, t) \exp(2\pi i n_x x/\mathcal{L}_x) \, T_{n_y}\left(\frac{2y}{h} - 1\right) \exp(2\pi i n_z z/\mathcal{L}_z). \qquad (9.18)$$

A fast transform (involving FFTs) is used to transform between the basis-function coefficients $\widehat{U}(n_x, n_y, n_z, t)$ and the velocity $U(x, y, z, t)$ in physical space on an $N_x \times N_y \times N_z$ grid. In the y direction the grid spacing is non-uniform with, conveniently, a finer grid spacing near the boundaries. The fast transform allows the use of a pseudo-spectral method to advance the velocity field according to the Navier–Stokes equations.

From specified initial conditions, the solution is advanced in time until the statistically stationary state is reached. The initial conditions specified may affect the number of steps M_0 required to reach stationarity, but they do not affect the statistics in the stationary state. The solution is then continued (for another M_T time steps) so that statistics can be time averaged. (One-point statistics are also averaged over the two homogeneous directions.) Table 9.2 shows numerical parameters used in DNS of channel flow at two different Reynolds numbers. Notice that a grid spacing of about $\frac{1}{20}\delta_\nu$ in the y direction is necessary at the wall, where $\delta_\nu \equiv \nu/u_\tau$ in the viscous lengthscale. (Notice also that the larger-Reynolds-number simulation is performed on a domain one quarter of the size and for a time duration one quarter as long as that in the lower-Reynolds-number simulation.) From the resolution requirement based on δ_ν, Reynolds (1990) estimates that the total number of modes required increases as $\mathrm{Re}^{2.7}$ (cf. $\mathrm{Re}_L^{2.25}$ for homogeneous turbulence).

9.2.2 Free shear flows

The temporally evolving mixing layer and plane wake are examples of statistically one-dimensional free shear flows that have been studied using DNS. For the mixing layer, Rogers and Moser (1994) used a pseudo-spectral method with Fourier modes in the x and z directions, and Jacobi polynomials in y. Because of the imposition of periodicity, the dimensions of the domain $\mathcal{L}_x$ and $\mathcal{L}_z$ must be large compared with the mixing-layer thickness $\delta(t)$, which increases with time. The Jacobi polynomials extend to $y = \pm\infty$, facilitating the specification of the boundary conditions in the free streams, namely uniform flow parallel to the x axis.

For temporally evolving flows, the specification of appropriate initial conditions is crucial. For the duration of the mixing-layer simulations, over which time $\delta(t)$ increases by a factor of three, the initial conditions are found to have a first-order effect on the evolution.

Table 9.2. *Numerical parameters for DNS of channel flow at* $Re_\tau = 180$ *(Kim et al. 1987) and* $Re_\tau = 595$ *(Dr N. N. Mansour, personal communication) (the variables not defined in the text are the total number of modes,* N_{xyz} *; maximum and minimum grid spacings in* $y, \Delta y_{max}$ *and* Δy_{min} *; and the time duration of the simulation,* T *)*

	Re_τ	
	180	595
N_x	192	384
N_y	129	256
N_z	160	384
N_{xyz}	4×10^6	38×10^6
$\mathcal{L}_x/h$	2π	π
$\mathcal{L}_z/h$	π	$\frac{1}{2}\pi$
$\Delta x/\delta_\nu$	12	9.7
$\Delta y_{min}/\delta_\nu$	0.05	0.04
$\Delta y_{max}/\delta_\nu$	4.4	4.9
$\Delta z/\delta_\nu$	6	7.3
Tu_τ/h	5	1.1
M	22,500	12,000
$N_{xyz}M$	9×10^{10}	45×10^{10}
Computer	Cray XMP	IBM SP2 (64 processors)
CPU time (h)	250	185

The spatially developing mixing layer is statistically stationary and two-dimensional. In DNS, inflow boundary conditions (at $x = 0$, say) and outflow conditions (at $x = \mathcal{L}_x$) are required. Roughly speaking, the inflow conditions here play the same role as that of the initial conditions in the temporal mixing layer. Not surprisingly, it is found that the flow is sensitive to the details of these inflow conditions (Buell and Mansour 1989).

9.2.3 Flow over a backward-facing step

DNS for this flow, sketched in Fig. 9.5, has been performed by Le, Moin, and Kim (1997). A turbulent boundary layer (of thickness $\delta \approx 1.2h$ and free-stream velocity U_0) enters at the left-hand boundary, it separates at the step ($x = 0$), and then reattaches downstream (at $x \approx 7h$). The flow is statistically stationary and two-dimensional. The Reynolds number considered $Re \equiv U_0 h/\nu = 5,100$, is quite low compared, for example, with $Re = 500,000$ in the experiments of Durst and Schmitt (1985).

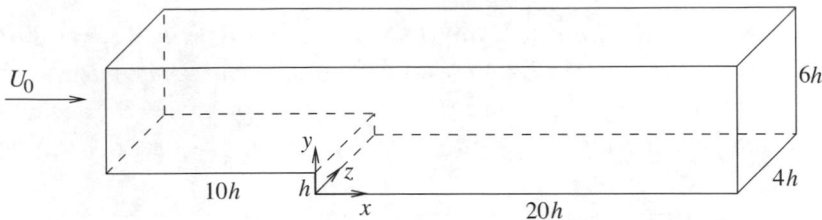

Fig. 9.5. A sketch of the solution domain used by Le *et al.* (1997) for DNS of flow over a backward-facing step. Dimensions are in units of the step height, *h*.

Table 9.3. *Numerical parameters for DNS of the flow over a backward-facing step (Le* et al. *1997)*

Number of nodes in x, N_x	786
Number of nodes in y, N_y	192
Number of nodes in z, N_z	64
Total number of nodes, N_{xyz}	8.3×10^6
Number of time steps, M	2.1×10^5
Node-steps, $N_{xyz}M$	1.8×10^{12}
Computer	Cray C-90
CPU time (h)	1,300

Some of the numerical characteristics of the DNS are given in Table 9.3. In spite of the low Reynolds number and the vast amount of computer time consumed (about 54 days), the resolution is marginal. According to the authors, it would be desirable to double the inlet length (upstream of the step), and to double the number of nodes N_z in the spanwise direction. In addition, the relatively short length $\mathcal{L}_z = 4h$ in the periodic spanwise direction results in the two-point velocity autocorrelation at some locations exceeding 0.7 for all spanwise separations r (rather than being negligible for $r = \frac{1}{2}\mathcal{L}_z$). The results of the simulations are in excellent agreement with the experiments of Jovic and Driver (1994) and include Reynolds-stress budgets.

9.3 Discussion

Where it can be applied, DNS provides a level of description and accuracy that cannot be equalled with other approaches. Highly accurate numerical methods have been developed to solve the Navier–Stokes equations, and the practitioners of DNS generally have high standards with respect to numerical accuracy.

DNS studies have proved extremely valuable in supplementing our knowledge from experiments of turbulence and turbulent flows. For example, DNS has been used to extract Lagrangian statistics (e.g., Yeung and Pope (1989)), and statistics of pressure fluctuations (e.g., the pressure–rate-of-strain tensor, Spalart (1988)), which are all but impossible to obtain experimentally. The details of near-wall flows, and homogeneous turbulence subjected to various deformations, are also more easily studied in DNS than in experiments. Some DNS results are presented in Chapters 5 and 7; but the reader is referred to Moin and Mahesh (1998) for a more comprehensive review of the contributions made by DNS studies to our understanding of turbulence and turbulent flows.

It should be appreciated that not all simulations based on the Navier–Stokes equations are 'direct.' Because of artificial modifications, incomplete resolution, or non-physical boundary conditions, a simulation need not directly correspond to a realizable turbulent flow. Non-physical simulations can, however, be used judiciously to isolate and study particular phenomena (see, e.g., Jiménez and Moin (1991) and Perot and Moin (1995)).

The drawback of DNS is of course its very large computational cost, and the fact that this cost increases rapidly with the Reynolds number (approximately as Re^3). Computer times are typically of order 200 h on a supercomputer, and then only flows with low or moderate Reynolds numbers can be simulated.

The observations about the computational cost of DNS signify more than the limitations of current computers. They signify also that there is a mismatch between DNS and the objective of determining the mean velocity and energy-containing motions in a turbulent flow. In DNS, over 99% of the effort is devoted to the dissipation range (see Fig. 9.4 and Exercise 9.2) and this effort increases strongly with the Reynolds number. By contrast, the mean flow and the statistics of the energy-containing motions exhibit only weak Reynolds-number dependences.

10

Turbulent-viscosity models

In this chapter and the next we consider RANS models in which the Reynolds equations are solved for the mean velocity field. The Reynolds stresses – which appear as unknowns in the Reynolds equations – are determined by a turbulence model, either via the turbulent viscosity hypothesis or more directly from modelled Reynolds-stress transport equations (Chapter 11).

Turbulent-viscosity models are based on the turbulent-viscosity hypothesis, which was introduced in Chapter 4 and has been used in subsequent chapters. According to the hypothesis, the Reynolds stresses are given by

$$\langle u_i u_j \rangle = \tfrac{2}{3} k \delta_{ij} - \nu_T \left(\frac{\partial \langle U_i \rangle}{\partial x_j} + \frac{\partial \langle U_j \rangle}{\partial x_i} \right), \tag{10.1}$$

or, in simple shear flow, the shear stress is given by

$$\langle uv \rangle = -\nu_T \frac{\partial \langle U \rangle}{\partial y}. \tag{10.2}$$

Given the turbulent viscosity field $\nu_T(x, t)$, Eq. (10.1) provides a most convenient closure to the Reynolds equations, which then have the same form as the Navier–Stokes equations (Eq. (4.46) on page 93). It is unfortunate, therefore, that for many flows the accuracy of the hypothesis is poor. The deficiencies of the turbulent-viscosity hypothesis – many of which have been mentioned above – are reviewed in Section 10.1.

If the turbulent-viscosity hypothesis is accepted as an adequate approximation, all that remains is to determine an appropriate specification of the turbulent viscosity $\nu_T(x, t)$. This can be written as the product of a velocity $u^*(x, t)$ and a length $\ell^*(x, t)$:

$$\nu_T = u^* \ell^*, \tag{10.3}$$

and the task of specifying ν_T is generally approached through specifications of u^* and ℓ^*. In algebraic models (Section 10.2) – the mixing-length model, for example – ℓ^* is specified on the basis of the geometry of the flow. In two-equation models (Section 10.4) – the k-ε model being the prime example – u^* and ℓ^* are related to k and ε, for which modelled transport equations are solved.

10.1 The turbulent-viscosity hypothesis

The turbulent-viscosity hypothesis can be viewed in two parts. First, there is the *intrinsic* assumption that (at each point and time) the Reynolds-stress anisotropy $a_{ij} \equiv \langle u_i u_j \rangle - \frac{2}{3}k\delta_{ij}$ is determined by the mean velocity gradients $\partial \langle U_i \rangle / \partial x_j$. Second, there is the *specific* assumption that the relationship between a_{ij} and $\partial \langle U_i \rangle / \partial x_j$ is

$$\langle u_i u_j \rangle - \tfrac{2}{3}k\delta_{ij} = -\nu_T \left(\frac{\partial \langle U_i \rangle}{\partial x_j} + \frac{\partial \langle U_j \rangle}{\partial x_i} \right), \tag{10.4}$$

or, equivalently,

$$a_{ij} = -2\nu_T \bar{S}_{ij}, \tag{10.5}$$

where $\bar{S}_{ij}$ is the mean rate-of-strain tensor. This is, of course, directly analogous to the relation for the viscous stress in a Newtonian fluid:

$$-(\tau_{ij} + P\delta_{ij})/\rho = -2\nu S_{ij}. \tag{10.6}$$

10.1.1 The intrinsic assumption

To discuss the intrinsic assumption we first describe a simple flow for which it is entirely incorrect. Then it is shown that, in a crucial respect, the physics of turbulence is vastly different than the physics of the molecular processes that lead to the viscous stress law (Eq. (10.6)). However, finally, it is observed that, for simple shear flows, the turbulent viscosity hypothesis is nevertheless quite reasonable.

Axisymmetric contraction

Figure 10.1 is a sketch of a wind-tunnel experiment, first performed by Uberoi (1956), to study the effect on turbulence of an axisymmetric contraction. The air flows through the turbulence-generating grid into the first straight section, in which the mean velocity $\langle U_1 \rangle$ is (ideally) uniform. In this section there is no mean straining ($\bar{S}_{ij} = 0$), and the turbulence (which is almost isotropic) begins to decay.

Following the first straight section there is an axisymmetric contraction, which is designed to produce a uniform extensive axial strain rate,

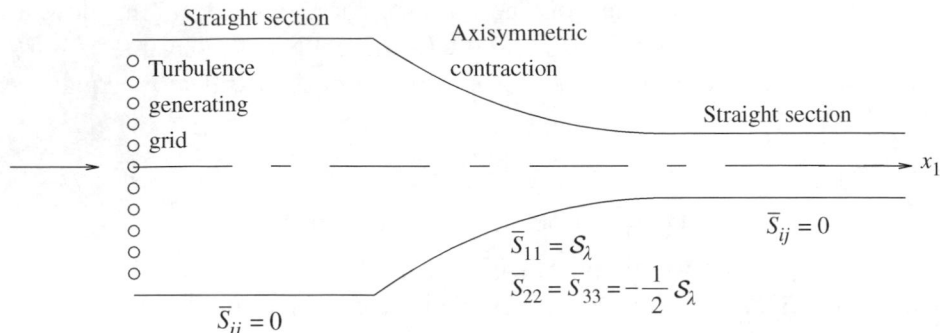

Fig. 10.1. A sketch of an apparatus, similar to that used by Uberoi (1956) and Tucker (1970), to study the effect of axisymmetric mean straining on grid turbulence.

$\bar{S}_{11} = \partial\langle U_1\rangle/\partial x_1 = \mathcal{S}_\lambda$, and hence uniform compressive lateral strain rates, $\bar{S}_{22} = \bar{S}_{33} = -\frac{1}{2}\mathcal{S}_\lambda$. The quantity $\mathcal{S}_\lambda k/\varepsilon$ (evaluated at the beginning of the contraction) measures the mean strain rate relative to the turbulence timescale. Figure 10.2 shows measurements of the normalized anisotropies ($b_{ij} \equiv \langle u_i u_j\rangle/\langle u_k u_k\rangle - \frac{1}{3}\delta_{ij} = \frac{1}{2}a_{ij}/k$) from the experiment of Tucker (1970) with $\mathcal{S}_\lambda k/\varepsilon = 2.1$. Also shown in Fig. 10.2 are DNS results for $\mathcal{S}_\lambda k/\varepsilon = 55.7$ obtained by Lee and Reynolds (1985). For this large value of $\mathcal{S}_\lambda k/\varepsilon$, rapid-distortion theory (RDT, see Section 11.4) accurately describes the evolution of the Reynolds stresses. According to RDT, the Reynolds stresses are determined not by the rate of strain, but by the total amount of mean strain experienced by the turbulence. In these circumstances the turbulence behaves not like a viscous fluid, but more like an elastic solid (Crow 1968): the turbulent viscosity hypothesis is qualitatively incorrect.

In the experiment depicted in Fig. 10.1, following the contraction there is a second straight section. Since there is no mean straining in this section, the turbulent-viscosity hypothesis inevitably predicts that the Reynolds-stress anisotropies are zero. However, the experimental data of Warhaft (1980) show instead that the anisotropies generated in the contraction decay quite slowly, on the turbulence timescale k/ε (see Fig. 10.2). These persisting anisotropies exist not because of the local mean strain rates (which are zero), but because of the prior history of straining to which the turbulence has been subjected.

Evidently, for this flow, both in the contraction section and in the downstream straight section, the intrinsic assumption of the turbulent-viscosity hypothesis is incorrect: the Reynolds-stress anisotropies are not determined by the local mean rates of strain.

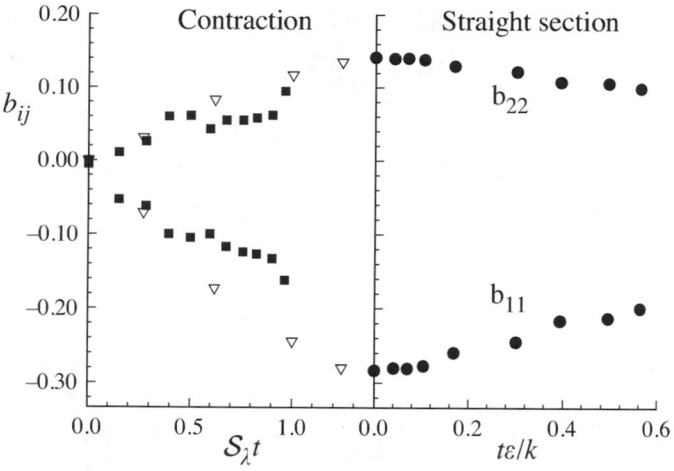

Fig. 10.2. Reynolds-stress anisotropies during and after axisymmetric straining. Contraction: experimental data of Tucker (1970), $\mathcal{S}_\lambda k/\varepsilon = 2.1$; $\triangle$ DNS data of Lee and Reynolds (1985), $\mathcal{S}_\lambda k/\varepsilon = 55.7$; the flight time t from the beginning of the contraction is normalized by the mean strain rate $\mathcal{S}_\lambda$. Straight section: experimental data of Warhaft (1980); the flight time from the beginning of the straight section is normalized by the turbulence timescale there.

Comparison with kinetic theory

Simple kinetic theory for ideal gases (see, e.g., Vincenti and Kruger (1965) and Chapman and Cowling (1970)) yields the Newtonian viscous stress law (Eq. (10.6)), with the kinematic viscosity given by

$$v \approx \tfrac{1}{2}\bar{C}\lambda, \qquad (10.7)$$

where $\bar{C}$ is the mean molecular speed, and λ is the mean free path. It is natural to seek to justify the turbulent-viscosity hypothesis through analogy with kinetic theory, and hence to give physical significance to u^* and ℓ^* by analogy to $\bar{C}$ and λ. However, a simple examination of the various timescales involved shows that such an analogy has no general validity.

In simple laminar shear flow (with shear rate $\partial U_1/\partial x_2 = \mathcal{S} = \mathcal{U}/\mathcal{L}$), the ratio of the molecular timescale $\lambda/\bar{C}$ and the shear timescale $\mathcal{S}^{-1}$ is

$$\frac{\lambda}{\bar{C}}\mathcal{S} = \frac{\lambda}{\mathcal{L}}\frac{\mathcal{U}}{\bar{C}} \sim \mathrm{KnMa}, \qquad (10.8)$$

which is typically very small (e.g., 10^{-10}, see Exercise 10.1). The significance of the molecular timescale being relatively minute is that the statistical state of the molecular motion rapidly adjusts to the imposed straining. By contrast, for turbulent shear flows, the ratio of the turbulence timescale $\tau = k/\varepsilon$ to the mean shear timescale $\mathcal{S}^{-1}$ is not small: in the self-similar round jet $\mathcal{S}k/\varepsilon$ is about 3 (Table 5.2 on page 131); in experiments on homogeneous turbulent shear flow it is typically 6 (Table 5.4 on page 157); and in turbulence subjected to rapid distortions it can be orders of magnitude larger. Consequently, as already observed, turbulence does not adjust rapidly to imposed mean straining, and so (in contrast to the case of molecular motion) there is no general basis for a local relationship between the stress and the rate of strain.

Simple shear flows

The example of rapid axisymmetric distortion and the timescale considerations given above show that, in general, the turbulent-viscosity hypothesis is incorrect. These general objections notwithstanding, there are important particular flows for which the hypothesis is more reasonable. In simple turbulent shear flows (e.g., the round jet, mixing layer, channel flow, and boundary layer) the turbulence characteristics and mean velocity gradients change relatively slowly (following the mean flow). As a consequence, the local mean velocity gradients characterize the history of the mean distortion to which the turbulence has been subjected; and the Reynolds-stress balance is dominated by local processes – production, dissipation, the pressure–rate-of-strain tensor – the non-local transport processes being small in comparison (see e.g., Figs. 7.35–7.38 on pages 316–317). In these circumstances, then, it is more reasonable to hypothesize that there is a relationship between the Reynolds stresses and the local mean velocity gradients.

An important observation is that in these particular flows (in which the turbulence characteristics change slowly following the mean flow), the production and dissipation of turbulent kinetic energy are approximately in balance, i.e., $\mathcal{P}/\varepsilon \approx 1$. By contrast, in the axisymmetric-contraction experiment (Fig. 10.1), in the contraction section $\mathcal{P}/\varepsilon$ is much greater than unity, whereas in the downstream straight section $\mathcal{P}/\varepsilon$ is zero: in both of these cases the turbulent-viscosity hypothesis is incorrect.

The gradient-diffusion hypothesis

Related to the turbulent-viscosity hypothesis is the gradient-diffusion hypothesis

$$\langle \boldsymbol{u}\phi' \rangle = -\Gamma_{\mathrm{T}} \nabla \langle \phi \rangle, \tag{10.9}$$

according to which the scalar flux $\langle \boldsymbol{u}\phi' \rangle$ is aligned with the mean scalar gradient (see Section 4.4). Most of the observations made above apply equally to the gradient-diffusion hypothesis. In homogeneous shear flow it is found that the direction of the scalar flux is significantly different than that of the mean gradient (Tavoularis and Corrsin 1985). However, in simple two-dimensional turbulent shear flows (in the usual coordinate system) the scalar equation

$$\langle v\phi' \rangle = -\Gamma_T \frac{\partial \langle \phi \rangle}{\partial y}, \tag{10.10}$$

can be used to define Γ_T, and thus no assumption is involved (for this component). The turbulent Prandtl number σ_T can be used to relate ν_T and Γ_T, i.e., $\Gamma_T = \nu_T/\sigma_T$; and for simple shear flows, σ_T is of order unity (see, e.g., Fig. 5.34 on page 162).

Both ν_T and Γ_T can be written as the product of a velocity scale and a lengthscale (Eq. (10.3)). They can also be expressed as the product of the square of a velocity scale and a timescale:

$$\Gamma_T = u^{*2} T^*. \tag{10.11}$$

As shown in Section 12.4, in ideal circumstances, Γ_T can be related to statistics of the turbulence: u^* is the r.m.s. velocity u', and T^* is the Lagrangian integral timescale T_L (see Eq. (12.158) on page 500).

EXERCISE _____

10.1   According to simple kinetic theory (see, e.g., Vincenti and Kruger (1965)) the kinematic viscosity of an ideal gas is

$$\nu \approx \tfrac{1}{2}\bar{C}\lambda, \tag{10.12}$$

and the mean molecular speed $\bar{C}$ is 1.35 times the speed of sound a. Show that the shear rate $S = U/\mathcal{L}$ normalized by the molecular timescale $\lambda/\bar{C}$ is

$$\frac{S\lambda}{\bar{C}} \approx 0.7\mathrm{MaKn}, \tag{10.13}$$

where the Mach number and Knudsen number are defined by $\mathrm{Ma} \equiv U/a$ and $\mathrm{Kn} \equiv \lambda/\mathcal{L}$.

Use the relation $a^2 = \gamma p/\rho$ (with $\gamma = 1.4$) to show that the ratio of the viscous shear stress τ_{12} to the normal stress (pressure) is

$$\frac{\tau_{12}}{p} \approx 0.9\mathrm{MaKn}. \tag{10.14}$$

Using the values $a = 332$ m s^{-1} and $\nu = 1.33 \times 10^{-5}$ m^2 s^{-1} (corresponding to air under atmospheric conditions) and $S = 1$ s^{-1}, obtain the following estimates:

$$\lambda = 5.9 \times 10^{-8} \text{ m}, \quad \lambda/\bar{C} = 1.3 \times 10^{-10} \text{ s},$$

$$\frac{S\lambda}{\bar{C}} = 1.3 \times 10^{-10}, \quad \frac{\tau_{12}}{p} = 1.7 \times 10^{-10}. \tag{10.15}$$

10.1.2 The specific assumption

We turn now to the *specific* assumption that the relationship between the Reynolds stresses and mean velocity gradients is that given by Eq. (10.1) (or, equivalently, Eq. (10.4) or (10.5)).

For simple shear flows, the single Reynolds stress of interest $\langle uv \rangle$ is related to the single significant mean velocity gradient $\partial \langle U \rangle / \partial y$ by Eq. (10.2). In essence, no assumption is involved, but rather the equation defines v_T. Examples of profiles of v_T thus obtained are given in Fig. 5.10 (on page 108) for the round jet, and in Fig. 7.30 (on page 307) for the boundary layer.

In general, the specific assumption in the turbulent-viscosity hypothesis is that the Reynolds-stress-anisotropy tensor a_{ij} is linearly related to the mean rate-of-strain tensor $\bar{S}_{ij}$ via the scalar turbulent viscosity, Eq. (10.5). Even for the simplest of flows, this is patently incorrect. In turbulent shear flow the normal strain rates are zero ($\bar{S}_{11} = \bar{S}_{22} = \bar{S}_{33} = 0$) and yet the normal Reynolds stresses are significantly different from each other (see Table 5.4 on page 157). An alternative perspective on the same observation is that the principal axes of a_{ij} are (by a significant amount) misaligned with those of $\bar{S}_{ij}$ (see Exercise 4.5 on page 90).

The reason that the simple linear stress law applies to the viscous stresses (Eq. (10.6)) but not to the Reynolds stresses can again be understood in terms of the timescale ratio, and in terms of the level of anisotropy. Compared with the molecular scales, the straining is very weak ($S\lambda/\bar{C} \ll 1$), and consequently it produces a very small departure from isotropy: in simple laminar shear flow the ratio of anisotropic and isotropic molecular stresses is

$$\frac{\tau_{12}}{p} \approx \frac{\frac{1}{2}\bar{C}\lambda \mathcal{U}}{P\mathcal{L}} \sim \text{KnMa}, \tag{10.16}$$

(see Exercise 10.1), which is typically very small. As a consequence, there is every reason to expect the anisotropic stresses to depend *linearly* on the velocity gradients. The Newtonian viscous stress law (Eq. (10.6)) is the most general possible *linear* relation consistent with the mathematical properties of the stress tensor. By contrast, in turbulent shear flow, the anisotropic-to-isotropic stress ratio $-\langle uv \rangle / (\frac{2}{3}k)$ is close to 0.5. In the turbulent case, then, the rate of straining is relatively large ($Sk/\varepsilon > 1$) and it leads to relatively large anisotropies. Consequently, there is no reason to suppose that the relationship is linear.

There are several classes of flows in which the mean velocity gradient tensor is more complex than that in simple shear flow, and in which the turbulent-viscosity hypothesis is known to fail significantly. Examples are strongly swirling flows (Weber, Visser, and Boysan 1990), flows with significant streamline curvature (Bradshaw 1973, Patel and Sotiropoulos 1997), and fully developed flow in ducts of non-circular cross-section (Melling and Whitelaw 1976, Bradshaw 1987).

In place of Eq. (10.5), a possible nonlinear turbulent viscosity hypothesis is

$$a_{ij} = -2v_{T1}\bar{S}_{ij} + v_{T2}(\bar{S}_{ik}\bar{\Omega}_{kj} - \bar{\Omega}_{ik}\bar{S}_{kj}) + v_{T3}(\bar{S}_{ik}\bar{S}_{kj} - \tfrac{1}{3}\bar{S}_{kk}^2\delta_{ij}), \quad (10.17)$$

where the coefficients v_{T1}, v_{T2} and v_{T3} may depend on the mean-velocity-gradient invariants such as $\bar{S}_{kk}^2$ (as well as on turbulence quantities). Note that a dependence of a_{ij} on the mean rate of rotation $\bar{\Omega}_{ij}$ (e.g., through the term in v_{T2}) is required so that the principal axes of a_{ij} are not aligned with those of $\bar{S}_{ij}$. Rational means of obtaining nonlinear turbulent viscosity laws have been developed, and are described in Section 11.9.

In summary: the *intrinsic* assumption of the turbulent-viscosity hypothesis – that a_{ij} is locally determined by $\partial\langle U_i\rangle/\partial x_j$ – has no general validity. However, for simple shear flows, in which the mean velocity gradients and turbulence characteristics evolve slowly (following the mean flow), the hypothesis is more reasonable. In such flows $\mathcal{P}/\varepsilon$ is close to unity, which is indicative of an approximate local balance in the Reynolds-stress equations between production by the mean shear and the other local processes – redistribution and dissipation.

10.2 Algebraic models

The algebraic models that have been introduced in previous chapters are the *uniform turbulent viscosity* and the *mixing-length model*. These models are now appraised relative to the criteria described in Chapter 8.

10.2.1 Uniform turbulent viscosity

In applications to a planar two-dimensional free shear flow, the uniform-turbulent-viscosity model can be written

$$v_T(x, y) = \frac{U_0(x)\delta(x)}{R_T}, \quad (10.18)$$

where $U_0(x)$ and $\delta(x)$ are the characteristic velocity scale and lengthscale of the mean flow, and R_T – which has the interpretation of a turbulent Reynolds number – is a flow-dependent constant. Thus the turbulent viscosity is taken to be constant across the flow (in the y direction), but it varies in the mean-flow direction.

The range of applicability of this model is extremely limited. In order to

Table 10.1. *Measured spreading rates S and corresponding values of the turbulent Reynolds number* R_T *for self-similar free shear flows*

Flow	Spreading rate S	Turbulent Reynolds number, R_T	Equation relating S to R_T
Round jet	0.094	35	5.84
Plane jet	0.10	31	5.200
Mixing layer	0.06–0.11	60–110	5.225
Plane wake	0.073–0.103	13–19	5.241
Axisymmetric wake	0.064–0.8	2–22	5.260

apply the model, it is necessary to define unambiguously the direction of flow, x; the characteristic flow width $\delta(x)$; and the characteristic velocity $U_0(x)$. This is possible only for the simplest of flows.

For the simple free shear flows to which it is applicable, the model is incomplete, in that R_T has to be specified. The appropriate value depends both upon the nature of the flow and on the definitions chosen for $\delta(x)$ and $U_0(x)$. In Chapter 5, it is shown that, for each self-similar free shear flow, there is an inverse relation between the rate of spreading S and the turbulent Reynolds number R_T. Table 10.1 summarizes the measured spreading rates and the corresponding values of R_T. (For each flow, the definitions of S, δ, and U_0 are given in Chapter 5.)

In self-similar free shear flows, the empirically determined turbulent viscosity is fairly uniform over the bulk of the flow, but it decreases to zero as the free stream is approached (see, e.g., Fig. 5.10 on page 108). Correspondingly, the mean velocity profile predicted by the uniform viscosity model agrees well with experimental data except at the edge of the flow (e.g., Fig. 5.15 on page 119).

In principle, the uniform-turbulent-viscosity model could be applied to simple wall-bounded flows. However, since the turbulent viscosity in fact varies significantly across the flow (see Fig. 7.30 on page 307), the resulting predicted mean velocity profile would be, for most purposes, uselessly inaccurate.

In summary, the uniform-turbulent-viscosity model provides a useful basic description of the mean velocity profiles in self-similar free shear flows. However, it is an incomplete model with a very limited range of applicability.

10.2.2 The mixing-length model

In application to two-dimensional boundary-layer flows, the mixing length $\ell_m(x, y)$ is specified as a function of position, and then the turbulent viscosity is obtained as

$$v_T = \ell_m^2 \left| \frac{\partial \langle U \rangle}{\partial y} \right|. \tag{10.19}$$

As shown in Section 7.1.7, in the log-law region, the appropriate specification of the mixing length is $\ell_m = \kappa y$, and then the turbulent viscosity is $v_T = u_\tau \kappa y$.

Several generalizations of Eq. (10.19) have been proposed in order to allow the application of the mixing-length hypothesis to all flows. On the basis of the mean rate of strain $\bar{S}_{ij}$ Smagorinsky (1963) proposed

$$v_T = \ell_m^2 (2\bar{S}_{ij}\bar{S}_{ij})^{1/2} = \ell_m^2 \mathcal{S}, \tag{10.20}$$

whereas, on the basis of the mean rate of rotation $\bar{\Omega}_{ij}$, Baldwin and Lomax (1978) proposed

$$v_T = \ell_m^2 (2\bar{\Omega}_{ij}\bar{\Omega}_{ij})^{1/2} = \ell_m^2 \Omega. \tag{10.21}$$

(Both of these formulae reduce to Eq. (10.19) in the case that $\partial \langle U_1 \rangle / \partial x_2$ is the only non-zero mean velocity gradient.)

In its generalized form, the mixing-length model is applicable to all turbulent flows, and it is arguably the simplest turbulence model. Its major drawback, however, is its incompleteness: the mixing length $\ell_m(x)$ has to be specified, and the appropriate specification is inevitably dependent on the geometry of the flow. For a complex flow that has not been studied before, the specification of $\ell_m(x)$ requires a large measure of guesswork, and consequently one should have little confidence in the accuracy of the resulting calculated mean velocity field. On the other hand, there are classes of technologically important flows that have been studied extensively, so that the appropriate specifications of $\ell_m(x)$ are well established. The prime example is boundary-layer flows in aeronautical applications. The Cebeci–Smith model (Smith and Cebeci 1967) and the Baldwin–Lomax model (Baldwin and Lomax 1978) provide mixing-length specifications that yield quite accurate calculations of attached boundary layers. Details of these models and their performance are provided by Wilcox (1993).

As illustrated in the following exercise, the mixing-length model can also be applied to free shear flows. The predicted mean velocity profile agrees well with experimental data (see, e.g., Schlichting (1979)). An interesting (though non-physical) feature of the solution is that the mixing layer has a definite edge at which the mean velocity goes to the free-stream velocity with zero slope but non-zero curvature.

10.2   Consider the self-similar temporal mixing layer in which the mean
lateral velocity $\langle V \rangle$ is zero, and the axial velocity $\langle U \rangle$ depends on
y and t only. The velocity difference is U_s, so that the boundary
conditions are $\langle U \rangle = \pm \frac{1}{2} U_s$ at $y = \pm \infty$. The thickness of the layer
$\delta(t)$ is defined (as in Fig. 5.21 on page 140) such that $\langle U \rangle = \pm \frac{2}{5} U_s$ at
$y = \pm \frac{1}{2}\delta$.

The mixing-length model is applied to this flow, with the mixing
length being uniform across the flow and proportional to the flow's
width, i.e., $\ell_m = \alpha\delta$, where α is a specified constant.

Starting from the Reynolds equations

$$\frac{\partial \langle U \rangle}{\partial t} = -\frac{\partial \langle uv \rangle}{\partial y}, \tag{10.22}$$

show that the mixing-length hypothesis implies that

$$\frac{\partial \langle U \rangle}{\partial t} = 2\alpha^2 \delta^2 \frac{\partial \langle U \rangle}{\partial y} \frac{\partial^2 \langle U \rangle}{\partial y^2}. \tag{10.23}$$

Show that this equation admits a self-similar solution of the form
$\langle U \rangle = U_s f(\xi)$, where $\xi = y/\delta$; and that $f(\xi)$ satisfies the ordinary
differential equation

$$-S\xi f' = 2\alpha^2 f' f'', \tag{10.24}$$

where $S \equiv U_s^{-1} \, d\delta/dt$ is the spreading rate.

Show that Eq. (10.24) admits two different solutions (denoted by
f_1 and f_2):

$$f_1 = -\frac{S}{12\alpha^2}\xi^3 + A\xi + B, \tag{10.25}$$

$$f_2 = C, \tag{10.26}$$

where A, B, and C are arbitrary constants.

The appropriate solution for f is made up of three parts. For $|\xi|$
greater than a particular value ξ^*, f is constant (i.e., f_2):

$$f = \begin{cases} -\frac{1}{2} & \text{for } \xi < -\xi^*, \\ \frac{1}{2} & \text{for } \xi > \xi^*. \end{cases} \tag{10.27}$$

Show that the appropriate solution for $-\xi^* < \xi < \xi^*$ satisfying
$f'(\pm\xi^*) = 0$ is

$$f = \frac{3}{4}\frac{\xi}{\xi^*} - \frac{1}{4}\left(\frac{\xi}{\xi^*}\right)^3. \tag{10.28}$$

Show that the spreading rate is related to the mixing-length constant by

$$S = 3\alpha^2/\xi^{*3},\tag{10.29}$$

and use the definition of δ (i.e., $f\left(\frac{1}{2}\right) = \frac{2}{5}$) to obtain

$$\xi^* \approx 0.8450.\tag{10.30}$$

How does ν_T vary across the flow?

10.3 Turbulent-kinetic-energy models

With the turbulent viscosity written as

$$\nu_T = \ell^* u^*,\tag{10.31}$$

in the mixing-length model the lengthscale is $\ell^* = \ell_m$ and the velocity scale is (in simple shear flow)

$$u^* = \ell_m \left| \frac{\partial \langle U \rangle}{\partial y} \right|.\tag{10.32}$$

The implication is that the velocity scale is locally determined by the mean velocity gradient; and, in particular, u^* is zero where $\partial \langle U \rangle / \partial y$ is zero. In fact, contrary to this implication, there are several circumstances in which the velocity gradient is zero and yet the turbulent velocity scale is non-zero. One example is decaying grid turbulence; another is the centerline of the round jet, where direct measurement shows ν_T to be far from zero (see Fig. 5.10 on page 108).

Independently, Kolmogorov (1942) and Prandtl (1945) suggested that it is better to base the velocity scale on the turbulent kinetic energy, i.e.,

$$u^* = ck^{1/2},\tag{10.33}$$

where c is a constant. If the lengthscale is again taken to be the mixing length, then the turbulent viscosity becomes

$$\nu_T = ck^{1/2}\ell_m.\tag{10.34}$$

As shown in Exercise 10.3, the value of the constant $c \approx 0.55$ yields the correct behavior in the log-law region.

In order for Eq. (10.34) to be used, the value of $k(x,t)$ must be known or estimated. Kolmogorov and Prandtl suggested achieving this by solving a model transport equation for k. This is called a *one-equation model*, because a model transport equation is solved for just one turbulence quantity, namely, k.

Before discussing the model transport equation for k, it is helpful to itemize all the components of the model:

(i) the mixing length $\ell_m(x, t)$ is specified;
(ii) a model transport equation is solved for $k(x, t)$;
(iii) the turbulent viscosity is defined by $v_T = ck^{1/2}\ell_m$;
(iv) the Reynolds stresses are obtained from the turbulent-viscosity hypothesis, Eq. (10.1); and
(v) the Reynolds equations are solved for $\langle U(x, t) \rangle$ and $\langle p(x, t) \rangle$.

Thus, from the specification of ℓ_m and from the solutions to the exact and model equations, the following fields are determined: $\langle U \rangle$, $\langle p \rangle$, ℓ_m, k, v_T, and $\langle u_i u_j \rangle$. These are referred to as 'knowns.'

We now consider the model transport equation for k. The exact equation (Eq. (5.132)) is

$$\frac{\bar{D}k}{\bar{D}t} \equiv \frac{\partial k}{\partial t} + \langle U \rangle \cdot \nabla k$$
$$= -\nabla \cdot T' + \mathcal{P} - \varepsilon, \qquad (10.35)$$

where the flux T' (Eq. (5.140)) is

$$T_i' = \tfrac{1}{2}\langle u_i u_j u_j \rangle + \langle u_i p' \rangle / \rho - 2v\langle u_j s_{ij} \rangle. \qquad (10.36)$$

In Eq. (10.35), any term that is completely determined by the 'knowns' is said to be 'in closed form.' Specifically, $\bar{D}k/\bar{D}t$ and $\mathcal{P}$ are in closed form. Conversely, the remaining terms (ε and $\nabla \cdot T'$) are 'unknown' and, in order to obtain a closed set of model equations, these terms must be modelled. That is, 'closure approximations' that model the unknowns in terms of the knowns are required.

As discussed extensively in Chapter 6, at high Reynolds number the dissipation rate ε scales as u_0^3/ℓ_0, where u_0 and ℓ_0 are the velocity scale and lengthscale of the energy-containing motions. Consequently, it is reasonable to model ε as

$$\varepsilon = C_D k^{3/2}/\ell_m, \qquad (10.37)$$

where C_D is a model constant. Indeed, an examination of the log-law region (Exercise 10.3) yields this relation with $C_D = c^3$.

Modelling assumptions such as Eq. (10.37) deserve close scrutiny. Equations (10.34) and (10.37) can be combined to eliminate ℓ_m to yield

$$v_T = cC_D k^2/\varepsilon, \qquad (10.38)$$

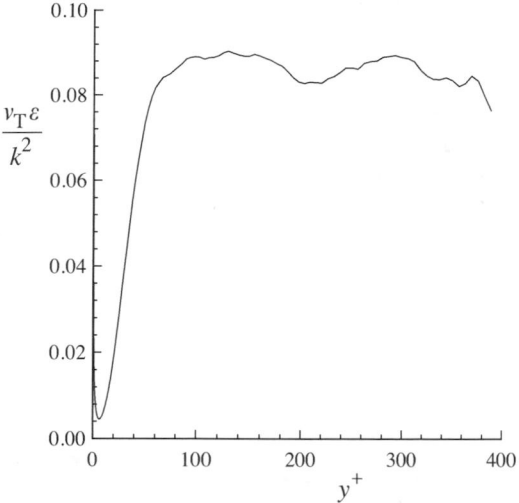

Fig. 10.3. The profile of $v_T\varepsilon/k^2$ (see Eq. (10.39)) from DNS of channel flow at Re = 13,750 (Kim *et al.* 1987).

or, equivalently,

$$\frac{v_T\varepsilon}{k^2} = cC_D. \tag{10.39}$$

For simple shear flows, k, ε, and $v_T = -\langle uv \rangle/(\partial\langle U \rangle/\partial y)$ can be measured, so that this modelling assumption can be tested directly. Figure 10.3 shows the left-hand side of Eq. (10.39) extracted from DNS data of fully developed turbulent channel flow. It may be seen that (except close to the wall, $y^+ < 50$) this quantity is indeed approximately constant, with a value around 0.09. Figure 10.4 shows the same quantity for the temporal-mixing layer: except near the edges, the value is everywhere close to 0.08.

The remaining unknown in the turbulent-kinetic-energy equation is the energy flux $\mathbf{T}'$ (Eq. (10.36)). This is modelled with a gradient-diffusion hypothesis as

$$\mathbf{T}' = -\frac{v_T}{\sigma_k}\nabla k, \tag{10.40}$$

where the 'turbulent Prandtl number' for kinetic energy[1] is generally taken to be $\sigma_k = 1.0$. Physically, Eq. (10.40) asserts that (due to velocity and pressure fluctuations) there is a flux of k down the gradient of k. Mathematically, the term ensures that the resulting model transport equation for k yields smooth solutions, and that a boundary condition can be imposed on k everywhere on the boundary of the solution domain.

---

[1] The symbol σ_k is standard notation. Note, however, that σ_k is a scalar, and that 'k' is not a suffix in the sense of Cartesian-tensor suffix notation.

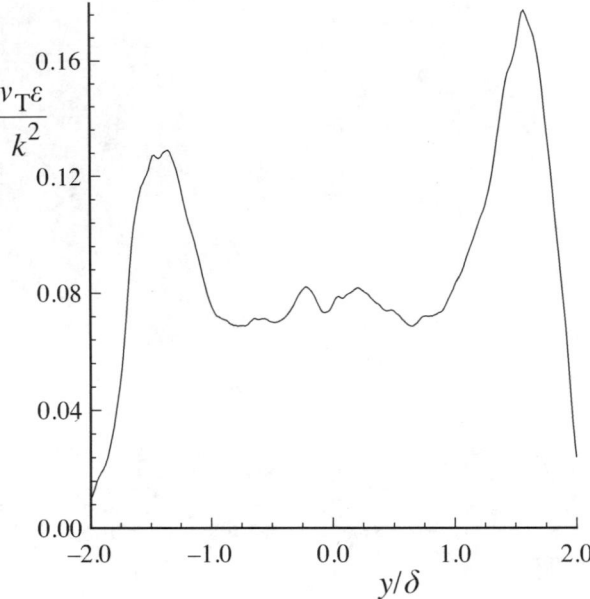

Fig. 10.4. The profile of $\nu_{\mathrm{T}}\varepsilon/k^2$ (see Eq. (10.39)) from DNS of the temporal mixing layer. (From data of Rogers and Moser (1994)).)

In summary, the one-equation model based on k consists of the model transport equation

$$\frac{\bar{\mathrm{D}}k}{\bar{\mathrm{D}}t} = \nabla \cdot \left(\frac{\nu_{\mathrm{T}}}{\sigma_k}\nabla k\right) + \mathcal{P} - \varepsilon, \qquad (10.41)$$

with $\nu_{\mathrm{T}} = ck^{1/2}\ell_{\mathrm{m}}$ and $\varepsilon = C_{\mathrm{D}}k^{3/2}/\ell_{\mathrm{m}}$, together with the turbulent-viscosity hypothesis (Eq. (10.1)) and the specification of ℓ_{m}.

A comparison of model predictions with experimental data (Wilcox 1993) shows that this one-equation model has a modest advantage in accuracy over mixing-length models. However, the major drawback of incompleteness remains: the length scale $\ell_{\mathrm{m}}(\boldsymbol{x})$ must be specified.

EXERCISES _____

10.3   Consider the log-law region of a wall-bounded flow. Use the log-law and the specification $\ell_{\mathrm{m}} = \kappa y$ to show that the appropriate value of the constant c (in the relation $\nu_{\mathrm{T}} = ck^{1/2}\ell_{\mathrm{m}}$) is

$$c = |\langle uv\rangle/k|^{1/2} \approx 0.55. \qquad (10.42)$$

Use the relation $\mathcal{P} = \varepsilon$ to show that

$$\varepsilon = c^3 k^{3/2}/\ell_{\mathrm{m}}, \qquad (10.43)$$

and hence

$$v_T = c^4 k^2 / \varepsilon. \tag{10.44}$$

10.4   For the one-equation model applied to simple shear flow, express the production $\mathcal{P}$ in terms of k, ℓ_m and $\partial \langle U \rangle / \partial y$. Hence (taking $C_D = c^3$ in Eq. (10.37)) show that the velocity scales u^* in the one-equation model and in the mixing-length model are related by

$$ck^{1/2} = \ell_m \left| \frac{\partial \langle U \rangle}{\partial y} \right| \left(\frac{\mathcal{P}}{\varepsilon} \right)^{-1/2} \tag{10.45}$$

Show that the corresponding relation for a general flow is

$$ck^{1/2} = \ell_m \mathcal{S} \left(\frac{\mathcal{P}}{\varepsilon} \right)^{-1/2} \tag{10.46}$$

(cf. Eq. (10.20)).

10.4 The k–ε model

10.4.1 An overview

The k–ε model belongs to the class of *two-equation models,* in which model transport equations are solved for two turbulence quantities – i.e., k and ε in the k–ε model. From these two quantities can be formed a lengthscale ($L = k^{3/2} / \varepsilon$), a timescale ($\tau = k / \varepsilon$), a quantity of dimension v_T (k^2 / ε), etc. As a consequence, two-equation models can be *complete* – flow-dependent specifications such as $\ell_m(x)$ are not required.

The k–ε model is the most widely used complete turbulence model, and it is incorporated in most commercial CFD codes. As is the case with all turbulence models, both the concepts and the details evolved over time; but Jones and Launder (1972) are appropriately credited with developing the 'standard' k–ε model, with Launder and Sharma (1974) providing improved values of the model constants. Significant earlier contributions are due to Davidov (1961), Harlow and Nakayama (1968), Hanjalić (1970), and others cited by Launder and Spalding (1972).

In addition to the turbulent viscosity hypothesis, the k–ε model consists of

  (i) the model transport equation for k (which is the same as that in the one-equation model, Eq. (10.41));

  (ii) the model transport equation for ε (which is described below); and

(iii) the specification of the turbulent viscosity as

$$v_T = C_\mu k^2 / \varepsilon, \tag{10.47}$$

where $C_\mu = 0.09$ is one of five model constants.

If it is supposed that v_T depends only on the turbulence quantities k and ε (independent of $\partial \langle U_i \rangle / \partial x_j$ etc.), then Eq. (10.47) is inevitable. The one-equation model implies the similar relation $v_T = c^4 k^2 / \varepsilon$ (see Exercise 10.3), so the model constants are related by $c = C_\mu^{1/4}$.

In simple turbulent shear flow, the k–ε model yields

$$\frac{|\langle uv \rangle|}{k} = \left(C_\mu \frac{\mathcal{P}}{\varepsilon} \right)^{1/2}, \tag{10.48}$$

(see Exercise 10.5) so that the specification $C_\mu = 0.09 = (0.3)^2$ stems from the empirical observation $|\langle uv \rangle|/k \approx 0.3$ in regions where $\mathcal{P}/\varepsilon$ is close to unity.

The quantity $v_T \varepsilon / k^2$ plotted in Figs. 10.3 and 10.4 is a 'measurement' of C_μ for channel flow and for the temporal mixing layer. As may be seen, $v_T \varepsilon / k^2$ is close to the value 0.09 everywhere except near the boundaries of the flows.

EXERCISE ──

10.5   Consider the k–ε model applied to a simple turbulent shear flow with $S = \partial \langle U \rangle / \partial y$ being the only non-zero mean velocity gradient. Obtain the relations

$$\frac{|\langle uv \rangle|}{k} = C_\mu \frac{Sk}{\varepsilon}, \tag{10.49}$$

$$\frac{\mathcal{P}}{\varepsilon} = C_\mu \left(\frac{Sk}{\varepsilon} \right)^2, \tag{10.50}$$

and hence verify Eq. (10.48).

Show that $\langle uv \rangle$ satisfies the Cauchy–Schwarz inequality (Eq. (3.100)) if, and only if, C_μ satisfies

$$C_\mu \le \frac{2/3}{Sk/\varepsilon}, \tag{10.51}$$

or, equivalently,

$$C_\mu \le \frac{4/9}{\mathcal{P}/\varepsilon}. \tag{10.52}$$

Show that Eq. (10.50) also holds for a general flow.

10.4.2 The model equation for ε

Quite different approaches are taken in developing the model transport equations for k and ε. The k equation amounts to the exact equation (Eq. (10.35)) with the turbulent flux T' modelled as gradient diffusion (Eq. (10.40)). The three other terms – $\bar{D}k/\bar{D}t$, $\mathcal{P}$, and ε – are in closed form (given the turbulent-viscosity hypothesis).

The exact equation for ε can also be derived, but it is not a useful starting point for a model equation. This is because (as discussed in Chapter 6) ε is best viewed as the energy-flow rate in the cascade, and it is determined by the large-scale motions, independent of the viscosity (at high Reynolds number). By contrast, the exact equation for ε pertains to processes in the dissipative range. Consequently, rather than being based on the exact equation, the standard model equation for ε is best viewed as being entirely empirical: it is

$$\frac{\bar{D}\varepsilon}{\bar{D}t} = \nabla \cdot \left(\frac{\nu_{\mathrm{T}}}{\sigma_\varepsilon} \nabla \varepsilon \right) + C_{\varepsilon 1} \frac{\mathcal{P}\varepsilon}{k} - C_{\varepsilon 2} \frac{\varepsilon^2}{k}. \tag{10.53}$$

The standard values of all the model constants due to Launder and Sharma (1974) are

$$C_\mu = 0.09, \; C_{\varepsilon 1} = 1.44, \; C_{\varepsilon 2} = 1.92, \; \sigma_k = 1.0, \; \sigma_\varepsilon = 1.3. \tag{10.54}$$

An understanding of the ε equation can be gained by studying its behaviors in various flows. We first examine homogeneous turbulence, for which the k and ε equations become

$$\frac{dk}{dt} = \mathcal{P} - \varepsilon, \tag{10.55}$$

$$\frac{d\varepsilon}{dt} = C_{\varepsilon 1} \frac{\mathcal{P}\varepsilon}{k} - C_{\varepsilon 2} \frac{\varepsilon^2}{k}. \tag{10.56}$$

Decaying turbulence

In the absence of mean velocity gradients, the production is zero, and the turbulence decays. The equations then have the solutions

$$k(t) = k_0 \left(\frac{t}{t_0} \right)^{-n}, \quad \varepsilon(t) = \varepsilon_0 \left(\frac{t}{t_0} \right)^{-(n+1)}, \tag{10.57}$$

where k and ε have the values k_0 and ε_0 at the reference time

$$t_0 = n \frac{k_0}{\varepsilon_0}, \tag{10.58}$$

and the decay exponent n is

$$n = \frac{1}{C_{\varepsilon 2} - 1}. \tag{10.59}$$

This power-law decay is precisely that observed in grid turbulence (see Section 5.4.6, Eqs. (5.274) and (5.277)), and so the behavior of the ε equation is correct for this flow.

The experimental values reported for the decay exponent n are generally in the range 1.15–1.45, and Mohamed and LaRue (1990) suggest that most of the data are consistent with $n = 1.3$. Equation (10.59) can be rearranged to give $C_{\varepsilon 2}$ in terms of n:

$$C_{\varepsilon 2} = \frac{n + 1}{n}, \tag{10.60}$$

and the values of $C_{\varepsilon 2}$ corresponding to $n = 1.15$, 1.3, and 1.45, are 1.87, 1.77, and 1.69. It may be seen, then, that the standard value ($C_{\varepsilon 2} = 1.92$) lies somewhat outside of the experimentally observed range. The reason for this is discussed below.

EXERCISES

10.6　Consider the k–ε model applied to decaying turbulence. Let $s(t)$ be the normalized time defined by

$$s(t) = \int_{t_0}^{t} \frac{\varepsilon(t')}{k(t')} \, \mathrm{d}t'.$$

(a) Obtain an explicit expression for $s(t)$.

(b) Derive and solve evolution equations in s for k and ε (i.e., $\mathrm{d}k/\mathrm{d}s = \ldots$).

(c) Grid turbulence is examined between $x/M = 40$ and $x/M = 200$. To what interval in normalized time does this correspond?

10.7　Show that the k–ε model gives the correct behavior for the final period of decay (see Exercise 6.11 on page 205), if $C_{\varepsilon 2}$ is modified to $C_{\varepsilon 2} = \frac{7}{5}$.

Homogeneous shear flow

As observed in Section 5.4.5, in homogeneous turbulent shear flow, the principal experimental observations are that the Reynolds stresses become self-similar, and that the non-dimensional parameters $\mathcal{S}k/\varepsilon$ and $\mathcal{P}/\varepsilon$ become constant: $\mathcal{S}k/\varepsilon \approx 6$ and $\mathcal{P}/\varepsilon \approx 1.7$. Since the imposed mean shear rate $\mathcal{S}$ is constant, the constancy of $\mathcal{S}k/\varepsilon$ implies that the turbulence timescale $\tau \equiv k/\varepsilon$ is also fixed. From the k and ε equations (Eqs. (10.55) and (10.56)) we obtain

$$\frac{\mathrm{d}}{\mathrm{d}t}\left(\frac{k}{\varepsilon}\right) = \frac{\mathrm{d}\tau}{\mathrm{d}t} = (C_{\varepsilon 2} - 1) - (C_{\varepsilon 1} - 1)\left(\frac{\mathcal{P}}{\varepsilon}\right). \tag{10.61}$$

Evidently the model predicts that τ does not change with time for the particular value of $\mathcal{P}/\varepsilon$,

$$\left(\frac{\mathcal{P}}{\varepsilon}\right)^* \equiv \frac{C_{\varepsilon 2} - 1}{C_{\varepsilon 1} - 1} \approx 2.1, \tag{10.62}$$

a considerably higher value than is observed in experiments and DNS.

Interpretation of the ε equation

We now offer an interpretation of the ε equation based on the relationship between the turbulence frequency $\omega \equiv \varepsilon/k$ and the characteristic mean strain rate $\mathcal{S}$.

An overly simple model is

$$\omega = \frac{\mathcal{S}}{\beta}, \tag{10.63}$$

where β is a constant; equal to 3, say. This model then predicts that $\mathcal{S}k/\varepsilon = \mathcal{S}/\omega$ is equal to the constant value $\beta = 3$ in all flows. In several shear flows this value of $\mathcal{S}k/\varepsilon$ is indeed measured (see Tables 5.2, 5.4, and 7.2 on pages 131, 157 and 283). However, for other flows the model is wrong: in grid turbulence ($\mathcal{S} = 0$) ω is not zero; and in homogeneous shear flow $\mathcal{S}k/\varepsilon$ is approximately 6.

Instead of setting ω *equal* to $\mathcal{S}/\beta$, consider instead a model that makes ω *relax* toward $\mathcal{S}/\beta$. Or (as a contrivance to produce the required result) consider the model that makes ω^2 relax toward $(\mathcal{S}/\beta)^2$. When it is applied to homogeneous turbulence, this model is described by the equation

$$\frac{\mathrm{d}\omega^2}{\mathrm{d}t} = -\alpha\omega\left(\omega^2 - \frac{\mathcal{S}^2}{\beta^2}\right), \tag{10.64}$$

where α is a constant and $\alpha\omega$ is the relaxation rate. It is a matter of algebra (see Exercise 10.8) to show that this equation is exactly equivalent to the ε equation (Eq. (10.56)) if the constants are specified as

$$\alpha = 2(C_{\varepsilon 2} - 1) \approx 1.84, \tag{10.65}$$

$$\beta = \left(\frac{C_{\varepsilon 2} - 1}{C_\mu[C_{\varepsilon 1} - 1]}\right)^{1/2} \approx 4.27. \tag{10.66}$$

Thus the ε equation can be interpreted through Eq. (10.64): at the rate $\alpha\omega$, the turbulence frequency (squared) relaxes toward $\mathcal{S}/\beta$ (squared).

EXERCISE

10.8   For homogeneous turbulence, from the model equations for k (Eq. (10.55)) and for ω^2 (Eq. (10.64)), show that the corresponding model equation for ε is

$$\frac{\mathrm{d}\varepsilon}{\mathrm{d}t} = \frac{\mathcal{P}\varepsilon}{k} + \frac{\alpha k \mathcal{S}^2}{2\beta^2} - (1 + \tfrac{1}{2}\alpha)\frac{\varepsilon^2}{k}. \tag{10.67}$$

By using Eq. (10.50) to eliminate $\mathcal{S}$, re-express the equation as

$$\frac{\mathrm{d}\varepsilon}{\mathrm{d}t} = \left(1 + \frac{\alpha}{2\beta^2 C_\mu}\right)\frac{\mathcal{P}\varepsilon}{k} - (1 + \tfrac{1}{2}\alpha)\frac{\varepsilon^2}{k}. \tag{10.68}$$

By comparing this result with the standard ε equation (Eq. (10.56)), verify the relationships between α and β, and $C_{\varepsilon 1}$ and $C_{\varepsilon 2}$ (Eqs. (10.65) and (10.66)).

The behavior in the log-law region

For inhomogeneous flows, the diffusion term in the ε equation (i.e., $\nabla \cdot [(\nu_\mathrm{T}/\sigma_\varepsilon)\nabla\varepsilon]$) has the same benefits as the analogous term in the k equation: it ensures that we obtain smooth solutions and it allows the specification of a boundary condition on ε everywhere on the boundary. As is now illustrated, the diffusion term plays an important role in near-wall flows.

Consider high-Reynolds-number, fully developed channel flow. The quantities of interest ($\langle U\rangle$, k, and ε) depend only on y, so that the k–ε equations reduce to

$$0 = \frac{\mathrm{d}}{\mathrm{d}y}\left(\frac{\nu_\mathrm{T}}{\sigma_k}\frac{\mathrm{d}k}{\mathrm{d}y}\right) + \mathcal{P} - \varepsilon, \tag{10.69}$$

$$0 = \frac{\mathrm{d}}{\mathrm{d}y}\left(\frac{\nu_\mathrm{T}}{\sigma_\varepsilon}\frac{\mathrm{d}\varepsilon}{\mathrm{d}y}\right) + C_{\varepsilon 1}\frac{\mathcal{P}\varepsilon}{k} - C_{\varepsilon 2}\frac{\varepsilon^2}{k}. \tag{10.70}$$

We now focus on the log-law region. Production and dissipation balance, both being equal to $u_\tau^3/(\kappa y)$. Hence in the k equation the diffusion term is zero, which implies that k is uniform, which is approximately correct. In the ε equation, the equality of $\mathcal{P}$ and ε leads to a net sink (equal to $-(C_{\varepsilon 2}-C_{\varepsilon 1})\varepsilon^2/k$) that varies as y^{-2}. This is balanced by the diffusion of ε away from the wall.

It is shown in Exercise 10.9 that the ε equation (Eq. (10.70)) is satisfied by

$$\varepsilon = \frac{C_\mu^{3/4} k^{3/2}}{\kappa y}, \tag{10.71}$$

and that the constants are related by

$$\kappa^2 = \sigma_\varepsilon C_\mu^{1/2} (C_{\varepsilon 2} - C_{\varepsilon 1}). \tag{10.72}$$

Equation (10.72) yields a value of the von Kármán constant implied by the k–ε model ($\kappa \approx 0.43$): or, alternatively, it can be used to adjust a model constant (e.g., σ_ε) to produce a particular value of κ.

EXERCISE

10.9 Consider the log-law region of a wall-bounded turbulent flow. Show that the turbulent viscosity hypothesis with $\nu_T = C_\mu k^2/\varepsilon$ and the log-law imply that

$$\varepsilon = \frac{C_\mu k^2}{u_\tau \kappa y}. \tag{10.73}$$

Given $\mathcal{P} = \varepsilon$, from the expression for $\mathcal{P}$ obtain the result

$$\varepsilon = \frac{C_\mu^{1/2} k u_\tau}{\kappa y}. \tag{10.74}$$

Hence verify Eq. (10.71) and compare it with Eq. (10.43). Substitute Eq. (10.71) for ε into the ε equation (Eq. (10.70), with k independent of y) to obtain Eq. (10.72). Verify that these equations yield the variation of the lengthscale

$$L \equiv \frac{k^{-3/2}}{\varepsilon} = C_\mu^{-3/4} \kappa y. \tag{10.75}$$

The behavior at the free-stream edge

As described in Section 5.5.2, there is an intermittent region between a turbulent flow and an irrotational, non-turbulent free stream or quiescent surroundings. In the non-turbulent fluid ($y \to \infty$), both k and ε are zero.

The k–ε model does not account for intermittency, and it yields solutions in which there is a sharp edge between the turbulent and non-turbulent regions (Cazalbou, Spalart, and Bradshaw 1994). For a statistically two-dimensional boundary layer or free shear flow, let $y_e(x)$ denote the location of the edge. Then, for $y > y_e(x)$, k and ε are zero, and the mean flow is irrotational. Just inside of the turbulent region, k, ε, and $\partial \langle U \rangle / \partial y$ vary as positive powers of the distance to the edge $y_e(x) - y$.

Because of entrainment, there is a mean flow through the edge from the non-turbulent side. The edge therefore corresponds to a turbulent front that propagates (relative to the fluid) by turbulent diffusion. The solutions to the k–ε model equations for such a propagating front are developed in Exercise 10.10.

EXERCISES

10.10 Consider a statistically stationary one-dimensional flow without turbulence production in which the k–ε equations reduce to

$$U_0 \frac{dk}{dx} = \frac{d}{dx}\left(\frac{\nu_T}{\sigma_k}\frac{dk}{dx}\right) - \varepsilon, \qquad (10.76)$$

$$U_0 \frac{d\varepsilon}{dx} = \frac{d}{dx}\left(\frac{\nu_T}{\sigma_\varepsilon}\frac{d\varepsilon}{dx}\right) - C_{\varepsilon 2}\frac{\varepsilon^2}{k}, \qquad (10.77)$$

where U_0 is the uniform mean velocity, which is positive. These equations admit a weak solution (with $k = 0$ and $\varepsilon = 0$ for $x \le 0$ and $k > 0$ and $\varepsilon > 0$ for $x > 0$) corresponding to a front between turbulent flow ($x > 0$) and non-turbulent flow ($x \le 0$). For small positive x the solutions are

$$k = k_0 \left(\frac{x}{\delta_0}\right)^p, \qquad (10.78)$$

$$\varepsilon = \varepsilon_0 \left(\frac{x}{\delta_0}\right)^q, \qquad (10.79)$$

where p, q, k_0, ε_0, and δ_0 are positive constants, with $\delta_0 \equiv k_0^{3/2}/\varepsilon_0$. By substituting Eqs. (10.78) and (10.79) into Eqs. (10.76) and (10.77), obtain the results

$$q = 2p - 1, \qquad (10.80)$$

$$\frac{U_0}{k_0^{1/2}} = \frac{C_\mu p}{\sigma_k} = \frac{C_\mu(2p-1)}{\sigma_\varepsilon}, \qquad (10.81)$$

and show that (for small x) convection and diffusion balance, with the dissipation terms being negligible in comparison. From Eq. (10.81) obtain

$$p = \frac{\sigma_k}{2\sigma_k - \sigma_\varepsilon} = \frac{1}{2 - \sigma}, \qquad (10.82)$$

where $\sigma \equiv \sigma_\varepsilon/\sigma_k$. Show that (for small x) k, ε, $L \equiv k^{3/2}/\varepsilon$, $\omega \equiv \varepsilon/k$, and ν_{T} vary as the following powers of x:

$$\frac{1}{2-\sigma}, \quad \frac{\sigma}{2-\sigma}, \quad \frac{\frac{3}{2}-\sigma}{2-\sigma}, \quad \frac{\sigma-1}{2-\sigma}, \quad 1, \tag{10.83}$$

respectively, and hence that all of these quantities are zero in the limit as x approaches zero provided that σ is between 1 and $1\frac{1}{2}$. Show that the speed of propagation of the turbulence front (relative to the mean flow) is

$$U_0 = \frac{1}{2\sigma_k - \sigma_\varepsilon} \left(\frac{\mathrm{d}\nu_{\mathrm{T}}}{\mathrm{d}x} \right)_{x=0_+}. \tag{10.84}$$

10.11   Show that the k–ε model equations applied to the self-similar temporal mixing layer can be written

$$\frac{\partial k}{\partial t} = \frac{\partial}{\partial y} \left(\frac{C_\mu k^2}{\sigma_k \varepsilon} \frac{\partial k}{\partial y} \right) - \varepsilon \left(1 - \frac{\mathcal{P}}{\varepsilon} \right), \tag{10.85}$$

$$\frac{\partial \varepsilon}{\partial t} = \frac{\partial}{\partial y} \left(\frac{C_\mu k^2}{\sigma_\varepsilon \varepsilon} \frac{\partial \varepsilon}{\partial y} \right) - \frac{\varepsilon^2}{k} \left(C_{\varepsilon 2} - C_{\varepsilon 1} \frac{\mathcal{P}}{\varepsilon} \right). \tag{10.86}$$

With $\delta(t)$ being the width of the layer, the similarity variables are defined by

$$\xi \equiv \frac{y}{\delta}, \quad \hat{k}(\xi) \equiv \frac{k}{U_{\mathrm{s}}^2}, \quad \hat{\varepsilon}(\xi) \equiv \frac{\varepsilon \delta}{U_{\mathrm{s}}^3}, \tag{10.87}$$

where U_{s} is the (constant) velocity difference. Transform Eqs. (10.85) and (10.86) to obtain

$$-S\xi \frac{\mathrm{d}\hat{k}}{\mathrm{d}\xi} = \frac{\mathrm{d}}{\mathrm{d}\xi} \left(\frac{C_\mu \hat{k}^2}{\sigma_k \hat{\varepsilon}} \frac{\mathrm{d}\hat{k}}{\mathrm{d}\xi} \right) - \hat{\varepsilon} \left(1 - \frac{\mathcal{P}}{\varepsilon} \right), \tag{10.88}$$

$$-S \left(\hat{\varepsilon} + \xi \frac{\mathrm{d}\hat{\varepsilon}}{\mathrm{d}\xi} \right) = \frac{\mathrm{d}}{\mathrm{d}\xi} \left(\frac{C_\mu \hat{k}^2}{\sigma_\varepsilon \hat{\varepsilon}} \frac{\mathrm{d}\hat{\varepsilon}}{\mathrm{d}\xi} \right) - \frac{\hat{\varepsilon}^2}{\hat{k}} \left(C_{\varepsilon 2} - C_{\varepsilon 1} \frac{\mathcal{P}}{\varepsilon} \right), \tag{10.89}$$

where S is the spreading rate $S = U_{\mathrm{s}}^{-1} \mathrm{d}\delta/\mathrm{d}t$.

Let ξ_{e} denote the edge of the turbulent region (so that k and ε are zero for $\xi \geq \xi_{\mathrm{e}}$) and let x be the distance from the edge,

$$x \equiv \xi_{\mathrm{e}} - \xi. \tag{10.90}$$

Show that the left-hand side of Eq. (10.88) can be written

$$-S\xi \frac{\mathrm{d}\hat{k}}{\mathrm{d}\xi} = S(\xi_{\mathrm{e}} - x) \frac{\mathrm{d}\hat{k}}{\mathrm{d}x}. \tag{10.91}$$

Hence show that (for small x) the equations have the power-law
solutions given in Exercise 10.10 ($k \sim x^p$ and $\varepsilon \sim x^q$).

10.4.3 Discussion

The k–ε model is arguably the simplest complete turbulence model, and
hence it has the broadest range of applicability[2]. It is incorporated in most
commercial CFD codes, and has been applied to a diverse range of problems
including heat transfer, combustion, and multi-phase flows.

A discussion of its accuracy is deferred to the next chapter (Section 11.10),
where its performance is compared with that of other turbulence models.
Briefly; although it is usually acceptably accurate for simple flows, it can be
quite inaccurate for complex flows, to the extent that the calculated mean
flow patterns can be qualitatively incorrect. The inaccuracies stem from the
turbulent-viscosity hypothesis and from the ε equation.

Also deferred to the next chapter (Section 11.7) is a discussion of near-wall
treatments. Modifications to the standard k–ε model are required in order to
apply it to the viscous near-wall region. For example, it may be seen from
Fig. 10.3 that the appropriate value of C_μ decreases as y^+ decreases below 50.

The values of the standard k–ε model constants (Eq. 10.54) represent
a compromise. For any particular flow it is likely that the accuracy of the
model calculations can be improved by adjusting the constants. For decaying
turbulence, $C_{\varepsilon 2} = 1.77$ (corresponding to the value of the decay exponent
$n = 1.3$) is more suitable than $C_{\varepsilon 2} = 1.92$ (corresponding to the value of the
decay exponent $n = 1.09$). A well-known deficiency of the k–ε model is that it
significantly overpredicts the rate of spreading for the round jet. This problem
can be remedied (for the round jet) by adjusting the value of $C_{\varepsilon 1}$ or $C_{\varepsilon 2}$.
However, such *ad hoc* flow-dependent adjustments are of limited value. For a
complete, generally applicable model, a single specification of the constants
is required; and the standard values represent a compromise chosen (with
subjective judgement) to give the 'best' performance for a range of flows.

Over the years, many 'modifications' to the standard k–ε model have been
proposed, the usual motivation being to remedy poor performance for a
particular class of flows. For example, Pope (1978) proposed an additional
source term in the ε equation of the form $\bar{S}_{ij}\bar{\Omega}_{jk}\bar{\Omega}_{ki}k^2/\varepsilon$, so that the modified
model yields the correct spreading rate for the round jet. Other modifications
to the ε equation (based on $\bar{\Omega}_{ij}$) have been proposed by, for example, Hanjalić
and Launder (1980) and Bardina, Ferziger, and Reynolds (1983). When these

---

[2] Recall from Section 8.3 that applicability does not imply accuracy.

modified models are applied to a range of flows, the general experience (Launder 1990, Hanjalić 1994) is that their overall performance is inferior to that of the standard model.

The model equation for ε has been presented here as being entirely empirical; but this is only one viewpoint. The renormalization group method (RNG) has been used to obtain the k–ε equation from the Navier–Stokes equations (see, e.g., Yakhot and Orszag (1986), Smith and Reynolds (1992), Smith and Woodruff (1998), and references therein). The values of the constants stemming from the RNG analysis are

$$C_{\mu} = 0.0845, \; C_{\varepsilon 1} = 1.42, \; C_{\varepsilon 2} = 1.68, \; \sigma_k = \sigma_{\varepsilon} = 0.72 \qquad (10.92)$$

(Orszag *et al.* 1996), cf. Eq. (10.54). In the RNG k–ε model there is also an additional term in the ε equation, which is an *ad hoc* model, not derived from RNG theory. It is this term which is largely responsible for the difference in performance of the standard and RNG models.

10.5 Further turbulent-viscosity models

10.5.1 *The* k–ω *model*

Historically, many two-equation models have been proposed. In most of these, k is taken as one of the variables, but there are diverse choices for the second. Examples are quantities with dimensions of kL (Rotta 1951), ω (Kolmogorov 1942), ω^2 (Saffman 1970), and τ (Speziale, Abid, and Anderson 1992).

For homogeneous turbulence, the choice of the second variable is immaterial, since there is an exact correspondence among the various equations, and their forms are essentially the same (see Exercise 10.12). For inhomogeneous flows, the difference lies in the diffusion term. Consider, for example, the following model equation for $\omega \equiv \varepsilon/k$:

$$\frac{\bar{D}\omega}{\bar{D}t} = \nabla \cdot \left(\frac{\nu_{\mathrm{T}}}{\sigma_{\omega}} \nabla \omega \right) + C_{\omega 1} \frac{\mathcal{P}\omega}{k} - C_{\omega 2}\, \omega^2. \qquad (10.93)$$

How does the k–ω model based on this equation differ from the k–ε model? One way to answer this question is to derive the ω equation implied by the k–ε model (see Exercise 10.13). Taking $\sigma_k = \sigma_{\varepsilon} = \sigma_{\omega}$ for simplicity, the result is

$$\frac{\bar{D}\omega}{\bar{D}t} = \nabla \cdot \left(\frac{\nu_{\mathrm{T}}}{\sigma_{\omega}} \nabla \omega \right) + (C_{\varepsilon 1} - 1)\frac{\mathcal{P}\omega}{k} - (C_{\varepsilon 2} - 1)\omega^2$$
$$+ \frac{2\nu_{\mathrm{T}}}{\sigma_{\omega}k}\nabla \omega \cdot \nabla k. \qquad (10.94)$$

Table 10.2. *Values of C_{Z1} and C_{Z2} for various specifications of $Z = C_Z k^p \varepsilon^q$*
(see Exercise 10.12)

Z	p	q	C_{Z1}	C_{Z2}
k	1	0	1.0	1.0
ε	0	1	1.44	1.92
$\omega = \varepsilon/k$	−1	1	0.44	0.92
$\tau = k/\varepsilon$	1	−1	−0.44	−0.92
$L = k^{3/2}/\varepsilon$	$\frac{3}{2}$	−1	0.06	−0.42
$kL = k^{5/2}/\varepsilon$	$\frac{5}{2}$	−1	1.06	0.58
$\nu_{\mathrm{T}} = C_\mu k^2/\varepsilon$	2	−1	0.56	0.08

Evidently, for homogeneous turbulence, the choices $C_{\omega1} = C_{\varepsilon1} - 1$ and $C_{\omega2} = C_{\varepsilon2} - 1$ make the models identical. However, for inhomogeneous flows, the k–ε model (written as a k–ω model) contains an additional term – the final term in Eq. (10.94).

The second most widely used two-equation model is the k–ω model that has been developed for over 20 years by Wilcox and others (see Wilcox (1993)). In this model, the expression for ν_{T} and the k equation are the same as those in the k–ε model. The difference lies in the use of Eq. (10.93) for ω rather than Eq. (10.53) for ε (or the implied ω equation, Eq. (10.94)). As described in detail by Wilcox (1993), for boundary-layer flows, the k–ω model is superior both in its treatment of the viscous near-wall region, and in its accounting for the effects of streamwise pressure gradients. However, the treatment of non-turbulent free-stream boundaries is problematic: a non-zero (non-physical) boundary condition on ω is required, and the calculated flow is sensitive to the value specified.

Menter (1994) proposed a two-equation model designed to yield the best behavior of the k–ε and k–ω models. It is written as a (non-standard) k–ω model, with the ω equation of the form of Eq. (10.94), but with the final term multiplied by a 'blending function.' Close to walls the blending function is zero (leading to the standard ω equation), whereas remote from walls the blending function is unity (corresponding to the standard ε equation). (The behavior of the k–ε model at a free-stream boundary has been analyzed by Cazalbou *et al.* (1994); see also Exercise 10.11.)

EXERCISES

10.12 Consider the quantity Z defined by

$$Z = C_Z k^p \varepsilon^q, \tag{10.95}$$

for given C_Z, p and q. From the standard k–ε model equations, show that, in homogeneous turbulence, the implied model equation for Z is

$$\frac{dZ}{dt} = C_{Z1}\frac{Z\mathcal{P}}{k} - C_{Z2}\frac{Z\varepsilon}{k}, \tag{10.96}$$

where

$$C_{Z1} = p + qC_{\varepsilon 1}, \tag{10.97}$$

$$C_{Z2} = p + qC_{\varepsilon 2}. \tag{10.98}$$

Hence verify the values of C_{Z1} and C_{Z2} in Table 10.2.

10.13   Given the standard k–ε model equations, show that the implied model equation for $\omega = \varepsilon/k$ is

$$\begin{aligned}
\frac{\bar{D}\omega}{\bar{D}t} = {}& \nabla \cdot \left(\frac{\nu_{\mathrm{T}}}{\sigma_\varepsilon}\nabla\omega\right) + (C_{\varepsilon 1} - 1)\frac{\mathcal{P}\omega}{k} - (C_{\varepsilon 2} - 1)\omega^2 \\
& + C_\mu\left(\frac{1}{\sigma_\varepsilon} + \frac{1}{\sigma_k}\right)\frac{1}{\omega}\nabla\omega \cdot \nabla k \\
& + C_\mu\left(\frac{1}{\sigma_\varepsilon} - \frac{1}{\sigma_k}\right)\left(\nabla^2 k + \frac{1}{k}\nabla k \cdot \nabla k\right).
\end{aligned} \tag{10.99}$$

10.5.2   The Spalart–Allmaras model

Spalart and Allmaras (1994) described a one-equation model developed for aerodynamic applications, in which a single model transport equation is solved for the turbulent viscosity ν_{T}. Earlier proposals for such a model are described by Nee and Kovasznay (1969) and Baldwin and Barth (1990).

It is useful at the outset to appreciate the context of the development of the Spalart–Allmaras model. There is a natural progression in the models described above – algebraic, one-equation, two-equation – on to the Reynolds-stress models described in the next chapter. Each successive level provides a fuller description of the turbulence, and thereby removes a qualitative deficiency of its predecessor. If accuracy were the only criterion in the selection of models for development and application, then the choice would naturally tend toward the models with the higher level of description. However, as discussed in Section 8.3, cost and ease of use are also important criteria that favor the simpler models. It is useful, therefore, for model developers to work toward the best possible model at each level of description.

Arguably, a one-equation model for ν_{T} is the lowest level at which a model can be complete. Spalart and Allmaras (1994) developed the model to

remove the incompleteness of algebraic and one-equation models based on
k, and yet have a model computationally simpler than two-equation models.
The model is designed for aerodynamic flows, such as transonic flow over
airfoils, including boundary-layer separation.

The model equation is of the form

$$\frac{\overline{D}\nu_T}{\overline{D}t} = \nabla \cdot \left(\frac{\nu_T}{\sigma_\nu} \nabla \nu_T \right) + S_\nu, \tag{10.100}$$

where the source term S_ν depends on the laminar and turbulent viscosities,
ν and ν_T; the mean vorticity (or rate of rotation) Ω; the turbulent viscosity
gradient $|\nabla \nu_T|$; and the distance to the nearest wall, ℓ_w. The details of the
model are quite complicated: the reader is referred to the original paper
which provides an enlightening account of the construction of the model to
achieve particular desired behaviors.

In applications to the aerodynamic flows for which it is intended, the model
has proved quite successful (see, e.g., Godin, Zingg, and Nelson (1997)). However, it has clear limitations as a general model. For example; it is incapable
of accounting for the decay of ν_T in isotropic turbulence, it implies that,
in homogeneous turbulence, ν_T is unaffected by irrotational mean straining;
and it overpredicts the rate of spreading of the plane jet by almost 40%.

EXERCISE

10.14 Consider the Spalart–Allmaras model applied to high-Reynolds-
number homogeneous turbulence. Argue that the laminar viscosity
ν and the distance to the nearest wall ℓ_w are not relevant quantities. Hence show that dimensional and other considerations reduce
Eq. (10.100) to

$$\frac{d\nu_T}{dt} = S_\nu(\nu_T, \Omega) = c_{b1}\nu_T\Omega, \tag{10.101}$$

where c_{b1} is a constant. Comment on the form of this equation for
irrotational mean straining.

For self-similar homogeneous turbulent shear flow (in which Ω
and S are equal), from the relation $-\langle uv \rangle = \nu_T \partial \langle U \rangle / \partial y$, show that
ν_T evolves by Eq. (10.101) with

$$c_{b1} = \left(\frac{\mathcal{P}}{\varepsilon} - 1 \right) \left(\frac{Sk}{\varepsilon} \right)^{-1}. \tag{10.102}$$

Use experimental data (Table 5.4 on page 157) to estimate c_{b1} according to Eq. (10.102) and compare it with the Spalart–Allmaras
value of $c_{b1} = 0.135$.

11

Reynolds-stress and related models

11.1 Introduction

Reynolds-stress closure

In Reynolds-stress models, model transport equations are solved for the individual Reynolds stresses $\langle u_i u_j \rangle$ and for the dissipation ε (or for another quantity, e.g., ω, that provides a length or time scale of the turbulence). Consequently, the turbulent-viscosity hypothesis is not needed; so one of the major defects of the models described in the previous chapter is eliminated.

The exact transport equation for the Reynolds stresses is obtained from the Navier–Stokes equations in Exercises 7.23–7.25 on pages 319–319: it is

$$\frac{\bar{D}}{\bar{D}t} \langle u_i u_j \rangle + \frac{\partial}{\partial x_k} T_{kij} = \mathcal{P}_{ij} + \mathcal{R}_{ij} - \varepsilon_{ij}. \qquad (11.1)$$

In a Reynolds-stress model, the 'knowns' are $\langle U \rangle$, $\langle p \rangle$, $\langle u_i u_j \rangle$, and ε. Thus in Eq. (11.1) both the mean-flow convection, $\bar{D}\langle u_i u_j \rangle / \bar{D}t$, and the production tensor, $\mathcal{P}_{ij}$ (Eq. (7.179)), are in closed form. However, models for the dissipation tensor ε_{ij} (Eq. (7.181)), the pressure–rate-of-strain tensor $\mathcal{R}_{ij}$ (Eq. (7.187)), and the Reynolds-stress flux T_{kij} (Eq. (7.195)) are required.

An outline of the chapter

By far the most important quantity to be modelled is the pressure–rate-of-strain tensor, $\mathcal{R}_{ij}$. This term is considered extensively in the next four sections in the context of homogeneous turbulence. This includes (in Section 11.4) a description of rapid-distortion theory (RDT), which applies to a limiting case, and provides useful insights. The extension to inhomogeneous flows is described in Section 11.6, and special near-wall treatments – both for k–ε and for Reynolds-stress models – are described in Section 11.7. There is a vast literature on Reynolds-stress models, with many different proposals and variants: the emphasis here is on the fundamental concepts and approaches.

In Reynolds-stress models, $\mathcal{R}_{ij}$ is modelled as a local function of $\langle u_i u_j \rangle$, ε, and $\partial \langle U_i \rangle / \partial x_j$. Elliptic relaxation models (Section 11.8) provide a higher level of closure and thereby allow the model for $\mathcal{R}_{ij}$ to be non-local.

Algebraic stress models and nonlinear turbulent-viscosity models are described in Section 11.9. These are simpler models that can be derived from Reynolds-stress closures. The chapter concludes with an appraisal of the relative merits of the range of models described.

Dissipation

The dissipation is treated summarily here. For high-Reynolds-number flows, a consequence of local isotropy is[1]

$$\varepsilon_{ij} = \tfrac{2}{3} \varepsilon \delta_{ij}. \tag{11.2}$$

This is taken as the model for ε_{ij}. For moderate-Reynolds-number flows, this isotropic relation is not completely accurate (see, e.g., Fig. 7.39 on page 318). However, to an extent this is of no consequence because the anisotropic component (i.e., $\varepsilon_{ij} - \tfrac{2}{3} \varepsilon \delta_{ij}$) has the same mathematical properties as $\mathcal{R}_{ij}$, and so can be absorbed into the model for $\mathcal{R}_{ij}$. As discussed in Section 11.7, close to walls the dissipation is anisotropic, and different models are appropriate.

The Reynolds number

Most Reynolds-stress models contain no Reynolds-number dependence (except for near-wall treatments), and therefore they implicitly assume that the terms being modelled are independent of the Reynolds number. For simplicity of exposition, we follow this expedient assumption. However, it is good to remember that, in moderate-Reynolds-number experiments, and especially in DNS, there can be (usually modest) Reynolds-number effects.

11.2 The pressure–rate-of-strain tensor

The fluctuating pressure appears in the Reynolds-stress equation (Eq. (7.178)) most directly as the velocity-pressure-gradient tensor

$$\Pi_{ij} \equiv -\frac{1}{\rho} \left\langle u_i \frac{\partial p'}{\partial x_j} + u_j \frac{\partial p'}{\partial x_i} \right\rangle. \tag{11.3}$$

---

[1] In this context it is not necessary to distinguish between the dissipation ε and the pseudo-dissipation $\tilde{\varepsilon}$.

This can be decomposed (see Exercise 7.24 on page 319) into the pressure-transport term $-\partial T_{kij}^{(p)}/\partial x_k$, Eq. (7.192), and the pressure–rate-of-strain tensor

$$\mathcal{R}_{ij} \equiv \left\langle \frac{p'}{\rho} \left(\frac{\partial u_i}{\partial x_j} + \frac{\partial u_j}{\partial x_i} \right) \right\rangle. \tag{11.4}$$

The trace of $\mathcal{R}_{ij}$ is zero ($\mathcal{R}_{ii} = 2\langle p'\nabla \cdot \boldsymbol{u}/\rho \rangle = 0$), and consequently the term does not appear in the kinetic-energy equation: it serves to *redistribute* energy among the Reynolds stresses.

As observed by Lumley (1975), the decomposition of Π_{ij} into a redistribution term and a transport term is not unique. For example, an alternative decomposition is

$$\Pi_{ij} = \mathcal{R}_{ij}^{(a)} - \frac{\partial}{\partial x_\ell} \left(\tfrac{2}{3} \delta_{ij} T_\ell^{(p)} \right), \tag{11.5}$$

with

$$\mathcal{R}_{ij}^{(a)} \equiv \Pi_{ij} - \tfrac{1}{3} \Pi_{\ell\ell} \delta_{ij}, \tag{11.6}$$

$$\boldsymbol{T}^{(p)} \equiv \langle \boldsymbol{u}p' \rangle / \rho. \tag{11.7}$$

The significance of $\boldsymbol{T}^{(p)}$ is that the source of kinetic energy due to pressure transport is

$$\frac{1}{2} \Pi_{ii} = -\nabla \cdot \boldsymbol{T}^{(p)}. \tag{11.8}$$

In homogeneous turbulence the pressure transport is zero, and all redistributive terms are equivalent (e.g., $\Pi_{ij} = \mathcal{R}_{ij} = \mathcal{R}_{ij}^{(a)}$). In examining such flows (in this and the next three sections) we focus on the pressure–rate-of-strain tensor $\mathcal{R}_{ij}$. For inhomogeneous flows it is a matter of convenience which decomposition to use; and in this case, as discussed in Section 11.6, there are reasons to favor the decomposition in terms of $\mathcal{R}_{ij}^{(a)}$, Eq. (11.5).

The importance of redistribution

It is worth recalling the behavior of Π_{ij} in the turbulent boundary layer. In the budget for $\langle u^2 \rangle$ (Fig. 7.35 on page 316), Π_{11} removes energy at about twice the rate of ε_{11}. These two sinks are (approximately) balanced by the production $\mathcal{P}_{11}$. In the budgets for $\langle v^2 \rangle$ and $\langle w^2 \rangle$ (Figs. 7.36 and 7.37) there is no production, but Π_{22} and Π_{33} are sources that approximately balance ε_{22} and ε_{33}. Thus energy is redistributed from the largest normal stress (which has all of the energy production) to the smaller normal stresses (which have no production). In the shear-stress budget (Fig. 7.38 on page 317), the production $\mathcal{P}_{12}$ is approximately balanced by $-\Pi_{12}$, the dissipation being small in comparison.

Evidently, along with production and dissipation, redistribution is a dominant process in the balance of the Reynolds stresses. Consequently, its modelling is crucial, and the subject of extensive research.

The Poisson equation for p'

Some insight into the pressure–rate-of-strain tensor can be gained by examining the Poisson equation for pressure (see Section 2.5). The Reynolds decomposition of this equation (performed in Exercise 11.1) leads to a Poisson equation for p' with two source terms:

$$\frac{1}{\rho} \nabla^2 p' = -2 \frac{\partial \langle U_i \rangle}{\partial x_j} \frac{\partial u_j}{\partial x_i} - \frac{\partial^2}{\partial x_i \, \partial x_j} (u_i u_j - \langle u_i u_j \rangle). \tag{11.9}$$

On the basis of this equation, the fluctuating pressure field can be decomposed into three contributions:

$$p' = p^{(r)} + p^{(s)} + p^{(h)}. \tag{11.10}$$

The *rapid pressure* $p^{(r)}$ satisfies

$$\frac{1}{\rho} \nabla^2 p^{(r)} = -2 \frac{\partial \langle U_i \rangle}{\partial x_j} \frac{\partial u_j}{\partial x_i}, \tag{11.11}$$

the *slow pressure* $p^{(s)}$ satisfies

$$\frac{1}{\rho} \nabla^2 p^{(s)} = - \frac{\partial^2}{\partial x_i \, \partial x_j} (u_i u_j - \langle u_i u_j \rangle), \tag{11.12}$$

and the *harmonic* contribution $p^{(h)}$ satisfies Laplace's equation $\nabla^2 p^{(h)} = 0$. Boundary conditions are specified on $p^{(h)}$ – dependent on those on $p^{(r)}$ and $p^{(s)}$ – so that p' satisfies the required boundary conditions.

The rapid pressure is so called because (unlike $p^{(s)}$) it responds immediately to a change in the mean velocity gradients. Also, in the rapid-distortion limit (i.e., $Sk/\varepsilon \to \infty$), the rapid pressure field $p^{(r)}$ has a leading-order effect, whereas $p^{(s)}$ is negligible (Section 11.4). Corresponding to $p^{(r)}$, $p^{(s)}$, and $p^{(h)}$, the pressure–rate-of-strain tensor can also be decomposed into three contributions, $\mathcal{R}_{ij}^{(r)}$, $\mathcal{R}_{ij}^{(s)}$, and $\mathcal{R}_{ij}^{(h)}$, with obvious definitions, e.g.,

$$\mathcal{R}_{ij}^{(r)} \equiv \left\langle \frac{p^{(r)}}{\rho} \left(\frac{\partial u_i}{\partial x_j} + \frac{\partial u_j}{\partial x_i} \right) \right\rangle. \tag{11.13}$$

As shown in the seminal work of Chou (1945), the Green-function solution to the Poisson equation (Eq. (2.48)) can be used to express the pressure–rate-of-strain tensor in terms of two-point velocity correlations. For example, in homogeneous turbulence, one contribution to $\mathcal{R}_{ij}^{(\mathrm{r})}$ is

$$\left\langle \frac{p^{(\mathrm{r})}}{\rho} \frac{\partial u_i}{\partial x_j} \right\rangle = 2 \frac{\partial \langle U_k \rangle}{\partial x_\ell} M_{i\ell jk}, \tag{11.14}$$

where the fourth-order tensor $\boldsymbol{M}$ is given by an integral of the two-point velocity correlation $R_{ij}(\boldsymbol{r}) = \langle u_i(\boldsymbol{x}) u_j(\boldsymbol{x} + \boldsymbol{r}) \rangle$:

$$M_{i\ell jk} = -\frac{1}{4\pi} \int \frac{1}{|\boldsymbol{r}|} \frac{\partial^2 R_{i\ell}}{\partial r_j \, \partial r_k} \, \mathrm{d}\boldsymbol{r} \tag{11.15}$$

(see Exercise 11.2).

The valuable conclusions from these considerations are that there are three qualitatively different contributions to $\mathcal{R}_{ij}$. The rapid pressure involves the mean velocity gradients, and in homogeneous turbulence $\mathcal{R}_{ij}^{(\mathrm{r})}$ is directly proportional to $\partial \langle U_k \rangle / \partial x_\ell$ (Eq. (11.14)). The slow pressure–rate-of-strain tensor $\mathcal{R}_{ij}^{(\mathrm{s})}$ can be expected to be significant in most circumstances (except rapid distortion); and, indeed, in decaying homogeneous anisotropic turbulence, $\mathcal{R}_{ij}^{(\mathrm{s})}$ is the only one of the three contributions that is non-zero. The harmonic component $\mathcal{R}_{ij}^{(\mathrm{h})}$ is zero in homogeneous turbulence, and is important only near walls: it is discussed in Section 11.7.5.

EXERCISES

11.1   Show that the Poisson equation for pressure (Eq. (2.42)) can alternatively be written

$$\frac{1}{\rho} \nabla^2 p = -\frac{\partial U_i}{\partial x_j} \frac{\partial U_j}{\partial x_i} = -\frac{\partial^2 U_i U_j}{\partial x_i \, \partial x_j}. \tag{11.16}$$

Hence show that the mean pressure satisfies

$$\frac{1}{\rho} \nabla^2 \langle p \rangle = -\frac{\partial \langle U_i \rangle}{\partial x_j} \frac{\partial \langle U_j \rangle}{\partial x_i} - \frac{\partial^2 \langle u_i u_j \rangle}{\partial x_i \, \partial x_j}, \tag{11.17}$$

and that the fluctuation pressure satisfies Eq. (11.9).

11.2   Consider homogeneous turbulence (in which the mean velocity gradient $\partial \langle U_k \rangle / \partial x_\ell$ is uniform). From Eqs. (2.48) and (11.11), show that the correlation at $\boldsymbol{x}$ between the rapid pressure and a random field $\phi(\boldsymbol{x})$ is given by

$$\frac{1}{\rho} \langle p^{(\mathrm{r})}(\boldsymbol{x}) \phi(\boldsymbol{x}) \rangle = \frac{1}{2\pi} \frac{\partial \langle U_k \rangle}{\partial x_\ell} \int\!\!\!\int\!\!\!\int_{-\infty}^{\infty} \left\langle \frac{\partial u_\ell(\boldsymbol{y})}{\partial y_k} \phi(\boldsymbol{x}) \right\rangle \frac{\mathrm{d}\boldsymbol{y}}{|\boldsymbol{x} - \boldsymbol{y}|}, \tag{11.18}$$

where integration is over all $\boldsymbol{y}$. Comment on the behavior of the two-point correlation necessary for the integral to converge. Show that a contribution to $\mathcal{R}_{ij}^{(r)}$ is

$$\left\langle \frac{p^{(r)}}{\rho} \frac{\partial u_i}{\partial x_j} \right\rangle = \frac{1}{2\pi} \frac{\partial \langle U_k \rangle}{\partial x_\ell} \int\!\!\!\int\!\!\!\int\limits_{-\infty}^{\infty} \frac{\partial^2}{\partial x_j \, \partial y_k} \langle u_i(x) u_\ell(y) \rangle \frac{\mathrm{d} y}{|x - y|}. \quad (11.19)$$

With the separation vector defined by $r \equiv y - x$, show that this equation can be rewritten in terms of the two-point velocity correlation $R_{ij}(r) \equiv \langle u_i(x) u_j(x + r) \rangle$ as

$$\left\langle \frac{p^{(r)}}{\rho} \frac{\partial u_i}{\partial x_j} \right\rangle = -\frac{1}{2\pi} \frac{\partial \langle U_k \rangle}{\partial x_\ell} \int\!\!\!\int\!\!\!\int\limits_{-\infty}^{\infty} \frac{1}{|r|} \frac{\partial^2 R_{i\ell}}{\partial r_j \, \partial r_k} \, \mathrm{d} r. \quad (11.20)$$

Hence verify Eq. (11.14).

11.3 Return-to-isotropy models

11.3.1 Rotta's model

The simplest situation in which to examine the slow pressure–rate-of-strain tensor is decaying homogeneous anisotropic turbulence. In this case, there is no production or transport, and $\mathcal{R}_{ij}^{(r)}$ and $\mathcal{R}_{ij}^{(h)}$ are zero, so that the exact Reynolds-stress equation is

$$\frac{\mathrm{d}}{\mathrm{d}t} \langle u_i u_j \rangle = \mathcal{R}_{ij}^{(s)} - \varepsilon_{ij}. \quad (11.21)$$

Since the trace of $\mathcal{R}_{ij}^{(s)}$ is zero, the term has no effect on the turbulent kinetic energy. Its effect is on the distribution of energy among the Reynolds stresses, which can be examined through the normalized anisotropy tensor

$$b_{ij} \equiv \frac{\langle u_i u_j \rangle}{\langle u_k u_k \rangle} - \tfrac{1}{3}\delta_{ij} = \frac{a_{ij}}{2k}. \quad (11.22)$$

Taking ε_{ij} to be isotropic (Eq. 11.2), the evolution equation for b_{ij} is

$$\frac{\mathrm{d}b_{ij}}{\mathrm{d}t} = \frac{\varepsilon}{k}\left(b_{ij} + \frac{\mathcal{R}_{ij}^{(s)}}{2\varepsilon} \right) \quad (11.23)$$

(see Exercise 11.3).

It is natural to suppose that the turbulence has a tendency to become less anisotropic as it decays, and indeed such a tendency to *return to isotropy* is evident in Fig. 10.2 on page 361. On the basis of this notion, Rotta (1951) proposed the model

$$\mathcal{R}_{ij}^{(s)} = -C_R \frac{\varepsilon}{k} \left(\langle u_i u_j \rangle - \tfrac{2}{3} k \delta_{ij} \right)$$
$$= -2 C_R \varepsilon b_{ij}, \tag{11.24}$$

where C_R is the 'Rotta constant.' Substituting this into Eq. (11.23) yields

$$\frac{\mathrm{d} b_{ij}}{\mathrm{d} t} = -(C_R - 1) \frac{\varepsilon}{k} b_{ij}, \tag{11.25}$$

showing that Rotta's model corresponds to a *linear* return to isotropy. Evidently a value of C_R greater than unity is required.

EXERCISE

11.3 From the definition of b_{ij} (Eq. (11.22)) and from the Reynolds-stress-evolution equation (Eq. (11.21)), show that the exact equation for b_{ij} in decaying turbulence is

$$\frac{\mathrm{d} b_{ij}}{\mathrm{d} t} = \frac{\varepsilon}{k} \left(b_{ij} + \frac{\mathcal{R}_{ij}^{(s)}}{2\varepsilon} + \frac{1}{3} \delta_{ij} - \frac{\varepsilon_{ij}}{2\varepsilon} \right). \tag{11.26}$$

Hence show that Eq. (11.23) follows from the assumed isotropy of ε_{ij}.

Show that, if, instead, ε_{ij} is taken to be proportional to $\langle u_i u_j \rangle$, then the resulting equation for b_{ij} is

$$\frac{\mathrm{d} b_{ij}}{\mathrm{d} t} = \frac{\mathcal{R}_{ij}^{(s)}}{2k}. \tag{11.27}$$

Show that, if Rotta's model is used in this equation, the result is the same as Eq. (11.25) but with C_R in place of $C_R - 1$.

11.3.2 The characterization of Reynolds-stress anisotropy

To examine the behavior of Rotta's model (and other return-to-isotropy models) it is useful to be able to characterize the Reynolds-stress anisotropy more simply than by the six components b_{ij} of the anisotropy tensor **b**. In fact, as now described, the state of anisotropy can be characterized by just two variables (denoted by ξ and η), which allows a simple graphical representation.

The anisotropy tensor has zero trace ($b_{ii} = 0$) and consequently it has just two independent invariants (see Appendix B). It is convenient to take as these two independent invariants the quantities ξ and η defined by

$$6\eta^2 = -2\mathrm{II}_b = b_{ii}^2 = b_{ij} b_{ji}, \tag{11.28}$$

Table 11.1. *Special states of the Reynolds-stress tensor in terms of the invariants ξ and η, the eigenvalues of* **b**, *and the shape of the Reynolds-stress ellipsoid defined in Exercise 11.8 (see also Fig. 11.1 and Exercise 11.4)*

State of turbulence	Invariants	Eigenvalues of **b**	Shape of Reynolds-stress ellipsoid	Designation in Fig. 11.1
Isotropic	$\xi = \eta = 0$	$\lambda_1 = \lambda_2 = \lambda_3 = 0$	Sphere	iso
Two-component axisymmetric	$\xi = -\frac{1}{6}, \eta = \frac{1}{6}$	$\lambda_1 = \lambda_2 = \frac{1}{6}$	Disk	2C, axi
One-component	$\xi = \frac{1}{3}, \eta = \frac{1}{3}$	$\lambda_1 = \frac{2}{3},$ $\lambda_2 = \lambda_3 = -\frac{1}{3}$	Line	1C
Axisymmetric (one large eigenvalue)	$\eta = \xi$	$-\frac{1}{3} \leq \lambda_1 = \lambda_2 \leq 0$	Prolate spheroid	axi, $\xi > 0$
Axisymmetric (one small eigenvalue)	$\eta = -\xi$	$0 \leq \lambda_1 = \lambda_2 \leq \frac{1}{6}$	Oblate spheroid	axi, $\xi < 0$
Two-component	$\eta = (\frac{1}{27} + 2\xi^3)^{1/2}$ $F(\xi, \eta) = 0$	$\lambda_1 + \lambda_2 = \frac{1}{3}$	Ellipse	2C

$$6\xi^3 = 3\text{III}_b = b_{ii}^3 = b_{ij}b_{jk}b_{ki}. \tag{11.29}$$

A related observation is that the sum of the eigenvalues (λ_1, λ_2, and λ_3) of **b** is zero, so that, in its principal axes, **b** is

$$\tilde{b}_{ij} = \begin{bmatrix} \lambda_1 & 0 & 0 \\ 0 & \lambda_2 & 0 \\ 0 & 0 & -\lambda_1 - \lambda_2 \end{bmatrix}. \tag{11.30}$$

Thus, in terms of ξ and η, II_b and III_b, or λ_1 and λ_2, the state of anisotropy of the Reynolds stresses can be characterized by two invariants.

At any point and time in any turbulent flow, ξ and η can be determined from the Reynolds stresses, and the result plotted as a point on the ξ–η plane. There are some special states of the Reynolds-stress tensor that correspond to particular points and curves in this plane. These states are determined in Exercise 11.4, summarized in Table 11.1, and plotted in Fig. 11.1 – which is known as the *Lumley triangle* (even though one side is not straight).[2]

Every Reynolds stress that can occur in a turbulent flow (i.e., that is

---

[2] In fact Lumley (1978) used III_b and $-\text{II}_b$ as coordinates (rather than ξ and η), in which case only one of the three sides of the triangle is straight.

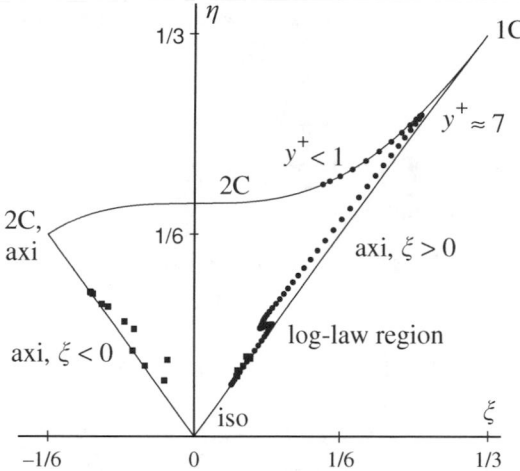

Fig. 11.1. The Lumley triangle on the plane of the invariants ξ and η of the Reynolds-stress anisotropy tensor. The lines and vertices correspond to special states (see Table 11.1). Circles: from DNS of channel flow (Kim *et al.* 1987). Squares: from experiments on a turbulent mixing layer (Bell and Mehta 1990). 1C, one-component; 2C, two-component.

realizable) corresponds to a point in the Lumley triangle. Points outside correspond to *non-realizable* Reynolds stresses – with negative or complex eigenvalues.

Also shown in Fig. 11.1 are values of (ξ, η) obtained in turbulent channel flow and in a turbulent mixing layer. Very close to the wall in channel flow ($y^+ < 5$, say) the turbulence is essentially two-component, $\langle v^2 \rangle$ being much smaller than $\langle u^2 \rangle$ and $\langle w^2 \rangle$. The anisotropy (measured by η) reaches a peak at $y^+ \approx 7$. Throughout the remainder of the channel the Reynolds stress is close to being axisymmetric with ξ positive. In the center of the turbulent mixing layer, the Reynolds stresses are close to axisymmetric with ξ positive, with somewhat lower anisotropy than in the log-region of the channel flow. At the edges of the mixing layer, the Reynolds stresses are close to axisymmetric with ξ negative.

EXERCISES

11.4   Show that ξ and η (Eqs. (11.28) and (11.29)) are related to the eigenvalues of **b** (Eq. (11.30)) by

$$\eta^2 = \tfrac{1}{3}(\lambda_1^2 + \lambda_1\lambda_2 + \lambda_2^2), \qquad (11.31)$$

$$\xi^3 = -\tfrac{1}{2}\lambda_1\lambda_2(\lambda_1 + \lambda_2). \qquad (11.32)$$

Let $\langle \tilde{u}_1^2 \rangle$, $\langle \tilde{u}_2^2 \rangle$ and $\langle \tilde{u}_3^2 \rangle$ be the eigenvalues of the Reynolds-stress tensor (i.e., the normal stresses in principal axes). Verify the entries in Table 11.1 for the following states of the Reynolds stresses:

(a) isotropic; $\langle \tilde{u}_1^2 \rangle = \langle \tilde{u}_2^2 \rangle = \langle \tilde{u}_3^2 \rangle$;

(b) two-component, axisymmetric; $\langle \tilde{u}_1^2 \rangle = \langle \tilde{u}_2^2 \rangle$, $\langle \tilde{u}_3^2 \rangle = 0$; and

(c) one-component; $\langle \tilde{u}_2^2 \rangle = \langle \tilde{u}_3^2 \rangle = 0$.

Show for axisymmetric Reynolds stresses (i.e., $\langle \tilde{u}_1^2 \rangle = \langle \tilde{u}_2^2 \rangle$, $\lambda_1 = \lambda_2$) that

$$\eta^2 = \lambda_1^2, \quad \xi^3 = -\lambda_1^3, \tag{11.33}$$

and hence verify the axisymmetric states given in Table 11.1.

The quantity F is defined as the determinant of the normalized Reynolds-stress tensor

$$F \equiv \det\left(\frac{\langle u_i u_j \rangle}{\frac{1}{3}\langle u_\ell u_\ell \rangle} \right), \tag{11.34}$$

so that it is unity in isotropic turbulence. Derive the following equivalent expressions for F

$$\begin{aligned} F &= \frac{27 \langle \tilde{u}_1^2 \rangle \langle \tilde{u}_2^2 \rangle \langle \tilde{u}_3^2 \rangle}{(\langle \tilde{u}_1^2 \rangle + \langle \tilde{u}_2^2 \rangle + \langle \tilde{u}_3^2 \rangle)^3} \\ &= (1 + 3\lambda_1)(1 + 3\lambda_2)(1 - 3\lambda_1 - 3\lambda_2) \\ &= 1 - \tfrac{9}{2}b_{ii}^2 + 9b_{ii}^3 \\ &= 1 - 27\eta^2 + 54\xi^3. \end{aligned} \tag{11.35}$$

Observe that F is zero in two-component turbulence (e.g., $\langle \tilde{u}_1^2 \rangle = 0$), and hence verify that the equation for the two-component curve in Fig. 11.1 is

$$\eta = (\tfrac{1}{27} + 2\xi^3)^{1/2}. \tag{11.36}$$

11.5   Evaluate ξ and η for homogeneous shear flow (i) from experimental data (see Table 5.4 on page 157), and (ii) according to the k–ε model. Locate the corresponding points on Fig. 11.1.

11.6   Show that, according to the Rotta model (Eq. (11.25)), the invariants b_{ii}^n (e.g., $b_{ii}^3 = b_{ij}b_{jk}b_{ki}$) evolve by

$$\frac{db_{ii}^n}{dt} = -n(C_R - 1)\frac{\varepsilon}{k}b_{ii}^n. \tag{11.37}$$

Hence show that ξ and η (Eqs. (11.28) and (11.29)) evolve by

$$\frac{d\xi}{dt} = -(C_R - 1)\frac{\varepsilon}{k}\xi, \tag{11.38}$$

$$\frac{d\eta}{dt} = -(C_R - 1)\frac{\varepsilon}{k}\eta. \tag{11.39}$$

How does the ratio ξ/η evolve? Show that the trajectories in the ξ-η plane generated by Eqs. (11.38) and (11.39) are straight lines directed toward the origin (for $C_R > 1$).

11.7 Show that

$$C_\mu \le \frac{2}{\sqrt{3}}\left(\frac{Sk}{\varepsilon}\right)^{-1}, \qquad C_\mu \le \frac{1}{\sqrt{3}}\left(\frac{Sk}{\varepsilon}\right)^{-1} \tag{11.40}$$

are necessary and sufficient conditions, respectively, for the Reynolds stresses given by the k-ε model to be realizable.

11.8 At a given point in a turbulent flow, and for given $\alpha \ge 0$, consider the quantity

$$E(V, \alpha) \equiv \frac{1}{\alpha^2}C_{ij}^{-1}(V_i - \langle U_i \rangle)(V_j - \langle U_j \rangle), \tag{11.41}$$

where C_{ij}^{-1} is the i-j component of the inverse of the Reynolds-stress tensor, $C_{ij} = \langle u_i u_j \rangle$. Verify that the equation

$$E(V, \alpha) \le 1 \tag{11.42}$$

defines an ellipsoid in velocity space, centered at $\langle U \rangle$, with principal axes aligned with those of $\langle u_i u_j \rangle$, and with the half-lengths of the principal axes being $\alpha \tilde{u}'_{(i)}$, where $\tilde{u}'^2_{(i)}$ ($i = 1, 2, 3$) are the eigenvalues of $\langle u_i u_j \rangle$. Verify the shapes of the Reynolds-stress ellipsoids given in Table 11.1.

If U is joint normally distributed (see Eq. (3.116)), show that the surface of the ellipsoid – which is given by $E(V, \alpha) = 1$ – is a surface of constant probability density. Show that the probability of U lying within the ellipsoid is

$$P(\alpha) = \sqrt{\frac{2}{\pi}}\int_0^\alpha s^2 e^{-s^2/2}\,ds = 1 - Q\left(\tfrac{3}{2}, \tfrac{1}{2}\alpha^2\right), \tag{11.43}$$

where Q is the incomplete gamma function defined by Eq. (12.202). (This is the χ^2 distribution with three degrees of freedom.)

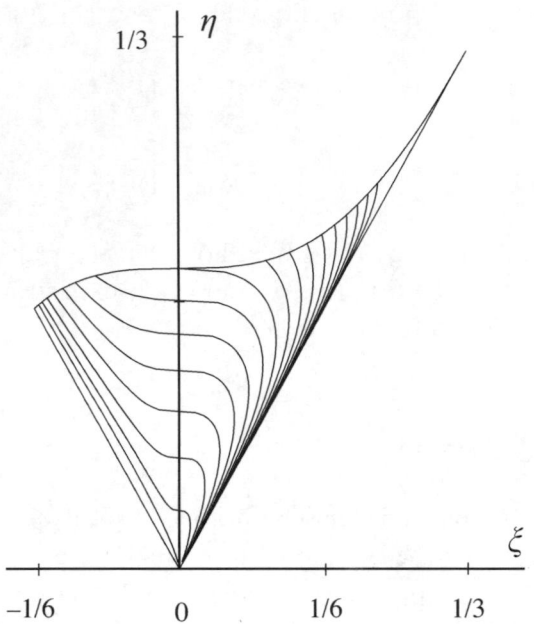

Fig. 11.2. Trajectories on the ξ–η plane given by the model of Sarkar and Speziale (1990) (Eqs. (11.51) and (11.57)).

11.3.3 Nonlinear return-to-isotropy models

When homogeneous anisotropic turbulence decays, its evolving state corresponds to a trajectory on the ξ–η plane. According to Rotta's model, these trajectories are straight lines directed toward the origin (see Exercise 11.6); and the normalized rate of return to isotropy

$$\frac{k}{\varepsilon}\frac{\mathrm{d}}{\mathrm{d}t}\ln b_{ii}^2 = -2(C_R - 1), \tag{11.44}$$

is independent of ξ and η. A careful examination of the available experimental data (see Chung and Kim (1995)) reveals a different behavior. The principal observations are the following.

(i) Rather than being straight lines, the return-to-isotropy trajectories are more like those shown in Fig. 11.2: there is a tendency toward the axisymmetric state with ξ positive.

(ii) The normalized rate of return to isotropy is not independent of ξ and η. In particular, the rate of return is typically larger (e.g., by a factor of two) for ξ negative than it is for ξ positive.

(iii) Over the Reynolds-number range of grid-turbulence experiments and DNS, the rate of return depends significantly on the Reynolds number.

The shortcomings of Rotta's model have motivated the consideration of more general models for the slow pressure–rate-of-strain tensor $\mathcal{R}_{ij}^{(s)}$ (e.g., Shih and Lumley (1985), Sarkar and Speziale (1990), and Chung and Kim (1995)). The development of these models illustrates, in a relatively simple setting, several of the general principles used in Reynolds-stress modelling. It is useful, therefore, to elaborate on the steps involved.

In a Reynolds-stress model applied to decaying anisotropic turbulence, the turbulence is characterized by $\langle u_i u_j \rangle$ and ε, and the only relevant material property is v. Hence, a *general model* can be written (in terms of an unspecified tensor function $\mathcal{F}_{ij}^{(1)}$) as

$$\mathcal{R}_{ij}^{(s)} = \mathcal{F}_{ij}^{(1)}(\langle u_k u_\ell \rangle, \varepsilon, v). \qquad (11.45)$$

In a wind-tunnel experiment on decaying anisotropic turbulence, the mean velocity $\langle U \rangle$ is also represented in a Reynolds-stress closure, and so the dependence of $\mathcal{R}_{ij}^{(s)}$ on $\langle U \rangle$ could also be considered. However, the principle of *Galilean invariance* precludes such a dependence: $\mathcal{R}_{ij}^{(s)}$ is the same in all inertial frames, whereas $\langle U \rangle$ obviously is not (see Section 2.9). Although it is not explicit in the notation, it should also be appreciated that Eq. (11.45) contains the assumption of *localness*: $\mathcal{R}_{ij}^{(s)}$ at position x and time t is supposed to depend on $\langle u_i u_j \rangle$ and ε at the same position and time.

Simple application of *dimensional analysis* allows Eq. (11.45) to be re-expressed in non-dimensional form as

$$\frac{\mathcal{R}_{ij}^{(s)}}{\varepsilon} = \mathcal{F}_{ij}^{(2)}(\mathbf{b}, \mathrm{Re}_L), \qquad (11.46)$$

where $\mathcal{F}_{ij}^{(2)}$ is a non-dimensional tensor function, and the Reynolds number is $\mathrm{Re}_L \equiv k^2/(\varepsilon v)$. It may be noted that Rotta's model (Eq. (11.24)) corresponds to the specification

$$\mathcal{F}_{ij}^{(2)} = -2C_R b_{ij}. \qquad (11.47)$$

The quantity $\mathcal{R}_{ij}^{(s)}/\varepsilon$ is a second-order tensor, and (as already implied by the notation) it is only reasonable for the model ($\mathcal{F}_{ij}^{(2)}$) also to be a second-order tensor. This may seem too obvious to mention. However, it is important to recognize that insistence on tensorially correct models guarantees that the behavior implied by the model is independent of the choice of coordinate system. Models not written in tensor form (e.g., those written for specific components, or that involve pseudovectors or the alternating symbol ε_{ijk}) do not guarantee *coordinate-system invariance* and should be viewed with suspicion. Additionally, to reproduce the properties of $\mathcal{R}_{ij}^{(s)}/\varepsilon$, the model $\mathcal{F}_{ij}^{(2)}$ has to be a symmetric tensor with zero trace.

The next issue is the *representation of tensor functions*. Going beyond Eq. (11.47), what is the most general representation of the tensor $\mathcal{F}^{(2)}$ in terms of the tensor $\mathbf{b}$ and the scalar $\mathrm{Re_L}$? Formally, we can write

$$\mathcal{F}_{ij}^{(2)} = \sum_n f^{(n)} \mathcal{T}_{ij}^{(n)}, \qquad (11.48)$$

where $\mathcal{T}_{ij}^{(n)}$, $n = 1, 2, 3, \ldots$ are all the symmetric, deviatoric tensors that can be formed from $\mathbf{b}$, and the scalar coefficients $f^{(n)}$ depend on $\mathrm{Re_L}$ and the invariants of $\mathbf{b}$. The only form of tensor that can be formed from $\mathbf{b}$ is powers of $\mathbf{b}$. Thus, the deviatoric tensors $\mathcal{T}_{ij}^{(n)}$ can be taken to be

$$\mathcal{T}_{ij}^{(n)} = b_{ij}^n - \tfrac{1}{3} b_{kk}^n \delta_{ij}. \qquad (11.49)$$

However, in view of the Cayley–Hamilton theorem (see Appendix B), it is sufficient to include just $\mathcal{T}^{(1)}$ and $\mathcal{T}^{(2)}$ in Eq. (11.48), because all the other tensors (i.e., $\mathcal{T}^{(n)}$, $n > 2$) can be expressed as linear combinations of these two. Thus Eq. (11.48) can be reduced to

$$\mathcal{F}_{ij}^{(2)}(\mathbf{b}, \mathrm{Re_L}) = f^{(1)}(\xi, \eta, \mathrm{Re_L}) b_{ij} + f^{(2)}(\xi, \eta, \mathrm{Re_L})(b_{ij}^2 - \tfrac{1}{3} b_{kk}^2 \delta_{ij}). \qquad (11.50)$$

This equation shows that $\mathcal{F}_{ij}^{(2)}$ (which is a tensor function of a tensor) is given by two *known* tensors, multiplied by scalar coefficients. Consequently, the modelling task has been reduced to the determination of these two scalar coefficients ($f^{(1)}$ and $f^{(2)}$). Clearly, this procedure (introduced by Robertson (1940)) of using the Cayley–Hamilton theorem to obtain such tensor representations is of great value in turbulence modelling.

In the present case, the procedure is straightforward. In other cases that arise in the study of turbulence it is far from simple. For such cases, there is a substantial body of work in the mechanics literature describing a range of tensor-representation theorems (e.g., Spencer (1971), Pennisi and Trovato (1987), Pennisi (1992), and references therein).

The steps taken so far have reduced the general model (Eq. (11.45)) to the more specific form (Eq. (11.50)), which (with Eq. (11.23)) leads to the modelled equation for the evolution of b_{ij}

$$\frac{k}{\varepsilon} \frac{\mathrm{d}b_{ij}}{\mathrm{d}t} = (1 + \tfrac{1}{2} f^{(1)}) b_{ij} + \tfrac{1}{2} f^{(2)}(b_{ij}^2 - \tfrac{1}{3} b_{kk}^2 \delta_{ij}). \qquad (11.51)$$

A particular model corresponds to a particular specification of $f^{(1)}$ and $f^{(2)}$ as functions of ξ, η and $\mathrm{Re_L}$. Rotta's model is given by $f^{(1)} = -2C_\mathrm{R}$, $f^{(2)} = 0$.

Considerations of *realizability* impose constraints on the specifications of $f^{(1)}$ and $f^{(2)}$. For a given specification, and for a given initial condition on b_{ij}, Eq. (11.51) defines the trajectory on the ξ–η plane followed by the

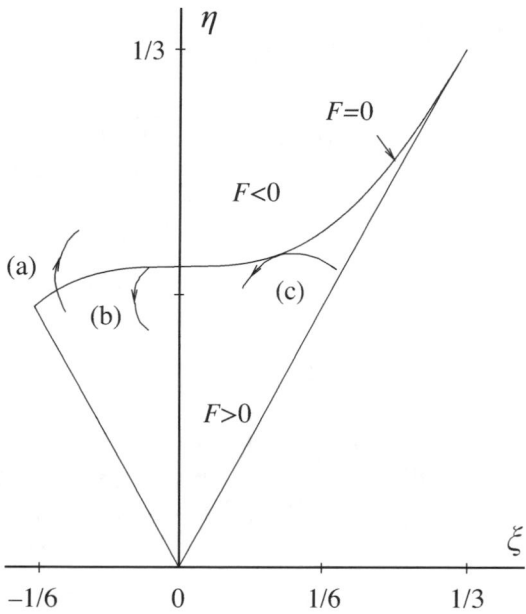

Fig. 11.3. The Lumley triangle showing trajectories of three types: (a) violates realizability; (b) satisfies weak realizability; (c) satisfies strong realizability. (Note that other types of trajectories are possible.)

Reynolds-stress anisotropy as the turbulence decays. With an inappropriate specification of $f^{(1)}$ and $f^{(2)}$ it is possible for a trajectory to pass from the realizable region (i.e., the Lumley triangle) into the non-realizable region. Trajectory (a) in Fig. 11.3 illustrates this non-realizable behavior.

A convenient quantity to consider in the present discussion of realizability is the determinant of the normalized Reynolds-stress tensor:

$$F \equiv \det\left(\frac{\langle u_i u_j \rangle}{\frac{1}{3}\langle u_k u_k \rangle}\right), \tag{11.52}$$

see Exercise 11.4. Within the Lumley triangle F is positive; on the two-component line F is zero; and in the non-realizable region, across the two-component line, F is negative.[3] As shown in Exercise 11.10, the model equation for b_{ij} (Eq. (11.51)) leads to a corresponding equation for F of the form

$$\frac{dF}{dt} = \frac{\varepsilon}{k}\mathcal{F}^{(3)}(\xi, \eta, \mathrm{Re_L}), \tag{11.53}$$

where the non-dimensional function $\mathcal{F}^{(3)}$ is determined by $f^{(1)}$ and $f^{(2)}$ (Eq. (11.64)).

---

[3] Above and to the right of the one-component point there is a non-realizable region with positive F, corresponding to a Reynolds-stress tensor with *two* negative eigenvalues.

The three trajectories shown in Fig. 11.3 are distinguished by the behavior of $\mathrm{d}F/\mathrm{d}t$ on the two-component line $(F = 0)$. Trajectory (a) has

$$\left(\frac{\mathrm{d}F}{\mathrm{d}t}\right)_{F=0} < 0, \tag{11.54}$$

leading to negative F and non-realizability. Trajectory (b) is characterized by

$$\left(\frac{\mathrm{d}F}{\mathrm{d}t}\right)_{F=0} > 0. \tag{11.55}$$

Clearly, if this condition is satisfied all along the two-component line, no trajectory can cross the line and realizability is guaranteed.[4] Trajectory (c) osculates the two-component line and is characterized by

$$\left(\frac{\mathrm{d}F}{\mathrm{d}t}\right)_{F=0} = 0, \qquad \left(\frac{\mathrm{d}^2 F}{\mathrm{d}t^2}\right)_{F=0} > 0. \tag{11.56}$$

Schumann (1977) and Lumley (1978) introduced the concept of realizability into Reynolds-stress modelling, and argued that models should yield trajectories of type (c). Many model developers have followed this idea, and accordingly have constructed models that satisfy Eq. (11.56). Such models are said to satisfy *strong realizability*. On the other hand, Pope (1985) argued that trajectories of type (b) do not violate realizability, and so (in general) the strong realizability conditions are too strong. Models with type (b) or (c) behavior, that satisfy either Eq. (11.55) or Eq. (11.56), are said to satisfy *weak realizability*.

For the nonlinear return-to-isotropy model (Eq. (11.50)), the constraints on $f^{(1)}$ and $f^{(2)}$ imposed by weak realizability are described by Sarkar and Speziale (1990), and those for strong realizability by Chung and Kim (1995). Realizability is described here for the particular case of the Reynolds stresses given by a return-to-isotropy model. It should be appreciated, however, that realizability is a general concept, applicable to all models and to all statistics.

The final step in the development of the return-to-isotropy model is the *specification of the model coefficients* $f^{(1)}(\xi, \eta, \mathrm{Re_L})$ and $f^{(2)}(\xi, \eta, \mathrm{Re_L})$. For this step, experimental data play a crucial role both in suggesting appropriate specifications, and in testing the resulting model. In the model of Sarkar and

---

[4] Excepting singular behavior at the one-component point.

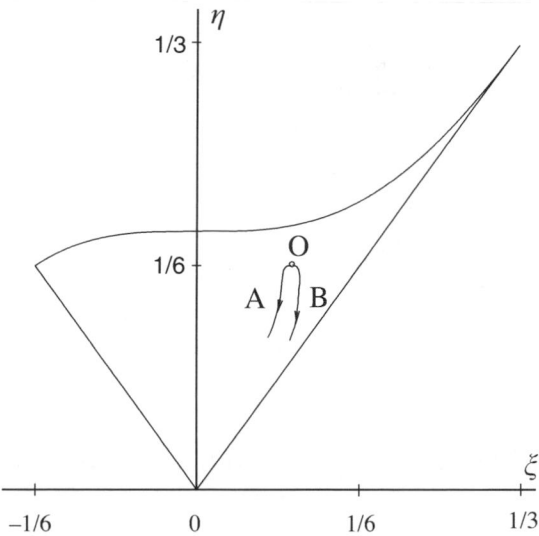

Fig. 11.4. A sketch of trajectories (A and B) on the ξ–η plane for two experiments (or DNS) in which the initial spectra are different, but the initial values of **b** are the same. A Reynolds-stress model yields a unique trajectory from an initial point O.

Speziale (1990), the specifications are simply

$$f^{(1)} = -3.4, \quad f^{(2)} = 4.2, \tag{11.57}$$

whereas in the model of Chung and Kim (1995) the specifications are complicated functions of ξ, η, and Re_L. The trajectories generated by the former model are shown in Fig. 11.2 on page 398.

We end this section with a note of caution. The seemingly flawless logic in the rational procedure of developing the general model can lead to an overestimation of the model's ability to describe the phenomenon at hand. Consider two direct numerical simulations of decaying homogeneous anisotropic turbulence. The initial conditions are chosen so that $\langle u_i u_j \rangle$, ε, and v are the same in the two simulations, but such that the spectra are radically different. Inevitably, the subsequent evolution of the turbulence is different in the two simulations, leading to different trajectories on the ξ–η plane, as sketched in Fig. 11.4. However, regardless of how $f^{(1)}$ and $f^{(2)}$ are specified, the general model (Eq. (11.50)) inevitably yields a unique trajectory from the initial condition, and hence is incapable of describing the different behaviors of the two simulations. The reason, of course, is that the state of the turbulence is only *partially* characterized by the variables represented in the model (i.e, $\langle u_i u_j \rangle$, ε, and v).

11.9   Show that the three principal invariants of $\mathbf{b}$ (see Eqs. (B.31)–(B.33)) are

$$I_b = 0, \tag{11.58}$$

$$II_b = -\tfrac{1}{2}b_{kk}^2 = -3\eta^2, \tag{11.59}$$

$$III_b = \tfrac{1}{3}b_{kk}^3 = 2\xi^3. \tag{11.60}$$

Use the Cayley–Hamilton theorem (Eq. (B.39)) to show that

$$b_{kk}^4 = \tfrac{1}{2}(b_{kk}^2)^2 = 18\eta^4. \tag{11.61}$$

11.10   Show that Eq. (11.51) implies the following evolution equations for η and ξ:

$$\frac{k}{\varepsilon}\eta\,\frac{d\eta}{dt} = (1 + \tfrac{1}{2}f^{(1)})\eta^2 + \tfrac{1}{2}f^{(2)}\xi^3, \tag{11.62}$$

$$\frac{k}{\varepsilon}\xi^2\,\frac{d\xi}{dt} = (1 + \tfrac{1}{2}f^{(1)})\xi^3 + \tfrac{1}{2}f^{(2)}\eta^4. \tag{11.63}$$

Hence show that F (Eq. (11.35)) evolves by

$$\frac{k}{\varepsilon}\frac{dF}{dt} = 54[(1 + \tfrac{1}{2}f^{(1)})(3\xi^3 - \eta^2) + \tfrac{1}{2}f^{(2)}(3\eta^4 - \xi^3)]. \tag{11.64}$$

11.4 Rapid-distortion theory

Homogeneous turbulence can be subjected to time-dependent uniform mean velocity gradients, the magnitude of which can be characterized by

$$\mathcal{S}(t) \equiv (2\bar{S}_{ij}\bar{S}_{ij})^{1/2} \tag{11.65}$$

(see Exercise 5.41 on page 158).[5] As observed above, in turbulent-shear flows, the turbulence-to-mean-shear time scale ratio $\tau\mathcal{S} = \mathcal{S}k/\varepsilon$ is typically in the range 3–6. In contrast, in this section we consider the *rapid-distortion limit* in which $\mathcal{S}k/\varepsilon$ is arbitrarily large. In this limiting case, the evolution of the turbulence is described *exactly* by rapid-distortion theory (RDT). RDT provides several useful insights, especially with regard to the rapid pressure–rate-of-strain tensor, models for which are considered in the next section.

---

[5] Obviously a different characterization, e.g., $(\bar{\Omega}_{ij}\bar{\Omega}_{ij})^{1/2}$, is needed for solid-body rotation in which $\bar{S}_{ij}$ is zero.

11.4.1 Rapid-distortion equations

In homogeneous turbulence, the fluctuating velocity evolves by (Eq. (5.138))

$$\frac{\bar{D}u_j}{\bar{D}t} = -u_i \frac{\partial \langle U_j \rangle}{\partial x_i} - u_i \frac{\partial u_j}{\partial x_i} + \nu \nabla^2 u_j - \frac{1}{\rho} \frac{\partial p'}{\partial x_j}, \tag{11.66}$$

and the Poisson equation for $p' = p^{(r)} + p^{(s)}$ is (Eq. (11.9))

$$\frac{1}{\rho} \nabla^2 (p^{(r)} + p^{(s)}) = -2 \frac{\partial \langle U_i \rangle}{\partial x_j} \frac{\partial u_j}{\partial x_i} - \frac{\partial^2 u_i u_j}{\partial x_i \partial x_j}. \tag{11.67}$$

On the right-hand sides of both of these equations, the first terms represent interactions between the turbulence field u and the mean velocity gradients; whereas the second terms represent turbulence–turbulence interactions. Given the turbulence field $u(x, t)$ at time t, the turbulence–turbulence terms are determined, and are independent of $\partial \langle U_i \rangle / \partial x_j$. On the other hand, the mean-velocity-gradient terms scale linearly with $\mathcal{S}$. Clearly, therefore, in the rapid-distortion limit (i.e., $\mathcal{S} \to \infty$), the terms that scale with $\mathcal{S}$ dominate, all others being negligible in comparison. Hence, in this limit, Eqs. (11.66) and (11.67) reduce to the *rapid-distortion equations*

$$\frac{\bar{D}u_j}{\bar{D}t} = -u_i \frac{\partial \langle U_j \rangle}{\partial x_i} - \frac{1}{\rho} \frac{\partial p^{(r)}}{\partial x_j}, \tag{11.68}$$

$$\frac{1}{\rho} \nabla^2 p^{(r)} = -2 \frac{\partial \langle U_i \rangle}{\partial x_j} \frac{\partial u_j}{\partial x_i}. \tag{11.69}$$

The deformation caused by the mean velocity gradients can be considered in terms of the *rate* $\mathcal{S}(t)$, the *amount* (from time 0 to t)

$$s(t) \equiv \int_0^t \mathcal{S}(t') \, dt', \tag{11.70}$$

and the *geometry* of the deformation

$$\mathcal{G}_{ij}(t) \equiv \frac{1}{\mathcal{S}(t)} \frac{\partial \langle U_i \rangle}{\partial x_j}. \tag{11.71}$$

Note that both s and $\mathcal{G}_{ij}$ are non-dimensional quantities. An interesting feature of rapid-distortion theory is that the turbulence field depends on the geometry and the amount of distortion, but it is independent of the rate $\mathcal{S}(t)$ – showing that the turbulent-viscosity hypothesis is qualitatively incorrect for rapid distortions (Crow 1968). To show this property of the rapid-distortion equations, we use s in place of t as an independent variable, and define

$$\tilde{u}(x, s) \equiv u(x, t), \quad \tilde{\mathcal{G}}_{ij}(s) \equiv \mathcal{G}_{ij}(t), \quad \tilde{p}(x, s) \equiv \frac{p^{(r)}(x, t)}{\rho \mathcal{S}(t)}. \tag{11.72}$$

Then (when divided by S) the rapid distortion equations (Eqs. (11.68) and (11.69)) become

$$\frac{\bar{D}\tilde{u}_j}{\bar{D}s} = -\tilde{u}_i\tilde{\mathcal{G}}_{ji} - \frac{\partial\tilde{p}}{\partial x_j}, \tag{11.73}$$

$$\nabla^2\tilde{p} = -2\tilde{\mathcal{G}}_{ij}\frac{\partial\tilde{u}_j}{\partial x_i}, \tag{11.74}$$

Given the initial turbulence field $u(x,0)$ and the distortion geometry $\tilde{\mathcal{G}}_{ij}(s)$, these equations can be integrated forward in s to determine the subsequent turbulence field as a function of the amount of distortion s (independent of $S(t)$ and t). (Having made this observation, we revert to the more familiar variables of Eqs. (11.68) and (11.69).)

To make progress analytically with the rapid distortion equations, it is necessary to circumvent or to solve the Poisson equation for $p^{(r)}$. In the first works on RDT, Prandtl (1933) and Taylor (1935b) considered the turbulent-vorticity equation (the curl of Eq. (11.68)), thus eliminating $p^{(r)}$. For irrotational mean distortions ($\bar{\Omega}_{ij} = 0$), this vorticity equation is

$$\frac{\bar{D}\omega_j}{\bar{D}t} = \omega_i\frac{\partial\langle U_j\rangle}{\partial x_i} = \omega_i\bar{S}_{ij}. \tag{11.75}$$

Thus, vortex lines (of the *fluctuating* vorticity field) move with, and are stretched by, the *mean* velocity field; and (as in inviscid flow) the vorticity $|\omega|$ increases in proportion to the amount of stretching.

In an axisymmetric contraction with $\bar{S}_{11} > 0$ and $\bar{S}_{22} = \bar{S}_{33} = -\frac{1}{2}\bar{S}_{11}$ (see Figs. 10.1 and 10.2 on pages 360 and 361), the vortex lines are tilted toward the x_1 axis, and are stretched in the x_1 direction, leading to an intensification of $|\omega_1|$. As a consequence $\langle u_2^2\rangle$ and $\langle u_3^2\rangle$ increase relative to $\langle u_1^2\rangle$, as is observed in Fig. 10.2.

11.4.2 The evolution of a Fourier mode

In place of the vorticity equation, an alternative approach to RDT, introduced by Taylor and Batchelor (1949) (and developed further by Batchelor and Proudman (1954) and Craya (1958)), is to consider Fourier modes. The initial turbulent velocity field $u(x,0)$ can be represented as the sum of Fourier modes (see Section 6.4). The rapid distortion equations (Eqs. (11.68) and (11.69)) are linear in u and $p^{(r)}$, and consequently each Fourier mode evolves independently. We start, therefore, by considering a single mode, which initially is specified to be

$$u(x,0) = \hat{u}(0)e^{i\kappa^\circ\cdot x}, \tag{11.76}$$

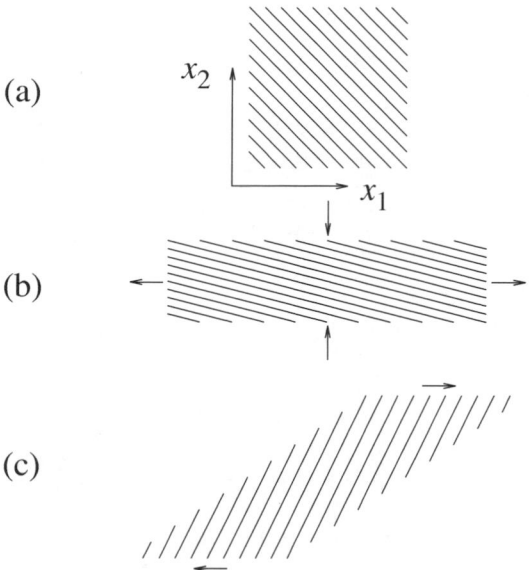

Fig. 11.5. Crests of the fields $\phi(x,t)$ evolving by $\bar{D}\phi/\bar{D}t = 0$: (a) the initial condition, $\hat{\phi}e^{i\kappa^\circ \cdot x}$, $\kappa_1^\circ = \kappa_2^\circ > 0$, $\kappa_3^\circ = 0$; (b) after plane straining ($\bar{S}_{11} = -\bar{S}_{22} > 0$); (c) after shearing $\partial \langle U_1 \rangle / \partial x_2 > 0$.

with $\kappa^\circ \cdot \hat{u}(0) = 0$ in order to satisfy the continuity equation, Eq. (6.128). (It should be understood that there is also a conjugate mode $\hat{u}^*(0)e^{-i\kappa^\circ \cdot x}$, so that the sum of these two modes results in a real velocity field.)

Before treating the rapid-distortion equations, in order to develop the necessary concepts, we first consider a much simpler equation. Specifically, let $\phi(x,t)$ be a scalar field fixed to the mean flow, i.e.,

$$\frac{\bar{D}\phi}{\bar{D}t} = 0, \tag{11.77}$$

and which is initially the Fourier mode

$$\phi(x,0) = \hat{\phi}e^{i\kappa^\circ \cdot x}, \tag{11.78}$$

for some specified $\hat{\phi}$ and κ°. Since ϕ does not change following the mean flow, the maximum value of ϕ remains equal to $\hat{\phi}$ for all time.

Figure 11.5(a) shows the crests of $\phi(x,0)$ (where $\phi = \hat{\phi}$) for $\kappa_1^\circ = \kappa_2^\circ > 0$, $\kappa_3^\circ = 0$. These are parallel planes (that appear as lines in Fig. 11.5(a)). Figure 11.5 also shows the crests of the field $\phi(x,t)$ after (b) plain straining and (c) shearing. The crests remain parallel, but their spacing and orientation are changed by the deformations.

On the basis of these observations it may be hypothesized that the field $\phi(x,t)$ evolves as a Fourier mode, but with a time-dependent wavenumber, i.e.,

$$\phi(x,t) = \hat{\phi}e^{i\hat{\kappa}(t)\cdot x}, \tag{11.79}$$

for some $\hat{\kappa}(t)$. It is readily shown (Exercise 11.11) that this is indeed a solution to $\bar{D}\phi/\bar{D}t = 0$, provided that the wavenumber evolves according to

$$\frac{d\hat{\kappa}_\ell}{dt} = -\hat{\kappa}_j \frac{\partial\langle U_j\rangle}{\partial x_\ell}, \tag{11.80}$$

with the initial condition $\hat{\kappa}(0) = \kappa°$.

The same type of solution is obtained for the rapid-distortion equations, namely

$$u(x,t) = \hat{u}(t)e^{i\hat{\kappa}(t)\cdot x}, \tag{11.81}$$

$$p^{(\mathrm{r})}(x,t) = \hat{p}(t)e^{i\hat{\kappa}(t)\cdot x}, \tag{11.82}$$

with the Fourier coefficient $\hat{u}(t)$ evolving by

$$\frac{d\hat{u}_j}{dt} = -\hat{u}_k \frac{\partial\langle U_\ell\rangle}{\partial x_k}\left(\delta_{j\ell} - 2\frac{\hat{\kappa}_j\hat{\kappa}_\ell}{\hat{\kappa}^2}\right), \tag{11.83}$$

(see Exercise 11.13). Thus, in the rapid-distortion limit, each Fourier mode of the turbulent velocity field (Eq. (11.81)) evolves independently according to the pair of ordinary differential equations for $\hat{\kappa}(t)$ and $\hat{u}(t)$, Eqs. (11.80) and (11.83).

The wavenumber vector $\hat{\kappa}(t)$ can be decomposed into its magnitude $\hat{\kappa}(t) \equiv |\hat{\kappa}(t)|$ and its direction

$$\hat{e}(t) \equiv \frac{\hat{\kappa}(t)}{\hat{\kappa}(t)}, \tag{11.84}$$

which is referred to as the *unit wavevector*. It is immediately apparent from Eq. (11.83) that the evolution of the Fourier coefficient $\hat{u}(t)$ is independent of the wavenumber's magnitude: the equation can be rewritten in terms of the unit wavevector as

$$\frac{d\hat{u}_j}{dt} = -\hat{u}_k \frac{\partial\langle U_\ell\rangle}{\partial x_k}(\delta_{j\ell} - 2\hat{e}_j\hat{e}_\ell). \tag{11.85}$$

Also, it follows from Eq. (11.80) that the unit wavenumber evolves by

$$\frac{d\hat{e}_\ell}{dt} = -\frac{\partial\langle U_\mathrm{m}\rangle}{\partial x_i}\hat{e}_\mathrm{m}(\delta_{i\ell} - \hat{e}_i\hat{e}_\ell). \tag{11.86}$$

This pair of equations fully determines the evolution of $\hat{u}(t)$, which, evidently, does not depend on the wavenumber's magnitude $\hat{\kappa}(t)$.

A unit vector can be represented as a point on the unit sphere, and correspondingly the evolution of $\hat{e}(t)$ given by Eq. (11.86) can be represented

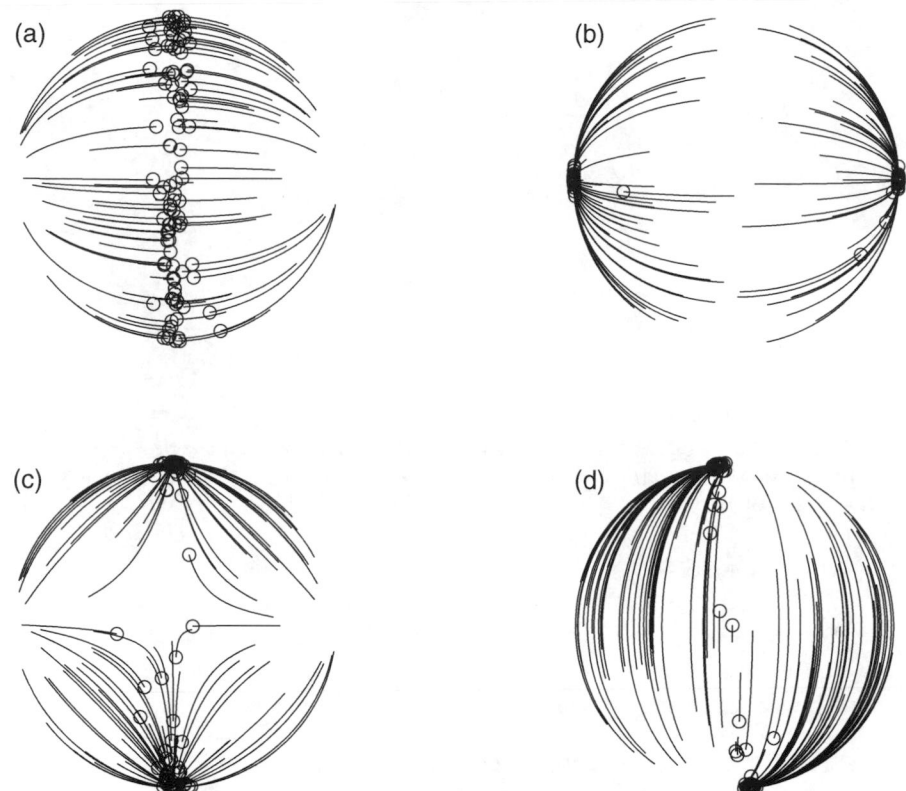

Fig. 11.6. Trajectories of the unit wavevector $\hat{e}(t)$ on the unit sphere from random initial conditions for (a) axisymmetric contraction, (b) axisymmetric expansion, (c) plane strain, and (d) shear. The $\hat{e}_1$ direction is horizontal, the $\hat{e}_2$ direction is vertical, and the $\hat{e}_3$ direction is into the page. The symbols mark the ends of the trajectories after distortion.

as a trajectory on the unit sphere. These trajectories for the various mean velocity gradients defined in Table 11.2 (on page 415) (from different initial conditions) are shown in Fig. 11.6.

The Fourier coefficient $\hat{u}(t)$ is orthogonal to $\hat{\kappa}(t)$, and hence to $\hat{e}(t)$ (because of continuity). Consequently, as depicted in Fig. 11.7, $\hat{u}(t)$ can be represented as a vector in the tangent plane of the unit sphere at $\hat{e}(t)$. As the point $\hat{e}(t)$ moves on the unit sphere, the coefficient $\hat{u}(t)$ evolves according to Eq. (11.85), and remains in the local tangent plane.

EXERCISES _____

11.11   Consider homogeneous turbulence with uniform mean velocity gradients. The mean velocity is zero at the origin, so that the mean

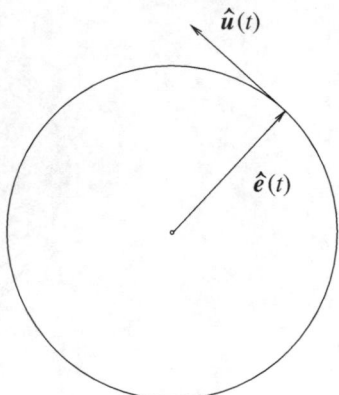

Fig. 11.7. A sketch of the unit sphere showing the unit wavevector $\hat{e}(t)$. The Fourier component of velocity $\hat{u}(t)$ is orthogonal to $\hat{e}(t)$, and so it is in the tangent plane of the unit sphere at $\hat{e}(t)$.

velocity field is

$$\langle U_j \rangle = \frac{\partial \langle U_j \rangle}{\partial x_\ell} x_\ell. \tag{11.87}$$

Show that $\phi(\boldsymbol{x}, t)$ given by Eq. (11.79) evolves by

$$\frac{\bar{D}\phi}{\bar{D}t} \equiv \frac{\partial \phi}{\partial t} + \langle U_j \rangle \frac{\partial \phi}{\partial x_j}$$
$$= i\phi x_\ell \left(\frac{d\hat{\kappa}_\ell}{dt} + \hat{\kappa}_j \frac{\partial \langle U_j \rangle}{\partial x_\ell} \right). \tag{11.88}$$

Hence show that $\bar{D}\phi/\bar{D}t$ is zero if $\hat{\boldsymbol{\kappa}}(t)$ evolves by Eq. (11.80).

11.12   Show that, if ϕ evolves by $\bar{D}\phi/\bar{D}t = 0$, then its gradient $\boldsymbol{g}$ evolves by

$$\frac{\bar{D}g_\ell}{\bar{D}t} = -g_j \frac{\partial \langle U_j \rangle}{\partial x_\ell}. \tag{11.89}$$

Comment on the connection between this result and Eq. (11.80).

11.13   With $\boldsymbol{u}$, $p^{(\mathrm{r})}$ and $\hat{\boldsymbol{\kappa}}(t)$ given by Eqs. (11.81), (11.82), and (11.80), show that the Poisson equation for $p^{(\mathrm{r})}$ (Eq. (11.69)) can be written in terms of the Fourier coefficients as

$$\hat{\kappa}^2 \hat{p} = i2\rho \hat{\kappa}_\ell \hat{u}_k \frac{\partial \langle U_\ell \rangle}{\partial x_k}. \tag{11.90}$$

Hence show that the rapid pressure gradient is

$$-\frac{1}{\rho} \frac{\partial p^{(\mathrm{r})}}{\partial x_j} = 2 \frac{\hat{\kappa}_j \hat{\kappa}_\ell}{\hat{\kappa}^2} u_k \frac{\partial \langle U_\ell \rangle}{\partial x_k}. \tag{11.91}$$

Show that

$$\frac{\bar{D}u_j}{\bar{D}t} = e^{i\hat{\kappa}(t)\cdot x}\frac{d\hat{u}_j}{dt}, \tag{11.92}$$

and hence that the rapid-distortion equation is satisfied if $\hat{u}(t)$ evolves by

$$\frac{d\hat{u}_j}{dt} = -\hat{u}_k\frac{\partial\langle U_j\rangle}{\partial x_k} + 2\frac{\hat{\kappa}_j\hat{\kappa}_\ell}{\hat{\kappa}^2}\hat{u}_k\frac{\partial\langle U_\ell\rangle}{\partial x_k}. \tag{11.93}$$

11.14   With $u(x,t)$ being the Fourier mode Eq. (11.81), show that the continuity equation requires that $\hat{\kappa}(t)$ and $\hat{u}(t)$ be orthogonal (i.e., $\hat{\kappa}(t)\cdot\hat{u}(t) = 0$). Given that this condition is satisfied initially (i.e., $\hat{\kappa}(0)\cdot\hat{u}(0) = 0$), show that the evolution equations for $\hat{\kappa}(t)$ and $\hat{u}(t)$ (Eqs. (11.80) and (11.83)) maintain this orthogonality.

11.4.3 The evolution of the spectrum

Rather than a single Fourier mode, we now consider a random fluctuating velocity field composed of many modes. The velocity-spectrum tensor $\Phi_{ij}(\kappa, t)$ provides a statistical description of the field; and, remarkably, the evolution of the spectrum is determined exactly by RDT.

The initial condition ($t = 0$) of the fluctuating velocity field can be expressed as the sum of Fourier modes

$$u(x,0) = \sum_{\kappa^\circ} e^{i\kappa^\circ\cdot x}\,\hat{u}(\kappa^\circ, 0), \tag{11.94}$$

where κ° denotes one of a set of wavenumbers (that can be random), and $\hat{u}(\kappa^\circ, 0)$ is the corresponding (random) Fourier coefficient. The requirements on these initial conditions are that κ° and $\hat{u}$ be orthogonal (to satisfy continuity), and that the modes be in complex conjugate pairs (so that $u(x,0)$ is real).

Each mode evolves independently according to the rapid-distortion equations, as described in the previous subsection. Consequently, the velocity field at subsequent times is simply

$$u(x,t) = \sum_{\kappa^\circ} e^{i\hat{\kappa}(t)\cdot x}\,\hat{u}(\kappa^\circ, t). \tag{11.95}$$

Recall that the wavenumber $\hat{\kappa}(t)$ evolves according to Eq. (11.80) from the initial condition $\hat{\kappa}(0) = \kappa^\circ$, and the Fourier coefficient $\hat{u}(\kappa^\circ, t)$ evolves by Eq. (11.83).

The velocity-spectrum tensor $\Phi_{ij}(\kappa, t)$ provides a statistical description of the turbulent velocity field (see Section 6.5). In terms of the Fourier representation of the velocity field (Eq. (11.95)) it is given explicitly by

$$\Phi_{ij}(\kappa, t) = \left\langle \sum_{\kappa^\circ} \delta(\kappa - \hat{\kappa}(t)) \hat{u}_i^*(\kappa^\circ, t) \hat{u}_j(\kappa^\circ, t) \right\rangle, \tag{11.96}$$

which is Eq. (6.155) adapted to the present context. In computations it is convenient also to consider the spectrum parametrized by the initial wavenumber:

$$\Phi_{ij}^\circ(\kappa, t) = \left\langle \sum_{\kappa^\circ} \delta(\kappa - \kappa^\circ) \hat{u}_i^*(\kappa^\circ, t) \hat{u}_j(\kappa^\circ, t) \right\rangle. \tag{11.97}$$

One of the important properties of the velocity-spectrum tensor is that it represents the Reynolds-stress density in wavenumber space: on integrating Eq. (11.96) or (11.97) over all κ we obtain

$$\langle u_i u_j \rangle = \int\!\!\!\int\!\!\!\int_{-\infty}^{\infty} \Phi_{ij}(\kappa, t)\, d\kappa = \int\!\!\!\int\!\!\!\int_{-\infty}^{\infty} \Phi_{ij}^\circ(\kappa, t)\, d\kappa$$

$$= \left\langle \sum_{\kappa^\circ} \hat{u}_i^*(\kappa^\circ, t) \hat{u}_j(\kappa^\circ, t) \right\rangle. \tag{11.98}$$

The equation for the evolution of the velocity-spectrum tensor can be derived (see Exercise 11.15) from its definition (Eq. (11.96)) and from the evolution equations for $\hat{\kappa}$ and $\hat{u}$. The result is

$$\frac{\partial \Phi_{ij}}{\partial t} = \frac{\partial \langle U_m \rangle}{\partial x_\ell} \kappa_m \frac{\partial \Phi_{ij}}{\partial \kappa_\ell} - \frac{\partial \langle U_i \rangle}{\partial x_k} \Phi_{kj} - \frac{\partial \langle U_j \rangle}{\partial x_k} \Phi_{ik}$$

$$+ 2 \frac{\partial \langle U_\ell \rangle}{\partial x_k} \left(\frac{\kappa_i \kappa_\ell}{\kappa^2} \Phi_{kj} + \frac{\kappa_j \kappa_\ell}{\kappa^2} \Phi_{ik} \right). \tag{11.99}$$

This is a closed statistical equation that describes exactly the evolution of the turbulence spectrum in the rapid-distortion limit. Apart from the specified mean velocity gradients and the independent variables (κ and t), the only quantity involved in the equation is the spectrum itself.

The first term on the right-hand side of Eq. (11.99) represents transport of the spectrum in wavenumber space, which is a result of the distortion of the wavenumber vector $\hat{\kappa}(t)$ by the mean velocity gradients. The evolution equation for $\Phi_{ij}^\circ(\kappa, t)$ (Eq. (11.97)) is the same as that for $\Phi_{ij}(\kappa, t)$ (Eq. (11.99)), except for the omission of this first term.

The next two terms in Eq. (11.99) represent production, and the remaining term stems from the rapid-pressure. Note that the rapid pressure term depends on the wavenumber's direction $\kappa/|\kappa|$, but not on its magnitude, $|\kappa|$.

When the equation for Φ_{ij} is integrated over wavenumber space, the transport term vanishes, leaving the Reynolds-stress equation in the form

$$\frac{d}{dt}\langle u_i u_j\rangle = -\langle u_j u_k\rangle \frac{\partial\langle U_i\rangle}{\partial x_k} - \langle u_i u_k\rangle \frac{\partial\langle U_j\rangle}{\partial x_k}$$
$$+ 2\frac{\partial\langle U_\ell\rangle}{\partial x_k}(M_{kji\ell} + M_{ikj\ell}) = \mathcal{P}_{ij} + \mathcal{R}_{ij}^{(r)}, \qquad (11.100)$$

where the fourth-order tensor M is

$$M_{ijk\ell} \equiv \int\!\!\!\int\!\!\!\int_{-\infty}^{\infty} \Phi_{ij}\frac{\kappa_k\kappa_\ell}{\kappa^2}\,d\kappa = \left\langle \sum_{\kappa^\circ} \hat{u}_i^*\hat{u}_j \frac{\hat{\kappa}_k\hat{\kappa}_\ell}{\hat{\kappa}^2}\right\rangle$$

$$= \left\langle \sum_{\kappa^\circ} \hat{u}_i^*\hat{u}_j\hat{e}_k\hat{e}_\ell\right\rangle. \qquad (11.101)$$

(It can be verified (see Exercise 11.16) that this definition of $M_{ijk\ell}$ is equivalent to that introduced above, Eq. (11.15).) The Reynolds-stress equation is *not* in closed form, because the tensor M involves information about the direction of the Fourier modes that is not contained in the Reynolds-stresses themselves.

EXERCISES

11.15 Differentiate Eq. (11.96) to obtain the result

$$\frac{\partial\Phi_{ij}}{\partial t} = \left\langle -\frac{\partial}{\partial\kappa_\ell}\left(\sum_{\kappa^\circ}\frac{d\hat{\kappa}_\ell}{dt}\delta(\kappa-\hat{\kappa})\hat{u}_i^*\hat{u}_j\right)\right.$$
$$\left. + \sum_{\kappa^\circ}\delta(\kappa-\hat{\kappa})\left(\hat{u}_j\frac{d\hat{u}_i^*}{dt} + \hat{u}_i^*\frac{d\hat{u}_j}{dt}\right)\right\rangle. \qquad (11.102)$$

Substitute Eq. (11.80) for $d\hat{\kappa}/dt$ in order to re-express the first term on the right-hand side as

$$\frac{\partial\langle U_m\rangle}{\partial x_\ell}\frac{\partial}{\partial\kappa_\ell}(\kappa_m\Phi_{ij}). \qquad (11.103)$$

(Hint: use the fact that $\hat{\kappa}$ is independent of κ, and the sifting property of the delta function.)

Substitute Eq. (11.83) for $d\hat{u}_j/dt$ in order to obtain the result

$$\left\langle\sum_{\kappa^\circ}\delta(\kappa-\hat{\kappa})\hat{u}_i^*\frac{d\hat{u}_j}{dt}\right\rangle = -\frac{\partial\langle U_\ell\rangle}{\partial x_k}\Phi_{ik}\left(\delta_{j\ell} - 2\frac{\kappa_j\kappa_\ell}{\kappa^2}\right). \qquad (11.104)$$

Use these results to verify Eq. (11.99).

11.16 Consider the function $m_{ijk\ell}(\mathbf{r})$ which is defined as the solution to the Poisson equation

$$\frac{\partial^2 m_{ijk\ell}(\mathbf{r})}{\partial r_q \partial r_q} = \frac{\partial^2 R_{ij}(\mathbf{r})}{\partial r_k \partial r_\ell}. \tag{11.105}$$

Using Eq. (2.48), write down the Green's function solution to this equation, and hence show that the solution at the origin is

$$m_{ijk\ell}(0) = -\frac{1}{4\pi} \int\limits_{-\infty}^{\infty}\!\!\!\int\!\!\!\int \frac{1}{|\mathbf{r}|} \frac{\partial^2 R_{ij}}{\partial r_k \partial r_\ell}\, d\mathbf{r} = M_{ijk\ell}. \tag{11.106}$$

Show that the Fourier transform of Eq. (11.105) is

$$-\kappa^2 \hat{m}_{ijk\ell}(\boldsymbol{\kappa}) = -\kappa_k \kappa_\ell \Phi_{ij}(\boldsymbol{\kappa}), \tag{11.107}$$

where $\hat{m}_{ijk\ell}(\boldsymbol{\kappa})$ denotes the Fourier transform of $m_{ijk\ell}(\mathbf{r})$. Hence obtain

$$m_{ijk\ell}(0) = \int\limits_{-\infty}^{\infty}\!\!\!\int\!\!\!\int \hat{m}_{ijk\ell}(\boldsymbol{\kappa})d\boldsymbol{\kappa} = \int\limits_{-\infty}^{\infty}\!\!\!\int\!\!\!\int \Phi_{ij}\frac{\kappa_k \kappa_\ell}{\kappa^2} d\boldsymbol{\kappa} = M_{ijk\ell}, \tag{11.108}$$

showing that the two definitions of $M_{ijk\ell}$ (Eqs. 11.15 and 11.101) are consistent.

11.17 From the expressions for $M_{ijk\ell}$, show that it satisfies the following conditions:

$$M_{ijk\ell} = M_{jik\ell}, \qquad M_{ijk\ell} = M_{ij\ell k},$$
$$M_{ijj\ell} = 0, \qquad M_{ijkk} = \langle u_i u_j \rangle. \tag{11.109}$$

11.18 For isotropic turbulence, the most general possible (isotropic) expression for $M_{ijk\ell}$ is

$$M_{ijk\ell} = k(\alpha \delta_{ij}\delta_{k\ell} + \beta \delta_{ik}\delta_{j\ell} + \gamma \delta_{i\ell}\delta_{jk}), \tag{11.110}$$

where α, β, and γ are constants. Show that Eq. (11.109) determines these constants to be

$$\alpha = \tfrac{4}{15}, \qquad \beta = \gamma = -\tfrac{1}{15}. \tag{11.111}$$

11.19 For isotropic turbulence, show that the Reynolds-stress production is

$$\mathcal{P}_{ij} = -\tfrac{4}{3}k\bar{S}_{ij}. \tag{11.112}$$

Use the results of Exercise 11.18 to show that the rapid pressure–rate-of-strain tensor (for isotropic turbulence) is

$$\mathcal{R}_{ij}^{(r)} = \tfrac{4}{5}k\bar{S}_{ij} = -\tfrac{3}{5}\mathcal{P}_{ij}. \tag{11.113}$$

Table 11.2. *Non-zero mean velocity gradients for simple deformations: S_λ is the largest eigenvalue of $\bar{S}_{ij}$*

	Axisymmetric contraction	Axisymmetric expansion	Plane strain	Shear
$\partial\langle U_1\rangle/\partial x_1$	S_λ	$-2S_\lambda$	S_λ	0
$\partial\langle U_2\rangle/\partial x_2$	$-\frac{1}{2}S_\lambda$	S_λ	$-S_\lambda$	0
$\partial\langle U_3\rangle/\partial x_3$	$-\frac{1}{2}S_\lambda$	S_λ	0	0
$\partial\langle U_1\rangle/\partial x_2$	0	0	0	S
$S \equiv (2\bar{S}_{ij}\bar{S}_{ij})^{1/2}$	$\sqrt{3}S_\lambda$	$2\sqrt{3}S_\lambda$	$2S_\lambda$	S

11.4.4 Rapid distortion of initially isotropic turbulence

We now describe the evolution of initially isotropic turbulence subjected to rapid distortion by the simple mean velocity gradients given in Table 11.2. Analytic solutions for the spectrum for some of these distortions are given in Townsend (1976), Reynolds (1987), Hunt and Carruthers (1990), and references therein. The results presented here are obtained from the numerical integration of the equation for $\Phi^\circ_{ij}(\boldsymbol{\kappa}, t)$ (which is similar to Eq. (11.99)).

The initial response

At the initial instant when the distortion is applied, the turbulence is isotropic and its relevant statistics can be determined explicitly (Crow 1968). The fourth-order tensor $M_{ijk\ell}$ (Eq. (11.101)) is

$$M_{ijk\ell} = \tfrac{1}{15}k(4\delta_{ij}\delta_{k\ell} - \delta_{ik}\delta_{j\ell} - \delta_{i\ell}\delta_{jk}), \tag{11.114}$$

and the rapid pressure–rate-of-strain tensor is

$$\mathcal{R}^{(r)}_{ij} = \tfrac{4}{5}k\bar{S}_{ij} = -\tfrac{3}{5}\mathcal{P}_{ij}, \tag{11.115}$$

(see Exercises 11.18 and 11.19). Thus, initially, the effect of the rapid pressure on the Reynolds stresses is to counteract 60% of the production. After the initial instant, the turbulence becomes anisotropic and these equations cease to be valid. For later reference, we note that the production of kinetic energy $\mathcal{P} = \tfrac{1}{2}\mathcal{P}_{ii}$ is zero at the initial instant, so that Eq. (11.115) can also be written

$$\mathcal{R}^{(r)}_{ij} = -\tfrac{3}{5}(\mathcal{P}_{ij} - \tfrac{2}{3}\mathcal{P}\delta_{ij}). \tag{11.116}$$

Axisymmetric contraction

For the axisymmetric contraction, the wavenumbers evolve (Eq. (11.80)) by

$$\frac{d\hat{\kappa}_1}{dt} = -\hat{\kappa}_1 S_\lambda, \quad \frac{d\hat{\kappa}_2}{dt} = \tfrac{1}{2}\hat{\kappa}_2 S_\lambda, \quad \frac{d\hat{\kappa}_3}{dt} = \tfrac{1}{2}\hat{\kappa}_3 S_\lambda. \tag{11.117}$$

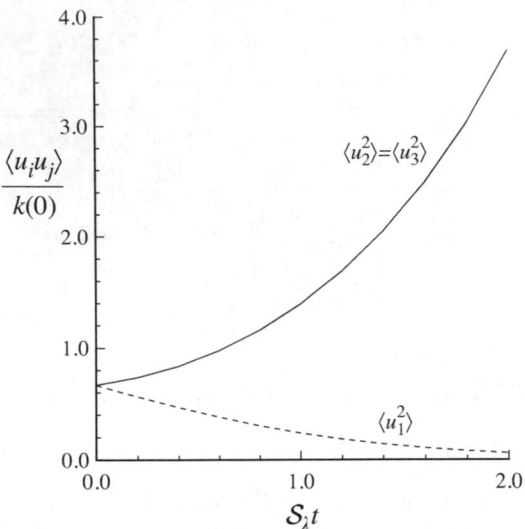

Fig. 11.8. The evolution of the Reynolds stresses for axisymmetric contraction rapid distortion. Dashed line $\langle u_1^2 \rangle / k(0)$, solid line $\langle u_2^2 \rangle / k(0) = \langle u_3^2 \rangle / k(0)$.

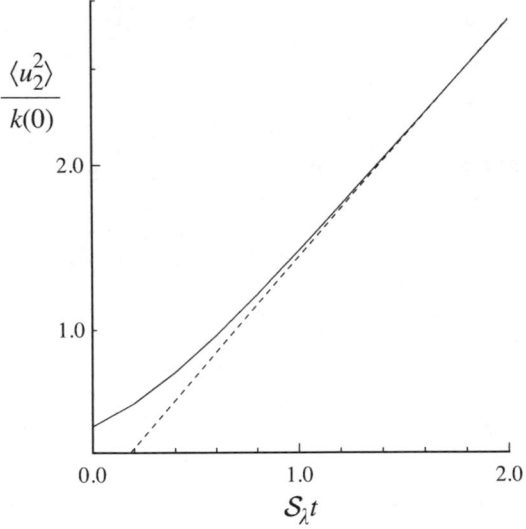

Fig. 11.9. The evolution of $\langle u_2^2 \rangle$ (on a logarithmic scale) for axisymmetric contraction rapid distortion (solid line). The dashed line is $\frac{1}{2} \exp(\mathcal{S}_\lambda t)$, indicating the asymptotic growth rate.

Evidently, $\hat{\kappa}_2$ and $\hat{\kappa}_3$ grow exponentially, while $\hat{\kappa}_1$ decreases. Consequently, as shown in Fig. 11.6(a), the trajectories of the unit wavevectors $\hat{e}(t)$ tend toward the prime meridian corresponding to $\hat{e}_1 = 0$.

The production of $\langle u_1^2 \rangle$ is negative ($\mathcal{P}_{11} = -2 \langle u_1^2 \rangle \mathcal{S}_\lambda$), while that of $\langle u_2^2 \rangle$ is positive ($\mathcal{P}_{22} = \langle u_2^2 \rangle \mathcal{S}_\lambda$): $\langle u_3^2 \rangle$ is statistically identical to $\langle u_2^2 \rangle$. The Reynolds-stress evolution (Fig. 11.8) is therefore as expected. As may be seen from Fig. 11.9, at large times $\langle u_2^2 \rangle$ increases as $\exp(\mathcal{S}_\lambda t)$. This is as a result of production: the rapid pressure–rate-of-stain tensor tends to zero in comparison. At early times, $\langle u_2^2 \rangle$ increases more slowly than exponentially, as the pressure–rate-of-strain tensor redistributes energy to $\langle u_1^2 \rangle$. At large times the wavevectors tend to the prime meridian ($\hat{e}_1 = 0$), and the Fourier components $\hat{\boldsymbol{u}}$ are aligned with the meridian ($\hat{u}_1 = 0$).

Axisymmetric expansion

Whereas rapid distortion in an axisymmetric contraction can be approximated in an experiment, rapid axisymmetric expansion in a diffuser is not possible because of flow separation. Nevertheless, the flow can be studied by DNS and RDT.

In some respects, this flow is the opposite of the axisymmetric contraction. The wavenumber $\hat{\kappa}_1$ grows exponentially, whereas $\hat{\kappa}_2$ and $\hat{\kappa}_3$ decrease, so that the trajectories of $\hat{e}(t)$ on the unit sphere (Fig. 11.6(b)) tend to the poles $\hat{e}_1 = \pm 1$. The production of $\langle u_1^2 \rangle$ is positive ($\mathcal{P}_{11} = 4\langle u_1^2 \rangle \mathcal{S}_\lambda$), whereas those of $\langle u_2^2 \rangle$ and $\langle u_3^2 \rangle$ are negative ($\mathcal{P}_{22} = \mathcal{P}_{33} = -2\langle u_2^2 \rangle \mathcal{S}_\lambda$). However, for this flow, the rapid pressure strongly suppresses the growth of $\langle u_1^2 \rangle$. As the wavevectors tend to $\hat{e}_1 = \pm 1$, the continuity equation (which is enforced by the rapid pressure) requires that $\hat{u}_1$ tend to zero.

Figure 11.10 shows the evolution of the Reynolds stresses. At large times these are in the proportions

$$\langle u_1^2 \rangle = 2\langle u_2^2 \rangle = 2\langle u_3^2 \rangle, \tag{11.118}$$

and each grows as $\exp(\mathcal{S}_\lambda t)$. It can be deduced, therefore, that the terms in the Reynolds-stress balance

$$\frac{\mathrm{d}}{\mathrm{d}t}\langle u_i u_j \rangle = \mathcal{P}_{ij} + \mathcal{R}_{ij}^{(\mathrm{r})}, \tag{11.119}$$

when they are divided by $\mathcal{S}_\lambda k$, are

$$\begin{bmatrix} 1 & 0 & 0 \\ 0 & \frac{1}{2} & 0 \\ 0 & 0 & \frac{1}{2} \end{bmatrix} = \begin{bmatrix} 4 & 0 & 0 \\ 0 & -1 & 0 \\ 0 & 0 & -1 \end{bmatrix} + \begin{bmatrix} -3 & 0 & 0 \\ 0 & \frac{3}{2} & 0 \\ 0 & 0 & \frac{3}{2} \end{bmatrix}. \tag{11.120}$$

The rapid pressure removes energy from $\langle u_1^2 \rangle$ at three quarters of the rate at which it is added by production.

The production not only adds energy, but also tends to increase the

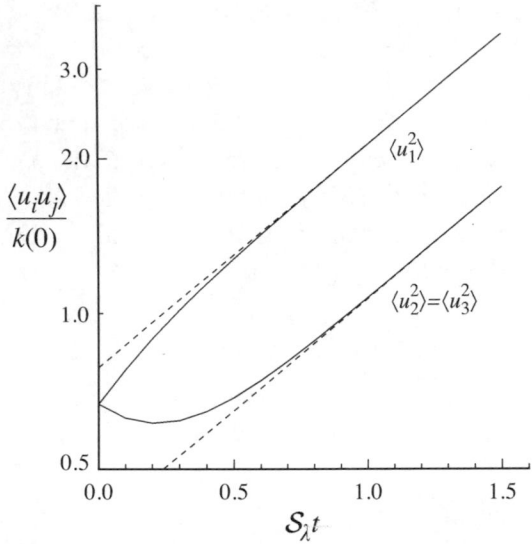

Fig. 11.10. The evolution of the Reynolds stresses for axisymmetric expansion rapid distortion. The dashed lines show the asymptotic growth as $\exp(\mathcal{S}_\lambda t)$.

anisotropy. For this flow (at large times), $\mathcal{P}_{ij}$ and $\mathcal{R}_{ij}^{(r)}$ given by Eq. (11.120) are related by

$$\mathcal{R}_{ij}^{(r)} = -\tfrac{9}{10}(\mathcal{P}_{ij} - \tfrac{2}{3}\mathcal{P}\delta_{ij}), \qquad (11.121)$$

showing that the rapid pressure counteracts 90% of the anisotropy production (cf. Eq. (11.116)).

Plane strain

For plane strain, the wavenumbers evolve by

$$\frac{d\hat{\kappa}_1}{dt} = -\hat{\kappa}_1 \mathcal{S}_\lambda, \quad \frac{d\hat{\kappa}_2}{dt} = \hat{\kappa}_2 \mathcal{S}_\lambda, \quad \frac{d\hat{\kappa}_3}{dt} = 0, \qquad (11.122)$$

leading to the trajectories shown in Fig. 11.6(c). At large times the wavevectors tend to the poles $\hat{e}_2 = \pm 1$.

The rates of Reynolds-stress production are

$$\mathcal{P}_{11} = -2\langle u_1^2 \rangle \mathcal{S}_\lambda, \quad \mathcal{P}_{22} = 2\langle u_2^2 \rangle \mathcal{S}_\lambda, \quad \mathcal{P}_{33} = 0. \qquad (11.123)$$

The decay of $\langle u_1^2 \rangle$ and the growth of $\langle u_2^2 \rangle$ observed in Fig. 11.11 are therefore to be expected. However, as the wavevectors tend to the poles $\hat{e}_2 = \pm 1$ (at which $\hat{u}_2$ is forced to be zero) the rapid pressure redistributes the energy from $\langle u_2^2 \rangle$ into $\langle u_3^2 \rangle$. Evidently, at large times, $\mathcal{R}_{22}$ extracts energy from $\langle u_2^2 \rangle$ at about 50% of the production rate $\mathcal{P}_{22}$, so that the net rate of gain in $\langle u_3^2 \rangle$ (i.e., $\mathcal{R}_{33}$) is slightly greater than that in $\langle u_2^2 \rangle$ (i.e., $\mathcal{P}_{22} + \mathcal{R}_{22}$).

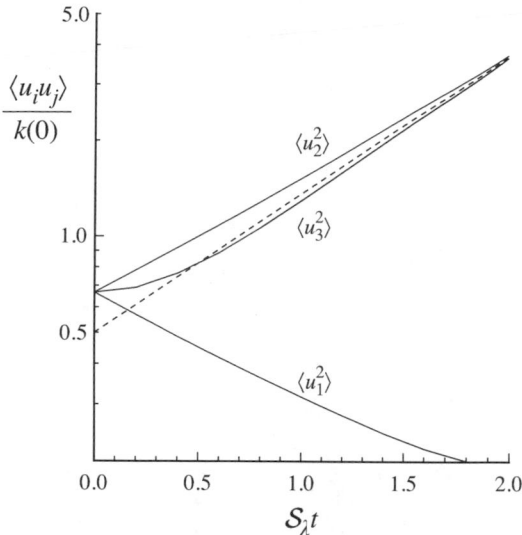

Fig. 11.11. The evolution of the Reynolds stresses for plane strain rapid distortion. The dashed line is $\frac{1}{2}\exp(\mathcal{S}_\lambda t)$.

Shear

In shear, the wavenumbers evolve by

$$\frac{\mathrm{d}\hat{\kappa}_1}{\mathrm{d}t} = 0, \quad \frac{\mathrm{d}\hat{\kappa}_2}{\mathrm{d}t} = -\hat{\kappa}_1\mathcal{S}, \quad \frac{\mathrm{d}\hat{\kappa}_3}{\mathrm{d}t} = 0. \qquad (11.124)$$

Consequently, as shown in Fig. 11.6, wavevectors in the western hemisphere ($\hat{e}_1 < 0$) move upward toward the north pole ($\hat{e}_2 = 1$); whereas the motion in the eastern hemisphere is downward toward the south pole.

The non-zero rates of Reynolds-stress production are

$$\mathcal{P}_{11} = -2\langle u_1 u_2 \rangle \mathcal{S}, \quad \mathcal{P}_{12} = -\langle u_2^2 \rangle \mathcal{S}. \qquad (11.125)$$

The evolution of the Reynolds-stress anisotropies b_{ij} is shown in Fig. 11.12, and the corresponding states are shown on the Lumley triangle in Fig. 11.13. Early on, the state is close to being axisymmetric with a negative third invariant ($\xi < 0$). This is in contrast to the state in non-rapid homogeneous shear, in which the third invariant is positive. At large times, the state moves toward the two-component limit ($\langle u_2^2 \rangle \approx 0$), and eventually to the one-component limit ($\langle u_1^2 \rangle > 0$).

As is the case for axisymmetric expansion and plane strain, at large times the rapid pressure has the effect of suppressing the rate of growth of the largest Reynolds stress. In fact, the asymptotic growth appears to be linear rather than exponential (Fig. 11.14). However, for shear, the suppression is

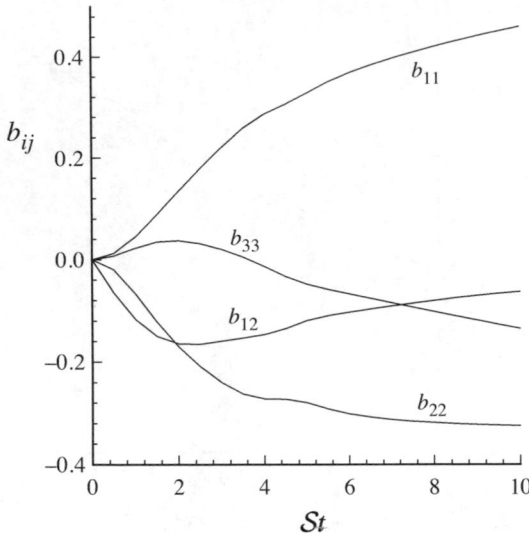

Fig. 11.12. The evolution of the Reynolds-stress anisotropies for shear rapid distortion.

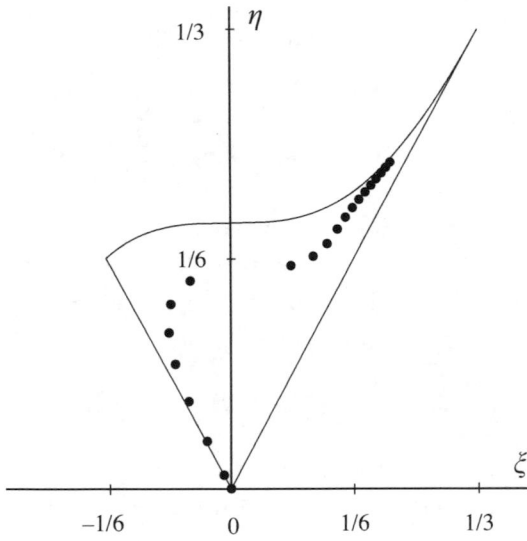

Fig. 11.13. The evolution of the Reynolds-stress invariants for shear rapid distortion. Starting from the origin (corresponding to isotropy), each symbol gives the state after an amount of shear $St = 0.5$.

indirect. At the poles $\hat{e}_2 = \pm 1$, the continuity equation does not preclude $\hat{u}_1$, and indeed the asymptotic one-component state is $\hat{e}_2 = \pm 1, |\hat{u}_1| > 0$. However, at these poles $\hat{u}_2$ is precluded, so the shear stress $\langle u_1 u_2 \rangle$ (and the production $\mathcal{P}_{11}$) are suppressed. At large times $\mathcal{R}_{22}$ is negative, constantly removing energy from the already small normal stress $\langle u_2^2 \rangle$.

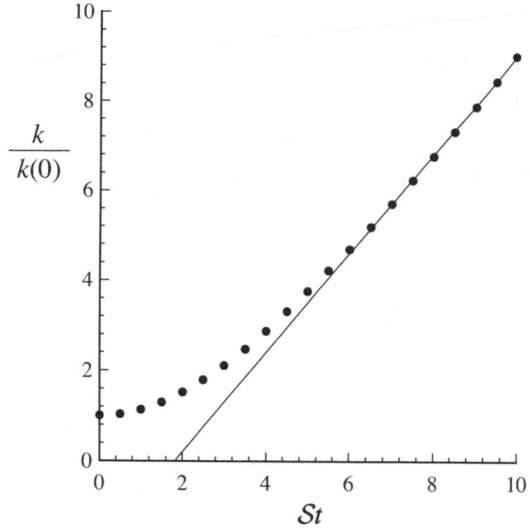

Fig. 11.14. The evolution of the turbulent kinetic energy for shearing with rapid distortion.

11.4.5 Final remarks

The principal conclusions on the application of RDT to homogeneous turbulence are as follows.

(i) RDT applies in the limit as the magnitude of the mean velocity gradients becomes very large compared with turbulence rates. The condition $S\tau = Sk/\varepsilon \gg 1$ is required in order to apply RDT to the energy-containing motions: to apply it to all scales of motion, the more stringent condition $S\tau_\eta \gg 1$ is required. How well RDT represents turbulence when these conditions are not met is discussed by Savill (1987) and Hunt and Carruthers (1990).

(ii) The evolution of the turbulence depends on the geometry and the amount of the deformation, but it is independent of the rate.

(iii) In the rapid-distortion equations (Eqs. (11.68) and (11.69)), the role of the rapid pressure is to enforce the continuity equation $\nabla \cdot \boldsymbol{u} = 0$.

(iv) The rapid-distortion equations are linear, and consequently Fourier modes (Eq. (11.71)) evolve independently: the wavenumber $\hat{\boldsymbol{\kappa}}(t)$ and coefficient $\hat{\boldsymbol{u}}(t)$ evolve by the ordinary differential equations Eqs. (11.80) and (11.83).

(v) The evolution of the Fourier coefficient depends on the unit wavevector $\hat{\boldsymbol{e}} = \hat{\boldsymbol{\kappa}}/|\hat{\boldsymbol{\kappa}}|$, but not on the wavenumber's magnitude $|\hat{\boldsymbol{\kappa}}|$.

(vi) In the rapid-distortion limit, the velocity-spectrum tensor $\Phi_{ij}(\boldsymbol{\kappa}, t)$ evolves by an exact closed equation (Eq. (11.99)).

(vii) The Reynolds-stress equation (Eq. (11.100)) is not in closed form, because the fourth-order tensor $M_{ijk\ell}$ (that determines the rapid pressure–rate-of-strain tensor, $\mathcal{R}_{ij}^{(r)}$) involves information about the direction of the Fourier modes (Eq. (11.101)). Two fields of homogeneous turbulence can have the same Reynolds-stress tensor

$$\langle u_i u_j \rangle = \sum_{\kappa^\circ} \langle \hat{u}_i^* \hat{u}_j \rangle, \tag{11.126}$$

but different values of

$$M_{ijk\ell} = \sum_{\kappa^\circ} \langle \hat{u}_i^* \hat{u}_j \hat{e}_k \hat{e}_\ell \rangle. \tag{11.127}$$

Consequently the Reynolds-stress evolution is not uniquely determined by the Reynolds stresses.

(viii) The solution of the rapid-distortion equations for four simple deformations reveals quite different behaviors depending on the geometry. For the axisymmetric expansion at large times, the rapid pressure counteracts 90% of the anisotropy production (Eq. (11.121)). At the initial instant when isotropic turbulence is subject to any distortion, the corresponding figure is 60%. However, for the axisymmetric contraction (at large times) it is zero: the rapid pressure–rate-of-strain tensor vanishes relative to production. In shear, the component $\langle u_2^2 \rangle$ has no production, and yet the rapid pressure removes energy from that component.

Some different approaches to RDT for homogeneous turbulence have been developed by Kassinos and Reynolds (1994) and Van Slooten and Pope (1997). As described by Hunt (1973) and Hunt and Carruthers (1990), RDT can also be applied to inhomogeneous flows.

11.5 Pressure–rate-of-strain models

The previous two sections consider homogeneous turbulence in the limiting cases of zero mean velocity gradients ($\mathcal{S}k/\varepsilon = 0$) and very large mean velocity gradients ($\mathcal{S}k/\varepsilon \to \infty$). In the first case, the pressure–rate-of-strain tensor is due entirely to the slow pressure (i.e., $\mathcal{R}_{ij} = \mathcal{R}_{ij}^{(s)}$); in the second case it is due entirely to the rapid pressure (i.e., $\mathcal{R}_{ij} = \mathcal{R}_{ij}^{(r)}$). Here we consider the case of more relevance to turbulent shear flows in which $\mathcal{S}k/\varepsilon$ is of order unity, and both the slow and the rapid pressure fields are significant.

For homogeneous turbulence, with the dissipation taken to be isotropic, the Reynolds-stress equation (Eq. (11.1)) is

$$\frac{d}{dt}\langle u_i u_j \rangle = \mathcal{P}_{ij} + \mathcal{R}_{ij} - \tfrac{2}{3}\varepsilon\delta_{ij}. \tag{11.128}$$

The pressure–rate-of-strain tensor $\mathcal{R}_{ij}$ is the only term that is not in closed form.

11.5.1 The basic model (LRR-IP)

The *basic model* for $\mathcal{R}_{ij}$ is

$$\mathcal{R}_{ij} = -C_R \frac{\varepsilon}{k}(\langle u_i u_j \rangle - \tfrac{2}{3}k\delta_{ij}) - C_2(\mathcal{P}_{ij} - \tfrac{2}{3}\mathcal{P}\delta_{ij}). \tag{11.129}$$

The first term is Rotta's model for $\mathcal{R}_{ij}^{(s)}$, and the second term is the *isotropization of production* (IP) model for $\mathcal{R}_{ij}^{(r)}$ proposed by Naot, Shavit, and Wolfshtein (1970). The IP model supposes that the rapid pressure partially counteracts the effect of production to increase the Reynolds-stress anisotropy. This is indeed the effect observed for rapid distortion axisymmetric expansion (see Eq. (11.121)). Also, if the model constant is taken to be $C_2 = \tfrac{3}{5}$, then the IP model yields the correct initial response of isotropic turbulence to all rapid distortions (see Eq. (11.116)).

This combination of Rotta's model and the IP model is the first of two models proposed by Launder, Reece, and Rodi (1975), and is also referred to as LRR-IP. Launder (1996) suggests the value $C_R = 1.8$ for the Rotta constant; and the IP constant is taken to be $C_2 = \tfrac{3}{5}$, consistent with RDT.

Figure 11.15 shows calculations of the LRR-IP model compared with DNS data for homogeneous shear flow. A closed set of ordinary differential equations is obtained from the Reynolds-stress equation (Eq. (11.128)) with the model for $\mathcal{R}_{ij}$ (Eq. (11.129)), together with the modelled ε equation (Eq. (10.56)). The only non-zero mean velocity gradient, $\partial\langle U_1 \rangle/\partial x_2 = \mathcal{S}$ is constant; the Reynolds stresses are initially isotropic; and so the single non-dimensional parameter characterizing the solutions to the equations is the initial value of $\mathcal{S}k/\varepsilon$, which is taken to be 2.36 to match the DNS.

It may be seen from Fig. 11.15 that the evolution of b_{12} and b_{11} is reasonably represented by the LRR-IP model, the discrepancies being typically less than 15%. On the other hand, the model predicts that b_{22} and b_{33} are equal, whereas the DNS data show that b_{22} is substantially smaller. (Because the ε equation predicts an asymptotic value of $\mathcal{P}/\varepsilon$ that is considerably higher than that observed (see Eq. (10.62)), the calculations of k are quite inaccurate: but the calculated values of b_{ij} are insensitive to this deficiency.) In

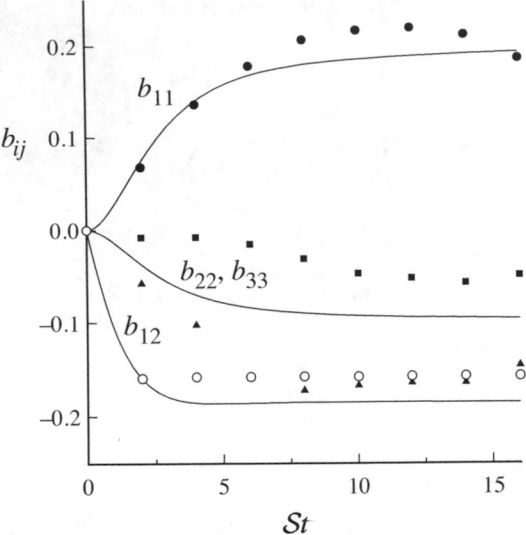

Fig. 11.15. Reynolds-stress anisotropies in homogeneous shear flow. Comparison of LRR-IP model calculations (lines) with the DNS data of Rogers and Moin (1987) (symbols): •, b_{11}; ○, b_{12}; triangles, b_{22}; squares, b_{33}.

Exercise 11.20 analytic solutions for the asymptotic values of the components of b_{ij} are obtained, Eqs. (11.133)–(11.134).

EXERCISE

11.20   In homogeneous turbulent shear flow at large times, the Reynolds-stress tensor becomes self-similar (i.e., $\langle u_i u_j \rangle / k$ tends to a constant tensor). In such a self-similar state, the rate of change of the Reynolds-stress tensor is

$$\frac{d}{dt}\langle u_i u_j \rangle = \frac{\langle u_i u_j \rangle}{k}\frac{dk}{dt} = \frac{\langle u_i u_j \rangle}{k}(\mathcal{P} - \varepsilon). \qquad (11.130)$$

Write down the Reynolds-stress equation (Eq. (11.128)) for homogeneous turbulence, with the rate of change approximated by Eq. (11.130), and $\mathcal{R}_{ij}$ modelled by LRR-IP (Eq. (11.129)). Manipulate this equation to obtain the result

$$\frac{\langle u_i u_j \rangle}{k} = \tfrac{2}{3}\delta_{ij} + \Theta\frac{(\mathcal{P}_{ij} - \tfrac{2}{3}\mathcal{P}\delta_{ij})}{\mathcal{P}}, \qquad (11.131)$$

where

$$\Theta \equiv \frac{(1 - C_2)\mathcal{P}/\varepsilon}{C_R - 1 + \mathcal{P}/\varepsilon}. \qquad (11.132)$$

Hence obtain

$$b_{ij} = \tfrac{1}{2}\Theta(\mathcal{P}_{ij} - \tfrac{2}{3}\mathcal{P}\delta_{ij})/\mathcal{P},$$
$$b_{11} = \tfrac{2}{3}\Theta, \quad b_{22} = b_{33} = -\tfrac{1}{3}\Theta, \tag{11.133}$$

$$b_{12} = -\sqrt{\tfrac{1}{6}\Theta(1 - \Theta)}. \tag{11.134}$$

11.5.2 Other pressure–rate-of-strain models

Many other pressure–rate-of-strain models have been proposed, including: the HL model of Hanjalić and Launder (1972); the quasi-isotropic model (LRR-QI) of Launder *et al.* (1975); the SL model of Shih and Lumley (1985); the JM model of Jones and Musonge (1988); the SSG model of Speziale, Sarkar, and Gatski (1991); and the FLT model of Fu, Launder, and Tselepidakis (1987). The first five of these (and the LRR-IP model) can be written

$$\frac{\mathcal{R}_{ij}}{\varepsilon} = \sum_{n=1}^{8} f^{(n)} \mathcal{T}_{ij}^{(n)}, \tag{11.135}$$

where the non-dimensional, symmetric, deviatoric tensors $\mathcal{T}_{ij}^{(n)}$ are given in Table 11.3, and are defined in terms of the normalized mean rate of strain

$$\widehat{S}_{ij} = \frac{k}{\varepsilon}\bar{S}_{ij} = \frac{1}{2}\frac{k}{\varepsilon}\left(\frac{\partial\langle U_i\rangle}{\partial x_j} + \frac{\partial\langle U_j\rangle}{\partial x_i}\right), \tag{11.136}$$

and the normalized mean rate of rotation

$$\widehat{\Omega}_{ij} = \frac{k}{\varepsilon}\bar{\Omega}_{ij} = \frac{1}{2}\frac{k}{\varepsilon}\left(\frac{\partial\langle U_i\rangle}{\partial x_j} - \frac{\partial\langle U_j\rangle}{\partial x_i}\right). \tag{11.137}$$

The coefficients $f^{(n)}$ defining some of the models are given in Table 11.4. (The FLT model involves additional tensors.)

The term in $\mathcal{T}_{ij}^{(1)}$ corresponds to the Rotta-model contribution, the Rotta *coefficient* being $-\tfrac{1}{2}f^{(1)}$. The LRR and JM models take the Rotta coefficient to be constant, whereas the SSG and SL models incorporate a dependence of the coefficient on $\mathcal{P}/\varepsilon$. Of the models shown in Table 11.4, only the SSG model has a non-zero value of $f^{(2)}$, which leads to the non-linear return to isotropy depicted in Fig. 11.2. The remaining tensors $\mathcal{T}_{ij}^{(3-8)}$ provide contributions to the model for the rapid pressure–rate-of-strain tensor.

The LRR-IP model is the first Reynolds-stress model to be widely used,

Table 11.3. *Definitions of the non-dimensional, symmetric, deviatoric tensors in Eq. (11.135)*

$$T_{ij}^{(1)} = b_{ij}$$

$$T_{ij}^{(2)} = b_{ij}^2 - \tfrac{1}{3}b_{kk}^2\delta_{ij}$$

$$T_{ij}^{(3)} = \widehat{S}_{ij}$$

$$T_{ij}^{(4)} = \widehat{S}_{ik}b_{kj} + b_{ik}\widehat{S}_{kj} - \tfrac{2}{3}\widehat{S}_{k\ell}b_{\ell k}\delta_{ij}$$

$$T_{ij}^{(5)} = \widehat{\Omega}_{ik}b_{kj} - b_{ik}\widehat{\Omega}_{kj}$$

$$T_{ij}^{(6)} = \widehat{S}_{ik}b_{kj}^2 + b_{ik}^2\widehat{S}_{kj} - \tfrac{2}{3}\widehat{S}_{k\ell}b_{\ell k}^2\delta_{ij}$$

$$T_{ij}^{(7)} = \widehat{\Omega}_{ik}b_{kj}^2 - b_{ik}^2\widehat{\Omega}_{kj}$$

$$T_{ij}^{(8)} = b_{ik}\widehat{S}_{k\ell}b_{\ell j} - \tfrac{1}{3}\widehat{S}_{k\ell}b_{\ell k}^2\delta_{ij}$$

but it should be appreciated that all of these models build on earlier contributions. For example, the LRR, JM, and SSG models are all particular cases of the general model proposed by Hanjalić and Launder (1972); and the particular form of the HL model includes the same dependence of the Rotta coefficient on $\mathcal{P}/\varepsilon$ as appears in the SSG model.

The *ideal* pressure–rate-of-strain model would be accurate in all circumstances, and would therefore conform with all known properties of $\mathcal{R}_{ij}$. These properties are as follows.

  (i) In the rapid-distortion limit ($\mathcal{S}k/\varepsilon \to \infty$), $\mathcal{R}_{ij}$ is linear in the mean velocity gradients.

  (ii) $\mathcal{R}_{ij}^{(r)}$ is determined by the tensor $M_{ijk\ell}$ (Eqs. (11.14)–(11.15)), which satisfies the exact relations given in Eq. (11.109).

 (iii) $\mathcal{R}_{ij}^{(r)}$ is linear in the Reynolds stresses, because $M_{ijk\ell}$ is linear in the spectrum (Eq. (11.101)).

 (iv) $\mathcal{R}_{ij}$ is such that the Reynolds-stress tensor remains realizable.

  (v) In the limit of two-dimensional turbulence, $\mathcal{R}_{ij}$ satisfies material-frame indifference (see Section (2.9) and Speziale (1981)).

To a large extent, research in this area has been motivated by the desire to develop models that conform to these properties. The basic model (LRR-IP) does not satisfy (ii), and LRR-QI was developed to do so. The Shih–Lumley model was developed to satisfy strong realizability; but, in doing so, terms nonlinear in b_{ij} were introduced (i.e., $T_{ij}^{(6)}$–$T_{ij}^{(8)}$) in opposition to (iii). In fact, the other models satisfy weak realizability in all but pathologic circumstances, and their coefficients can be modified to satisfy it in all circumstances (Durbin and Speziale 1994). None of the models mentioned satisfies (v), but models

Table 11.4. *Specifications of the coefficients $f^{(n)}$ in Eq. (11.135) for various pressure–rate-of-strain models*

	LRR-IP Launder et al. (1975)	LRR-QI Launder et al. (1975)	JM Jones and Musonge (1988)	SSG Speziale et al. (1991)	SL Shih and Lumley (1985)
$f^{(1)}$	$-2C_R$	$-2C_R$	$-2C_1$	$-C_1 - C_1^* \dfrac{\mathcal{P}}{\varepsilon}$	$-\beta + \dfrac{6}{5}\dfrac{\mathcal{P}}{\varepsilon}$
$f^{(2)}$	0	0	0	C_2	0
$f^{(3)}$	$\frac{4}{3}C_2$	$\frac{4}{5}$	$2(C_4 - C_2)$	$C_3 - \sqrt{6}\,C_3^*\eta$	$\frac{4}{5}$
$f^{(4)}$	$2C_2$	$\frac{6}{11}(2 + 3C_2)$	$-3C_2$	C_4	$12C_2$
$f^{(5)}$	$2C_2$	$\frac{2}{11}(10 - 7C_2)$	$3C_2 + 4C_3$	C_5	$\frac{4}{3}(2 - 7C_2)$
$f^{(6)}$	0	0	0	0	$\frac{4}{5}$
$f^{(7)}$	0	0	0	0	$\frac{4}{5}$
$f^{(8)}$	0	0	0	0	$-\frac{8}{5}$
Values of constants	$C_R = 1.8,$ $C_2 = 0.6$	$C_R = 1.5,$ $C_2 = 0.4$	$C_1 = 1.5,$ $C_2 = -0.53,$ $C_3 = 0.67,$ $C_4 = -0.12$	$C_1 = 3.4,$ $C_1^* = 1.8,$ $C_2 = 4.2,$ $C_3 = 0.8,$ $C_3^* = 1.3,$ $C_4 = 1.25,$ $C_5 = 0.4$	β – see Shih and Lumley (1985), $C_2 = \frac{1}{10}\left(1 + \frac{4}{5}F^{1/2}\right)$

that do have been proposed by Haworth and Pope (1986) and Ristorcelli, Lumley, and Abid (1995).

A subtle but important point to appreciate is that the properties enumerated above pertain to the *ideal* pressure–rate-of-strain model. However, no such model exists. As already observed (see Fig. 11.4), the decay of anisotropic turbulence is not uniquely determined by the 'knowns' in a Reynolds-stress closure. Similarly, the tensor $M_{ijk\ell}$ is not uniquely determined by $\langle u_i u_j \rangle$ (see Eqs. (11.126) and (11.127)). In the absence of the ideal model, it is natural to seek the 'best' model – 'best' in the sense that the model is as accurate as possible when it is applied to the range of flows of interest. Many of the properties enumerated above pertain to extreme situations – the rapid-distortion limit, two-dimensional turbulence, two-component turbulence – that have limited relevance to the flows of interest. Arguably, therefore, in the specification of the form of the model and of the coefficients $f^{(n)}$, the quantitative performance of the model when applied to flows of interest is of more importance than the satisfaction of exact constraints in extreme limits.

In comparison to LRR-IP, the LRR-QI model has the theoretical advantage of conforming to property (ii). However, experience shows that, in practice, LRR-IP is distinctly superior (Launder 1996). The SSG model conforms to yet fewer of the properties than does LRR-IP – it is nonlinear in b_{ij}.

EXERCISES _____

11.21

    (a) Which of the models given in Table 11.4 is consistent with properties (i) and (iii)?

    (b) Which of the models is correct for the initial rapid distortion of isotropic turbulence?

    (c) Which of the models is correct for arbitrary rapid distortions?

11.22 Show that LRR-QI is the most general model that is linear in b_{ij} and $\partial \langle U_i \rangle / \partial x_j$ and that is consistent with the properties of $M_{ijk\ell}$ (Eq. (11.101)).

11.6 Extension to inhomogeneous flows

The development in the previous sections has focused on homogeneous turbulence, whereas the principal application of Reynolds-stress models is to inhomogeneous turbulent flows. Described in this section are the remaining steps leading to complete Reynolds-stress models for inhomogeneous flows. Additional treatments for near-wall regions are described in the next section.

    The terms to be modelled in the exact Reynolds-stress equations (Eq. (11.1)) are the dissipation tensor ε_{ij}, the pressure–rate-of-strain tensor $\mathcal{R}_{ij}$, and the transport T_{kij}. As discussed in Section 11.1, the dissipation ε_{ij} can be taken to be isotropic, $\frac{2}{3} \varepsilon \delta_{ij}$ (except close to walls). The form of the model dissipation equation used is described later in this section.

11.6.1 Redistribution

As discussed in Section 11.2, there are different decompositions of the velocity–pressure-gradient tensor Π_{ij} into redistributive and transport terms; for example, Eq. (7.187) in terms of $\mathcal{R}_{ij}$, and Eq. (11.5) in terms of $\mathcal{R}_{ij}^{(a)}$. The analysis of the channel-flow DNS data performed by Mansour, Kim, and Moin (1988) suggests strongly that, for such near-wall flows, it is preferable to use the decomposition in terms of $\mathcal{R}_{ij}^{(a)}$ rather than that in terms of $\mathcal{R}_{ij}$. It is found that the pressure transport of kinetic energy $\frac{1}{2}\Pi_{ii}$ is quite small (see Figs. 7.18 and 7.34 on pages 285 and 314), so that in fact Π_{ij} is itself almost redistributive. The profiles of Π_{ij} (Figs. 7.35–7.38 on pages 316–317) exhibit simple behavior with Π_{ij} being zero at the wall (because $\boldsymbol{u}$ is zero there). In contrast, $\mathcal{R}_{22}$ and $\partial T_{222}^{(p)}/\partial y$ exhibit much more complicated behavior, including several sign changes. Although both terms are zero at the wall, within the viscous sublayer they are large, but almost cancel. In view of these considerations, we henceforth use the decomposition Eq. (11.5) in terms of $\mathcal{R}_{ij}^{(a)}$ and $\boldsymbol{T}^{(p)}$.

For inhomogeneous flows, the redistribution $\mathcal{R}_{ij}^{(a)}$ is modelled in terms of local quantities. That is, $\mathcal{R}_{ij}^{(a)}(\boldsymbol{x},t)$ is modelled in terms of $\langle u_i u_j \rangle$, $\partial \langle U_i \rangle / \partial x_j$, and ε, evaluated at $(\boldsymbol{x},t)$, just as in homogeneous turbulence (Eq. (11.135)). In terms of the ease of solution of the resulting model equations, this is certainly an expedient assumption compared with the alternative of including non-local quantities. However, it should be recognized that $p'(\boldsymbol{x},t)$ is governed by a Poisson equation (Eq. (11.9)), so that it is influenced by quantities such as $\partial \langle U_i \rangle / \partial x_j$ some distance from $\boldsymbol{x}$. As a consequence, the modelling of $\mathcal{R}_{ij}^{(a)}$ is less secure for inhomogeneous flows than it is for homogeneous turbulence. (In contrast, in elliptic relaxation models – Section 11.8 – the modelling of $\mathcal{R}_{ij}^{(a)}$ is non-local.)

11.6.2 Reynolds-stress transport

With the velocity–pressure-gradient tensor Π_{ij} decomposed according to Eq. (11.5), the exact equation for the evolution of Reynolds stresses is

$$\frac{\bar{D}}{\bar{D}t} \langle u_i u_j \rangle + \frac{\partial}{\partial x_k} \left(T_{kij}^{(v)} + T_{kij}^{(p')} + T_{kij}^{(u)} \right) = \mathcal{P}_{ij} + \mathcal{R}_{ij}^{(a)} - \varepsilon_{ij}, \qquad (11.138)$$

where the three fluxes are viscous diffusion

$$T_{kij}^{(v)} = -v \frac{\partial \langle u_i u_j \rangle}{\partial x_k}, \qquad (11.139)$$

pressure transport

$$T_{kij}^{(p')} = \tfrac{2}{3} \delta_{ij} \langle u_k p' \rangle / \rho, \qquad (11.140)$$

and turbulent convection

$$T_{kij}^{(u)} = \langle u_i u_j u_k \rangle. \qquad (11.141)$$

Viscous diffusion is negligible except in the viscous wall region; and, since the term is in closed form, it requires no further discussion.

Pressure transport

Before the advent of DNS, there was little reliable information about pressure correlations. On the basis of an analysis of nearly homogeneous turbulence, Lumley (1978) proposed the model

$$\frac{1}{\rho}\langle u_i p'\rangle = -\tfrac{1}{5}\langle u_i u_j u_j\rangle. \tag{11.142}$$

Since the pressure transport given by Eq. (11.140) is isotropic, it can be examined through the corresponding term in the kinetic-energy equation

$$\tfrac{1}{2}T_{kii}^{(p')} = \langle u_k p'\rangle/\rho, \tag{11.143}$$

for which Lumley's model is

$$\tfrac{1}{2}T_{kii}^{(p')} = -\tfrac{1}{5}\langle u_k u_i u_i\rangle. \tag{11.144}$$

This is $-\tfrac{2}{5}$ of the convective flux

$$\tfrac{1}{2}T_{kii}^{(u)} = \tfrac{1}{2}\langle u_k u_i u_i\rangle. \tag{11.145}$$

The kinetic-energy budgets for channel flow (Fig. 7.18 on page 285) and for the boundary layer (Fig. 7.34 on page 314) show that pressure transport is not very significant close to the wall, and that Lumley's model is qualitatively incorrect. At the edge of the boundary layer, however, the pressure transport is more important, and Lumley's model is at least qualitatively correct.

To examine the pressure transport for free shear flows, the kinetic-energy budget for the self-similar temporal-mixing layer is shown in Fig. 11.16. It may be seen from Fig. 11.16(a) that the pressure transport is relatively small over most of the layer, and that Lumley's model is quite reasonable.

The edge of the layer is examined in more detail in Fig. 11.16(b). The rotational turbulent fluctuations within the layer induce irrotational fluctuations in the non-turbulent region ($y/\delta > 1$, say). This transfer of energy is effected by the fluctuating pressure field, and hence it appears in the kinetic-energy budget as pressure transport. As may be seen from Fig. 11.16(b), at the edge of the layer this pressure transport becomes dominant.

Demuren *et al.* (1996) used DNS data to examine separately the pressure transport due to the slow and the rapid pressure, and proposed models for each contribution. In most Reynolds-stress models the pressure transport is either neglected or (implicitly or explicitly) it is modelled together with the turbulent convection by a gradient-diffusion assumption.

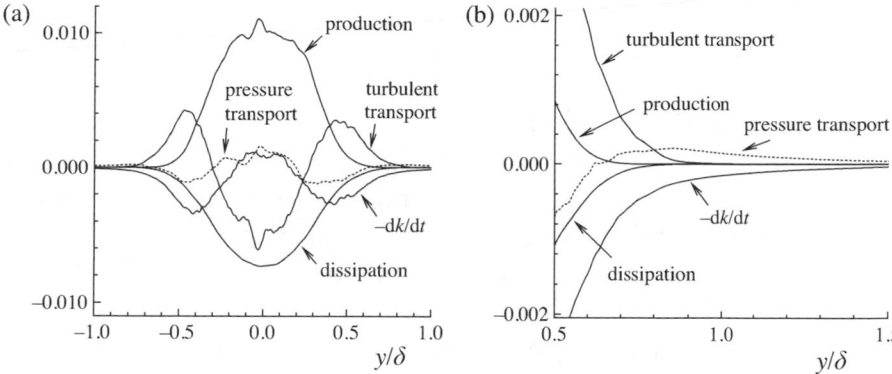

Fig. 11.16. The kinetic energy budget in the temporal mixing layer from the DNS data of Rogers and Moser (1994): (a) across the whole flow; (b) an expanded view of the edge of the layer. The contributions to the budget are production $\mathcal{P}$, dissipation $-\varepsilon$, rate of change $-dk/dt$, turbulent transport, and pressure transport (dashed line). All quantities are normalized by the velocity difference and the layer thickness δ (see Fig. 5.21 on page 140).

Gradient-diffusion models

The simplest gradient-diffusion model for $T'_{kij} = T^{(u)}_{kij} + T^{(p')}_{kij}$, due to Shir (1973), is

$$T'_{kij} = -C_s \frac{k^2}{\varepsilon} \frac{\partial \langle u_i u_j \rangle}{\partial x_k}, \tag{11.146}$$

where C_s is a model constant. In more general use is the model of Daly and Harlow (1970), which uses the Reynolds-stress tensor to define an anisotropic diffusion coefficient:

$$T'_{kij} = -C_s \frac{k}{\varepsilon} \langle u_k u_\ell \rangle \frac{\partial \langle u_i u_j \rangle}{\partial x_\ell}. \tag{11.147}$$

For this model Launder (1990) suggests the value of the constant $C_s = 0.22$.

If $T^{(u)}_{kij} \equiv \langle u_k u_i u_j \rangle$ is to be modelled separately, then a consistent model is required to be symmetric with respect to all three indices. Such symmetric models, necessarily involving cross-diffusion, have been proposed by Mellor and Herring (1973),

$$\langle u_i u_j u_k \rangle = -C_s \frac{k^2}{\varepsilon} \left(\frac{\partial \langle u_j u_k \rangle}{\partial x_i} + \frac{\partial \langle u_i u_k \rangle}{\partial x_j} + \frac{\partial \langle u_i u_j \rangle}{\partial x_k} \right), \tag{11.148}$$

and by Hanjalić and Launder (1972),

$$\langle u_i u_j u_k \rangle = -C_s \frac{k}{\varepsilon} \left(\langle u_i u_\ell \rangle \frac{\partial \langle u_j u_k \rangle}{\partial x_\ell} + \langle u_j u_\ell \rangle \frac{\partial \langle u_i u_k \rangle}{\partial x_\ell} + \langle u_k u_\ell \rangle \frac{\partial \langle u_i u_j \rangle}{\partial x_\ell} \right). \tag{11.149}$$

However, consistent models for T'_{kij} are required to be symmetric with respect to i and j only – a requirement that is satisfied by all four of the above models.

Examination of the transport equation for the triple correlation $\langle u_i u_j u_k \rangle$ can motivate yet more elaborate models, involving the mean velocity gradients. However, the general experience of practitioners is that the modelling of T'_{kij} is not a critical ingredient in the overall model, and that the relatively simple Daly–Harlow model is adequate (Launder 1990). This view is questioned by Parneix, Laurence, and Durbin (1998b), who suggest that deficiencies in the transport model are responsible for inaccuracies in the calculation of the flow over a backward-facing step. Direct tests of the models for $T^{(u)}_{kij}$ against experimental data can be found in Schwarz and Bradshaw (1994) and references therein.

11.6.3 The dissipation equation

The standard model equation for ε used in Reynolds-stress models is that proposed by Hanjalić and Launder (1972),

$$\frac{\bar{D}\varepsilon}{\bar{D}t} = \frac{\partial}{\partial x_i}\left(C_\varepsilon \frac{k}{\varepsilon} \langle u_i u_j \rangle \frac{\partial \varepsilon}{\partial x_j} \right) + C_{\varepsilon 1}\frac{\mathcal{P}\varepsilon}{k} - C_{\varepsilon 2}\frac{\varepsilon^2}{k}, \tag{11.150}$$

with $C_\varepsilon = 0.15$, $C_{\varepsilon 1} = 1.44$, and $C_{\varepsilon 2} = 1.92$ (Launder 1990). There are two differences between this equation and that used in the k–ε model (Eq. (10.53)). First, the production $\mathcal{P}$ is evaluated directly from the Reynolds stresses rather than as $2\nu_T \bar{S}_{ij}\bar{S}_{ij}$. Second, the diffusion term involves an anisotropic diffusivity.

As mentioned in Section 10.4.3, several modifications to the dissipation equation have been proposed. In a Reynolds-stress model, the invariants of the anisotropy tensor b_{ij} can be used in such modifications. For example, Launder (1996) suggested modifying the coefficients to

$$C_{\varepsilon 1} = 1.0, \quad C_{\varepsilon 2} = 1.92/(1 + 3.4\eta F), \tag{11.151}$$

where the invariants η and F are defined by Eqs. (11.28) and (11.34). Other modifications were suggested by Hanjalić, Jakirlić, and Hadžić (1997).

EXERCISE _____

11.23 Show that the dissipation equation Eq. (11.150) is consistent with the log-law, with the Kármán constant given by

$$\kappa^2 = \frac{4(C_{\varepsilon 2} - C_{\varepsilon 1})|b_{12}|^3}{C_\varepsilon(b_{22} + \frac{1}{3})}. \tag{11.152}$$

Evaluate κ for b_{ij} given by the LRR-IP model (Eq. (11.133) with $\mathcal{P}/\varepsilon = 1$).

11.7 Near-wall treatments

11.7.1 Near-wall effects

In a turbulent flow, the presence of a wall causes a number of different effects. For a boundary layer in the usual coordinate system, for example, some of these effects are:

(i) *low Reynolds number* – the turbulence Reynolds number $\mathrm{Re_L} \equiv k^2/(\varepsilon v)$ tends to zero as the wall is approached;

(ii) *high shear rate* – the highest mean shear rate $\partial\langle U\rangle/\partial y$ occurs at the wall (see Exercise 11.24);

(iii) *two-component turbulence* – for small y, $\langle v^2\rangle$ varies as y^4 whereas $\langle u^2\rangle$ and $\langle w^2\rangle$ vary as y^2 (Eqs. (7.60)–(7.63)), so that, as the wall is approached, the turbulence tends to the two-component limit (Fig. 11.1); and

(iv) *wall blocking* – through the pressure field ($p^{(\mathrm{h})}$, Eq. (11.10)), the impermeability condition $V = 0$ (at $y = 0$) affects the flow up to (of order) an integral scale from the wall (Hunt and Graham 1978).

These effects necessitate modifications to the basic k–ε and Reynolds-stress turbulence models. (It is informative also to study turbulent flows from which one or both of (i) and (ii) are absent, e.g., turbulence at a free surface (Brumley 1984) and shear-free turbulence near a wall (Thomas and Hancock 1977, Perot and Moin 1995, Aronson, Johansson, and Löfdahl 1997).)

The forms of the basic k–ε and Reynolds-stress models have not changed since the 1970s. The same is not true for the near-wall treatments. In the late 1980s, detailed DNS data for the viscous near-wall region, which is very difficult to study experimentally, became available. These data showed existing models to be qualitatively incorrect, and led to new and continuing developments.

EXERCISE

11.24 Consider $\tau_\eta \partial\langle U\rangle/\partial y$ as a non-dimensional measure of the shear rate. Show that its value at the wall is

$$\tau_\eta \frac{\partial\langle U\rangle}{\partial y} = (\varepsilon_0^+)^{-1/2} \approx 2, \qquad (11.153)$$

(see Exercise 7.8 on page 287); whereas in the log-law region it is

$$\tau_\eta \frac{\partial \langle U \rangle}{\partial y} = (\kappa y^+)^{-1/2}, \tag{11.154}$$

which is less than 0.3 (for $y^+ > 30$). An alternative non-dimensional measure of the shear rate is $\mathcal{S}k/\varepsilon$. How does this vary close to the wall?

11.7.2 Turbulent viscosity

It may be recalled that the mixing-length specification $\ell_m = \kappa y$ is too large in the near-wall region ($y^+ < 30$, say), and a much improved specification is obtained from the van Driest damping function, Eq. (7.145). So also, the standard k–ε specification $\nu_T = C_\mu k^2/\varepsilon$ yields too large a turbulent viscosity in the near-wall region, as may be seen from Fig. 10.3.

The original k–ε model of Jones and Launder (1972) includes various damping functions to allow the model to be used within the viscous near-wall region. The turbulent viscosity is given by

$$\nu_T = f_\mu C_\mu \frac{k^2}{\varepsilon}, \tag{11.155}$$

where the damping function f_μ depends on the turbulence Reynolds number according to

$$f_\mu = \exp\left(\frac{-2.5}{1 + \mathrm{Re}_L/50}\right). \tag{11.156}$$

At the wall f_μ is small, and it tends to unity at large Re_L, so that the standard k–ε formula is recovered.

The quantity $\nu_T \varepsilon/k^2$ plotted in Fig. 10.3 (on page 371) is a 'measurement' of $f_\mu C_\mu$ for channel flow. Rodi and Mansour (1993) assessed a number of proposals for f_μ on the basis of similar DNS data: most early proposals, such as Eq. (11.156), are quite inaccurate. Rodi and Mansour (1993) suggested instead the empirical relation

$$f_\mu = 1 - \exp(-0.0002y^+ - 0.00065y^{+2}). \tag{11.157}$$

A tenable view on this approach is that the damping function f_μ is devoid of physical justification, and it is used to compensate for the incorrect physics in the basic model. There is no reason to suppose that Eq. (11.157) is accurate in application to other flows – near a separation or reattachment point, for example.

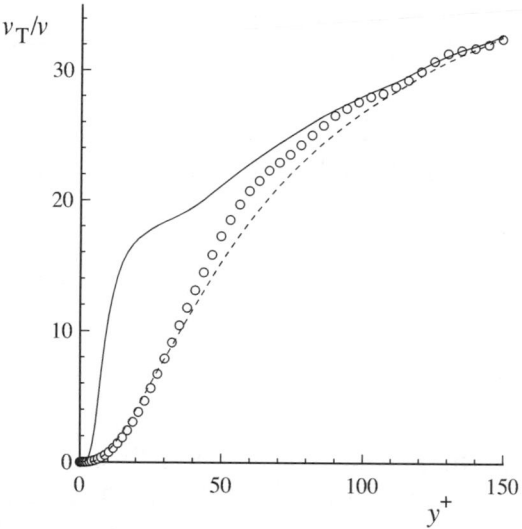

Fig. 11.17. The turbulent viscosity against y^+ for channel flow at Re = 13,750. Symbols, DNS data of Kim *et al.* (1987); solid line, $0.09k^2/\varepsilon$; dashed line, $0.22\langle v^2 \rangle k/\varepsilon$.

Durbin (1991) argued that it is the cross-stream velocity v that is responsible for turbulent transport in a boundary-layer flow, and that the appropriate expression for the turbulent viscosity is

$$\nu_{\mathrm{T}} = C'_{\mu} \frac{\langle v^2 \rangle k}{\varepsilon}, \tag{11.158}$$

where C'_{μ} is a constant. (This is also the cross-stream diffusion coefficient in the Daly–Harlow model, Eq. (11.147).) Figure 11.17 shows that Eq. (11.158) does indeed provide an excellent representation of ν_{T} throughout the near-wall region. This implies – as noted by Launder (1986) – that the shapes of the profiles of f_{μ} and $\langle v^2 \rangle / k$ are similar.

The strong suppression of $\langle v^2 \rangle$ (relative to k) in the near-wall region (see Figs. 7.14 and 7.33 on pages 282 and 313) is due to the inviscid wall-blocking effect, not to viscous effects. It is inappropriate, therefore, to refer to f_{μ} as a 'low-Reynolds number correction.'

11.7.3 Model equations for k and ε

Most proposed model equations for k and ε have the same form as the original suggestions of Jones and Launder (1972). The equation for k, which includes transport by the molecular viscosity, is

$$\frac{\bar{\mathrm{D}}k}{\bar{\mathrm{D}}t} = \nabla \cdot \left[\left(\nu + \frac{\nu_{\mathrm{T}}}{\sigma_k} \right) \nabla k \right] + \mathcal{P} - \varepsilon. \tag{11.159}$$

At the wall, the value of the dissipation is

$$\varepsilon_\mathrm{o} = v\left(\frac{\partial^2 k}{\partial y^2}\right)_{y=0} = 2v\left(\frac{\partial k^{1/2}}{\partial y}\right)^2_{y=0}, \qquad (11.160)$$

(see Eq. (7.73)). Jones and Launder (1972) choose as the second variable

$$\tilde{\varepsilon} \equiv \varepsilon - \varepsilon_\mathrm{o}, \qquad (11.161)$$

so that the simpler boundary condition $\tilde{\varepsilon}_{y=0} = 0$ can be applied. The model equation for $\tilde{\varepsilon}$ is

$$\frac{\bar{\mathrm{D}}\tilde{\varepsilon}}{\bar{\mathrm{D}}t} = \nabla \cdot \left[\left(v + \frac{v_\mathrm{T}}{\sigma_\varepsilon}\right)\nabla\tilde{\varepsilon}\right] + C_{\varepsilon 1}f_1\frac{\tilde{\varepsilon}\mathcal{P}}{k} - C_{\varepsilon 2}f_2\frac{\tilde{\varepsilon}^2}{k} + E. \qquad (11.162)$$

Different specifications of the damping functions f_1 and f_2 and of the additional term E are given in Patel, Rodi, and Scheuerer (1985) and Rodi and Mansour (1993). The DNS data reveal substantial discrepancies in all of the proposed models (Rodi and Mansour 1993).

The turbulence timescale $\tau = k/\varepsilon$ tends to zero at the wall, and yet of course all relevant physical timescales are strictly positive. Durbin (1991) argued that (a constant times) the Kolmogorov timescale $\tau_\eta = (v/\varepsilon)^{1/2}$ is the smallest relevant scale, and thereby defined the modified timescale

$$T \equiv \max(\tau, C_T\tau_\eta), \qquad (11.163)$$

taking $C_T = 6$. (T is determined by $C_T\tau_\eta$ up to $y^+ \approx 5$, and by τ for larger y^+.) The standard dissipation equation written with T replacing τ is

$$\frac{\bar{\mathrm{D}}\varepsilon}{\bar{\mathrm{D}}t} = \nabla \cdot \left[\left(v + \frac{v_\mathrm{T}}{\sigma_\varepsilon}\right)\nabla\varepsilon\right] + C_{\varepsilon 1}\frac{\mathcal{P}}{T} - C_{\varepsilon 2}\frac{\varepsilon}{T}. \qquad (11.164)$$

The solution to this equation – without damping functions or additional terms – yields profiles of ε that are in good agreement with DNS data (Durbin 1991).

11.7.4 The dissipation tensor

In Reynolds-stress models, the dissipation tensor ε_{ij} and the pressure–rate-of-strain tensor $\mathcal{R}_{ij}$ are the quantities primarily affected by the presence of walls. At high Reynolds number and remote from walls, the dissipation is taken to be isotropic,

$$\varepsilon_{ij} = \tfrac{2}{3}\varepsilon\delta_{ij}. \qquad (11.165)$$

However, as observed in Chapter 7 (e.g., Fig. 7.39 on page 318), as the wall is approached, ε_{ij} becomes distinctly anisotropic.

From the definition of ε_{ij} (Eq. (7.181)) and from the series expansions for u_i (Eqs. (7.56)–(7.58)), it is straightforward to show that, as the wall is approached ($y \to 0$), ε_{ij} is given by

$$\frac{\varepsilon_{ij}}{\varepsilon} = \frac{\langle u_i u_j \rangle}{k}, \qquad \text{for } i \neq 2, \; j \neq 2,$$

$$\frac{\varepsilon_{i2}}{\varepsilon} = 2\frac{\langle u_i u_2 \rangle}{k}, \qquad \text{for } i \neq 2, \qquad\qquad (11.166)$$

$$\frac{\varepsilon_{22}}{\varepsilon} = 4\frac{\langle u_2^2 \rangle}{k}.$$

Rotta (1951) proposed the simple model

$$\varepsilon_{ij} = \frac{\langle u_i u_j \rangle}{k}\varepsilon \qquad\qquad (11.167)$$

for the limit of low-Reynolds number turbulence. At a wall, this is consistent with Eq. (11.166), since $\langle v^2 \rangle/k$ and $\langle uv \rangle/k$ are zero at $y = 0$. However, for small y, Eq. (11.167) underestimates ε_{12} and ε_{22} by factors of 2 and 4, respectively.

To provide a more accurate approximation to ε_{ij} for small y, consistent with Eq. (11.166), Launder and Reynolds (1983) and Kebede, Launder, and Younis (1985) introduced the quantity

$$\varepsilon_{ij}^* = \frac{\varepsilon\left(\langle u_i u_j \rangle + n_j n_\ell \langle u_\ell u_i \rangle + n_i n_\ell \langle u_\ell u_j \rangle + \delta_{ij} n_\ell n_{\mathrm{m}} \langle u_\ell u_{\mathrm{m}} \rangle\right)/k}{1 + \frac{5}{2} n_\ell n_{\mathrm{m}} \langle u_\ell u_{\mathrm{m}} \rangle/k}, \qquad (11.168)$$

where $\boldsymbol{n}$ is the unit normal to the wall. The denominator is unity at the wall, and is specified so that the trace of ε_{ij}^* is 2ε. It is readily verified that, for a wall at $y = 0$ (i.e., $n_j = \delta_{j2}$), Eq. (11.168) reverts to Eq. (11.166).

With Eq. (11.168) giving the value of ε_{ij} at the wall, and Eq. (11.165) holding remote from the wall, it is natural to model ε_{ij} as a blending of these two functions:

$$\varepsilon_{ij} = f_{\mathrm{s}}\varepsilon_{ij}^* + (1 - f_{\mathrm{s}})\tfrac{2}{3}\varepsilon\delta_{ij}, \qquad\qquad (11.169)$$

where the blending function f_{s} decreases from unity at the wall to zero far from the wall. Lai and So (1990) propose

$$f_{\mathrm{s}} = \exp\left[-\left(\frac{\mathrm{Re}_{\mathrm{L}}}{150}\right)^2\right]. \qquad\qquad (11.170)$$

Figure 11.18 shows the profiles of $\varepsilon_{ij}/(\tfrac{2}{3}\varepsilon)$ from the boundary-layer data of Spalart (1988) compared with Rotta's model (Eq. (11.167)) and Eq. (11.169).

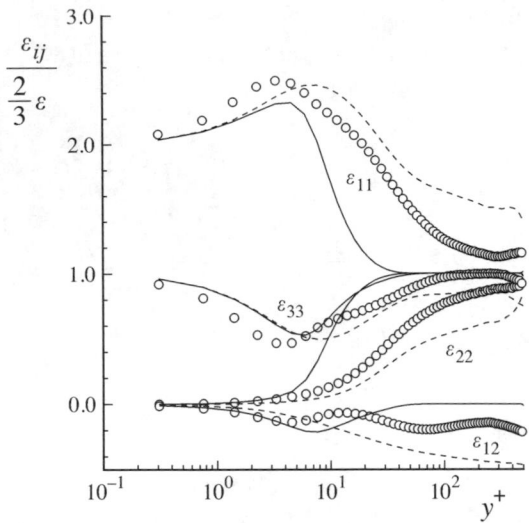

Fig. 11.18. Normalized dissipation components in a turbulent boundary layer at $Re_\theta = 1{,}410$: symbols, DNS data of Spalart (1988); dashed lines, Rotta's model, Eq. (11.167); solid lines, Eq. (11.169).

It may be seen that both models are reasonably accurate for the viscous sublayer, the more complicated model (Eq. (11.169)) having no perceptible advantage. Further out in the flow ($y^+ > 200$), the dissipation is less anisotropic than is predicted by Rotta's model. On the other hand, for this low-Reynolds-number flow, it is evident that the blending function in Eq. (11.169) causes the modelled ε_{ij} to become isotropic more rapidly than the data as y increases. (There are more complicated proposals for f_s due to Hanjalić and Jakirlić (1993) and Launder (1996), which yield somewhat better agreement with the data.)

Some of the near-wall treatments described above involve the distance from the wall y (e.g., Eq. (11.157)) or the normal to the wall n (e.g., Eq. (11.168)). For geometrically simple flows, such as the flat-plate boundary layer, it is trivial to define and evaluate these quantities. For the general case $\ell_w(x)$ – defined as the distance from the point x to the closest point on a wall – can be used in place of y; and the vector

$$n^w(x) \equiv \nabla \ell_w(x) \tag{11.171}$$

can be used in place of n. The properties of n^w are that it equals n at every point on a smooth surface, its magnitude $|n^w|$ is less than or equal to unity, and it can vary discontinuously with x.

There are differing opinions about the desirability of using ℓ_w and n^w in near-wall treatments. An alternative is to use instead local quantities with similar behaviors. For example, at a wall $\nabla k^{1/2}$ is a vector in the direction of n, and Eq. (11.160) can be used to give

$$n_i n_j = \frac{2\nu}{\varepsilon} \frac{\partial k^{1/2}}{\partial x_i} \frac{\partial k^{1/2}}{\partial x_j}. \tag{11.172}$$

This relation is exact at the wall, and provides a good approximation through the viscous sublayer. Craft and Launder (1996) use this approximation to $n_i n_j$ in Eq. (11.168) so that Eq. (11.169) becomes a local model for ε_{ij}.

11.7.5 Fluctuating pressure

Harmonic pressure

To make precise the decomposition of the fluctuating pressure p' into rapid $p^{(r)}$, slow $p^{(s)}$, and harmonic $p^{(h)}$ contributions (Eqs. (11.10)–(11.12)), it is necessary to specify boundary conditions. The boundary condition on p' is obtained from the equation for the evolution of the fluctuating velocity, Eq. (5.138). In the usual coordinate system, the normal component of this equation evaluated at the wall ($y = 0$) yields the Neumann condition

$$\frac{1}{\rho} \frac{\partial p'}{\partial y} = \nu \frac{\partial^2 v}{\partial y^2}. \tag{11.173}$$

Evidently the normal pressure gradient at the wall is due entirely to viscous effects: for inviscid flow $\partial p'/\partial y$ would be zero.

The source term in the Poisson equation for p' (Eq. (11.9)) is independent of the viscosity. It is natural, therefore, to specify zero-normal-gradient (inviscid) boundary conditions on $p^{(r)}$ and $p^{(s)}$, and to specify the viscous condition (Eq. (11.173)) on the harmonic contribution $p^{(h)}$. (Mansour et al. (1988) refer to $p^{(h)}$ as the *Stokes pressure*.)

An examination of DNS data of channel flow (Mansour et al. 1988) reveals that the harmonic pressure plays a minor role in the Reynolds-stress budgets, and is insignificant beyond $y^+ \approx 15$. This confirms the fact that the significant effect of the wall on the fluctuating pressure field is an inviscid blocking effect arising from the impermeability condition, rather than a viscous effect due to the no-slip condition.

EXERCISE

11.25 Consider a semi-infinite body of turbulence above an infinite plane wall at $y = 0$. The turbulence is statistically homogeneous in the x and z directions. A Fourier mode of the normal velocity derivative at the wall is

$$\left(\frac{\partial^2 v}{\partial y^2}\right)_{y=0} = \frac{u_\tau}{\delta_v^2} \hat{v} \exp\left(i\kappa_1 \frac{x}{\delta_v} + i\kappa_3 \frac{z}{\delta_v}\right), \tag{11.174}$$

where κ_1 and κ_3 are normalized wavenumbers and $\hat{v}$ is the normalized Fourier coefficient. Determine the corresponding harmonic pressure field $p^{(h)}$, and hence show that it decays exponentially with y.

Wall reflection

The 'inertial' pressure is defined by

$$p^{(i)} \equiv p^{(r)} + p^{(s)} = p' - p^{(h)}. \tag{11.175}$$

The Green's function solution for $p^{(i)}$ provides some insight into the blocking effect of a wall. Consider a turbulent flow above an infinite plane wall at $y = 0$. The inertial pressure is governed by the same Poisson equation as p' (Eq. (11.9)), which is written

$$\nabla^2 p^{(i)} = S. \tag{11.176}$$

The source $S(x,t)$ is given by the right-hand side of Eq. (11.9) (multiplied by ρ). For simplicity, it is assumed that, far from the region of interest ($|x| \to \infty$), the flow is non-turbulent so that $p^{(i)}$ is zero there. The boundary condition at the wall is

$$\left(\frac{\partial p^{(i)}}{\partial y}\right)_{y=0} = 0. \tag{11.177}$$

The source is defined in the upper half-space $y \geq 0$. This definition is extended by reflecting S about the wall:

$$S(x,y,z,t) \equiv S(x,|y|,z,t), \quad \text{for } y < 0. \tag{11.178}$$

Since S is symmetric about $y = 0$, the solution $p^{(i)}$ to the Poisson equation in the whole domain is also symmetric about $y = 0$, and so satisfies the boundary condition, Eq. (11.177). This solution is

$$p^{(i)}(x,t) = -\frac{1}{4\pi} \int\!\!\!\int\!\!\!\int_{-\infty}^{\infty} S(x',t) \frac{dx'}{|x - x'|}, \tag{11.179}$$

(cf. Eq. (2.49)).

As shown in Fig. 11.19, for every point $x' = (x',y',z')$ in the upper half-space, the image point $x'' = (x'',y'',z'')$ is defined by

$$(x'',y'',z'') = (x',-y',z'). \tag{11.180}$$

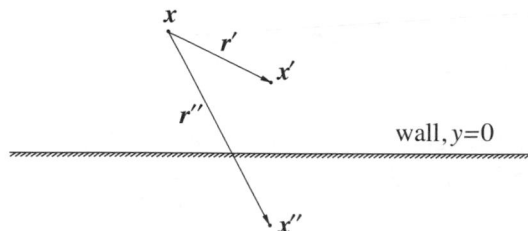

Fig. 11.19. A sketch of the point x' and its image x'', showing the vectors r' and r'' that appear in the Green's function solutions, Eqs. (11.181) and (11.182).

Then with the definitions $r' \equiv x' - x$ and $r'' \equiv x'' - x$, the solution can be re-written

$$p^{(i)}(x,t) = -\frac{1}{4\pi} \int_{-\infty}^{\infty} \int_{0}^{\infty} \int_{-\infty}^{\infty} S(x',t) \left(\frac{1}{|r'|} + \frac{1}{|r''|} \right) dx'. \qquad (11.181)$$

So the correlation at x between $p^{(i)}$ and a random field $\phi(x,t)$ is

$$\langle p^{(i)}(x,t)\phi(x,t) \rangle = -\frac{1}{4\pi} \int_{-\infty}^{\infty} \int_{0}^{\infty} \int_{-\infty}^{\infty} \langle S(x',t)\phi(x,t) \rangle \left(\frac{1}{|r'|} + \frac{1}{|r''|} \right) dx'. \qquad (11.182)$$

Substituting the fluctuating rate of strain s_{ij} for ϕ yields the pressure–rate-of-strain tensor $\mathcal{R}_{ij}$, and a similar formula can be obtained for the velocity–pressure-gradient tensor Π_{ij}.

Equation (11.182) shows that pressure correlations can be viewed as having two contributions: one due to the free-space Green's function $|r'|^{-1}$; the second due to $|r''|^{-1}$, which is called the *wall reflection* contribution (or *wall echo*). It is a simple matter to estimate that the relative magnitudes of these two contributions are in the ratio of L_s^{-1} to y^{-1}, where L_s is the characteristic correlation lengthscale of S and ϕ. Thus remote from a wall ($L_s/y \ll 1$) the wall reflection is negligible, but it may be significant when the turbulence lengthscale is comparable to the distance from the wall – as it is throughout the log-law region.

In some Reynolds-stress closures additional redistribution terms are used to account for wall reflections. With the LRR-IP model, Gibson and Launder (1978) propose additional slow $\mathcal{R}_{ij}^{(s,w)}$ and rapid $\mathcal{R}_{ij}^{(r,w)}$ terms, the former being

$$\mathcal{R}_{ij}^{(s,w)} = 0.2 \frac{\varepsilon}{k} \frac{L}{y} \left(\langle u_\ell u_m \rangle n_\ell n_m \delta_{ij} - \tfrac{3}{2} \langle u_i u_\ell \rangle n_j n_\ell - \tfrac{3}{2} \langle u_j u_\ell \rangle n_i n_\ell \right), \qquad (11.183)$$

where the turbulence lengthscale is $L = k^{3/2}/\varepsilon$. The factor L/y is consistent with the preceding Green's function analysis, and it is essentially uniform in the log-law region, $L/y \approx 2.5$ (see Eq. (7.84)). The effect of this term is to reduce $\langle v^2 \rangle$ and $\langle uv \rangle$ and to increase $\langle u^2 \rangle$ and $\langle w^2 \rangle$ (see Exercise 11.26).

The SSG model (without wall-reflection terms) performs quite well in the log-law region; and Speziale (1996) suggests that the need for wall reflection terms in other models is due to deficiencies in their pressure–rate-of-strain models $\mathcal{R}_{ij}^{(s)}$ and $\mathcal{R}_{ij}^{(r)}$.

EXERCISE _____

11.26   For a near-wall flow in the usual coordinate system, write in matrix form the components of $\mathcal{R}_{ij}^{(s,w)}$ given by Eq. (11.183).

For the log-law region of a boundary layer, estimate L/y and compare the relative magnitudes of the terms $\mathcal{R}_{22}^{(s)}$, $\mathcal{R}_{22}^{(s,w)}$, and $-\frac{2}{3}\varepsilon$ in the LRR-IP model.

_____

11.7.6 Wall functions

The near-wall region adds complication and expense to the task of performing turbulence-model calculations of turbulent flows. As described in the previous subsections, near-wall effects require additions or modifications to the basic models; and the need to resolve the steep profiles (of $\langle U \rangle$ and ε for $y^+ < 30$, for example) can lead to a substantial fraction of the computational effort being devoted to the near-wall region. At the same time, if the mean flow is approximately parallel to the wall, then the log-law relations apply, giving simple algebraic relations for turbulence-model variables in the log-law region.

The idea of the 'wall-function' approach (Launder and Spalding 1972) is to apply boundary conditions (based on log-law relations) some distance away from the wall, so that the turbulence-model equations are not solved close to the wall (i.e., between the wall and the location at which boundary conditions are applied).

The implementation of wall functions for a statistically-two-dimensional flow in the usual coordinate system is now described. The wall-function boundary conditions are applied at a location $y = y_p$ in the log-law region (e.g., where y^+ is around 50). The subscript 'p' indicates quantities evaluated at y_p, e.g., $\langle U \rangle_p$, k_p and ε_p.

For a high-Reynolds-number zero-pressure-gradient boundary layer, the log-law (Eq. (7.43)) is

$$\langle U \rangle = u_\tau \left(\frac{1}{\kappa} \ln y^+ + B \right), \tag{11.184}$$

the balance of production and dissipation yields

$$\varepsilon = \frac{u_\tau^3}{\kappa y}, \tag{11.185}$$

and the k–ε expression for the turbulent viscosity yields (Eq. (10.48))

$$-\langle uv \rangle = u_\tau^2 = C_\mu^{1/2} k. \tag{11.186}$$

Rather than using these equations directly, wall functions are designed to provide 'robust' boundary conditions under all circumstances, and to revert to the above relations under the ideal conditions.

Equation (11.186) is used to define the nominal friction velocity as

$$u_\tau^* \equiv C_\mu^{1/4} k_p^{1/2}, \tag{11.187}$$

and the corresponding estimate of y_p^+ is

$$y_p^* \equiv \frac{y_p u_\tau^*}{\nu}. \tag{11.188}$$

The nominal mean velocity is then obtained from the log-law:

$$\langle U \rangle_p^* = u_\tau^* \left(\frac{1}{\kappa} \ln y_p^* + B \right). \tag{11.189}$$

In the mean-momentum equation, the boundary condition is applied not by specifying $\langle U \rangle_p$, but by specifying the shear stress as

$$-\langle uv \rangle_p = u_\tau^{*2} \frac{\langle U \rangle_p}{\langle U \rangle_p^*}. \tag{11.190}$$

This is a robust condition in that the shear stress is of the opposite sign to the velocity $\langle U \rangle_p$; at a separation or reattachment point ($\langle U \rangle_p = 0$) everything is well defined (even where u_τ is zero); if $\langle U \rangle_p$ exceeds the nominal value $\langle U \rangle_p^*$, then $-\langle uv \rangle$ exceeds u_τ^{*2} and thus provides a 'restoring force.'

On the basis of Eq. (11.185), the boundary condition on ε is

$$\varepsilon_p = \frac{u_\tau^{*3}}{\kappa y_p}, \tag{11.191}$$

while zero-normal gradient conditions are applied to k and to the normal stresses.

It is usual for wall functions to be implemented in a more complicated way than that described above, the treatment being incorporated within the finite-volume equations used in the numerical solution procedure (Jones 1994). The location y_p is taken to be the first grid node away from the wall.

The simplifications and economies provided by wall functions are very attractive, and they are widely used in the application of commercial CFD codes to complex turbulent flows. However, under many flow conditions – e.g., strong pressure gradients, separated and impinging flows – their physical basis is uncertain and their accuracy is poor (see, e.g., Wilcox (1993)).

Wall functions introduce y_p as an artificial parameter. For boundary-layer flows for which the log-law relations are accurate, the overall solution is insensitive to the choice of y_p (within the log-law region). However, in other flows it is found that the overall solution is sensitive to this choice. As a consequence it might not be possible to obtain numerically accurate, grid-independent solutions (since refining the grid generally implies reducing y_p).

EXERCISES

11.27 The k–ε or Reynolds-stress model equations are to be solved by a finite-difference method for fully developed channel flow without the use of wall functions. To provide adequate spatial resolution, the grid spacing Δy is specified as

$$\Delta y = \min[\max(a\delta_v, by), c\delta], \qquad (11.192)$$

where δ is the channel half-height, and a, b, and c are positive constants (with $b > c$). Show that the number of grid points N_y required is

$$N_y \approx \int_0^\delta \frac{\mathrm{d}y}{\Delta y} = \frac{1}{c} + \frac{1}{b} \ln\left(\frac{c\delta}{a\delta_v}\right). \qquad (11.193)$$

Use the relation $\mathrm{Re}_\tau = \delta/\delta_v \approx 0.09\mathrm{Re}^{0.88}$ (see Section 7.1.5 on page 278) to estimate the high-Reynolds-number asymptote

$$N_y \sim \frac{0.88}{b} \ln(\mathrm{Re}). \qquad (11.194)$$

11.28 Further to Exercise 11.27, consider turbulence-model calculations of fully developed channel flow, with wall functions applied at $y = y_p$. The finite-difference grid – which extends from $y = y_p$ to $y = \delta$ – has the spacing

$$\Delta y = \min(by, c\delta), \qquad (11.195)$$

with $0 < c < b$. Show that the number N_y of grid points required is

$$N_y \approx \int_{y_p}^\delta \frac{\mathrm{d}y}{\Delta y} = \frac{1}{c} + \frac{1}{b}\left[\ln\left(\frac{c\delta}{by_p}\right) - 1\right]. \qquad (11.196)$$

Show that, if y_p is specified as a fraction of δ, then N_y is independent of the Reynolds number; whereas, if y_p is specified as a multiple of δ_v (e.g., $y_p^+ \equiv y_p/\delta_v = 50$), then N_y increases as $\ln(\text{Re})$.

11.8 Elliptic relaxation models

From a theoretical viewpoint, the modelling of the redistribution $\mathcal{R}_{ij}^{(a)}$ in terms of local quantities is dubious for strongly inhomogeneous flows. From a practical viewpoint, for near-wall flows, the various damping functions and wall-reflection terms are not very satisfactory. The damping functions are essentially *ad hoc* corrections that are often found to be inaccurate beyond the flow used in their development (e.g., Rodi and Scheuerer (1986)). The use of the distance from the wall ℓ_w and the normal direction n is an imperfect way to incorporate non-local information into otherwise local models. In the model equations, it would be preferable for the presence of walls to be felt solely through boundary conditions imposed at the walls.

Durbin (1993) proposed a higher level of closure to address these issues. The essential innovation is that – rather than being local – the pressure–rate-of-strain model is based on the solution of an elliptic equation (similar to the Poisson equation[6]). This, and related *elliptic relaxation models* (Durbin 1991, Dreeben and Pope 1997a), which use neither damping functions nor wall-reflection terms, have met with considerable success in the calculation of a variety of near-wall flows.

Durbin's (1993) model is now described. The starting point is the exact Reynolds-stress equation written

$$\frac{\bar{D}\langle u_i u_j \rangle}{\bar{D}t} + \frac{\partial}{\partial x_k}\left(T_{kij}^{(v)} + T_{kij}^{(p')} + T_{kij}^{(u)}\right) = \mathcal{P}_{ij} + \mathcal{R}_{ij}^{(e)} - \frac{\langle u_i u_j \rangle}{k}\varepsilon, \qquad (11.197)$$

(cf. Eq. (11.138)) with

$$\mathcal{R}_{ij}^{(e)} \equiv (\Pi_{ij} - \tfrac{1}{3}\Pi_{kk}\delta_{ij}) - \left(\varepsilon_{ij} - \frac{\langle u_i u_j \rangle}{k}\varepsilon\right). \qquad (11.198)$$

The primary quantity to be modelled is the redistribution term $\mathcal{R}_{ij}^{(e)}$, which consists of two contributions: the deviatoric part of the velocity–pressure-gradient tensor $\mathcal{R}_{ij}^{(a)}$ (which is zero at a wall); and the difference between ε_{ij} and the simple anisotropic (Rotta) model $\varepsilon\langle u_i u_j \rangle/k$. The dissipation is decomposed in this way so that, at a wall, the dissipation in Eq. (11.197) is exact, and correspondingly $\mathcal{R}_{ij}^{(e)}$ is zero.

---

[6] In fact, the elliptic equation solved (Eq. (11.203)) is closer to the modified Helmholtz equation, which is easier to solve numerically than the Poisson equation.

The turbulent-transport term is modelled as

$$\frac{\partial}{\partial x_k}\left(C_s T \langle u_k u_\ell \rangle \frac{\partial \langle u_i u_j \rangle}{\partial x_\ell}\right), \tag{11.199}$$

where the timescale is $T = \max(k/\varepsilon, 6\tau_\eta)$, as previously defined (see Eq. (11.163)). Except in the viscous sublayer, this transport model is identical to the Daly–Harlow model (Eq. (11.147)).

The quantity $\bar{\mathcal{R}}_{ij}$ is defined as

$$\bar{\mathcal{R}}_{ij} = -\frac{(C_R - 1)}{T}(\langle u_i u_j \rangle - \tfrac{2}{3}k\delta_{ij}) - C_2(\mathcal{P}_{ij} - \tfrac{2}{3}\mathcal{P}\delta_{ij}). \tag{11.200}$$

This is the same as the basic LRR-IP model for $\mathcal{R}_{ij}$ (Eq. (11.129)), except that $1/T$ is used in place of ε/k, and $C_R - 1$ is used in place of C_R to account for the anisotropic dissipation in Eq. (11.197) (see Exercise 11.3).

Outside of the viscous sublayer (so that $T = k/\varepsilon$), if the local model

$$\mathcal{R}_{ij}^{(e)} = \bar{\mathcal{R}}_{ij}, \tag{11.201}$$

were used, then the resulting Reynolds-stress model would be identical to the basic LRR-IP model. However, in place of Eq. (11.201), $\mathcal{R}_{ij}^{(e)}$ is determined by the equations

$$\mathcal{R}_{ij}^{(e)} = k f_{ij}, \tag{11.202}$$

$$(I - L_D^2 \nabla^2)f_{ij} = \frac{\bar{\mathcal{R}}_{ij}}{k}, \tag{11.203}$$

where the lengthscale L_D is obtained from $L = k^{3/2}/\varepsilon$ and the Kolmogorov scale η as

$$L_D = C_L \max(L, C_\eta \eta), \tag{11.204}$$

with $C_L = 0.2$ and $C_\eta = 80$. Equation (11.203) is the *elliptic relaxation equation* for the quantity f_{ij}. The elliptic operator consists of the identity operator I and the Laplacian scaled by L_D. In homogeneous turbulence (where $\nabla^2 f_{ij}$ is zero), or if L_D is taken to be zero, then Eq. (11.203) simply yields $f_{ij} = \bar{\mathcal{R}}_{ij}/k$, so that the local model Eq. (11.201) is recovered. In general, however, because of the Laplacian in Eq. (11.203), $f_{ij}(x, t)$ is determined non-locally by $\bar{\mathcal{R}}_{ij}(x', t)$, and by the boundary conditions imposed on f_{ij}.

For inhomogeneous flows, it is rarely possible to provide convincing justification for a model, either from the Navier–Stokes equations, or from experimental observations; and certainly there is no such justification for Eq. (11.203). However, some understanding is provided by a manipulation of the Poisson equation for pressure (Durbin 1991, 1993). If the Poisson equation is written

$$\nabla^2 p'(x) = S(x), \qquad (11.205)$$

and $\phi(x)$ is a random field, then a solution for the correlation $\langle p'(x)\phi(x)\rangle$ is

$$\langle p'(x)\phi(x)\rangle = -\frac{1}{4\pi} \iiint \langle S(y)\phi(x)\rangle \frac{dy}{|x-y|}, \qquad (11.206)$$

cf. Eqs. (11.179) and (11.182). (The integration is over the flow domain, and a harmonic component $\langle p^{(h)}(x)\phi(x)\rangle$ can be added to satisfy the appropriate boundary conditions.) Now suppose that the two-point correlation can be approximated by

$$\langle S(y)\phi(x)\rangle = \langle S(y)\phi(y)\rangle e^{-|x-y|/L_D}. \qquad (11.207)$$

Then, Eq. (11.206) becomes

$$\langle p'(x)\phi(x)\rangle = \iiint \langle S(y)\phi(y)\rangle \left\{ -\frac{1}{4\pi} \frac{e^{-|x-y|/L_D}}{|x-y|} \right\} dy. \qquad (11.208)$$

When it is multiplied by $-L_D^2$, the term in braces ($\{\ \}$) is the Green's function for the operator in Eq. (11.203). Thus, a loose connection between the Poisson equation for pressure and the elliptic relaxation equation is established.

Perhaps more important than the specific form of the elliptic relaxation equation is the fact that it allows additional boundary conditions to be applied. In a local Reynolds-stress model, only one boundary condition per Reynolds stress can be applied at a wall (i.e., $\langle u_i u_j \rangle = 0$), and then the asymptotic variation of the Reynolds stresses with distance is determined by the model. In an elliptic relaxation model, on the other hand, two boundary conditions can be applied to each $\langle u_i u_j \rangle$ and f_{ij} pair. The additional boundary condition can be used to ensure the correct asymptotic Reynolds-stress variation.

Reynolds-stress models using elliptic relaxation have been used with success to calculate boundary layers with adverse pressure gradients and convex curvatures (Durbin 1993). For channel flow (for which DNS data are available) the Reynolds-stress budgets are reproduced quite accurately (Dreeben and Pope 1997a).

The k–ε–$\overline{v^2}$ model with elliptic relaxation

Prior to the elliptic-relaxation Reynolds-stress model, Durbin (1991) introduced elliptic relaxation for a k–ε–$\overline{v^2}$ model. For the boundary-layer flows considered, $\overline{v^2}$ is a model for $\langle v^2 \rangle$, and the turbulent viscosity model is used with

$$\nu_T = C_\mu' \frac{\overline{v^2} k}{\varepsilon} \qquad (11.209)$$

(see Eq. (11.158) and Fig. 11.17).

This model has also been applied successfully to a number of complex flows (e.g., Durbin (1995) and Parneix, Durbin, and Behnia (1998a)). There $\overline{v^2}$ is interpreted as a scalar which, adjacent to a wall, is a model for the variance of the normal velocity, i.e., $\overline{v^2} = \langle u_i u_j \rangle n_i n_j$.

EXERCISE

11.29   As a generalization of the elliptic relaxation equation (Eq. (11.203)), consider the equation

$$\frac{\mathcal{R}_{ij}^{(e)}}{k} - L_D^{(2-p-q)} \nabla \cdot \left[L_D^p \nabla \left(L_D^q \frac{\mathcal{R}_{ij}^{(e)}}{k} \right) \right] = \frac{\bar{\mathcal{R}}_{ij}}{k}. \qquad (11.210)$$

Durbin's model has exponents $p = q = 0$, whereas Wizman *et al.* (1996) suggest a 'neutral' model with $p = 0$ and $q = 2$, and another model with $p = -2$ and $q = 2$. Consider the log-law region in the usual coordinates, and assume that the turbulence statistics are self-similar (with $L_D \sim y$, $k \sim y^0$, $\mathcal{R}_{ij}^{(e)} \sim y^{-1}$, and $\bar{\mathcal{R}}_{ij} \sim y^{-1}$). Show that a solution to Eq. (11.210) is

$$\mathcal{R}_{ij}^{(e)} = \bar{\mathcal{R}}_{ij} \Big/ \left(1 - \frac{C_L^2 \kappa^2}{C_\mu^3} (q-1)(p+q-2) \right). \qquad (11.211)$$

Comment on the effects of the three models mentioned above.

11.9 Algebraic stress and nonlinear viscosity models

11.9.1 Algebraic stress models

By the introduction of an approximation for the transport terms, a Reynolds-stress model can be reduced to a set of algebraic equations. These equations form an *algebraic stress model* (ASM), which implicitly determines the Reynolds stresses (locally) as functions of k, ε, and the mean velocity gradients. Because of the approximation involved, algebraic stress models are inherently less general and less accurate than Reynolds-stress models. However, because of their relative simplicity, they have been used as turbulence models (in conjunction with the model equations for k and ε). In addition, an algebraic stress model provides some insights into the Reynolds-stress model from which it is derived; and it can also be used to obtain a nonlinear turbulent-viscosity model.

A standard model Reynolds-stress transport equation is

$$\mathcal{D}_{ij} \equiv \frac{\bar{D}\langle u_i u_j \rangle}{\bar{D}t} - \frac{\partial}{\partial x_k}\left(\frac{C_s k}{\varepsilon}\langle u_k u_\ell \rangle \frac{\partial}{\partial x_\ell}\langle u_i u_j \rangle\right) = \mathcal{P}_{ij} + \mathcal{R}_{ij} - \tfrac{2}{3}\varepsilon\delta_{ij}. \quad (11.212)$$

This is a coupled set of six partial differential equations. The terms on the right-hand side are local, algebraic functions of $\partial\langle U_i \rangle/\partial x_j$, $\langle u_i u_j \rangle$, and ε – they do not involve derivatives of the Reynolds stresses. In algebraic stress models, the transport terms $\mathcal{D}_{ij}$ (on the left-hand side of Eq. (11.212)) are *approximated* by an algebraic expression, so that the entire equation becomes algebraic. Specifically, Eq. (11.212) becomes a set of six algebraic equations that implicitly determines the Reynolds stresses as functions of k, ε, and the mean velocity gradients.

In some circumstances (e.g., the log-law region of high-Reynolds-number fully developed channel flow), the transport terms in Eq. (11.212) are negligible, so that (in a sense) the Reynolds stresses are in local equilibrium with the imposed mean velocity gradient. However, the complete neglect of the transport terms is inconsistent unless $\mathcal{P}/\varepsilon$ is unity, since half the trace of Eq. (11.212) is

$$\tfrac{1}{2}\mathcal{D}_{\ell\ell} = \mathcal{P} - \varepsilon. \quad (11.213)$$

Rodi (1972) introduced the more general *weak-equilibrium assumption*. The Reynolds stress can be decomposed as

$$\langle u_i u_j \rangle = k\frac{\langle u_i u_j \rangle}{k} = k(2b_{ij} + \tfrac{2}{3}\delta_{ij}), \quad (11.214)$$

and so spatial and temporal variations in $\langle u_i u_j \rangle$ can be considered to be due to variations in k and b_{ij}. In the weak-equilibrium assumption, the variations in $\langle u_i u_j \rangle/k$ (or equivalently in b_{ij}) are neglected, but the variations in $\langle u_i u_j \rangle$ due to those in k are retained. For the mean convection term this leads to the approximation

$$\frac{\bar{D}}{\bar{D}t}\langle u_i u_j \rangle = \frac{\langle u_i u_j \rangle}{k}\frac{\bar{D}k}{\bar{D}t} + k\frac{\bar{D}}{\bar{D}t}\left(\frac{\langle u_i u_j \rangle}{k}\right)$$

$$\approx \frac{\langle u_i u_j \rangle}{k}\frac{\bar{D}k}{\bar{D}t}. \quad (11.215)$$

The same approximation applied to the entire transport term yields

$$\mathcal{D}_{ij} \approx \frac{\langle u_i u_j \rangle}{k}\frac{1}{2}\mathcal{D}_{\ell\ell} = \frac{\langle u_i u_j \rangle}{k}(\mathcal{P} - \varepsilon), \quad (11.216)$$

where the last step follows from Eq. (11.213).

The use of the weak-equilibrium assumption (Eq. (11.216)) in the model Reynolds-stress equation (Eq. (11.212)) leads to the algebraic stress model

$$\frac{\langle u_i u_j \rangle}{k}(\mathcal{P} - \varepsilon) = \mathcal{P}_{ij} + \mathcal{R}_{ij} - \tfrac{2}{3}\varepsilon\delta_{ij}. \tag{11.217}$$

This comprises five independent algebraic equations (since the trace contains no information), which can be used to determine $\langle u_i u_j \rangle/k$ (or equivalently b_{ij}) in terms of k, ε, and $\partial\langle U_i \rangle/\partial x_j$.

As an example of the insights that an ASM can provide, if $\mathcal{R}_{ij}$ is given by the LRR-IP model, then Eq. (11.217) can be manipulated to yield

$$b_{ij} = \frac{\tfrac{1}{2}(1 - C_2)}{C_R - 1 + \mathcal{P}/\varepsilon} \frac{(\mathcal{P}_{ij} - \tfrac{2}{3}\mathcal{P}\delta_{ij})}{\varepsilon}, \tag{11.218}$$

see Exercise 11.20. Consequently it may be seen that an implication of the model is that the Reynolds stress anisotropy is directly proportional to the production anisotropy.

For simple shear flow, Eq. (11.218) is readily solved to obtain the anisotropies b_{ij} as functions of $\mathcal{P}/\varepsilon$ (see Exercise 11.20): these are plotted in Fig. 11.20. For large $\mathcal{P}/\varepsilon$, $|b_{12}|$ tends to the asymptote $\sqrt{\tfrac{1}{6}C_2(1 - C_2)} = \tfrac{1}{5}$, whereas the value given by the k–ε model continually increases and becomes non-realizable.

Again, for simple shear flow, if the relation

$$-\langle uv \rangle = \frac{C_\mu k^2}{\varepsilon}\frac{\partial\langle U \rangle}{\partial y} \tag{11.219}$$

is used to define C_μ, then it can be deduced (Exercise 11.31) that the ASM (Eq. (11.218)) yields

$$C_\mu = \frac{\tfrac{2}{3}(1 - C_2)(C_R - 1 + C_2\mathcal{P}/\varepsilon)}{(C_R - 1 + \mathcal{P}/\varepsilon)^2}. \tag{11.220}$$

Consequently, as shown in Fig. 11.21, the value of C_μ implied by the LRR-IP model decreases with $\mathcal{P}/\varepsilon$, corresponding to 'shear-thinning' behavior – C_μ decreases with increasing shearing, $\mathcal{S}k/\varepsilon$.

With respect to the mean flow convection, the weak equilibrium assumption (Eq. (11.215)) amounts to

$$\frac{\bar{\mathrm{D}}}{\bar{\mathrm{D}}t}b_{ij} = 0, \tag{11.221}$$

where b_{ij} are the components of the Reynolds-stress anisotropy tensor $\mathbf{b}$, referred to an inertial frame. Part of the assumption embodied in Eq. (11.221) is, therefore, that the principal axes of $\mathbf{b}$ do not rotate (relative to an inertial

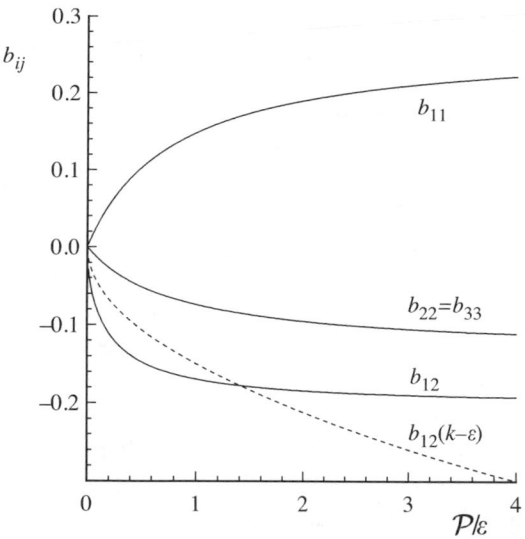

Fig. 11.20. Reynolds-stress anisotropies as functions of $\mathcal{P}/\varepsilon$ according to the LRR-IP algebraic stress model. The dashed line shows b_{12} according to the k–ε model.

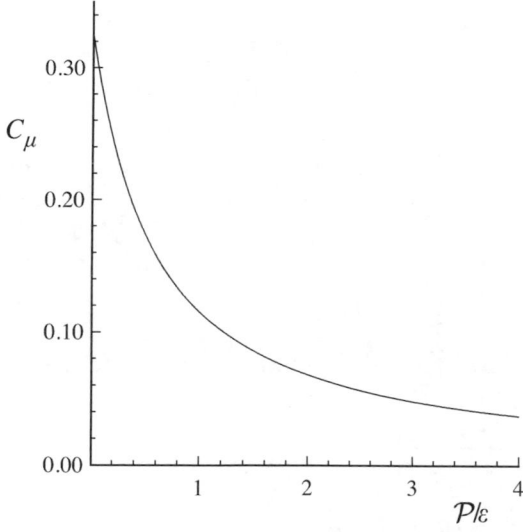

Fig. 11.21. The value of C_μ as a function of $\mathcal{P}/\varepsilon$ given by the LRR-IP algebraic stress model (Eq. (11.220)).

frame, following the mean flow). For some flows, for example those with significant mean streamline curvature, Girimaji (1997) argues that a better assumption is that the components of **b** are fixed relative to a particular rotating frame.

Table 11.5. *Coefficients in Eq. (11.223) for various models*

Coefficient	General	LRR-IP	LRR-QI
g^{-1}	$-\frac{1}{2}f^{(1)} - 1 + \dfrac{P}{\varepsilon}$	$C_R - 1 + \dfrac{P}{\varepsilon}$	$C_R - 1 + \dfrac{P}{\varepsilon}$
γ_1	$\frac{2}{3} - \frac{1}{2}f^{(3)}$	$\frac{2}{3}(1 - C_2)$	$\frac{4}{15}$
γ_2	$1 - \frac{1}{2}f^{(4)}$	$1 - C_2$	$\frac{1}{11}(5 - 9C_2)$
γ_3	$1 - \frac{1}{2}f^{(5)}$	$1 - C_2$	$\frac{1}{11}(1 + 7C_2)$

EXERCISES

11.30 Consider the algebraic stress model based on LRR-IP applied to a simple shear flow (Eqs. (11.130)–(11.134)), and obtain expressions for b_{ij} in the limit $P/\varepsilon \to \infty$. Are these values realizable? Are they consistent with RDT?

11.31 Manipulate Eq. (11.219) to obtain

$$C_\mu = \frac{4b_{12}^2}{P/\varepsilon}. \tag{11.222}$$

Hence use Eq. (11.134) to verify the expression for C_μ, Eq. (11.220).

11.32 Consider a general model for the pressure–rate-of-strain tensor that is linear in b_{ij} and in the mean velocity gradients, i.e., Eq. (11.135) with $f^{(2)} = f^{(6-8)} = 0$. Show that the corresponding algebraic stress model (Eq. (11.217)) can be written

$$\begin{aligned} b_{ij} = -g[\gamma_1 \widehat{S}_{ij} + \gamma_2(\widehat{S}_{ik} b_{kj} + b_{ik} \widehat{S}_{kj} - \tfrac{2}{3} \widehat{S}_{k\ell} b_{\ell k} \delta_{ij}) \\ + \gamma_3(\widehat{\Omega}_{ik} b_{kj} - b_{ik} \widehat{\Omega}_{kj})], \end{aligned} \tag{11.223}$$

where the coefficients (in general, and for the LRR models) are those given in Table 11.5.

11.9.2 Nonlinear turbulent viscosity

The algebraic stress model equation Eq. (11.217) is an *implicit* equation for $\langle u_i u_j \rangle / k$, or equivalently for the anisotropy b_{ij}. Clearly, there is benefit in obtaining an *explicit* relation of the form

$$b_{ij} = \mathcal{B}_{ij}(\widehat{\mathbf{S}}, \widehat{\mathbf{\Omega}}), \tag{11.224}$$

where $\widehat{\mathbf{S}}$ and $\widehat{\mathbf{\Omega}}$ are the normalized mean rate-of-strain and rotation tensors (Eqs. (11.136) and (11.137)).

Table 11.6. *The complete set of independent, symmetric, deviatoric functions* $\widehat{\mathcal{T}}^{(n)}$ *of a deviatoric symmetric tensor* $\widehat{\mathbf{S}}$ *and an antisymmetric tensor* $\widehat{\mathbf{\Omega}}$, *shown in matrix notation: braces denote traces, e.g.,* $\{\widehat{\mathbf{S}}^2\} = \widehat{S}_{ij}\widehat{S}_{ji}$

$$\widehat{\mathcal{T}}^{(1)} = \widehat{\mathbf{S}}, \qquad\qquad \widehat{\mathcal{T}}^{(6)} = \widehat{\mathbf{\Omega}}^2\widehat{\mathbf{S}} + \widehat{\mathbf{S}}\widehat{\mathbf{\Omega}}^2 - \tfrac{2}{3}\{\widehat{\mathbf{S}}\widehat{\mathbf{\Omega}}^2\}\mathbf{I},$$
$$\widehat{\mathcal{T}}^{(2)} = \widehat{\mathbf{S}}\widehat{\mathbf{\Omega}} - \widehat{\mathbf{\Omega}}\widehat{\mathbf{S}}, \qquad \widehat{\mathcal{T}}^{(7)} = \widehat{\mathbf{\Omega}}\widehat{\mathbf{S}}\widehat{\mathbf{\Omega}}^2 - \widehat{\mathbf{\Omega}}^2\widehat{\mathbf{S}}\widehat{\mathbf{\Omega}},$$
$$\widehat{\mathcal{T}}^{(3)} = \widehat{\mathbf{S}}^2 - \tfrac{1}{3}\{\widehat{\mathbf{S}}^2\}\mathbf{I}, \qquad \widehat{\mathcal{T}}^{(8)} = \widehat{\mathbf{S}}\widehat{\mathbf{\Omega}}\widehat{\mathbf{S}}^2 - \widehat{\mathbf{S}}^2\widehat{\mathbf{\Omega}}\widehat{\mathbf{S}},$$
$$\widehat{\mathcal{T}}^{(4)} = \widehat{\mathbf{\Omega}}^2 - \tfrac{1}{3}\{\widehat{\mathbf{\Omega}}^2\}\mathbf{I}, \qquad \widehat{\mathcal{T}}^{(9)} = \widehat{\mathbf{\Omega}}^2\widehat{\mathbf{S}}^2 + \widehat{\mathbf{S}}^2\widehat{\mathbf{\Omega}}^2 - \tfrac{2}{3}\{\widehat{\mathbf{S}}^2\widehat{\mathbf{\Omega}}^2\}\mathbf{I},$$
$$\widehat{\mathcal{T}}^{(5)} = \widehat{\mathbf{\Omega}}\widehat{\mathbf{S}}^2 - \widehat{\mathbf{S}}^2\widehat{\mathbf{\Omega}}, \qquad \widehat{\mathcal{T}}^{(10)} = \widehat{\mathbf{\Omega}}\widehat{\mathbf{S}}^2\widehat{\mathbf{\Omega}}^2 - \widehat{\mathbf{\Omega}}^2\widehat{\mathbf{S}}^2\widehat{\mathbf{\Omega}}.$$

The most general possible expression of the form Eq. (11.224) can be written

$$\mathcal{B}_{ij}(\widehat{\mathbf{S}}, \widehat{\mathbf{\Omega}}) = \sum_{n=1}^{10} G^{(n)}\widehat{\mathcal{T}}_{ij}^{(n)}, \qquad (11.225)$$

where the tensors $\widehat{\mathcal{T}}^{(n)}$ are given in Table 11.6, and the coefficients can depend upon the five invariants $\widehat{S}_{ii}^2$, $\widehat{\Omega}_{ii}^2$, $\widehat{S}_{ii}^3$, $\widehat{\Omega}_{ij}^2\widehat{S}_{ji}$, and $\widehat{\Omega}_{ij}^2\widehat{S}_{ji}^2$ (Pope 1975). Like b_{ij}, each of the tensors $\widehat{\mathcal{T}}^{(n)}$ is non-dimensional, symmetric, and deviatoric. As a set they form an *integrity basis*, meaning that every symmetric deviatoric second-order tensor formed from $\widehat{\mathbf{S}}$ and $\widehat{\mathbf{\Omega}}$ can be expressed as a linear combination of these ten. (The proof of this is based on the Cayley–Hamilton theorem, Pope (1975).)

With the specification $G^{(1)} = -C_\mu$, $G^{(n)} = 0$ for $n > 1$, Eq. (11.225) reverts to the linear k–ε turbulent-viscosity formula

$$b_{ij} = -C_\mu \widehat{S}_{ij}, \qquad (11.226)$$

or, equivalently,

$$\langle u_i u_j \rangle - \tfrac{2}{3}k\delta_{ij} = -C_\mu \frac{k^2}{\varepsilon}\left(\frac{\partial \langle U_i \rangle}{\partial x_j} + \frac{\partial \langle U_j \rangle}{\partial x_i}\right). \qquad (11.227)$$

A non-trivial specification of $G^{(n)}$ for $n > 1$ yields a nonlinear turbulent-viscosity model, i.e., an explicit formula for $\langle u_i u_j \rangle$ that is nonlinear in the mean velocity gradients.

For flows that are statistically two-dimensional, the situation is considerably simpler. The tensors $\widehat{\mathcal{T}}^{(1)}$, $\widehat{\mathcal{T}}^{(2)}$, and $\widehat{\mathcal{T}}^{(3)}$ form an integrity basis, and there are just two independent invariants $\widehat{S}_{kk}^2$ and $\widehat{\Omega}_{kk}^2$ (Pope 1975, Gatski and Speziale 1993). Consequently $G^{(4)} - G^{(10)}$ can be set to zero. Furthermore, the term in $\widehat{\mathcal{T}}^{(3)}$ can be absorbed into the modified pressure (see Exercise 11.33) so that the value of $G^{(3)}$ has no effect on the mean velocity field. With $G^{(3)} = 0$, the nonlinear viscosity model for statistically two-dimensional flows is

$$b_{ij} = G^{(1)}\widehat{\mathcal{T}}_{ij}^{(1)} + G^{(2)}\widehat{\mathcal{T}}_{ij}^{(2)}, \tag{11.228}$$

or, equivalently,

$$\langle u_i u_j \rangle - \tfrac{2}{3}k\delta_{ij} = 2G^{(1)}\frac{k^2}{\varepsilon}\bar{S}_{ij} + 2G^{(2)}\frac{k^3}{\varepsilon^2}(\bar{S}_{ik}\bar{\Omega}_{kj} - \bar{\Omega}_{ik}\bar{S}_{kj}). \tag{11.229}$$

One way to obtain a suitable specification of the coefficients $G^{(n)}$ is from an algebraic stress model. Since the nonlinear turbulent-viscosity formula Eq. (11.225) provides a completely general expression for b_{ij} in terms of mean velocity gradients, it follows that to every algebraic stress model there corresponds a nonlinear viscosity model. It is a matter of algebra to determine the corresponding coefficients $G^{(n)}$. For example, for statistically two-dimensional flows, the coefficients $G^{(n)}$ corresponding to the LRR-IP algebraic stress model are

$$G^{(1)} = -C_\mu, \quad G^{(2)} = -\lambda C_\mu, \quad G^{(3)} = 2\lambda C_\mu, \quad G^{(4-10)} = 0, \tag{11.230}$$

where

$$\lambda \equiv \frac{1 - C_2}{C_R - 1 + \mathcal{P}/\varepsilon}, \tag{11.231}$$

$$C_\mu \equiv \frac{\tfrac{2}{3}\lambda}{1 - \tfrac{2}{3}\lambda^2 \widehat{S}_{ii}^2 - 2\lambda^2 \widehat{\Omega}_{ii}^2}, \tag{11.232}$$

see Exercise 11.36. Figure 11.22 shows $-G^{(1)} = C_\mu$ and $-G^{(2)} = \lambda C_\mu$ as functions of $\mathcal{S}k/\varepsilon$ and $\Omega k/\varepsilon$ (where $\Omega = (2\bar{\Omega}_{ij}\bar{\Omega}_{ij})^{1/2}$).

The nonlinear viscosity model defined by Eqs. (11.230)–(11.232) is not completely explicit, because the definition of λ contains $\mathcal{P}/\varepsilon = -2b_{ij}\widehat{S}_{ij}$. Girimaji (1996) gives fully explicit formulae, obtained by solving the cubic equation for λ, see Exercise 11.36. Taulbee (1992) and Gatski and Speziale (1993) extend this approach to three-dimensional flows, where, in general, all ten coefficients $G^{(n)}$ are non-zero.

Nonlinear viscosity models, not based on algebraic stress models, have been proposed by Yoshizawa (1984), Speziale (1987), Rubinstein and Barton (1990), Craft, Launder, and Suga (1996), and others. The first three mentioned are quadratic in the mean velocity gradients, and so $G^{(1)} - G^{(4)}$ are non-zero. In the model of Craft et al. $G^{(5)}$ is also non-zero. In addition to mean velocity gradients, the models of Yoshizawa and Speziale also involve $\bar{D}\bar{S}_{ij}/\bar{D}t$.

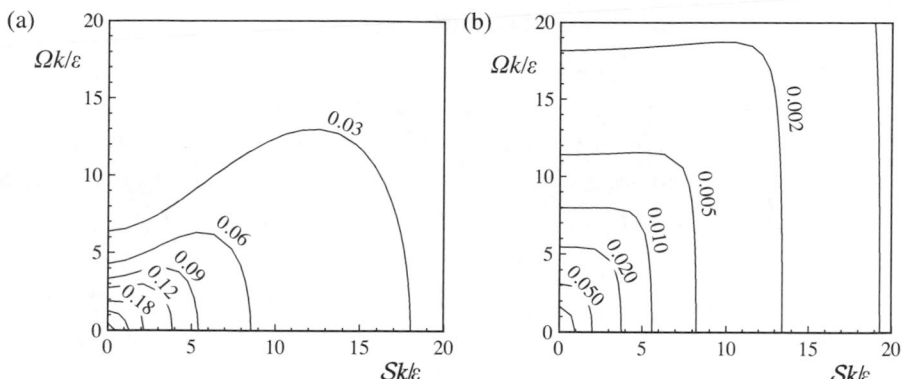

Fig. 11.22. Contour plots of (a) $C_\mu = -G^{(1)}$, and (b) $-G^{(2)}$, for the LRR-IP nonlinear viscosity model (Eqs. (11.230)–(11.232)).

EXERCISES

11.33 Consider a statistically two-dimensional turbulent flow in the x_1–x_2 plane (so that $\langle U_3 \rangle = 0$ and $\partial \langle U_i \rangle / \partial x_3 = 0$). By evaluating each component, show that

$$\widehat{S}_{ij}^2 = \tfrac{1}{2}\delta_{ij}^{(2)}\widehat{S}_{kk}^2, \tag{11.233}$$

where

$$\boldsymbol{\delta}^{(2)} \equiv \begin{bmatrix} 1 & 0 & 0 \\ 0 & 1 & 0 \\ 0 & 0 & 0 \end{bmatrix}. \tag{11.234}$$

Hence show that

$$\widehat{T}_{ij}^{(3)} = -\widehat{S}_{kk}^2 \, \widehat{T}_{ij}^{(0)}, \tag{11.235}$$

where

$$\widehat{T}_{ij}^{(0)} \equiv \tfrac{1}{3}\delta_{ij} - \tfrac{1}{2}\delta_{ij}^{(2)}. \tag{11.236}$$

11.34 In the nonlinear viscosity model (Eq. (11.225)), the contribution from $\widehat{\boldsymbol{\mathcal{T}}}^{(3)}$ in the mean-momentum equation is

$$\frac{\partial}{\partial x_i}\langle u_i u_j \rangle^{(3)} = \frac{\partial}{\partial x_i}\left(2kG^{(3)}\widehat{T}_{ij}^{(3)}\right). \tag{11.237}$$

For the two-dimensional flows being considered, show that this can be re-expressed as

$$\frac{\partial}{\partial x_i}\langle u_i u_j \rangle^{(3)} = \frac{\partial}{\partial x_j}\left(\tfrac{1}{3}kG^{(3)}\widehat{S}_{kk}^2\right). \tag{11.238}$$

(This term can be absorbed into the modified pressure, so that $G^{(3)}$ has no effect on the mean velocity field.)

11.35   Evaluate $\widehat{\mathcal{T}}^{(2)}$ for simple shear flow in which $\partial\langle U_1\rangle/\partial x_2 = \mathcal{S}$ is the only non-zero mean velocity gradient. Hence show that (for this flow) the shear stress $\langle u_1 u_2\rangle$ given by the nonlinear viscosity model (Eq. (11.229)) depends on $G^{(1)}$, but is independent of $G^{(2)}$.

11.36   Consider a pressure–rate-of-strain model that is linear both in the mean velocity gradients and in the anisotropy. The corresponding algebraic stress model is Eq. (11.223). For turbulent flows that are statistically two-dimensional, the corresponding nonlinear viscosity model is given by (Pope 1975)

$$G^{(1)} = -C_\mu, \qquad G^{(2)} = -C_\mu g \gamma_3,$$

$$G^{(3)} = 2C_\mu g \gamma_2, \qquad G^{(4-10)} = 0,$$

(11.239)

where g, γ_1, γ_2, and γ_3 are defined in Table 11.5. Use this result to verify Eq. (11.230).

By using the relation $\mathcal{P}/\varepsilon = -2b_{ij}\widehat{S}_{ij}$, show that λ (Eq. (11.231)) satisfies a cubic equation.

11.37   The model of Rubinstein and Barton (1990) can be written

$$\langle u_i u_j\rangle - \tfrac{2}{3}k\delta_{ij} = -2\nu_{\mathrm{T}}\bar{S}_{ij} + \frac{k^3}{\varepsilon}\left[C_{\tau 1}\frac{\partial\langle U_i\rangle}{\partial x_k}\frac{\partial\langle U_j\rangle}{\partial x_k}\right.$$

$$+ C_{\tau 2}\left(\frac{\partial\langle U_i\rangle}{\partial x_k}\frac{\partial\langle U_k\rangle}{\partial x_j} + \frac{\partial\langle U_j\rangle}{\partial x_k}\frac{\partial\langle U_k\rangle}{\partial x_i}\right)$$

$$\left. + C_{\tau 3}\frac{\partial\langle U_k\rangle}{\partial x_i}\frac{\partial\langle U_k\rangle}{\partial x_j}\right] - Q\delta_{ij},$$

(11.240)

where Q is defined such that the trace of the right-hand side vanishes, and the constants are $C_{\tau 1} = 0.034$, $C_{\tau 2} = 0.104$, and $C_{\tau 3} = -0.014$. Re-express this model in the form of Eq. (11.225) and determine the coefficients $G^{(n)}$.

11.38   Craft *et al.* (1996) propose a nonlinear viscosity model involving the tensors $\widehat{\mathcal{T}}^{(1-5)}$ and also

$$\widehat{\mathcal{T}}_{ij}^* \equiv \widehat{S}_{ij}^3 - \tfrac{1}{3}\widehat{S}_{kk}^3 \delta_{ij}.$$

(11.241)

Re-express $\widehat{\mathcal{T}}_{ij}^*$ in terms of $\widehat{\mathcal{T}}_{ij}^{(n)}$.

Table 11.7. *Attributes of various RANS turbulence models (the first four are turbulent-viscosity models)*

Model	Specified fields	Fields from differential equations used in modelling	Primary quantities modelled
Mixing length	ℓ_{m}	$\dfrac{\partial \langle U_i \rangle}{\partial x_j}$	$\langle u_i u_j \rangle$
One-equation (ν_{T})	–	$\nu_{\mathrm{T}},\ \dfrac{\partial \langle U_i \rangle}{\partial x_j}$	$\langle u_i u_j \rangle,\ \dfrac{\bar{\mathrm{D}} \nu_{\mathrm{T}}}{\bar{\mathrm{D}} t}$
One-equation (k–ℓ_{m})	ℓ_{m}	$k,\ \dfrac{\partial \langle U_i \rangle}{\partial x_j}$	$\langle u_i u_j \rangle,\ \varepsilon$
Two-equation (k–ε) isotropic viscosity nonlinear viscosity algebraic stress	–	$k,\ \varepsilon,\ \dfrac{\partial \langle U_i \rangle}{\partial x_j}$	$\langle u_i u_j \rangle,\ \dfrac{\bar{\mathrm{D}} \varepsilon}{\bar{\mathrm{D}} t}$
Reynolds stress	–	$\langle u_i u_j \rangle,\ \varepsilon,\ \dfrac{\partial \langle U_i \rangle}{\partial x_j}$	$\mathcal{R}_{ij},\ \dfrac{\bar{\mathrm{D}} \varepsilon}{\bar{\mathrm{D}} t}$
Reynolds stress/ elliptic relaxation	–	$\langle u_i u_j \rangle,\ f_{ij},\ \varepsilon,\ \dfrac{\partial \langle U_i \rangle}{\partial x_j}$	$\bar{\mathcal{R}}_{ij},\ \dfrac{\bar{\mathrm{D}} \varepsilon}{\bar{\mathrm{D}} t}$

11.10 Discussion

Table 11.7 shows attributes of the principal turbulence models introduced in this and the previous chapter. All of these are RANS models, in that the turbulence-model equations are solved together with the Reynolds-averaged Navier–Stokes equations, to which they provide closure. These models – especially the k–ε and Reynolds-stress models – are discussed in this section with respect to the criteria for appraising models described in Chapter 8.

The level of description

In Table 11.7, the models are listed in order of increasing level of description. The first two models listed are intended solely to determine a turbulent viscosity ν_{T}, and they provide little information about the turbulence. In the k–ε model, the dependent variables provide a description of the turbulent scales of velocity ($k^{1/2}$), length ($L = k^{3/2}/\varepsilon$), and time ($\tau = k/\varepsilon$). The one-equation k–ℓ_{m} model provides the same level of description, but the lengthscale ℓ_{m} is prescribed rather than calculated.

The models mentioned thus far (the first four in Table 11.7) are turbulent-viscosity models. The all-important Reynolds-stress tensor is not represented directly, but is modelled as a local function of k, ε and $\partial\langle U_i\rangle/\partial x_j$. However this is done – by an isotropic or nonlinear viscosity model, or by an algebraic stress model – the level of description is the same; and the assumption that $\langle u_i u_j\rangle$ is locally determined by $\partial\langle U_i\rangle/\partial x_j$ is unavoidable.

In Reynolds-stress models, the direct representation of the Reynolds stresses removes this intrinsic assumption of turbulent-viscosity models – that $\langle u_i u_j\rangle$ is locally determined by $\partial\langle U_i\rangle/\partial x_j$. In the Reynolds-stress equation, the mean convection $\bar{D}\langle u_i u_j\rangle/\bar{D}t$ is balanced by four processes – production, dissipation, redistribution, and turbulent transport – of which the first three are usually dominant (and comparable). At this level of closure, production and dissipation (at least its isotropic part) are in closed form, so the principal quantity to be modelled is the redistribution due to the fluctuating pressure field, $\mathcal{R}_{ij}$.

In regions of strong inhomogeneity, especially near walls, the assumption that $\mathcal{R}_{ij}$ is locally determined (by $\langle u_i u_j\rangle$, ε, and $\partial\langle U_i\rangle/\partial x_j$) is questionable. The fluctuating pressure field p' is governed by a Poisson equation, so that there can be substantial non-local effects such as wall-blocking. The higher level of description in elliptic relaxation models allows non-local effects to be incorporated into the modelling of redistribution (as $\mathcal{R}_{ij}^{(e)} = k f_{ij}$, Eq. (11.202)). In particular, near a wall, $\mathcal{R}_{ij}^{(e)}$ is strongly affected by the boundary conditions imposed on f_{ij} at the wall.

In the last three models listed, the lengthscales and timescales used in the modelling are obtained from ε (through $L = k^{3/2}/\varepsilon$ and $\tau = k/\varepsilon$). In some situations – for example, when the turbulence is approximately self-similar – a single scale may be adequate to parametrize the state of the turbulence. However, in general, especially for rapidly changing flows, more than one parameter may be required to characterize the states and responses of the various scales of turbulent motion. Models that introduce additional scales have been proposed by Hanjalić, Launder, and Schiestel (1980), Wilcox (1988), and Lumley (1992), among others. These models have not found widespread use.

EXERCISE

11.39   Under what circumstances, or with what assumptions or choice of parameters, does each model in the following list reduce to its successor?

(a) An elliptic-relaxation Reynolds-stress model.

(b) A Reynolds-stress model.

(c) An algebraic stress model (using k–ε).

(d) A nonlinear viscosity model (using k–ε).

(e) The standard (isotropic viscosity) k–ε model.

(f) The one-equation k–ℓ_m model.

(g) The mixing-length model.

Completeness

The mixing-length and one-equation k–ℓ_m models are not complete, because they require the specification of $\ell_\mathrm{m}(\boldsymbol{x}, t)$. The remaining models are complete.

The Spalart–Allmaras model for ν_T involves the distance to the wall ℓ_w; and some near-wall treatments in other models involve ℓ_w and the normal $\boldsymbol{n}$. Since ℓ_w and $\boldsymbol{n}$ can be defined unambiguously for general flows (e.g., Eq. (11.171)), their use does not render the models incomplete.

Cost and ease of use

In this discussion we take the k–ε model as a reference: the model is incorporated into most commercial CFD codes, and it is generally regarded as being easy to use and computationally inexpensive when it is used in conjunction with wall functions.[*]

If wall functions are not employed, the task of performing k–ε calculations for the viscous near-wall region is significantly more difficult and expensive. This is due to the need to resolve k and ε (which vary strongly in the near-wall region); and also to the fact that the source terms in these equations become very large close to the wall. (In the log-law region, the term $C_{\varepsilon 2}\varepsilon^2/k$ varies as u_τ^4/y^2.)

The Spalart–Allmaras model is – by design – much simpler and less expensive for near-wall aerodynamic flows. This is because, compared with k and ε, the turbulent viscosity ν_T behaves benignly in the near-wall region, and is more easily resolved.

In comparison with the k–ε model, Reynolds-stress models are somewhat more difficult and costly because

(i) in general there are seven turbulence equations to be solved (for $\langle u_i u_j \rangle$ and ε) instead of two (for k and ε);

(ii) the model Reynolds-stress equation is substantially more complicated than the k equation (and hence requires coding effort); and

(iii) in the mean-momentum equation, the term

$$-\frac{\partial}{\partial x_i}\langle u_i u_j \rangle \tag{11.242}$$

results in a less favorable numerical coupling between the flow and turbulence equations than does the corresponding term

$$\frac{\partial}{\partial x_i}\left[v_\mathrm{T}\left(\frac{\partial \langle U_i \rangle}{\partial x_j} + \frac{\partial \langle U_j \rangle}{\partial x_i} \right) \right] \tag{11.243}$$

in the k–ε model.

Typically, the CPU time required for a Reynolds-stress-model calculation can be more than that for a k–ε calculation by a factor of two.

The primary motivation for the use of algebraic stress models is to avoid the cost and difficulty of solving the Reynolds-stress model equations. However, the general experience is that these benefits are not realized. The algebraic stress model equations are coupled nonlinear equations, often with multiple roots, which are non-trivial to solve economically. In addition, with respect to item (iii) above, algebraic stress models have the same disadvantage as Reynolds-stress models. As discussed in the previous section, algebraic stress models can be recast as nonlinear viscosity models. These add little cost and difficulty to k–ε-model calculations.

The further six fields $f_{ij}(x,t)$ introduced by elliptic relaxation models clearly result in additional computational cost. The elliptic equation (Eq. (11.203)) is not particularly difficult or expensive to solve, but it is of a different structure than the convective–diffusive equations that CFD codes are usually designed to solve.

Range of applicability

The basic k–ε and Reynolds-stress models can be applied to any turbulent flow. They also provide lengthscale and timescale information that can be used in the modelling of additional processes. Consequently, they provide a basis for the modelling of turbulent reactive flows, multi-phase flows, etc. Model transport equations for the scalar flux can be solved in conjunction with a Reynolds-stress model to provide closure to the mean scalar equations. Such so-called *second-moment closures* have successfully been extended to atmospheric flows in which buoyancy effects are significant (e.g., Zeman and Lumley (1976)). Although it can, in principle, be applied to any turbulent flow (in the class considered), the Spalart–Allmaras model is intended only for aerodynamic applications.

Accuracy

The ideal way to assess their accuracies is to compare the performances of various turbulence models for a broad range of test flows. Extraneous errors (Fig. 8.2) must be shown to be small; the test flows should be different than those used in development of the model; and the tests should not be performed by the model's developers. Even though they do not conform to this ideal in all respects, the following works are useful in assessing the accuracies of various turbulence models: Bradshaw, Launder, and Lumley (1996), Luo and Lakshminarayana (1997), Kral, Mani, and Ladd (1996), Behnia, Parneix, and Durbin (1998), Godin *et al.* (1997), Menter (1994), and Kline, Cantwell, and Lilley (1982). Also valuable are the reviews and assessments of Hanjalić (1994) and Wilcox (1993).

The works cited support the following conclusions.

(i) The k–ε model performs reasonably well for two-dimensional thin shear flows in which the mean streamline curvature and mean pressure gradient are small.

(ii) For boundary layers with strong pressure gradients the k–ε model performs poorly. However, the k–ω model performs satisfactorily, and indeed its performance is superior for many flows.

(iii) For flows far removed from simple shear (e.g., the impinging jet and three-dimensional flows), the k–ε model can fail profoundly.

(iv) The use of nonlinear viscosity models is beneficial and allows the calculation of secondary flows (which cannot be calculated using the isotropic viscosity hypothesis).

(v) Reynolds-stress models can be successful (whereas turbulent viscosity models are not) in calculating flows with significant mean streamline curvature, flows with strong swirl or mean rotation, secondary flows in ducts, and flows with rapid variations in the mean flow.

(vi) Reynolds-stress-model calculations are sensitive to the details of the modelling of the pressure–rate-of-strain tensor, including wall-reflection terms.

(vii) The elliptic relaxation models (both Reynolds-stress and k–ε–$\overline{v^2}$) have been quite successful in application to a number of challenging two-dimensional flows, including the impinging jet, and separated boundary layers.

(viii) The dissipation equation is frequently blamed for poor performance of a model. For many flows, much improved performance can be obtained by altering the model constants ($C_{\varepsilon1}$ or $C_{\varepsilon2}$) or by adding correction terms. No correction to the dissipation equation that is effective in all flows has been found.

In summary, especially for complex flows, Reynolds-stress models have been demonstrated to be superior to two-equation models. Until around 1990 this superiority was not well established, because of deficiencies in the models used, and because there had been little testing on complex flows. RANS models have improved considerably in the time since Bradshaw (1987) wrote 'the best modern methods allow almost all flows to be calculated to higher accuracy than the best-informed guess, which means that the methods are genuinely useful even if they cannot replace experiments.'

12

PDF methods

The mean velocity $\langle U(x,t)\rangle$ and the Reynolds stresses $\langle u_i u_j\rangle$ are the first and second moments of the Eulerian PDF of velocity $f(V;x,t)$ (Eq. (3.153)). In PDF methods, a model transport equation is solved for a PDF such as $f(V;x,t)$.

The exact transport equation for $f(V;x,t)$ is derived from the Navier–Stokes equations in Appendix H, and discussed in Section 12.1. In this equation, all convective transport is in closed form – in contrast to the term $\partial\langle u_i u_j\rangle/\partial x_i$ in the mean-momentum equation, and $\partial\langle u_i u_j u_k\rangle/\partial x_i$ in the Reynolds-stress equation. A closed model equation for the PDF – based on the *generalized Langevin model* (GLM) – is given in Section 12.2, and it is shown how this is closely related to models for the pressure–rate-of-strain tensor, $\mathcal{R}_{ij}$.

Central to PDF methods are *stochastic Lagrangian models*, which involve new concepts and require additional mathematical tools. The necessary background on *diffusion processes* and *stochastic differential equations* is given in Appendix J. The simplest stochastic Lagrangian model is the *Langevin equation*, which provides a model for the velocity following a fluid particle. This model is introduced and examined in Section 12.3.

A closure cannot be based on the PDF of velocity alone, because this PDF contains no information on the turbulence timescale. One way to obtain closure is to supplement the PDF equation with the model dissipation equation. A superior way, described in Section 12.5, is to consider the joint PDF of velocity and a turbulence frequency.

In practice, model PDF equations are solved by *Lagrangian particle methods* (Section 12.6), which themselves provide valuable insights. Further models (beyond GLM) are described in Section 12.7.

463

12.1 The Eulerian PDF of velocity

12.1.1 Definitions and properties

As defined by Eq. (3.153), $f(V;x,t)$ is the one-point, one-time, Eulerian PDF of the velocity $U(x,t)$, where $V = \{V_1, V_2, V_3\}$ is the independent variable in the sample space – *velocity space*. Integration over the entire velocity space $\int_{-\infty}^{\infty} \int_{-\infty}^{\infty} \int_{-\infty}^{\infty} (\quad) \, dV_1 \, dV_2 \, dV_3$ is abbreviated to $\int (\quad) \, dV$, so that the normalization condition is written

$$\int f(V;x,t)\,dV = 1. \tag{12.1}$$

For any function $Q(U(x,t))$, its mean (or expectation) is defined by

$$\langle Q(U(x,t)) \rangle \equiv \int Q(V) f(V;x,t)\,dV, \tag{12.2}$$

so that the mean velocity and Reynolds stresses are given by

$$\langle U(x,t) \rangle = \int V f(V;x,t)\,dV, \tag{12.3}$$

and (with abbreviated notation)

$$\langle u_i u_j \rangle = \int (V_i - \langle U_i \rangle)(V_j - \langle U_j \rangle) f \, dV. \tag{12.4}$$

The issue of realizability for PDFs is much simpler than that for Reynolds stresses. For $f(V;x,t)$ to be a realizable PDF, it has to be non-negative for all V, and to satisfy the normalization condition, Eq. (12.1). In addition, since momentum and energy are bounded, $f(V;x,t)$ must be such that $\langle U \rangle$ and $\langle u_i u_j \rangle$ (Eqs. (12.3) and (12.4)) are bounded.

In exact PDF transport equations, the 'unknowns' appear as conditional expectations, the definitions and properties of which are now reviewed. Let $\phi(x,t)$ be a random field, and let $f_{U\phi}(V,\psi;x,t)$ be the (one-point, one-time) Eulerian joint PDF of $U(x,t)$ and $\phi(x,t)$, where ψ is the sample-space variable corresponding to ϕ. The PDF of $\phi(x,t)$ conditional on $U(x,t) = V$ is defined by

$$f_{\phi|U}(\psi|V,x,t) \equiv \frac{f_{U\phi}(V,\psi;x,t)}{f(V;x,t)}, \tag{12.5}$$

(cf. Eq. (3.95)); and the conditional mean of ϕ is defined by

$$\langle \phi(x,t)|U(x,t) = V \rangle \equiv \int_{-\infty}^{\infty} \psi f_{\phi|U}(\psi|V,x,t)\,d\psi, \tag{12.6}$$

which may be abbreviated to $\langle \phi|V \rangle$. The unconditional mean is then obtained as

$$\langle \phi(x,t) \rangle = \iint_{-\infty}^{\infty} \psi f_{U\phi}(V,\psi;x,t) \, d\psi \, dV$$

$$= \int \langle \phi(x,t) | U(x,t) = V \rangle f(V;x,t) \, dV$$

$$= \int \langle \phi|V \rangle f \, dV. \tag{12.7}$$

12.1.2 The PDF transport equation

The exact transport equation for $f(V;x,t)$ is derived in Appendix H. The equation can be expressed in several informative ways, two of which are

$$\frac{\partial f}{\partial t} + V_i \frac{\partial f}{\partial x_i} = -\frac{\partial}{\partial V_i} \left[f \left\langle \frac{DU_i}{Dt} \middle| V \right\rangle \right], \tag{12.8}$$

and

$$\frac{\partial f}{\partial t} + V_i \frac{\partial f}{\partial x_i} = \frac{1}{\rho} \frac{\partial \langle p \rangle}{\partial x_i} \frac{\partial f}{\partial V_i} - \frac{\partial}{\partial V_i} \left[f \left\langle \nu \nabla^2 U_i - \frac{1}{\rho} \frac{\partial p'}{\partial x_i} \middle| V \right\rangle \right], \tag{12.9}$$

and a third is Eq. (H.25) on page 706.

The terms on the left-hand sides represent the rate of change and convection. They involve only the subject of the equation, f, and the independent variables x, V and t: consequently these terms are in closed form.

The first equation, Eq. (12.8), is a mathematical identity and so contains no physics (except that incompressibility is assumed). It indicates, however, that the fundamental quantity determining the evolution of the PDF is the conditional acceleration $\langle DU/Dt|V \rangle$. The physics enters when the Navier–Stokes equations are used to replace the acceleration (DU/Dt) by the specific force $(\nu \nabla^2 U - \nabla p/\rho)$ causing the acceleration. This, together with the Reynolds decomposition of the pressure, leads to the second equation, Eq. (12.9).

Also in closed form is the mean pressure gradient term – the first term or the right-hand side of Eq. (12.9). (This is so because; given $f(V;x,t)$, the mean velocity and Reynolds-stress fields are known (from Eqs. (12.3) and (12.4)); hence the source in the Poisson equation for $\langle p \rangle$ (Eq. (4.13)) is known; and so $\partial \langle p \rangle / \partial x_i$ is determined from the solution to the Poisson equation.)

Exercises 12.3 and 12.4 illustrate how the equations for the evolution of the moments of f (e.g., $\langle U \rangle$ and $\langle u_j u_k \rangle$) can be obtained from the PDF transport equation. An important technical point that arises is the evaluation of quantities of the form

$$Q_{ij} \equiv \int \frac{\partial}{\partial V_i} [f A_j(V)] \, dV. \tag{12.10}$$

In all circumstances encountered here, quantities of this form are zero: as shown in Exercise 12.1, a sufficient condition for Q_{ij} to be zero is that the mean $\langle |\mathbf{A}| \rangle$ exists.

<hr />

EXERCISES

12.1   With Q_{ij} defined by Eq. (12.10), use Gauss's theorem to show that

$$Q_{ij} = \lim_{V \to \infty} \oint f A_j \frac{V_i}{V} \, dS(V), \tag{12.11}$$

where $\oint (\;) \, dS(V)$ denotes integration over the surface of a sphere of radius V in velocity space. Hence obtain

$$|Q_{ij}| \leq \lim_{V \to \infty} 4\pi V^2 \{f|\mathbf{A}|\}_{S(V)}, \tag{12.12}$$

where $\{ \; \}_{S(V)}$ denotes the area average over the sphere of radius V. Show that $\langle |\mathbf{A}| \rangle$ is given by

$$\langle |\mathbf{A}| \rangle = \int_0^\infty 4\pi V^2 \{f|\mathbf{A}|\}_{S(V)} \, dV, \tag{12.13}$$

and hence that the existence of $\langle |\mathbf{A}| \rangle$ is a sufficient condition for Q_{ij} to be zero.

12.2   Obtain the general result

$$\int B_k(V) \frac{\partial}{\partial V_i} [f A_j(V)] \, dV = -\langle A_j(U(x,t)) \, B_{k,i}(U[x,t]) \rangle, \tag{12.14}$$

where $B_{k,i}$ denotes the derivative

$$B_{k,i}(V) \equiv \frac{\partial B_k(V)}{\partial V_i}. \tag{12.15}$$

12.3   Multiply Eq. (12.9) by V_j and integrate over velocity space to obtain the mean-momentum equation. Show that this is identical to Eq. (4.12) on page 85. What is the result when the same procedure is applied to Eq. (12.8)?

12.4   The exact Reynolds-stress equation (for $\langle u_j u_k \rangle$) can be derived by multiplying Eq. (12.9) by $v_j v_k$ and integrating over velocity space, where v is defined by

$$v(V, x, t) \equiv V - \langle U(x,t) \rangle. \tag{12.16}$$

When this is done, show that

(a) the term $\partial f/\partial t$ leads to $\partial \langle u_j u_k \rangle / \partial t$;

(b) the convection term can be re-expressed as

$$V_i v_j v_k \frac{\partial f}{\partial x_i} = \frac{\partial}{\partial x_i}(V_i v_j v_k f) + f V_i \left(v_k \frac{\partial \langle U_j \rangle}{\partial x_i} + v_j \frac{\partial \langle U_k \rangle}{\partial x_i} \right),$$
(12.17)

and hence that this term accounts for mean convection, turbulent transport, and production; and

(c) the mean pressure-gradient term vanishes.

12.1.3 The PDF of the fluctuating velocity

The PDF of the fluctuating velocity $u(x,t)$ is denoted by $g(v;x,t)$, where v is the sample-space variable. The first moment of g is zero,

$$\int v_i g(v;x,t)\, dv = \langle u_i(x,t) \rangle = 0,$$
(12.18)

and g contains no information about the mean velocity. However, the information contained in $\langle U(x,t) \rangle$ and $g(v;x,t)$ together is identical to that contained in $f(V;x,t)$.

The transport equation for $g(v;x,t)$, derived in Appendix H, is

$$\frac{\partial g}{\partial t} + (\langle U_i \rangle + v_i) \frac{\partial g}{\partial x_i} + \frac{\partial \langle u_i u_j \rangle}{\partial x_i} \frac{\partial g}{\partial v_j} - v_i \frac{\partial \langle U_j \rangle}{\partial x_i} \frac{\partial g}{\partial v_j}$$

$$= -\frac{\partial}{\partial v_i}\left[g \left\langle \nu \nabla^2 u_i - \frac{1}{\rho} \frac{\partial p'}{\partial x_i} \middle| v \right\rangle \right]$$

$$= \nu \nabla^2 g + \frac{\partial^2}{\partial x_i \partial v_i}\left[g \left\langle \frac{p'}{\rho} \middle| v \right\rangle \right] + \frac{1}{2} \frac{\partial^2}{\partial v_i \partial v_j}\left[g \left(\mathcal{R}^c_{ij} - \varepsilon^c_{ij} \right) \right], \quad (12.19)$$

where the *conditional pressure–rate-of-strain tensor* is

$$\mathcal{R}^c_{ij}(v,x,t) \equiv \left\langle \frac{p'}{\rho}\left(\frac{\partial u_i}{\partial x_j} + \frac{\partial u_j}{\partial x_i} \right) \middle| v \right\rangle,$$
(12.20)

and the *conditional dissipation tensor* is

$$\varepsilon^c_{ij}(v,x,t) \equiv 2\nu \left\langle \frac{\partial u_i}{\partial x_k} \frac{\partial u_j}{\partial x_k} \middle| v \right\rangle.$$
(12.21)

It may be seen that the left-hand side of the equation for g (Eq. (12.19)) is considerably more complicated than that for f (Eq. (12.9)). The final term (on the left-hand side) corresponds to production in the Reynolds-stress equation. The first form given for the right-hand side is similar to that in the equation for f. The second form shows a decomposition of the terms corresponding to viscous diffusion, pressure transport, redistribution, and dissipation in the Reynolds-stress equations.

In homogeneous turbulence, Eq. (12.19) simplifies considerably to

$$\frac{\partial g}{\partial t} = v_i \frac{\partial \langle U_j \rangle}{\partial x_i} \frac{\partial g}{\partial v_j} + \frac{1}{2} \frac{\partial^2}{\partial v_i \, \partial v_j} \left[g \left(\mathcal{R}_{ij}^c - \varepsilon_{ij}^c \right) \right]. \tag{12.22}$$

EXERCISES

12.5 For a non-symmetric tensor function $H_{ij}(v)$, integrate by parts to show that

$$\int v_k \frac{\partial^2}{\partial v_i \, \partial v_j} [g H_{ij}(v)] \, dv = 0, \tag{12.23}$$

$$\int v_k v_\ell \frac{\partial^2}{\partial v_i \, \partial v_j} \left[g H_{ij}(v) \right] dv = \int g(H_{k\ell}(v) + H_{\ell k}(v)) \, dv$$
$$= \langle H_{k\ell}(u) \rangle + \langle H_{\ell k}(u) \rangle. \tag{12.24}$$

12.6 Show that consistent results are obtained when Eq. (12.19) is integrated over velocity space, and when it is multiplied by v_k and integrated over velocity space.

12.7 Show that

$$\langle \mathcal{R}_{ij}^c (u(x,t), x, t) \rangle = \mathcal{R}_{ij}(x, t), \tag{12.25}$$

and similarly for ε_{ij}^c.

12.8 Multiply Eq. (12.19) by $v_k v_\ell$ and integrate over velocity space to obtain the exact transport equation for the Reynolds-stress tensor $\langle u_k u_\ell \rangle$. Perform a term-by-term comparison with Eq. (7.194) on page 319.

12.2 The model velocity PDF equation

In the PDF transport equation (Eq. (12.9)) derived from the Navier–Stokes equations, convection and the mean pressure appear in closed form, whereas models are required for the conditional expectations of the viscous term and the fluctuating pressure. Different types of models have been proposed by Lundgren (1969) and Pope (1981b); but since its introduction (Pope 1983b) the prevalent type of model has been the generalized Langevin model (GLM). In this section, the behavior and properties of the PDF equation with GLM are discussed. The physical basis for the model is discussed in Section 12.3.

12.2.1 The generalized Langevin model

The PDF transport equation incorporating the generalized Langevin model is

$$
\frac{\partial f}{\partial t} + V_i \frac{\partial f}{\partial x_i} - \frac{1}{\rho} \frac{\partial \langle p \rangle}{\partial x_i} \frac{\partial f}{\partial V_i}
$$

$$
= -\frac{\partial}{\partial V_i} [f G_{ij} (V_j - \langle U_j \rangle)] + \tfrac{1}{2} C_0 \varepsilon \frac{\partial^2 f}{\partial V_i \, \partial V_i}, \qquad (12.26)
$$

where $C_0(x,t)$ and $G_{ij}(x,t)$ are coefficients that define the particular model: C_0 is non-dimensional, whereas G_{ij} has dimensions of inverse time. For homogeneous turbulence, the corresponding equation for the PDF of the fluctuating velocity is

$$
\frac{\partial g}{\partial t} - v_i \frac{\partial \langle U_j \rangle}{\partial x_i} \frac{\partial g}{\partial v_j} = -\frac{\partial}{\partial v_i} (g G_{ij} v_j) + \tfrac{1}{2} C_0 \varepsilon \frac{\partial^2 g}{\partial v_i \, \partial v_i}. \qquad (12.27)
$$

Before discussing the specification of the coefficients and the behavior of these equations, we make the following general observations.

(i) When the mean-momentum equation is formed from Eq. (12.26) (by multiplying by V and integrating), the modelled terms vanish, correctly leaving the Reynolds equations. (The viscous term $v \nabla^2 \langle U \rangle$ has been omitted for simplicity, but can readily be included.)

(ii) Because of the appearance of the dissipation $\varepsilon(x,t)$, Eq. (12.26) by itself is not closed, but, together, the model equations for f and ε are closed.

(iii) In general, the coefficients $C_0(x,t)$ and $G_{ij}(x,t)$ depend on the local values of $\langle u_i u_j \rangle$, ε, and $\partial \langle U_i \rangle / \partial x_j$, but they are independent of V.

(iv) The two terms in G_{ij} and C_0 *jointly* model the exact terms in v and p'. Each model term does not correspond to one of the exact terms.

(v) As is shown later in this section, the coefficients are subject to the constraint

$$
(1 + \tfrac{3}{2} C_0) \varepsilon + G_{ij} \langle u_i u_j \rangle = 0, \qquad (12.28)
$$

which ensures that the kinetic energy evolves correctly in homogeneous turbulence.

(vi) A very important observation (Pope 1985, Durbin and Speziale 1994) is that the GLM equation (Eq. (12.26)) ensures realizability, provided only that C_0 is non-negative and that C_0 and G_{ij} are bounded. (This result is explained in Section 12.6.)

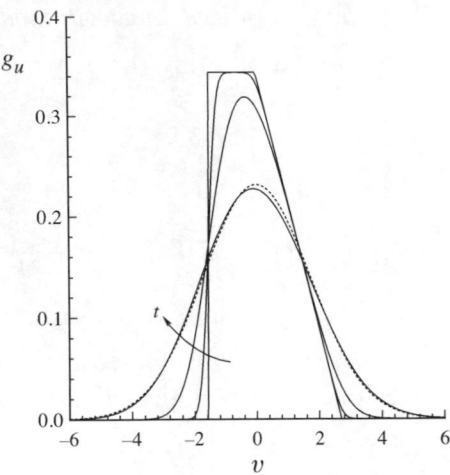

Fig. 12.1. The effect of diffusion on the shape of the PDF: the solutions to Eq. (12.29) for $\mathcal{D}t = 0$, 0.02, 0.2, and 1. The dashed line is the Gaussian with the same mean (0) and variance (3) as those of the PDF at $\mathcal{D}t = 1$.

(vii) The generalized Langevin model is, in fact, a class of models: a particular model corresponds to a particular specification of C_0 and G_{ij}.

12.2.2 The evolution of the PDF

The GLM PDF equation (Eq. (12.26)) consists of the *drift* term (involving G_{ij}) and the *diffusion* term (involving C_0). The latter causes the PDF to diffuse in velocity space. To illustrate this effect, we consider the simpler equation

$$\frac{\partial g_u}{\partial t} = \mathcal{D}\frac{\partial^2 g_u}{\partial v^2}, \tag{12.29}$$

where $g_u(v; t)$ is the PDF of a scalar-valued random process $u(t)$, and $\mathcal{D}$ is the (constant) diffusion coefficient which, like ε, has dimensions of velocity squared per time. It is simply shown that Eq. (12.29) causes the mean $\langle u(t) \rangle$ to be conserved and the variance to increase at the rate $2\mathcal{D}$.

Figure 12.1 shows the evolution of $g_u(v; t)$ given by this equation from the non-physical initial condition given in Exercise 3.13 on page 53. Even at a very early time ($\mathcal{D}t = 0.02$), the effect of diffusion smoothing the PDF is evident. At the latest time shown ($\mathcal{D}t = 1$) the PDF is close to Gaussian. In fact, the analytic solution to Eq. (12.29) (obtained in Exercise 12.9) shows that, from any initial condition, diffusion causes the PDF to tend to a Gaussian.

For several PDF equations, such as Eq. (12.29), analytic solutions can be obtained for the corresponding characteristic function. In Appendix I, the properties of characteristic functions are reviewed, and Table I.1 on page 710 provides a summary of useful results.

The simplest specification of the drift coefficient is that it is isotropic, i.e.,

$$G_{ij} = -\frac{\delta_{ij}}{T_{\mathrm{L}}}, \tag{12.30}$$

where the constraint Eq. (12.28) determines the timescale T_{L} to be

$$T_{\mathrm{L}}^{-1} = (\tfrac{1}{2} + \tfrac{3}{4}C_0)\frac{\varepsilon}{k}. \tag{12.31}$$

This specification, with C_0 a constant, is called the *simplified Langevin model* (SLM).

To illustrate the effect of the drift term, we consider the equation

$$\frac{\partial g_u}{\partial t} = \frac{1}{T_{\mathrm{L}}}\frac{\partial}{\partial v}(g_u v), \tag{12.32}$$

for the PDF of a zero-mean random process $u(t)$, taking T_{L} to be constant. It is readily shown that the zero mean is preserved, and that the variance decreases exponentially as

$$\sigma(t)^2 \equiv \langle u(t)^2 \rangle = \langle u(0)^2 \rangle e^{-2t/T_{\mathrm{L}}}, \tag{12.33}$$

so that the standard deviation is $\sigma(t) = \sigma(0)e^{-t/T_{\mathrm{L}}}$.

The distinctive behavior of Eq. (12.32) is that it preserves the shape of the PDF. As may readily be verified, the solution at time t is given in terms of the initial condition (at $t = 0$) by

$$g_u(v, t) = \frac{\sigma(0)}{\sigma(t)} g_u \left(v\frac{\sigma(0)}{\sigma(t)}, 0 \right). \tag{12.34}$$

For the non-physical initial condition considered previously, this solution is shown in Fig. 12.2 at $t/T_{\mathrm{L}} = 0, \tfrac{1}{2}$ and 1.

For decaying homogeneous turbulence (with $\partial \langle U_i \rangle / \partial x_j = 0$), the simplified Langevin model (SLM) reduces to

$$\frac{\partial g}{\partial t} = (\tfrac{1}{2} + \tfrac{3}{4}C_0)\frac{\varepsilon}{k}\frac{\partial}{\partial v_i}(g v_i) + \tfrac{1}{2}C_0\varepsilon\frac{\partial^2 g}{\partial v_i \, \partial v_i}. \tag{12.35}$$

This can be integrated over v_2 and v_3 to yield the equation for the marginal PDF of u_1, denoted by $g_u(v; t)$:

$$\frac{\partial g_u}{\partial t} = (\tfrac{1}{2} + \tfrac{3}{4}C_0)\frac{\varepsilon}{k}\frac{\partial}{\partial v}(g_u v) + \tfrac{1}{2}C_0\varepsilon\frac{\partial^2 g_u}{\partial v^2}. \tag{12.36}$$

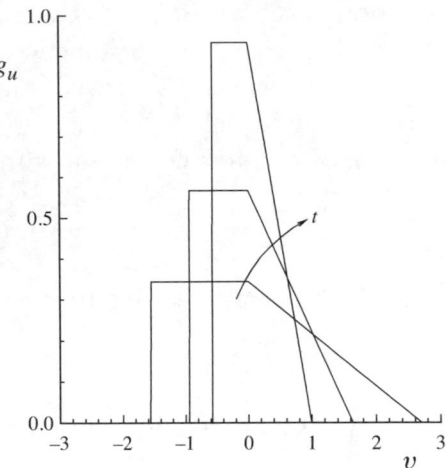

Fig. 12.2. Solutions (Eq. (12.34)) to Eq. (12.32) for $t/T_L = 0, \frac{1}{2}, 1$.

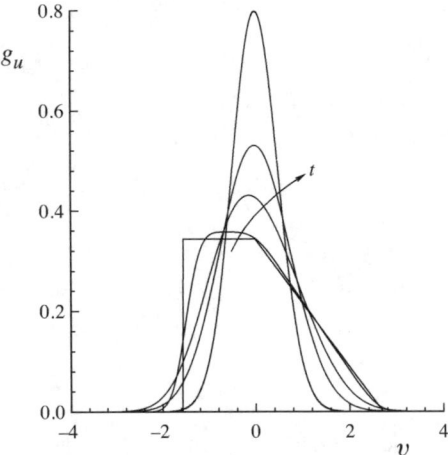

Fig. 12.3. The PDF $g_u(v;t)$ evolving according to the simplified Langevin model, Eq. (12.36). The PDF is shown at times at which the standard deviations are 1, 0.99, 0.9, 0.75, and 0.5. (The constant C_0 is taken to be 2.1.)

Both terms preserve the mean $\langle u_1 \rangle = 0$. The diffusion term causes the variance $\langle u_1^2 \rangle$ to increase at the rate $C_0 \varepsilon$, while the drift term causes it to decrease at the rate $(C_0 + \frac{2}{3})\varepsilon$ – assuming that $\langle u_1^2 \rangle = \frac{2}{3}k$. The net effect is to produce the correct rate of decay of the variance $\langle u_1^2 \rangle$, namely $\frac{2}{3}\varepsilon$.

Figure 12.3 shows the solution to Eq. (12.36) from the non-physical initial condition. As may be seen, as time proceeds the PDF tends to a Gaussian and becomes narrower. The analytic solution to Eq. (12.36) is developed in Exercise 12.11.

For the general case, the qualitative behavior of the generalized Langevin model is the same as that in the examples above. The drift term deforms the PDF $g(v;t)$ without qualitatively affecting its shape; while the diffusion term makes it tend toward an isotropic joint normal. For homogeneous turbulence, as shown in Exercise 12.12, the GLM admits joint-normal solutions, and the solution tends to a joint normal from any initial condition. This is the correct physical behavior, since measurements in homogeneous turbulence indeed show the one-point PDF of velocity to be joint normal (see Section 5.5.3 and Fig. 5.46 on page 175).

EXERCISES

12.9 By integrating Eq. (12.29) show that it (a) satisfies the normalization condition, (b) conserves the mean of $u(t)$, and (c) causes the variance to increase at the rate $2\mathcal{D}$.

Show that, corresponding to Eq. (12.29), the characteristic function

$$\Psi_u(s,t) \equiv \langle e^{iu(t)s} \rangle, \tag{12.37}$$

evolves by

$$\frac{\partial \Psi_u}{\partial t} = -\mathcal{D}s^2 \Psi_u. \tag{12.38}$$

(Hint: refer to Table I.1 on page 710.) By differentiating this equation with respect to s, verify the properties (a), (b), and (c). Show that the solution to Eq. (12.38) is

$$\Psi_u(s,t) = \Psi_u(s,0)e^{-\mathcal{D}s^2 t}. \tag{12.39}$$

Show that (for $t > 0$) the characteristic function of the random variable

$$\tilde{u} \equiv \frac{u}{\sqrt{2\mathcal{D}t}} \tag{12.40}$$

is

$$\Psi_{\tilde{u}}(\hat{s},t) = \exp(-\tfrac{1}{2}\hat{s}^2)\, \Psi_u\!\left(\frac{\hat{s}}{\sqrt{2\mathcal{D}t}}, 0\right). \tag{12.41}$$

Use this result to show that, from any initial PDF, the solution to the diffusion equation Eq. (12.29) tends to a Gaussian at large time.

12.10 Show that, if the PDF $g_u(v;t)$ evolves by Eq. (12.32), then correspondingly the characteristic function (Eq. (12.37)) evolves by

$$\left(\frac{\partial}{\partial t} + \frac{s}{T_L}\frac{\partial}{\partial s}\right)\Psi_u = 0. \tag{12.42}$$

Hence obtain the solution

$$\Psi_u(s,t) = \Psi_u(se^{-t/T_{\mathrm{L}}}, 0).\tag{12.43}$$

12.11 Let $\hat{g}(\hat{v}; \hat{t})$ be the standardized PDF of $u(t)$, i.e.,

$$\hat{g}(\hat{v}; \hat{t}) = g_u(\hat{v}\sigma(t), t)\sigma(t),\tag{12.44}$$

where $\sigma(t)^2 = \langle u(t)^2 \rangle$, and

$$d\hat{t} \equiv \tfrac{3}{4}C_0 \frac{\varepsilon}{k}\,dt.\tag{12.45}$$

Show that, if g_u evolves according to the SLM (Eq. (12.36)), then $\hat{g}$ evolves by

$$\frac{\partial \hat{g}}{\partial \hat{t}} = \frac{\partial}{\partial \hat{v}}(\hat{g}\hat{v}) + \frac{\partial^2 \hat{g}}{\partial \hat{v}^2}.\tag{12.46}$$

With $\hat{\Psi}_u(\hat{s}, \hat{t})$ being the characteristic function of $u(t)/\sigma(t)$, show that it evolves by

$$\frac{\partial \hat{\Psi}_u}{\partial \hat{t}} = -\hat{s}\frac{\partial \hat{\Psi}_u}{\partial \hat{s}} - \hat{s}^2 \hat{\Psi}_u.\tag{12.47}$$

Use the method of characteristics to obtain the solution

$$\hat{\Psi}_u(\hat{s}, \hat{t}) = \hat{\Psi}_u(\hat{s}e^{-\hat{t}}, 0)\exp[-\tfrac{1}{2}\hat{s}^2(1 - e^{-2\hat{t}})].\tag{12.48}$$

Comment on the long-time behavior. Determine σ as a function of $\hat{t}$, and hence obtain the solution for $\Psi_u(s,t)$.

12.12 With the definition

$$A_{ji} = -G_{ij} + \frac{\partial \langle U_i \rangle}{\partial x_j},\tag{12.49}$$

show that the GLM for $g(v,t)$ (Eq. (12.27)) can be rewritten

$$\frac{\partial g}{\partial t} = A_{ji}\frac{\partial}{\partial v_i}(gv_j) + \tfrac{1}{2}C_0\varepsilon\frac{\partial^2 g}{\partial v_i\,\partial v_i},\tag{12.50}$$

and that the corresponding equation for the characteristic function $\Psi(s,t)$ is

$$\left(\frac{\partial}{\partial t} + A_{ji}s_i\frac{\partial}{\partial s_j}\right)\Psi = -\tfrac{1}{2}C_0\varepsilon s^2\Psi,\tag{12.51}$$

where $s^2 = s_i s_i$.

Show that, along the trajectories $\hat{s}(s_0, t)$ given by

$$\hat{s}(s_0, t) = \mathbf{B}(t)s_0,\tag{12.52}$$

with

$$\mathbf{B}(t) \equiv \exp \left(\int_0^t \mathbf{A}(t') \, dt' \right), \tag{12.53}$$

Ψ evolves by

$$\frac{d \ln \Psi}{dt} = -\tfrac{1}{2} C_0 \varepsilon \hat{s}^2. \tag{12.54}$$

Hence show that the solution to Eq. (12.51) is

$$\Psi(s,t) = \Psi(\mathbf{B}^{-1} s, 0) e^{-s^{\mathrm{T}} \widehat{\mathbf{C}} s / 2}, \tag{12.55}$$

where

$$\widehat{\mathbf{C}}(t) = (\mathbf{B}^{-1})^{\mathrm{T}} \int_0^t C_0 \varepsilon \mathbf{B} \mathbf{B}^{\mathrm{T}} \, dt' \, \mathbf{B}^{-1}. \tag{12.56}$$

Show that, if $g(\boldsymbol{v}, t)$ is initially joint normal, then it remains joint normal for all time. Show also that, from any initial condition, the distribution tends to a joint normal (with covariance $\widehat{\mathbf{C}}$), provided that $\mathbf{B}^{-1}$ tends to zero.

12.2.3 Corresponding Reynolds-stress models

From the GLM equation for the PDF (Eq. (12.26)), it is straightforward to derive the corresponding equations for the first and second moments $\langle \boldsymbol{U}(\boldsymbol{x}, t) \rangle$ and $\langle u_i u_j \rangle$. As already observed, the first moment equation is the Reynolds equation (Eq. (4.12) with the viscous term neglected). The second-moment equation is the partially modelled Reynolds-stress equation

$$\frac{\bar{\mathrm{D}}}{\bar{\mathrm{D}} t} \langle u_i u_j \rangle + \frac{\partial}{\partial x_k} \langle u_i u_j u_k \rangle = \mathcal{P}_{ij} + G_{ik} \langle u_j u_k \rangle + G_{jk} \langle u_i u_k \rangle + C_0 \varepsilon \delta_{ij}. \tag{12.57}$$

The first three terms – mean convection, turbulent transport, and production – are exact.

The above equation can be compared with the exact Reynolds-stress equation Eq. (7.194) on page 319. If viscous and pressure transport are neglected, then the model terms (those in G_{ij} and C_0 in Eq. (12.57)) correspond to redistribution and dissipation, i.e.,

$$\mathcal{R}_{ij} - \varepsilon_{ij} = G_{ik} \langle u_j u_k \rangle + G_{jk} \langle u_i u_k \rangle + C_0 \varepsilon \delta_{ij}. \tag{12.58}$$

Thus, taking dissipation to be isotropic, the pressure–rate-of-strain model implied by the GLM is

$$\mathcal{R}_{ij} = G_{ik} \langle u_j u_k \rangle + G_{jk} \langle u_i u_k \rangle + (\tfrac{2}{3} + C_0) \varepsilon \delta_{ij}. \tag{12.59}$$

The requirement that $\mathcal{R}_{jk}$ be redistributive (i.e., $\mathcal{R}_{ii} = 0$) leads to the constraint Eq. (12.28) on the GLM coefficients.

The simplified Langevin model (SLM)

This model is defined by the simplest possible choice of G_{ij}, namely

$$G_{ij} = -(\tfrac{1}{2} + \tfrac{3}{4}C_0)\frac{\varepsilon}{k}\delta_{ij}. \tag{12.60}$$

Substituting this specification into Eq. (12.59) yields

$$\mathcal{R}_{ij} = -(1 + \tfrac{3}{2}C_0)\varepsilon\left(\frac{\langle u_i u_j\rangle}{k} - \tfrac{2}{3}\delta_{ij}\right), \tag{12.61}$$

which is Rotta's model (Eq. (11.24) on page 393) with the coefficient

$$C_{\mathrm{R}} = 1 + \tfrac{3}{2}C_0. \tag{12.62}$$

Thus, the simplified Langevin model corresponds (at the Reynolds-stress level) to Rotta's model, with $C_{\mathrm{R}} = (1 + \tfrac{3}{2}C_0)$. When it is applied to homogeneous turbulence, the SLM yields the same Reynolds-stress evolution as does Rotta's model.

The isotropization-of-production model

What specification of G_{ij} and C_0 corresponds to the LRR-IP model? The GLM equation for $g(\boldsymbol{v}, t)$ in homogeneous turbulence (Eq. (12.27)) can be written

$$\frac{\partial g}{\partial t} = -\frac{\partial}{\partial v_i}\left[g\left(G_{ij} - \frac{\partial\langle U_i\rangle}{\partial x_j}\right)v_j\right] + \tfrac{1}{2}C_0\varepsilon\frac{\partial^2 g}{\partial v_i\,\partial v_i}. \tag{12.63}$$

The term in the mean velocity gradient leads to production $\mathcal{P}_{ij}$ in the Reynolds-stress equations. Obviously, therefore, a contribution to G_{ij} of

$$G'_{ij} = C_2\frac{\partial\langle U_i\rangle}{\partial x_j} \tag{12.64}$$

leads to the term $-C_2\mathcal{P}_{ij}$ in the Reynolds-stress equations, i.e., the counteraction of production according to the IP model. Thus the specification of G_{ij} corresponding to the LRR-IP model is the sum of this IP contribution (Eq. (12.64)) and an isotropic contribution corresponding to Rotta's model:

$$G_{ij} = -\tfrac{1}{2}C_{\mathrm{R}}\frac{\varepsilon}{k}\delta_{ij} + C_2\frac{\partial\langle U_i\rangle}{\partial x_j}. \tag{12.65}$$

The constraint Eq. (12.28) then yields

$$C_{\mathrm{R}} = 1 + \tfrac{3}{2}C_0 - C_2\frac{\mathcal{P}}{\varepsilon}, \tag{12.66}$$

or, rearranging,

$$C_0 = \tfrac{2}{3}\left(C_R - 1 + C_2\frac{\mathcal{P}}{\varepsilon}\right).$$ (12.67)

To summarize: for given values of C_R and C_2, the specification of G_{ij} and C_0 according to Eqs. (12.65) and (12.67) produces the generalized Langevin model corresponding to LRR-IP. In homogeneous turbulence, the evolution of the Reynolds stresses given by the two models is identical.

The Haworth–Pope model

Haworth and Pope (1986) proposed the model[1]

$$G_{ij} = \frac{\varepsilon}{k}(\alpha_1\delta_{ij} + \alpha_2 b_{ij} + \alpha_3 b_{ij}^2) + H_{ijk\ell}\frac{\partial\langle U_k\rangle}{\partial x_\ell}$$ (12.68)

where the fourth-order tensor **H** is

$$\begin{aligned}
H_{ijk\ell} = &\,\beta_1\delta_{ij}\delta_{k\ell} + \beta_2\delta_{ik}\delta_{j\ell} + \beta_3\delta_{i\ell}\delta_{jk} \\
&+ \gamma_1\delta_{ij}b_{k\ell} + \gamma_2\delta_{ik}b_{j\ell} + \gamma_3\delta_{i\ell}b_{jk} \\
&+ \gamma_4 b_{ij}\delta_{k\ell} + \gamma_5 b_{ik}\delta_{j\ell} + \gamma_6 b_{i\ell}\delta_{jk}.
\end{aligned}$$ (12.69)

The terms in $\alpha_{(i)}$ are the most general possible involving the Reynolds-stress anisotropy b_{ij} alone, while the term in **H** is linear both in the mean velocity gradients and in the anisotropy. Values of the coefficients (taken to be constants) are given by Haworth and Pope (1986) and Haworth and Pope (1987).

The corresponding model for the pressure–rate-of-strain tensor is readily deduced from Eq. (12.59). It is of the general form Eq. (11.135), with the coefficients $f^{(n)}$ determined by $\alpha_{(i)}, \beta_{(i)}$, and $\gamma_{(i)}$ (see Pope (1994b) and Exercise 12.13, Eq. (12.72)).

The Lagrangian IP model (LIPM)

A useful model in this class is LIPM, proposed by Pope (1994b). Three constants are specified: $C_0 = 2.1$, $C_2 = 0.6$, and $\alpha_2 = 3.5$. The remaining coefficients are then given by $\alpha_3 = -3\alpha_2$, $\beta_1 = -\tfrac{1}{5}$, $\beta_2 = \tfrac{1}{2}(1 + C_2)$, $\beta_3 = -\tfrac{1}{2}(1 - C_2)$, $\gamma_{1-4} = 0$, $\gamma_5 = -\gamma_6 = \tfrac{3}{2}(1 - C_2)$, and

$$\alpha_1 = -(\tfrac{1}{2} + \tfrac{3}{4}C_0) + \tfrac{1}{2}C_2\frac{\mathcal{P}}{\varepsilon} + 3\alpha_2 b_{ii}^3.$$ (12.70)

The performance of this model is little different than that of the LRR-IP model (Eqs. (12.65)–(12.67)), but it has the virtue of having a constant value

---

[1] It is useful to include the term in α_3 which is not in the original proposal of Haworth and Pope (1986).

of C_0. The implied value of the Rotta constant is

$$C_R = 1 + \tfrac{3}{2}C_0 - C_2\frac{\mathcal{P}}{\varepsilon} - \tfrac{2}{3}\alpha_2 F, \qquad (12.71)$$

(cf. Eq. (12.66)), where F is the determinant of the normalized Reynolds-stress tensor, Eq. (11.34).

The general case

Equation (12.59) shows that to each GLM there corresponds a unique model for the redistribution $\mathcal{R}_{ij}$. Furthermore, the model is realizable (subject only to the conditions that the coefficients are bounded, and that C_0 is non-negative).

The converse is more subtle. We have determined specifications of C_0 and G_{ij} corresponding to Rotta's model and to the LRR-IP model, but these specifications are not unique: note that $\mathcal{R}_{ij}$ has only five independent components, whereas G_{ij} has nine. These issues are discussed in more depth by Pope (1994b).

EXERCISES _____

12.13   From Eq. (12.59) show that the pressure–rate-of-strain model corresponding to the Haworth–Pope model Eq. (12.68) is Eq. (11.135) with the coefficients $f^{(n)}$ given by

$$
\begin{aligned}
f^{(1)} &= 4\alpha_1 + \tfrac{4}{3}\alpha_2 + 2b_{ii}^2\alpha_3, \\
f^{(2)} &= 4\alpha_2 + \tfrac{4}{3}\alpha_3, \\
f^{(3)} &= \tfrac{4}{3}(\beta_2 + \beta_3), \\
f^{(4)} &= 2(\beta_2 + \beta_3) + \tfrac{2}{3}(\gamma_2 + \gamma_3 + \gamma_5 + \gamma_6), \\
f^{(5)} &= 2(\beta_2 - \beta_3) + \tfrac{2}{3}(\gamma_2 - \gamma_3 - \gamma_5 + \gamma_6), \\
f^{(6)} &= 2(\gamma_2 + \gamma_3), \\
f^{(7)} &= 2(\gamma_2 - \gamma_3), \\
f^{(8)} &= 4(\gamma_5 + \gamma_6).
\end{aligned}
\qquad (12.72)
$$

12.14   Show that, for arbitrary γ_5, the remaining coefficients (β_i, γ_i) in the Haworth–Pope model can be specified so that the corresponding pressure–rate-of-strain model is LRR-IP.

12.15   Show that the coefficients $f^{(n)}$ (Eq. (12.72)) corresponding to the Haworth–Pope model satisfy

$$\tfrac{3}{2}f^{(3)} - f^{(4)} + \tfrac{1}{3}f^{(6)} + \tfrac{1}{6}f^{(8)} = 0. \qquad (12.73)$$

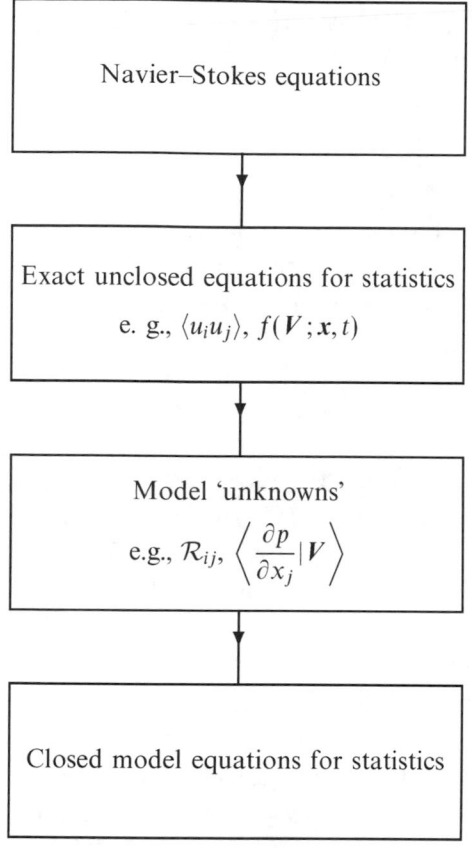

Fig. 12.4. The Eulerian modelling approach.

Discuss the significance of this linear dependence for the coefficients $f^{(n)}$.

12.2.4 Eulerian and Lagrangian modelling approaches

Up to this point, in considering Reynolds-stress and PDF closures, we have followed the Eulerian approach to modelling, which is summarized in Fig. 12.4. Starting from the Navier–Stokes equations, exact transport equations for statistics such as $\langle u_i u_j \rangle$ and $f(V; x, t)$ are derived. These equations are unclosed: they contain, as unknowns, Eulerian statistics such as $\mathcal{R}_{ij}$ and $\langle \partial p / \partial x_j | V \rangle$. The 'modelling' consists in approximating these unknown statistics by functions of the knowns, which leads to a closed set of modelled equations.

An alternative approach is to perform the modelling directly on the

Navier–Stokes equations. Consider, for example, the following model equation for the evolution of velocity following a fluid particle:

$$\frac{DU_i}{Dt} = -\frac{1}{\rho}\frac{\partial \langle p \rangle}{\partial x_i} + G_{ij}(U_j - \langle U_j \rangle), \tag{12.74}$$

where the coefficient $G_{ij}(x,t)$ is a local function of $\langle u_i u_j \rangle$, $\partial \langle U_i \rangle / \partial x_j$, and ε. The term in G_{ij} models the specific force (due to the fluctuating pressure gradient and to viscosity) as a linear function of the velocity fluctuation.

The equation for the evolution of the Eulerian PDF $f(V;x,t)$ derived from this model equation is

$$\frac{\partial f}{\partial t} + V_i \frac{\partial f}{\partial x_i} = \frac{1}{\rho}\frac{\partial \langle p \rangle}{\partial x_i}\frac{\partial f}{\partial V_i} - \frac{\partial}{\partial V_i}[f G_{ij}(V_j - \langle U_j \rangle)], \tag{12.75}$$

which follows directly from Eq. (12.8). This equation and the model dissipation equation form a closed set. Indeed, Eq. (12.75) is identical to the generalized Langevin model equation (Eq. (12.26)) with the diffusion coefficient set to zero (i.e., $C_0 = 0$). Thus, a closed model PDF equation can also be obtained from a model for the fluid particle velocity such as Eq. (12.74).

The right-hand side of Eq. (12.74) is a deterministic model for the specific forces on a fluid particle. As it stands, it is not a satisfactory model, since there is no diffusion term in Eq. (12.75) causing the shape of the PDF to relax to a joint normal. In the next section, Langevin equations are introduced. These are *stochastic models* for the specific forces on a fluid particle that result in satisfactory behavior of the PDF. This alternative Lagrangian modelling approach is summarized in Fig. 12.5.

EXERCISES

12.16   What are the necessary conditions for the mean velocity given by Eq. (12.75) to be solenoidal? Write down the steps in the argument showing that the equations for f and ε (Eqs. (12.75) and (10.53)) form a closed set.

12.17   Derive the mean-momentum and Reynolds-stress equations from Eq. (12.75). Show that the equations for $\langle u_i u_j \rangle$ and ε are closed for homogeneous turbulence, but not for the general case.

12.2.5 Relationships between Lagrangian and Eulerian PDFs

Although the models described in the subsequent sections are Lagrangian, it is of course essential to understand their implications for Eulerian quantities. This can be achieved through consideration of the relationships between

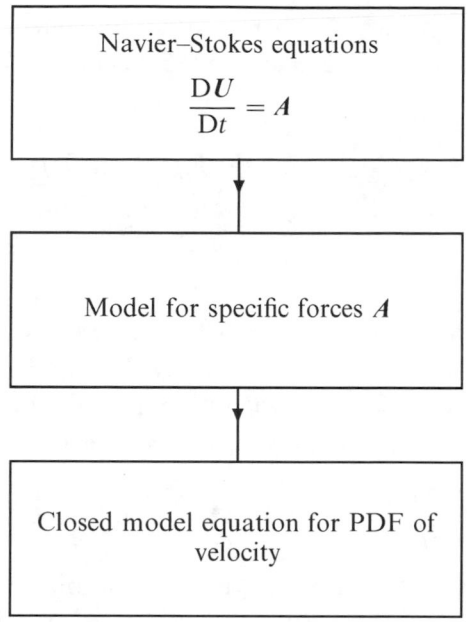

Fig. 12.5. The Lagrangian modelling approach.

Lagrangian and Eulerian PDFs which are established here, Eqs. (12.80) and (12.87).

It may be recalled from Section 2.2 that $X^+(t, Y)$ and $U^+(t, Y)$ denote the position and velocity of the fluid particle originating from position Y at the reference time t_0. These are Lagrangian quantities, and their joint PDF is accordingly called the Lagrangian PDF and is denoted by $f_L(V, x; t|Y)$. This can be expressed in terms of the fine-grained Lagrangian PDF f_L' as

$$f_L(V, x; t|Y) = \langle f_L'(V, x; t|Y)\rangle, \tag{12.76}$$

where

$$f_L'(V, x; t|Y) \equiv \delta(U^+(t, Y) - V)\delta(X^+(t, Y) - x). \tag{12.77}$$

(The properties of fine-grained PDF's such as f_L' are described in Appendix H.)

Now consider the integral of f_L' over all initial points, i.e., $\int f_L' \, dY$. Since there is a one-to-one mapping between points Y and $X^+(t, Y)$, this integral can be re-expressed as

$$\int f_L' \, dY = \int f_L' J^{-1} \, dX^+, \tag{12.78}$$

where J is the determinant of the Jacobian $\partial X_i^+/\partial Y_j$ (Eq. (2.23)). For the incompressible flows being considered, a consequence of the continuity equation is that J is unity (see Exercise 2.5 on page 16). Thus Eq. (12.78) becomes

$$\int f'_{\rm L}\,{\rm d}Y = \int \delta(U^+(t,Y)-V)\delta(X^+(t,Y)-x)\,{\rm d}X^+$$
$$= \delta(U^+(t,Y))-V)|_{X^+(t,Y)=x}$$
$$= \delta(U(x,t)-V). \tag{12.79}$$

The sifting property of the delta function singles out the fluid particle located at $X^+(t,Y)=x$, which has velocity $U(x,t)$. It may be recognized that $\delta(U(x,t)-V)$ is the fine-grained Eulerian PDF (Eq. (H.1)), so that the expectation of Eq. (12.79) leads to the required result:

$$\int f_{\rm L}(V,x;t|Y)\,{\rm d}Y = f(V;x,t). \tag{12.80}$$

It is usual to think of Y as the fixed initial position, so that (in a turbulent flow) $X^+(t,Y)$ is random. However, we can instead consider X^+ to be fixed (i.e., $X^+(t,Y)=x$), in which case Y is the random initial position of the fluid particle that is at x at time t. From this viewpoint it may be seen that $f_{\rm L}(V,x;t|Y)$ has a second interpretation: namely, for fixed x and t, $f_{\rm L}$ is the joint PDF of the velocity $U^+(t,Y)=U(x,t)$ and initial position Y of fluid particles that are at $X^+(t,Y)=x$ at time t. Thus the left-hand side of Eq. (12.80) represents the marginal PDF of velocity of the fluid particle at x (irrespective of its initial position), which is identical to the Eulerian PDF.

The result Eq. (12.80) applies equally to the PDFs of the fluctuating velocities. The PDF $g(v;x,t)$ of the Eulerian velocity fluctuation $u(x,t)$ is related to f simply by

$$g(v;x,t) = f\left(\langle U(x,t)\rangle + v;x,t\right). \tag{12.81}$$

The joint PDF of $X^+(t,Y)$ and the fluctuation

$$u^+(t,Y) \equiv U^+(t,Y) - \langle U(X^+(t,Y),t)\rangle \tag{12.82}$$

is denoted by $g^+(v,x;t|Y)$ and is related to $f_{\rm L}$ by

$$g^+(v,x;t|Y) = f_{\rm L}(\langle U(x,t)\rangle + v,x;t|Y). \tag{12.83}$$

It follows straightforwardly from Eqs. (12.80)–(12.83) that g^+ and g are related by

$$\int g^+(v,x;t|Y)\,{\rm d}Y = g(v;x,t). \tag{12.84}$$

Homogeneous turbulence

By definition, in homogeneous turbulence the fluctuating Eulerian velocity field $u(x, t)$ is statistically homogeneous: its PDF $g(v; t)$ is independent of x.

The appropriate Lagrangian quantities to consider are the velocity fluctuation $u^+(t, Y)$ and the displacement

$$R^+(t, Y) \equiv X^+(t, Y) - Y. \tag{12.85}$$

Because of statistical homogeneity, the joint PDF of $u^+(t, Y)$ and $R^+(t, Y)$ – denoted by $g_R^+(v, r; t)$ – is independent of Y. The marginal PDF of $u^+(t, Y)$

$$g_L(v; t) = \int g_R^+(v, r; t) \, dr \tag{12.86}$$

is also independent of Y.

With these definitions, for homogeneous turbulence, Eq. (12.84) can be re-expressed as

$$
\begin{aligned}
g(v; t) &= \int g^+(v, x; t | Y) \, dY \\
&= \int g_R^+(v, x - Y; t) \, dY \\
&= g_L(v; t).
\end{aligned}
\tag{12.87}
$$

Thus, in homogeneous turbulence, the PDFs of $u(x, t)$ and $u^+(t, Y)$ are the same, independent of x and Y. As far as one-point, one-time statistics are concerned, all points – whether fixed (x) or moving (X^+) – are statistically equivalent. In the simpler case of homogeneous turbulence with zero mean velocity, the one-point, one-time PDFs of $U(x, t)$, $u(x, t)$, $U^+(t, Y)$, and $u^+(t, Y)$ are all equal:

$$
\begin{aligned}
f(V; x, t) &= g(v; t) = g_L(v; t) = \int f_L(V, x; t | Y) \, dx \\
&= \int f_L(V, x; t | Y) \, dY,
\end{aligned}
\tag{12.88}
$$

for $V = v$.

12.3 Langevin equations

The Langevin equation was originally proposed (Langevin 1908) as a stochastic model for the velocity of a microscopic particle undergoing Brownian motion. In this section the equation is described and shown to provide a good model for the velocity of a fluid particle in turbulence.

The stochastic process $U^*(t)$ generated by the Langevin equation is called the *Ornstein–Uhlenbeck (OU) process*, and its PDF evolves by the *Fokker–Planck* equation. In the terminology of stochastic processes, $U^*(t)$ is a *diffusion process*, and the Langevin equation is a *stochastic differential equation (SDE)*. The necessary mathematics connected with diffusion processes is summarized in Appendix J, where the properties of the OU process are described.

12.3.1 Stationary isotropic turbulence

We begin by considering the simplest case of homogeneous isotropic turbulence, made statistically stationary by artificial forcing. The mean velocity is zero, and the values of k and ε are constant. In these circumstances all fluid particles are statistically identical, and their three components of velocity are also statistically identical. It is sufficient, therefore, to consider one component of the fluid particle velocity, which is denoted[2] by $U^+(t)$. The model for $U^+(t)$ given by the Langevin equation is denoted by $U^*(t)$.

The Langevin equation is the stochastic differential equation

$$dU^*(t) = -U^*(t)\frac{dt}{T_L} + \left(\frac{2\sigma^2}{T_L}\right)^{1/2} dW(t), \tag{12.89}$$

where T_L and σ^2 are positive constants. For the reader unfamiliar with stochastic differential equations, this equation can be understood through the finite-difference equation

$$U^*(t + \Delta t) = U^*(t) - U^*(t)\frac{\Delta t}{T_L} + \left(\frac{2\sigma^2 \Delta t}{T_L}\right)^{1/2} \xi(t), \tag{12.90}$$

in the limit as Δt tends to zero. Here $\xi(t)$ is a standardized Gaussian random variable ($\langle\xi(t)\rangle = 0$, $\langle\xi(t)^2\rangle = 1$) which is independent of itself at different times ($\langle\xi(t)\xi(t')\rangle = 0$, for $t \neq t'$), and which is independent of $U^*(t)$ at past times (e.g., $\langle\xi(t)U^*(t')\rangle = 0$ for $t' \leq t$). In Eq. (12.89), the deterministic drift term $(-U^* dt/T_L)$ causes the velocity to relax toward zero on the timescale T_L, whereas the diffusion term adds a zero-mean random increment of standard deviation $\sigma\sqrt{2\,dt/T_L}$.

Figure 12.6 shows realizations, or *sample paths*, of the OU process $U^*(t)$ generated by the Langevin equation. Because Eq. (12.89) is stochastic (i.e., it contains randomness), two sample paths from the same initial condition are different.

---

[2] Here the superscript $+$ denotes a Lagrangian quantity, and is not to be confused with normalization by viscous wall scales.

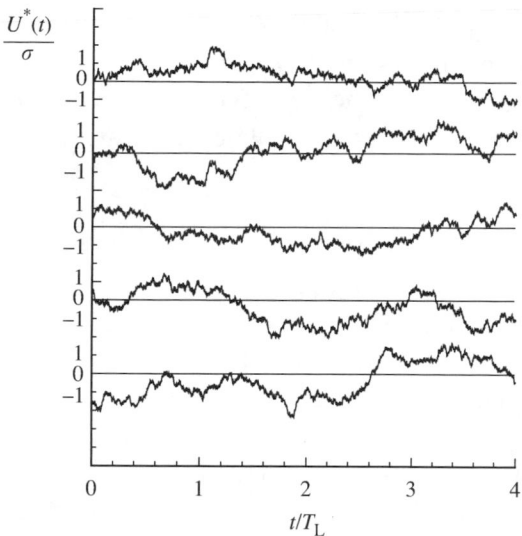

Fig. 12.6. Sample paths of the Ornstein–Uhlenbeck process generated by the Langevin equation, Eq. (12.89).

The PDF $f_L^*(V;t)$ of $U^*(t)$ evolves by the *Fokker–Planck* equation

$$\frac{\partial f_L^*}{\partial t} = \frac{1}{T_L}\frac{\partial}{\partial V}(Vf_L^*) + \frac{\sigma^2}{T_L}\frac{\partial^2 f_L^*}{\partial V^2}. \tag{12.91}$$

The properties of the OU process are described in Appendix J. Briefly, $U^*(t)$ is a statistically stationary, Gaussian, Markov process, with continuous sample paths that are nowhere differentiable. As a stationary Gaussian process, it is completely characterized by its mean (zero), its variance σ^2, and its autocorrelation function, which is

$$\rho(s) = e^{-|s|/T_L}. \tag{12.92}$$

If $\rho(s)$ is the Lagrangian velocity autocorrelation function, then the *Lagrangian integral timescale* is defined by

$$T_L \equiv \int_0^\infty \rho(s)\,ds. \tag{12.93}$$

Evidently, Eq. (12.92) is consistent with this definition, so the coefficient T_L in the Langevin equation (Eq. (12.89)) is indeed the integral timescale of the process.

Comparison with observations

To what extent does the OU process $U^*(t)$ model the behavior of the velocity $U^+(t)$ of a fluid particle in turbulent flow? The Langevin equation is correct in yielding a Gaussian PDF of velocity. In isotropic turbulence, the one-time PDF of the Lagrangian velocity $U^+(t)$ is identical to the one-point, one-time Eulerian PDF (Eq. (12.88)). The clear evidence from experiments and DNS (e.g., Yeung and Pope (1989)) is that these PDFs are very close to Gaussian. The specification

$$\sigma^2 = \tfrac{2}{3}k \qquad (12.94)$$

produces the correct velocity variance.

A limitation of the Langevin equation is that $U^+(t)$ is differentiable, whereas $U^*(t)$ is not. Hence the model is qualitatively incorrect if $U^*(t)$ is examined on an infinitesimal timescale. However, consider high-Reynolds-number turbulence in which there is a large separation between the integral timescale T_L and the Kolmogorov timescale τ_η; and let us examine $U^+(t)$ on inertial-range timescales s, $T_L \gg s \gg \tau_\eta$. This is best done through the Lagrangian structure function

$$D_L(s) \equiv \langle [U^+(t+s) - U^+(t)]^2 \rangle. \qquad (12.95)$$

The Kolmogorov hypotheses (both original, from 1941, and refined, in 1962) predict

$$D_L(s) = C_0 \varepsilon s, \quad \text{for } \tau_\eta \ll s \ll T_L, \qquad (12.96)$$

where C_0 is a universal constant; whereas the Langevin equation yields

$$D_L^*(s) \equiv \langle [U^*(t+s) - U^*(t)]^2 \rangle$$
$$= \frac{2\sigma^2}{T_L} s, \quad \text{for } s \ll T_L, \qquad (12.97)$$

(see Eq. (J.46)). Thus the Langevin equation is consistent with the Kolmogorov hypotheses in yielding a linear dependence of D_L on s in the inertial range.

In place of σ^2 and T_L, the two parameters in the Langevin equation can be expressed in terms of k and ε. Equation (12.94) gives σ^2 in terms of k, and we introduce the model coefficient[3] C_0 through the relation

$$\frac{2\sigma^2}{T_L} = C_0 \varepsilon, \qquad (12.98)$$

so that the timescale is given by

$$T_L^{-1} = \frac{C_0 \varepsilon}{2\sigma^2} = \tfrac{3}{4} C_0 \frac{\varepsilon}{k}. \qquad (12.99)$$

---

[3] Note the distinction between the Kolmogorov constant C_0 and the Langevin-model constant C_0.

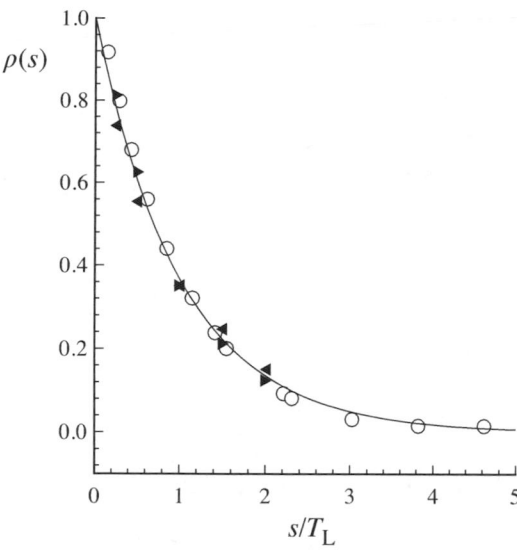

Fig. 12.7. The Lagrangian velocity autocorrelation function. Line, Langevin model $\rho(s) = \exp(-s/T_L)$; solid symbols, experimental data of Sato and Yamamoto (1987) $\triangleright$, $R_\lambda = 46$ and $\triangleleft$, $R_\lambda = 66$; open symbols, DNS data of Yeung and Pope (1989), $R_\lambda = 90$.

With the coefficients re-expressed in this way, the Langevin equation becomes

$$dU^*(t) = -\tfrac{3}{4}C_0\frac{\varepsilon}{k}U^*(t)\,dt + (C_0\varepsilon)^{1/2}\,dW(t). \qquad (12.100)$$

The Langevin equation is quantitatively consistent with the Kolmogorov hypotheses if the model coefficient C_0 is taken to be the Kolmogorov constant C_0. For then $D_L(s)$ (given by Eq. (12.96)) and $D_L^*(s)$ (given by Eq. (12.97)) are identical (for $\tau_\eta \ll s \ll T_L$). The value of C_0 and the choice of C_0 are discussed in Section 12.4.

Lagrangian statistics in high-Reynolds-number flows have proven inaccessible both to experiment and to DNS. However, at low or moderate Reynolds number both techniques have been used to measure the Lagrangian autocorrelation function $\rho(s)$. Figure 12.7 shows measurements of $\rho(s)$ in isotropic turbulence compared with the exponential (Eq. (12.92)) given by the Langevin equation. At very small times ($s/T_L \ll 1$) the exponential arising from the Langevin equation is qualitatively incorrect in that it has a negative slope – which is a consequence of $U^*(t)$ not being differentiable. However, for larger times, the exponential provides a very reasonable approximation to the observed autocorrelations.

12.18   In the finite-difference form of the Langevin equation (Eq. (12.90)), the random variable $\xi(t)$ is defined to be independent of $U^*(t')$ for $t' \leq t$. Show that $\xi(t)$ and $U^*(t + \Delta t)$ are correlated.

12.19   By taking the mean of Eq. (12.90), show that the mean $\langle U^*(t) \rangle$ evolves by

$$\frac{\mathrm{d}\langle U^* \rangle}{\mathrm{d}t} = -\frac{\langle U^* \rangle}{T_\mathrm{L}}. \tag{12.101}$$

Show that the mean of the square of Eq. (12.90) is

$$\langle U^*(t + \Delta t)^2 \rangle = \langle U^*(t)^2 \rangle \left(1 - \frac{\Delta t}{T_\mathrm{L}} \right)^2 + \frac{2\sigma^2 \, \Delta t}{T_\mathrm{L}}. \tag{12.102}$$

Hence show that the variance $\langle u^{*2} \rangle \equiv \langle U^{*2} \rangle - \langle U^* \rangle^2$ evolves by

$$\frac{\mathrm{d}}{\mathrm{d}t} \langle u^{*2} \rangle = \frac{2}{T_\mathrm{L}} (\sigma^2 - \langle u^{*2} \rangle). \tag{12.103}$$

From the initial conditions $\langle U^* \rangle = V_1$ and $\langle u^{*2} \rangle = 0$ at $t = t_1$, obtain solutions to Eqs. (12.101) and (12.103). Show that these are consistent with Eq. (J.41) on page 720.

12.20   For $t_0 < t$, multiply Eq. (12.90) by $U^*(t_0)$ and take the mean to obtain

$$\frac{\mathrm{d}}{\mathrm{d}t} \langle U^*(t)U^*(t_0) \rangle = -\frac{\langle U^*(t)U^*(t_0) \rangle}{T_\mathrm{L}}. \tag{12.104}$$

Hence show that, for stationary $U^*(t)$, the autocovariance is

$$R(s) = \sigma^2 e^{-s/T_\mathrm{L}}, \quad \text{for } s \geq 0. \tag{12.105}$$

How can $R(s)$ be determined for $s < 0$?

12.21   With $\Psi(s, t)$ being the characteristic function of $U^*(t)$, show that Eq. (12.90) corresponds to

$$\Psi(s, t + \Delta t) = \Psi \left[\left(1 - \frac{\Delta t}{T_\mathrm{L}} \right) s, t \right] \exp \left(-\frac{s^2 \sigma^2 \Delta t}{T_\mathrm{L}} \right). \tag{12.106}$$

By expanding the right-hand side to first order in Δt, show that $\Psi(s, t)$ evolves by

$$\frac{\partial \Psi}{\partial t} = -\frac{s}{T_\mathrm{L}} \frac{\partial \Psi}{\partial s} - \frac{s^2 \sigma^2}{T_\mathrm{L}} \Psi. \tag{12.107}$$

Show that the Fourier transform of this equation is the Fokker–Planck equation, Eq. (12.91). (See Exercise J.4 on page 725 for a solution to this equation.)

12.3.2 The generalized Langevin model

The rudimentary extension of the Langevin equation (Eq. (12.100)) to inhomogeneous turbulent flows is now described, the result being the *simplified Langevin model* (SLM). The *generalized Langevin model* (GLM) is then obtained as an extension of the SLM.

The subject of these Langevin equations is the velocity $U^*(t)$ of a particle with position $X^*(t)$. The particle models the behavior of fluid particles and consequently it moves with its own velocity, i.e.,

$$\frac{dX^*(t)}{dt} = U^*(t). \tag{12.108}$$

The simplified Langevin model is written as the stochastic differential equation

$$dU_i^*(t) = -\frac{1}{\rho}\frac{\partial\langle p\rangle}{\partial x_i}\,dt - (\tfrac{1}{2} + \tfrac{3}{4}C_0)\frac{\varepsilon}{k}\left(U_i^*(t) - \langle U_i\rangle\right)dt$$
$$+ (C_0\varepsilon)^{1/2}\,dW_i(t), \tag{12.109}$$

where $W(t)$ is a vector-valued Wiener process (with the property $\langle dW_i\,dW_j\rangle = dt\,\delta_{ij}$, and the coefficients ($\partial\langle p\rangle/\partial x_i$, k, ε, and $\langle U_i\rangle$) are evaluated at the particle's location $X^*(t)$. Compared with the scalar Langevin equation (Eq. (12.100)) for stationary isotropic turbulence, the differences in the SLM model are as follows.

(i) The SLM applies to the particle vector velocity $U^*(t)$.
(ii) The drift term in the mean pressure gradient is added: this is an exact term from the Navier–Stokes equations (written in terms of $\langle p\rangle + p'$).
(iii) In the second drift term, the particle velocity $U^*(t)$ relaxes toward the local Eulerian mean $\langle U\rangle$. (In the stationary isotropic turbulence, to which Eq. (12.100) pertains, the mean velocity is zero.)
(iv) The additional $\frac{1}{2}$ in the coefficient ($\frac{1}{2} + \frac{3}{4}C_0$) leads to the correct energy-dissipation rate ε, as is shown below. (Or, conversely, this $\frac{1}{2}$ is omitted from Eq. (12.100) because the artificial forcing exactly balances ε.)

These modifications (relative to Eq. (12.100)) are the minimum necessary for consistency with the mean momentum and kinetic energy equations.

The generalized Langevin model (Pope 1983b) is

$$dU_i^*(t) = -\frac{1}{\rho}\frac{\partial\langle p\rangle}{\partial x_i}\,dt + G_{ij}(U_j^*(t) - \langle U_j\rangle)\,dt + (C_0\varepsilon)^{1/2}\,dW_i(t), \tag{12.110}$$

where the coefficient $G_{ij}(x,t)$ (which is evaluated at $X^*(t)$) depends on the local values of $\langle u_i u_j \rangle$, ε, and $\partial \langle U_i \rangle / \partial x_j$. As mentioned above, GLM is a class of models; a particular model corresponds to a particular specification of G_{ij} and C_0. Examples of particular models are SLM for which G_{ij} is given by

$$G_{ij} = -(\tfrac{1}{2} + \tfrac{3}{4}C_0)\frac{\varepsilon}{k}\delta_{ij}, \tag{12.111}$$

and the IP model for which G_{ij} and C_0 are given by Eqs. (12.65) and (12.67).

The GLM is characterized by the drift term being linear in U^*, and the diffusion coefficient being isotropic, i.e., $(C_0\varepsilon)^{1/2}\delta_{ij}$. Consider instead the general diffusion coefficient denoted by $\mathcal{B}_{ij}$, so that the diffusion term is $\mathcal{B}_{ij}\,dW_j$. In this case, the second-order Lagrangian structure function given by the diffusion process is

$$D^*_{ij}(s) \equiv \langle [U^*_i(t+s) - U^*_i(t)][U^*_j(t+s) - U^*_j(t)] \rangle$$
$$= \mathcal{B}_{ik}\mathcal{B}_{jk}\,s, \quad \text{for } s \ll \frac{k}{\varepsilon} \tag{12.112}$$

(cf. Eq. (12.97)); whereas the Kolmogorov hypotheses yield

$$D^{\mathrm{L}}_{ij}(s) \equiv \langle [U^+_i(t+s) - U^+_i(t)][U^+_j(t+s) - U^+_j(t)] \rangle$$
$$= C_0\varepsilon s\delta_{ij}, \quad \text{for } \tau_\eta \ll s \ll T_{\mathrm{L}}, \tag{12.113}$$

(cf. Eq. (12.96)). Thus, the choice of an isotropic diffusion coefficient in GLM is necessary for consistency with local isotropy; and, as is evident from a comparison of these equations for $D^*_{ij}(s)$ and $D^{\mathrm{L}}_{ij}(s)$, consistency with the Kolmogorov hypotheses is achieved by the specification

$$\mathcal{B}_{ij} = (C_0\varepsilon)^{1/2}\delta_{ij}, \tag{12.114}$$

with $C_0 = \mathcal{C}_0$.

Given the form of the diffusion term, the linearity of G_{ij} in U^* ensures that the GLM yields joint-normal velocity distributions in homogeneous turbulence, in accord with observations. (Durbin and Speziale (1994) consider further the use of anisotropic diffusion in Langevin models.)

The Lagrangian PDF equation

In the generalized Langevin model, $X^*(t)$ and $U^*(t)$ model the fluid-particle properties and hence their joint PDF f^*_{L} is a model for f_{L}. Since the fluid particle is defined to originate from Y at time t_0 (i.e., $X^+(t_0, Y) = Y$), the appropriate definition of $f^*_{\mathrm{L}}(V, x; t | Y)$ is the joint PDF of $X^*(t)$ and $U^*(t)$ conditional on $X^*(t_0) = Y$.

The evolution equation for f^*_{L} is simply the Fokker–Planck equation

(Eq. (J.56)) obtained from the equations for $X^*(t)$ and $U^*(t)$ (Eqs. (12.108) and (12.110)). It is

$$\frac{\partial f_{\mathrm{L}}^*}{\partial t} = -V_i \frac{\partial f_{\mathrm{L}}^*}{\partial x_i} + \frac{1}{\rho} \frac{\partial f_{\mathrm{L}}^*}{\partial V_i} \frac{\partial \langle p \rangle}{\partial x_i} - G_{ij} \frac{\partial}{\partial V_i} \left[f_{\mathrm{L}}^* \left(V_j - \langle U_j \rangle \right) \right] + \tfrac{1}{2} C_0 \varepsilon \frac{\partial^2 f_{\mathrm{L}}^*}{\partial V_i \partial V_i},$$
(12.115)

where all the coefficients are evaluated at (x, t) (see Exercise 12.22).

The Eulerian PDF equation

On the basis of the relationship between Lagrangian and Eulerian PDFs (Eq. (12.80)), the model Eulerian PDF corresponding to the generalized Langevin model is

$$f^*(V; x, t) = \int f_{\mathrm{L}}^*(V, x; t | Y) \, \mathrm{d}Y.$$
(12.116)

The evolution equation for f^* is readily obtained by integrating the equation for f_{L}^* (Eq. (12.115)) over all Y. Since this equation contains no dependence on Y (other than in f_{L}^*), the result is simply

$$\frac{\partial f^*}{\partial t} + V_i \frac{\partial f^*}{\partial x_i} - \frac{1}{\rho} \frac{\partial f^*}{\partial V_i} \frac{\partial \langle p \rangle}{\partial x_i} = -G_{ij} \frac{\partial}{\partial V_i} [f^*(V_j - \langle U_j \rangle)] + \tfrac{1}{2} C_0 \varepsilon \frac{\partial^2 f^*}{\partial V_i \partial V_i}.$$
(12.117)

This equation is precisely the model Eulerian PDF equation examined in Section 12.2, namely Eq. (12.26) (although here the PDF is denoted by f^* to emphasize that it originates from the stochastic Lagrangian model equations for $X^*(t)$ and $U^*(t)$, Eqs. (12.108) and (12.110)). As a consequence, all of the deductions made in Section 12.2 stemming from Eq. (12.26) reflect the properties of the stochastic Lagrangian models.

Figure 12.8 summarizes the connections between the stochastic Lagrangian model (GLM) and the evolution equations for various statistics. In particular, the GLM equation for $U^*(t)$ (Eq. (12.110)), implies that the Reynolds stresses $\langle u_i u_j \rangle$ evolve according to Eq. (12.57).

Realizability

The stochastic Lagrangian models $X^*(t)$ and $U^*(t)$ are realizable, subject only to the minimal conditions that the coefficients in the GLM equation (Eq. (12.110)) be real and bounded. Given that $X^*(t)$ and $U^*(t)$ are realizable, it follows that all of their statistics are also realizable, including the Lagrangian and Eulerian PDFs f_{L}^* and f^*, and the Reynolds stresses obtained from them.

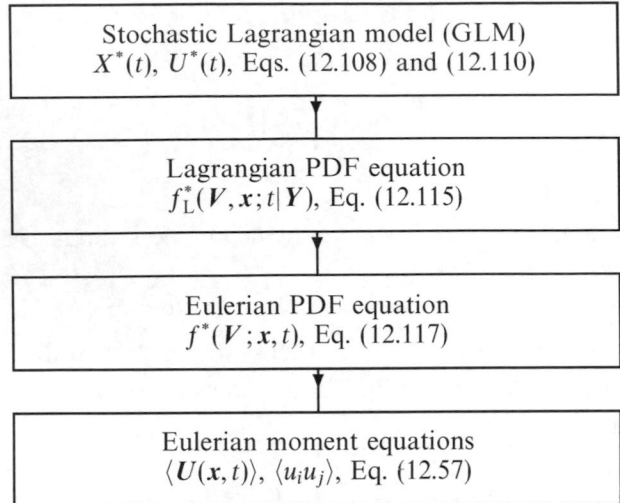

Fig. 12.8. Evolution equations for statistics that follow from stochastic Lagrangian models.

Thus, not only is realizability readily achieved in PDF methods, but the route from stochastic models to Reynolds-stress equations depicted in Fig. 12.8 provides a simple way to address realizability issues in Reynolds-stress closures. In particular, a pressure–rate-of-strain model that can be written as the right-hand side of Eq. (12.59) is realizable providing that G_{ij} is bounded and that C_0 is positive and bounded. These issues are considered further by Pope (1994b) and Durbin and Speziale (1994).

EXERCISES

12.22 Consider the six-dimensional diffusion process $\mathbf{Z}^*(t)$, where the components of $\mathbf{Z}^*$ are $\{Z_1^*, Z_2^*, Z_3^*, Z_4^*, Z_5^*, Z_6^*\} = \{X_1^*, X_2^*, X_3^*, U_1^*, U_2^*, U_3^*\}$. With $\mathbf{X}^*(t)$ and $\mathbf{U}^*(t)$ evolving according to Eqs. (12.108) and (12.110), show that $\mathbf{Z}^*(t)$ evolves by the general diffusion equation

$$dZ_i^* = a_i(\mathbf{Z}^*, t)\, dt + \mathcal{B}_{ij}(\mathbf{Z}^*, t)\, dW_j, \qquad (12.118)$$

and write down explicit expressions for the drift vector a_i, the diffusion matrix $\mathcal{B}_{ij}$, and the diffusion coefficient $B_{ij} = \mathcal{B}_{ik}\mathcal{B}_{jk}$. Hence use the general form of the Fokker–Planck equation (Eq. (J.56)) to verify Eq. (12.115).

12.23 In place of the Navier–Stokes equations, and as a generalization of Eq. (12.74), consider the equation

$$\frac{DU_i}{Dt} = -\frac{1}{\rho}\frac{\partial\langle p\rangle}{\partial x_i} + \bar{A}_i(\mathbf{U}, \mathbf{x}, t), \qquad (12.119)$$

where $\bar{A}_i$ is a differentiable function. Show that the corresponding Eulerian PDF of $U(x,t)$, $f^*(V;x,t)$, evolves by

$$\frac{\partial f^*}{\partial t} + V_i \frac{\partial f^*}{\partial x_i} - \frac{1}{\rho} \frac{\partial f^*}{\partial V_i} \frac{\partial \langle p \rangle}{\partial x_i} = -\frac{\partial}{\partial V_i}[f^* \bar{A}_i(V,x,t)]. \tag{12.120}$$

Show that the Eulerian PDF equation (Eq. (12.117)) stemming from the GLM can be written in the same form with

$$\bar{A}_i = G_{ij}(V_j - \langle U_j \rangle) - \tfrac{1}{2}C_0\varepsilon \frac{\partial \ln f^*}{\partial V_i}. \tag{12.121}$$

Show that, if f^* is joint normal, then $\bar{A}$ is given by

$$\bar{A}_i = (G_{ij} + \tfrac{1}{2}C_0\varepsilon C_{ij}^{-1})(V_j - \langle U_j \rangle), \tag{12.122}$$

where C_{ij}^{-1} is the i–j component of the inverse of the Reynolds-stress tensor $C_{ij} = \langle u_i u_j \rangle$. Hence argue that, for homogeneous turbulence, in which f^* is joint normal, Eqs. (12.119) and (12.122) provide a deterministic model with the same Eulerian PDF evolution as that of the GLM. For general flows, what are the shortcomings of this deterministic model?

12.24 By comparing the Navier–Stokes equations (Eq. (2.35)) with the GLM equation (Eq. (12.110)), show that the generalized Langevin model amounts to modelling the fluctuating pressure gradient and viscous terms as

$$-\frac{1}{\rho} \frac{\partial p'}{\partial x_i} + v \nabla^2 U_i = \tilde{A}_i(U,x,t) \equiv G_{ij}(U_j - \langle U_j \rangle) + (C_0\varepsilon)^{1/2}\dot{W}_i, \tag{12.123}$$

where $\dot{W}$ is white noise (see Eq. (J.29)).

12.25 Let $u^*(t)$ be the fluctuating component of velocity following the particle, which is defined by

$$u^*(t) \equiv U^*(t) - \langle U(X^*(t),t) \rangle. \tag{12.124}$$

For $X^*(t)$ evolving by Eq. (12.108), and $U^*(t)$ evolving by a diffusion process, show that $u^*(t)$ evolves by

$$du_i^*(t) = dU_i^*(t) - \left(\frac{\partial \langle U_i \rangle}{\partial t} + \frac{dX_j^*}{dt} \frac{\partial \langle U_i \rangle}{\partial x_j} \right) dt$$

$$= dU_i^*(t) - u_j^* \frac{\partial \langle U_i \rangle}{\partial x_j} dt - \frac{\bar{D}\langle U_i \rangle}{\bar{D}t} dt. \tag{12.125}$$

For homogeneous turbulence, and for $U^*(t)$ evolving by the GLM

(Eq. (12.110)), show that $\boldsymbol{u}^*(t)$ evolves by

$$\mathrm{d}u_i^* = -u_j^* \frac{\partial \langle U_i \rangle}{\partial x_j}\,\mathrm{d}t + G_{ij}u_j^*\,\mathrm{d}t + (C_0\varepsilon)^{1/2}\,\mathrm{d}W_i(t). \qquad (12.126)$$

Verify that the equation for the evolution of the Reynolds-stress tensor $\langle u_i^* u_j^* \rangle$ obtained from Eq. (12.126) is consistent with Eq. (12.57). In particular, show that the first term on the right-hand side of Eq. (12.126) leads to the production $\mathcal{P}_{ij}$.

Obtain from Eq. (12.126) the evolution equation for $g^*(\boldsymbol{v};t)$, the PDF of $\boldsymbol{u}^*(t)$. Verify that this equation is identical to Eq. (12.27).

12.4 Turbulent dispersion

In the preceding sections, stochastic Lagrangian models (such as the GLM) have been used as a basis for turbulence modelling – specifically to obtain a closed model equation for the Eulerian PDF of velocity. Historically, stochastic Lagrangian models were used much earlier and more extensively in the context of turbulent dispersion. In this section, the simplest case of dispersion behind a line source in grid turbulence is considered, both because of its historical significance and because it sheds light on the performance of the Langevin equation.

The problem considered

A simple statement of the usual problem addressed in the application of turbulence modelling is this: given the geometry and initial and boundary conditions of a flow, determine the mean velocity field $\langle U(x,t) \rangle$ and some turbulence characteristics (e.g., $\langle u_i u_j \rangle$ and ε). Studies in turbulent dispersion address a different problem, namely: given a turbulent flow field (in terms of $\langle U(x,t) \rangle$, $\langle u_i u_j \rangle$ and ε, say), determine the mean field $\langle \phi(x,t) \rangle$ of a conserved passive scalar that originates from a specified source. The principal applications of such studies concern the dispersion of pollutants in the atmosphere and in rivers, lakes, and oceans (see, e.g., Fischer *et al.* (1979) and Hunt (1985)). In these applications the source may be continuous in time (e.g., an effluent discharging from a pipe into the ocean) or it may be confined to a short duration of time (e.g., an accidental release of toxic gases). The size of the source is usually very small compared with the size of the domain studied, and so may be approximated as a point source.

The Eulerian approach

One approach to determining the mean scalar field $\langle\phi(x,t)\rangle$ is to solve the transport equation for $\langle\phi\rangle$ (Eq. (4.41)), with a model for the scalar flux $\langle u\phi'\rangle$. For example, if the turbulent diffusion hypothesis (Eq. (4.42)) is used, the resulting model equation is

$$\frac{\bar{D}\langle\phi\rangle}{\bar{D}t} = \nabla \cdot [(\Gamma + \Gamma_{\mathrm{T}})\nabla\langle\phi\rangle], \qquad (12.127)$$

where Γ and Γ_{T} are the molecular and turbulent diffusivities, respectively.

The Lagrangian approach

The alternative, Lagrangian, approach to turbulent dispersion originated with G. I. Taylor's classic (1921) paper 'Diffusion by continuous movements.' Taylor argued that (at high Reynolds number) the spatial transport of ϕ due to molecular diffusion is negligible compared with the convective transport by the mean flow and turbulent motions. With complete neglect of molecular diffusion, ϕ is conserved following a fluid particle ($D\phi/Dt = 0$), and consequently the evolution of the mean field $\langle\phi\rangle$ can be determined from the statistics of the motion of fluid particles. The relevant statistics are now considered.

PDFs of particle position

Recall that $X^+(t, Y)$ denotes the position at time t of the fluid particle originating from position Y at time t_0. The PDF of $X^+(t, Y)$ is called the *forward PDF* of particle position, and can be expressed as

$$f_X(x;t|Y) = \langle f_X'(x;t|Y)\rangle, \qquad (12.128)$$

where the fine-grained PDF is

$$f_X'(x;t|Y) = \delta(X^+(t, Y) - x). \qquad (12.129)$$

It is related to the Lagrangian PDF f_{L} (Eq. (12.76)) by

$$f_X(x;t|Y) = \int f_{\mathrm{L}}(V, x;t|Y)\,\mathrm{d}V. \qquad (12.130)$$

In these definitions, Y is considered to be a (non-random) independent variable, and then $X^+(t, Y)$ gives the random trajectory of the fluid particle forward in time, for $t > t_0$.

It is useful also to consider fluid particle trajectories backward in time. To do so we define $Y^+(t, x)$ to be the position at time t_0 of the fluid particle that is at x at time t (for $t > t_0$). Here x is the non-random forward position, and

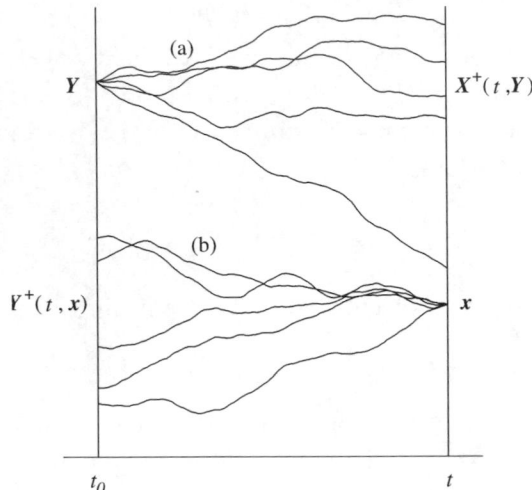

Fig. 12.9. Sketches of forward (a) and backward (b) fluid-particle trajectories (on different realizations of the turbulent flow). (a) Forward trajectories – fluid-particle paths originating at Y at time t_0. (b) Backward trajectories – fluid-particle paths that reach x at time t.

$Y^+(t, x)$ is the random initial position. Forward and backward fluid-particle trajectories are illustrated in Fig. 12.9.

The function $X^+(t, Y)$ provides a one-to-one mapping between initial points Y and final points X^+; and similarly, $Y^+(t, x)$ provides a one-to-one mapping between final points x and initial points Y^+. These functions are, of course, inverses of each other; that is

$$X^+(t, Y^+(t, x)) = x, \qquad (12.131)$$

$$Y^+(t, X^+(t, Y)) = Y. \qquad (12.132)$$

For the constant-density flows being considered, the Jacobian of the mapping $X^+(t, Y)$ is unity:

$$\det\left(\frac{\partial X_i^+(t, Y)}{\partial Y_j}\right) = J(t, Y) = 1 \qquad (12.133)$$

(see Exercise 2.5). Consequently, the Jacobian of the inverse is also unity:

$$\det\left(\frac{\partial Y_j^+(t, x)}{\partial x_i}\right) = J(t, Y^+(t, x))^{-1} = 1. \qquad (12.134)$$

The PDF of $Y^+(t, x)$ is called the *backward PDF* of fluid particle position, and can be written

$$f_Y(Y; t|x) = \langle f_Y'(Y; t|x)\rangle, \qquad (12.135)$$

where the fine-grained PDF is

$$f'_Y(Y;t|x) = \delta(Y^+(t,x) - Y).$$ (12.136)

A simple and valuable – though by no means obvious – relationship is that the backward and forward PDFs are equal:

$$f_X(x;t|Y) = f_Y(Y;t|x).$$ (12.137)

This is shown by the development

$$
\begin{aligned}
f'_X(x;t,Y) &\equiv \delta(X^+(t,Y) - x) \\
&= \int \delta(Y - y)\,\delta(X^+(t,y) - x)\,dy \\
&= \int \delta(Y - y)\,\delta(X^+(t,y) - x)J(t,y)^{-1}\,dX^+ \\
&= \delta(Y - Y^+(t,x)) = f'_Y(Y;t|x).
\end{aligned}
$$ (12.138)

In the third line, the integral over all y is re-expressed as an integral over all X^+. The Jacobian is unity, so that the next line follows from the sifting property of the delta function; where $X^+(t,y)$ equals x and y equals $Y^+(t,x)$.

The Lagrangian formulation

Returning to the problem of turbulent dispersion, we consider an unbounded turbulent flow in which a source at the time t_0 initializes the scalar field deterministically to some value

$$\phi(x,t_0) = \phi_0(x).$$ (12.139)

If molecular diffusion is completely neglected, then ϕ is conserved following a fluid particle. Thus, ϕ has the same value at the initial and final points of a fluid-particle trajectory:

$$\phi(X^+(t,Y),t) = \phi(Y,t_0) = \phi_0(Y),$$ (12.140)

or conversely

$$\phi(x,t) = \phi(Y^+(t,x),t_0) = \phi_0(Y^+(t,x)).$$ (12.141)

The expectation of this last equation yields the required result

$$
\begin{aligned}
\langle \phi(x,t) \rangle = \langle \phi_0(Y^+(t,x)) \rangle &= \int f_Y(Y;t|x)\phi_0(Y)\,dY \\
&= \int f_X(x;t|Y)\phi_0(Y)\,dY.
\end{aligned}
$$ (12.142)

In the particular case of a unit point source at a location Y_0, i.e.,

$$\phi_0(x) = \delta(x - Y_0), \tag{12.143}$$

Eq. (12.142) yields

$$\langle \phi(x, t) \rangle = f_X(x; t | Y_0). \tag{12.144}$$

Thus, the mean conserved scalar field resulting from a unit point source is given by the PDF of position of the fluid particles originating at the source.

It is important to appreciate that the neglect of molecular diffusion in this development is justifiable only insofar as spatial transport of ϕ is concerned. Molecular diffusion has a first-order effect on the scalar variance $\langle \phi'^2 \rangle$ and the scalar PDF $f_\phi(\psi; x, t)$, for example, even at high Reynolds numbers. Consequently, even though expressions for these quantities can be obtained from Eq. (12.141), they have no physical validity. (The effects of molecular diffusion are considered by Saffman (1960) and Pope (1998b).)

Dispersion from a point source

The simplest case to consider is dispersion from a point source in statistically stationary isotropic turbulence. The unit source (Eq. (12.143)) is located at the origin ($Y_0 = 0$), and the release occurs at time $t_0 = 0$. The isotropic turbulent velocity field has zero mean ($\langle U(x, t) \rangle = 0$), and is maintained statistically stationary (by artificial forcing). The r.m.s. velocity is u' so that the Reynolds stresses are $\langle u_i u_j \rangle = u'^2 \delta_{ij}$.

For this case, the mean scalar field $\langle \phi(x, t) \rangle$ is equal to the PDF $f_X(x; t | 0)$ of the position $X^+(t, 0)$ of fluid particles originating from the source (Eq. (12.144)). The dispersion can therefore be characterized by the first and second moments of this PDF, namely the mean and variance of $X^+(t, 0)$. (If f_X is joint normal – as the Langevin model predicts – then these moments fully characterize the dispersion.)

The equation for fluid-particle motion $\partial X^+ / \partial t = U^+$ can be integrated,

$$X^+(t, 0) = \int_0^t U^+(t', 0) \, dt', \tag{12.145}$$

so that statistics of X^+ can be expressed in terms of Lagrangian velocity statistics. The mean is zero

$$\langle X^+(t, 0) \rangle = \int_0^t \langle U^+(t', 0) \rangle \, dt' = 0, \tag{12.146}$$

simply because the mean velocity is taken to be zero.

The covariance of the fluid-particle position is

$$\langle X_i^+(t,0)X_j^+(t,0)\rangle = \int_0^t \int_0^t \langle U_i^+(t',0)U_j^+(t'',0)\rangle \, dt' \, dt''. \qquad (12.147)$$

Given that the turbulence is stationary and isotropic, the two-time Lagrangian velocity correlation can be written

$$\langle U_i^+(t',0)U_j^+(t'',0)\rangle = u'^2 \rho(t'-t'')\delta_{ij}, \qquad (12.148)$$

where $\rho(s)$ is the Lagrangian velocity autocorrelation function. Thus the covariance of position is also isotropic,

$$\langle X_i^+(t,0)X_j^+(t,0)\rangle = \sigma_X^2(t)\delta_{ij}, \qquad (12.149)$$

and is characterized by the standard derivation $\sigma_X(t)$. From these three equations we obtain

$$\sigma_X^2(t) = u'^2 \int_0^t \int_0^t \rho(t'-t'') \, dt' \, dt''. \qquad (12.150)$$

By a non-trivial manipulation (see Exercise 12.26), the double integral can be reduced to a single integral to yield

$$\sigma_X^2(t) = 2u'^2 \int_0^t (t-s)\rho(s) \, ds. \qquad (12.151)$$

For very short times ($t \ll T_L$, so that $\rho(s)$ is adequately approximated by $\rho(0) = 1$) the integral in Eq. (12.151) is $\frac{1}{2}t^2$, which leads to the result

$$\sigma_X(t) \approx u't, \quad \text{for } t \ll T_L. \qquad (12.152)$$

For the short times considered, there is a negligible change in the fluid-particle velocity ($\rho(s) \approx 1$), so that there is essentially straight-line motion

$$X^+(t,0) \approx U^+(0,0)t, \quad \text{for } t \ll T_L \qquad (12.153)$$

(from which Eq. (12.152) follows immediately).

At the other extreme, for very large times ($t \gg T_L$), the integral in Eq. (12.151) can be approximated by

$$\int_0^t (t-s)\rho(s) \, ds \approx t \int_0^\infty \rho(s) \, ds = t\, T_L, \quad \text{for } t \gg T_L, \qquad (12.154)$$

which leads to the square-root spreading

$$\sigma_X(t) \approx \sqrt{2u'^2 T_L t}, \quad \text{for } t \gg T_L. \qquad (12.155)$$

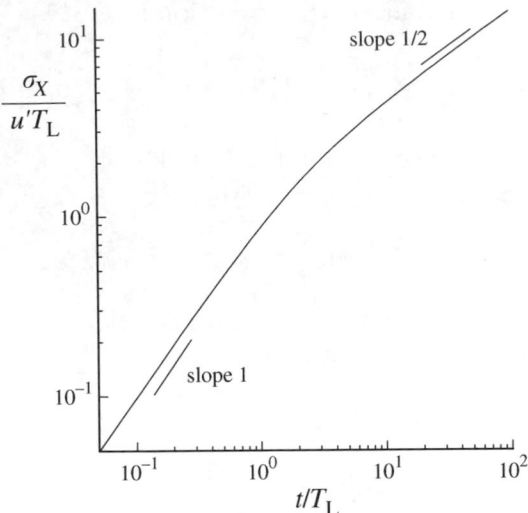

Fig. 12.10. The standard deviation σ_X of dispersion from a point source given by the Langevin model (Eq. (12.159)).

This is precisely the spreading given by the diffusion equation with the constant turbulent diffusivity

$$\Gamma_T = u'^2 T_L, \tag{12.156}$$

see Exercise 12.27.

In fact, for all times, the dispersion can be expressed in terms of a diffusivity $\widehat{\Gamma}_T(t)$ defined by

$$\widehat{\Gamma}_T(t) \equiv \frac{\mathrm{d}}{\mathrm{d}t}\left(\tfrac{1}{2}\sigma_X^2\right), \tag{12.157}$$

and from Eq. (12.151) we obtain

$$\widehat{\Gamma}_T(t) = u'^2 \int_0^t \rho(s)\,\mathrm{d}s. \tag{12.158}$$

To summarize: fluid particles disperse from the origin isotropically (Eqs. (12.146) and (12.149)). In each coordinate direction, the standard deviation of the particle's displacement is $\sigma_X(t)$, which is given in terms of the Lagrangian velocity autocorrelation function $\rho(s)$ by Eq. (12.151). At large times ($t \gg T_L$) the dispersion corresponds to diffusion with the constant turbulent diffusivity $\Gamma_T = u'^2 T_L$, so that σ_X increases as the square-root of time (Eq. (12.155)). For small times ($t \ll T_L$), straight-line fluid particle motion leads to the linear increase $\sigma_X \approx u't$, which corresponds to a time-dependent diffusivity $\widehat{\Gamma}_T(t) \approx u'^2 t$.

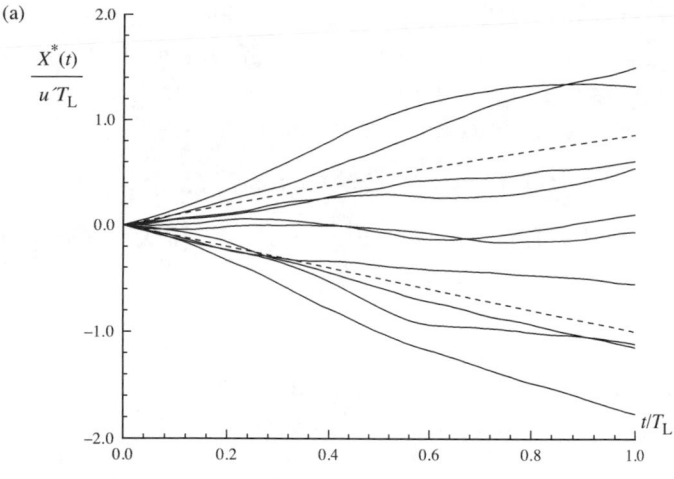

(a)

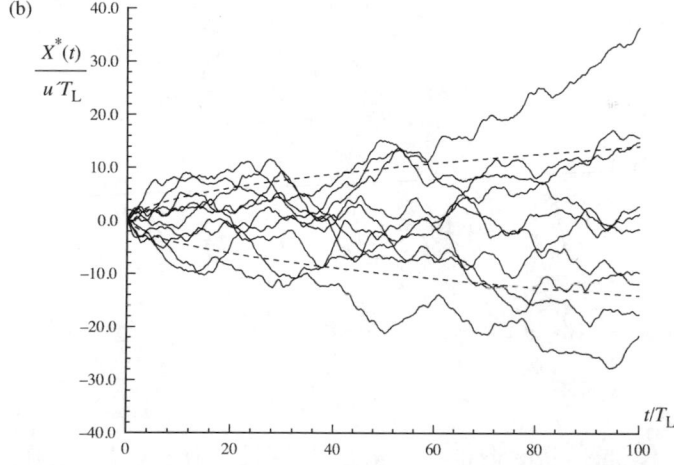

(b)

Fig. 12.11. Samples of fluid-particle paths given by the Langevin model, shown for (a) moderate times and (b) long times. The dashed lines show $\pm\sigma_X(t)$.

The Langevin equation

The Langevin model for the fluid-particle velocity yields a complete prediction for turbulent dispersion. According to the model, the Lagrangian velocity autocorrelation function is the exponential $\rho(s) = \exp(-|s|/T_L)$. With this expression for $\rho(s)$, Eq. (12.151) can be integrated to yield

$$\sigma_X^2(t) = 2u'^2 T_L[t - T_L(1 - e^{-t/T_L})]. \tag{12.159}$$

Figure 12.10 shows $\sigma_X(t)$ given by this formula: the linear and square-root behaviors at small and large times, respectively, are clearly visible.

According to the Langevin model, each component of the fluid-particle velocity is an Ornstein–Uhlenbeck process – which is a Gaussian process. It follows that the fluid particle position (i.e., the integral of the OU process) is also a Gaussian process. Thus, the mean scalar field predicted by the Langevin model is the Gaussian distribution

$$\langle\phi(x,t)\rangle = (\sigma_X\sqrt{2\pi})^{-3}\exp(-\tfrac{1}{2}x_ix_i/\sigma_X^2),\qquad(12.160)$$

with $\sigma_X(t)$ given by Eq. (12.159). Figure 12.11 shows fluid particle paths generated by the Langevin equation.

EXERCISES

12.26 From Eq. (12.150), with the substitutions $r = t''$ and $s = t'' - t'$, and recalling that $\rho(s)$ is an even function, obtain

$$\sigma_X^2(t) = 2u'^2\int_0^t\int_0^r\rho(s)\,ds\,dr.\qquad(12.161)$$

Manipulate the expression

$$\int_0^t\frac{d}{dr}\left(r\int_0^r\rho(s)\,ds\right)dr,$$

to show that

$$\int_0^t\int_0^r\rho(s)\,ds\,dr = t\int_0^t\rho(s)\,ds - \int_0^t r\rho(r)\,dr.\qquad(12.162)$$

Hence verify Eq. (12.151).

12.27 Show that the diffusion equation

$$\frac{\partial\langle\phi\rangle}{\partial t} = \widehat{\Gamma}_T(t)\nabla^2\langle\phi\rangle\qquad(12.163)$$

admits the Gaussian solution, Eq. (12.160), with $\sigma_X(t)$ varying according to Eq. (12.157).

12.28 Show that the exponential approximation $\rho(s) = \exp(-|s|/T_L)$ yields

$$\widehat{\Gamma}_T(t) = u'^2 T_L(1 - e^{-t/T_L})\qquad(12.164)$$

and Eq. (12.159) for $\sigma_X(t)^2$. Verify the short-time and long-time behaviors of these expressions.

12.29 Let $X^*(t)$ and $U^*(t)$ be components of position and velocity of a fluid particle evolving according to the Langevin model, Eq. (12.89). Show that their covariance is

$$\langle X^*U^*\rangle = u'^2 T_L(1 - e^{-t/T_L}).\qquad(12.165)$$

Determine the correlation coefficient ρ_{XU}. How does this behave at small and large times?

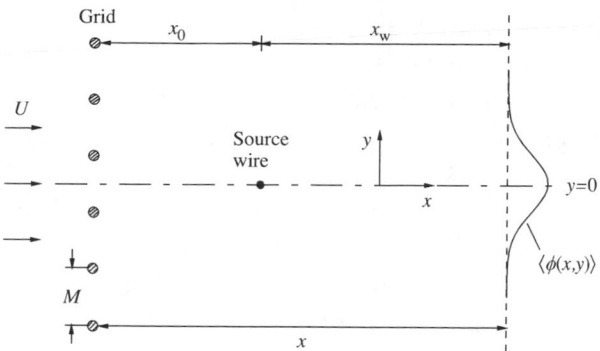

Fig. 12.12. A sketch of a thermal line-source experiment, showing a heated wire downstream of a turbulence-generating grid.

12.30 Taylor (1921) proposed a stochastic model for one component of position of a fluid particle $X^*(t)$. According to the model, $X^*(t)$ is a Markov process, and over a small time interval Δt the increment $\Delta_{\Delta t} X^*(t)$ has the properties

$$\langle \Delta_{\Delta t} X^*(t)\rangle = 0, \quad \langle [\Delta_{\Delta t} X^*(t)]^2\rangle = d^2,$$
$$\langle \Delta_{\Delta t} X^*(t+\Delta t)\,\Delta_{\Delta t} X^*(t)\rangle = cd^2, \tag{12.166}$$

where the variance d^2 and correlation coefficient c depend on Δt. Show that, with a suitable specification of d^2 and c, in the limit $\Delta t \to 0$ this model is equivalent to the Langevin equation, Eq. (12.89). (Hint: show that $U^*(t) \equiv \Delta_{\Delta t} X^*(t)/\Delta t$ tends to a diffusion process.)

A line source in grid turbulence

The simplest experiment used to study turbulent dispersion is the *thermal wake* sketched in Fig. 12.12. A fine heated wire is placed normal to the mean flow, a distance x_0 downstream of a turbulence-generating grid of mesh size M. The wire is intended to be sufficiently fine that it does not significantly affect the velocity field. Although the wire may be quite hot, the excess temperature of the air passing by it rapidly falls to a few degrees. Consequently, except in the immediate vicinity of the wire, the excess temperature $\phi(\boldsymbol{x}, t)$ is a conserved passive scalar.

In experiments (e.g., Warhaft (1984) and Stapountzis *et al.* (1986)) the mean profile is found to be Gaussian,

$$\langle \phi(\boldsymbol{x})\rangle = \frac{1}{\sigma_Y \sqrt{2\pi}} \exp\left(\frac{-\frac{1}{2}y^2}{\sigma_Y^2}\right), \tag{12.167}$$

with a characteristic width $\sigma_Y(x_w)$ that increases with distance downstream of the wire, x_w. (The integral of $\langle \phi \rangle$ across the flow is conserved, and, as implied by Eq. (12.167), ϕ is normalized so that this integral is unity.)

In the laboratory frame, the thermal wake is statistically stationary and two-dimensional: statistics depend solely on x_w and y. To a good approximation, in a frame moving with the mean velocity, the temperature field is statistically one-dimensional: statistics depend solely on y and t – the mean flight time from the source. This temporally evolving problem corresponds to dispersion from a plane source (at $y = 0$ and $t = 0$), and can be analyzed in much the same way as dispersion from a point source. Indeed, if the turbulence were non-decaying, the expressions obtained in the previous subsection for σ_X (Eqs. (12.151) and (12.159)) would apply to the thickness of the thermal wake σ_Y.

Anand and Pope (1985) applied the Langevin model to the thermal wake, taking into account the decay of the turbulence, and the non-negligible effects of molecular diffusion close to the source. Their result is

$$\frac{\sigma_Y^2}{L_0^2} = 2\frac{\Gamma x_w}{U} + \frac{4n^2}{3}\left(\frac{(1+x_w/x_0)^{r-s}}{r(r-s)} + \frac{(1+x_w/x_0)^{-s}}{rs} - \frac{1}{s(r-s)}\right), \quad (12.168)$$

where L_0 is the lengthscale $k^{3/2}/\varepsilon$ at the wire, Γ is the molecular diffusivity, n is the turbulence-decay exponent, and the constants r and s are given in terms of the Langevin model constant C_0 by

$$r = \tfrac{1}{2}n\left(\tfrac{3}{2}C_0 - 1\right) + 1, \quad (12.169)$$

$$s = \tfrac{1}{2}n\left(\tfrac{3}{2}C_0 + 1\right) - 1. \quad (12.170)$$

Figure 12.13 compares the Langevin model prediction of σ_Y (Eq. (12.168)) with the experimental data. Even though the value $C_0 = 2.1$ was selected with respect to these data, the agreement over a substantial range may be seen to be excellent.

The values of C_0 and $\mathcal{C}_0$

Given the consistency between the Langevin equation and the Kolmogorov hypotheses, it might be thought that the value $C_0 = 2.1$ obtained from thermal-wake data would provide a good estimate for the Kolmogorov constant $\mathcal{C}_0$. This is not the case, because the experiments are at fairly low Reynolds numbers, and the predictions of σ_Y^2 depend on the autocorrelation function $\rho(s)$ for time intervals s outside the inertial subrange. Although $\mathcal{C}_0$ is difficult to determine, there are indications that its value is greater than 4, possibly around 6 (Pope 1994a). It is for this reason that the notation

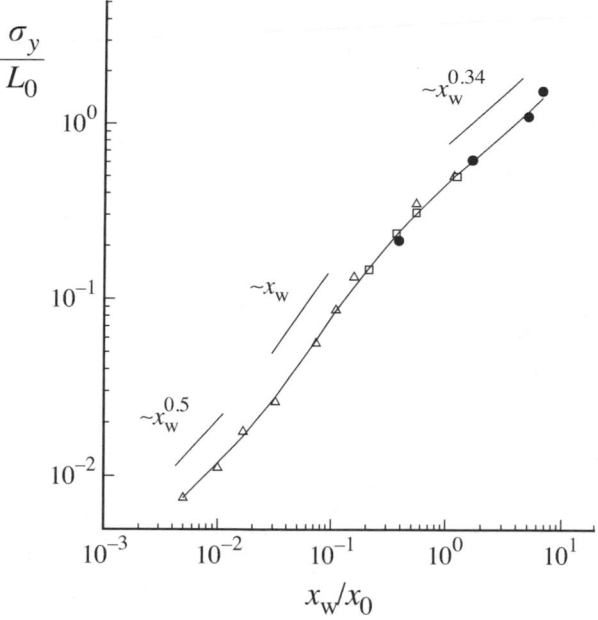

Fig. 12.13. The thickness of the thermal wake σ_Y (normalized by the turbulence lengthscale L_0) as a function of the distance x_w downstream of the wire (normalized by the distance from the grid to the wire x_0). Line, Langevin model Eq. (12.168); symbols, experimental data of Warhaft (1984), $x_0/M = 20\,(\bullet)$, $52\,(\square)$, and $60\,(\triangle)$.

distinguishes between the Langevin-model constant C_0 and the Kolmogorov constant $\mathcal{C}_0$. The extent of the consistency between the Langevin model and the Kolmogorov hypotheses (at high Reynolds number) is an open question.

EXERCISE _____

12.31   Consider the gradient-diffusion hypothesis applied to the thermal wake, with the turbulent diffusivity given by

$$\Gamma_T = \frac{C_\mu\, k^2}{\sigma_T\; \varepsilon}. \qquad (12.171)$$

Show that the predicted thickness of the thermal wake is

$$\frac{\sigma_Y^2}{L_0^2} = \frac{2\Gamma x_w}{U} + \frac{2nC_\mu}{(2-n)\sigma_T}\left[\left(1 + \frac{x_w}{x_0}\right)^{2-n} - 1\right], \qquad (12.172)$$

(cf. Eq. (12.168)). Show that, both qualitatively and quantitatively, this result is significantly at variance with the experimental data (Fig. 12.13).

12.5 The velocity–frequency joint PDF

12.5.1 Complete PDF closure

The generalized Langevin model provides a model equation for the PDF of velocity $f(V; x, t)$, Eq. (12.26). By itself, this equation does not provide a complete closure because the coefficients in the equation involve quantities (e.g., ε and $\tau \equiv k/\varepsilon$) that are not known in terms of f. Together, the model equations for f and ε (or f and $\omega \equiv \varepsilon/k$) do provide a complete closure; and the calculations of Haworth and El Tahry (1991) provide an example of the application of this approach.

The alternative approach (described here) to obtaining a complete closure is to incorporate information about the distribution of the turbulent scales within the PDF framework. Specifically, we consider the joint PDF of velocity and the turbulent frequency (defined below). A model equation for this joint PDF provides a complete closure.

Beyond providing a complete closure, there are additional benefits to describing the distribution of turbulent frequency. Large fluctuations in dissipation rate (i.e., internal intermittency) can be accounted for; and, in inhomogeneous flows, fuller account can be taken of the different behavior of turbulent fluid depending on its origin and history. For non-turbulent fluid the turbulence frequency is zero, whereas for turbulent fluid it is strictly positive. By exploiting this property, it is possible to obtain (from the velocity-frequency joint PDF) the intermittency factor $\gamma(x, t)$, and the turbulent and non-turbulent conditional PDFs.

We redefine $\omega(x, t)$ to be the *instantaneous* turbulent frequency

$$\omega(x, t) \equiv \varepsilon_0(x, t)/k(x, t), \tag{12.173}$$

where $\varepsilon_0(x, t)$ is the *instantaneous* dissipation rate (Eq. (6.312)). Its mean is

$$\langle \omega(x, t) \rangle = \frac{\varepsilon(x, t)}{k(x, t)} = \frac{1}{\tau(x, t)}, \tag{12.174}$$

which is identical to the second variable in the k–ω model. The Eulerian PDF of $\omega(x, t)$ is denoted by $f_\omega(\theta; x, t)$ (where θ is the sample-space variable corresponding to ω) so that the mean is also given by

$$\langle \omega(x, t) \rangle = \int_0^\infty \theta f_\omega(\theta; x, t) \, d\theta. \tag{12.175}$$

Because ω is, by definition, non-negative, the lower limit of the sample space is zero.

The velocity–frequency joint PDF – the Eulerian joint PDF of $\mathbf{U}(x, t)$

and $\omega(\boldsymbol{x}, t)$ – is denoted by $\bar{f}(V, \theta; \boldsymbol{x}, t)$. As usual, the marginal PDFs are obtained as

$$f(V, \boldsymbol{x}, t) = \int_0^\infty \bar{f}(V, \theta; \boldsymbol{x}, t) \, \mathrm{d}\theta, \qquad (12.176)$$

$$f_\omega(\theta; \boldsymbol{x}, t) = \int \bar{f}(V, \theta; \boldsymbol{x}, t) \, \mathrm{d}V. \qquad (12.177)$$

As with the velocity, models for the turbulent frequency are developed from the Lagrangian viewpoint. In the next two subsections, stochastic models $\omega^*(t)$ for the turbulent frequency following a fluid particle, $\omega^+(t, \boldsymbol{Y})$, are described. The model proposed by Pope and Chen (1990) yields a log-normal PDF of ω in homogeneous turbulence, whereas that proposed by Jayesh and Pope (1995) yields a gamma distribution. Stochastic Lagrangian models for $\boldsymbol{U}^*(t)$ and $\omega^*(t)$ lead to a closed model equation for the Eulerian velocity–frequency joint PDF, $\bar{f}(V, \theta; \boldsymbol{x}, t)$.

12.5.2 The log-normal model for the turbulence frequency

As discussed in Section 6.7, the instantaneous dissipation exhibits intermittent behavior, and is approximately log-normally distributed. Pope and Chen (1990) designed a stochastic model for $\omega^*(t)$ to incorporate these properties.

To begin with, we consider statistically stationary homogeneous turbulence, so that the mean turbulence frequency $\langle \omega \rangle$ is constant. (For this case, the Eulerian mean $\langle \omega \rangle$ and Lagrangian mean $\langle \omega^* \rangle$ are equal.) If ω^* is log-normally distributed, then (by definition) the quantity

$$\chi^*(t) \equiv \ln[\omega^*(t)/\langle \omega \rangle] \qquad (12.178)$$

is normally distributed. Hence a log-normal model for $\omega^*(t)$ is obtained by taking $\chi^*(t)$ to be an Ornstein–Uhlenbeck process (since this yields a normal distribution for $\chi^*(t)$).

The OU process for $\chi^*(t)$ is written

$$\mathrm{d}\chi^* = -\left(\chi^* + \tfrac{1}{2}\sigma^2\right) \frac{\mathrm{d}t}{T_\chi} + \left(\frac{2\sigma^2}{T_\chi}\right)^{1/2} \mathrm{d}W, \qquad (12.179)$$

where the coefficients to be specified are the variance of χ^*, σ^2, and its autocorrelation timescale, T_χ. By reference to moderate-Reynolds-number DNS data, these coefficients are taken to be

$$\sigma^2 = 1.0, \qquad T_\chi^{-1} = C_\chi \langle \omega \rangle, \qquad (12.180)$$

with $C_\chi = 1.6$. It follows from its definition (Eq. (12.178)) and the assumption

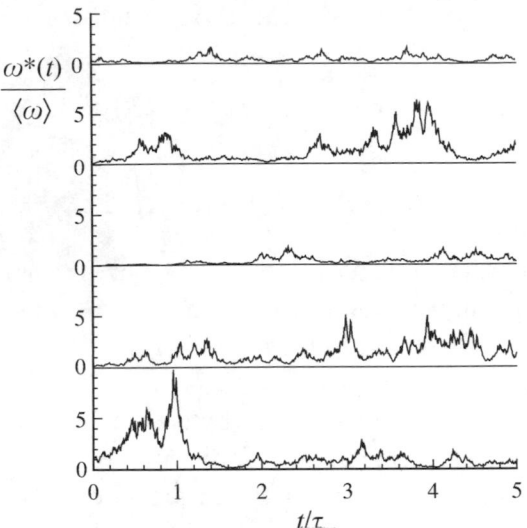

Fig. 12.14. Sample paths of the log-normal stochastic model for the turbulence frequency, Eq. (12.181).

of normality that the mean of χ^* is $-\frac{1}{2}\sigma^2$ (see Exercise 12.32). The factor of $\frac{1}{2}\sigma^2$ in the drift term in Eq. (12.179) ensures that $\langle\chi^*\rangle$ has the correct value (in the stationary state).

The corresponding stochastic differential equation for $\omega^*(t) = \langle\omega\rangle\exp[\chi^*(t)]$, obtained from Eq. (12.179) by using the Ito transformation (Eq. (J.52)), is

$$\mathrm{d}\omega^* = -C_\chi\omega^*\langle\omega\rangle\left[\ln\left(\frac{\omega^*}{\langle\omega\rangle}\right) - \frac{1}{2}\sigma^2\right]\mathrm{d}t + \omega^*(2C_\chi\langle\omega\rangle\sigma^2)^{1/2}\,\mathrm{d}W. \quad (12.181)$$

Sample paths of this process are shown in Fig. 12.14. It may be seen that the process is indeed highly variable and intermittent. There are sustained tranquil periods of small amplitude; and also periods of large fluctuations, with ω^* achieving ten times its mean value.

When it is extended to non-stationary and inhomogeneous flows (Pope 1991a), the stochastic differential equation for $\omega^*(t)$ is

$$\mathrm{d}\omega^* = -\omega^*\langle\omega\rangle\left\{S_\omega + C_\chi\left[\ln\left(\frac{\omega^*}{\langle\omega\rangle}\right) - \left\langle\frac{\omega^*}{\langle\omega\rangle}\ln\left(\frac{\omega^*}{\langle\omega\rangle}\right)\right\rangle\right]\right\}\mathrm{d}t$$
$$+\langle\omega\rangle^2 h\,\mathrm{d}t + \omega^*(2C_\chi\langle\omega\rangle\sigma^2)^{1/2}\,\mathrm{d}W. \quad (12.182)$$

The diffusion term is unaltered; the drift term is re-expressed (so that its mean is zero even if ω^* is not log-normal); and additional drift terms involving the coefficients h and S_ω are included. According to Eq. (12.182), if h is zero, then a non-turbulent particle (i.e., $\omega^* = 0$) remains non-turbulent for all time. The term in h is therefore included in order to allow non-turbulent fluid to become turbulent: the details are given by Pope (1991a).

For homogeneous turbulence, the term in h is negligible, and the evolution of the mean $\langle \omega \rangle$ implied by Eq. (12.182) is

$$\frac{d\langle \omega \rangle}{dt} = -\langle \omega \rangle^2 S_\omega. \tag{12.183}$$

Thus, S_ω determines the evolution of the mean, and, for consistency with the k–ε and k–ω models, this can be specified as

$$S_\omega = -C_{\omega 1}\frac{\mathcal{P}}{\varepsilon} + C_{\omega 2}, \tag{12.184}$$

with $C_{\omega 1} = C_{\varepsilon 1} - 1$ and $C_{\omega 2} = C_{\varepsilon 2} - 1$, cf. Eq. (10.94).

In addition to the model for $\omega^*(t)$ described above, Pope and Chen (1990) proposed a *refined* Langevin model (for the velocity $U^*(t)$) in which the coefficients depend on $\omega^*(t)$ (among other quantities). In particular, in place of $C_0\varepsilon = C_0 k\langle \omega \rangle$, the diffusion coefficient is $C_0 k\omega^*$: that is, it is based on the instantaneous dissipation, consistent with ideas of internal intermittency and the refined similarity hypotheses.

The log-normal model for ω^*, together with the refined Langevin model for $U^*(t)$, yields a closed model equation for the velocity–frequency joint PDF. This model has been applied to a number of flows by Pope (1991a), Anand, Pope, and Mongia (1993), and Minier and Pozorski (1995). As an example, Fig. 12.15 shows profiles of the skewness and kurtosis of velocity in the self-similar plane mixing layer. Clearly these higher moments are calculated quite accurately.

EXERCISES _____

12.32 Consider ω^* to be log-normally distributed, and χ^* defined by Eq. (12.178) to have variance σ^2. Show that the normalization condition

$$\langle e^{\chi^*} \rangle = 1 \tag{12.185}$$

(stemming from Eq. 12.178) implies that the mean of χ^* is

$$\langle \chi^* \rangle = -\tfrac{1}{2}\sigma^2. \tag{12.186}$$

Also obtain the result

$$\left\langle \frac{\omega^*}{\langle \omega \rangle} \ln\left(\frac{\omega^*}{\langle \omega \rangle}\right) \right\rangle = \langle \chi^* e^{\chi^*} \rangle = \tfrac{1}{2}\sigma^2. \tag{12.187}$$

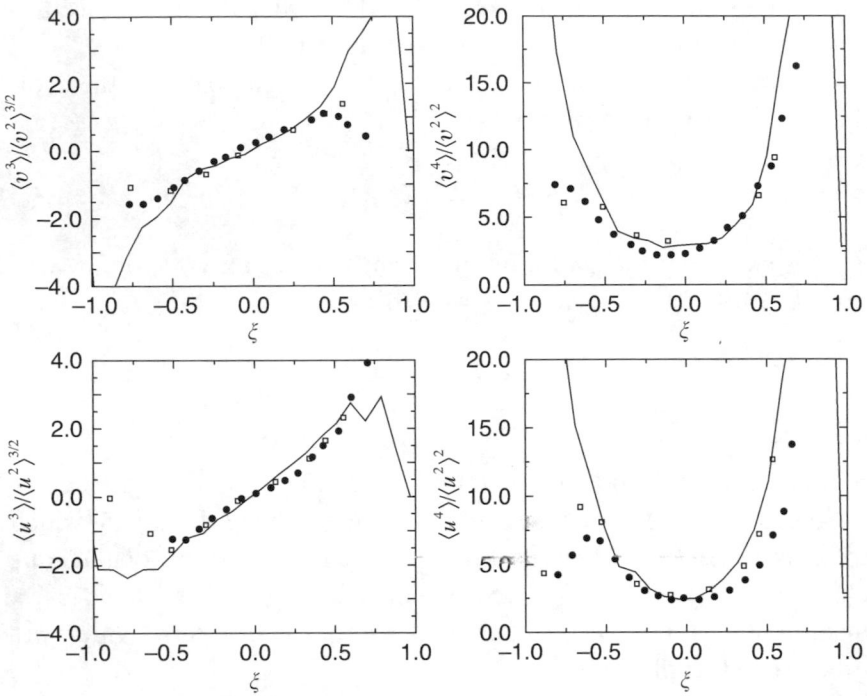

Fig. 12.15. Profiles of skewness and flatness of the axial (u) and lateral (v) velocities in the self-similar plane mixing layer. Lines, calculations by Minier and Pozorski (1995) based on the log-normal/refined Langevin model of Pope (1991a); symbols, experimental data of Wygnanski and Fiedler (1970) ($\bullet$) and of Champagne *et al.* (1976) ($\square$). The abscissa is a normalized cross-stream coordinate. (From Minier and Pozorski (1995).)

12.33   Given N independent samples $\omega^{(n)}$, $n = 1, 2, \ldots, N$, of ω^* drawn from the same distribution $f_\omega(\theta)$, the mean $\langle \omega \rangle$ is estimated by taking the ensemble average

$$\langle \omega \rangle \approx \langle \omega \rangle_N \equiv \frac{1}{N} \sum_{n=1}^{N} \omega^{(n)}. \qquad (12.188)$$

Show that the normalized r.m.s. error ϵ_N in this approximation

$$\epsilon_N^2 \equiv \left\langle \left(\frac{\langle \omega \rangle_N - \langle \omega \rangle}{\langle \omega \rangle} \right)^2 \right\rangle, \qquad (12.189)$$

is given by

$$\epsilon_N^2 = \frac{1}{N} \, \text{var} \left(\frac{\omega^*}{\langle \omega \rangle} \right). \qquad (12.190)$$

How many samples N are needed in order for the error ϵ_N to be 1% if

(a) $f_\omega(\theta)$ is the log-normal distribution with $\mathrm{var}[\ln(\omega^*/\langle\omega\rangle)] = \sigma^2 = 1$, and

(b) $f_\omega(\theta)$ is the gamma distribution with $\mathrm{var}(\omega^*/\langle\omega\rangle) = \sigma^2 = \frac{1}{4}$?

12.5.3 The gamma-distribution model

Although it is physically accurate (compared with DNS data), the log-normal model has several undesirable features: the *ad hoc* term in h (in Eq. (12.182)) is required in order to entrain non-turbulent fluid; the stochastic differential equation is complicated, making it difficult to analyze and implement numerically; and the very long tails of the underlying log-normal distribution lead to substantial statistical fluctuations in numerical implementations (see Exercise 12.33). Jayesh and Pope (1995) developed an alternative model for $\omega^*(t)$, designed to avoid these difficulties and to be as simple as possible.

Beginning again with statistically stationary isotropic turbulence, the Jayesh–Pope model is

$$d\omega^* = -(\omega^* - \langle\omega\rangle)\frac{dt}{T_\omega} + \left(\frac{2\sigma^2\langle\omega\rangle\omega^*}{T_\omega}\right)^{1/2} dW, \qquad (12.191)$$

where the timescale is taken to be $T_\omega^{-1} = C_3\langle\omega\rangle$ with $C_3 = 1.0$, and σ^2 is the variance of $\omega^*/\langle\omega\rangle$, which is taken to be $\sigma^2 = \frac{1}{4}$. Because the drift term is linear in ω^*, the autocorrelation function is the exponential $\exp(-|s|/T_\omega)$.

The stationary distribution of ω^* (see Exercise 12.34) is

$$f_\omega(\theta) = \frac{1}{\Gamma(\alpha)}\left(\frac{\alpha}{\langle\omega\rangle}\right)^\alpha \theta^{\alpha-1} \exp\left(-\frac{\alpha\theta}{\langle\omega\rangle}\right), \qquad (12.192)$$

where $\alpha = 1/\sigma^2 = 4$, and $\Gamma(\alpha)$ is the gamma function. This is indeed the gamma distribution with mean $\langle\omega\rangle$ and variance $\langle\omega\rangle^2\sigma^2$ (see Exercise 12.35).

Sample paths of $\omega^*(t)$ generated by Eq. (12.191) are shown in Fig. 12.16. Clearly these do not exhibit intermittency, and excursions beyond $3\langle\omega\rangle$ are rare.

The log-normal and gamma PDFs are compared in Fig. 12.17. At first sight it may be surprising that these distributions have the same mean and that the log-normal's variance is over six times that of the gamma distribution. The explanation lies in the long tail of the log-normal: in the integrals of $\theta f_\omega(\theta)$ and $\theta^2 f_\omega(\theta)$, 13% and 45%, respectively, of the contributions are from the tails $\theta > 5\langle\omega\rangle$ (i.e., beyond the range shown in Fig. 12.17).

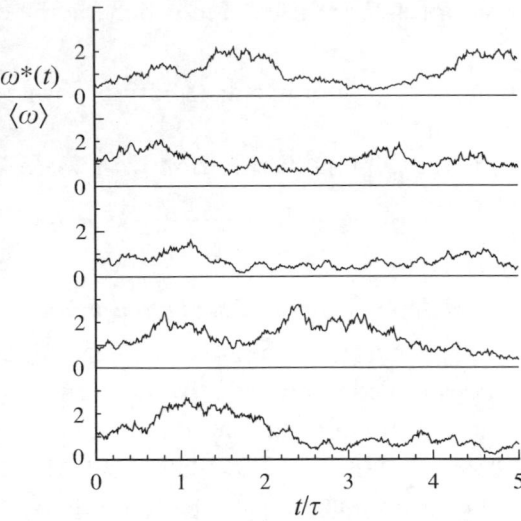

Fig. 12.16. Sample paths of the gamma-distribution model for the turbulence frequency, Eq. (12.191).

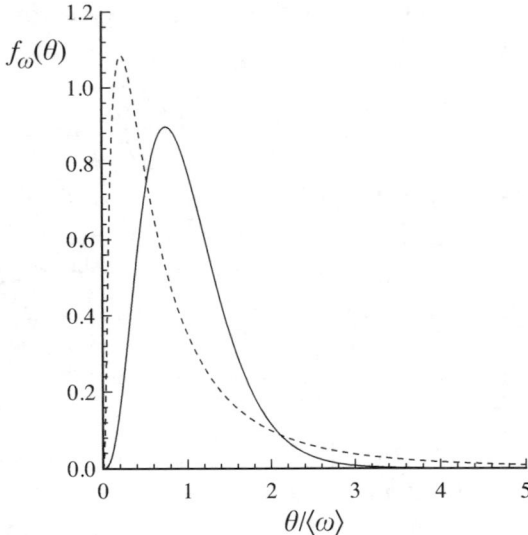

Fig. 12.17. Stationary PDFs of the turbulence frequency given by the log-normal model (dashed line) and the gamma-distribution model (solid line).

The extension of the gamma-distribution model to inhomogeneous flows involves the straightforward addition of the source S_ω (just as in the log-normal model, Eq. (12.182)), and the redefinition of the timescale T_ω in terms of a *conditional mean turbulence frequency* Ω.

To explain the definition of Ω and the rationale for its use, we show in

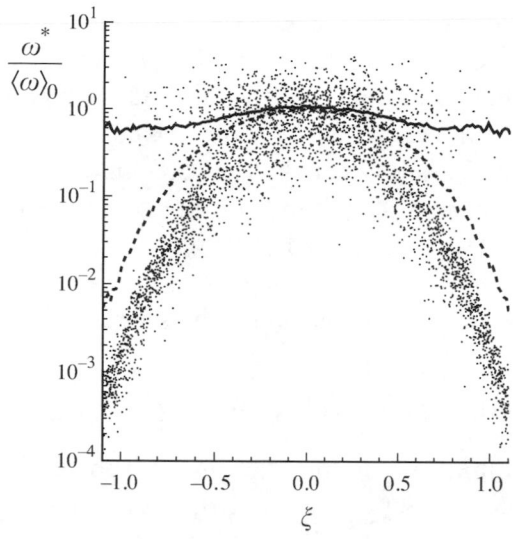

Fig. 12.18. A scatter plot of the turbulence frequency ω^* (normalized by $\langle\omega\rangle$ at $\xi = 0$) against the normalized lateral distance in the self-similar temporal shear layer. The dashed line is the unconditional mean, $\langle\omega\rangle$. The solid line is the conditional mean, Ω, Eq. (12.193). (From Van Slooten, Jayesh, and Pope (1998).)

Fig. 12.18 a scatter plot of ω^* obtained from an application of the model to the temporal mixing layer. Note that ω^* is shown on a logarithmic scale, and that the range of the cross-stream coordinate ξ extends well into the intermittent turbulent/non-turbulent region. In the center of the layer, the PDF of ω^* is found, as expected, to be a gamma distribution: there are no samples with ω^* less than $\frac{1}{10}\langle\omega\rangle$. At the edges of the layer (e.g., $\xi = \pm 0.8$), the situation is quite different. There is a band of particles with very small values of ω^*, which correspond to non-turbulent fluid being entrained, and there is a scattering of particles with higher values of ω^*, comparable to those in the center of the layer, which correspond to fully turbulent fluid. It may be seen that the profile of the mean $\langle\omega\rangle$ provides a separation between the non-turbulent and turbulent samples.

Jayesh and Pope (1995) argued that the appropriate rate Ω for modelling turbulence processes in intermittent regions is that of the turbulent fluid. (In contrast, the unconditional mean $\langle\omega\rangle$ is based on turbulent and non-turbulent fluid.) Accordingly, Ω is defined by

$$\Omega \equiv C_\Omega \langle \omega^* | \omega^* \geq \langle\omega\rangle \rangle$$

$$= C_\Omega \int_{\langle\omega\rangle}^{\infty} \theta f_\omega(\theta)\, d\theta \bigg/ \int_{\langle\omega\rangle}^{\infty} f_\omega(\theta)\, d\theta, \qquad (12.193)$$

where $C_\Omega \approx 0.69$ is a constant, explained below.

In the intermittent region, the condition $\omega^* \geq \langle\omega\rangle$ excludes the non-turbulent particles. Consequently, as may be seen in Fig. 12.18, at the edge of the layer Ω is about half of its peak value, and is two orders of magnitude greater than $\langle\omega\rangle$. In the center of the layer, the condition $\omega^* \geq \langle\omega\rangle$ excludes some turbulent particles, so that, if C_Ω were unity, Ω would exceed $\langle\omega\rangle$. Instead, C_Ω is specified so that Ω and $\langle\omega\rangle$ are equal when the PDF of ω^* is the gamma distribution (see Exercise 12.36).

The model for inhomogeneous flows (Van Slooten and Pope 1999) is obtained from Eq. (12.191) by including the source term S_ω, and redefining $T_\omega^{-1} = C_3\Omega$:

$$d\omega^* = -C_3(\omega^* - \langle\omega\rangle)\Omega\,dt - \Omega\omega^* S_\omega\,dt + (2C_3\sigma^2\langle\omega\rangle\Omega\omega^*)^{1/2}\,dW. \quad (12.194)$$

Because the gamma distribution for ω^* is substantially different than the observed approximately log-normal distribution for the instantaneous dissipation (see Fig. 12.17), it is best to abandon the precise definition of ω^* (Eq. (12.173)). Instead, ω^* can be considered an imprecisely defined characteristic rate of turbulence processes, which is related to the mean dissipation by

$$\varepsilon = k\Omega. \quad (12.195)$$

12.5.4 The model joint PDF equation

The generalized Langevin model for $U^*(t)$, Eq. (12.110), and Eq. (12.194) for $\omega^*(t)$ yield the following equation for the velocity–frequency joint PDF $\bar{f}(V, \theta; x, t)$:

$$\frac{\partial \bar{f}}{\partial t} + V_i \frac{\partial \bar{f}}{\partial x_i} = \frac{1}{\rho} \frac{\partial \langle p\rangle}{\partial x_i} \frac{\partial \bar{f}}{\partial V_i} - G_{ij} \frac{\partial}{\partial V_i}[\bar{f}(V_j - \langle U_j\rangle)]$$

$$+ \frac{1}{2} C_0 \Omega k \frac{\partial^2 \bar{f}}{\partial V_i \partial V_i} + \frac{\partial}{\partial \theta}\{\bar{f}[C_3(\theta - \langle\omega\rangle)\Omega + \Omega\theta S_\omega]\}$$

$$+ C_3\sigma^2\langle\omega\rangle\,\Omega\frac{\partial^2(\bar{f}\theta)}{\partial \theta^2}. \quad (12.196)$$

From the joint PDF $\bar{f}(V, \theta; x, t)$ the fields of $\langle U\rangle$, $\langle\omega\rangle$, Ω, k, and $\langle u_i u_j\rangle$ can be determined, so that all the coefficients in Eq. (12.196) are known in terms of $\bar{f}$. Thus this single model equation is closed. Turbulent-flow calculations based on this equation have been performed by, for example, Anand, Hsu, and Pope (1997) and Van Slooten et al. (1998).

It may be recalled from Section 10.5 on page 383 that the k–ω model has difficulty in treating non-turbulent free-streams. Because of the use of the conditional mean Ω, the velocity–frequency joint PDF model experiences no such difficulties.

EXERCISES

12.34   Identify the coefficients $a(\theta)$ and $b(\theta)^2$ defined by writing the Jayesh–Pope model (Eq. (12.191)) in the general form

$$d\omega^* = a(\omega^*)\,dt + b(\omega^*)\,dW. \qquad (12.197)$$

Use Eq. (J.21) to show that the stationary distribution of ω^* is

$$f_\omega(\theta) = C\theta^{\alpha-1}\exp\left(-\frac{\alpha\theta}{\langle\omega\rangle}\right), \qquad (12.198)$$

where $\alpha = 1/\sigma^2$ and C is a constant. Use the substitution $x = \alpha\theta/\langle\omega\rangle$ to show that the normalization condition yields

$$\int_0^\infty f_\omega(\theta)\,d\theta = 1 = C\left(\frac{\langle\omega\rangle}{\alpha}\right)^\alpha\int_0^\infty x^{\alpha-1}e^{-x}\,dx = C\left(\frac{\langle\omega\rangle}{\alpha}\right)^\alpha\Gamma(\alpha), \qquad (12.199)$$

where $\Gamma(\alpha)$ is the gamma function. Hence verify Eq. (12.192).

12.35   Use the results of Exercise 3.10 on page 51 to show that the moments of the gamma distribution (Eq. (12.192)) are

$$\frac{\langle\omega^n\rangle}{\langle\omega\rangle^n} = \frac{\Gamma(n+\alpha)}{\alpha^n\Gamma(\alpha)} = \frac{(n+\alpha-1)!}{\alpha^n(\alpha-1)!}. \qquad (12.200)$$

Compare the values of the higher moments (e.g., $n = 4$) with those of the log-normal distribution (see Eq. (6.333)).

12.36   With $f_\omega(\theta)$ being the gamma distribution, Eq. (12.192), obtain the result

$$\int_z^\infty \theta^n f_\omega(\theta)\,d\theta = \left(\frac{\langle\omega\rangle}{\alpha}\right)^n \frac{\Gamma(n+\alpha)}{\Gamma(\alpha)}Q\left(n+\alpha, \frac{\alpha z}{\langle\omega\rangle}\right), \qquad (12.201)$$

where $Q(a, z)$ is the incomplete gamma function defined by

$$Q(a, z) = \frac{1}{\Gamma(a)}\int_z^\infty x^{a-1}e^{-x}\,dx. \qquad (12.202)$$

Hence show that $\langle\omega\rangle$ and Ω (Eq. (12.193)) are equal if C_Ω is specified as

$$C_\Omega = \frac{Q(\alpha, \alpha)}{Q(\alpha+1, \alpha)}. \qquad (12.203)$$

(For $\alpha = 4$, this equation yields $C_\Omega \approx 0.6893$.)

12.6 The Lagrangian particle method

In the previous section, a model equation (Eq. (12.196)) for the velocity–frequency joint PDF $\tilde{f}(V, \theta; x, t)$ based on stochastic models for the evolution of fluid-particle properties was obtained. For a general three-dimensional flow, at a given time t, this joint PDF is a function of seven independent variables (i.e., $V_1, V_2, V_3, \theta, x_1, x_2,$ and x_3). It is not feasible, computationally, to represent the joint PDF accurately through a discretization of the seven-dimensional V–θ–x space, as is required in finite-difference, finite-volume, and finite-element methods. Instead, the model PDF equations are solved by *particle methods*. (The book by Hockney and Eastwood (1998) provides an introduction to the use of particle methods in various applications.)

Some aspects of the particle method are simple and obvious. For example, there is a large number of particles, each of which evolves according to the stochastic model equations. However, there are other important aspects that are less obvious. The purpose of this section is to describe the conceptual basis for the particle methods. For the most part, this is most simply done at the level of the velocity PDF $f(V; x, t)$ (as opposed to the velocity–frequency PDF $\tilde{f}$).

12.6.1 Fluid and particle systems

It is important to make a clear distinction between the turbulent flow and the particle method used to model it. We refer to these as the *fluid system* and the *particle system*, respectively.

The fluid system

The basic representation of the turbulent flow is in terms of the Eulerian velocity field $U(x, t)$. This is governed by the continuity and Navier–Stokes equations, and the pressure field $p(x, t)$ is determined from it by the Poisson equation, Eq. (2.42). The basic probabilistic description considered is the one-point one-time Eulerian PDF $f(V; x, t)$, the first two moments of which are the fields of the mean velocity $\langle U(x, t) \rangle$ and the Reynolds stresses $\langle u_i u_j \rangle$. For simplicity we consider the flow in a fixed closed region of volume $\mathcal{V}$, with no flow through the boundary of the region, so the mass of fluid $\mathcal{M} = \rho \mathcal{V}$ is constant.

The particle system

The basic representation in the particle system is in terms of the properties of a large number (N) of particles. Each particle represents a mass $m \equiv \mathcal{M}/N$ of fluid. The nth particle has position $X^{(n)}(t)$ and velocity $U^{(n)}(t)$. The particles are statistically identical, so that in some analyses it is sufficient to consider a single particle, whose properties are denoted by $X^*(t)$, $U^*(t)$, etc.

The one-time joint PDF of $X^*(t)$ and $U^*(t)$ is denoted by $f_L^*(V, x; t)$. The marginal PDF of position is then

$$f_X^*(x; t) = \int f_L^*(V, x; t) \, dV, \qquad (12.204)$$

and the PDF of $U^*(t)$ conditional on $X^*(t) = x$ is

$$f^*(V|x; t) = f_L^*(V, x; t)/f_X^*(x; t). \qquad (12.205)$$

The first moment of this PDF is the conditional mean particle velocity

$$\langle U^*(t)|X^*(t) = x \rangle = \int V f^*(V|x; t) \, dV, \qquad (12.206)$$

which is abbreviated to $\langle U^*|x \rangle$. The fluctuating particle velocity is defined by

$$u^*(t) \equiv U^*(t) - \langle U^*(t)|X^*(t) \rangle, \qquad (12.207)$$

so that the conditional covariance of the particle velocity is

$$\langle u_i^* u_j^*|x \rangle = \langle u_i^*(t) u_j^*(t)|X^*(t) = x \rangle$$
$$= \int (V_i - \langle U_i^*|x \rangle)(V_j - \langle U_j^*|x \rangle) f^*(V|x; t) \, dV. \qquad (12.208)$$

Correspondence

It is important to recognize that there are fundamental differences between the fluid and particle systems. In the fluid system there is an underlying instantaneous velocity field $U(x, t)$, from which velocity gradients (e.g., $\partial^2 U_i/(\partial x_j \, \partial x_k)$), the pressure field $p(x, t)$, and multi-point, multi-time statistics (e.g., $\langle U_i(x, t) U_j(x + r, t + s) \rangle$) can be determined. In the particle system, on the other hand, there is no underlying instantaneous velocity field. The only fields pertaining to the particles are mean fields such as $\langle U^*(t)|x \rangle$.

Given these inherent differences, only in a limited way can the particle system model the fluid system. Specifically, we seek a *correspondence* between the systems at the level of one-point, one-time statistics. That is, if the models were perfect, then the conditional PDF of the particle velocity $f^*(V|x; t)$ would be equal to the PDF of the fluid velocity $f(V; x, t)$, and consequently the moments of the particle velocity, $\langle U^*|x \rangle$ and $\langle u_i^* u_j^*|x \rangle$, would be equal to the mean velocity $\langle U(x, t) \rangle$ and the Reynolds stresses $\langle u_i u_j \rangle$. Some of the distinctions and correspondences between the fluid and particle systems are shown in Table 12.1.

Table 12.1. *A comparison between the fluid system and the particle system showing: the fundamentally different basic representations; the correspondence at the level of one-point, one-time statistics; and the correspondence between the governing equations for these statistics*

	Fluid system		Particle system
Basic representation	$U(x,t)$		$X^*(t)$, $U^*(t)$
Governing equations	Navier–Stokes		$dX^*/dt = U^*$, Stochastic model for U^*
Corresponding velocity statistics	$f(V;x,t)$	$\Longleftrightarrow$	$f^*(V|x;t)$
	$\langle U(x,t)\rangle$	$\Longleftrightarrow$	$\langle U^*(t)|X^*(t) = x\rangle$
	$\langle u_i u_j\rangle$	$\Longleftrightarrow$	$\langle u_i^* u_j^*|x\rangle$
Pressure	$\langle p(x,t)\rangle$	$\Longleftrightarrow$	$P(x,t)$
PDF equation	Eq. (12.9)	$\Longleftrightarrow$	Eq. (12.221)
Mean continuity	$\nabla \cdot \langle U\rangle = 0$	$\Longleftrightarrow$	$\nabla \cdot \langle U^*|x\rangle = 0$
Mean momentum	Eq. (12.224)	$\Longleftrightarrow$	Eq. (12.222)
Poisson equation	Eq. (4.13)	$\Longleftrightarrow$	Eq. (12.227)
Reynolds-stress equation	Eq. (7.194)	$\Longleftrightarrow$	Eq. (12.229)

Some fluid statistics (e.g., $\langle \partial U_i/\partial x_j\, \partial U_k/\partial x_\ell\rangle$) have no equivalents in the particle system. For others (for example, the two-point, two-time correlation $\langle U_i(x,t)U_j(x+r,t+s)\rangle$) an equivalent quantity in the particle system can be defined (i.e., $\langle U_i^{(n)}(t)U_j^{(m)}(t+s)|X^{(n)}(t) = x, X^{(m)}(t+s) = x+r\rangle$, where n and m denote two different particles). However, equality of such multi-point, multi-time statistics is not required – they are not intended to correspond.

In most particle methods, the particles are statistically independent (or, more precisely, they have a weak dependence that vanishes in the limit $N \to \infty$). Consequently, the particles are best conceived as being (models for) fluid particles *on different realizations* of the turbulent flow. Two particles at (or very close to) the same location x are likely to have completely different velocities, i.e., independent samples from $f(V;x,t)$. In view of this independence, the two-particle correlation introduced above is

$$\langle U_i^{(n)}(t)U_j^{(m)}(t+s)|X^{(n)}(t) = x,\ X^{(m)}(t+s) = x+r\rangle$$
$$= \langle U_i^*(t)|x\rangle\langle U_j^*(t+s)|x+r\rangle, \tag{12.209}$$

which clearly does not correspond to the two-point, two-time correlation in the fluid system, $\langle U_i(x,t)U_j(x+r,t+s)\rangle$.

12.6.2 Corresponding equations

In order for fluid and particle statistics to correspond, their governing equations must also correspond. For example, the mean continuity equation is

$$\nabla \cdot \langle U(x,t) \rangle = 0, \tag{12.210}$$

and therefore a necessary condition for $\langle U(x,t) \rangle$ and $\langle U^*(t)|x \rangle$ to correspond is

$$\nabla \cdot \langle U^*(t)|x \rangle = 0. \tag{12.211}$$

The satisfaction of equations such as this imposes constraints on the particle system, which are now discussed.

Particle equations

Each particle – like a fluid particle – moves with its own velocity

$$\frac{\mathrm{d}X^*(t)}{\mathrm{d}t} = U^*(t), \tag{12.212}$$

while its velocity is taken to evolve by the diffusion process

$$\mathrm{d}U^*(t) = a(U^*(t), X^*(t), t)\,\mathrm{d}t + b(X^*(t), t)\,\mathrm{d}W, \tag{12.213}$$

where $a(V,x,t)$ and $b(x,t)$ are the drift and diffusion coefficients. This form is sufficiently general to include the generalized Langevin model; and the results obtained below are unchanged if a completely general diffusion term is used instead. The Fokker–Planck equation for $f_{\mathrm{L}}^*(V,x;t)$ obtained from these particle equations is

$$\frac{\partial f_{\mathrm{L}}^*}{\partial t} + V_i \frac{\partial f_{\mathrm{L}}^*}{\partial x_i} = -\frac{\partial}{\partial V_i}[f_{\mathrm{L}}^* a_i(V,x,t)] + \tfrac{1}{2}b(x,t)^2 \frac{\partial^2 f_{\mathrm{L}}^*}{\partial V_i \partial V_i}. \tag{12.214}$$

The particle-position density

The correspondence between $f(V;x,t)$ and $f^*(V|x;t) = f_{\mathrm{L}}^*(V,x;t)/f_X^*(x;t)$ is not sufficient to determine the particle position density in physical space f_X^*: the condition $f = f^*$ is satisfied by $f_{\mathrm{L}}^* = f f_X^*$ for any positive specification of f_X^*. However, as is now shown, the evolution equations dictate that $f_X^*(x,t)$ must be constant and uniform, i.e.,

$$f_X^*(x;t) = \mathcal{V}^{-1}. \tag{12.215}$$

When the equation for f^* is derived from the Fokker–Planck equation for f_{L}^*, the term

$$(V_i - \langle U_i^*|x \rangle)\frac{f^*}{f_X^*}\frac{\partial f_X^*}{\partial x_i} \tag{12.216}$$

arises, see Exercise 12.38. There is no equivalent of f_X^* in the fluid system, so, clearly, in order for the equations for f and f^* to correspond, this term must be zero. This occurs if, and only if, f_X^* is uniform.

Also obtained from the Fokker–Planck equation (see Exercise 12.37) is the equation for the evolution of f_X^*:

$$\left(\frac{\partial}{\partial t} + \langle U_i^* | x \rangle \frac{\partial}{\partial x_i}\right) \ln f_X^* = -\frac{\partial \langle U_i^* | x \rangle}{\partial x_i}. \tag{12.217}$$

For the particle-mean-continuity equation (Eq. (12.211)) to be satisfied, it is necessary and sufficient for the left-hand side of Eq. (12.217) to be zero. Given that f_X^* is uniform, this condition then requires that f_X^* also be constant. (This result is also implied by the normalization condition $\int f_X^* \, dx = 1$.)

EXERCISES

12.37   From the definitions Eqs. (12.204)–(12.206) obtain the results

$$\int V f_L^*(V, x; t) \, dV = f_X^*(x; t) \langle U^*(t) | x \rangle. \tag{12.218}$$

By integrating the Fokker–Planck equation for f_L^* Eq. (12.214) over all V, show that the particle-position density $f_X^*(x; t)$ evolves by

$$\frac{\partial f_X^*}{\partial t} + \frac{\partial}{\partial x_i}\left(\langle U_i^* | x \rangle f_X^*\right) = 0. \tag{12.219}$$

Divide by f_X^* to verify Eq. (12.217).

12.38   The equation for the evolution of the conditional PDF of the particle velocity $f^*(V | x; t)$ is obtained from the Fokker–Planck equation for $f_L^*(V, x; t)$ (Eq. (12.214)) by dividing throughout by the PDF of position f_X^*. Assuming that the particle-mean continuity equation (Eq. (12.211)) is satisfied, obtain the result

$$\frac{1}{f_X^*}\left(\frac{\partial f_L^*}{\partial t} + V_i \frac{\partial f_L^*}{\partial x_i}\right) = \frac{\partial f^*}{\partial t} + V_i \frac{\partial f^*}{\partial x_i} + (V_i - \langle U_i^* | x \rangle) \frac{f^*}{f_X^*} \frac{\partial f_X^*}{\partial x_i}. \tag{12.220}$$

PDF equations

The equation for the evolution of the conditional PDF of the particle velocity $f^*(V | x; t)$ is obtained by dividing the Fokker–Planck equation (Eq. (12.214)) by f_X^*. Given that f_X^* is uniform, the result is simply

$$\frac{\partial f^*}{\partial t} + V_i \frac{\partial f^*}{\partial x_i} = -\frac{\partial}{\partial V_i}[f^* a_i(V, x, t)] + \frac{1}{2}b(x, t)^2 \frac{\partial^2 f^*}{\partial V_i \partial V_i}. \tag{12.221}$$

Since f^* corresponds to the Eulerian velocity PDF $f(V; x, t)$, this equation for f^* corresponds to the evolution equation for f, Eq. (12.9) or equivalently Eq. (H.25) on page 706. If the diffusion model for $U^*(t)$ were perfect, then the right-hand sides of these equations would be equal. This condition provides a relationship between the model coefficients a and b and conditional statistics of the fluid velocity field – which in general are not known. More definite information is provided by comparing the moment equations.

The mean momentum equation

The particle-mean-momentum equation is obtained by multiplying Eq. (12.221) by V_j and integrating:

$$\frac{\partial}{\partial t}\langle U_j^*|x\rangle + \frac{\partial}{\partial x_i}\langle U_i^* U_j^*|x\rangle = A_j(x,t),$$ (12.222)

where $A(x,t)$ is the conditional mean drift

$$A(x,t) = \langle a(U^*(t), X^*(t), t)|X^*(t) = x\rangle$$
$$= \int a(V, x, t)f^*(V; x, t)\,dV.$$ (12.223)

This particle-mean-momentum equation corresponds to the Reynolds equations, which (with neglect of the viscous term) can be written

$$\frac{\partial}{\partial t}\langle U_j\rangle + \frac{\partial}{\partial x_i}\langle U_i U_j\rangle = -\frac{1}{\rho}\frac{\partial\langle p\rangle}{\partial x_j}.$$ (12.224)

(The incorporation of viscous effects in the near-wall region is described in Section 12.7.2.)

The right-hand side of the Reynolds equations (i.e., $-\nabla\langle p\rangle/\rho$) is the irrotational vector field given by the gradient of the scalar $-\langle p\rangle/\rho$. For the particle- and fluid-mean-momentum equations to correspond, the conditional drift A must also be irrotational, and hence can be written as the gradient of a scalar. When multiplied by $-\rho$, this scalar field is denoted by $P(x,t)$ and is called the *particle-pressure field*, so that A is given by

$$A = -\frac{1}{\rho}\nabla P.$$ (12.225)

Clearly then, the particle- and fluid-mean-momentum equations correspond if, and only if, the mean fluid pressure $\langle p(x,t)\rangle$ and the particle pressure $P(x,t)$ correspond.

The Poisson equation for pressure

With $A(x, t)$ given by Eq. (12.225), the divergence of Eq. (12.222) is

$$\frac{\partial}{\partial t} \frac{\partial \langle U_j^* | x \rangle}{\partial x_j} + \frac{\partial^2 \langle U_i^* U_j^* | x \rangle}{\partial x_i \, \partial x_j} = -\frac{1}{\rho} \frac{\partial^2 P}{\partial x_j \, \partial x_j}. \qquad (12.226)$$

In view of the particle-mean-continuity equation (Eq. (12.211)), the first term is zero, so that $P(x, t)$ is determined by the Poisson equation:

$$\nabla^2 P = -\rho \frac{\partial^2 \langle U_i^* U_j^* | x \rangle}{\partial x_i \, \partial x_j}. \qquad (12.227)$$

Both sides of this equation correspond directly to the Poisson equation for $\langle p(x, t) \rangle$, Eq. (4.13).

In the particle system, there are no underlying instantaneous fields, but there are mean fields (e.g., $\langle U^*(t) | X^*(t) = x \rangle$) defined as conditional means of the particle properties. For pressure, neither is there an underlying instantaneous field nor is pressure a particle property from which a conditional mean field can be obtained. Thus the word 'mean' is avoided in describing the particle-pressure field $P(x, t)$, because it is not the mean of an underlying random object. It is instead obtained as the solution to the Poisson equation.

It is clear from Eq. (12.226) that the satisfaction of the Poisson equation is a necessary condition for the particle mean continuity equation to be satisfied. Thus the conditional mean of the drift coefficient ($A = \langle a | x \rangle = -\nabla P / \rho$) is determined by this condition.

The generalized Langevin model

As previously written (Eq. (12.110)), the generalized Langevin model contains quantities pertaining both to the fluid system and to the particle system. Written consistently in terms of particle properties, it is

$$dU_i^*(t) = -\frac{1}{\rho} \frac{\partial P}{\partial x_i} \, dt + G_{ij}(U_j^*(t) - \langle U_j^*(t) | X^*(t) \rangle) \, dt$$

$$+ [C_0 \varepsilon(X^*(t), t)]^{1/2} \, dW_i. \qquad (12.228)$$

EXERCISES

12.39 For the fluid system, the instantaneous continuity equation is $\nabla \cdot U = 0$ and, stemming from this, the Jacobian $J(t, Y)$ (Eq. (2.23)) is unity. Is there an equation corresponding to $\nabla \cdot U = 0$ and a quantity corresponding to $J(t, Y)$ for the particle system?

Compare the steps involved in relating Lagrangian and Eulerian PDFs for the fluid and particle systems.

12.40   Identify the drift $a(V, x, t)$ and diffusion $b(x, t)$ coefficients corresponding to the GLM, Eq. (12.228). Verify that the conditional mean drift is $-\nabla P/\rho$. Write down the evolution equation for $f^*(V|x, t)$ and verify that the first moment equation for $\langle U^*(t)|x \rangle$ corresponds to the Reynolds equation. Show that the second moment equation is

$$
\left(\frac{\partial}{\partial t} + \langle U_k^*|x \rangle \frac{\partial}{\partial x_k} \right) \langle u_i^* u_j^* |x \rangle + \frac{\partial}{\partial x_k} \langle u_i^* u_j^* u_k^* |x \rangle
$$

$$
+ \langle u_i^* u_k^* |x \rangle \frac{\partial \langle U_j^*|x \rangle}{\partial x_k} + \langle u_j^* u_k^* |x \rangle \frac{\partial \langle U_i^*|x \rangle}{\partial x_k}
$$

$$
= G_{ik} \langle u_j^* u_k^* |x \rangle + G_{jk} \langle u_i^* u_k^* |x \rangle + C_0 \varepsilon \delta_{ij}. \qquad (12.229)
$$

Compare this equation with the exact Reynolds-stress equation, Eq. (7.194). Which terms correspond?

12.6.3 Estimation of means

The development so far in this section has involved conditional means, such as $\langle U^*|x \rangle$, and $\langle u_i^* u_j^* |x \rangle$ which appear in the generalized Langevin model. In a numerical implementation, with a large number N of particles, these conditional means have to be *estimated* from the particle properties, $X^{(n)}(t)$, $U^{(n)}(t)$, $n = 1, 2, \ldots, N$. Usually a *kernel estimator* (described below) is used. The purpose here is not to describe the details of a particular numerical method; but the estimation of conditional means is, conceptually, an important ingredient in the particle implementation of PDF methods.

Kernel estimation

With $Q(V)$ being a given function of velocity, we consider the estimation of the conditional mean $\langle Q(U^*[t])|X^*(t) = x \rangle$ from the properties of the ensemble of N particles. The particles' positions $X^{(n)}(t)$ are random, uniformly distributed in the flow domain. Since (with probability one) there are no particles at the point of interest x, it is inevitable that the estimate must involve particles in the vicinity of x. A kernel estimate is a weighted mean over the particles in the vicinity of x, the weight being proportional to a specified kernel function, $K(r, h)$. A simple example of a kernel function in D dimensions is

$$
K(r, h) = \begin{cases} \alpha_D h^{-D} \left(1 - \dfrac{r}{h} \right), & \text{for } \dfrac{r}{h} \le 1 \\[2mm] 0, & \text{for } \dfrac{r}{h} > 1, \end{cases} \qquad (12.230)
$$

where h is the specified (positive) bandwidth and $r = |r|$. In general, a kernel function is required to integrate to unity:

$$\int K(r, h)\, dr = 1. \tag{12.231}$$

For the kernel given by Eq. (12.230), this condition determines the coefficient α_D for the cases $D = 1, 2$, and 3 (see Exercise 12.41). It is usually also required that K be non-negative, and that it have bounded support (i.e., $K = 0$ for $|r| > h$).

For simplicity we consider a point x that is at least a distance h from the boundary, and we consider a fixed time t (which is not indicated explicitly in the notation). The kernel estimate for

$$\langle Q|x \rangle \equiv \langle Q(U^*[t]) | X^*(t) = x \rangle \tag{12.232}$$

is

$$\langle Q|x \rangle_{N,h} \equiv \frac{(\mathcal{V}/N) \sum_{n=1}^{N} K(x - X^{(n)}, h) Q^{(n)}}{(\mathcal{V}/N) \sum_{n=1}^{N} K(x - X^{(n)}, h)}, \tag{12.233}$$

where $Q^{(n)}$ is written for $Q(U^{(n)}[t])$. It may be seen that $\langle Q|x \rangle_{N,h}$ is the ensemble average of the particle values $Q^{(n)}$, weighted by the kernel which is centered at x. Indeed, Eq. (12.233) can alternatively be written

$$\langle Q|x \rangle_{N,h} = \sum_{n=1}^{N} w^{(n)} Q^{(n)}, \tag{12.234}$$

where the weights are

$$w^{(n)} \equiv \frac{(\mathcal{V}/N) K(x - X^{(n)}, h)}{(\mathcal{V}/N) \sum_{m=1}^{N} K(x - X^{(m)}, h)}, \tag{12.235}$$

and evidently they sum to unity, $\sum_{n=1}^{N} w^{(n)} = 1$.

Estimation errors

The error ϵ_Q in the kernel estimate

$$\epsilon_Q \equiv \langle Q|x \rangle_{N,h} - \langle Q|x \rangle, \tag{12.236}$$

can be decomposed into a deterministic part – called the *bias* –

$$B_Q \equiv \langle \langle Q|x \rangle_{N,h} \rangle - \langle Q|x \rangle = \langle \epsilon_Q \rangle, \tag{12.237}$$

and a random part, i.e.,

$$\epsilon_Q = B_Q + \epsilon_s \xi, \tag{12.238}$$

where ξ is a standardized random variable ($\langle\xi\rangle = 0, \langle\xi^2\rangle = 1$) and ϵ_s is the r.m.s. error

$$\epsilon_s^2 = \text{var}(\langle Q|x\rangle_{N,h}), \tag{12.239}$$

see Exercise 12.42. The bias arises because the estimate is based on particles at $x + r$ (for $|r| < h$) where the mean $\langle Q|x + r\rangle$ differs from $\langle Q|x\rangle$. As shown in Exercise 12.43, to leading order the bias varies as $h^2 \nabla^2 \langle Q|x\rangle$. The statistical error arises because of the finite number N_h of samples with non-zero weights. As shown in Exercise 12.44, to a first approximation the variance of $\langle Q|x\rangle_{N,h}$ varies as

$$\epsilon_s^2 \sim \frac{1}{N_h} \text{var}(Q) \sim \frac{1}{Nh^D} \text{var}(Q). \tag{12.240}$$

The bias increases with h while the statistical error decreases with h. Consequently, as shown in Exercise 12.45, for a given N there is an optimal choice of h, and this varies as $N^{-1/(D+4)}$. With this choice, the total error varies as

$$\langle\epsilon_Q^2\rangle^{1/2} \sim N^{-2/(4+D)}. \tag{12.241}$$

The conclusion is that kernel estimation can be used to estimate mean fields from particle values. As the number of particles N tends to infinity, the error involved tends to zero, but it does so rather slowly. Further information about kernel estimation can be found in Eubank (1988) and Härdle (1990), where better kernel functions than Eq. (12.230) are discussed.

Particle and particle-mesh methods

In order to integrate the generalized Langevin equation (Eq. (12.228)), means such as $\langle U(t)^*|X^{(n)}(t)\rangle$ must be evaluated at every particle location ($n = 1, 2, \ldots, N$). The direct approach is to perform a kernel estimation at every particle location. It appears at first sight that to perform this estimation for every particle requires on the order of N^2 operations. However, it is possible to perform the task in on the order of N operations (Welton and Pope 1997, Welton 1998). The resulting method is a pure particle method, resembling smoothed particle hydrodynamics (SPH, see, e.g., Monaghan (1992)).

The generally used alternative is to use a particle-mesh method, resembling the cloud-in-cell method (Birdsall and Fuss 1969). The solution domain is covered with a mesh, and kernel estimates of mean fields are formed at the mesh nodes. On the basis of these values, the mean fields over the whole field are represented by linear splines, so that the values at the particle locations are simply obtained by linear interpolation.

12.6.4 Summary

The following points summarize the theoretical basis of particle methods for the numerical solution of model PDF equations.

(i) An ensemble of N particles with properties $X^{(n)}(t)$, $U^{(n)}(t)$, $n = 1, 2, \dots, N$ can consistently represent a turbulent flow. The particles can be thought of as (models for) fluid particles on different realizations of the flow.

(ii) The fluid system and the particle system correspond at the level of one-point, one-time Eulerian PDFs (see Table 12.1). Multi-particle statistics do not correspond to multi-point fluid statistics.

(iii) Particle methods are used to obtain numerical solutions to the model PDF equations. In these methods, model equations (e.g., Eqs. (12.212), (12.213), and (12.228)) are integrated forward in time to determine the evolution of the particle properties.

(iv) Several important properties of the particle system are determined from the correspondence between the evolution equations for the fluid and particle PDFs, $f(V; x, t)$ and $f^*(V|x, t)$.

  (a) The PDF of the particle position f_X^* is uniform. An initially uniform distribution remains uniform if the particle mean continuity equation is satisfied (see Eq. (12.217)).

  (b) In the stochastic model equation for velocity (Eq. (12.213)), the conditional mean drift A (Eq. (12.223)) is given by $A = -\nabla P / \rho$, where $P(x, t)$ is the particle pressure.

  (c) The particle pressure (which corresponds to the mean fluid pressure $\langle p(x, t) \rangle$) is determined by a Poisson equation, Eq. (12.227). The satisfaction of the Poisson equation is necessary to the continued satisfaction of the particle mean continuity equation.

(v) Equation (12.228) gives the generalized Langevin model written exclusively in terms of particle properties.

(vi) Conditional means (e.g., $\langle U^*(t)|x \rangle$) can be determined using kernel estimation. The errors in such estimates tend to zero as the number of particles N tends to infinity, but they do so quite slowly (Eq. (12.241)).

EXERCISES _____

12.41 For the kernel $K(r, h)$ given by Eq. (12.230), show that the normalization condition Eq. (12.231) determines the coefficients α_D to be

$$\alpha_1 = 1, \quad \alpha_2 = \alpha_3 = \left(\frac{3}{\pi}\right). \tag{12.242}$$

Show that the first moments of the kernel are zero,

$$K_i \equiv \int r_i K(r,h)\,dr = 0, \tag{12.243}$$

and that the second moments are

$$K_{ij} \equiv \int r_i r_j K(r,h)\,dr = \beta_D h^2 \delta_{ij}, \tag{12.244}$$

where

$$\beta_1 = \tfrac{1}{6}, \quad \beta_2 = \tfrac{3}{20}, \quad \beta_3 = \tfrac{2}{15}. \tag{12.245}$$

Obtain the result

$$\int K(r,h)^2\,dr = h^{-D}\gamma_D, \tag{12.246}$$

where

$$\gamma_1 = \frac{2}{3}, \quad \gamma_2 = \frac{3}{4\pi}, \quad \gamma_3 = \frac{3}{5\pi}. \tag{12.247}$$

12.42 From the definition of ϵ_Q (Eq. (12.236)), show that the mean-square error in the kernel estimate is

$$\langle \epsilon_Q^2 \rangle = B_Q^2 + \epsilon_s^2. \tag{12.248}$$

Hence confirm the validity of the decomposition $\epsilon_Q = B_Q + \epsilon_s \xi$ (Eq. (12.238)).

12.43 Recalling that the joint PDF of $X^{(n)}(t)$ and $U^{(n)}(t)$ is $f_L^*(V, x, t) = f_X^*(x;t)f^*(V|x;t)$, and that f_X^* is the uniform distribution $\mathcal{V}^{-1}$, show that the expectation of the numerator in the kernel estimate (Eq. (12.233)) is

$$\left\langle \frac{\mathcal{V}}{N} \sum_{n=1}^{N} K(x - X^{(n)}(t), h)\, Q(U^{(n)}[t]) \right\rangle =$$

$$\iint K(r,h)Q(V)f^*(V|x - r;t)\,dV\,dr. \tag{12.249}$$

Show that the expectation of the denominator in Eq. (12.233) is unity. By expanding $f^*(V|x - r;t)$ in a Taylor series about x, obtain

$$\iint K(r,h)Q(V)f^*(V|x - r;t)\,dV\,dr =$$

$$\langle Q|x \rangle - K_i \frac{\partial \langle Q|x \rangle}{\partial x_i} + \frac{1}{2!} K_{ij} \frac{\partial^2 \langle Q|x \rangle}{\partial x_i\,\partial x_j} \dots, \tag{12.250}$$

where the moments of the kernel, K_i and K_{ij}, are defined by Eqs. (12.243) and (12.244). Hence show that, in the limit $N \to \infty$, the bias in the kernel estimate is

$$
\begin{aligned}
B_Q &= \langle Q|x\rangle_{\infty,h} - \langle Q|x\rangle \\
&= \tfrac{1}{2}\beta_D h^2 \nabla^2 \langle Q|x\rangle + \mathcal{O}(h^4),
\end{aligned}
\tag{12.251}
$$

where the coefficient β_D is defined by Eq. (12.244).

12.44   Let w^* denote any one of the weights defined by Eq. (12.235). Show that the mean is

$$
\langle w^{(n)}\rangle = \langle w^*\rangle = \frac{1}{N}.
\tag{12.252}
$$

Noting that, because of the normalization condition $\sum_{n=1}^{N} w^{(n)} = 1$, the weights are not mutually independent, obtain the result

$$
\langle w^{(n)} w^{(m)}\rangle = \frac{1}{N^2} + \frac{(N\delta_{nm} - 1)}{N - 1}\,\mathrm{var}(w^*).
\tag{12.253}
$$

Consider the statistically homogeneous case in which $\langle Q|x\rangle$ is independent of x, and in which $w^{(n)}$ and $Q^{(n)}$ are independent. Obtain the result

$$
\begin{aligned}
\frac{\epsilon_s^2}{\mathrm{var}(Q)} &= \frac{\mathrm{var}(\langle Q|x\rangle_{N,h})}{\mathrm{var}(Q)} = \frac{1}{N} + N\,\mathrm{var}(w^*) \\
&= N\langle w^{*2}\rangle = \frac{1}{N}\left\langle \left(\frac{w^*}{\langle w^*\rangle}\right)^2\right\rangle.
\end{aligned}
\tag{12.254}
$$

From Eq. (12.235), argue that, for large N, the weight w^* (which is a random variable) can be written

$$
w^* = \frac{\mathcal{V}}{N} K(r^*, h),
\tag{12.255}
$$

where r^* is a uniformly distributed random variable with density $1/\mathcal{V}$. Hence obtain

$$
\langle w^{*2}\rangle = \frac{\mathcal{V}}{N^2}\int K(r, h)^2\, dr = \frac{\mathcal{V}\gamma_D}{N^2 h^D},
\tag{12.256}
$$

where the coefficient γ_D is defined by Eq. (12.246).

With $\mathcal{L} \equiv \mathcal{V}^{1/D}$ being a characteristic dimension of the solution domain, show that the variance of the kernel estimate is

$$
\epsilon_s^2 = \frac{\gamma_D}{N}\left(\frac{\mathcal{L}}{h}\right)^D \mathrm{var}(Q).
\tag{12.257}
$$

Show that this can be re-expressed as

$$\epsilon_s^2 = \frac{\gamma_D'}{N_h} \text{var}(Q), \tag{12.258}$$

where N_h is the expected number of particles within the support of the kernel, and γ_D' is a constant.

12.45   The results above show that, to leading order, the mean square error in the kernel estimate is of the form

$$\langle \epsilon_Q^2 \rangle = C_b h^4 + \frac{C_s}{N h^D}, \tag{12.259}$$

where C_b and C_s are positive coefficients (independent of N and h) describing the bias and statistical errors. Show that the choice of h that minimizes the error is

$$h = \left(\frac{D C_s}{4 N C_b} \right)^{1/(D+4)}. \tag{12.260}$$

Determine $\langle \epsilon_Q^2 \rangle$ for this choice of h, and hence confirm Eq. (12.241).

12.7 Extensions

In this section we describe various extensions to the model based on the joint PDF of velocity and the turbulence frequency. As is the case with other turbulence models, in PDF methods walls can be treated either by using wall functions or by solving the model equations (with some modifications) through the viscous wall region, with boundary conditions imposed at the wall. These two approaches are described in the first two subsections. By adding a random unit vector – the wavevector – to the PDF formulation (Section 12.7.3), it is possible to provide a much more accurate representation of rapid distortions. In Section 12.7.4 we briefly describe the extension of PDF methods to flows with chemical reactions. This involves adding composition variables to the PDF formulation.

12.7.1 Wall functions

We consider a statistically two-dimensional flow in the usual coordinate system, so that U and V are the components of velocity parallel and normal to the wall located at $y = 0$. As described in Section 11.7.6, the idea of the wall-function approach is to impose boundary conditions at a location $y = y_p$ in the log-law region (e.g., where y^+ is around 50), so that the

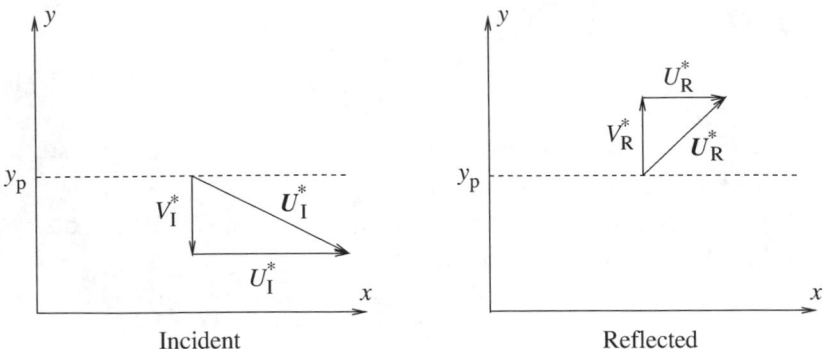

Fig. 12.19. Incident and reflected particle velocities for wall functions imposed at $y = y_p$.

model equations do not have to be solved in the viscous near-wall region. For the velocity–frequency joint-PDF equation, a satisfactory wall-function treatment has been developed by Dreeben and Pope (1997b). This is best described in terms of the behavior of the particles used in the numerical solution of the model PDF equation.

The general particle has position $X^*(t)$, velocity $U^*(t)$, and frequency $\omega^*(t)$; its distance from the wall is $Y^*(t) = X_2^*(t) \geq y_p$. The wall-function boundary conditions are imposed by *reflecting* the particle off the boundary $y = y_p$, as depicted in Fig. 12.19.

Suppose that the particle hits the boundary at time $t = \bar{t}$, (i.e., $Y^*(\bar{t}) = y_p$). The particle properties immediately prior to $\bar{t}$ are called the *incident* properties, and are indicated by the subscript I (e.g., U_I^*, V_I^* and ω_I^*). Note that V_I^* is negative, since the particle hits the boundary ($y = y_p$) by moving downwards from above ($y > y_p$). The boundary conditions are imposed by specifying the particle properties immediately after $\bar{t}$. These *reflected* properties are denoted by the subscript R.

Every incident particle results in a reflected particle so that, in an infinitesimal region at the wall, incident and reflected particles occur with equal probability. Hence, mean properties at the wall are the average of incident and reflected properties. For example, the mean normal velocity is

$$\langle V \rangle_p = \tfrac{1}{2}\big[\langle V_I^* \rangle + \langle V_R^* \rangle\big]. \tag{12.261}$$

(Recall that the subscript p denotes quantities evaluated at $y = y_p$.)

The particle position is continuous (i.e., $X_R^* = X_I^*$), while the normal component of velocity is specified by

$$V_R^* = -V_I^*. \tag{12.262}$$

This condition appropriately ensures that V_R^* is positive, and that the mean normal velocity is zero.

In standard wall functions, the boundary condition for the axial momentum equation amounts to a specification of the shear stress $\langle uv \rangle_p$ – according to Eq. (11.190), for example. In PDF methods, the same effect is achieved through the specification

$$U_R^* = U_I^* + \alpha V_I^*. \tag{12.263}$$

A straightforward analysis of the resulting Reynolds stresses (Exercise 12.46) shows that the appropriate specification of the coefficient α is

$$\alpha = -\frac{2\langle uv \rangle_p}{\langle v^2 \rangle_p}. \tag{12.264}$$

The shear stress arises because, in being reflected from the wall, the particle's specific axial momentum decreases by $\alpha|V_I^*|$. As shown in Exercise 12.47, these boundary conditions are consistent with $\boldsymbol{U}^*$ being joint-normally distributed. For the statistically two-dimensional flows considered, the spanwise velocity is conserved on reflection, i.e.,

$$W_R^* = W_I^*. \tag{12.265}$$

The wall-function boundary condition imposed on the turbulence frequency ω^* is based on similar ideas, but the details are less straightforward. In the log-law region, the mean $\langle \omega \rangle$ increases (as y^{-1}) as the wall is approached, and there is a positive flux $\langle v\omega \rangle$ away from the wall. Consequently, an appropriate boundary condition causes ω^* to increase on reflection ($\omega_R^* > \omega_I^*$), at least in the mean. The condition proposed by Dreeben and Pope (1997b) is

$$\omega_R^* = \omega_I^* \exp\left(\frac{\beta V_I^*}{y_p \langle \omega \rangle}\right), \tag{12.266}$$

where β is a non-dimensional (negative) coefficient that is specified to yield the appropriate flux $\langle v\omega \rangle_p$. The exponential form is consistent with a joint-normal distribution of $\boldsymbol{U}^*$ and $\ln \omega^*$.

The wall-function boundary conditions have been described as they are implemented in the particle method used to solve the model PDF transport equation. In terms of the joint PDF of $\boldsymbol{U}$ and ω, $\bar{f}(\boldsymbol{V}, \theta; \boldsymbol{x}, t)$, the boundary condition is (see Exercise 12.48) that for $V_2 > 0$,

$$\bar{f}(V_1, V_2, V_3, \theta; x_1, y_p, x_3, t) = \exp\left(\frac{\beta V_2}{y_p \langle \omega \rangle}\right)$$

$$\times \bar{f}\left[V_1 + \alpha V_2, -V_2, V_3, \theta \exp\left(\frac{\beta V_2}{y_p \langle \omega \rangle}\right); x_1, y_p, x_3, t\right]. \tag{12.267}$$

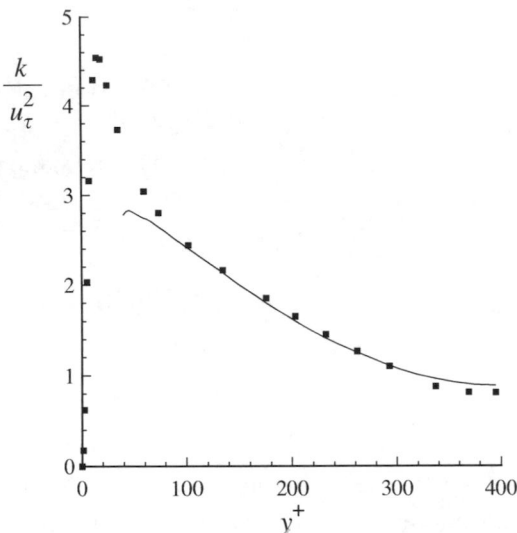

Fig. 12.20. The turbulent-kinetic-energy profile (in wall units) for fully developed channel flow at Re = 13,750. Symbols, DNS data of Kim *et al.* (1987); line, velocity-frequency joint PDF calculation using wall functions (from Dreeben and Pope (1997b)).

Figures 12.20 and 12.21 illustrate velocity–frequency joint-PDF model calculations of channel flow using wall functions (imposed at $y^+ \approx 40$). It may be seen from Fig 12.20 that the turbulent kinetic energy is well represented in the outer region ($y^+ > 50$), and that no calculation is performed for the viscous near-wall region. The main requirement on wall functions is that they yield the correct wall shear stress. Figure 12.21 shows that, for channel flow, these wall functions are successful in that respect.

The advantages and disadvantages of the use of wall functions in PDF methods are the same as with other turbulence models. Compared with solving the model equations through the viscous wall region, there is a very considerable saving in the amount of computation required. However, in near-wall flow regions where the profiles differ from the log-law, the wall functions have little physical basis, and may be quite inaccurate.

EXERCISES

12.46   From Eqs. (12.261)–(12.263) obtain the results

$$\langle V \rangle_p = 0, \tag{12.268}$$

$$\langle v^2 \rangle_p = \langle V_I^{*2} \rangle = \langle V_R^{*2} \rangle, \tag{12.269}$$

$$\langle uv \rangle_p = -\tfrac{1}{2}\alpha \langle v^2 \rangle_p. \tag{12.270}$$

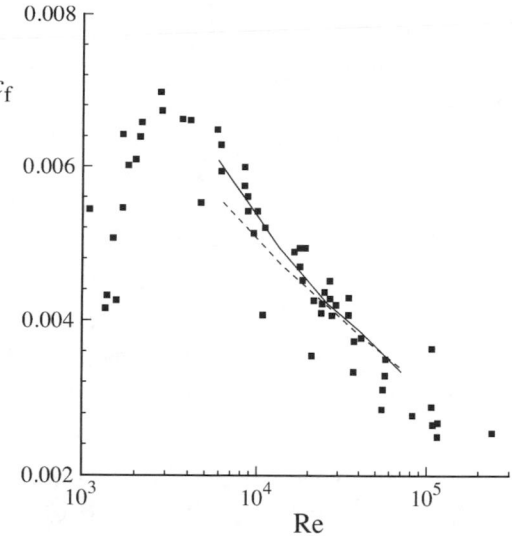

Fig. 12.21. The skin-friction coefficient $c_f \equiv \tau_w/(\frac{1}{2}\rho U_0^2)$ against the Reynolds number ($\text{Re} = 2\overline{U}\delta/\nu$) for channel flow: symbols, experimental data compiled by Dean (1978); solid line, velocity–frequency joint-PDF calculations using wall functions (Dreeben and Pope 1997b); dashed line, near-wall joint-PDF calculations using elliptic relaxation (Section 12.7.2, Dreeben and Pope (1998)).

If V^* is normally distributed, show that

$$-\langle V_I^* \rangle = \langle V_R^* \rangle = \left(\frac{2\langle v^2 \rangle}{\pi} \right)^{1/2}. \qquad (12.271)$$

12.47 Let U and V be correlated, jointly normal random variables with $\langle V \rangle = 0$, and for specified α let U^* and V^* be defined by

$$U^* = U + \alpha V, \qquad (12.272)$$

$$V^* = -V. \qquad (12.273)$$

Show that, for a particular choice of α, the random variables $\{U^*, V^*\}$ and $\{U, V\}$ are identically distributed. Argue on the basis of this result that the wall functions Eqs. (12.262)–(12.264) are consistent with the particle velocity being joint normally distributed.

A sample of $\{U, V\}$ corresponds to a point in the sample space, which lies on an ellipse of constant probability density. Show that the point corresponding to $\{U^*, V^*\}$ (obtained from $\{U, V\}$ through Eqs. (12.272) and (12.273) with the appropriate choice of α) lies on the same ellipse. What is the significance of this observation?

12.48 Write down expressions for U_I^*, V_I^*, W_I^*, and ω_I^* in terms of U_R^*,

V_R^*, W_R^*, and ω_R^*. Show that the determinant of the Jacobian of the transformation between these variables is $\exp[\beta V_R/(y_p\langle\omega\rangle)]$. Hence verify Eq. (12.267).

12.7.2 The near-wall elliptic-relaxation model

The alternative to wall functions is to solve the model PDF equation through the viscous near-wall region, so that boundary conditions are imposed at the wall. We describe here the extensions to the velocity–frequency joint-PDF model developed by Dreeben and Pope (1998) that account for near-wall effects. The three extensions are to incorporate the direct effects of viscosity, to determine the GLM coefficients C_0 and G_{ij} via an elliptic relaxation equation, and to modify the defining equation relating $\langle\omega\rangle$ to k and ε.

Viscous transport

The exact evolution equation for the PDF of velocity $f(V; x, t)$ deduced from the Navier–Stokes equations is

$$\frac{\partial f}{\partial t} + V_i\frac{\partial f}{\partial x_i} = \nu\,\nabla^2 f + \frac{1}{\rho}\frac{\partial\langle p\rangle}{\partial x_i}\frac{\partial f}{\partial V_i} + \dot{f}_R, \qquad (12.274)$$

where $\dot{f}_R$ is written for the pressure-fluctuation and dissipative terms (identified in Eq. (H.25)).

The GLM model described in previous sections is appropriate to high-Reynolds-number regions, and the viscous transport term $\nu\,\nabla^2 f$ (on the right-hand side of Eq. (12.274)) is absent from the corresponding model PDF equation, Eq. (12.26). However, this viscous term is, of course, of prime importance in the viscous sublayer, and must be included in a successful near-wall model.

From the microscopic viewpoint, viscous effects arise from the random motion of molecules; so a natural and convenient way to include viscous effects in the particle implementation of PDF methods is to add a random motion to the particles. Specifically, the particle position $X^*(t)$ is redefined to evolve by the diffusion process

$$\mathrm{d}X^*(t) = U^*(t)\,\mathrm{d}t + \sqrt{2\nu}\,\mathrm{d}W', \qquad (12.275)$$

where $W'(t)$ is a Wiener process (independent of that appearing in GLM). With this definition, $X^*(t)$ is no longer the position of a fluid particle: it is instead a model for the position of a molecule that diffuses with diffusion coefficient ν (see Exercise 12.49).

As before, $U^*(t)$ is a model for the continuum fluid velocity following the particle, i.e., it is a model for[4]

$$U^\star(t) \equiv U(X^*(t), t).$$ (12.276)

Even though $U(x, t)$ is differentiable, $U^\star(t)$ is not, because the argument $X^*(t)$ is not differentiable. Instead, $U^\star(t)$ is a diffusion process with infinitesimal increment

$$\begin{aligned}
dU_i^\star(t) &= \frac{\partial U_i}{\partial t}\,dt + \frac{\partial U_i}{\partial x_j}\,dX_j^* + \frac{1}{2}\frac{\partial^2 U_i}{\partial x_j\,\partial x_k}\,dX_j^*\,dX_k^* \\
&= \frac{DU_i}{Dt}\,dt + v\,\nabla^2 U_i\,dt + \sqrt{2v}\,\frac{\partial U_i}{\partial x_j}\,dW_j' \\
&= -\frac{1}{\rho}\frac{\partial p}{\partial x_i}\,dt + 2v\,\nabla^2 U_i\,dt + \sqrt{2v}\,\frac{\partial U_i}{\partial x_j}\,dW_j'.
\end{aligned}$$ (12.277)

(The first line is the increment obtained from Eq. (12.276), the second line is obtained by substituting Eq. (12.275) for dX^*, and the final expression is obtained by using the Navier–Stokes equation to substitute for DU_i/Dt.)

On the right-hand side of Eq. (12.277), the velocity can be decomposed into its mean and fluctuation. The GLM for $U^*(t)$ is extended by including the resulting mean terms, i.e., the extended model is

$$\begin{aligned}
dU_i^*(t) &= -\frac{1}{\rho}\frac{\partial \langle p \rangle}{\partial x_i}\,dt + 2v\,\nabla^2 \langle U_i \rangle\,dt + \sqrt{2v}\,\frac{\partial \langle U_i \rangle}{\partial x_j}\,dW_j' \\
&\quad + G_{ij}(U_j^* - \langle U_j \rangle)\,dt + (C_0\varepsilon)^{1/2}\,dW_i.
\end{aligned}$$ (12.278)

(Note that this equation is written in the relatively simple notation used in Sections 12.1–12.5. In the more precise but cumbersome notation of Section 12.6, $\langle U \rangle$ is replaced by $\langle U^*(t)|X^*(t)\rangle$, and $\langle p \rangle$ is replaced by the particle pressure, P.)

The model equation for the PDF of velocity (deduced from the stochastic models for X^* and U^*) is precisely of the form of that obtained from the Navier–Stokes equations, Eq. (12.274). In particular, it contains the exact viscous transport term $v\,\nabla^2 f^*$ (see Exercise 12.50, Eq. (12.282)). As a consequence, both the model mean-momentum and the Reynolds-stress equations contain the exact viscous-transport terms (see Exercise 12.51).

EXERCISES _____

12.49   Consider a mixture of ideal gases in laminar flow, so that the velocity $U(x, t)$ is non-random. The position $X^*(t)$ of a molecule of a

---

[4] Note the distinction between the fluid velocity $U^\star(t)$ obtained from $U(x, t)$ and which evolves by Eq. (12.277), and the stochastic process $U^*(t)$ which evolves by the model Eq. (12.278).

particular species is modelled by the stochastic differential equation

$$dX^*(t) = U(X^*(t), t)\, dt + \sqrt{2\Gamma}\, dW(t), \qquad (12.279)$$

where Γ is a positive constant. The concentration of the species is denoted by $\phi(x, t)$. This is simply the expected number density of the molecules of the species, which is the product of the total number of such molecules and the PDF of $X^*(t)$. From the Fokker–Planck equation for this PDF, show that $\phi(x, t)$ evolves by

$$\frac{\partial \phi}{\partial t} + U_i \frac{\partial \phi}{\partial x_i} = \Gamma \frac{\partial^2 \phi}{\partial x_i\, \partial x_i}. \qquad (12.280)$$

Comment on the validity of the model of molecular motion, Eq. (12.279), and on the coefficient Γ. How can the same result be obtained for turbulent flow?

12.50   Consider the six-dimensional diffusion process $Z^*(t)$, where the components of Z^* are $\{Z_1^*, Z_2^*, Z_3^*, Z_4^*, Z_5^*, Z_6^*\} = \{X_1^*, X_2^*, X_3^*, U_1^*, U_2^*, U_3^*\}$. With $X^*(t)$ and $U^*(t)$ evolving according to Eqs. (12.275) and (12.278), show that $Z^*(t)$ evolves by the general diffusion equation

$$dZ_i^* = a_i(Z^*, t)\, dt + \mathcal{B}_{ij}(Z^*, t)\, dW_j'', \qquad (12.281)$$

where $\{W_1'', W_2'', W_3'', W_4'', W_5'', W_6''\} = \{W_1', W_2', W_3', W_1, W_2, W_3\}$.

Write down explicit expressions for the drift vector a_i, the diffusion matrix $\mathcal{B}_{ij}$, and the diffusion coefficient $B_{ij} = \mathcal{B}_{ik}\mathcal{B}_{jk}$. Hence use the general form of the Fokker–Planck equation (Eq. (J.56)) to show that the joint PDF of $X^*(t)$ and $U^*(t)$, $f_L^*(V, x; t)$, evolves by

$$\frac{\partial f_L^*}{\partial t} + V_i \frac{\partial f_L^*}{\partial x_i} = \nu\, \nabla^2 f_L^* + \frac{1}{\rho} \frac{\partial \langle p \rangle}{\partial x_i} \frac{\partial f_L^*}{\partial V_i}$$

$$- \frac{\partial}{\partial V_i} [G_{ij}(V_j - \langle U_j \rangle) f_L^*] + 2\nu \frac{\partial \langle U_j \rangle}{\partial x_i} \frac{\partial^2 f_L^*}{\partial x_i\, \partial V_j}$$

$$+ \nu \frac{\partial \langle U_i \rangle}{\partial x_k} \frac{\partial \langle U_j \rangle}{\partial x_k} \frac{\partial^2 f_L^*}{\partial V_i\, \partial V_j} + \frac{1}{2} C_0 \varepsilon \frac{\partial^2 f_L^*}{\partial V_i\, \partial V_i}. \qquad (12.282)$$

Integrate this equation over all V to obtain the equation for the evolution of the PDF of position, $f_X^*(x; t)$. Show that this equation is consistent with f_X^* being uniform, and hence that the PDF of velocity conditional on $X^*(t) = x$, $f^*(V|x; t)$, also evolves by Eq. (12.282).

12.51   From the model PDF equation, Eq. (12.282) written for $f^*(V|x; t)$, show that the model mean-momentum equation is identical to the Reynolds equations (including the viscous term).

Similarly, show that the model equation for the Reynolds stress

$\langle u_i u_j \rangle$ is the same as that given by the GLM (Eq. (12.57)), but with the addition of the exact viscous-transport term $v \nabla^2 \langle u_i u_j \rangle$ on the right-hand side.

Elliptic relaxation

Dreeben and Pope (1998) developed an elliptic relaxation model to determine the GLM coefficients G_{ij} and C_0 on the basis of Durbin's (1991, 1993) approach (section 11.8). The elliptic equation solved for the tensor field $g_{ij}(x,t)$ is

$$g_{ij} - L_D \nabla^2 (L_D g_{ij}) = k\left(\bar{G}_{ij} + \tfrac{1}{2}\langle \omega \rangle \delta_{ij}\right), \tag{12.283}$$

where $\bar{G}_{ij}$ is a local model for G_{ij} (very similar to LIPM), and the lengthscale L_D is defined by Eq. (11.204). The solution g_{ij} is then used to specify G_{ij} as

$$G_{ij} = \left(g_{ij} - \tfrac{1}{2}\varepsilon \delta_{ij}\right)/k. \tag{12.284}$$

It may be seen that, for homogeneous turbulence, and with $\varepsilon = k\langle \omega \rangle$, these two equations revert to the local model, i.e., $G_{ij} = \bar{G}_{ij}$.

Let $\bar{C}_0$ denote the value of C_0 used in the local model. By construction, like every consistent GLM, the local model satisfied the constraint Eq. (12.28) relating its coefficients $\bar{C}_0$ and $\bar{G}_{ij}$. However, with $C_0 = \bar{C}_0$, the model G_{ij} most likely does not satisfy the constraint (except in homogeneous turbulence). This difficulty is overcome by determining the value of C_0 used in the GLM (Eq. (12.278)) via the constraint

$$C_0 = -\frac{2}{3}\left(\frac{G_{ij}\langle u_i u_j \rangle}{\varepsilon} + 1\right) = -\frac{2}{3}\frac{g_{ij}\langle u_i u_j \rangle}{k\varepsilon}. \tag{12.285}$$

At the wall, boundary conditions are required on each component of g_{ij}. Satisfactory conditions are found to be

$$g_{ij} = -\gamma_0 \varepsilon_0 n_i n_j, \tag{12.286}$$

where ε_0 is the dissipation rate at the wall, n is the unit normal, and the constant is taken to be $\gamma_0 = 4.5$. In the usual coordinate system in which the wall is at $y = 0$, this condition yields $g_{22} = -\gamma_0 \varepsilon_0$, and all other components are zero.

An analysis of the resulting model Reynolds stress equations shows that the Reynolds-stress components very close to the wall scale as $\langle u^2 \rangle \sim y^2$, $\langle v^2 \rangle \sim y^3$, $\langle w^2 \rangle \sim y^2$, and $\langle uv \rangle \sim y^3$ (providing that γ_0 is at least 2). These scalings are correct (see Eqs. (7.60)–(7.63)) except for $\langle v^2 \rangle$, for which the correct scaling is $\langle v^2 \rangle \sim y^4$.

The turbulence frequency and dissipation

As discussed in Section 11.7, very close to a wall, the Kolmogorov timescale $\tau_\eta = (\nu/\varepsilon)^{1/2}$ is the relevant physical timescale, whereas $\tau = k/\varepsilon$ is zero at the wall. It is appropriate, therefore, to modify the definition of the turbulence frequency so that $\langle\omega\rangle$ is of order $1/\tau_\eta$ at the wall (rather than being infinite according to the relation $\langle\omega\rangle = \varepsilon/k$). To this end, Dreeben and Pope (1998) relate $\langle\omega\rangle$ and ε by

$$\varepsilon = \langle\omega\rangle(k + \nu C_T^2\langle\omega\rangle), \tag{12.287}$$

with $C_T = 6$. In high-Reynolds-number regions this reverts to the original relation ($\langle\omega\rangle = \varepsilon/k$), while at a wall it yields $\langle\omega\rangle^{-1} = C_T\tau_\eta$ – just as in Durbin's model, Eq. (11.163). The value of $\langle\omega\rangle$ at the wall, denoted by $\langle\omega\rangle_o$, is known from Eqs. (11.160) and (12.287) :

$$\langle\omega\rangle_o = \frac{1}{C_T}\left(\frac{\partial(2k)^{1/2}}{\partial y}\right)_{y=0}. \tag{12.288}$$

Boundary conditions

At the wall ($y = 0$), boundary conditions on the joint PDF of U and ω, $\bar{f}(V,\theta;x,t)$ are required (in addition to the boundary conditions on g_{ij}, Eq. (12.286), which have already been discussed). The specified boundary condition is

$$\bar{f}(V,\theta;x_1,0,x_3,t) = \delta(V)f_\omega^o(\theta), \tag{12.289}$$

where f_ω^o is the gamma distribution with mean $\langle\omega\rangle_o$ (Eq. (12.288)) and variance $\sigma^2\langle\omega\rangle_o^2$. The term $\delta(V)$ corresponds to the deterministic no-slip and impermeability conditions $U(x_1,0,x_3,t) = 0$.

It is instructive to examine the behavior of particles very close to the wall, for $y^+ \ll 1$, say. The evolution of the particle position $X^*(t)$ (Eq. (12.275)) is dominated by diffusion, the convection by the fluid velocity being small in comparison (for $y^+ \ll 1$).

We consider the distance of the particle from the wall in wall units

$$Y^+(t^+) \equiv X_2^*(t)/\delta_\nu, \tag{12.290}$$

as a function of time in wall units,

$$t^+ \equiv t\nu/\delta_\nu^2. \tag{12.291}$$

With the neglect of the fluid velocity U_2^*, Eq. (12.275) transforms into

$$\mathrm{d}Y^+(t^+) = \sqrt{2}\,\mathrm{d}W^+(t^+), \tag{12.292}$$

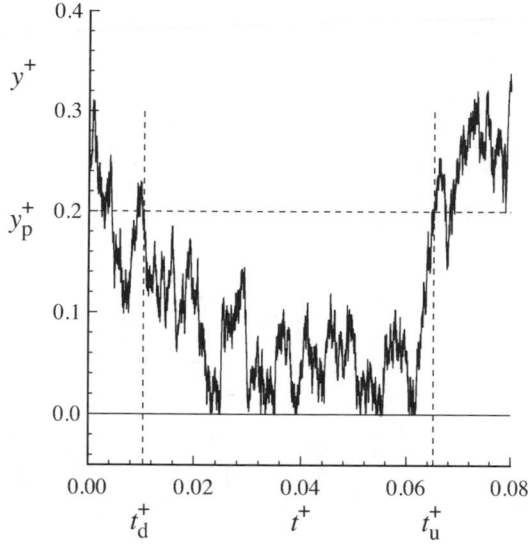

Fig. 12.22. The distance $Y^+(t^+)$ of a particle from the wall (in wall units) as a function of time: a sample path of reflected Brownian motion, Eq. (12.293). For the given level y_p^+, there is a down-crossing at t_d^+ and the subsequent up-crossing is at t_u^+.

where the Wiener process W^+ is such that $\langle dW^+(t^+)^2 \rangle = dt^+$. The particle is reflected from the wall, so that the stochastic process for Y^+ can be written

$$Y^+(t^+ + dt^+) = |Y^+(t^+) + \sqrt{2}\, dW^+(t^+)|. \qquad (12.293)$$

A sample path of this process, which is called *reflected Brownian motion*, is shown in Fig. 12.22.

For a given distance from the wall $y_p^+ = 0.2$, Fig. 12.22 shows a down-crossing at t_d^+ (i.e., where the sample path $Y^+(t^+)$ crosses the line $y^+ = y_p^+$ from above) and the subsequent up-crossing at t_u^+. It may be observed that in the time interval (t_d^+, t_u^+), the particle hits the wall numerous times. In fact, a remarkable property of reflected Brownian motion is that, if the particle hits the wall in the time interval (t_d^+, t_u^+), then (with probability one) it does so an infinite number of times (see Exercise J.2 on page 724). This property holds for all positive y_p^+, no matter how small.

The *incident* properties U_I^* and ω_I^* of the particle are those immediately before the wall is hit, and the *reflected* properties, U_R^* and ω_R^*, are those immediately after. The reflected velocity is specified to be zero ($U_R^* = 0$), while ω_R^* is sampled from the appropriate gamma distribution. What is known about the incident properties, U_I^* and ω_I^*? Because of the behavior of reflected Brownian motion, U_I^* and ω_I^* are statistically identical to U_R^* and ω_R^*. For, if the particle hits the wall at some time t_h^+, with probability one

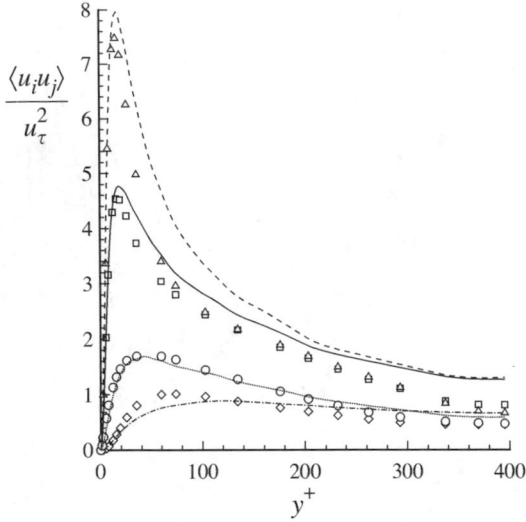

Fig. 12.23. Reynolds stresses in fully developed turbulent channel flow at Re = 13,750. Symbols, DNS data of Kim *et al.* (1987), $\triangle \langle u^2 \rangle$, $\diamond \langle v^2 \rangle$, $\bigcirc \langle w^2 \rangle$, $\square\ k$; lines, near-wall velocity–frequency joint-PDF calculations (from Dreeben and Pope (1998)).

it also hit the wall in the arbitrarily small preceding time interval, at which time the particle properties had been set to U_R^* and ω_R^*. Consequently, the boundary condition on the joint PDF at the wall (Eq. (12.289)) is enforced by the specification of the reflected properties $U_R^* = 0$ and ω_R^*. (Note that, in the wall-function treatment, the joint PDF at y_p is different than the joint PDF of the reflected-particle properties.)

Application to channel flow

Figure 12.23 illustrates the application of the near-wall velocity–frequency joint-PDF model to fully developed turbulent channel flow. It may be seen that the Reynolds stresses are calculated reasonably accurately throughout the channel, and that there is excellent agreement with the DNS data in the viscous near-wall region. The skin-friction coefficient given by the model is shown in Fig. 12.21.

12.7.3 *The wavevector model*

As discussed in Chapter 11, it is not possible to construct a Reynolds-stress model that is accurate for arbitrary rapid distortions of homogeneous turbulence. This is because the rapid pressure–rate-of-strain tensor $\mathcal{R}_{ij}^{(r)}$ depends on the orientation of the velocity correlations – information that is not contained in the Reynolds stresses. On the basis of ideas advanced by Kassinos and Reynolds (1994) and Reynolds and Kassinos (1995), Van Slooten and Pope (1997) developed a PDF model that yields the exact evolution of the Reynolds stresses in rapidly distorted homogeneous turbulence. The model is for the joint PDF of velocity, the turbulence frequency, and a unit vector – the wavevector – which contains the directional information needed to describe rapid distortions. The development of the model is now described, first for rapid distortions, and then for the general case.

Rapid distortion of homogeneous turbulence

We recall from Section 11.4 that, for the case being considered, the fluctuating velocity field can be expressed as the sum of Fourier modes:

$$u(x,t) = \sum_{\kappa^o} \hat{u}(\kappa^o, t) e^{i\hat{\kappa}(t)\cdot x}. \tag{12.294}$$

Each Fourier mode evolves independently: the time-dependent wavenumber vector $\hat{\kappa}(t)$ evolves by the ordinary differential equation Eq. (11.80) from the initial condition $\hat{\kappa}(0) = \kappa^o$; and the corresponding Fourier coefficient $\hat{u}(\kappa^o, t)$ evolves by Eq. (11.83).

The evolution of $\hat{u}(\kappa^o, t)$ depends on the direction of the wavenumber vector, given by the unit wavevector

$$\hat{e}(t) \equiv \frac{\hat{\kappa}(t)}{\hat{\kappa}(t)}, \tag{12.295}$$

but it does not depend on its magnitude $\hat{\kappa}(t) = |\hat{\kappa}(t)|$. The two quantities $\hat{u}(\kappa^o, t)$ and $\hat{e}(t)$ evolve according to the closed pair of ordinary differential equations.

$$\frac{d\hat{u}_i}{dt} = -\hat{u}_j \frac{\partial \langle U_\ell \rangle}{\partial x_j} (\delta_{i\ell} - 2\hat{e}_i \hat{e}_\ell) \tag{12.296}$$

and

$$\frac{d\hat{e}_i}{dt} = -\hat{e}_\ell \frac{\partial \langle U_\ell \rangle}{\partial x_j} (\delta_{ij} - \hat{e}_i \hat{e}_j) \tag{12.297}$$

(see Eqs. (11.85) and (11.86)). Because of the continuity equation, $\hat{u}$ is orthogonal to $\hat{\kappa}$ and e:

$$\hat{e} \cdot \hat{u} = 0. \tag{12.298}$$

The Reynolds-stress tensor is given by

$$\langle u_i u_j \rangle = \left\langle \sum_{\kappa^o} \hat{u}_i^*(\kappa^o, t) \hat{u}_j(\kappa^o, t) \right\rangle. \tag{12.299}$$

Thus, given the initial Fourier modes ($\hat{u}(\kappa^o, 0)$ and κ^o), the evolution of the Reynolds stresses can be determined from the solution of the ordinary differential equations for $\hat{u}$ and $\hat{e}$.

This description of the effect of rapid distortions is in wavenumber space, whereas PDF methods are based in physical space. Consequently, the velocity–wavevector PDF method of Van Slooten and Pope (1997) is not a direct implementation of the preceding development: it is instead an exact mathematical analogy, based on Eqs. (12.296)–(12.299).

In the particle implementation of the PDF method, the general particle has a fluctuating velocity $u^*(t)$ and a unit vector (called the wavevector) $e^*(t)$.[5] In accord with Eq. (12.298), these vectors are orthogonal,

$$e^*(t) \cdot u^*(t) = 0, \tag{12.300}$$

and they are prescribed to evolve according to Eqs. (12.296) and (12.297):

$$\frac{du_i^*}{dt} = -u_j^* \frac{\partial \langle U_\ell \rangle}{\partial x_j} (\delta_{i\ell} - 2e_i^* e_\ell^*), \tag{12.301}$$

$$\frac{de_i^*}{dt} = -e_\ell^* \frac{\partial \langle U_\ell \rangle}{\partial x_j} (\delta_{ij} - e_i^* e_j^*). \tag{12.302}$$

It follows from the exact analogy between Eqs. (12.296)–(12.298) for $\hat{u}(\kappa^\circ, t)$ and $\hat{e}(t)$ and Eqs. (12.300)–(12.302) for $u^*(t)$ and $e^*(t)$ that (given appropriate initial conditions) the Reynolds-stress tensor determined from the particles (i.e., $\langle u_i^* u_j^* \rangle$) is identical to that obtained from the Fourier coefficients (Eq. (12.299)). The evolution of the Reynolds stresses $\langle u_i^* u_j^* \rangle$ given by the model is therefore exact for arbitrary rapid distortions of homogeneous turbulence.

Equation (12.301) can be rewritten

$$\frac{du_i^*}{dt} = -u_j^* \frac{\partial \langle U_i \rangle}{\partial x_j} + G_{ij}^* u_j^*, \tag{12.303}$$

with

$$G_{ij}^* \equiv 2e_i^* e_\ell^* \frac{\partial \langle U_\ell \rangle}{\partial x_j}. \tag{12.304}$$

In the rapid-distortion limit, the diffusion in the generalized Langevin model vanishes, so that the GLM equation for $u^*(t)$, Eq. (12.126), becomes

$$\frac{du_i^*}{dt} = -u_j^* \frac{\partial \langle U_i \rangle}{\partial x_j} + G_{ij} u_j^*. \tag{12.305}$$

Evidently, the wavevector model and GLM have similar forms. However,

---

[5] Here, and for the remainder of the section, the superscript * indicates a particle property. In Eq. (12.299) it denotes the complex conjugate.

the coefficient G_{ij}^*, that yields the correct behavior for rapid distortions, depends on $e^*(t)$; whereas, in GLM, G_{ij} is modelled in terms of $\langle u_i u_j \rangle$, ε, and $\partial \langle U_i \rangle / \partial x_j$.

Homogeneous turbulence

For the case of homogeneous turbulence (not in the rapid-distortion limit) the velocity model proposed by Van Slooten and Pope (1997) can be written

$$du_i^*(t) = \left(G_{ij}^* - \frac{\partial \langle U_i \rangle}{\partial x_j} \right) u_j^* \, dt + (C_0 \varepsilon)^{1/2} \, dW_i, \qquad (12.306)$$

$$G_{ij}^* = \frac{\varepsilon}{k} \left[-\left(\tfrac{1}{2} + \tfrac{3}{4} C_0 \right) \delta_{ij} + \alpha_2 (b_{ij} - b_{\ell\ell}^2 \delta_{ij}) \right] + 2 e_i^* e_\ell^* \frac{\partial \langle U_\ell \rangle}{\partial x_j}, \qquad (12.307)$$

with $C_0 = 2.1$ and $\alpha_2 = 2.0$. In Eq. (12.306), the term in the mean velocity gradient is exact, and corresponds to production. In the expression for G_{ij}^*, the term in the mean velocity gradient represents the rapid pressure (cf. Eq. (12.304)); while the remaining terms represent the slow pressure, and correspond to a nonlinear return-to-isotropy model (for non-zero α_2).

In the absence of mean velocity gradients, Eq. (12.306) becomes a Langevin equation with no dependence on $e^*(t)$; whereas in the rapid-distortion limit, all terms involving ε are negligible, and the equation correctly reverts to Eq. (12.303).

A general diffusion process for $e^*(t)$ can be written

$$de_i^*(t) = E_i \, dt + B_{ij} \, dW_j + B_{ij}' \, dW_j', \qquad (12.308)$$

where the Wiener process $W'(t)$ is independent of that in the velocity equation, $W(t)$, and the coefficients E_i, B_{ij} and B_{ij}' can depend both on $u^*(t)$ and on $e^*(t)$. The model of Van Slooten and Pope (1997) corresponds to the simplest specifications of the coefficients that is consistent with $e^*(t)$ being a unit vector orthogonal to $u^*(t)$. The resulting coefficients (see Exercise 12.52) are

$$E_i = -\frac{e_i^* \varepsilon}{2k} \left(C_e + C_0 \frac{k}{u_j^* u_j^*} \right) - \frac{u_i^* u_k^* e_\ell^*}{u_m^* u_m^*} \left(G_{\ell k}^* - \frac{\partial \langle U_\ell \rangle}{\partial x_k} \right), \qquad (12.309)$$

$$B_{ij} = -\frac{u_i^* e_j^*}{u_k^* u_k^*} (C_0 \varepsilon)^{1/2}, \qquad (12.310)$$

$$B_{ij}' = \left(\delta_{ij} - e_i^* e_j^* - \frac{u_i^* u_j^*}{u_\ell^* u_\ell^*} \right) \left(\frac{C_e \varepsilon}{k} \right)^{1/2}, \qquad (12.311)$$

where the new model constant is taken to be $C_e = 0.03$.

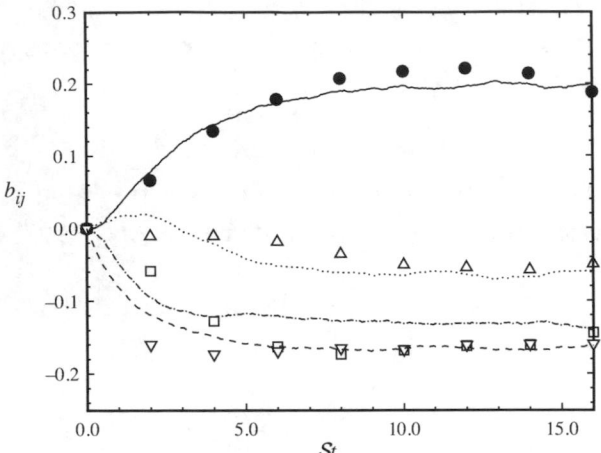

Fig. 12.24. The evolution of Reynolds-stress anisotropies in homogeneous shear flow with $(Sk/\varepsilon)_0 = 2.36$. Velocity–wavevector PDF model calculations of Van Slooten and Pope (1997) (lines) compared with the DNS data of Rogers and Moin (1987) (symbols): ($-$,●), b_{11}; (---,$\triangledown$), b_{12}; (-··-,□), b_{22}; (···,$\triangle$), b_{33}.

Van Slooten and Pope (1997) applied this model to a wide range of homogeneous turbulent flows, and found good agreement with experimental and DNS data. As an example, Fig. 12.24 shows the Reynolds-stress anisotropies calculated in a homogeneous shear flow.

Inhomogeneous flows

In a straightforward way, Van Slooten *et al.* (1998) extend the model to inhomogeneous flows and combine it with the turbulent frequency to obtain a model equation for the joint PDF of velocity, wavevector, and the turbulence frequency. The model has been applied to a number of flows, including swirling jets (Van Slooten and Pope 1999). While the performance of the model for these flows is quite good, so also is that of the simpler LIPM. Nevertheless, the wavevector has the advantage of being exact for the rapid distortion of homogeneous turbulence, and its performance can be expected to be superior for flows closer to this limit.

EXERCISES _____

12.52 Consider the general model for $e^*(t)$, Eq. (12.308), and the model for $u^*(t)$, Eq. (12.306), rewritten (for simplicity) as

$$\mathrm{d}u_i^*(t) = A_{ij}u_j^*\,\mathrm{d}t + b\,\mathrm{d}W_i. \tag{12.312}$$

Show that the requirement that $e^*(t)$ remain of unit length leads to the following constraints on the coefficients:

$$e_i^* B_{ij} = 0, \quad e_i^* B_{ij}' = 0, \tag{12.313}$$

$$2e_i^* E_i + B_{ij}B_{ij} + B_{ij}'B_{ij}' = 0. \tag{12.314}$$

Similarly, show that the orthogonality of $u^*(t)$ and $e^*(t)$ leads to the constraints

$$u_i^* B_{ij}' = 0, \quad u_i^* B_{ij} + be_j^* = 0, \tag{12.315}$$

$$u_i^* E_i + e_i^* u_j^* A_{ij} + bB_{ii} = 0. \tag{12.316}$$

Show that the specifications of B_{ij} and B_{ij}' by Eqs. (12.310) and (12.311) satisfy Eqs. (12.313) and (12.315). From these specifications, obtain the results

$$B_{ij}B_{ij} = \frac{C_0\varepsilon}{u_k^* u_k^*}, \tag{12.317}$$

$$B_{ij}'B_{ij}' = \frac{C_e\varepsilon}{k}, \tag{12.318}$$

and $B_{ii} = 0$. Hence show that the constraints Eqs. (12.314) and (12.316) can be rewritten

$$e_i^* E_i = -\frac{1}{2}\left(\frac{C_0\varepsilon}{u_k^* u_k^*} + \frac{C_e\varepsilon}{k}\right), \tag{12.319}$$

$$u_i^* E_i = -e_i^* u_j^* A_{ij}, \tag{12.320}$$

and that these are satisfied by the specification of E_i (Eq. (12.309)).

12.53 In the general diffusion process for $e^*(t)$, Eq. (12.308), take $B_{ij} = 0$, and determine the coefficients E_i and B_{ij}' corresponding to $e^*(t)$ being an isotropic diffusion on the unit sphere.

12.7.4 Mixing and reaction

PDF methods are particularly attractive for dealing with turbulent reactive flows because nonlinear chemical reactions appear in closed form in the PDF equations. Reviews of PDF methods for reactive flows are provided by Pope (1985), Kuznetsov and Sabel'nikov (1990), and Dopazo (1994). The principal ideas are presented here first for the PDF of a single scalar $\phi(x,t)$, and then for the joint PDF of velocity and a set of scalars.

The composition PDF

In the simplest case, the conservation equation for a reactive scalar $\phi(x,t)$ is

$$\frac{D\phi}{Dt} = \Gamma\nabla^2\phi + S(\phi(x,t)), \tag{12.321}$$

where S is the source due to chemical reaction, which is a known function of ϕ. In many applications, especially combustion, $S(\phi)$ is a highly nonlinear function, so that the source based on the mean $S(\langle\phi\rangle)$ does not provide a realistic approximation to the mean source $\langle S(\phi)\rangle$. As a consequence, the chemical source term results in a severe closure problem in mean-flow and second-moment closures.

The one-point, one-time Eulerian PDF of $\phi(x,t)$ is denoted by $f_\phi(\psi;x,t)$, where ψ is the sample-space variable. The evolution equation for f_ϕ deduced from Eq. (12.321) is

$$\frac{\partial f_\phi}{\partial t} + \frac{\partial}{\partial x_i}\left[f_\phi(\langle U_i\rangle + \langle u_i|\psi\rangle)\right] = -\frac{\partial}{\partial\psi}\left(f_\phi\left\langle\frac{D\phi}{Dt}\middle|\psi\right\rangle\right)$$

$$= -\frac{\partial}{\partial\psi}\{f_\phi[\langle\Gamma\nabla^2\phi|\psi\rangle + S(\psi)]\}, \tag{12.322}$$

or alternatively

$$\frac{\bar{D}f_\phi}{\bar{D}t} = \Gamma\nabla^2 f_\phi - \frac{\partial}{\partial x_i}(f_\phi\langle u_i|\psi\rangle)$$

$$- \frac{\partial^2}{\partial\psi^2}\left(f_\phi\left\langle\Gamma\frac{\partial\phi}{\partial x_i}\frac{\partial\phi}{\partial x_i}\middle|\psi\right\rangle\right) - \frac{\partial}{\partial\psi}[f_\phi S(\psi)], \tag{12.323}$$

where $\langle u_i|\psi\rangle$ denotes $\langle u_i(x,t)|\phi(x,t) = \psi\rangle$, etc. (see Exercise 12.54). The most important observation is that, because of the relation

$$\langle S(\phi)|\phi = \psi\rangle = S(\psi), \tag{12.324}$$

the chemical source appears in closed form.

The fluctuating velocity field leads to the turbulent convective flux $-f_\phi\langle u|\psi\rangle$, which is not in closed form. A gradient-diffusion model (with the usual deficiencies) is

$$-f_\phi\langle u|\psi\rangle = \Gamma_T\nabla f_\phi, \tag{12.325}$$

where $\Gamma_T(x,t)$ is the turbulent diffusivity.

The effects of molecular diffusion, which are also not in closed form, appear in terms of the *conditional Laplacian* $\langle\Gamma\nabla^2\phi|\psi\rangle$, or alternatively in terms of the *conditional scalar dissipation* $\langle 2\Gamma\nabla\phi\cdot\nabla\phi|\psi\rangle$. Models for the effects of molecular diffusion are called *mixing models*[6] and are discussed

---

[6] They are also called *molecular-mixing models* and *micromixing models*.

below. An early and simple mixing model is

$$\langle \Gamma \nabla^2 \phi | \psi \rangle = -\tfrac{1}{2} C_\phi \frac{\varepsilon}{k} (\psi - \langle \phi \rangle), \tag{12.326}$$

where the model constant C_ϕ – generally taken to be 2.0 – is the mechanical-to-scalar timescale ratio τ/τ_ϕ (see Section 5.5, and Exercise 12.55). This linear deterministic model is called the IEM model (interaction by exchange with the mean, Villermaux and Devillon (1972)) or the LMSE model (linear mean-square estimation, Dopazo and O'Brien (1974)).

With the incorporation of these models, the composition-PDF equation (Eq. (12.322)) becomes

$$\frac{\bar{D} f_\phi}{\bar{D} t} = \frac{\partial}{\partial x_i} \left(\Gamma_{\rm T} \frac{\partial f_\phi}{\partial x_i} \right) + \frac{\partial}{\partial \psi} \left[f_\phi \left(\tfrac{1}{2} C_\phi \frac{\varepsilon}{k} (\psi - \langle \phi \rangle) - S(\psi) \right) \right]. \tag{12.327}$$

Assuming that $\langle U \rangle$, k, ε and $\Gamma_{\rm T}$ are known from a turbulence-model calculation, then Eq. (12.327) is a closed model equation that can be solved to determine $f_\phi(\psi; x, t)$.

This equation is readily extended to a set of compositions, and many reactive-flow calculations have been based on it, e.g., Chen, Dibble, and Bilger (1990), Roekaerts (1991), Hsu, Anand, and Razdan (1990), and Jones and Kakhi (1997). These calculations employ a node-based particle method (Pope (1981a)), in which there is an ensemble of particles at each node of a finite-difference grid. Alternatively, a distributed-particle algorithm (Pope (1985), see Exercise 12.56) – similar to that described in Section 12.6 – has also been used, e.g., Tsai and Fox (1996) and Colucci *et al.* (1998).

12.54 With

$$f'_\phi(\psi; x, t) \equiv \delta(\phi(x, t) - \psi) \tag{12.328}$$

being the fine-grained PDF of composition, obtain the results

$$\frac{\partial f'_\phi}{\partial t} = -\frac{\partial}{\partial \psi} \left(f'_\phi \frac{\partial \phi}{\partial t} \right), \tag{12.329}$$

$$\frac{\partial f'_\phi}{\partial x_i} = -\frac{\partial}{\partial \psi} \left(f'_\phi \frac{\partial \phi}{\partial x_i} \right), \tag{12.330}$$

$$\frac{D f'_\phi}{D t} = -\frac{\partial}{\partial \psi} \left(f'_\phi \frac{D \phi}{D t} \right). \tag{12.331}$$

(Hint: see Eq. (H.8) on page 703.) Similar to Eq. (12.3) on page 464, obtain the result

$$\langle f'_\phi U \rangle = f_\phi(\langle U \rangle + \langle u|\psi \rangle), \tag{12.332}$$

where $\langle u|\psi \rangle$ is an abbreviation for $\langle u(x,t)|\phi(x,t) = \psi \rangle$. Hence verify Eq. (12.322). Differentiate Eq. (12.330) with respect to x_i to obtain

$$\nabla^2 f'_\phi = -\frac{\partial}{\partial \psi}\left(f'_\phi \nabla^2 \phi\right) + \frac{\partial^2}{\partial \psi^2}(f'_\phi \nabla\phi \cdot \nabla\phi). \tag{12.333}$$

By taking the mean of this equation, verify Eq. (12.323).

12.55 Show that the equations for the evolution of the mean $\langle \phi \rangle$ and variance $\langle \phi'^2 \rangle$ obtained from the model PDF equation (Eq. (12.327)) are

$$\frac{\bar{D}\langle \phi \rangle}{\bar{D}t} = \frac{\partial}{\partial x_i}\left(\Gamma_T \frac{\partial \langle \phi \rangle}{\partial x_i}\right)$$

$$\frac{\bar{D}\langle \phi'^2 \rangle}{\bar{D}t} = \frac{\partial}{\partial x_i}\left(\Gamma_T \frac{\partial \langle \phi'^2 \rangle}{\partial x_i}\right) + 2\Gamma_T \frac{\partial \langle \phi \rangle}{\partial x_i}\frac{\partial \langle \phi \rangle}{\partial x_i}$$
$$-C_\phi \frac{\varepsilon}{k}\langle \phi'^2 \rangle + 2\langle \phi' S(\phi) \rangle. \tag{12.335}$$

12.56 Consider a particle method in which the position $X^*(t)$ and composition $\phi^*(t)$ of a particle evolve by

$$dX^* = a(X^*, t)\,dt + b(X^*, t)\,dW, \tag{12.336}$$

$$d\phi^* = c(\phi^*, X^*, t)\,dt, \tag{12.337}$$

where a, b, and c are coefficients, and $W(t)$ is an isotropic Wiener process. Show that the joint PDF of X^* and $\phi^* - f^*_{\phi X}(\psi, x; t)$ – evolves by

$$\frac{\partial f^*_{\phi X}}{\partial t} = -\frac{\partial}{\partial x_i}[f^*_{\phi X} a_i(x, t)] + \frac{1}{2}\frac{\partial^2}{\partial x_i\,\partial x_i}[f^*_{\phi X} b(x, t)^2]$$
$$-\frac{\partial}{\partial \psi}[f^*_{\phi X} c(\psi, x, t)]. \tag{12.338}$$

Integrate this equation over all ψ to obtain the evolution equation for the particle-position density $f^*_X(x; t)$. Show that, if f^*_X is initially uniform, then it will remain uniform, provided that the coefficients satisfy

$$\nabla \cdot a = \tfrac{1}{2}\nabla^2 b^2. \tag{12.339}$$

The PDF of $\phi^*(t)$ conditional on $X^*(t) = x$ is $f^*_\phi(\psi; x, t) = f^*_{\phi X}(\psi, x; t)/$

$f_X^*(x;t)$. Take f_X^* to be uniform, and show that f_ϕ^* evolves by the model PDF equation Eq. (12.327) provided that the coefficients are specified by

$$a(x,t) = \langle U \rangle + \nabla \Gamma_T, \qquad (12.340)$$

$$b(x,t)^2 = 2\Gamma_T, \qquad (12.341)$$

$$c(\psi, x, t) = -\tfrac{1}{2} C_\phi \frac{\varepsilon}{k}(\psi - \langle \phi^* | x \rangle) + S(\psi). \qquad (12.342)$$

The conserved scalar PDF in isotropic turbulence

The central issue in composition-PDF approaches is the modelling of the effects of molecular diffusion. The simplest case to consider is the decay (from a specified initial condition) of the statistically homogeneous field of a single conserved passive scalar in homogeneous isotropic turbulence. A DNS study of this case is described by Eswaran and Pope (1988) for an initial condition corresponding to statistically identical blobs of fluid with $\phi = 0$ and $\phi = 1$, for which the initial PDF is

$$f_\phi(\psi;0) = \tfrac{1}{2}[\delta(\psi) + \delta(1 - \psi)]. \qquad (12.343)$$

As the field $\phi(x,t)$ evolves, the mean is conserved, the variance decays according to

$$\frac{\mathrm{d}\langle \phi'^2 \rangle}{\mathrm{d}t} = -\varepsilon_\phi = -2\Gamma \langle \nabla \phi' \cdot \nabla \phi' \rangle, \qquad (12.344)$$

(cf. Eq. (5.281) on page 163), and the shapes adopted by the evolving PDF $f_\phi(\psi, t)$ are those shown in Fig. 12.25(a). The DNS data show that the PDF $f_\phi(\psi, t)$ tends to a Gaussian at large times.

For this case of a statistically homogeneous conserved scalar field, the composition PDF equations (Eqs. (12.322) and (12.323)) reduce to

$$\frac{\partial f_\phi}{\partial t} = -\frac{\partial}{\partial \psi}(f_\phi \langle \Gamma \nabla^2 \phi | \psi \rangle) = -\frac{1}{2} \frac{\partial^2}{\partial \psi^2}[f_\phi \varepsilon_\phi^c(\psi, t)], \qquad (12.345)$$

where ε_ϕ^c is the conditional scalar dissipation

$$\varepsilon_\phi^c(\psi, t) \equiv \langle 2\Gamma \nabla \phi' \cdot \nabla \phi' | \phi = \psi \rangle. \qquad (12.346)$$

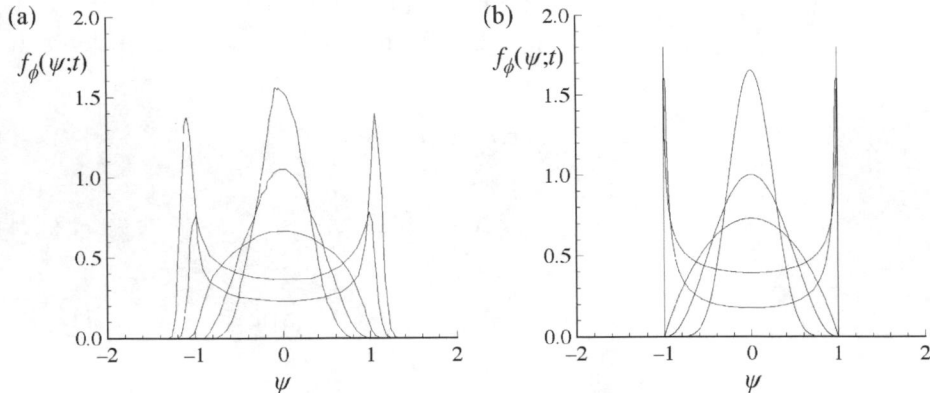

Fig. 12.25. The evolution of the PDF $f_\phi(\psi;t)$ of a conserved passive scalar in isotropic turbulence from a double-delta-function initial condition: (a) DNS of Eswaran and Pope (1988); (b) calculated from the mapping closure (Pope 1991b).

Gaussian-field models

It is very informative to study three different models based on Gaussian fields. First, the assumption that $\phi(x,t)$ is a Gaussian field implies that the conditional Laplacian is precisely that given by the IEM model, Eq. (12.326) – see Exercise 12.57. This model is directly analogous to the degenerate Langevin equation in which the diffusion coefficient is zero (Eq. (12.32) on page 471). We have seen that this model has the serious deficiency of preserving the shape of the PDF. From a Gaussian initial condition it gives correctly a Gaussian evolution. However, from a double-delta-function initial condition, there is no relaxation to a Gaussian. Instead, the delta functions persist and move toward each other in composition space.

The second model considered, which also follows from the assumption that $\phi(x,t)$ is a Gaussian field (see Exercise 12.57), is that the conditional and unconditional scalar dissipations are equal. The PDF equation (Eq. (12.345)) then reduces to

$$\frac{\partial f_\phi}{\partial t} = -\tfrac{1}{2}\varepsilon_\phi \frac{\partial^2 f_\phi}{\partial \psi^2}, \tag{12.347}$$

which is the diffusion equation, but with a negative diffusion coefficient, (i.e., $-\tfrac{1}{2}\varepsilon_\phi$). This is a classic example of an unstable differential equation: the analytic solution to Eq. (12.347) (see Exercise 12.9 on page 473) shows that the Fourier coefficients of the solution diverge exponentially in time, most rapidly for the highest modes. Except when one starts from the particular initial condition of a Gaussian, the PDF $f_\phi(\psi;t)$ becomes negative, and hence non-realizable.

The assumption of a Gaussian field for $\phi(x,t)$ leads to different flawed mixing models based on the conditional Laplacian and on the conditional dissipation. For a Gaussian initial condition, both models correctly predict a Gaussian evolution. However, for other initial conditions, the two models display different kinds of unsatisfactory behavior.

EXERCISE _____

12.57 Let $\phi(x)$ be a statistically homogeneous Gaussian field. Then, at every location, the quantities ϕ, $\partial\phi/\partial x_i$, and $\nabla^2\phi$ are jointly normal random variables. Show that the covariance between ϕ and $\Gamma\nabla^2\phi$ is

$$\langle\phi\,\Gamma\,\nabla^2\phi\rangle = -\langle\Gamma\,\nabla\phi\cdot\nabla\phi\rangle = -\tfrac{1}{2}\varepsilon_\phi. \qquad (12.348)$$

Hence, using Eq. (3.119) on page 63, show that the conditional Laplacian is

$$\langle\Gamma\,\nabla^2\phi|\psi\rangle = -\frac{1}{2}\frac{\varepsilon_\phi}{\langle\phi'^2\rangle}(\psi - \langle\phi\rangle). \qquad (12.349)$$

Show that this is identical to the IEM model, Eq. (12.326). Show that ϕ and $\partial\phi/\partial x_i$ are uncorrelated and hence independent (because they are jointly normal). Hence show that the conditional scalar dissipation $\varepsilon_\phi^c(\psi)$ (Eq. (12.346)) is equal to the unconditional scalar dissipation ε_ϕ. (It is stressed that these results are specific to a Gaussian field, and do not apply in general.)

The mapping closure

For a non-Gaussian initial PDF $f_\phi(\psi;0)$ the assumption that $\phi(x,t)$ is a Gaussian field is clearly inconsistent. In contrast, the *mapping closure* (Chen, Chen, and Kraichnan 1989, Pope 1991b, Gao and O'Brien 1991) provides a consistent Gaussian assumption and leads to a satisfactory model.

Let $\theta(x)$ be a statistically homogeneous Gaussian field, and let η be the sample-space variable corresponding to θ. At every time t there is a uniquely-defined non-decreasing function $X(\eta,t)$ – the 'mapping' – such that the *surrogate field*

$$\phi^s(x,t) \equiv X(\theta(x),t) \qquad (12.350)$$

has the same one-point PDF as that of the composition field $\phi(x,t)$. The closure is achieved through the assumption that the statistics of the composition field $\phi(x,t)$ are the same as those of the (non-Gaussian) surrogate field $\phi^s(x,t)$: in particular that $\langle\Gamma\,\nabla^2\phi|\psi\rangle$ is equal to $\langle\Gamma\,\nabla^2\phi^s|\psi\rangle$ – which is known in terms of the mapping. This assumption leads to an equation for the evolution of the mapping $X(\eta,t)$, which implicitly determines the PDF.

For the test case of a double-delta-function initial condition, Fig. 12.25(b) shows the evolution of the PDF's shape predicted by the mapping closure. Clearly there is excellent agreement with the DNS data. At large times, the PDF given by the mapping closure tends to a Gaussian. For a single composition, a particle implementation of the mapping closure is described by Pope (1991b).

The velocity–composition PDF

The thermochemical composition of the fluid in chemically reacting turbulent flows can be described by scalars $\phi = \{\phi_1, \phi_2, \ldots, \phi_{n_\phi}\}$. Depending on the chemistry involved, the number of scalars n_ϕ can be two or three, about ten, or even on the order of 100. We now extend the previous considerations to the joint PDF of velocity and a set of scalars. For simplicity we take the scalars to evolve according to the conservation equation

$$\frac{D\phi_\alpha}{Dt} = \Gamma \nabla^2 \phi_\alpha + S_\alpha(\phi), \tag{12.351}$$

where the source due to reaction S is a known function of the compositions. The molecular diffusion Γ is taken to be constant, uniform, and the same for each composition. (In practice, the *differential diffusion* of species can be important; see, e.g., Yeung (1998), Nilsen and Kosály (1999), and references therein.)

The one-point, one-time Eulerian joint PDF of velocity $U(x,t)$ and composition $\phi(x,t)$ is denoted by $\hat{f}(V,\psi;x,t)$, where $\psi = \{\psi_1, \psi_2, \ldots, \psi_{n_\phi}\}$ are the sample-space variables for composition. The evolution equation for $\hat{f}(V,\psi;x,t)$ deduced from the Navier–Stokes equations and Eq. (12.351) is

$$\frac{\partial \hat{f}}{\partial t} + V_i \frac{\partial \hat{f}}{\partial x_i} - \frac{1}{\rho}\frac{\partial \langle p \rangle}{\partial x_i}\frac{\partial \hat{f}}{\partial V_i} + \frac{\partial}{\partial \psi_\alpha}[\hat{f}S_\alpha(\psi)]$$
$$= -\frac{\partial}{\partial V_i}\left(\hat{f}\left\langle \nu \nabla^2 U_i - \frac{1}{\rho}\frac{\partial p'}{\partial x_i}\middle| V,\psi\right\rangle\right) - \frac{\partial}{\partial \psi_\alpha}[\hat{f}\langle \Gamma \nabla^2 \phi_\alpha | V,\psi\rangle]. \tag{12.352}$$

The terms on the left-hand side are in closed form, most notably convection and reaction. Thus, the gradient-diffusion model used in the composition PDF model is obviated, while the benefit of the reaction being in closed form is retained.

For the unclosed terms on the right-hand side, the expectations are conditioned on $U(x,t) = V$ and $\phi(x,t) = \psi$. Since, by assumption, the compositions are passive, their values have no effect on ρ, v, U, and p'. Hence the first term on the right-hand side can be modelled, independent of ψ, by the generalized Langevin model.

The simplest mixing model is again IEM

$$\langle \Gamma \nabla^2 \phi_\alpha | V, \psi \rangle = -\tfrac{1}{2} C_\phi \frac{\varepsilon}{k} (\psi_\alpha - \langle \phi_\alpha \rangle), \qquad (12.353)$$

cf. Eq. (12.326). According to this model, the conditional Laplacian is independent of V – which is incorrect (Pope 1998b). Several other models containing a dependence on V have been proposed (e.g., Pope (1985) and Fox (1996)).

This approach is readily extended to the joint PDF of velocity, turbulence frequency, and composition to produce a simple closed model PDF equation for turbulent reactive flows. The particle method described in Section 12.6 allows this equation to be solved, even though the joint PDF is a function of many independent variables. For example, Saxena and Pope (1998) describe calculations of a (statistically stationary and axisymmetric) methane–air jet flame based on the joint PDF of velocity, turbulence frequency, and 14 compositions, i.e., $f_{U\omega\phi}(V_1, V_2, V_3, \theta, \psi_1, \psi_2, \ldots, \psi_{14}; x, r)$.

EXERCISE _____

12.58   Write down the velocity-composition joint-PDF equation (Eq. 12.352)), with the generalized Langevin model (see Eq. (12.26)) and the IEM model (Eq. (12.353)) substituted for the two unclosed terms on the right-hand side. Multiply by the appropriate sample-space variables and integrate to obtain the mean evolution equations

$$\frac{\bar{D}\langle U_j \rangle}{\bar{D}t} + \frac{\partial \langle u_i u_j \rangle}{\partial x_i} + \frac{1}{\rho} \frac{\partial \langle p \rangle}{\partial x_j} = 0, \qquad (12.354)$$

$$\frac{\bar{D}\langle \phi_\beta \rangle}{\bar{D}t} + \frac{\partial \langle u_i \phi'_\beta \rangle}{\partial x_i} - \langle S_\beta(\boldsymbol{\phi}) \rangle = 0. \qquad (12.355)$$

Multiply the model joint-PDF equation by $V_j \psi_\beta$ and integrate to obtain

$$\frac{\partial \langle U_j \phi_\beta \rangle}{\partial t} + \frac{\partial \langle U_i U_j \phi_\beta \rangle}{\partial x_i} + \frac{\langle \phi_\beta \rangle}{\rho} \frac{\partial \langle p \rangle}{\partial x_j} - \langle U_j S_\beta \rangle$$
$$= G_{j\ell} \langle u_\ell \phi_\beta \rangle - \tfrac{1}{2} C_\phi \frac{\varepsilon}{k} \langle U_j \phi'_\beta \rangle. \qquad (12.356)$$

Hence obtain the model scalar flux equation

$$\frac{\bar{D}\langle u_j \phi'_\beta \rangle}{\bar{D}t} + \frac{\partial \langle u_i u_j \phi'_\beta \rangle}{\partial x_i} + \langle u_i u_j \rangle \frac{\partial \langle \phi_\beta \rangle}{\partial x_i} + \langle u_i \phi'_\beta \rangle \frac{\partial \langle U_j \rangle}{\partial x_i} - \langle u_j S_\beta \rangle$$

$$= \left(G_{j\ell} - \tfrac{1}{2} C_\phi \frac{\varepsilon}{k} \delta_{j\ell} \right) \langle u_\ell \phi'_\beta \rangle. \tag{12.357}$$

Mixing models

Because of the deficiencies of IEM and other simple models, there have been many attempts to develop better mixing models (most of which are reviewed by Dopazo (1994)). Although improvements have been made, there is no model that is satisfactory in all respects.

Some of the behaviors and properties of an ideal mixing model can be deduced from the conservation equations, and others from experimental and DNS data. For a set of conserved scalars with equal diffusivities the following should hold.

(i) The means $\langle \phi_\alpha \rangle$ are not directly affected by mixing.

(ii) Mixing causes the variances $\langle \phi'_{(\alpha)} \phi'_{(\alpha)} \rangle$ to decrease with time. (A stronger condition is that it causes the eigenvalues of the covariance matrix $\langle \phi'_\alpha \phi'_\beta \rangle$ to decrease.)

(iii) A generalization of the boundedness condition (Eq. (2.55) on page 21) is that the convex region in the sample space occupied by the scalars decreases with time. Let $C(t)$ denote the smallest convex region in the sample space occupied by the compositions at time t, so that the PDF is zero for all compositions outside $C(t)$. Then, for $t_2 > t_1$, $C(t_2)$ is contained within $C(t_1)$.

(iv) An ideal mixing model satisfies the invariance properties of linearity and independence implied by the conservation equation for ϕ (Pope 1983a).

(v) For statistically homogeneous scalar fields in homogeneous turbulence, the joint PDF $f_\phi(\psi; t)$ tends to a joint normal (Juneja and Pope 1996).

In more general circumstances, an ideal mixing model would accurately account for

(vi) differential diffusion

(vii) the influence of the lengthscales of the scalar fields, and

(viii) the influence of reaction on mixing.

Table 12.2. *Different levels of PDF models*

Model	PDF	Additional fields required for closure
Composition PDF	$f_\phi(\psi; x, t)$	$k, \varepsilon, \langle U \rangle$
Velocity PDF	$f(V; x, t)$	ε
Velocity–frequency PDF	$\bar{f}(V, \theta; x, t)$	–
Velocity–frequency PDF with elliptic relaxation	$\bar{f}(V, \theta; x, t)$	g_{ij}
Velocity–frequency– wavevector PDF	$f_{U\omega e}(V, \theta, \eta; x, t)$	–

As previously noted, the IEM model preserves the shape of the PDF, in opposition to (v). This deficiency could be remedied by adding a diffusion term, so that $\phi^+(t)$ is modelled by a Langevin equation, but then the boundedness condition (iii) would be violated. For simple bounds, a diffusion process that satisfies boundedness can be constructed (e.g., Valiño and Dopazo (1991) and Fox (1992,1994)) but the general case remains problematic. A satisfactory, tractable extension of the closure of mapping for the general case has also proved elusive.

The particle implementation of PDF methods allows the use of models based on particle interactions (Curl 1963, Janicka, Kolbe, and Kollmann 1977, Pope 1985, Subramaniam and Pope 1998). These models readily satisfy the general boundedness condition (iii), and provide a relaxation of the PDF's shape. To address (vii), Fox (1997) has developed a model that incorporates spectral information about the scalar field.

12.8 Discussion

Table 12.2 shows some of the PDF models introduced in this chapter. For simplicity, compositions are shown only in the first of these models, but it should be appreciated that all of the other models are readily extended also to include compositions. These PDF models are discussed in this section with respect to the criteria for appraising models described in Chapter 8. In this discussion it is natural to compare PDF methods with Reynolds-stress closures.

Level of description

The various joint PDFs provide complete one-point, one-time, statistical descriptions of the flow quantities considered. In general, this is much more information than is contained in the means and covariances. For some flows (e.g., homogeneous turbulence) this higher level of description may provide little benefit over second-moment closures. However, in general, the fuller description is beneficial in allowing more processes to be treated exactly, and in providing more information that can be used in the construction of closure models. Specifically,

(i) in PDF models that include velocity, turbulent convection is in closed form and so gradient-diffusion models (e.g., for $\langle u_i u_j u_k \rangle$ and $\langle u_i \phi' \rangle$) are avoided;

(ii) with the inclusion of composition, the reaction is in closed form, in marked contrast to the closure problem encountered in moment closures;

(iii) in the wavevector model, the pressure–rate-of-strain tensor is treated exactly in the rapid-distortion limit (for homogeneous turbulence); and

(iv) from the *distribution* of the turbulence frequency, the *conditional* mean turbulence frequency Ω can be obtained and used in the modelling, thereby removing the free-stream boundary-condition problem encountered in the k–ω model.

Completeness

All the PDF models considered are complete.

Cost and ease of use

Compared with Reynolds-stress models, PDF methods are here at a disadvantage. The particle methods described in Section 12.6 are less well developed – both in terms of algorithms and in terms of availability of codes – than the finite-volume methods generally used for k–ε and Reynolds-stress calculations. In a typical particle-method calculation, the number of particles may be more by a factor of 100 than the number of cells in a typical finite-volume calculation. Not surprisingly, therefore, the CPU time requirements of particle methods are larger – but not by a factor of 100. For example, Hsu *et al.* (1990) quote a factor of five in comparing a composition PDF calculation with a k–ε calculation.

Compared with LES, however, the computational cost of PDF methods is modest. Just as in all one-point statistical models, in PDF methods scales much below the integral scale do not need to be resolved, and statistical symmetries can be exploited to reduce the dimensionality of the problem.

The range of applicability

Like k–ε and Reynolds-stress models, PDF methods can be applied to any turbulent flow (within the class considered in this book). PDF methods have found most application in reactive flows because nonlinear chemical reactions can be treated without closure approximations.

Accuracy

PDF methods for treating homogeneous turbulence can be constructed to be equivalent to any realizable Reynolds-stress closure, and consequently (for these flows) the accuracies of the two methods are comparable. For inhomogeneous flows, velocity-PDF methods have the theoretical advantage that turbulent convective transport is in closed form; and the wavevector model has theoretical advantages for flows approaching rapid distortions. However, in the absence of a comparative study of Reynolds-stress and PDF-model calculations, the gains in accuracy achieved by PDF methods in treating various flows are difficult to assess.

13

Large-eddy simulation

13.1 Introduction

In large-eddy simulation (LES), the larger three-dimensional unsteady turbulent motions are directly represented, whereas the effects of the smaller-scale motions are modelled. In computational expense, LES lies between Reynolds-stress models and DNS, and it is motivated by the limitations of each of these approaches. Because the large-scale unsteady motions are represented explicitly, LES can be expected to be more accurate and reliable than Reynolds-stress models for flows in which large-scale unsteadiness is significant – such as the flow over bluff bodies, which involves unsteady separation and vortex shedding.

As discussed in Chapter 9, the computational cost of DNS is high, and it increases as the cube of the Reynolds number, so that DNS is inapplicable to high-Reynolds-number flows. Nearly all of the computational effort in DNS is expended on the smallest, dissipative motions (see Fig 9.4 on page 351), whereas the energy and anisotropy are contained predominantly in the larger scales of motion. In LES, the dynamics of the larger-scale motions (which are affected by the flow geometry and are not universal) are computed explicitly, the influence of the smaller scales (which have, to some extent, a universal character) being represented by simple models. Thus, compared with DNS, the vast computational cost of explicitly representing the small-scale motions is avoided.

There are four conceptual steps in LES.

(i) A *filtering* operation is defined to decompose the velocity $U(x,t)$ into the sum of a filtered (or resolved) component $\overline{U}(x,t)$ and a residual (or subgrid-scale, SGS) component $u'(x,t)$. The filtered velocity field $\overline{U}(x,t)$ – which is three-dimensional and time-dependent – represents the motion of the large eddies.

(ii) The equations for the evolution of the filtered velocity field are derived from the Navier–Stokes equations. These equations are of the standard form, with the momentum equation containing the *residual-stress tensor* (or SGS stress tensor) that arises from the residual motions.

(iii) Closure is obtained by modelling the residual-stress tensor, most simply by an eddy-viscosity model.

(iv) The model filtered equations are solved numerically for $\overline{U}(x,t)$, which provides an approximation to the large-scale motions in one realization of the turbulent flow.

There are two viewpoints on the separation of the modelling issues (i)–(iii) from the numerical solution (iv). In one viewpoint, expressed by Reynolds (1990), the issues are quite separate. The filtering and modelling are independent of the numerical method, and in particular they are independent of the grid employed. Hence 'filtered' and 'residual' are the appropriate terms, not 'resolved' or 'subgrid.' The numerical method is then supposed to provide an accurate solution to the filtered equations. In practice, the modelling and numerical issues are interwoven to some extent; and (as discussed in Section 13.5) even in principle they are connected. (The alternative viewpoint, discussed in Section 13.6.4, is that modelling and numerical issues should deliberately be combined.)

Much of the pioneering work on LES (e.g., Smagorinsky (1963), Lilly (1967), and Deardorff (1974)) was motivated by meteorological applications, and atmospheric boundary layers remain a focus of LES activities (e.g., Mason (1994)). The development and testing of LES methodologies has focused primarily on isotropic turbulence (e.g., Kraichnan (1976) and Chasnov (1991)), and on fully developed turbulent channel flow (e.g., Deardorff (1970), Schumann (1975), Moin and Kim (1982) and Piomelli (1993)). A primary goal of work in this area is to apply LES to flows in complex geometries that occur in engineering applications (e.g., Akselvoll and Moin (1996) and Haworth and Jansen (2000)). The collection of works compiled by Galperin and Orszag (1993) provides an overview of the history of LES and its range of applications.

The LES methodologies that have been developed depend to some extent on the type of flow considered and on the numerical method used. It is useful, therefore, to keep in mind the following examples of LES which span the range of flows and numerical methods:

(i) isotropic turbulence using a pseudo-spectral method (i.e., in wavenumber space),

Table 13.1. *Resolution in DNS and in some variants of LES*

Model	Acronym	Resolution
Direct numerical simulation	DNS	Turbulent motions of all scales are fully resolved
Large-eddy simulation with near-wall resolution	LES-NWR	The filter and grid are sufficiently fine to resolve 80% of the energy everywhere
Large-eddy simulation with near-wall modelling	LES-NWM	The filter and grid are sufficiently fine to resolve 80% of the energy remote from the wall, but not in the near-wall region
Very-large-eddy simulation	VLES	The filter and grid are too coarse to resolve 80% of the energy

(ii) isotropic turbulence using a finite-difference method (in physical space),

(iii) a free shear flow using a uniform rectangular grid,

(iv) fully developed turbulent channel flow using a non-uniform rectangular grid,

(v) an atmospheric boundary layer using a rectangular grid,

(vi) the flow over a bluff body using a structured grid, and

(vii) flow in a complex geometry using an unstructured grid.

There are important distinctions among variants of LES, which are summarized in Table 13.1. Considering, first, the flow remote from walls, we make a distinction between LES and VLES (very-large-eddy simulation). In LES, the filtered velocity field accounts for the bulk (say 80%) of the turbulent kinetic energy everywhere in the flow field. In VLES the grid and filter are too large to resolve the energy-containing motions, and instead a substantial fraction of the energy resides in the residual motions. Although VLES can be performed on coarser grids, and therefore is less expensive, the simulation is more strongly dependent on the modelling of the residual motions. In practice, the fraction of energy resolved is seldom estimated, so that it is not always clear whether a particular simulation is LES or VLES.

For wall-bounded flows, the same distinction between LES and VLES applies with respect to the flow remote from the wall; and a further distinction depends on the treatment of the near-wall motions. For smooth walls, the near-wall motions scale with the viscous lengthscale δ_ν (which decreases with the Reynolds number compared with the flow scale δ). If the filter and grid are chosen so that the bulk (say 80%) of the energy in these motions is resolved, then the result is a large-eddy simulation with near-wall resolution, LES-NWR. This requires a very fine grid near the wall, and the computational cost increases as a power of the Reynolds number (Section 13.4.5), so that, like DNS, LES-NWR is infeasible for high-Reynolds-number flows. The alternative is large-eddy simulation with near-wall modelling, LES-NWM, in which the filter and grid are too coarse to resolve the near-wall motions, so that their influence is modelled (either explicitly or implicitly). Meteorological applications of LES are either LES-NWM or VLES.

An outline of the chapter

In the next two sections, the LES formulation is developed by examining the filtering operation and its application to the Navier–Stokes equations. For LES performed in physical space, the basic residual-stress model is the eddy-viscosity model proposed by Smagorinsky (1963), which is described and examined in Section 13.4. Further residual-stress models are described in Section 13.6, following a consideration of LES in wavenumber space (Section 13.5). The chapter concludes with an appraisal of LES and its variants.

13.2 Filtering

In DNS, the velocity field $U(x,t)$ has to be resolved on lengthscales down to the Kolmogorov scale η. In LES, a low-pass filtering operation is performed so that the resulting filtered velocity field $\overline{U}(x,t)$ can be adequately resolved on a relatively coarse grid. Specifically, the required grid spacing h is proportional to the specified filter width Δ. In the ideal case the filter width is somewhat smaller than ℓ_{EI} – the size of the smallest energy-containing motions (see Fig. 6.2 on page 188). For then the grid spacing is as large as possible, subject to the condition that the energy-containing motions are resolved.

13.2.1 The general definition

The general filtering operation (introduced by Leonard (1974)) is defined by

$$\overline{U}(x,t) = \int G(r,x)U(x-r,t)\,dr, \qquad (13.1)$$

where integration is over the entire flow domain, and the specified filter function G satisfies the normalization condition

$$\int G(r,x)\,dr = 1. \qquad (13.2)$$

In the simplest case, the filter function is homogeneous, i.e., independent of x.

The residual field is defined by

$$u'(x,t) \equiv U(x,t) - \overline{U}(x,t), \tag{13.3}$$

so that the velocity field has the decomposition

$$U(x,t) = \overline{U}(x,t) + u'(x,t). \tag{13.4}$$

This appears analogous to the Reynolds decomposition. Important differences, however, are that $\overline{U}(x,t)$ is a random field, and that (in general) the filtered residual is not zero:

$$\overline{u'}(x,t) \neq 0. \tag{13.5}$$

EXERCISE _____

13.1   Show from Eq. (13.1) that the operations of filtering and differentiating with respect to time commute, i.e.,

$$\frac{\partial \overline{U}}{\partial t} = \overline{\left(\frac{\partial U}{\partial t}\right)}. \tag{13.6}$$

Show that the operations of filtering and taking the mean commute, i.e.,

$$\overline{(\langle U \rangle)} = \langle \overline{U} \rangle. \tag{13.7}$$

Differentiate Eq. (13.1) with respect to x_j to obtain the result

$$\frac{\partial \overline{U}_i}{\partial x_j} = \overline{\left(\frac{\partial U_i}{\partial x_j}\right)} + \int U_i(x-r,t)\frac{\partial G(r,x)}{\partial x_j}\,dr, \tag{13.8}$$

showing that the operations of filtering and differentiation with respect to position do not commute in general, but do so for homogeneous filters.

13.2.2 Filtering in one dimension

The properties of the various filters are most simply examined in one dimension. We consider, therefore, a random scalar function $U(x)$ defined for all x $(-\infty < x < \infty)$. The extension to the three-dimensional vector case is straightforward; and, with this in mind, we refer to $U(x)$ as the 'velocity field.' With $G(r)$ being a homogeneous filter, the filtered velocity field is given

Table 13.2. *Filter functions and transfer functions for one-dimensional filters:
the box filter function has the same second moment as the Gaussian* ($\frac{1}{12}\Delta^2$); *the
other filters have the same value of the transfer function at the characteristic
wavenumber* $\kappa_c \equiv \pi/\Delta$, *i.e.,* $\widehat{G}(\kappa_c) = \exp(-\pi^2/24)$

Name	Filter function	Transfer function
General	$G(r)$	$\widehat{G}(\kappa) \equiv \displaystyle\int_{-\infty}^{\infty} e^{-i\kappa r} G(r)\, dr$
Box	$\dfrac{1}{\Delta} H(\tfrac{1}{2}\Delta - \|r\|)$	$\dfrac{\sin(\tfrac{1}{2}\kappa\Delta)}{\tfrac{1}{2}\kappa\Delta}$
Gaussian	$\left(\dfrac{6}{\pi\Delta^2}\right)^{1/2} \exp\left(-\dfrac{6r^2}{\Delta^2}\right)$	$\exp\left(-\dfrac{\kappa^2\Delta^2}{24}\right)$
Sharp spectral	$\dfrac{\sin(\pi r/\Delta)}{\pi r}$	$H(\kappa_c - \|\kappa\|),$ $\kappa_c \equiv \pi/\Delta$
Cauchy	$\dfrac{a}{\pi\Delta[(r/\Delta)^2 + a^2]},\quad a = \dfrac{\pi}{24}$	$\exp(-a\Delta\|\kappa\|)$
Pao		$\exp\left(-\dfrac{\pi^{2/3}}{24}(\Delta\|\kappa\|)^{4/3}\right)$

by the convolution

$$\overline{U}(x) \equiv \int_{-\infty}^{\infty} G(r) U(x - r)\, dr. \tag{13.9}$$

The most commonly used filters (which are defined in Table 13.2 and
shown in Fig. 13.1) are the box filter, the Gaussian filter; and the sharp
spectral filter. With the box filter, $\overline{U}(x)$ is simply the average of $U(x')$ in the
interval $x - \tfrac{1}{2}\Delta < x' < x + \tfrac{1}{2}\Delta$. The Gaussian filter function is the Gaussian
distribution with mean zero and variance $\sigma^2 = \tfrac{1}{12}\Delta^2$. This value of σ^2 was
chosen by Leonard (1974) so as to match the second moments $\int_{-\infty}^{\infty} r^2 G(r)\, dr$
of the Gaussian and box filters. (The other filters defined in Table 13.2 are
discussed below.)

Figure 13.2 shows a sample velocity field $U(x)$ and the corresponding
filtered field $\overline{U}(x)$ obtained using the Gaussian filter with $\Delta \approx 0.35$. It is
evident that $\overline{U}(x)$ follows the general trends of $U(x)$, but that the short
lengthscale fluctuations have been removed. These appear in the residual

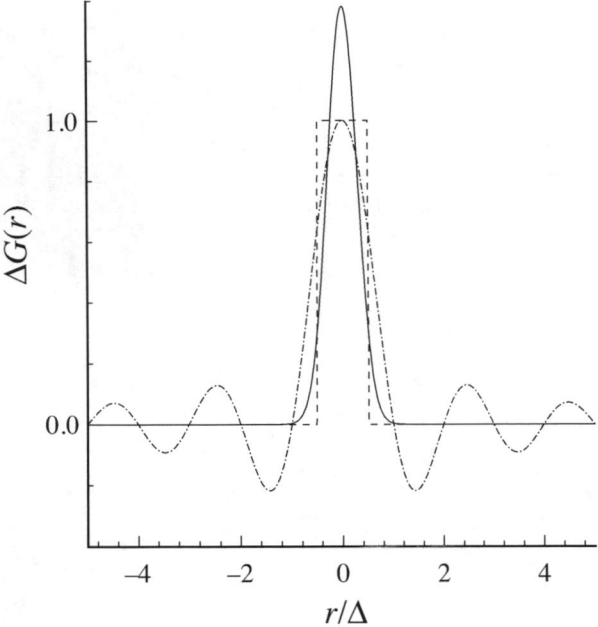

Fig. 13.1. Filters $G(r)$: box filter, dashed line; Gaussian filter, solid line; sharp spectral filter, dot–dashed line.

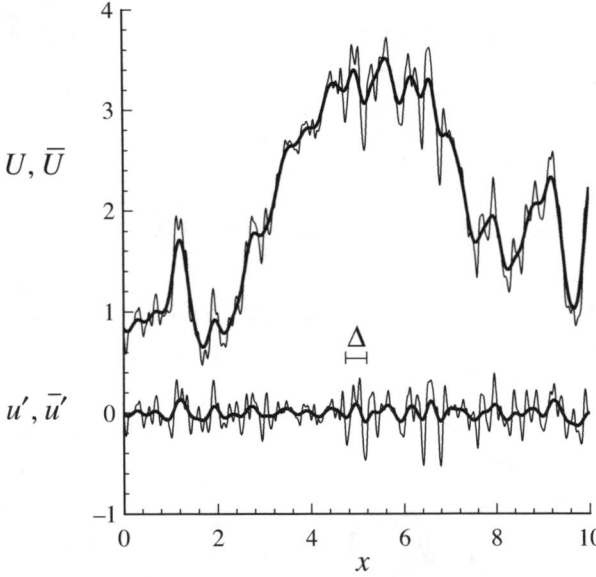

Fig. 13.2. Upper curves: a sample of the velocity field $U(x)$ and the corresponding filtered field $\overline{U}(x)$ (bold line), using the Gaussian filter with $\Delta \approx 0.35$. Lower curves: the residual field $u'(x)$ and the filtered residual field $\overline{u'(x)}$ (bold line).

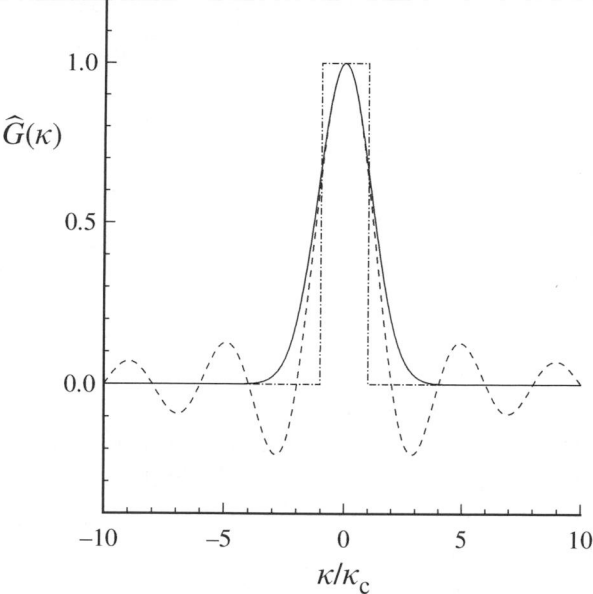

Fig. 13.3. Filter transfer functions $\widehat{G}(\kappa)$: box filter, dashed line; Gaussian filter, solid line; sharp spectral filter, dot–dashed line.

field $u'(x)$ which is also shown in Fig. 13.2. As previously noted, the filtered residual is non-zero – as may be observed.

13.2.3 Spectral representation

The effects of the filter are most clearly revealed in wavenumber space. Suppose that $U(x)$ has a Fourier transform

$$\hat{U}(\kappa) \equiv \mathcal{F}\{U(x)\}. \tag{13.10}$$

Then the Fourier transform of the filtered velocity is

$$\hat{\overline{U}}(\kappa) \equiv \mathcal{F}\{\overline{U}(x)\}$$
$$= \widehat{G}(\kappa)\hat{U}(\kappa), \tag{13.11}$$

where the *transfer function* $\widehat{G}(\kappa)$ is 2π times the Fourier transform of the filter:

$$\widehat{G}(\kappa) \equiv \int_{-\infty}^{\infty} G(r)e^{-i\kappa r}\, \mathrm{d}r = 2\pi\mathcal{F}\{G(r)\}. \tag{13.12}$$

This result follows directly from the filtering equation (Eq. (13.9)) and the convolution theorem (Eq. (D.15) on page 681).

The transfer functions for the various filters are given in Table 13.2 and are shown in Fig. 13.3. Since the filter functions $G(r)$ considered are real and even, so also are the transfer functions. At the origin the transfer functions are unity, because the normalization condition is

$$\int_{-\infty}^{\infty} G(r)\,dr = \widehat{G}(0) = 1. \tag{13.13}$$

The significance of the sharp spectral filter is now apparent: it annihilates all Fourier modes of wavenumber $|\kappa|$ greater than the cutoff wavenumber

$$\kappa_c \equiv \frac{\pi}{\Delta}, \tag{13.14}$$

whereas it has no effect on lower wavenumber modes. However, although it is sharp in wavenumber space, the sharp spectral filter is decidedly non-local in physical space. The converse is true for the box filter. Of the filters considered, only the Gaussian is reasonably compact both in physical space and in wavenumber space.

(The Cauchy and Pao filters defined in Table 13.2 are not used in practice, but they have a theoretical significance that is revealed in Section 13.4. Their transfer functions and filter functions are compared with those of the Gaussian in Figs. 13.8 and 13.9, below.)

EXERCISES _____

13.2   Let $U(x)$ have the Fourier transform $\widehat{U}(\kappa)$ (Eq. (13.10)), so that $\overline{U}(x)$ has the Fourier transform $\widehat{\overline{U}}(\kappa) = \widehat{G}(\kappa)\widehat{U}(\kappa)$ (Eq. (13.11)). Show that the Fourier transform of the residual $u'(x)$ is

$$\widehat{u'}(\kappa) \equiv \mathcal{F}\{u'(x)\} = [1 - \widehat{G}(\kappa)]\widehat{U}(\kappa), \tag{13.15}$$

that the Fourier transform of the filtered residual $\overline{u'}$ is

$$\widehat{\overline{u'}}(\kappa) \equiv \mathcal{F}\{\overline{u'}(x)\} = \widehat{G}(\kappa)[1 - \widehat{G}(\kappa)]\widehat{U}(\kappa), \tag{13.16}$$

and that the Fourier transform of the doubly filtered field $\overline{\overline{U}}(x)$ is

$$\widehat{\overline{\overline{U}}}(\kappa) \equiv \mathcal{F}\{\overline{\overline{U}}(x)\} = \widehat{G}(\kappa)^2 \widehat{U}(\kappa). \tag{13.17}$$

Show that both Eq. (13.4) and the above equations lead to the result

$$\overline{u'}(x) = \overline{U}(x) - \overline{\overline{U}}(x). \tag{13.18}$$

13.3   For the sharp spectral filter, show that

$$\widehat{G}(\kappa)^2 = \widehat{G}(\kappa). \tag{13.19}$$

Hence obtain the results (for the sharp spectral filter)

$$\overline{\overline{U}}(x) = \overline{U}(x), \tag{13.20}$$

$$\overline{u'}(x) = 0. \tag{13.21}$$

More generally, show that the filtered residual is zero (Eq. (13.21)) if, and only if, the filtering operation is a *projection*, i.e., it yields Eq. (13.20).

13.4   Let $\langle\ \rangle_\Delta$ denote the operation of applying the Gaussian filter of width Δ, i.e.,

$$\langle U(x)\rangle_\Delta \equiv \int_{-\infty}^{\infty} \left(\frac{6}{\pi\Delta^2}\right)^{1/2} \exp\left(\frac{-6r^2}{\Delta^2}\right) U(x-r)\,\mathrm{d}r, \tag{13.22}$$

and let $\widehat{G}(\kappa;\Delta)$ denote the corresponding transfer function, i.e.,

$$\widehat{G}(\kappa;\Delta) \equiv \exp\left(-\frac{\kappa^2\Delta^2}{24}\right). \tag{13.23}$$

Obtain the result

$$\widehat{G}(\kappa;\Delta_\mathrm{a})\widehat{G}(\kappa;\Delta_\mathrm{b}) = \widehat{G}(\kappa;\Delta_\mathrm{c}), \tag{13.24}$$

where

$$\Delta_\mathrm{c} = (\Delta_\mathrm{a}^2 + \Delta_\mathrm{b}^2)^{1/2}. \tag{13.25}$$

Hence show that

$$\langle\langle U(x)\rangle_\Delta\rangle_\Delta = \langle U(x)\rangle_{\sqrt{2}\Delta}. \tag{13.26}$$

13.5   Show that the second moment of the filter function (if it exists) is

$$\int_{-\infty}^{\infty} r^2 G(r)\,\mathrm{d}r = -\left(\frac{\mathrm{d}^2\widehat{G}(\kappa)}{\mathrm{d}\kappa^2}\right)_{\kappa=0}. \tag{13.27}$$

Hence verify that the second moments of the box and Gaussian filters are $\Delta^2/12$. Comment on the implications for the sharp spectral and Cauchy filters.

13.6   A set of N basis functions $\varphi_n(x)$, $n = 1, 2, \ldots, N$, is defined on the interval $0 \le x \le \mathcal{L}$. Given the velocity $U(x)$, the filtered velocity $\overline{U}(x)$ is defined as the least-square projection of $U(x)$ onto the basis functions. That is, $\overline{U}(x)$ is given by

$$\overline{U}(x) = \sum_{n=1}^{N} a_n\varphi_n(x), \tag{13.28}$$

where the basis-function coefficients a_n are determined by the condition that they minimize the mean-square residual

$$\chi \equiv \frac{1}{\mathcal{L}} \int_0^{\mathcal{L}} [\overline{U}(x) - U(x)]^2 \, dx. \tag{13.29}$$

By substituting Eq. (13.28) into Eq. (13.29) and differentiating with respect to a_m, show that the coefficients satisfy the matrix equation

$$\sum_{n=1}^{N} B_{mn} a_n = v_m, \tag{13.30}$$

with

$$B_{mn} \equiv \frac{1}{\mathcal{L}} \int_0^{\mathcal{L}} \varphi_m(x) \varphi_n(x) \, dx, \tag{13.31}$$

$$v_m \equiv \frac{1}{\mathcal{L}} \int_0^{\mathcal{L}} \varphi_m(x) U(x) \, dx. \tag{13.32}$$

Show that the matrix **B** is symmetric positive definite. Show that $\overline{U}(x)$ can be expressed as the result of a filtering operation

$$\overline{U}(x) = \int_{x-\mathcal{L}}^{x} G(r, x) U(x - r) \, dr, \tag{13.33}$$

where the filter is

$$G(r, x) = \frac{1}{\mathcal{L}} \sum_{n=1}^{N} \sum_{m=1}^{N} B_{mn}^{-1} \varphi_n(x) \varphi_m(x - r), \tag{13.34}$$

and B_{mn}^{-1} denotes the m–n element of the inverse of the matrix **B**. Argue, both from Eq. (13.33) and from the equation for $\partial \chi / \partial a_m$, that the filtered residual $\overline{u}(x)$ is zero.

13.2.4 The filtered energy spectrum

In order to examine the important issue of the effect of filtering on the spectrum, we now take $U(x)$ to be statistically homogeneous and consider the fluctuating field $u(x) \equiv U(x) - \langle U \rangle$. We can think of $u(x)$ as a component of the velocity $\mathbf{u}(\mathbf{x})$ along a line in statistically homogeneous turbulence, e.g., $u(x) = u_1(e_1 x)$. The autocovariance is

$$R(r) \equiv \langle u(x + r) u(x) \rangle, \tag{13.35}$$

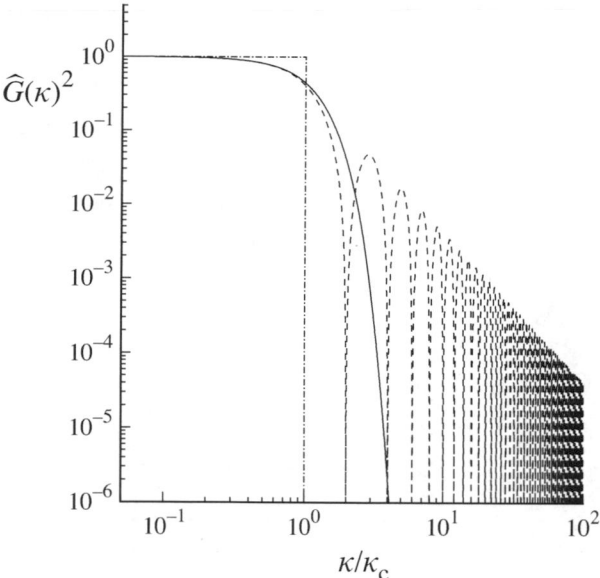

Fig. 13.4. Attenuation factors $\widehat{G}(\kappa)^2$: box filter, dashed line; Gaussian filter, solid line; sharp spectral filter, dot–dashed line.

and, consistent with the notation in Chapter 6, the spectrum $E_{11}(\kappa)$ is defined as twice its Fourier transform:

$$E_{11}(\kappa) = \frac{1}{\pi} \int_{-\infty}^{\infty} R(r)e^{-i\kappa r}\, dr. \tag{13.36}$$

It follows that the filtered fluctuation $\bar{u}(x)$ is also statistically stationary; and, as shown in Exercise 13.7, its spectrum is

$$\overline{E}_{11}(\kappa) = |\widehat{G}(\kappa)|^2 E_{11}(\kappa). \tag{13.37}$$

The attenuation factors $\widehat{G}(\kappa)^2$ for the various filters are shown in Fig. 13.4. Clearly, the box filter is not very effective at attenuating high wavenumber modes.

To illustrate the effect of filtering on a turbulence spectrum, Fig. 13.5 shows the one-dimensional spectrum $E_{11}(\kappa)$ obtained from the model spectrum (Eq. (6.246)) at Reynolds number $R_\lambda = 500$, and the corresponding filtered spectrum $\overline{E}_{11}(\kappa)$. This is obtained from Eq. (13.37) by using the Gaussian filter with $\Delta = \ell_{EI} = \frac{1}{6}L_{11}$, where L_{11} is the integral length scale

$$L_{11} = \frac{1}{\langle u^2 \rangle} \int_0^\infty R(r)\, dr. \tag{13.38}$$

With this specification of Δ, the energy in the filtered velocity field $\frac{1}{2}\langle \bar{u}^2 \rangle$

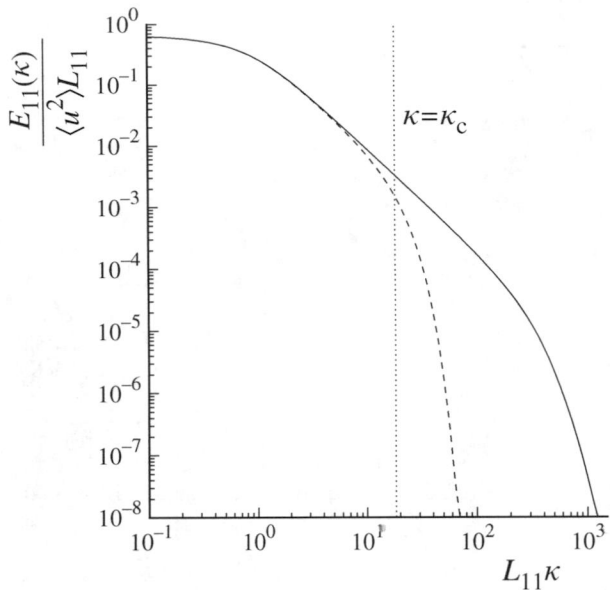

Fig. 13.5. The one-dimensional spectrum $E_{11}(\kappa)$ (solid line) obtained for the model spectrum at $R_\lambda = 500$; and the filtered spectrum $\overline{E}_{11}(\kappa)$ (dashed line) obtained using the Gaussian filter with $\Delta = \frac{1}{6}L_{11}$.

is 92% of the total energy $\frac{1}{2}\langle u^2 \rangle$. (Note that this result is based on one-dimensional filtering and on the one-dimensional spectrum. In the more relevant three-dimensional case, with the corresponding three-dimensional filter of width $\Delta = \ell_{\mathrm{EI}}$ approximately 80% of the energy is resolved, see Exercise 13.10.)

EXERCISE

13.7   Show that the autocovariance of the filtered fluctuation is

$$\overline{R}(r) \equiv \langle \overline{u}(x+r)\overline{u}(x) \rangle$$
$$= \int_{-\infty}^{\infty} \int_{-\infty}^{\infty} G(y)G(z)R(r+z-y)\,\mathrm{d}y\,\mathrm{d}z. \qquad (13.39)$$

Show that the spectrum of $\overline{u}(x)$ can be written

$$\overline{E}_{11}(\kappa) \equiv \frac{1}{\pi}\int_{-\infty}^{\infty} \overline{R}(r)e^{-i\kappa r}\,\mathrm{d}r$$
$$= \frac{1}{\pi}\int_{-\infty}^{\infty} \int_{-\infty}^{\infty} \int_{-\infty}^{\infty} G(y)e^{-i\kappa y}G(z)e^{i\kappa z}$$
$$R(r+z-y)e^{-i\kappa(r+z-y)}\,\mathrm{d}y\,\mathrm{d}z\,\mathrm{d}r. \qquad (13.40)$$

For fixed y and z show

$$\frac{1}{\pi} \int_{-\infty}^{\infty} R(r + z - y)e^{-i\kappa(r+z-y)} \, dr = E_{11}(\kappa), \qquad (13.41)$$

and hence obtain the result

$$\overline{E}_{11}(\kappa) = \widehat{G}(\kappa)\widehat{G}^*(\kappa)E_{11}(\kappa) = |\widehat{G}(\kappa)|^2 E_{11}(\kappa). \qquad (13.42)$$

13.2.5 The resolution of filtered fields

Consider the statistically homogeneous field $u(x)$ in the interval $0 \le x < \mathcal{L}$, and let the filtered field $\overline{u}(x)$ be represented discretely by its values at the N nodes of a uniform grid of spacing $h = \mathcal{L}/N$. What grid spacing h is required (or equivalently how many nodes are required) to resolve $\overline{u}(x)$ adequately? This is, of course, an important question because the cost of a LES calculation increases (at least linearly) with the number of nodes. The answer depends both on the choice of filter and on the information to be extracted from $\overline{u}(x)$.

To address the issues of filtering and resolution, we suppose $u(x)$ to be periodic (with period $\mathcal{L}$) and consider the Fourier series of $u(x)$ and $\overline{u}(x)$. The Fourier series for $u(x)$ is written in the form of an inverse discrete Fourier transform (see Eq. (F.6) on page 693)

$$u(x) = \sum_{n=1-\frac{1}{2}N_{max}}^{\frac{1}{2}N_{max}} c_n e^{i\kappa_n x}, \qquad (13.43)$$

where c_n is the Fourier coefficient corresponding to the nth wavenumber

$$\kappa_n = \frac{2\pi n}{\mathcal{L}}. \qquad (13.44)$$

The even integer N_{max} determines the maximum resolved wavenumber,

$$\kappa_{max} = \kappa_{N_{max}/2} = \frac{\pi N_{max}}{\mathcal{L}}, \qquad (13.45)$$

which is chosen to be sufficiently large to resolve $u(x)$. (As discussed in Chapter 9, isotropic turbulence is adequately resolved with $\kappa_{max}\eta \ge 1.5$.)

The discrete Fourier transform (see Appendix F) provides a one-to-one mapping between these Fourier coefficients and the values of $u(x)$ on a uniform grid with N_{max} nodes, i.e., $u(nh_{max})$, $n = 0, 1, \ldots, N_{max} - 1$, where

$$h_{max} \equiv \frac{\mathcal{L}}{N_{max}} = \frac{\pi}{\kappa_{max}}. \qquad (13.46)$$

Thus h_{max} is the largest grid spacing that can resolve $u(x)$.

The Fourier series of the filtered field $\bar{u}(x)$ can be written

$$\bar{u}(x) = \sum_{n=1-\frac{1}{2}N_{\max}}^{\frac{1}{2}N_{\max}} \bar{c}_n e^{i\kappa_n x}, \tag{13.47}$$

where the coefficients $\bar{c}_n$ are those of $u(x)$ attenuated by the transfer function of the filter:

$$\bar{c}_n = \widehat{G}(\kappa_n) c_n, \tag{13.48}$$

see Eq. (13.11).

The sharp spectral filter

We consider first the sharp spectral filter with cutoff wavenumber $\kappa_c < \kappa_{\max}$. For simplicity we suppose that κ_c is chosen so that

$$N \equiv \frac{\kappa_c \mathcal{L}}{\pi} \tag{13.49}$$

is an even integer. Then the coefficients $\bar{c}_n$ produced by the sharp spectral filter with $\widehat{G}(\kappa) = H(\kappa_c - |\kappa|)$ are

$$\bar{c}_n = c_n, \qquad \text{for } |n| < \tfrac{1}{2}N,$$

$$\bar{c}_n = 0, \qquad \text{for } |n| \geq \tfrac{1}{2}N. \tag{13.50}$$

Thus the Fourier series for $\bar{u}(x)$ (Eq. (13.47)) can be re-expressed as

$$\bar{u}(x) = \sum_{n=-\frac{1}{2}N+1}^{\frac{1}{2}N} c_n e^{i\kappa_n x}. \tag{13.51}$$

Correspondingly, without loss of information, the discrete Fourier transform allows $\bar{u}(x)$ to be represented in physical space on a grid of spacing

$$h \equiv \frac{\mathcal{L}}{N} = \frac{\pi}{\kappa_c}. \tag{13.52}$$

It is for this reason that the characteristic filter width is defined as $\Delta \equiv \pi/\kappa_c$.

To summarize: the filtered field given by the sharp spectral filter can be represented *exactly* by $N = \kappa_c \mathcal{L}/\pi$ Fourier modes, or equivalently by the N values $\bar{u}(nh)$, $n = 0, 1, \dots, N-1$, on a grid of spacing $h = \Delta = \pi/\kappa_c$. More Fourier modes, or more grid points, provide no further information.

The Gaussian filter

We turn our attention now to the Gaussian filter, for which the transfer function $\widehat{G}(\kappa)$ is strictly positive for all κ. As a consequence, the Fourier series for $u(x)$ and $\bar{u}(x)$ (Eq. (13.47)) contain the same information: given the Fourier coefficients of $\bar{u}(x)$, those of $u(x)$ can be recovered as

$$c_n = \bar{c}_n / \widehat{G}(\kappa). \tag{13.53}$$

This is a well-conditioned operation. On the other hand, if $\bar{u}(x)$ is represented in physical space by the N_{max} values of $\bar{u}(nh_{\mathrm{max}})$, $n = 0, 1, \ldots, N_{\mathrm{max}} - 1$, then the process of recovering $u(nh_{\mathrm{max}})$ (or c_n) is poorly conditioned.

In LES, the objective of filtering is to allow the filtered field $\bar{u}(x)$ to be adequately resolved with fewer modes or nodes than are required to resolve $u(x)$. To examine the issues involved, we consider the modified series

$$\bar{u}(x) \approx \tilde{u}(x) \equiv \sum_{n=1-\frac{1}{2}N}^{\frac{1}{2}N} \tilde{c}_n e^{i\kappa_n x}, \tag{13.54}$$

for some even N ($N \leq N_{\mathrm{max}}$), and the corresponding representation in physical space on a grid of spacing $h = \mathcal{L}/N$. The wavenumber κ_{r} of the highest resolved mode is

$$\kappa_{\mathrm{r}} = \kappa_{N/2} = \frac{\pi}{h}. \tag{13.55}$$

The appropriate coefficients $\tilde{c}_n$ to consider depend on whether $\bar{u}(x)$ is known in wavenumber space or in physical space. If $\bar{u}(x)$ is known in wavenumber space, the best approximation is obtained with $\tilde{c}_n = \bar{c}_n$ for $|n| < \frac{1}{2}N$ and $\tilde{c}_{N/2} = 0$ (to ensure that $\tilde{u}(x)$ is real). Then the spectrum of $\tilde{u}(x)$, $\tilde{E}_{11}(\kappa)$, is identical to that of $\bar{u}(x)$, $\bar{E}_{11}(\kappa)$, for $\kappa < \kappa_{\mathrm{r}}$ and is zero at higher wavenumbers.

The resolution is characterized by $h/\Delta = \kappa_{\mathrm{c}}/\kappa_{\mathrm{r}}$. For the model spectrum, Fig. 13.6 shows the spectrum $\kappa^2 \bar{E}_{11}(\kappa)$ of the filtered velocity derivative $d\bar{u}(x)/dx$. Resolving up to wavenumber κ_{r} means neglecting all contributions with $\kappa > \kappa_{\mathrm{r}}$. As shown in Exercise 13.8, the resolutions $\kappa_{\mathrm{c}}/\kappa_{\mathrm{r}} = h/\Delta = \frac{1}{2}$ and 1 lead to the neglect of 2% and 28%, respectively, of the contributions to $\langle (d\bar{u}/dx)^2 \rangle$. Hence $h = \frac{1}{2}\Delta$ is deemed to yield good resolution, whereas $h = \Delta$ yields poor resolution.

If, on the other hand, $\bar{u}(x)$ is known on the grid in physical space, then the coefficients $\tilde{c}_n$ can be obtained from the discrete Fourier transform of $\tilde{u}(nh) = \bar{u}(nh)$ for $n = 0, 1, \ldots, N - 1$. In this case, Fourier modes in $\bar{u}(x)$ at higher wavenumbers than are resolved by the grid ($|\kappa| \geq \kappa_{\mathrm{r}}$) are *aliased* to resolved wavenumbers (see Appendix F). For the model spectrum, Fig. 13.6

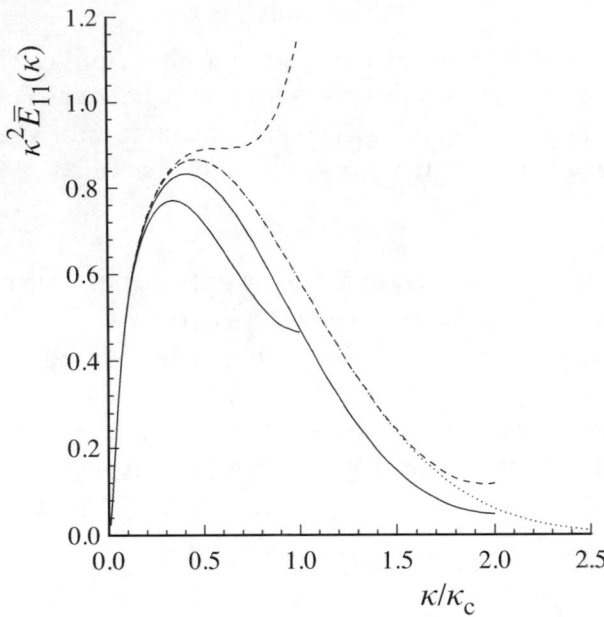

Fig. 13.6. The spectrum of $d\overline{u}(x)/dx$, $\kappa^2\overline{E}_{11}(\kappa)$ (dotted line); aliased spectra $\kappa^2\tilde{E}_{11}(\kappa)$ for $h/\Delta = \frac{1}{2}$ and 1 (dashed lines); and the spectra $\tilde{D}_h(\kappa)$ (Eq. (13.57)) of the finite-difference approximation to $d\tilde{u}/dx$ (Eq. (13.56)) for $h/\Delta = \frac{1}{2}$ and 1 (solid lines), for the model spectrum ($R_\lambda = 500$) and the Gaussian filter. (The spectra are scaled so that $\kappa^2\overline{E}_{11}(\kappa)$ integrates to unity.)

shows the aliased spectra $\tilde{E}_{11}(\kappa)$ (multiplied by κ^2) for two choices of N corresponding to $h/\Delta = \kappa_c/\kappa_r = \frac{1}{2}$ and 1. Clearly, for $h/\Delta = 1$, the aliased spectrum $\tilde{E}_{11}(\kappa)$ is a poor approximation to $\overline{E}_{11}(\kappa)$.

If the LES equations are solved in physical space, then spatial derivatives have to be approximated, and the accuracy of the approximation depends on the resolution. An indication of the errors involved is obtained by studying the approximation of the derivative $d\overline{u}(x)/dx$ by the simple finite-difference formula

$$\mathcal{D}_h\tilde{u}(x) \equiv [\tilde{u}(x + \tfrac{1}{2}h) - \tilde{u}(x - \tfrac{1}{2}h)]/h. \tag{13.56}$$

The spectrum of $d\overline{u}(x)/dx$ is $\kappa^2\overline{E}_{11}(\kappa)$, whereas that of $\mathcal{D}_h\tilde{u}(x)$ is

$$\tilde{D}_h(\kappa) \equiv \left(\frac{\sin(\tfrac{1}{2}\kappa h)}{\tfrac{1}{2}\kappa h}\right)^2 \kappa^2\tilde{E}_{11}(\kappa) \tag{13.57}$$

(see Exercise 13.9). For the model spectrum, Fig. 13.6 shows the spectrum of $d\overline{u}(x)/dx$, $\kappa^2\overline{E}_{11}(\kappa)$, and its approximation by $\tilde{D}_h(\kappa)$ for $h/\Delta = \frac{1}{2}$ and $h/\Delta = 1$. It may clearly be seen that the error in the approximation is greatest at high wavenumber, and that it decreases as h/Δ decreases. A quantitative indicator of the accuracy is the variance of $\mathcal{D}_h \tilde{u}(x)$ relative to that of $d\overline{u}(x)/dx$ (which is the ratio of the integrals of the spectra). For $h/\Delta = 1$, $\frac{1}{2}$, and $\frac{1}{4}$, the values of this ratio are 0.60, 0.86, and 0.96, respectively.

In an LES calculation of a turbulent flow with a fixed filter width Δ, the computational cost is roughly proportional to $(h/\Delta)^{-4}$ – since the time step and number of nodes in each direction scale as h^{-1}. Good resolution therefore comes at a high price: doubling the resolution – say from $h/\Delta = 1$ to $h/\Delta = \frac{1}{2}$ – increases the cost by a factor of 16.

EXERCISES

13.8   Suppose that $u(x)$ has the Kolmogorov spectrum $E_{11}(\kappa)$ given by Eq. (6.240) on page 231. Show that, for the Gaussian filter, the spectrum of $d\overline{u}(x)/dx$ is

$$\kappa^2 \overline{E}_{11}(\kappa) = C_1 \varepsilon^{2/3} \kappa^{1/3} \exp\left(-\frac{\pi^2 \kappa^2}{12\kappa_c^2}\right). \qquad (13.58)$$

If $u(x)$ is represented by its Fourier coefficients up to wavenumber κ_r, show that the fraction of $[d\overline{u}(x)/dx]^2$ resolved is

$$\frac{\int_0^{\kappa_r} \kappa^2 \overline{E}_{11}(\kappa)\, d\kappa}{\int_0^{\infty} \kappa^2 \overline{E}_{11}(\kappa)\, d\kappa} = \frac{\int_0^{(\pi^2/12)(\kappa_r/\kappa_c)^2} t^{-1/3} e^{-t}\, dt}{\int_0^{\infty} t^{-1/3} e^{-t}\, dt} = P\left(\frac{2}{3}, \frac{\pi^2}{12}\left(\frac{\kappa_r}{\kappa_c}\right)^2\right), \qquad (13.59)$$

where P is the incomplete gamma function. Hence show that, for $\kappa_c/\kappa_r = h/\Delta = \frac{1}{2}$ and 1, this fraction is 0.98 and 0.72, respectively.

13.9   Show that the Fourier transform of $\tilde{u}(x + \frac{1}{2}h)$ is

$$\mathcal{F}\{\tilde{u}(x + \tfrac{1}{2}h)\} = e^{i\kappa h/2} \mathcal{F}\{\tilde{u}(x)\}. \qquad (13.60)$$

Hence show that the Fourier transform of $\mathcal{D}_h \tilde{u}(x)$ (Eq. (13.56)) is

$$\mathcal{F}\{\mathcal{D}_h \tilde{u}(x)\} = \frac{i \sin(\tfrac{1}{2}\kappa h)}{\tfrac{1}{2} h} \mathcal{F}\{\tilde{u}(x)\}$$

$$= \frac{\sin(\tfrac{1}{2}\kappa h)}{\tfrac{1}{2}\kappa h} \mathcal{F}\left\{\frac{d\tilde{u}(x)}{dx}\right\}, \qquad (13.61)$$

and verify Eq. (13.57).

13.2.6 Filtering in three dimensions

For the general case, the filtered velocity field $\overline{U}(x,t)$ is defined by Eq. (13.1) in terms of the filter function $G(r, x)$. The filter may be uniform (i.e., $G(r, x)$

does not depend on x) or it may be non-uniform. It may be isotropic (i.e., $G(r, x)$ depends on r only through $r = |r|$) or it may be anisotropic.

All of the filters introduced in one dimension can be used to generate uniform, isotropic three-dimensional filters. The box filter becomes a volume average over the sphere of radius $\frac{1}{2}\Delta$; the Gaussian filter function is the joint normal with mean zero and covariance $\delta_{ij}\Delta^2/12$; and the sharp spectral filter annihilates all Fourier modes with $|\kappa| \geq \kappa_c$. Most of the results obtained in one dimension carry through straightforwardly for these three-dimensional filters. In particular, for homogeneous turbulence, the filtered energy spectrum function is

$$\overline{E}(\kappa) = \widehat{G}(\kappa)^2 E(\kappa). \tag{13.62}$$

This relation can be used (see Exercise 13.10) to estimate that, in high-Reynolds-number isotropic turbulence, 80% of the kinetic energy is resolved if the filter width is taken to be $\Delta \approx 1.2\ell_{EI}$ for the sharp spectral filter, or $\Delta \approx 0.8\ell_{EI}$ for the Gaussian filter.

For some inhomogeneous flows, the LES equations may be solved on a uniform anisotropic rectangular grid, with grid spacings h_1, h_2, and h_3 in the three coordinate directions. Then, an anisotropic filter is used, with the characteristic filter widths Δ_1, Δ_2, and Δ_3 generally being proportional to h_1, h_2, and h_3. For example, the anisotropic Gaussian filter function is joint normal with covariance $\delta_{ij}\Delta_{(i)}^2/12$. Some of the models described below involve a characteristic filter width Δ that, with anisotropic filters, is generally taken to be $\Delta = (\Delta_1\Delta_2\Delta_3)^{1/3}$, as suggested by Deardorff (1970). Scotti, Meneveau, and Lilly (1993) provide some theoretical support for this choice, and the issue is addressed further by Scotti, Meneveau, and Fatica (1997).

Historically, before the general filtering formalism was introduced by Leonard (1974), one view of LES (e.g., Deardorff (1970)) was that $\overline{U}(x, t)$ represents the velocity $U(x, t)$ averaged over a cell of a rectangular grid. That is, for the cell centered at x, $\overline{U}(x, t)$ is

$$\overline{U}(x, t) = \frac{1}{h_1 h_2 h_3} \int_{x_3-h_3/2}^{x_3+h_3/2} \int_{x_2-h_2/2}^{x_2+h_2/2} \int_{x_1-h_1/2}^{x_1+h_1/2} U(x', t)\, dx_1'\, dx_2'\, dx_3'. \tag{13.63}$$

The corresponding filter is

$$G(r) = \prod_{i=1}^{3} \frac{1}{\Delta_i} H\left(\tfrac{1}{2}\Delta_{(i)} - |r_{(i)}|\right), \tag{13.64}$$

with $\Delta = h$, which is anisotropic even if the grid spacings are the same (i.e., $h_1 = h_2 = h_3$).

In LES-NWR for wall-bounded flows, the grid spacing is taken to be finer close to the wall, so that the near-wall structures can be resolved (see, e.g., Piomelli (1993)). This leads to a non-uniform filter (if the filter width is taken to be proportional to the grid spacing). For non-uniform filters, the operations of filtering and spatial differentiation do not commute (see Exercise 13.1), an issue which is discussed further by Ghosal and Moin (1995) and Vasilyev, Lund, and Moin (1998). These authors also address the issues of filtering near boundaries, and the discrete implementation of the filtering operation. To avoid these difficulties, in LES of channel flow the usual practice is to have a non-uniform grid in the wall-normal (y) direction, but to filter only in x–z planes (see Exercise 13.13).

EXERCISES

13.10 Consider high-Reynolds-number homogeneous isotropic turbulence and an isotropic filter of width Δ and characteristic wavenumber $\kappa_c = \pi/\Delta$. The kinetic energy of the residual motions is

$$\langle k_r \rangle = \int_0^\infty [1 - \widehat{G}(\kappa)^2] E(\kappa) \, d\kappa. \tag{13.65}$$

For the sharp spectral filter with cutoff κ_c in the inertial subrange, use the Kolmogorov spectrum to obtain the estimate for the fraction of the energy in the residual motions

$$\frac{\langle k_r \rangle}{k} = \tfrac{3}{2} C (\kappa_c L)^{-2/3}, \tag{13.66}$$

where the lengthscale is $L \equiv k^{3/2}/\varepsilon$. Show that 80% of the energy is resolved (i.e., $\langle k_r \rangle / k = 0.2$) if κ_c is chosen by

$$\kappa_c L = (\tfrac{15}{2} C)^{3/2} \approx 38. \tag{13.67}$$

Using the relations $\ell_{EI} = \tfrac{1}{6} L_{11}$ and $L_{11} = 0.43 L$ from Section 6.5, show that the corresponding filter width is

$$\frac{\Delta}{\ell_{EI}} = \frac{6\pi}{0.43 \kappa_c L} \approx 1.16. \tag{13.68}$$

Perform the same analysis for the Gaussian filter to obtain

$$\frac{\langle k_r \rangle}{k} = C I_0 \, 96^{-1/3} (\Delta/L)^{2/3}, \tag{13.69}$$

where I_0 is given by

$$I_0 \equiv \int_0^\infty (1 - e^{-x}) x^{-4/3} \, dx \approx 4.062, \tag{13.70}$$

and show that 80% of the energy is resolved for $\kappa_c \equiv \pi/\Delta$ given by

$$\kappa_c L = \pi \left(\frac{5}{96^{1/3}} C I_0 \right)^{3/2} \approx 54. \qquad (13.71)$$

13.11 Consider solving the LES equations by a pseudo-spectral method in which wavenumbers with $|\boldsymbol{\kappa}| < \kappa_r$ are resolved. Argue that, for the Gaussian filter and for $\kappa_r = \kappa_c$ (corresponding to poor spatial resolution), a factor of $(54/38)^3 \approx 2.9$ more nodes is required than would be required if the sharp spectral filter were used (with κ_c chosen to resolve 80% of the energy in each case). Show that the corresponding factor is 23 if the better resolution $\kappa_r = 2\kappa_c$ ($h = \frac{1}{2}\Delta$) is used.

13.12 Consider the LES of high-Reynolds-number isotropic turbulence using a pseudo-spectral method and the sharp spectral filter with cutoff wavenumber $\kappa_c = \kappa_r$ in the inertial subrange. Show that, if 90% of the energy is to be resolved, then a factor of

$$2^{\frac{9}{2}} \approx 23$$

more modes is required than would be needed if only 80% of the energy were to be resolved.

13.13 Consider isotropic turbulence with energy-spectrum function $E(\kappa)$. A sharp spectral filter with cutoff wavenumber κ_c is used in the 1–2 plane, but no filtering is performed in the 3 direction. That is, the filter annihilates modes with $\kappa_1^2 + \kappa_2^2 \geq \kappa_c^2$. Show that the energy-spectrum function $\overline{E}(\kappa)$ of the filtered field is

$$\overline{E}(\kappa) = \left\{ 1 - H(\kappa - \kappa_c) \left[1 - \left(\frac{\kappa_c}{\kappa} \right)^2 \right]^{1/2} \right\} E(\kappa). \qquad (13.72)$$

Show that the attenuation factor has the high-wavenumber asymptote $\frac{1}{2}(\kappa_c/\kappa)^2$. Show that, if filtering is performed in one direction only, the corresponding asymptote is (κ_c/κ).

13.2.7 The filtered rate of strain

From the filtered velocity field $\overline{U}(\boldsymbol{x},t)$ we can form the filtered velocity gradients $\partial \overline{U}_i/\partial x_j$ and the filtered rate-of-strain tensor[1]

$$\overline{S}_{ij} \equiv \frac{1}{2} \left(\frac{\partial \overline{U}_i}{\partial x_j} + \frac{\partial \overline{U}_j}{\partial x_i} \right), \qquad (13.73)$$

---

[1] Note that $\overline{S}_{ij}$ (with a long overbar) is based on $\overline{U}$, whereas $\bar{S}_{ij}$ (with a short overbar) is based on $\langle U \rangle$.

and, on this basis, the characteristic filtered rate of strain is defined by

$$\overline{S} \equiv (2\overline{S}_{ij}\overline{S}_{ij})^{1/2}. \tag{13.74}$$

Both of these quantities are prominent in the SGS models described below, and it is instructive to examine their dependences on the type of filter and filter width.

We consider high-Reynolds-number isotropic turbulence so that there is an inertial subrange in which the energy-spectrum function $E(\kappa)$ adopts the Kolmogorov form $C\varepsilon^{2/3}\kappa^{-5/3}$ (Eq. (6.239)). The mean-square of $\overline{S}$ is determined by the spectrum $E(\kappa)$ and the filter transfer function $\widehat{G}(\kappa)$ by

$$\begin{aligned}
\langle\overline{S}^2\rangle &= 2\langle\overline{S}_{ij}\overline{S}_{ij}\rangle \\
&= 2\int_0^\infty \kappa^2 \overline{E}(\kappa)\,d\kappa \\
&= 2\int_0^\infty \kappa^2 \widehat{G}(\kappa)^2 E(\kappa)\,d\kappa.
\end{aligned} \tag{13.75}$$

The second line is the standard relationship between the mean-square rate of strain of an isotropic velocity field and its spectrum $\overline{E}(\kappa)$; the final expression follows from Eq. (13.62).

The factor of κ^2 weights the integral toward high wavenumbers, whereas the factor $\widehat{G}(\kappa)^2$ attenuates it beyond the cutoff wavenumber κ_c. Consequently, the contributions to $\langle\overline{S}^2\rangle$ are predominantly from wavenumbers around κ_c.

If the cutoff wavenumber κ_c is in the inertial subrange, the integral in Eq. (13.75) can be well approximated by substituting the Kolmogorov spectrum for $E(\kappa)$. This leads to the estimate

$$\begin{aligned}
\langle\overline{S}^2\rangle &\approx 2\int_0^\infty \kappa^2 \widehat{G}(\kappa)^2 C\varepsilon^{2/3}\kappa^{-5/3}\,d\kappa \\
&= a_{\mathrm{f}} C\varepsilon^{2/3}\Delta^{-4/3},
\end{aligned} \tag{13.76}$$

where the constant a_{f} is defined by

$$a_{\mathrm{f}} \equiv 2\int_0^\infty (\kappa\Delta)^{1/3}\widehat{G}(\kappa)^2\Delta\,d\kappa, \tag{13.77}$$

and depends on the filter type, but is independent of Δ. (Values of a_{f} are deduced in Exercise 13.14.)

This type of analysis originated with Lilly (1967). The most important deduction is that the r.m.s. of $\overline{S}$ scales as $\Delta^{-2/3}$ (for Δ in the inertial subrange).

13.14 Show that, for the sharp spectral filter, the constant a_f defined by Eq. (13.77) is

$$a_f = \tfrac{3}{2}\pi^{4/3} \approx 6.90 \qquad (13.78)$$

and that for the Gaussian filter it is

$$a_f = 12^{2/3}\Gamma(\tfrac{2}{3}) \approx 7.10. \qquad (13.79)$$

13.15 Consider

$$E(\kappa) = C\varepsilon^{2/3}\kappa^{-5/3}\exp(-\beta_o\kappa\eta), \qquad (13.80)$$

as a model for the high-wavenumber spectrum (see Eq. (6.249) and Exercise 6.33 on page 237). For this spectrum, and for the Cauchy filter (see Table 13.2), show that Eq. (13.75) leads to

$$\langle \bar{S}^2 \rangle = 2\Gamma(\tfrac{4}{3})\left(\frac{\pi}{12} + \beta_o\frac{\eta}{\Delta}\right)^{-4/3} C\varepsilon^{2/3}\Delta^{-3/4}$$
$$\approx \frac{\varepsilon^{2/3}}{(\eta + \tfrac{1}{8}\Delta)^{4/3}}, \qquad (13.81)$$

(for $C = 1.5$). Show that for the Cauchy filter

$$a_f = 2\left(\frac{12}{\pi}\right)^{4/3}\Gamma(\tfrac{4}{3}) \approx 10.66. \qquad (13.82)$$

13.16 Show that, with the Pao spectrum

$$E(\kappa) = C\varepsilon^{2/3}\kappa^{-5/3}\exp[-\tfrac{3}{2}C(\kappa\eta)^{4/3}], \qquad (13.83)$$

and the Pao filter

$$\hat{G}(\kappa) = \exp\left(-\frac{\pi^{2/3}}{24}(\kappa\Delta)^{4/3}\right), \qquad (13.84)$$

Eq. (13.75) leads to

$$\langle \bar{S}^2 \rangle = \varepsilon^{2/3}\left(\eta^{4/3} + \frac{\pi^{2/3}}{18C}\Delta^{4/3}\right)^{-1}$$
$$\approx \frac{\varepsilon^{2/3}}{\eta^{4/3} + (\tfrac{1}{7}\Delta)^{4/3}}, \qquad (13.85)$$

(for $C = 1.5$). Show that, for the Pao filter,

$$a_f = 18\pi^{-2/3} \approx 8.39. \qquad (13.86)$$

13.3 Filtered conservation equations

The conservation equations governing the filtered velocity field $\overline{U}(x,t)$ are obtained by applying the filtering operation to the Navier–Stokes equations. We consider spatially uniform filters, so that filtering and differentiation commute.

The filtered continuity equation is

$$\overline{\left(\frac{\partial U_i}{\partial x_i}\right)} = \frac{\partial \overline{U}_i}{\partial x_i} = 0, \tag{13.87}$$

from which we obtain

$$\frac{\partial u_i'}{\partial x_i} = \frac{\partial}{\partial x_i}(U_i - \overline{U}_i) = 0. \tag{13.88}$$

Thus both the filtered field $\overline{U}$ and the residual field u' are solenoidal.

13.3.1 Conservation of momentum

The filtered momentum equation (written in conservative form) is simply

$$\frac{\partial \overline{U}_j}{\partial t} + \frac{\partial \overline{U_i U_j}}{\partial x_i} = v\frac{\partial^2 \overline{U}_j}{\partial x_i \partial x_i} - \frac{1}{\rho}\frac{\partial \overline{p}}{\partial x_j}, \tag{13.89}$$

where $\overline{p}(x,t)$ is the filtered pressure field. This equation differs from the Navier–Stokes equations because the filtered product $\overline{U_i U_j}$ is different than the product of the filtered velocities $\overline{U}_i \overline{U}_j$. The difference is the *residual-stress tensor* defined by

$$\tau_{ij}^{R} \equiv \overline{U_i U_j} - \overline{U}_i \overline{U}_j, \tag{13.90}$$

which is analogous to the Reynolds-stress tensor

$$\langle u_i u_j \rangle = \langle U_i U_j \rangle - \langle U_i \rangle \langle U_j \rangle. \tag{13.91}$$

(Strictly, the stress tensors are $-\rho\tau_{ij}^{R}$ and $-\rho\langle u_i u_j \rangle$.) The *residual kinetic energy* is

$$k_{\mathrm{r}} \equiv \tfrac{1}{2}\tau_{ii}^{R}, \tag{13.92}$$

and the *anisotropic residual-stress tensor* is defined by

$$\tau_{ij}^{r} \equiv \tau_{ij}^{R} - \tfrac{2}{3}k_{\mathrm{r}}\delta_{ij}. \tag{13.93}$$

The isotropic residual stress is included in the modified filtered pressure

$$\bar{p} \equiv \overline{p} + \tfrac{2}{3}\rho k_{\mathrm{r}}. \tag{13.94}$$

With these definitions, the filtered momentum equation (Eq. (13.89)) can be rewritten

$$\frac{\overline{D}\,\overline{U}_j}{\overline{D}t} = \nu\,\frac{\partial^2 \overline{U}_j}{\partial x_i\,\partial x_i} - \frac{\partial \tau^r_{ij}}{\partial x_i} - \frac{1}{\rho}\frac{\partial \bar{p}}{\partial x_j},\tag{13.95}$$

where the substantial derivative based on the filtered velocity is

$$\frac{\overline{D}}{\overline{D}t} \equiv \frac{\partial}{\partial t} + \overline{U}\cdot\nabla.\tag{13.96}$$

As usual, the divergence of the momentum equation yields a Poisson equation for the modified pressure $\bar{p}$.

Like the Reynolds equations for $\langle U\rangle$, the filtered equations for $\overline{U}$ (Eqs. (13.87) and (13.95)) are unclosed. Closure is achieved by modelling the residual (or SGS) stress tensor τ^r_{ij}: this is the topic of Sections 13.4 and 13.6.

In several important respects, however, the filtered equations are quite different than the Reynolds equations. The fields involved – $\overline{U}(x,t)$, $\bar{p}(x,t)$, and $\tau^r_{ij}(x,t)$ – are random, three-dimensional, and unsteady, even if the flow is statistically stationary or homogeneous. Also the stress tensor to be modelled depends on the specification of the type and width of the filter. (In fact, only for positive filters, $G \geq 0$, is it guaranteed that k_r and τ^R_{ij} are positive semi-definite, see Exercise 13.21.)

With $\tau^r_{ij}(x,t)$ being given by a residual-stress model, Eqs. (13.87) and (13.95) can be solved to determine $\overline{U}(x,t)$ and $\bar{p}(x,t)$. The filtered velocity field depends on the type of filter and the filter width Δ, yet these quantities do not appear directly in the equations – they appear only indirectly through the model for $\tau^r_{ij}(x,t)$.

EXERCISE _____

13.17   Consider high-Reynolds-number homogeneous turbulence with the sharp spectral filter in the inertial subrange. Use the Kolmogorov spectrum to obtain the estimate for the mean residual kinetic energy

$$\langle k_r\rangle = \int_{\kappa_c}^{\infty} E(\kappa)\,\mathrm{d}\kappa \approx \tfrac{3}{2}C\left(\frac{\varepsilon\Delta}{\pi}\right)^{2/3}.\tag{13.97}$$

13.3.2 Decomposition of the residual stress

The closure problem in LES arises from the nonlinear convective term in the Navier–Stokes equations, which (via Eqs. (13.90) and (13.93)) is decomposed as

$$\overline{U_i U_j} = \overline{U_i U_j} + \tau_{ij}^R = \overline{U}_i \overline{U}_j + \tau_{ij}^r + \tfrac{2}{3} k_r \delta_{ij}. \tag{13.98}$$

Other decompositions can be considered. Leonard (1974) introduced a decomposition of τ_{ij}^R into three component stresses, as described in Exercise 13.18. However, since two of these component stresses are not Galilean-invariant (Speziale (1985), Exercise 13.19), a preferred Galilean-invariant decomposition is that proposed by Germano (1986):

$$\tau_{ij}^R = \mathcal{L}_{ij}^\circ + \mathcal{C}_{ij}^\circ + \mathcal{R}_{ij}^\circ, \tag{13.99}$$

where the *Leonard stresses* are

$$\mathcal{L}_{ij}^\circ \equiv \overline{\overline{U}_i \overline{U}_j} - \overline{\overline{U}}_i \overline{\overline{U}}_j, \tag{13.100}$$

the *cross stresses* are

$$\mathcal{C}_{ij}^\circ \equiv \overline{\overline{U}_i u_j'} + \overline{u_i' \overline{U}_j} - \overline{\overline{U}}_i \overline{u_j'} - \overline{u_i'} \, \overline{\overline{U}}_j, \tag{13.101}$$

and the *SGS Reynolds stresses* are

$$\mathcal{R}_{ij}^\circ \equiv \overline{u_i' u_j'} - \overline{u_i'} \, \overline{u_j'}. \tag{13.102}$$

The significance of these stresses is discussed in Section 13.5.2.

With the sharp spectral filter, a preferable decomposition is

$$\overline{U_i U_j} = \overline{\overline{U}_i \overline{U}_j} + \tau_{ij}^\kappa, \tag{13.103}$$

with

$$\begin{aligned}
\tau_{ij}^\kappa &= \overline{u_i \overline{U}_j} + \overline{u_j \overline{U}_i} + \overline{u_i u_j} \\
&= \mathcal{C}_{ij}^\circ + \mathcal{R}_{ij}^\circ.
\end{aligned} \tag{13.104}$$

This is preferable because each component stress is a filtered quantity and hence can be represented exactly in terms of the resolved modes, i.e., the Fourier modes with wavenumber $|\boldsymbol{\kappa}| < \kappa_c$. This decomposition is also preferable (for the same reason) for other projections, such as the least-square projection (Exercise 13.6).

EXERCISES

13.18 From the decomposition $\boldsymbol{U} = \overline{\boldsymbol{U}} + \boldsymbol{u}'$, show that (as originally proposed by Leonard (1974)) the residual stress tensor can be decomposed as

$$\tau_{ij}^R \equiv L_{ij} + C_{ij} + R_{ij}, \tag{13.105}$$

where the *Leonard stresses* are

$$L_{ij} \equiv \overline{\overline{U_i}\,\overline{U_j}} - \overline{U_i}\,\overline{U_j}, \tag{13.106}$$

the *cross stresses* are

$$C_{ij} \equiv \overline{\overline{U_i}u_j'} + \overline{u_i'\overline{U_j}}, \tag{13.107}$$

and the SGS *Reynolds stresses* are

$$R_{ij} \equiv \overline{u_i'u_j'}. \tag{13.108}$$

(Note that, although the same names are used, these three stresses are different than those defined by Eqs. (13.100)–(13.102).) Show that τ_{ij}^{R} and τ_{ij}^{κ} differ by the Leonard stress, i.e.,

$$\tau_{ij}^{\mathrm{R}} - \tau_{ij}^{\kappa} = L_{ij}. \tag{13.109}$$

13.19 In order to consider the Galilean invariance of the various stresses in the original Leonard decomposition, Eq. (13.105), consider the transformed velocity field

$$\boldsymbol{W}(\boldsymbol{x},t) \equiv \boldsymbol{U}(\boldsymbol{x},t) + \boldsymbol{V}, \tag{13.110}$$

where $\boldsymbol{V}$ is a constant velocity difference. Obtain the result

$$\overline{W_i W_j} - \overline{W}_i\,\overline{W}_j = \overline{U_i U_j} - \overline{U}_i\,\overline{U}_j, \tag{13.111}$$

showing that the residual stress τ_{ij}^{R} is Galilean invariant; and also

$$\overline{\overline{W}_i\overline{W}_j} - \overline{W}_i\,\overline{W}_j = \overline{\overline{U}_i\overline{U}_j} - \overline{U}_i\,\overline{U}_j - V_i\overline{u_j'} - V_j\overline{u_i'}, \tag{13.112}$$

showing that (in general) the Leonard stresses are not Galilean-invariant. For the other stresses in Eq. (13.105), show that R_{ij} is Galilean invariant whereas (in general) C_{ij} is not.

13.20 Verify the validity of Germano's decomposition, Eq. (13.99). Show that the Leonard stresses, cross stresses and SGS Reynolds stresses defined by Eqs. (13.100)–(13.102) are Galilean invariant. Show that, if the filter is a projection, then Germano's decomposition (Eqs. (13.99)–(13.102)) is identical to Leonard's (Eqs. (13.105)–(13.108)).

13.21 For a general filter $G(\boldsymbol{r}, \boldsymbol{x})$ satisfying the normalization condition Eq. (13.2), the *filtered density function* (Pope 1990) is defined by

$$\bar{f}(\boldsymbol{V}; \boldsymbol{x}, t) \equiv \int G(\boldsymbol{r}, \boldsymbol{x})\,\delta[\boldsymbol{U}(\boldsymbol{x} - \boldsymbol{r}, t) - \boldsymbol{V}]\,\mathrm{d}\boldsymbol{r}. \tag{13.113}$$

Obtain the results

$$\overline{\boldsymbol{U}} = \int \boldsymbol{V}\bar{f}(\boldsymbol{V}; \boldsymbol{x}, t)\,\mathrm{d}\boldsymbol{V}, \tag{13.114}$$

$$\overline{U_i U_j} = \int V_i V_j \overline{f}(V; x, t) \, dV, \qquad (13.115)$$

$$\tau_{ij}^R = \int (V_i - \overline{U}_i)(V_j - \overline{U}_j) \overline{f}(V; x, t) \, dV, \qquad (13.116)$$

where integration is over all V; and $\overline{U}$, $\overline{U_i U_j}$, and τ_{ij}^R are evaluated at x, t. Show that $\overline{f}$ satisfies the normalization condition Eq. (12.1) and that, if the filter is everywhere non-negative, then $\overline{f}$ is also non-negative, and hence has the properties of a joint PDF. Argue that, for such positive filters, the residual stress τ_{ij}^R is positive semi-definite (Gao and O'Brien 1993). Use similar reasoning to show that $\mathcal{L}_{ij}^o$ and $\mathcal{R}_{ij}^o$ (Eqs. (13.100) and (13.102)) are also positive semi-definite for positive filters.

13.3.3 Conservation of energy

An important issue is the transfer of kinetic energy between the filtered velocity field and the residual motions. The filtered kinetic energy $\overline{E}(x, t)$ is obtained by filtering the kinetic-energy field $E(x, t) \equiv \frac{1}{2} U \cdot U$, i.e.,

$$\overline{E} \equiv \frac{1}{2} \overline{U \cdot U}. \qquad (13.117)$$

This can be decomposed as

$$\overline{E} = E_f + k_r, \qquad (13.118)$$

where

$$E_f \equiv \frac{1}{2} \overline{U} \cdot \overline{U}, \qquad (13.119)$$

is the kinetic energy of the filtered velocity field, and k_r is the residual kinetic energy

$$k_r \equiv \frac{1}{2} \overline{U \cdot U} - \frac{1}{2} \overline{U} \cdot \overline{U} = \frac{1}{2} \tau_{ii}^R. \qquad (13.120)$$

The conservation equation for E_f is obtained by multiplying Eq. (13.95) by $\overline{U}_j$. The result can be written

$$\frac{\overline{D} E_f}{\overline{D} t} - \frac{\partial}{\partial x_i} \left[\overline{U}_j \left(2 \nu \overline{S}_{ij} - \tau_{ij}^r - \frac{\overline{p}}{\rho} \delta_{ij} \right) \right] = -\varepsilon_f - \mathcal{P}_r, \qquad (13.121)$$

where ε_f and $\mathcal{P}_r$ are defined by

$$\varepsilon_f \equiv 2 \nu \overline{S}_{ij} \overline{S}_{ij}, \qquad (13.122)$$

$$\mathcal{P}_r \equiv -\tau_{ij}^r \overline{S}_{ij}. \qquad (13.123)$$

The terms on the left-hand side of Eq. (13.121) represent transport; but of most interest are the source or sink terms on the right-hand side. The sink $-\varepsilon_f$ represents viscous dissipation directly from the filtered velocity field. For a high-Reynolds-number flow, with the filter width much larger than the Kolmogorov scale, this term is relatively small (see Exercise 13.23).

The final term in Eq. (13.121), $\mathcal{P}_r$, is the rate of production of residual kinetic energy. The term appears as a sink $(-\mathcal{P}_r)$ in the equation for E_f and as a source $(+\mathcal{P}_r)$ in the equation for k_r. It represents, therefore, the rate of transfer of energy from the filtered motions to the residual motions. Sometimes $\mathcal{P}_r$ is called the SGS dissipation and denoted by ε_s. This terminology is inappropriate because, unlike true dissipation, $\mathcal{P}_r$ is due entirely to inviscid, inertial processes, and it can be negative.

At high Reynolds number, with the filter in the inertial subrange, the filtered velocity field accounts for nearly all of the kinetic energy, i.e.,

$$\langle E_f \rangle \approx \langle E \rangle. \tag{13.124}$$

The dominant sink in the equation for $\langle E_f \rangle$ is $\langle \mathcal{P}_r \rangle$, whereas that in the equation for $\langle E \rangle$ (Eq. (5.125) on page 124) is the rate of dissipation of kinetic energy, ε. Consequently, in the circumstances considered, these two quantities are nearly equal:

$$\langle \mathcal{P}_r \rangle \approx \varepsilon. \tag{13.125}$$

An equivalent view of this result (which is due to Lilly (1967)) is that, in the equation for the mean residual kinetic energy $\langle k_r \rangle$, there is a close balance between production $\langle \mathcal{P}_r \rangle$ and dissipation ε.

While, in the mean, energy is transferred from the large scales, $\langle \mathcal{P}_r \rangle > 0$, nevertheless, locally there can be *backscatter* – i.e., transfer of energy from the residual motions to the filtered velocity field, $\mathcal{P}_r < 0$. This is discussed in Section 13.5.

EXERCISES

13.22 For isotropic turbulence, verify that $\langle \overline{E} \rangle$, $\langle E_f \rangle$, $\langle k_r \rangle$, and $\langle \varepsilon_f \rangle$ are given by integrals over all κ of the spectrum $E(\kappa)$ multiplied by 1, $\widehat{G}(\kappa)^2$, $1 - \widehat{G}(\kappa)^2$, and $2\nu\kappa^2\widehat{G}(\kappa)^2$, respectively.

13.23 Consider LES of a high-Reynolds-number flow with the filter width Δ less than ℓ_{EI}. Use the results of Exercise 13.15 to estimate that

$$\frac{\langle \varepsilon_f \rangle}{\varepsilon} \approx \left(1 + \frac{\Delta}{8\eta}\right)^{-4/3}. \tag{13.126}$$

13.4 The Smagorinsky model

In order to close the equations for the filtered velocity, a model for the anisotropic residual stress tensor τ_{ij}^r is needed. The simplest model is that proposed by Smagorinsky (1963), which also forms the basis for several of the more advanced models described in Section 13.6.

13.4.1 The definition of the model

The model can be viewed in two parts. First, the linear eddy-viscosity model

$$\tau_{ij}^r = -2v_r \overline{S}_{ij}, \tag{13.127}$$

is used to relate the residual stress to the filtered rate of strain. The coefficient of proportionality $v_r(\boldsymbol{x}, t)$ is the eddy viscosity of the residual motions. Second, by analogy to the mixing-length hypothesis (Eq. (10.20)), the eddy viscosity is modelled as

$$\begin{aligned} v_r &= \ell_S^2 \overline{S} \\ &= (C_S \Delta)^2 \overline{S}, \end{aligned} \tag{13.128}$$

where $\overline{S}$ is the characteristic filtered rate of strain (Eq. (13.74)), ℓ_S is the *Smagorinsky lengthscale* (analogous to the mixing length) which, through the *Smagorinsky coefficient* C_S, is taken to be proportional to the filter width Δ.

According to the eddy-viscosity model (Eq. (13.127)), the rate of transfer of energy to the residual motions is

$$\mathcal{P}_r \equiv -\tau_{ij}^r \overline{S}_{ij} = 2v_r \overline{S}_{ij} \overline{S}_{ij} = v_r \overline{S}^2. \tag{13.129}$$

For the Smagorinsky model (or for any other eddy-viscosity model with $v_r > 0$), this energy transfer is everywhere from the filtered motions to the residual motions: there is no backscatter.

13.4.2 Behavior in the inertial subrange

It is informative to study the Smagorinsky model applied to high-Reynolds-number turbulence, with the filter width Δ in the inertial subrange (i.e., $\ell_{EI} > \Delta > \ell_{DI}$). For this case, in the mean, the transfer of energy to the residual motions $\langle \mathcal{P}_r \rangle$ is balanced by the dissipation ε:

$$\varepsilon = \langle \mathcal{P}_r \rangle = \langle v_r \overline{S}^2 \rangle = \ell_S^2 \langle \overline{S}^3 \rangle. \tag{13.130}$$

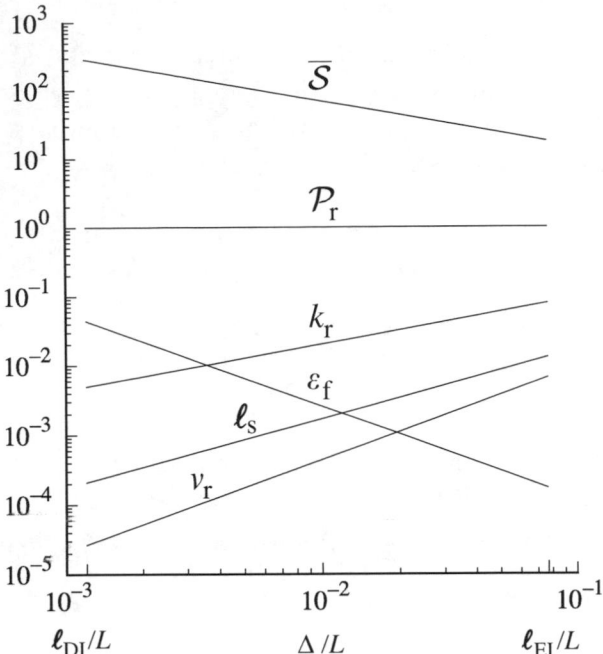

Fig. 13.7. Estimates of normalized filtered and residual quantities as functions of the filter width Δ for the sharp spectral filter in the inertial subrange of high-Reynolds-number turbulence. The normalizations and estimates are given in Table 13.3.

Using the estimate of $\langle \overline{S}^2 \rangle$ obtained from the Kolmogorov spectrum (Eq. (13.76)), this equation can be solved for the Smagorinsky length:

$$\ell_S = \frac{\Delta}{(Ca_f)^{3/4}} \left(\frac{\langle \overline{S}^3 \rangle}{\langle \overline{S}^2 \rangle^{3/2}} \right)^{-1/2}, \tag{13.131}$$

where C is the Kolmogorov constant, and a_f is the filter-dependent constant defined by Eq. (13.77). This analysis originated with Lilly (1967), who used the sharp spectral filter (see Eq. (13.78)) and the approximation $\langle \overline{S}^3 \rangle \approx \langle \overline{S}^2 \rangle^{3/2}$ to obtain the result

$$C_S = \frac{\ell_S}{\Delta} = \frac{1}{\pi} \left(\frac{2}{3C} \right)^{3/4} \approx 0.17. \tag{13.132}$$

Thus, the supposition of the Smagorinsky model that ℓ_S scales with Δ is confirmed for Δ in the inertial subrange.

From the scalings $\ell_S \sim \Delta$ and $\overline{S} \sim \varepsilon^{1/3}\Delta^{-2/3}$, the scalings of other quantities in the inertial subrange are readily deduced. For example, the eddy viscosity scales as $\nu_r \sim \ell_S^2 \overline{S} \sim \varepsilon^{1/3}\Delta^{4/3}$ and the residual stresses as $\tau_{ij}^r \sim \ell_S^2 \overline{S}^2 \sim \varepsilon^{2/3}\Delta^{2/3}$.

Table 13.3. *Estimates of filtered and residual quantities for the sharp spectral filter in the inertial subrange of high-Reynolds-number turbulence. Quantities are normalized by k, ε, and $L \equiv k^{3/2}/\varepsilon$. The estimates of the last three quantities are based on the Smagorinsky model*

Quantity	Normalized quantity	Estimate of normalized quantity	Equation used
Residual kinetic energy	$\dfrac{\langle k_r \rangle}{k}$	$\frac{3}{2}C\left(\dfrac{\Delta}{\pi L}\right)^{2/3}$	13.97
Rate of transfer of energy to residual motions	$\dfrac{\langle \mathcal{P}_r \rangle}{\varepsilon}$	1	13.125
Dissipation from filtered motions	$\dfrac{\langle \varepsilon_f \rangle}{\varepsilon}$	$\frac{3}{2}\pi^{4/3}C\left(\dfrac{\Delta}{\eta}\right)^{-4/3}$	13.122
Filtered rate of strain	$\dfrac{\langle \bar{S}^2 \rangle^{1/2} k}{\varepsilon}$	$\pi^{2/3}\left(\frac{3}{2}C\right)^{1/2}\left(\dfrac{\Delta}{L}\right)^{-2/3}$	13.76 13.78
Residual stress	$\dfrac{\langle \tau_{ij}^r \tau_{ij}^r \rangle^{1/2}}{k}$	$\dfrac{2}{(3C)^{1/2}}\left(\dfrac{\Delta}{\pi L}\right)^{2/3}$	13.127 13.128
Residual eddy viscosity	$\dfrac{\langle v_r \rangle \varepsilon}{k^2}$	$\dfrac{2}{3\pi^{2/3}}C\left(\dfrac{\Delta}{L}\right)^{4/3}$	13.129
Smagorinsky lengthscale	$\dfrac{\ell_S}{L}$	$\dfrac{1}{\pi}\left(\dfrac{2}{3C}\right)^{3/4}\dfrac{\Delta}{L}$	13.132

A summary of these scalings is given in Table 13.3 and they are depicted in Fig. 13.7.

EXERCISES

13.24 Suppose that $\bar{S}$ is log-normally distributed with $\mathrm{var}[\ln(\bar{S}^2/\langle \bar{S}^2 \rangle)] = \sigma^2$. Obtain the result

$$\frac{\langle \bar{S}^3 \rangle}{\langle \bar{S}^2 \rangle^{3/2}} = \exp(\tfrac{3}{8}\sigma^2). \qquad (13.133)$$

Show that, if this estimate is used with $\sigma^2 = 1$, then the Smagorinsky constant given by Lilly's analysis is reduced by about 20%. Discuss the dependence of σ^2 on Δ/L, and hence the possibility of a weak dependence of C_S on Δ/L due to internal intermittency (Novikov 1990).

13.25 Show that the Lilly analysis based on the Pao filter (see Table 13.2) yields

$$C_S = \pi^{1/2}(18C)^{-3/4} \approx 0.15. \qquad (13.134)$$

13.4.3 The Smagorinsky filter

With a specification of the Smagorinsky lengthscale ℓ_S, the LES equations using the Smagorinsky model can be solved to determine the filtered velocity field $\overline{U}(x,t)$. This can be done without the explicit specification of the filter: for the filter appears neither in the LES equations (Eqs. (13.87) and (13.95)) nor in the Smagorinsky model with the residual viscosity specified as $\nu_r = \ell_S^2 \overline{S}$.

For homogeneous isotropic turbulence there is a unique implied filter – the *Smagorinsky filter* – for which the calculated spectra are consistent (Pope 1998a). Suppose that the isotropic turbulence is computed using DNS, which yields (at a particular time) the energy-spectrum function $E(\kappa)$; and the same flow is computed using LES (with the Smagorinsky model and specified ℓ_S), which yields the energy spectrum of the filtered field $\overline{E}(\kappa)$. Thus, in principle (and in practice at moderate Reynolds numbers) the spectra $E(\kappa)$ and $\overline{E}(\kappa)$ can be computed. These spectra are related by the filter-transfer function $\widehat{G}(\kappa)$ according to Eq. (13.62), which can be inverted to yield

$$\widehat{G}(\kappa) = \left(\frac{\overline{E}(\kappa)}{E(\kappa)} \right)^{1/2}. \qquad (13.135)$$

This is the transfer function of the Smagorinsky filter: only with this filter does the filtered DNS velocity field have the same spectrum as that of the filtered velocity field obtained from LES. (While with this filter the Smagorinsky model produces the correct spectrum, there is no implication that it is accurate with respect to other statistics.)

The Smagorinsky filter can be roughly estimated for high-Reynolds-number isotropic turbulence and for ℓ_S smaller than the integral lengthscale ($\ell_S/L_{11} \ll 1$). This is accomplished by estimating $E(\kappa)$ and $\overline{E}(\kappa)$. For $E(\kappa)$ we assume the model spectrum with Pao's form (Eq. (6.254)) for the dissipation range, i.e.,

$$E(\kappa) = f_L(\kappa L) C \varepsilon^{2/3} \kappa^{-5/3} \exp\left[-\tfrac{3}{2} C (\kappa \eta)^{4/3}\right]. \qquad (13.136)$$

(Pao's form, rather than Eq. (6.248), is used to simplify the algebra.)

In order to estimate $\overline{E}(\kappa)$, in the Smagorinsky model $\overline{S}$ is approximated by $\langle \overline{S}^2 \rangle^{1/2}$. Then, the residual viscosity

$$v_r = \ell_S^2 \langle \overline{S}^2 \rangle^{1/2}, \tag{13.137}$$

is non-random and uniform, and the LES momentum equation (Eq. (13.95)) becomes

$$\frac{\overline{D}\,\overline{U}_j}{\overline{D}t} = (v + v_r)\frac{\partial^2 \overline{U}_j}{\partial x_i\,\partial x_i} - \frac{1}{\rho}\frac{\partial \overline{p}}{\partial x_j}. \tag{13.138}$$

This is identical to the Navier–Stokes equation except that the effective viscosity is $v + v_r$, so that the LES solution of Eq. (13.138) is the same as the DNS solution of the Navier–Stokes equations at a Reynolds number – lower by a factor of $(1 + v_r/v)$. The effective Kolmogorov scale is defined by

$$\overline{\eta} \equiv \left(\frac{(v + v_r)^3}{\varepsilon}\right)^{1/4} = \ell_S\left(1 + \frac{v}{v_r}\right)^{1/2}, \tag{13.139}$$

(see Exercise 13.26). Provided that $\overline{\eta}$ is small compared with the integral scale, it is consistent with Eq. (13.136) to suppose that the LES yields the spectrum

$$\overline{E}(\kappa) = f_L(\kappa L)C\varepsilon^{2/3}\kappa^{-5/3}\exp[-\tfrac{3}{2}C(\kappa\overline{\eta})^{4/3}]. \tag{13.140}$$

With these estimates of $E(\kappa)$ and $\overline{E}(\kappa)$, the transfer function of the Smagorinsky filter deduced from Eq. (13.135) is

$$\widehat{G}(\kappa) = \exp[-\tfrac{3}{4}C\kappa^{4/3}(\overline{\eta}^{4/3} - \eta^{4/3})]. \tag{13.141}$$

This corresponds to the Pao filter defined in Table 13.2, with filter width $\Delta = \ell_S/C_S$, where

$$C_S = \frac{\pi^{1/2}}{(18C)^{3/4}}\left[1 + \frac{18C}{\pi^{2/3}}\left(\frac{\eta}{\Delta}\right)^{4/3}\right]^{1/4}$$

$$\approx 0.15\left[1 + \left(\frac{7\eta}{\Delta}\right)^{4/3}\right]^{1/4}, \tag{13.142}$$

see Exercise 13.26.

The important point is that there is a unique filter given by Eq. (13.135). The particular filter deduced from the preceding argument is an approximation (because fluctuations in $\overline{S}$ are neglected), and it depends on the form assumed for the dissipation spectrum. As shown here, the Pao spectrum leads to the Pao filter, whereas the exponential spectrum $\exp(-\beta_0\kappa\eta)$ leads to the Cauchy filter, and the model spectrum leads to the filter Eq. (13.150) deduced in Exercise 13.27. These filter-transfer functions are shown in Fig. 13.8, and the corresponding filter functions (obtained as inverse Fourier transforms)

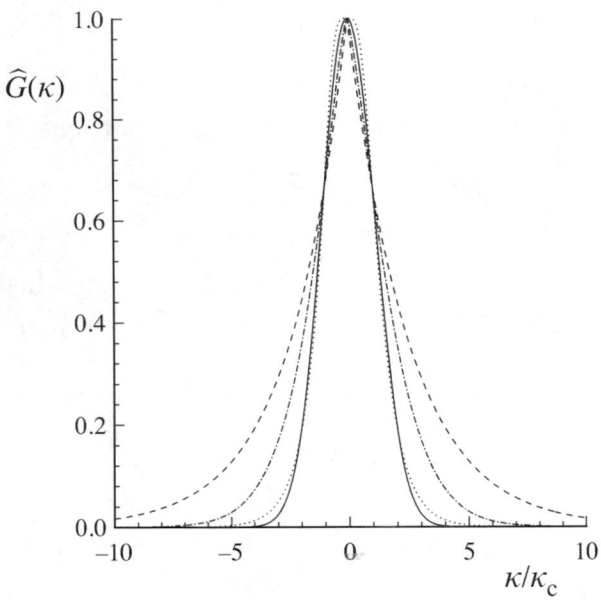

Fig. 13.8. Filter-transfer functions: Gaussian, solid line; Cauchy, dashed line; Pao, dot–dashed line; implied by model spectrum (Eq. (13.150)), dotted line.

are shown in Fig. 13.9. The model spectrum provides the most accurate representation of the dissipation spectrum (see Fig. 6.15 on page 236), and so the filter deduced from it is the best estimate of the Smagorinsky filter. It is interesting to observe from Figs. 13.8 and 13.9 that, in many respects, this filter is quite similar to the Gaussian filter.

(In this discussion it is implicitly assumed that the LES equations are solved accurately. It is common practice, however, to perform LES of isotropic turbulence using the sharp spectral filter with cutoff κ_c equal to the largest represented wavenumber κ_{max} (see, e.g., Meneveau, Lund, and Cabot (1996)). Such simulations are under-resolved since the true spectrum extends beyond κ_{max}. As a consequence, the computed spectrum exhibits a cusp at κ_{max} rather than a dissipation range of the form of Eq. (13.140).)

EXERCISES

13.26 Show that $\widehat{G}(\kappa)$ given by Eq. (13.141) is the transfer function of the Pao filter with

$$\Delta^{4/3} = 18C\pi^{-2/3}(\overline{\eta}^{4/3} - \eta^{4/3}) = \frac{18\nu_r}{\pi^{2/3}\varepsilon^{1/3}}. \qquad (13.143)$$

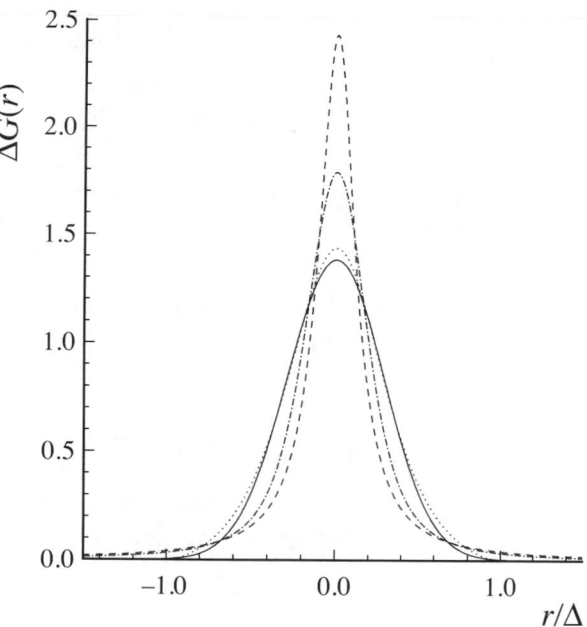

Fig. 13.9. Filter functions: Gaussian, solid line; Cauchy, dashed line; Pao, dot–dashed line; implied by model spectrum (Eq. (13.150)), dotted line.

Show that the value of $\langle \overline{\mathcal{S}}^2 \rangle$ obtained from Eq. (13.140) is

$$\langle \overline{\mathcal{S}}^2 \rangle = \varepsilon^{2/3} / \overline{\eta}^{4/3}, \tag{13.144}$$

and hence obtain the result

$$\varepsilon = (\nu + \nu_{\mathrm{r}}) \langle \overline{\mathcal{S}}^2 \rangle. \tag{13.145}$$

Starting from Eq. (13.137) with $\ell_{\mathrm{S}} = C_{\mathrm{S}} \Delta$, i.e.,

$$\nu_{\mathrm{r}} = (C_{\mathrm{S}} \Delta)^2 \langle \overline{\mathcal{S}}^2 \rangle^{1/2}, \tag{13.146}$$

use Eqs. (13.143) and (13.145) to eliminate ν_{r} and $\langle \overline{\mathcal{S}}^2 \rangle$, and hence to obtain

$$C_{\mathrm{S}}^2 = \gamma \left[\gamma + \left(\frac{\eta}{\Delta} \right)^{4/3} \right]^{1/2}, \tag{13.147}$$

where γ is defined by

$$\gamma \equiv \frac{\pi^{2/3}}{18C}. \tag{13.148}$$

Hence verify Eq. (13.142). From the definition of $\overline{\eta}$ (Eq. (13.139)), from Eq. (13.145), and from $v_r = \ell_S^2 \langle \overline{S}^2 \rangle^{1/2}$, obtain the result

$$\overline{\eta} = \ell_S \left(1 + \frac{v}{v_r} \right)^{1/2}. \tag{13.149}$$

13.27 Consider LES of homogeneous isotropic turbulence using the Smagorinsky model with the filter in the inertial subrange so that ℓ_S is much larger than the Kolmogorov scale ($\ell_S \approx \overline{\eta} \gg \eta$). Show that, if $\overline{E}(\kappa)$ is approximated by the model spectrum (Eqs. (6.246) and (6.248)), then the implied Smagorinsky filter has the transfer function

$$\widehat{G}(\kappa) = \exp\{-\tfrac{1}{2}\beta\{[(\kappa\ell_S)^4 + c_\eta^4]^{1/4} - c_\eta\}\}. \tag{13.150}$$

If the filter is scaled such that $\widehat{G}(\kappa_c) = \exp(-\pi^2/24)$ (as it is for the Gaussian, Cauchy, and Pao filters), show that the implied value of the Smagorinsky constant is

$$C_S = \frac{c_\eta}{\pi} \left[\left(1 + \frac{\pi^2}{12\beta c_\eta} \right)^4 - 1 \right]^{1/4} \approx 0.16, \tag{13.151}$$

for $\beta = 5.2$ and $c_\eta = 0.4$.

13.4.4 Limiting behaviors

An ideal application of LES is to high-Reynolds-number turbulent flow with the filter in the inertial subrange. However, in practice, departures from this ideal are the norm; and some insights are gained by considering extreme departures. Specifically we consider the three cases of the filter being in the dissipative range; the filter being large compared with the integral scale; and laminar flow.

A very small filter width

If the filter is in the far dissipation range (i.e., $\Delta/\eta \ll 1$), a Taylor-series analysis can be performed to show that the residual-stress tensor is (to leading order)

$$\tau_{ij}^R \equiv \overline{U_i U_j} - \overline{U}_i \overline{U}_j = \frac{\Delta^2}{12} \frac{\partial \overline{U}_i}{\partial x_k} \frac{\partial \overline{U}_j}{\partial x_k} \tag{13.152}$$

(see Exercise 13.28). This result applies to filters with finite moments such as the Gaussian, but not to the sharp spectral filter. Like the Smagorinsky model, this expression for τ_{ij}^{R} is quadratic in Δ and in the filtered velocity gradients, but the tensorial form is different.

Despite this difference, a value of the Smagorinsky coefficient can be determined such that the mean rate of transfer of energy to the residual scales $\langle \mathcal{P}_r \rangle$ is consistent with Eq. (13.152) (see Exercise 13.28 and Eq. (13.160)). For high-Reynolds-number turbulence, the resulting estimate of $C_S \approx 0.13$ is somewhat lower than the inertial-range value of $C_S \approx 0.17$. Presumably, around $\Delta/\eta = 1$, there is a transition of the Smagorinsky coefficient between the constant values in the inertial and far-dissipation ranges.

In the limit considered ($\Delta/\eta \to 0$), the filtered rate of strain $\overline{S}_{ij}$ tends to S_{ij}, and hence $\langle \overline{S}^2 \rangle$ tends to $\langle S_{ij} S_{ij} \rangle \approx \langle s_{ij} s_{ij} \rangle = \varepsilon/\nu$. Thus, the Smagorinsky model yields

$$\frac{\langle v_r^2 \rangle^{1/2}}{\nu} = \left(\frac{\ell_S}{\eta}\right)^2 = C_S^2 \left(\frac{\Delta}{\eta}\right)^2, \qquad (13.153)$$

in the far-dissipation range – a more rapid decrease with Δ than $v_r \sim \Delta^{4/3}$ in the inertial subrange.

In contrast to the result obtained here, the analysis in Section 13.4.3 predicts that C_S increases with decreasing Δ/η (Eqs. (13.142) and (13.147)). On the other hand, Lilly-type analyses (using the sharp spectral filter) performed by Voke (1996) and Meneveau and Lund (1997) suggest that C_S decreases with Δ/η. These differences can be attributed, in part, to the fact that the Smagorinsky model provides a poor description of the residual stresses at a detailed level – as discussed below. As a consequence, different results can be obtained when different aspects of the model's behavior are used to estimate C_S. From a practical viewpoint, the uncertainty in this limiting behavior of the Smagorinsky model is not of great concern since, for small Δ/η, the viscous stresses dominate the residual stresses.

EXERCISE

13.28   Consider an isotropic filter (such as the Gaussian) which has the second moments

$$\int r_i r_j G(r)\, dr = \frac{\Delta^2}{12}\delta_{ij}, \qquad (13.154)$$

and finite higher moments. This is applied to a turbulent flow, with the filter width Δ being much smaller than the Kolmogorov scale η. For given x, and for r on the order of Δ (i.e., $|r| \ll \eta$), the velocity

$U(x+r)$ can be expanded in the Taylor series

$$U_i(x+r) = U_i(x) + \frac{\partial U_i}{\partial x_k} r_k + \frac{1}{2!} \frac{\partial^2 U_i}{\partial x_k \, \partial x_\ell} r_k r_\ell \dots, \qquad (13.155)$$

where the velocity gradients are evaluated at x. By multiplying Eq. (13.155) by $G(r)$ and integrating, show that the filtered velocity field is (to leading order)

$$\overline{U}_i(x) = U_i(x) + \frac{\Delta^2}{24} \frac{\partial^2 U_i}{\partial x_k \, \partial x_k} + \mathcal{O}(\Delta^4). \qquad (13.156)$$

Follow a similar procedure to show that the residual-stress tensor is

$$\tau_{ij}^{\mathrm{R}} \equiv \overline{U_i U_j} - \overline{U}_i \overline{U}_j = \frac{\Delta^2}{12} \frac{\partial \overline{U}_i}{\partial x_k} \frac{\partial \overline{U}_j}{\partial x_k} + \mathcal{O}(\Delta^4). \qquad (13.157)$$

Show that the mean rate of transfer of energy to the residual motions is

$$\langle \mathcal{P}_\mathrm{r} \rangle = - \left\langle \frac{\partial \overline{U}_i}{\partial x_j} \tau_{ij}^{\mathrm{R}} \right\rangle = -\frac{\Delta^2}{12} \left\langle \frac{\partial \overline{U}_i}{\partial x_j} \frac{\partial \overline{U}_i}{\partial x_k} \frac{\partial \overline{U}_j}{\partial x_k} \right\rangle + \mathcal{O}(\Delta^4). \qquad (13.158)$$

Use the results of Exercise 6.11 on page 205 to show that, in high-Reynolds-number, locally isotropic turbulence (and for $\Delta/\eta \ll 1$),

$$\langle \mathcal{P}_\mathrm{r} \rangle = \frac{7}{72\sqrt{15}} \Delta^2 (-S) \left(\frac{\varepsilon}{\nu} \right)^{3/2}, \qquad (13.159)$$

where S is the velocity-derivative skewness (which is negative). By comparing this result with Eq. (13.130), show that it is consistent with the Smagorinsky model with

$$c_\mathrm{S}^2 = \frac{7}{72\sqrt{15}} (-S) \frac{\langle \overline{\mathcal{S}}^2 \rangle^{3/2}}{\langle \overline{\mathcal{S}}^3 \rangle}. \qquad (13.160)$$

Take $\langle \overline{\mathcal{S}}^3 \rangle \approx \langle \overline{\mathcal{S}}^2 \rangle^{3/2}$ and $S \approx -0.7$ (see Fig. 6.33 on page 262) to obtain

$$C_\mathrm{S} \approx 0.13. \qquad (13.161)$$

A very large filter width

The opposite extreme, of the filter width Δ being large compared with the turbulence integral scale L, can be analysed for the case of homogeneous turbulence. In the limit as Δ/L tends to infinity, the filtered velocity $\overline{U}(x,t)$ tends to the mean $\langle U(x,t) \rangle$, so the residual velocity $u'(x,t)$ tends to the fluctuation $u(x, t)$. Consequently, the residual stress tensor τ_{ij}^{R} tends to the Reynolds-stress tensor $\langle u_i u_j \rangle$. For homogeneous turbulent shear flow, the residual shear stress given by the Smagorinsky model is

$$\tau_{12}^{R} = -v_r \frac{\partial \overline{U}_1}{\partial x_2} = -\ell_S^2 \left| \frac{\partial \overline{U}_1}{\partial x_2} \right| \frac{\partial \overline{U}_1}{\partial x_2}, \tag{13.162}$$

whereas the Reynolds shear stress given by the mixing-length model is

$$\langle u_1 u_2 \rangle = -v_T \frac{\partial \langle U_1 \rangle}{\partial x_2} = -\ell_m^2 \left| \frac{\partial \langle U_1 \rangle}{\partial x_2} \right| \frac{\partial \langle U_1 \rangle}{\partial x_2}. \tag{13.163}$$

Evidently, in the limit considered ($\Delta/L \to \infty$), the residual eddy viscosity v_r is the turbulent viscosity v_T, and the Smagorinsky length ℓ_S is the mixing length ℓ_m. The mixing length can be evaluated as

$$\ell_m = |\langle u_1 u_2 \rangle|^{1/2} / \left| \frac{\partial \langle U_1 \rangle}{\partial x_2} \right|, \tag{13.164}$$

and it is, of course, independent of Δ. Hence, as Δ/L tends to infinity, the Smagorinsky coefficient tends to zero as

$$C_S = \frac{\ell_m}{\Delta}. \tag{13.165}$$

Laminar flow

For laminar flow, the Reynolds equations revert to the Navier–Stokes equations, and the Reynolds stresses are zero. In contrast, in general, the residual-stress tensor is non-zero in laminar flow. For filter widths Δ that are small compared with the lengthscales of the laminar velocity field, the Taylor-series analysis of Exercise 13.28 is valid, and it leads to Eq. (13.157) for the residual stress tensor τ_{ij}^{R}.

This general result notwithstanding, for several important flows the residual stresses are essentially zero. Consider, for example, unidirectional flow in which the sole component of velocity, U_1, depends only on x_2 and x_3. It is evident from Eq. (13.157) that the only non-zero residual stress is τ_{11}^{R}, whereas the shear stresses that effect momentum transport are zero. The same is true for two- and three-dimensional boundary-layer flows (to within the boundary-layer approximations).

For the laminar shear flows in which the residual shear stresses are zero, the appropriate value of the Smagorinsky coefficient is $C_S = 0$: a non-zero value of C_S would lead, incorrectly, to residual shear stresses on the order of Δ^2. Consequently, the Smagorinsky model with a constant non-zero value of C_S is incorrect for laminar flow. In some implementations (e.g., Schumann (1975) and Moin and Kim (1982)) the Smagorinsky model is based on $\overline{S}_{ij} - \langle \overline{S}_{ij} \rangle$ rather than on $\overline{S}_{ij}$, so that ν_r is zero in laminar flow. Note, however, that means such as $\langle \overline{S}_{ij} \rangle$ are not readily accessible in LES, but have to be estimated by calculating an appropriate type of average.

13.4.5 Near-wall resolution

As discussed in the introduction to the chapter, in LES there are two distinct ways to treat near-wall regions (see Table 13.1 on page 560). In LES-NWR (LES with near-wall resolution) the filter and grid are sufficiently fine to resolve 80% of the energy everywhere, including in the viscous wall region; whereas in LES-NWM (LES with near-wall modelling) the near-wall motions are not resolved. In this section, we examine the computational costs of these two approaches, and we consider the use of the Smagorinsky model in LES-NWR.

First, we recall from Chapter 7 that, in boundary-layer-type flows, the viscous wall region is very important: the production, dissipation, kinetic energy, and Reynolds-stress anisotropy all achieve their peak values at y^+ less than 20, i.e., within 20 viscous lengthscales δ_ν of the wall. Relative to the flow lengthscale δ, the viscous lengthscale δ_ν is small, and it decreases with the Reynolds number approximately as $\delta_\nu/\delta \sim \mathrm{Re}^{-0.88}$.

In LES-NWR, in order to resolve the near-wall motions, the filter width and grid spacing in the viscous near-wall region must be on the order of δ_ν. From this it can be estimated that the number of grid nodes required increases as $\mathrm{Re}^{1.76}$ (see Chapman (1979) and Exercise 13.29). As a consequence, LES-NWR is infeasible for high-Reynolds-number flows such as occur in aeronautical and meteorological applications.

In contrast, in LES-NWM, the filter width and grid spacing scale with the flow lengthscale δ, and consequently the computational requirements are independent of the Reynolds number (see Exercise 13.29). However, then the important near-wall processes are not resolved, and instead are modelled. Some of the near-wall treatments used in LES-NWM are described in Section 13.6.5.

In the near-wall region (just like elsewhere) the residual stresses τ_{ij}^R depend on the type and width of the filter. In LES of channel flow, the usual practice is to filter only in x–z planes. If this is done with a uniform filter (in particular with Δ independent of y) then it can be shown that, very close to the wall, the components of τ_{ij}^R vary with the same powers of y as do the

Reynolds stresses $\langle u_i u_j \rangle$ (see Exercise 13.30). In particular, the shear stress τ_{12}^R varies as y^3 (to leading order).

In the Smagorinsky model, the specification $\ell_S = C_S \Delta$ (for constant C_S) is justifiable for Δ in the inertial subrange of high-Reynolds-number turbulence. These circumstances do not pertain to the viscous wall region, and this specification of ℓ_S leads, incorrectly, to a non-zero residual viscosity and shear stress at the wall (see Exercise 13.31). Instead, Moin and Kim (1982) use a van Driest damping function to specify ℓ_S as

$$\ell_S = C_S \Delta [1 - \exp(-y^+/A^+)] \qquad (13.166)$$

(cf. Eq. (7.145)).

EXERCISES

13.29   The purpose of this exercise is to investigate the computational cost of LES-NWR and LES-NWM for wall-bounded flows, in particular to determine how the required number of grid points N_{xyz} scales with the Reynolds number. Consider LES of fully developed turbulent channel flow using four different grids.

(A)   Conventional LES-NWR using a structured grid with

$$\Delta x = a_x \delta_v, \quad \Delta z = a_z \delta_v, \qquad (13.167)$$

$$\Delta y = \min[\max(a\delta_v, by), c\delta], \qquad (13.168)$$

where a_x, a_z, a, b, and c are positive constants, with $b > c$. Note that the scaling of Δx and Δz with δ_v allows resolution of near-wall structures, but leads to over-resolution away from the wall.

(B)   Optimal-resolution LES-NWR using an irregular grid (Chapman 1979): Δy is again specified by Eq. (13.168), but Δx and Δz are taken to scale with Δy as

$$\Delta x = a_x \Delta y, \quad \Delta z = a_z \Delta y. \qquad (13.169)$$

(C)   LES-NWM with resolution of near-wall mean profiles: Δy is again specified by Eq. (13.168) so that the profiles of mean velocity, Reynolds stresses, etc. are resolved in the viscous near-wall region, but Δx and Δz are specified as

$$\Delta x = a_x \delta, \quad \Delta z = a_z \delta, \qquad (13.170)$$

so that, for fixed positive values of a_x and a_z (no matter how small), as the Reynolds number tends to infinity ($\delta_v/\delta \to 0$), the near-wall structures are not resolved.

(D) LES-NWM without resolution of near-wall mean profiles: Δx and Δz are specified by Eq. (13.170), and Δy is specified by

$$\Delta y = \min[\max(a\delta, by), c\delta], \tag{13.171}$$

for $a < c < b$.

For each of these four specifications, evaluate the approximate number of grid points as

$$N_{xyz} = \int_0^\delta \int_0^\delta \int_0^\delta \frac{dx\,dy\,dz}{\Delta x\,\Delta y\,\Delta z}, \tag{13.172}$$

expressing the result in terms of $\mathrm{Re}_\tau = \delta/\delta_v$. Using the empirical relation (for high Reynolds number) $\mathrm{Re}_\tau \approx 0.09\,\mathrm{Re}^{0.88}$, obtain the following scalings for the four grids:

$$N_{xyz}^A \sim \mathrm{Re}_\tau^2 \ln \mathrm{Re}_\tau \sim \mathrm{Re}^{1.76} \ln \mathrm{Re}, \tag{13.173}$$

$$N_{xyz}^B \sim \mathrm{Re}_\tau^2 \sim \mathrm{Re}^{1.76}, \tag{13.174}$$

$$N_{xyz}^C \sim \ln \mathrm{Re}_\tau \sim \ln \mathrm{Re}, \tag{13.175}$$

$$N_{xyz}^D \quad \text{independent of Re.} \tag{13.176}$$

(Note that the biggest difference is between B and C, which results from the different near-wall x–z resolutions, which scale with δ_v for B, and with δ for C.)

13.30 Consider turbulent channel flow in the usual coordinate system, so that $y = x_2$ is the normal distance from the wall. Very close to the wall, to leading order in y, the velocity field can be written

$$U_1(\boldsymbol{x}, t) = a(x, z, t)y, \tag{13.177}$$

$$U_2(\boldsymbol{x}, t) = b(x, z, t)y^2, \tag{13.178}$$

$$U_3(\boldsymbol{x}, t) = c(x, z, t)y, \tag{13.179}$$

where a, b, and c are random functions, cf. Eqs. (7.56)–(7.58). Filtering is applied in x–z planes only, using a uniform filter of width Δ. Show that the filtered velocity field is

$$\overline{U}_1 = \overline{a}y, \quad \overline{U}_2 = \overline{b}y^2, \quad \overline{U}_3 = \overline{c}y, \tag{13.180}$$

and that the residual stresses are

$$\tau_{11}^R = (\overline{a^2} - \overline{a}^2)\,y^2, \quad \tau_{22}^R = (\overline{b^2} - \overline{b}^2)\,y^4,$$
$$\tau_{12}^R = (\overline{ab} - \overline{a}\overline{b})\,y^3, \quad \text{etc.} \tag{13.181}$$

Confirm that (for the filter considered) the residual stresses τ_{ij}^{R} vary with the same powers of y as do the Reynolds stresses $\langle u_i u_j \rangle$.

Argue that, in an LES, since $\overline{U}$ is solenoidal and vanishes at the wall, its leading-order behavior is given by Eq. (13.180), whereas the behavior of τ_{ij}^{R} depends on the residual-stress model employed.

13.31 Further to Exercise 13.30, consider the performance of the Smagorinsky model in an LES of turbulent channel flow. Show that $\partial \overline{U}_1 / \partial x_2 = \overline{a}$ and $\partial \overline{U}_3 / \partial x_2 = \overline{c}$ are the only leading-order velocity gradients. Show that the modelled residual shear stress is

$$\tau_{12}^{r} = -\ell_{S}^2 (\overline{a}^2 + \overline{c}^2)^{1/2} \overline{a}, \qquad (13.182)$$

and hence the variation $\ell_{S} \sim y^{3/2}$ is required in order to produce the correct behavior. In contrast, show that the standard specification $\ell_{S} = C_{S}\Delta$ leads to ν_r and τ_{12}^{r} being independent of y (to leading order). By neglecting the fluctuations in $\overline{a}$ and $\overline{c}$, obtain the estimate

$$\frac{\nu_r}{\nu} \approx \frac{C_{S}\Delta}{\delta_{\nu}}, \qquad (13.183)$$

valid for $\Delta \gg \delta_{\nu} \equiv \nu / (\tau_w / \rho)^{1/2}$.

13.4.6 Tests of model performance

In general, a model can be tested in two rather different ways. In the context of LES, these two ways are known as *a priori* and *a posteriori* testing. An *a priori* test uses experimental or DNS data to measure directly the accuracy of a modelling assumption, for example, the relation for the residual-stress tensor τ_{ij}^{r} given by the Smagorinsky model (Eqs. (13.127) and (13.128)). In an *a posteriori* test, the model is used to perform a calculation for a turbulent flow, and the accuracy of calculated statistics (e.g., $\langle U \rangle$ and $\langle u_i u_j \rangle$) is assessed, again by reference to experimental or DNS data. It is natural and appropriate to perform *a priori* tests to assess directly the validity and accuracy of approximations being made. However, for the LES approach to be useful, it is success in *a posteriori* tests that is needed.

For homogeneous turbulence, *a priori* tests of the Smagorinsky model are reported by Clark, Ferziger, and Reynolds (1979) and McMillan and Ferziger (1979). For a 64^3 DNS with $R_\lambda \approx 38$, the filtered velocity field $\overline{U}(x, t)$ is extracted, and the Smagorinsky prediction of the residual stresses $\tau_{ij}^{r,\mathrm{Smag}}$ is determined from it. This is then compared with the residual stresses $\tau_{ij}^{r,\mathrm{DNS}}$ obtained directly from the DNS velocity field. It is found by McMillan and Ferziger (1979) that the correlation between $\tau_{ij}^{\mathrm{r,Smag}}$ and $\tau_{ij}^{\mathrm{r,DNS}}$ is greatest when the Smagorinsky coefficient is taken to be $C_{\mathrm{S}} = 0.17$, in agreement with the Lilly analysis. However, the correlation coefficient is quite small, around $\frac{1}{3}$. A higher correlation coefficient, around $\frac{1}{2}$, is found for the scalar quantity

$$\overline{U}_i \frac{\partial \tau_{ij}^{\mathrm{r}}}{\partial x_j} = -\frac{\partial \overline{U}_i}{\partial x_j} \tau_{ij}^{\mathrm{r}} + \frac{\partial}{\partial x_j} (\overline{U}_i \tau_{ij}^{\mathrm{r}}), \qquad (13.184)$$

which is related to the transfer of energy to the residual motions, $\mathcal{P}_{\mathrm{r}}$.

Subsequent to these early works, there have been numerous studies in which *a priori* testing of the Smagorinsky model has been performed, which support the conclusions drawn. For example, from experiments in a round jet at $\mathrm{R}_\lambda \approx 310$, Liu, Meneveau, and Katz (1994) show that the correlation coefficient between the Smagorinsky-model stresses and measured residual stresses is no greater than $\frac{1}{4}$. As discussed in Section 13.5.6, the significance of *a priori* testing (especially the measurement of correlation coefficients) is not as obvious as it may appear.

In *a posteriori* testing, statistics from an LES calculation are compared with those obtained from experiment or from DNS. For LES of a statistically stationary flow, long-time averaging can be performed to obtain estimates of the mean filtered velocity field $\langle \overline{U} \rangle$, and of the *resolved Reynolds stresses*

$$R_{ij}^{\mathrm{f}} \equiv \langle (\overline{U}_i - \langle \overline{U}_i \rangle)(\overline{U}_j - \langle \overline{U}_j \rangle) \rangle. \qquad (13.185)$$

In *a posteriori* testing, it is common for $\langle \overline{U} \rangle$ and R_{ij}^{f} from the LES to be compared with $\langle \overline{U} \rangle$ and $\langle u_i u_j \rangle$ from experiments or DNS. It should be recognized, however, that these quantities are not directly equivalent (see Exercise 13.32), although the difference between them decreases with decreasing filter width. DNS data can be filtered to obtain a direct comparison between filtered quantities such as R_{ij}^{f}.

For inhomogeneous flows, the general conclusion from *a posteriori* testing is that the Smagorinsky model is too dissipative – that is, it transfers too much energy to the residual motions. In channel-flow calculations, the Smagorinsky coefficient is generally decreased, e.g., to $C_{\mathrm{S}} = 0.1$ (Deardorff 1970, Piomelli, Moin, and Ferziger (1988)) or to $C_{\mathrm{S}} = 0.065$ (Moin and Kim 1982). Because the Smagorinsky model yields a spurious residual stress in laminar flow, the model is, again, much too dissipative in transitional flows. A detailed *a posteriori* study of the turbulent mixing layer by Vreman, Geurts, and Kuerten (1997) reveals similar deficiencies.

It is important to recognize that the performance of a residual-stress model depends on the Reynolds number and on the choice of the type and width of the filter. When LES is tested against DNS, inevitably the Reynolds numbers are low. Consequently there is considerable overlap between the energy-containing and dissipative ranges of scales, and the filter is likely to be placed in this overlap region: the residual viscosity v_r is typically about twice the molecular viscosity v. This situation is quite different than the ideal of a high-Reynolds-number flow, with the filter being placed toward the beginning of a substantial inertial subrange. For such an ideal application of LES to high-Reynolds-number free shear flow, there is every reason to suppose that the Smagorinsky model is satisfactory; for the resulting equations are similar to the Navier–Stokes equations at a lower Reynolds number, and the one-point statistics of free shear flows are known to be insensitive to the Reynolds number.

In LES-NWM of atmospheric boundary layers, the filter width and grid spacing scale with the boundary-layer thickness, and so, at the ground, they are large compared with the energy-containing turbulent motions. As in LES-NWR, ℓ_S is attenuated close to the wall (but in a different way). Mason and Thomson (1992) investigated various blendings of $C_S \Delta$ and the mixing length, ℓ_m, but concluded that the Smagorinsky model is inherently incapable of yielding the correct logarithmic velocity profiles.

In LES, a primary function of the residual-stress model is to remove energy from the resolved scales at the appropriate rate, this rate being $\mathcal{P}_r = -\tau_{ij}^r \bar{S}_{ij}$ (Eq. (13.123)). According to the Smagorinsky model, $-\tau_{ij}^r$ and $\bar{S}_{ij}$ are almost perfectly correlated (see Exercise 13.33), whereas *a priori* testing shows that in fact the correlation is substantially weaker. Consequently, as observed by Jiménez and Moser (1998), if the Smagorinsky coefficient is set to yield the appropriate value of $\langle \mathcal{P}_r \rangle$, then the magnitudes of the modelled residual stresses are too low. Close to a wall, if the filter width is not small compared with the energy-containing scales, then a substantial fraction of the shear stress arises from the residual-stress model. It is understandable, therefore, that in such circumstances the Smagorinsky model performs poorly, since no choice of C_S can produce the correct levels both of $\langle \mathcal{P}_r \rangle$ and of $\langle \tau_{12}^r \rangle$.

EXERCISES

13.32 Mean quantities $\langle Q(x,t) \rangle$ (e.g., $\langle U \rangle$ and $\langle u_i u_j \rangle$) vary on a lengthscale L comparable to that of the energy-containing motions. Argue that, for $\Delta \ll L$, such means are little affected by filtering, i.e.,

$$\langle Q \rangle \approx \overline{\langle Q \rangle} = \langle \bar{Q} \rangle. \tag{13.186}$$

Hence argue that, in LES of a statistically stationary flow, $\langle U \rangle$ can be accurately approximated as the long-time average of $\bar{U}$.

Write down an exact expression for the Reynolds stresses $\langle u_i u_j \rangle$

in terms of means of $\overline{U}$ and u'. Use Eq. (13.186) to obtain the approximation

$$\langle u_i u_j \rangle \approx R_{ij}^{\mathrm{f}} + \langle C_{ij} \rangle + \langle R_{ij} \rangle, \qquad (13.187)$$

where R_{ij}^{f} are the resolved Reynolds stresses, Eq. (13.185), and C_{ij} and R_{ij} are the cross stresses and SGS Reynolds stresses given by Eqs. (13.107) and (13.108), respectively.

13.33 Let ρ_{r} denote the correlation coefficient between the residual stresses and the filtered rate of strain:

$$\rho_{\mathrm{r}} \equiv \langle \tau_{ij}^{\mathrm{r}} \overline{S}_{ij} \rangle / (\langle \tau_{k\ell}^{\mathrm{r}} \tau_{k\ell}^{\mathrm{r}} \rangle \langle \overline{S}_{mn} \overline{S}_{mn} \rangle)^{1/2}. \qquad (13.188)$$

Show that the Smagorinsky model yields

$$\rho_{\mathrm{r}} = -\langle \overline{S}^3 \rangle / (\langle \overline{S}^2 \rangle \langle \overline{S}^4 \rangle)^{1/2}. \qquad (13.189)$$

By taking $\overline{S}$ to be log-normal with $\mathrm{var}[\ln(\overline{S}^2/\langle \overline{S}^2 \rangle)] = \sigma^2 = 1$, obtain the estimate (for the Smagorinsky model)

$$\rho_{\mathrm{r}} = -\exp(-\tfrac{1}{8}\sigma^2) \approx -0.88. \qquad (13.190)$$

13.5 LES in wavenumber space

Like DNS, LES is used as a research tool to study homogeneous turbulence. In this application, the modelling and numerical solution are usually performed in wavenumber space. The residual-stress models used in wavenumber space are different from those used in physical space, and they cannot be used for inhomogeneous turbulent flows. The rudiments of LES in wavenumber space are given here, since they provide a useful alternative viewpoint on several issues.

For the most part we consider the sharp spectral filter, which is the natural choice for LES in wavenumber space. However, in the final subsection, the Gaussian filter is considered in order to examine further the issues of resolution and modelling – which are qualitatively different for this type of filter.

13.5.1 Filtered equations

As in Section 6.4, we consider homogeneous turbulence with zero mean velocity, in which the velocity field is periodic (with period $\mathcal{L}$) in each of the three coordinate directions. Then the velocity field $u(x, t)$ has the Fourier series

$$u(x, t) = \sum_{\kappa} e^{i\kappa \cdot x} \hat{u}(\kappa, t), \tag{13.191}$$

where the wavenumber vectors κ are integer multiples of $\kappa_0 \equiv 2\pi/\mathcal{L}$ (Eqs. (6.105)–(6.109)). The Fourier coefficients $\hat{u}(\kappa, t)$ satisfy conjugate symmetry (Eq. (6.121)) and are orthogonal to κ.

For the sharp spectral filter with cutoff wavenumber κ_c, the Fourier coefficients of the filtered velocity field are

$$\hat{\bar{u}}(\kappa, t) = H(\kappa_c - \kappa)\hat{u}(\kappa, t), \tag{13.192}$$

where H is the Heaviside function and $\kappa = |\kappa|$ is the magnitude of the wavenumber. The Fourier series for the filtered velocity is then

$$\bar{u}(x, t) = \sum_{\kappa} e^{i\kappa \cdot x} \hat{\bar{u}}(\kappa, t) = \sum_{\substack{\kappa \\ \kappa < \kappa_c}} e^{i\kappa \cdot x} \hat{u}(\kappa, t). \tag{13.193}$$

In an LES computation, the N^3 wavenumbers κ represented lie on a uniform grid within a cube of side $2\kappa_c$ in wavenumber space. In each coordinate direction, the wavenumbers represented are $-(\frac{1}{2}N-1)\kappa_0, -(\frac{1}{2}N-2)\kappa_0, \ldots, -\kappa_0, 0, \kappa_0, \ldots, (\frac{1}{2}N-1)\kappa_0, \frac{1}{2}N\kappa_0$. The cutoff wavenumber κ_c and the number of modes N^3 are related by

$$\kappa_c = \tfrac{1}{2}N\kappa_0. \tag{13.194}$$

Note that the filter is taken to be isotropic, so that the second sum in Eq. (13.193) is restricted to wavenumbers within the sphere of radius κ_c (which is within the cube of side $2\kappa_c$). Equivalently, the fraction of the represented coefficients $\hat{\bar{u}}(\kappa, t)$ that are non-zero is approximately $\frac{4}{3}\pi\kappa_c^3/(2\kappa_c)^3 = \pi/6$.

The Navier–Stokes equations in wavenumber space (Eq. (6.145) on page 214) can be written

$$\left(\frac{d}{dt} + \nu\kappa^2\right)\hat{u}_j(\kappa, t) = -i\kappa_\ell P_{jk}(\kappa) \sum_{\kappa', \kappa''} \delta_{\kappa, \kappa' + \kappa''}\hat{u}_k(\kappa', t)\hat{u}_\ell(\kappa'', t), \tag{13.195}$$

where P_{jk} is the projection tensor (Eq. (6.133)). The Kronecker delta is unity for triadic interactions, i.e.,

$$\kappa = \kappa' + \kappa'', \tag{13.196}$$

and is zero otherwise. Equation (13.195) is a coupled set of ordinary differential equations, one equation for each of the infinite number of modes.

The filtered equation obtained from Eq. (13.195) is, for the non-zero coefficients ($\kappa < \kappa_c$),

$$\left(\frac{d}{dt} + \nu\kappa^2\right)\hat{\bar{u}}_j(\boldsymbol{\kappa}, t) = -i\kappa_\ell P_{jk}(\boldsymbol{\kappa})\sum_{\boldsymbol{\kappa}',\boldsymbol{\kappa}''}\delta_{\boldsymbol{\kappa},\boldsymbol{\kappa}'+\boldsymbol{\kappa}''}H(\kappa_c - \kappa)\hat{u}_k(\boldsymbol{\kappa}', t)\hat{u}_\ell(\boldsymbol{\kappa}'', t).$$

(13.197)

This is a finite set of (approximately $\pi N^3/6$) ordinary differential equations, but it is unclosed. The closure problem arises because the nonlinear term (on the right-hand side of Eq. (13.197)) includes unknown Fourier coefficients, namely $\hat{u}(\boldsymbol{\kappa}')$ and $\hat{u}(\boldsymbol{\kappa}'')$ for $\kappa' \equiv |\boldsymbol{\kappa}'| \geq \kappa_c$ and $\kappa'' \equiv |\boldsymbol{\kappa}''| \geq \kappa_c$.

EXERCISE

13.34   Consider LES of high-Reynolds-number homogeneous isotropic turbulence using the sharp spectral filter and a pseudo-spectral method with largest resolved wavenumber $\kappa_{max} = \kappa_c$. As in DNS, adequate resolution of the large scales requires the size $\mathcal{L}$ of the domain to be eight integral lengthscales, i.e., $\mathcal{L} = 8L_{11}$ (see Eq. (9.5)). So that no more than 20% of the energy is in the residual motions, the cutoff wavenumber is taken to be $\kappa_c L_{11} = 15$ (see Table 6.2 on page 240). Show that these requirements determine

$$\frac{\kappa_{max}}{\kappa_0} = \frac{15}{\pi/4} \approx 19,$$

(13.198)

as the ratio of the resolved wavenumbers, corresponding to a 38^3 simulation.

13.5.2 Triad interactions

There is, of course, a direct connection between the nonlinear term in Eq. (13.197), the velocity product in physical space $u_k u_\ell$, and various stresses formed from it (e.g., $\overline{u_k u_\ell}$). From Eq. (13.191) we obtain

$$u_k(\boldsymbol{x}, t)u_\ell(\boldsymbol{x}, t) = \sum_{\boldsymbol{\kappa}'}\sum_{\boldsymbol{\kappa}''}e^{i(\boldsymbol{\kappa}'+\boldsymbol{\kappa}'')\cdot\boldsymbol{x}}\,\hat{u}_k(\boldsymbol{\kappa}', t)\hat{u}_\ell(\boldsymbol{\kappa}'', t)$$

$$= \sum_{\boldsymbol{\kappa}',\boldsymbol{\kappa}''}\delta_{\boldsymbol{\kappa},\boldsymbol{\kappa}'+\boldsymbol{\kappa}''}\,e^{i\boldsymbol{\kappa}\cdot\boldsymbol{x}}\,\hat{u}_k(\boldsymbol{\kappa}', t)\hat{u}_\ell(\boldsymbol{\kappa}'', t).$$

(13.199)

Figure 13.10 is a sketch of qualitatively different triad interactions. These are referred to as types (a)–(d), and are defined in Table 13.4. In terms of the modes represented in the LES (i.e., $|\boldsymbol{\kappa}| < \kappa_c$), only interactions of type (a) can be represented exactly. Consider the summation in Eq. (13.199) for

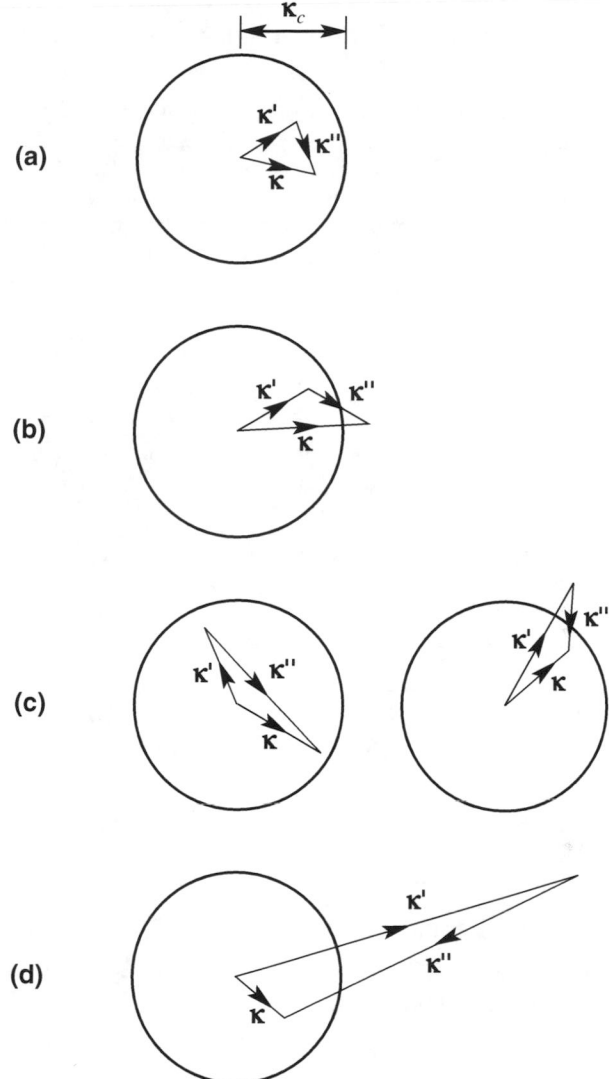

Fig. 13.10. Sketches of the various types of triad interactions defined in Table 13.4: (a) resolved, (b) Leonard, (c) cross, (d) SGS.

$u_k u_\ell$ restricted to these wavenumbers. The restrictions $\kappa' < \kappa_c$ and $\kappa'' < \kappa_c$ amount to replacing u_k by $\bar{u}_k$ and u_ℓ by $\bar{u}_\ell$, and the restriction $\kappa < \kappa_c$ amounts to filtering the result. Thus, the sum of the interactions of type (a) yields

$$\overline{\bar{u}_k \bar{u}_\ell} = \sum_{\kappa', \kappa''} \delta_{\kappa, \kappa' + \kappa''} H(\kappa_c - \kappa) H(\kappa_c - \kappa') H(\kappa_c - \kappa'') \hat{u}_k(\kappa') \hat{u}_\ell(\kappa'')$$

$$= \sum_{\kappa', \kappa''} \delta_{\kappa, \kappa' + \kappa''} H(\kappa_c - \kappa) \hat{\bar{u}}_k(\kappa') \hat{\bar{u}}_\ell(\kappa''). \tag{13.200}$$

Table 13.4. *Definitions of the types of triad interactions sketched in Fig. 13.10, and their contributions to $u_k u_\ell$ (Eq. (13.199)) (all possible interactions with $\kappa < \kappa_c$ are included, but there are additional interactions with $\kappa \geq \kappa_c$)*

Type designation	Defining wavenumber ranges	Contribution to $u_k u_\ell$
Type (a), resolved	$\kappa < \kappa_c$ $\kappa' < \kappa_c$ $\kappa'' < \kappa_c$	$\overline{\overline{u}_k \overline{u}_\ell}$
Type (b), Leonard	$\kappa_c \leq \kappa < 2\kappa_c$ $\kappa' < \kappa_c$ $\kappa'' < \kappa_c$	$\overline{u}_k \overline{u}_\ell - \overline{\overline{u}_k \overline{u}_\ell}$
Type (c), cross	$\kappa < \kappa_c$ $\kappa_c \leq \max(\kappa',\kappa'') < 2\kappa_c$ $\min(\kappa',\kappa'') < \kappa_c$	$\overline{\overline{u}_k u'_\ell} + \overline{u'_k \overline{u}_\ell}$
Type (d), SGS	$\kappa < \kappa_c$ $\kappa' \geq \kappa_c$ $\kappa'' \geq \kappa_c$	$\overline{u'_k u'_\ell}$

Given the Fourier coefficients $\hat{\overline{u}}(\boldsymbol{\kappa}, t)$, the triad interactions of type (b) are known, but they yield contributions of wavenumber κ beyond the filter cutoff ($\kappa \geq \kappa_c$). The sum of the contributions of types (a) and (b) to $u_k u_\ell$ in Eq. (13.199) is $\overline{u}_k \overline{u}_\ell$: hence the contribution from type (b) alone is

$$\overline{u}_k \overline{u}_\ell - \overline{\overline{u}_k \overline{u}_\ell} = -\mathcal{L}^{\circ}_{k\ell}, \tag{13.201}$$

i.e., the negative of the Leonard stress (Eq. (13.100)). (Recall that, with the sharp spectral filter, the two different definitions of the Leonard, cross, and SGS Reynolds stresses are equivalent.)

As previously mentioned, for LES in wavenumber space, the residual-stress tensor is best defined as

$$\tau^{\kappa}_{k\ell} \equiv \overline{u_k u_\ell} - \overline{\overline{u}_k \overline{u}_\ell}, \tag{13.202}$$

so that the filtered product is decomposed as

$$\overline{u_k u_\ell} = \overline{\overline{u}_k \overline{u}_\ell} + \tau^{\kappa}_{k\ell} = \overline{\overline{u}_k \overline{u}_\ell} + \mathcal{C}^{\circ}_{k\ell} + \mathcal{R}^{\circ}_{k\ell}. \tag{13.203}$$

In this way, each of the terms in Eq. (13.203) is a filtered quantity, and hence has no contribution from wavenumbers greater than κ_c. In contrast, in the decomposition

$$\overline{u_k u_\ell} = \overline{u}_k \overline{u}_\ell + \tau^{R}_{k\ell}, \tag{13.204}$$

the term $\overline{u}_k \overline{u}_\ell$ includes contributions of type (b) from wavenumber $\kappa_c \leq \kappa <$

$2\kappa_c$; and (since $\overline{u_k u_\ell}$ contains no such modes), the residual-stress model is required to cancel these contributions exactly – a tall order!

Triad interactions of type (c) are interactions between a represented mode (e.g., $\kappa' < \kappa_c$) and a residual mode ($\kappa'' \geq \kappa_c$) that produce contributions with $\kappa < \kappa_c$. Such interactions give rise to the cross stresses (Eq. (13.101)).

In interactions of type (d), two residual modes ($\kappa' \geq \kappa_c$ and $\kappa'' \geq \kappa_c$) give rise to a represented contribution ($\kappa < \kappa_c$). These interactions produce the SGS Reynolds stresses $\overline{u'_k u'_\ell}$. Together, the triad interactions of types (c) and (d) produce $\tau^\kappa_{k\ell}$ (Eq. (13.202)).

The filtered Navier–Stokes equations (Eq. (13.197)) can be rewritten on the basis of these different types of triad interactions (for $\kappa < \kappa_c$) as

$$\left(\frac{d}{dt} + \nu\kappa^2\right)\hat{\bar{u}}_j(\boldsymbol{\kappa}, t) = F_j^<(\boldsymbol{\kappa}, t) + F_j^>(\boldsymbol{\kappa}, t), \tag{13.205}$$

where $\boldsymbol{F}^<$ arises from the resolved interactions of type (a),

$$F_j^<(\boldsymbol{\kappa}) \equiv -i\kappa_\ell P_{jk}(\boldsymbol{\kappa}) \sum_{\boldsymbol{\kappa}',\boldsymbol{\kappa}''} \delta_{\boldsymbol{\kappa},\boldsymbol{\kappa}'+\boldsymbol{\kappa}''} \, \hat{\bar{u}}_k(\boldsymbol{\kappa}')\hat{\bar{u}}_\ell(\boldsymbol{\kappa}''), \tag{13.206}$$

and is in closed form; whereas $F_j^>(\boldsymbol{\kappa}, t)$ arises from interactions of types (c) and (d),

$$F_j^>(\boldsymbol{\kappa}) \equiv -i\kappa_\ell P_{jk}(\boldsymbol{\kappa}) \sum_{\max(\kappa',\kappa'')\geq\kappa_c} \delta_{\boldsymbol{\kappa},\boldsymbol{\kappa}'+\boldsymbol{\kappa}''} H(\kappa_c - \kappa)\hat{u}_k(\boldsymbol{\kappa}')\hat{u}_\ell(\boldsymbol{\kappa}''), \tag{13.207}$$

and has to be modelled.

13.5.3 The spectral energy balance

Various energy and spectral equations can be derived from Eq. (13.205). Most simply, let

$$\check{E}(\boldsymbol{\kappa}, t) \equiv \tfrac{1}{2}\hat{\bar{u}}_j^*(\boldsymbol{\kappa}, t)\hat{\bar{u}}_j(\boldsymbol{\kappa}, t) \tag{13.208}$$

denote the instantaneous energy of the mode of wavenumber $\boldsymbol{\kappa}$. Then, from Eq. (13.205), we have

$$\frac{d\check{E}}{dt} = -2\nu\kappa^2\check{E} + T_f + T_r, \tag{13.209}$$

where $T_f(\boldsymbol{\kappa}, t)$ and $T_r(\boldsymbol{\kappa}, t)$ are defined by

$$T_f \equiv \tfrac{1}{2}(\hat{\bar{u}}_j^* F_j^< + \hat{\bar{u}}_j F_j^{<*}), \tag{13.210}$$

$$T_r \equiv \tfrac{1}{2}(\hat{\bar{u}}_j^* F_j^> + \hat{\bar{u}}_j F_j^{>*}). \tag{13.211}$$

The first term on the right-hand side of Eq. (13.209) is molecular dissipation from the resolved motions, and is always negative. The second term, $T_f(\kappa, t)$, represents the rate of transfer to wavenumber κ of energy from other resolved modes by triad interactions of type (a). The total energy is conserved in such interactions (see Exercise 6.17 on page 214), so the term vanishes when it is summed over all resolved modes:

$$\sum_\kappa T_f(\kappa, t) = 0. \qquad (13.212)$$

The final term in Eq. (13.209), $T_r(\kappa, t)$, represents the rate of gain of energy from the residual motions via triad interactions of types (c) and (d). Positive values ($T_r > 0$) correspond to *backscatter*, whereas the term is predominantly negative (*forward scatter*). The expected rate of transfer of energy to the residual motions is

$$\langle \mathcal{P}_r \rangle = -\left\langle \sum_\kappa T_r(\kappa, t) \right\rangle. \qquad (13.213)$$

13.5.4 The spectral eddy viscosity

In order to perform a large-eddy simulation, the unclosed term $F^>(\kappa, t)$ in Eq. (13.205) has to be modelled. The simplest type of model, and that which is generally used, is based on a *spectral eddy viscosity*. The net effect of the residual motions is supposed to be similar to that of molecular dissipation, but with a spectral viscosity $\nu_e(\kappa|\kappa_c)$ that depends both on the wavenumber κ and on the cutoff κ_c. Thus the model is

$$F^>(\kappa, t) = -\nu_e(\kappa|\kappa_c)\kappa^2 \hat{\bar{u}}(\kappa, t). \qquad (13.214)$$

Starting with the work of Kraichnan (1976), various turbulence theories have been used to estimate $\nu_e(\kappa|\kappa_c)$. Chollet and Lesieur (1981) proposed the form

$$\nu_e(\kappa|\kappa_c) = \nu_e^+(\kappa/\kappa_c)\sqrt{\frac{E(\kappa_c, t)}{\kappa_c}}, \qquad (13.215)$$

which is based on the value of the energy-spectrum function at the cutoff wavenumber $E(\kappa_c, t)$; and the form of the non-dimensional function ν_e^+ proposed by Chollet (1984) is

$$\nu_e^+(\kappa/\kappa_c) = C^{-3/2}\left[0.441 + 15.2\exp\left(-3.03\frac{\kappa_c}{\kappa}\right)\right], \qquad (13.216)$$

which is shown in Fig. 13.11. As first predicted by Kraichnan (1976), for κ_c in the inertial subrange, the appropriate specification of $\nu_e(\kappa|\kappa_c)$ rises sharply

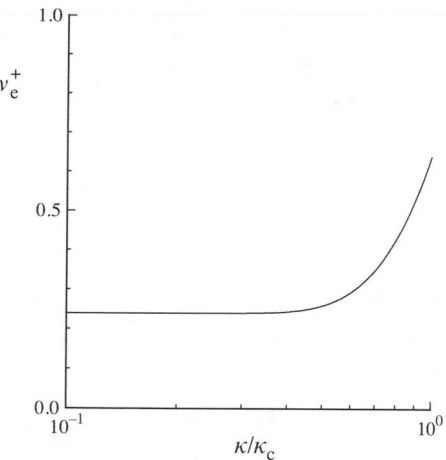

Fig. 13.11. The non-dimensional spectral viscosity $v_e^+(\kappa/\kappa_c)$, Eq. (13.216).

as κ approaches κ_c. According to the spectral-eddy-viscosity model, there is only forward scatter, with $\check{E}(\kappa, t)$ decreasing at the rate

$$-T_r = 2v_e\kappa^2\check{E}(\kappa, t), \tag{13.217}$$

due to this modelled term.

13.5.5 Backscatter

The expected rate of transfer of energy $\langle\mathcal{P}_r\rangle$ can be decomposed into forward-scatter and backscatter contributions:

$$\langle\mathcal{P}_r\rangle = \langle\mathcal{P}_{rf}\rangle - \langle\mathcal{P}_{rb}\rangle, \tag{13.218}$$

with

$$\langle\mathcal{P}_{rf}\rangle = -\left\langle \sum_{\kappa} T_r(\boldsymbol{\kappa}, t)H(-T_r(\boldsymbol{\kappa}, t)) \right\rangle, \tag{13.219}$$

$$\langle\mathcal{P}_{rb}\rangle = \left\langle \sum_{\kappa} T_r(\boldsymbol{\kappa}, t)H(T_r(\boldsymbol{\kappa}, t)) \right\rangle. \tag{13.220}$$

Leslie and Quarini (1979) found that backscatter can be very significant, with the ratio $\langle\mathcal{P}_{rb}\rangle/\langle\mathcal{P}_r\rangle$ being greater than unity.

Backscatter can be included in the model for $F^>(\boldsymbol{\kappa}, t)$ by adding a white-noise contribution (Chasnov 1991). Then $\hat{\bar{u}}(\boldsymbol{\kappa}, t)$ becomes a diffusion process, and its governing stochastic differential equation can be written

$$d\hat{\breve{u}}_j(\boldsymbol{\kappa}, t) = -[v + v_{\mathrm{e}}(\kappa|\kappa_{\mathrm{c}})]\kappa^2\hat{\breve{u}}_j\,dt + F_j^<\,dt$$
$$+ \dot{E}_{\mathrm{b}}(\kappa|\kappa_{\mathrm{c}})^{1/2}P_{jk}(\boldsymbol{\kappa})\,dW_k(\boldsymbol{\kappa}, t), \qquad (13.221)$$

where the isotropic Wiener processes $W(\boldsymbol{\kappa}, t)$ are independent for each mode,

$$\langle dW_i(\boldsymbol{\kappa}, t)\,dW_j(\boldsymbol{\kappa}', t)\rangle = \delta_{\boldsymbol{\kappa}, \boldsymbol{\kappa}'}\,\delta_{ij}\,dt, \qquad (13.222)$$

and $\dot{E}_{\mathrm{b}}$ is a specified coefficient. The expected energy of the mode $\langle \breve{E}(\boldsymbol{\kappa}, t)\rangle$ deduced from Eq. (13.221) evolves by

$$\frac{d\langle\breve{E}\rangle}{dt} = -2v\kappa^2(v + v_{\mathrm{e}})\langle\breve{E}\rangle + \langle T_{\mathrm{f}}\rangle + \dot{E}_{\mathrm{b}}, \qquad (13.223)$$

so $\dot{E}_{\mathrm{b}}$ is identified as the mean rate of addition of energy by backscatter (see Exercise 13.35).

EXERCISE

13.35   Verify that Eq. (13.221) is consistent with the continuity equation in that $\kappa_j\,d\hat{\breve{u}}_j(\boldsymbol{\kappa}, t)$ is identically zero. Recalling the definition of P_{ij} (Eq. (6.133)), obtain the result

$$\langle P_{jk}(\boldsymbol{\kappa})\,dW_k(\boldsymbol{\kappa}, t)P_{j\ell}(\boldsymbol{\kappa})\,dW_\ell(\boldsymbol{\kappa}, t)\rangle = 2\,dt. \qquad (13.224)$$

Hence verify Eq. (13.223).

13.5.6 A statistical view of LES

What is the ideal LES model for the residual motions? That is, in physical space, what is the ideal model for the residual stresses τ_{ij}^{r}? Or, in wavenumber space, what is the ideal model for $F^>$?

The corresponding question is easy to answer for the turbulence models considered in the previous chapters. For example, the ideal turbulent-viscosity model yields the correct Reynolds-stress field (that which occurs in the flow considered). However, for LES the question is more difficult to answer.

To examine the issues involved, consider an accurate DNS of homogeneous isotropic turbulence performed with a pseudo-spectral method with N_{DNS}^3 modes. We also consider a LES of the same flow using a cutoff wavenumber κ_{c} so that the number of modes represented N_{LES}^3 is significantly decreased ($N_{\mathrm{LES}}^3 \ll N_{\mathrm{DNS}}^3$). As is often done in practice, the initial conditions for the DNS (at $t = 0$) are set by specifying the Fourier amplitudes $|\hat{\boldsymbol{u}}(\boldsymbol{\kappa}, 0)|$ deterministically to obtain a specified spectrum $E(\kappa, 0)$, while the Fourier

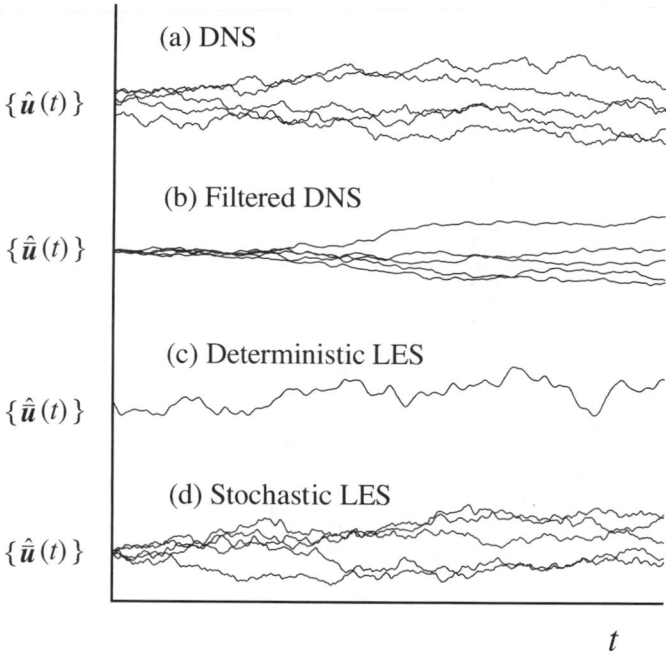

Fig. 13.12. Sketches of DNS and LES sample paths for simulations of isotropic turbulence: (a) DNS, $\{\hat{u}(\kappa, t)\}$; (b) filtered DNS, $\{\hat{\bar{u}}(\kappa, t)\}$; (c) LES with a deterministic residual-stress model, $\{\hat{\bar{u}}(\kappa, t)\}$; (d) LES with a stochastic backscattering model, $\{\hat{\bar{u}}(\kappa, t)\}$. The paths in (a) are in the N_{DNS}^3-dimensional state space; those in (b)–(d) are in the N_{LES}^3-dimensional state space.

phases are set at random. We consider different realizations of the DNS, in which the initial conditions $\hat{u}(\kappa, 0)$, are identical for $|\kappa| < \kappa_{\mathrm{c}}$, but different for $|\kappa| \geq \kappa_{\mathrm{c}}$ (as suggested by Piomelli and Chasnov (1996)).

Figure 13.12(a) is a sketch of the evolution of the flow (in the N_{DNS}^3-dimensional state space) on different realizations of the DNS. The initial states differ (because of the different phases for $|\kappa| \geq \kappa_{\mathrm{c}}$), and consequently the sample paths are different. (The state $\{\hat{u}(\kappa, t)\}$ evolves deterministically according to the Navier–Stokes equations, hence the sample paths do not cross in the N_{DNS}^3-dimensional state space.)

Figure 13.12(b) shows the evolution of the filtered DNS fields in the N_{LES}^3-dimensional state space. These sample paths are the projections of the DNS sample paths shown in Fig. 13.12(a) onto the lower-dimensional LES state space. The initial states $\{\hat{\bar{u}}(\kappa, 0)\}$ are identical – by construction. However, as time evolves, the states of the different realizations disperse.

For a deterministic LES model – such as that of the spectral eddy viscosity, Eq. (13.214) – there is a unique evolution of the state $\{\hat{\bar{u}}(\kappa, t)\}$ from the given

initial condition, as depicted in Fig. 13.12(c). For a stochastic LES model –
such as Eq. (13.221) – we have that states of different realizations disperse
(Fig. 13.12(d)); but the cause of the dispersion (i.e., white noise) is entirely
different than that in the case of the filtered DNS fields.

The following conclusions can be drawn from these considerations.

(i) The filtered DNS fields $\{\hat{\bar{u}}(\kappa, t)\}$ are not uniquely determined by the
filtered initial condition $\{\hat{\bar{u}}(\kappa, 0)\}$.

(ii) Hence, it is impossible to construct a LES model that produces filtered
velocity fields that match those from DNS realization by realization.

(iii) Instead, the best that can be achieved is a statistical correspondence
between the LES and DNS filtered fields.

(iv) Correspondence can (in principle) be achieved at the level of one-time
statistics by the deterministic model in which $F^{>}(\kappa, t)$ is replaced by
its conditional expectation

$$\langle F^{>}(\kappa, t) | \hat{\bar{u}}(\kappa', t) \rangle. \tag{13.225}$$

(This result follows directly from the equation for the evolution of
the one-time joint PDF of the Fourier coefficients $\{\hat{\bar{u}}(\kappa, t)\}$, and it
depends on the initial joint PDF being continuous – as opposed to
the sharp initial condition on $\{\hat{\bar{u}}(\kappa, 0)\}$ considered above: see Adrian
(1990) and Exercise 13.36.)

(v) *A priori* testing of the correlation coefficient between modelled and
measured residual stresses is of questionable significance. With the
ideal model, Eq. (13.225), the correlation coefficient for $F^{>}$ obtained
from DNS and from the model may be significantly less than unity.
The matching is only for conditional means – that is, means conditioned on the entire field. More meaningful, therefore, is *a priori*
testing based on conditional statistics, as performed by Piomelli, Yu,
and Adrian (1996), for example.

EXERCISE _____

13.36 Consider a random vector process $u(t) = \{u_1(t), u_2(t), ..., u_N(t)\}$ that
evolves from a random initial condition $u(0)$ according to the ordinary
differential equation

$$\frac{\mathrm{d}u(t)}{\mathrm{d}t} = A(t), \tag{13.226}$$

where $A(t)$ is a differentiable random vector. Show that the PDF of

$u(t), f(v; t)$, evolves by

$$\frac{\partial f(v; t)}{\partial t} = -\frac{\partial}{\partial v_i} [f(v; t) B_i(v, t)], \qquad (13.227)$$

with

$$B(v, t) \equiv \langle A(t) | u(t) = v \rangle. \qquad (13.228)$$

Consider a second vector process $\hat{u}(t) = \{\hat{u}_1(t), \hat{u}_2(t), ..., \hat{u}_N(t)\}$ that evolves from the same initial condition as u (i.e., $\hat{u}(0) = u(0)$), according to the deterministic equation

$$\frac{d\hat{u}}{dt} = B(\hat{u}(t), t). \qquad (13.229)$$

Show that the one-time PDF of $\hat{u}(t)$ is identical to that of $u(t)$.

Argue that this observation justifies the claim made in item (iv) on page 614.

13.5.7 Resolution and modelling

With the sharp spectral filter (with cutoff wavenumber κ_c), the issues of resolution and modelling are clear and separate. All wavenumbers $|\kappa| < \kappa_c$ have to be represented, and there is no benefit in representing higher-wavenumber modes. The filtered velocity field is represented with complete accuracy by the Fourier modes $\bar{\hat{u}}(\kappa, t)$ for $|\kappa| < \kappa_c$. The filtered velocity field contains no direct information[2] about the residual motions $|\kappa| \geq \kappa_c$, hence the effect of these residual motions has to be modelled.

For the Gaussian filter – or for any other filter with a strictly positive transfer function $\hat{G}(\kappa)$ – the issues of resolution and modelling are less clear and separate. Consider DNS and LES of homogeneous turbulence (with $\langle U \rangle = 0$) using a pseudo-spectral method. In the DNS, all wavenumbers up to κ_{DNS} are represented, whereby, to ensure adequate resolution of the instantaneous velocity field, κ_{DNS} is chosen so that $\kappa_{DNS}\eta$ is greater than 1.5, say. From a given initial condition, the Navier–Stokes equations (Eq. (13.195)) are integrated forward in time to determine the Fourier coefficients $\hat{u}(\kappa, t)$ for $|\kappa| < \kappa_{DNS}$.

The LES is performed using the Gaussian filter with characteristic wave-

---

[2] Direct information, as opposed to information that may be inferred from a statistical hypothesis.

number κ_c, for which the transfer function is

$$\widehat{G}(\kappa) = \exp\left[-\frac{1}{24}\left(\frac{\pi\kappa}{\kappa_c}\right)^2\right]. \tag{13.230}$$

As discussed in Section 13.2, the resulting filtered velocity field $\bar{u}(x,t)$ has spectral content beyond κ_c. Consequently, in the LES, modes are represented up to a higher wavenumber κ_{LES}, e.g., $\kappa_{LES} = 2\kappa_c$.

The form of the LES equations depends on the choice of decomposition of $\overline{u_i u_j}$. With the decomposition $\overline{u_i u_j} = \bar{u}_i \bar{u}_j + \tau_{ij}^R$, the equation is (for $|\kappa| < \kappa_{LES}$)

$$\left(\frac{d}{dt} + \nu\kappa^2\right)\hat{\bar{u}}_j(\kappa,t) = -i\kappa_\ell P_{jk}(\kappa)\sum_{\kappa',\kappa''}\delta_{\kappa,\kappa'+\kappa''}\hat{\bar{u}}_k(\kappa',t)\hat{\bar{u}}_\ell(\kappa'',t) + F_j^>(\kappa,t).$$

$$\tag{13.231}$$

This is the same as the Navier–Stokes equations (Eq. (13.195)) except that it is written for the Fourier coefficients $\hat{\bar{u}}$ of the filtered field; the wavenumbers κ, κ', and κ'' are restricted to be within the sphere of radius κ_{LES}; and the term $F_j^>(\kappa,t)$ representing the residual model has been added.

In a typical application, the relative magnitudes of the wavenumbers might be $\kappa_c = \frac{1}{8}\kappa_{DNS}$ and $\kappa_{LES} = 2\kappa_c = \frac{1}{4}\kappa_{DNS}$, so that the LES requires fewer modes by a factor of $4^3 = 64$. This saving comes at the cost of the uncertainties involved in the modelling of $F_j^>(\kappa,t)$. It should be observed, however, that the apparent closure problem – the need to model $F^>$ – can alternatively be viewed as a resolution problem. For, if the filtered field is fully resolved, $F^>$ can be determined – as is now shown.

The Fourier coefficients of the instantaneous $u(x,t)$ and filtered $\bar{u}(x,t)$ velocity fields are related by

$$\hat{\bar{u}}(\kappa,t) = \widehat{G}(\kappa)\hat{u}(\kappa,t). \tag{13.232}$$

Since (for the Gaussian filter) $\widehat{G}(\kappa)$ is strictly positive, this relation can be inverted to yield

$$\hat{u}(\kappa,t) = \frac{\hat{\bar{u}}(\kappa,t)}{\widehat{G}(\kappa)}. \tag{13.233}$$

An exact closed set of equations for the coefficients $\hat{\bar{u}}(\kappa,t)$ is obtained by multiplying the Navier–Stokes equations (Eq. (13.195)) by $\widehat{G}(\kappa)$:

$$\left(\frac{d}{dt} + \nu\kappa^2\right)\hat{\bar{u}}_j(\kappa,t) = -i\kappa_\ell P_{jk}(\kappa)\sum_{\kappa',\kappa''}\delta_{\kappa,\kappa'+\kappa''}\frac{\widehat{G}(\kappa)\hat{\bar{u}}_k(\kappa',t)\hat{\bar{u}}_\ell(\kappa'',t)}{\widehat{G}(\kappa')\widehat{G}(\kappa'')}. \tag{13.234}$$

By comparing this equation with the LES equation (Eq. (13.231)), the following exact equation for the residual motions is obtained:

$$F_j^>(\kappa, t) = -i\kappa_\ell P_{jk}(\kappa) \sum_{\kappa', \kappa''} \delta_{\kappa, \kappa' + \kappa''} \left(\frac{\widehat{G}(\kappa)}{\widehat{G}(\kappa')\widehat{G}(\kappa'')} - 1 \right) \widehat{\overline{u}}_k(\kappa', t)\widehat{\overline{u}}_\ell(\kappa'', t). \quad (13.235)$$

These results are exact only if the filtered velocity fields are fully resolved, so that the wavenumbers considered extend to κ_{DNS} and beyond. If, instead, the filtered velocity coefficients are represented only up to $\kappa_{LES} < \kappa_{DNS}$, then $F^>(\kappa, t)$ (for $|\kappa| < \kappa_{LES}$) can be decomposed as

$$F_j^>(\kappa, t) = F_j^f(\kappa, t) + F_j^{>>}(\kappa, t), \quad (13.236)$$

where $F_j^f(\kappa, t)$ is defined by the right-hand side of Eq. (13.235) but with the sum restricted to κ' and κ'' being within in the sphere of radius κ_{LES}. Thus F^f is known in terms of the represented coefficients whereas $F^{>>}$ has to be modelled. As κ_{LES} increases, the relative contribution of $F^{>>}$ decreases, and becomes negligible for $\kappa_{LES} = \kappa_{DNS}$.

For large values of $|\kappa|/\kappa_c$, the transfer function $\widehat{G}(\kappa)$ is very small, which raises questions of ill-conditioning of Eqs. (13.234) and (13.235). If the filtered velocity field is represented by its Fourier coefficients $\widehat{\overline{u}}(\kappa, t)$, then ill-conditioning can be avoided (see Exercise 13.38), but if $\overline{u}(x, t)$ is represented on a grid in physical space – as it would be in the LES of an inhomogeneous flow – then the task of determining τ_{ij}^R from $\overline{u}(x, t)$ is certainly ill-conditioned, however fine the grid. Consequently, if, by analogy to Eq. (13.236), τ_{ij}^R is decomposed as

$$\tau_{ij}^R = \tau_{ij}^f + \tau_{ij}^{>>}, \quad (13.237)$$

then the contribution τ_{ij}^f cannot be reliably determined for motions of wavelength κ_{LES} much larger than κ_c (see Exercises 13.37 and 13.39).

We now draw the important conclusions from these considerations. For LES using the sharp spectral filter (or other projection) the conclusions are as follows.

(i) The filtered velocity field is represented with complete accuracy by a finite set of modes.

(ii) The filtered velocity field provides no direct information about the residual motions.

(iii) The issues of modelling and resolution are separate.

For LES using the Gaussian filter (or other invertible filter, $\widehat{G}(\kappa) > 0$) the conclusions are as follows.

(iv) The filtered velocity field cannot be represented with complete accuracy by a finite set of modes. The accuracy increases with $\kappa_{\text{LES}}/\kappa_{\text{c}}$.

(v) The filtered velocity field contains information about the residual motions. The extent to which this information can be used in a LES to represent the effects of the residual motions depends on the resolution of the filtered fields, and on whether they are known on a grid in physical space or in wavenumber space.

(vi) Resolution and modelling are inherently connected: if the filtered fields are fully resolved in wavenumber space (at the same expense as DNS) then the effects of the residual motions are known, and no modelling is required.

EXERCISES

13.37 Show that, for the Gaussian filter (Eq. (13.230)), the transfer function in Eq. (13.234) can be re-expressed as

$$\frac{\delta_{\kappa,\kappa'+\kappa''}\widehat{G}(\kappa)}{\widehat{G}(\kappa')\widehat{G}(\kappa'')} = \delta_{\kappa,\kappa'+\kappa''}\exp\left(-\frac{\pi^2\boldsymbol{\kappa}'\cdot\boldsymbol{\kappa}''}{12\kappa_{\text{c}}^2}\right). \tag{13.238}$$

If $\boldsymbol{\kappa}'$ and $\boldsymbol{\kappa}''$ lie within the sphere of radius $2\kappa_{\text{c}}$, show that the exponential term is greater than 0.037; and show that, for a sphere of radius $4\kappa_{\text{c}}$, the corresponding bound is 1.9×10^{-6}.

13.38 Computations involving variables of greatly different magnitudes can be ill-conditioned. This problem can be alleviated by scaling the variables so that they are comparable in magnitude. Show that solving the LES equation (Eq. (13.234)) for the scaled variables $\widehat{\bar{u}}(\boldsymbol{\kappa},t)/\widehat{G}(\boldsymbol{\kappa})$ is equivalent to solving the DNS equation, Eq. (13.231).

13.39 Given the fully resolved filtered velocity field $\bar{u}(x,t)$ obtained with the Gaussian filter, show that the residual stress τ_{ij}^{R} can (in principle) be obtained as

$$\tau_{ij}^{\text{R}} = \mathcal{F}^{-1}\{\widehat{G}\mathcal{F}[\mathcal{F}^{-1}\{\widehat{G}^{-1}\mathcal{F}(\bar{u}_i)\}\mathcal{F}^{-1}\{\widehat{G}^{-1}\mathcal{F}(\bar{u}_j)\}]\} - \bar{u}_i\bar{u}_j. \tag{13.239}$$

For $\kappa_{\text{LES}} > \kappa_{\text{c}}$, $\widehat{H}(\boldsymbol{\kappa})$ is defined by

$$\widehat{H}(\boldsymbol{\kappa}) \equiv H(\kappa_{\text{LES}} - |\boldsymbol{\kappa}|). \tag{13.240}$$

Show that τ_{ij}^{f} (Eq. (13.237)) – the contribution to τ_{ij}^{R} from modes of wavenumber less than κ_{LES} – can be obtained from $\bar{u}(x,t)$ as

$$\tau_{ij}^{\text{f}} = \mathcal{F}^{-1}\{\widehat{H}\widehat{G}\mathcal{F}[\mathcal{F}^{-1}\{\widehat{H}\widehat{G}^{-1}\mathcal{F}(\bar{u}_i)\}\mathcal{F}^{-1}\{\widehat{H}\widehat{G}^{-1}\mathcal{F}(\bar{u}_j)\}]\} - \bar{u}_i\bar{u}_j. \tag{13.241}$$

Comment on the conditioning of these operations.

13.6 Further residual-stress models

We now return to physical space to consider residual-stress models applicable to inhomogeneous turbulent flows. These models are motivated by the shortcomings of the Smagorinsky model.

13.6.1 The dynamic model

One of the problems with the Smagorinsky model is that the appropriate value of the coefficient C_S is different in different flow regimes. In particular, it is zero in laminar flow, and it is attenuated near walls compared with its value ($C_S \approx 0.15$) in high-Reynolds-number free turbulent flows. The dynamic model provides a methodology for determining an appropriate *local* value of the Smagorinsky coefficient. The model was proposed by Germano *et al.* (1991), with important modifications and extensions provided by Lilly (1992) and Meneveau *et al.* (1996).

Grid and test filters

The dynamic model involves filters of different filter widths. For simplicity, we consider homogeneous isotropic filters, and denote by $G(r;\Delta)$ and $\widehat{G}(\kappa;\Delta)$ the filter of width Δ and its transfer function.

The *grid filter* has filter width $\overline{\Delta}$, which is proportional to the grid spacing h (e.g., $\overline{\Delta} = h$, or, for better resolution, $\overline{\Delta} = 2h$). The operation of grid filtering is denoted by an overbar, e.g.,

$$\overline{U}(x,t) \equiv \int U(x-r,t)G(|r|;\overline{\Delta})\,dr. \tag{13.242}$$

The LES equations are deemed to be solved for $\overline{U}(x,t)$, although this filtering operation is not explicitly performed.

The *test filter* has filter width $\widetilde{\Delta}$, which is typically taken to be twice $\overline{\Delta}$, and test filtering is denoted by a tilde, e.g.,

$$\widetilde{U}(x,t) \equiv \int U(x-r,t)G(|r|;\widetilde{\Delta})\,dr. \tag{13.243}$$

In fact, since U is unknown in a LES calculation, it is more relevant to consider the test filter applied to $\overline{U}$, to yield the doubly filtered quantity $\widetilde{\overline{U}}$.

Note that, if the sharp spectral filter is used, $\widetilde{\overline{U}}$ is the same as $\widetilde{U}$, since the product of the transfer functions is

$$\widehat{G}(\kappa;\widetilde{\Delta})\widehat{G}(\kappa;\overline{\Delta}) = H(\pi/\widetilde{\Delta} - |\kappa|)H(\pi/\overline{\Delta} - |\kappa|)$$
$$= H(\pi/\widetilde{\Delta} - |\kappa|) = \widehat{G}(\kappa;\widetilde{\Delta}) \tag{13.244}$$

(for $\widetilde{\Delta} \geq \overline{\Delta}$). On the other hand, for the Gaussian filter we have

$$\widehat{G}(\kappa;\widetilde{\Delta})\widehat{G}(\kappa;\overline{\Delta}) = \widehat{G}(\kappa;(\widetilde{\Delta}^2 + \overline{\Delta}^2)^{1/2}), \qquad (13.245)$$

see Exercise 13.4 on page 567. Thus, for both filters, the double-filtering operation can be written

$$\overline{\widetilde{U}}(x,t) \equiv \int \overline{U}(x-r,t)G(|r|;\widetilde{\Delta})\,dr$$

$$= \int U(x-r,t)G(|r|;\widetilde{\overline{\Delta}})\,dr, \qquad (13.246)$$

where the effective double-filter width is

$$\widetilde{\overline{\Delta}} = \begin{cases} \widetilde{\Delta}, & \text{for the sharp spectral filter} \\[2mm] (\widetilde{\Delta}^2 + \overline{\Delta}^2)^{1/2}, & \text{for the Gaussian filter.} \end{cases} \qquad (13.247)$$

The smallest resolved motions

The decomposition of the velocity $U = \overline{U} + u'$ is readily extended to

$$U = \overline{\widetilde{U}} + (\overline{U} - \overline{\widetilde{U}}) + u', \qquad (13.248)$$

where the three contributions represent (loosely) motions larger than $\widetilde{\overline{\Delta}}$, motions between $\overline{\Delta}$ and $\widetilde{\overline{\Delta}}$; and motions smaller than $\overline{\Delta}$. The smallest resolved motions $\overline{U} - \overline{\widetilde{U}}$ can be determined from $\overline{U}$ and are conceptually important in several models, including the dynamic model.

In physical space, $\overline{U} - \overline{\widetilde{U}}$ is given by the convolution of U with the filter difference $G(r;\overline{\Delta}) - G(r;\widetilde{\overline{\Delta}})$. For homogeneous turbulence, the Fourier coefficients of $\overline{U} - \overline{\widetilde{U}}$ are those of U multiplied by the transfer-function difference $\widehat{G}(\kappa;\overline{\Delta}) - \widehat{G}(\kappa;\widetilde{\overline{\Delta}})$. This difference is shown in Fig. 13.13 for the sharp spectral and Gaussian filters with $\widetilde{\Delta} = 2\overline{\Delta}$: as may be seen, these correspond to perfect and imperfect band-pass filters, respectively.

If the test and grid filters are taken to be the same (i.e., $\widetilde{\Delta} = \overline{\Delta}$), then we have

$$\overline{U} - \overline{\widetilde{U}} = \overline{U} - \overline{\overline{U}} = \overline{u'}. \qquad (13.249)$$

For the sharp spectral filter $\overline{u'}$ is zero; but, for the Gaussian filter, the corresponding transfer-function difference is similar in shape to that for $\widetilde{\Delta} = 2\overline{\Delta}$ (see Fig. 13.13).

We have interpreted $\overline{U} - \overline{\widetilde{U}}$ as the smallest resolved motions – resolved by a grid of spacing $\overline{\Delta}$. Equally, $\overline{U} - \overline{\widetilde{U}}$ represents the largest motions that are not resolved on a grid of spacing $\widetilde{\overline{\Delta}}$.

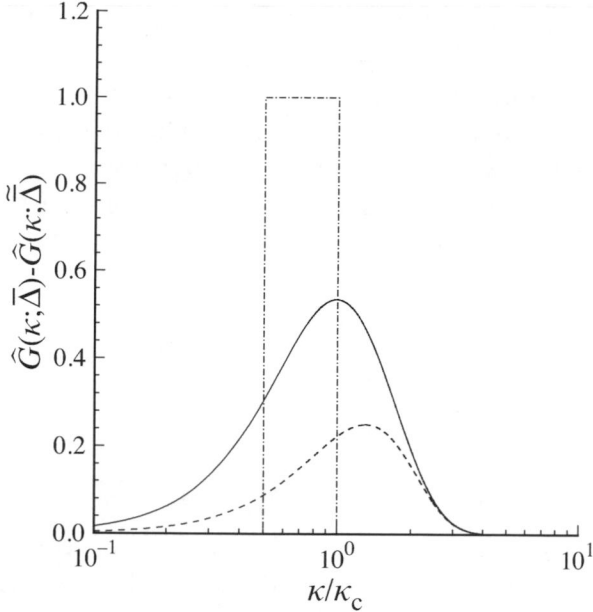

Fig. 13.13. Filter-transfer-function differences corresponding to the smallest resolved motions. Solid line, Gaussian filter with $\widetilde{\Delta} = 2\overline{\Delta}$; dot–dashed line, sharp spectral filter with $\widetilde{\Delta} = 2\overline{\Delta}$; dashed line, Gaussian filter with $\widetilde{\Delta} = \overline{\Delta}$.

EXERCISE

13.40   Consider high-Reynolds-number isotropic turbulence with the sharp spectral grid and test filters applied within the inertial subrange. Use the Kolmogorov spectrum to estimate that the ratio of the energy in the smallest resolved scales to the residual kinetic energy is $(\widetilde{\Delta}/\overline{\Delta})^{2/3} - 1$. Show that this ratio is 0.59 and 1 for $\widetilde{\Delta}/\overline{\Delta} = 2$ and 2.8, respectively.

Germano's identity

The residual stresses based on the single- and double-filtering operations are defined by

$$\tau_{ij}^{R} \equiv \overline{U_i U_j} - \overline{U}_i \overline{U}_j, \tag{13.250}$$

$$T_{ij} \equiv \widetilde{\overline{U_i U_j}} - \widetilde{\overline{U}}_i \widetilde{\overline{U}}_j. \tag{13.251}$$

An identity due to Germano (1992) is obtained by applying the test filter to Eq. (13.250) and subtracting the result from Eq. (13.251):

$$\mathcal{L}_{ij} \equiv T_{ij} - \widetilde{\tau}_{ij}^{R} = \widetilde{\overline{U}_i \overline{U}_j} - \widetilde{\overline{U}}_i \widetilde{\overline{U}}_j. \tag{13.252}$$

The significance of this identity is that $\mathcal{L}_{ij}$ – which is called the *resolved stress* – is known in terms of $\overline{U}$, whereas T_{ij} and τ_{ij}^{R} are not.

The right-hand side of Eq. (13.252) is in the form of a residual stress (relative to the test filter) based on the grid-filtered velocity. Hence, with respect to the test filter, $\mathcal{L}_{ij}$ can loosely be interpreted as the contribution to the residual stress from the largest unresolved motions. Also, for $\widetilde{\Delta} = \overline{\Delta}$, $\mathcal{L}_{ij}$ defined here by Eq. (13.252) is identical to the Leonard stress $\mathcal{L}_{ij}^{\mathrm{o}}$ defined by Eq. (13.100).

The Smagorinsky coefficient

The Smagorinsky model (Eqs. (13.127) and (13.128)) for the deviatoric part of τ_{ij}^{R} can be written

$$\tau_{ij}^{\mathrm{r}} \equiv \tau_{ij}^{\mathrm{R}} - \tfrac{1}{3}\tau_{kk}^{\mathrm{R}}\delta_{ij} = -2c_{\mathrm{S}}\overline{\Delta}^2\overline{S}\,\overline{S}_{ij}. \qquad (13.253)$$

The coefficient has been redefined as c_{S} (in place of C_{S}^2) to allow for the possibility of negative values, corresponding to backscatter. The same model equation written for the filter of width $\widetilde{\overline{\Delta}}$ is

$$T_{ij}^{\mathrm{d}} \equiv T_{ij} - \tfrac{1}{3}T_{kk}\delta_{ij} = -2c_{\mathrm{S}}\widetilde{\overline{\Delta}}^2\,\widetilde{\overline{S}}\,\widetilde{\overline{S}}_{ij}, \qquad (13.254)$$

where $\widetilde{\overline{S}}_{ij}$ and $\widetilde{\overline{S}}$ are defined straightforwardly based on $\widetilde{\overline{U}}$, similarly to Eqs. (13.73) and (13.74). Taking c_{S} to be uniform, and defining

$$M_{ij} \equiv 2\widetilde{\overline{\Delta}^2\overline{S}\,\overline{S}_{ij}} - 2\widetilde{\overline{\Delta}}^2\,\widetilde{\overline{S}}\,\widetilde{\overline{S}}_{ij}, \qquad (13.255)$$

Eqs. (13.253) and (13.254) lead to

$$\mathcal{L}_{ij}^{\mathrm{S}} \equiv T_{ij}^{\mathrm{d}} - \widetilde{\tau}_{ij}^{\mathrm{r}} = c_{\mathrm{S}}M_{ij}, \qquad (13.256)$$

which is the Smagorinsky model for the deviatoric part of $\mathcal{L}_{ij}$ (Eq. (13.252)),

$$\mathcal{L}_{ij}^{\mathrm{d}} \equiv \mathcal{L}_{ij} - \tfrac{1}{3}\mathcal{L}_{kk}\delta_{ij}. \qquad (13.257)$$

In a LES, both M_{ij} and $\mathcal{L}_{ij}^{\mathrm{d}}$ (which are functions of x and t) are known in terms of $\overline{U}(x,t)$. This information can be used to determine the value of the Smagorinsky coefficient c_{S} for which the model $\mathcal{L}_{ij}^{\mathrm{S}}$ provides the best approximation to $\mathcal{L}_{ij}^{\mathrm{d}}$. Of course, the single coefficient c_{S} cannot be chosen to match the five independent components of $\mathcal{L}_{ij}^{\mathrm{S}}$ and $\mathcal{L}_{ij}^{\mathrm{d}}$. However, as shown by Lilly (1992), the mean-square error is minimized by the specification

$$c_{\mathrm{S}} = M_{ij}\mathcal{L}_{ij}/M_{k\ell}M_{k\ell}, \qquad (13.258)$$

see Exercise 13.41. (Note that, since M_{ij} is deviatoric, $M_{ij}\mathcal{L}_{ij}$ is equal to $M_{ij}\mathcal{L}_{ij}^{\mathrm{d}}$.)

13.41 The mean-square error between the deviatoric stress $\mathcal{L}_{ij}^{d}$ and the Smagorinsky model prediction $\mathcal{L}_{ij}^{S}$ (Eq. (13.256)) is defined by

$$\epsilon = (\mathcal{L}_{ij}^{S} - \mathcal{L}_{ij}^{d})(\mathcal{L}_{ij}^{S} - \mathcal{L}_{ij}^{d}). \tag{13.259}$$

Differentiate with respect to c_S to obtain

$$\frac{1}{2}\frac{\partial \epsilon}{\partial c_S} = M_{ij}(\mathcal{L}_{ij}^{S} - \mathcal{L}_{ij}^{d}) = c_S M_{ij}M_{ij} - M_{ij}\mathcal{L}_{ij}^{d}, \tag{13.260}$$

$$\frac{1}{2}\frac{\partial^2 \epsilon}{\partial c_S^2} = M_{ij}M_{ij}. \tag{13.261}$$

Hence argue that the specification of c_S according to Eq. (13.258) minimizes ϵ.

13.42 Consider the application of the dynamic model to a laminar flow. Over distances on the order of the filter width $\widetilde{\overline{\Delta}}$, the velocity can be assumed to vary linearly with x. Use the results of Exercise 13.28 to show that the various stresses are

$$\tau_{ij}^{R} = \tilde{\tau}_{ij}^{R} = \frac{\overline{\Delta}^2}{12}\frac{\partial \overline{U}_i}{\partial x_k}\frac{\partial \overline{U}_j}{\partial x_k},$$
$$= T_{ij}/\alpha^2 = \mathcal{L}_{ij}/(\alpha^2 - 1), \tag{13.262}$$

where α is the filter width ratio $\alpha \equiv \widetilde{\overline{\Delta}}/\overline{\Delta}$. Show that the tensor M_{ij} (Eq. (13.255)) is

$$M_{ij} = -2\overline{\Delta}^2(\alpha^2 - 1)\overline{S}\,\overline{S}_{ij}. \tag{13.263}$$

Hence show from Eq. (13.258) that the dynamic model yields

$$c_S = -\frac{\overline{S}_{ij}}{12\overline{S}^3}\frac{\partial \overline{U}_i}{\partial x_k}\frac{\partial \overline{U}_j}{\partial x_k}. \tag{13.264}$$

Verify the values of c_S for the following homogeneous laminar deformations (see Table 11.2 on page 415): axisymmetric contraction, $c_S = -0.012$; axisymmetric expansion, $c_S = 0.012$; plane strain, $c_S = 0$; and shear, $c_S = 0$.

Application to channel flow

Given the low correlation between the stress and the rate of strain observed in *a priori* tests, it is not surprising that the values of c_S obtained from Eq. (13.258) exhibit very large fluctuations, including negative values. As a consequence, LES calculations using Eq. (13.258) are found to be unstable.

To overcome this problem, in their calculations of fully developed turbulent channel flow, Germano *et al.* (1991) and Piomelli (1993) averaged the numerator and denominator in Eq. (13.258).[3] Thus, denoting this averaging by ( )$_{\text{ave}}$, the Smagorinsky coefficient is obtained as

$$c_S = (M_{ij}\mathcal{L}_{ij})_{\text{ave}}/(M_{k\ell}M_{k\ell})_{\text{ave}}. \tag{13.265}$$

With this implementation, the dynamic model leads to good calculations of transitional and fully turbulent channel flow, provided that the grid is fine enough to resolve the near-wall energy-containing motions. Thus the need for damping functions or other special near-wall treatments is obviated. In part, the success of the model can be attributed to the fact that it correctly yields $c_S = 0$ for laminar shear flow (see Exercise 13.42), and that, close to the wall, the mean residual shear stress $\langle \tau_{12}^R \rangle$ correctly varies with the cube of the distance from the wall (see Exercise 13.43).

The same methodology has been applied successfully to temporally evolving free shear flows (e.g., Ghosal and Rogers (1997) and Vreman *et al.* (1997)), and also to statistically axisymmetric flows (Akselvoll and Moin 1996), in which case the averaging in Eq. (13.265) is performed in the circumferential direction. In some implementations 'clipping' is performed by setting negative values of c_S to zero, so that backscatter is eliminated (Vreman *et al.* 1997, Zang, Street, and Koseff 1993).

EXERCISE _____

13.43　Consider turbulent channel flow in the usual coordinate system so that $y = x_2$ is the normal distance from the wall. As shown in Exercise 13.30 on page 600, the LES velocity field very close to the wall (to leading order in y) can be written

$$\overline{U}_1 = \overline{a}y, \quad \overline{U}_2 = \overline{b}y^2, \quad \overline{U}_3 = \overline{c}y, \tag{13.266}$$

where $\overline{a}$, $\overline{b}$, and $\overline{c}$ are random functions of x, z, and t. Throughout this exercise we consider the region very close to the wall such that Eq. (13.266) is valid, and only the leading terms in y need be retained. Filtering is applied in x–z planes only, using uniform filters of width $\overline{\Delta}$ and $\widetilde{\Delta}$. From Eq. (13.252) obtain the result

$$\mathcal{L}_{12} = (\widetilde{\overline{a}\,\overline{b}} - \widetilde{\overline{a}}\,\widetilde{\overline{b}})\,y^3, \tag{13.267}$$

and show that the components of $\mathcal{L}_{ij}$ vary with the same power of y as do the Reynolds stresses and the true residual stresses,

---

[3] The equation for c_S used by Germano *et al.* (1991) is not exactly the same as Eq. (13.258).

Eq. (13.181). Show that the leading-order velocity gradients $\partial \overline{U}_i / \partial x_j$ are independent of y, and therefore so also is M_{ij}, Eq. (13.255). Hence show that, according to the dynamic model, c_S, v_r, and τ^r_{12} all vary correctly as y^3. Compare the modelled and correct behaviors of the other components of τ^r_{ij}.

The localized dynamic model

For general flows (without directions of statistical homogeneity) an alternative approach is needed in order to obtain stable fields of the Smagorinsky coefficient c_S. Ghosal *et al.* (1995) developed such an approach, starting from the observation that there is an inconsistency in the standard derivation of the dynamic model. Specifically, since c_S varies in space, there is little justification for assuming it to be uniform in order to evaluate $\widetilde{\tau}^r_{ij}$ (see Eqs. (13.253)–(13.256)).

If the spatial variation of c_S is taken into account in the analysis, then the difference (error) between $\mathcal{L}^d_{ij}$ and $\mathcal{L}^S_{ij}$ is no longer determined by the local value of c_S. Instead, a constrained variational problem can be solved to determine the non-negative field $c_S(x)$ that minimizes the global difference between $\mathcal{L}^d_{ij}$ and $\mathcal{L}^S_{ij}$. The Smagorinsky model with $c_S(x)$ determined in this way is called the *localized dynamic model*. Piomelli and Liu (1995) describe a simpler model based on similar ideas.

The Lagrangian dynamic model

For general inhomogeneous flows, one might consider using the dynamic model with the averaging in Eq. (13.265) for $c_S(x)$ being performed over some volume around x. However, there is some arbitrariness and difficulty in specifying the volume (especially near walls), and the evaluation of a volume average at each grid point incurs a computational cost. Instead, Meneveau *et al.* (1996) developed the Lagrangian dynamic model, in which weighted averages are formed backward in time along fluid-particle paths (based on the filtered velocity field).

According to this model, the Smagorinsky coefficient is evaluated as

$$c_S = \mathcal{J}_{\mathrm{LM}} / \mathcal{J}_{\mathrm{MM}}, \tag{13.268}$$

where $\mathcal{J}_{\mathrm{LM}}$ and $\mathcal{J}_{\mathrm{MM}}$ represent the averages $(M_{ij}\mathcal{L}_{ij})_{\mathrm{ave}}$ and $(M_{ij}M_{ij})_{\mathrm{ave}}$. The simple relaxation equation

$$\frac{\overline{\mathrm{D}}\mathcal{J}_{\mathrm{MM}}}{\overline{\mathrm{D}}t} = -(\mathcal{J}_{\mathrm{MM}} - M_{ij}M_{ij})/T \tag{13.269}$$

is solved for $\mathcal{J}_{MM}$, where T is a specified relaxation time. This is equivalent to averaging along the particle path, with relative weight $\exp[-(t-t')/T]$ at the earlier time t'. The similar equation that is solved for $\mathcal{J}_{LM}$ is

$$\frac{\overline{D}\mathcal{J}_{LM}}{\overline{D}t} = -I_0(\mathcal{J}_{LM} - M_{ij}\mathcal{L}_{ij})/T, \qquad (13.270)$$

where the indicator function

$$I_0 \equiv 1 - H(-\mathcal{J}_{LM})H(-M_{ij}\mathcal{L}_{ij}) \qquad (13.271)$$

prevents $\mathcal{J}_{LM}$ (and hence c_S) from becoming negative.

The Lagrangian dynamic model has been applied by its originators to transitional and fully turbulent channel flow with similar success to that obtained with the original plane-averaged dynamic model. A clear demonstration of its general applicability is provided by the LES calculations of the flow in the cylinder of a spark-ignition engine performed by Haworth and Jansen (2000).

Discussion

The general conclusion of *a posteriori* tests is that the dynamic model is quite successful, provided that the energy-containing motions are resolved. This proviso necessitates resolution on the order of the viscous lengthscale δ_ν close to walls (i.e., LES-NWR).

In view of this success, it is useful to review the physical basis of the two components of the model. The first component is the basic Smagorinsky-model assumption that the residual stress τ_{ij}^r is aligned with the filtered rate of strain $\overline{S}_{ij}$, Eq. (13.253). As discussed in Section 13.4.6, *a priori* tests reveal that there is a much weaker correlation between τ_{ij}^r and $\overline{S}_{ij}$ than the near-perfect correlation implied by the Smagorinsky model. Hence no value of the Smagorinsky coefficient c_S can yield the correct levels both of τ_{ij}^r and the rate of energy transfer of $\mathcal{P}_r = -\tau_{ij}^r \overline{S}_{ij}$.

The second component of the model – the dynamic aspect – is based on the assumption that the Smagorinsky coefficient is independent of the filter width (see Eqs. (13.253) and (13.254)). Equivalently, the assumption is that the Smagorinsky lengthscale ℓ_S is proportional to the filter width Δ. This assumption is well founded for the case of the filter being within the inertial subrange of high-Reynolds-number turbulence, as demonstrated by Eq. (13.131). However, the success of the model is attributed to its behavior for flow regimes for which the assumption lacks justification – laminar flow, transitional flow, and flow in the viscous wall region. Such considerations lead Jiménez and Moser (1998) to conclude that 'the physical basis for the good *a posteriori* performance of the dynamic-Smagorinsky subgrid models in LES [...] appears to be only weakly related to their ability to correctly represent the subgrid physics.'

Porté-Agel, Meneveau, and Parlange (2000) describe a scale-dependent dynamic model in which the Smagorinsky coefficient c_S is assumed to vary as an unknown power of the filter width Δ. By filtering at three levels, it is possible to deduce the exponent in the assumed power law, and hence to determine c_S.

13.6.2 *Mixed models and variants*

Over the years, many residual-stress models based on gradients of the filtered velocity $\partial \overline{U}_i/\partial x_j$ and on the doubly filtered velocity $\widetilde{\overline{U}}$ or $\overline{\overline{U}}$ have been proposed. Often these models are used in combination with the Smagorinsky model, to form so-called mixed models. The dynamic procedure can be used to determine the model coefficients. *A priori* tests of several of these models and their variants have been performed by Liu *et al.* (1994).

The Bardina model

A model proposed by Bardina, Ferziger, and Reynolds (1980) – also called the scale-similarity model – is most simply interpreted as an explicit incorporation of the Leonard stresses (although different physical interpretations of the model have been suggested by its originators and others). As discussed in Section 13.3, the residual stress can be decomposed in different ways into the Leonard stress, the cross stress, and the SGS Reynolds stress (see Eqs. (13.105) and (13.99)). Because each component of the decomposition is Galilean invariant, the preferred decomposition is that proposed by Germano (1986);

$$\tau_{ij}^R = \mathcal{L}_{ij}^o + \mathcal{C}_{ij}^o + \mathcal{R}_{ij}^o, \tag{13.272}$$

in which the Leonard stress is

$$\mathcal{L}_{ij}^o \equiv \overline{\overline{U}_i \overline{U}_j} - \overline{\overline{U}}_i \overline{\overline{U}}_j. \tag{13.273}$$

In an LES, $\overline{U}$ is known, and hence $\mathcal{L}_{ij}^o$ can be evaluated.

Previously, we have taken the Smagorinsky model as a model for τ_{ij}^r, the deviatoric part of τ_{ij}^R (Eq. (13.127)). If, instead, we take the Smagorinsky model as a model for $\mathcal{C}_{ij}^o + \mathcal{R}_{ij}^o$ and include the Leonard stresses explicitly, then the resulting model is

$$\tau_{ij}^r = (\mathcal{L}_{ij}^o - \tfrac{1}{3}\mathcal{L}_{kk}^o \delta_{ij}) - 2c_S \overline{\Delta}^2 \overline{S}\, \overline{S}_{ij}. \tag{13.274}$$

The term in $\mathcal{L}^o$ is the Bardina (or scale-similarity) model.

Successful LES calculations using this mixed model, in which the dynamic procedure is used to determine c_S, are reported by Zang *et al.* (1993) and Vreman *et al.* (1997). Indeed, for the turbulent-mixing layer, this is the most successful of six models evaluated by Vreman *et al.* (1997).

The Clark model

As observed by Clark *et al.* (1979), for a filter such as the Gaussian, an approximation for the Leonard stress is

$$\mathcal{L}_{ij}^c \equiv \frac{\overline{\Delta}^2}{12} \frac{\partial \overline{U}_i}{\partial x_k} \frac{\partial \overline{U}_j}{\partial x_k}. \tag{13.275}$$

(This can be shown by an adaptation of the Taylor-series analysis performed in Exercise 13.28 on page 595.) If this approximation is used in Eq. (13.274), the result is the mixed model

$$\tau_{ij}^r = \frac{\overline{\Delta}^2}{12} \left(\frac{\partial \overline{U}_i}{\partial x_k} \frac{\partial \overline{U}_j}{\partial x_k} - \frac{1}{3} \frac{\partial \overline{U}_\ell}{\partial x_k} \frac{\partial \overline{U}_\ell}{\partial x_k} \delta_{ij} \right) - 2c_S \overline{\Delta}^2 \overline{\mathcal{S}} S_{ij}. \tag{13.276}$$

The first term on the right-hand side is the Clark model. Vreman *et al.* (1997) applied this model (with use of the dynamic procedure to determine c_S) to the turbulent-mixing layer and obtained good results, comparable to those with the Bardina mixed model.

Alternative residual velocity scales

The eddy viscosity ν_r given by the Smagorinsky model can be written

$$\nu_r = \overline{\Delta} q_r, \tag{13.277}$$

where the velocity scale is

$$q_r = c_S \overline{\Delta} \, \overline{\mathcal{S}}. \tag{13.278}$$

Several eddy-viscosity models based on different definitions of the velocity scale q_r have been proposed.

Arguably, the velocity scale q_r should be representative of the residual motions. Using a filter such as the Gaussian, an estimate of the residual velocity is provided by the identity

$$\overline{u'} = \overline{U} - \overline{\overline{U}}. \tag{13.279}$$

On the basis of this observation, Bardina *et al.* (1980) define the velocity scale q_r as

$$q_r = c_q \left| \overline{U} \cdot \overline{U} - \overline{\overline{U}} \cdot \overline{\overline{U}} \right|^{1/2}, \tag{13.280}$$

with $c_q = 0.126$. LES calculations using a Galilean-invariant version of this model are described by Colucci *et al.* (1998).

The structure-function model

In the structure-function model of Métais and Lesieur (1992) the velocity scale is obtained from the second-order structure function defined as

$$\overline{F}_2(x) = \frac{1}{4\pi\overline{\Delta}^2} \oint |\overline{U}(x+r) - \overline{U}(x)|^2 \, \mathrm{d}\mathcal{S}(\overline{\Delta}). \qquad (13.281)$$

This is the area average, over the surface of the sphere of radius $\overline{\Delta}$ centered at x, of the square of the difference in velocity over the distance $\overline{\Delta}$. The velocity scale used to define v_r (Eq. 13.277) is then

$$q_r = c_q \overline{F}_2^{1/2}, \qquad (13.282)$$

with $c_q = 0.063$. A description of the theoretical basis for the structure-function model, as well as a review of extensions and applications, is provided by Lesieur and Métais (1996).

EXERCISE

13.44   Show that a first-order estimate of $\overline{F}_2$ (Eq. (13.281)) based on a Taylor-series expansion of $\overline{U}(x+r)$ is

$$\overline{F}_2 \approx \tfrac{1}{3}\overline{\Delta}^2(\overline{S}_{ij}\overline{S}_{ij} + \overline{\Omega}_{ij}\overline{\Omega}_{ij}), \qquad (13.283)$$

where $\overline{\Omega}_{ij}$ is the rate of rotation of the filtered velocity field. If the rate-of-strain and rate-of-rotation invariants are equal, show that the structure-function model (based on Eq. (13.283)) reduces to the Smagorinsky model with

$$C_S = 3^{-1/4}c_q^{1/2} \approx 0.19. \qquad (13.284)$$

13.6.3 Transport-equation models

All of the models for the residual-stress tensor considered above relate $\tau_{ij}^r(x,t)$ to the filtered velocity field $\overline{U}(x',t)$ at the same time t and in the neighborhood of x. In the pursuit of more accurate models, it is natural to incorporate history and non-local effects through transport equations for τ_{ij}^r and other quantities related to the residual motions. Such approaches parallel conventional turbulence modelling: just as the Smagorinsky model is analogous to the mixing-length model, transport-equation models that are analogous to one-equation models, algebraic stress models, Reynolds-stress models, and PDF models have been proposed. A difference, however, is that, in each case, the filter width Δ is taken as the characteristic lengthscale of the residual motions, so that a scale equation (analogous to the model equation for ε or ω) is not used.

One of the few examples of LES in which a model transport equation is solved for the residual stress τ_{ij}^R is the early work of Deardorff (1974). The model equation is of the same form as a Reynolds-stress closure: production is in closed form; redistribution is modelled by Rotta's model and by a simple rapid model; transport is by gradient diffusion; and the dissipation is taken to be isotropic. The instantaneous rate of dissipation ε_r of the residual kinetic energy $k_r \equiv \frac{1}{2}\tau_{ii}^R$ is related to k_r and the filter width Δ by

$$\varepsilon_r = \frac{C_E k_r^{3/2}}{\Delta}, \tag{13.285}$$

where the constant C_E is taken to be $C_E = 0.7$. As shown in Exercise 13.45, under the ideal conditions of a high Reynolds number and the sharp spectral filter in the inertial subrange, this form can be justified, and the constant C_E can be related to the Kolmogorov constant C.

In order to decrease the computational cost, subsequently Deardorff (1980) reverted to an isotropic eddy-viscosity model, with the eddy viscosity v_r given by

$$v_r = C_v k_r^{1/2}\Delta, \tag{13.286}$$

and with a model transport equation used to determine the residual kinetic energy. Again, under ideal conditions, this form for v_r can be justified, and the value of the constant $C_v \approx 0.1$ can be estimated (see Exercise 13.45). This type of model has been used extensively by the meteorological community, for example, Mason (1989), Schmidt and Schumann (1989) and Sullivan, McWilliams, and Moeng (1994). Rather than using an eddy-viscosity model, Schmidt and Schumann (1989) obtain the residual stress from an algebraic approximation to its transport equation – analogous to the algebraic stress model.

Wong (1992) describes (but does not demonstrate) a method in which equations are solved for the residual kinetic energies $\overline{k}_r$ and $\widetilde{k}_r$, corresponding to two different filter widths, $\overline{\Delta}$ and $\widetilde{\Delta}$. The dynamic procedure can be used to determine both coefficients, C_E and C_v. A different use of the equation for k_r in combination with the localized dynamic model is described by Ghosal et al. (1995).

Analogous to the joint PDF of velocity, the *filtered density function* (FDF) is defined by

$$\overline{f}(V;x,t) = \int G(r)\delta(U(x-r,t)-V)\,dr. \qquad (13.287)$$

For positive filters ($G \geq 0$), $\overline{f}$ has all of the properties of a PDF (see Exercise 13.21 on page 584). The first moments of $\overline{f}$ are the components of the filtered velocity $\overline{U}$; the second moments are the residual stresses τ_{ij}^{R}. Gicquel *et al.* (1998) performed LES calculations using a model equation for $\overline{f}(V;x,t)$, analogous to the Langevin model.

LES approaches incorporating a model equation for k_r have proved advantageous in meteorological applications in which other effects (e.g., buoyancy) are present. Approaches based on the filtered density function are advantageous also for reactive flows in which the important processes of mixing and reaction occur on subgrid scales. However, with these exceptions, the general experience is that, in LES, the additional computational cost and complexity involved in solving additional transport equations is not justified by an assured increase in accuracy. The benefits of using additional transport equations are more evident in VLES, in which a greater fraction of the energy is in the subgrid-scale motions.

EXERCISE _____

13.45 Consider high-Reynolds-number homogeneous turbulence with the sharp spectral filter in the inertial subrange. Use the Kolmogorov spectrum (with $C = 1.5$) and the relation $\langle \varepsilon_r \rangle \approx \varepsilon$ to obtain the estimate for C_E (Eq. (13.285))

$$C_E \approx \pi\left(\tfrac{3}{2}C\right)^{-3/2} \approx 0.93. \qquad (13.288)$$

With the eddy viscosity given by Eq. (13.286), show that the approximations

$$\varepsilon \approx \langle \mathcal{P}_r \rangle = \langle v_r \overline{S}^2 \rangle \approx C_v \Delta \langle k_r \rangle^{1/2} \langle \overline{S}^2 \rangle \qquad (13.289)$$

lead to the estimate

$$C_v \approx \left(\tfrac{3}{2}C\right)^{-3/2}/\pi \approx 0.094. \qquad (13.290)$$

13.6.4 Implicit numerical filters

In the numerical solution of the LES momentum equation, Eq. (13.95), various numerical errors are incurred, the most important being the spatial-truncation error. One way to express this error is through the *modified equation*, which is the partial differential equation satisfied by the *numerical* solution. The modified equation corresponding to the LES momentum equation can be written

$$\frac{\overline{D}\,\overline{U}_j}{\overline{D}t} = \nu\,\frac{\partial^2\overline{U}_j}{\partial x_i\,\partial x_i} - \frac{\partial}{\partial x_i}(\tau_{ij}^{\mathrm{r}} + \tau_{ij}^{h}) - \frac{1}{\rho}\,\frac{\partial\bar{p}}{\partial x_j}, \qquad (13.291)$$

so that the spatial-truncation error appears as an additional numerical stress, τ_{ij}^{h}, which depends on the grid spacing h. If the spatial discretization is pth-order accurate, then τ_{ij}^{h} is of order h^p. (A simple example of a first-order accurate scheme is given in Exercise 13.46, in which the numerical stress can be expressed as an additional viscous term, with the numerical viscosity ν_{num} being proportional to h.)

There are differing viewpoints on the role of the numerical stress in LES. The simplest (which is implicitly assumed in the preceding discussion) is that the LES equations should be solved accurately. That is, for a given filter width Δ, the grid spacing h should be chosen to be sufficiently small that the numerical stress τ_{ij}^{h} is negligible compared with the modelled residual stress τ_{ij}^{r}.

The opposite viewpoint, advocated by Boris *et al.* (1992), is that no explicit filtering should be performed and no explicit residual stress model should be used ($\tau_{ij}^{\mathrm{r}} = 0$). Instead, an appropriate numerical method is used to attempt to solve the Navier–Stokes equations for $\overline{U}$. Because the grid is not fine enough to resolve the solution to the Navier–Stokes equations, significant numerical stresses τ_{ij}^{h} arise. Thus, filtering and residual-stress modelling are performed implicitly by the numerical method.

The numerical stress τ_{ij}^{h} depends on the type of numerical method used, and hence this choice is crucial. The ideal scheme is accurate (i.e., τ_{ij}^{h} is small) for the well-resolved contributions to $\overline{U}(x,t)$ (with long wavelength relative to h), while it attenuates poorly resolved shorter-wavelength contributions (and prevents aliasing errors from contaminating the well-resolved modes). Energy is removed from the resolved motions by numerical dissipation at the rate

$$\varepsilon_{\mathrm{num}} \equiv -\tau_{ij}^{h}\overline{S}_{ij}. \qquad (13.292)$$

Boris *et al.* (1992) use so-called monotone numerical methods, and refer to their approach as MILES – monotone integrated large-eddy simulation. In addition to the works cited by Boris *et al.* (1992), representative LES calculations using this approach are described by Tamura and Kuwahara (1989), Knight *et al.* (1998), and Okong'o and Knight (1998).

Compared with explicit modelling of the residual stresses, the MILES approach has advantages and disadvantages, advocates and detractors. The advantages are that (for a given grid size) as much as possible of the turbulent motion is represented explicitly by the LES velocity field $\overline{U}(x,t)$, and that energy is removed from $\overline{U}$ only where and when it is necessary to do so. It is argued that the details of how energy is removed are unimportant, just so long as there is a mechanism to remove energy from the smallest resolved scales without contaminating the larger scales. (This is similar to the argument used to justify the use of residual-stress models that perform poorly in *a priori* tests.) A further advantage is that the time and effort required to develop and test a residual-stress model are eliminated.

The primary disadvantage is that the modelling and the numerics are inseparably coupled. Sometimes the approach is referred to as 'no model,' but it should be appreciated that this is an inadequate description: for a given flow the simulation results depend both on the numerical method and on the grid used. It is not possible to refine the grid to obtain grid-independent solutions (short of performing DNS). Another disadvantage is that there is no representation or estimation of the subgrid-scale motions that can be used for defiltering or in models for other subgrid-scale processes.

EXERCISE

13.46 Consider the numerical solution of the Burgers equation (Burgers 1940)

$$\frac{\partial u}{\partial t} + u\,\frac{\partial u}{\partial x} = v\,\frac{\partial^2 u}{\partial x^2}, \qquad (13.293)$$

for the positive velocity $u(x,t) > 0$. A crude finite-difference method is used, in which the spatial derivatives at x are approximated by

$$\frac{\partial u}{\partial x} \approx \mathcal{D}_h^- u(x) \equiv \frac{u(x) - u(x-h)}{h}, \qquad (13.294)$$

$$\frac{\partial^2 u}{\partial x^2} \approx \mathcal{D}_h^2 u(x) \equiv \frac{u(x+h) - 2u(x) + u(x-h)}{h^2}, \qquad (13.295)$$

where h is the grid spacing. Use a Taylor-series analysis to determine the leading-order truncation errors in these approximations. Hence show that the equation solved, i.e.,

$$\frac{\partial u}{\partial t} + u\mathcal{D}_h^- u = v\mathcal{D}_h^2 u, \qquad (13.296)$$

is equivalent to the modified equation

$$\frac{\partial u}{\partial t} + u\,\frac{\partial u}{\partial x} = (v + v_{\text{num}})\,\frac{\partial^2 u}{\partial x^2} + \mathcal{O}(h^2), \qquad (13.297)$$

where the numerical viscosity is

$$\nu_{\text{num}} = \tfrac{1}{2}uh. \tag{13.298}$$

(Thus the leading-order truncation error is equivalent to an additional viscous term.)

13.6.5 Near-wall treatments

In LES with near-wall resolution, no-slip and impermeability boundary conditions are applied to the filtered velocity at the wall. The correct behavior of the Smagorinsky lengthscale ℓ_S near the wall is $\ell_S \approx y^{3/2}$, which leads to $\nu_r \approx y^3$ and $\tau_{12}^r \approx y^3$ (see Exercise 13.31 on page 601). In contrast, for constant C_S, the Smagorinsky model yields, incorrectly, constant values of ℓ_S, ν_r, and τ_{12}^r. One approach to overcome this deficiency is to damp ℓ_S by, for example, the van Driest damping function (Eq. (13.166)). More satisfactory is to determine c_S from the dynamic model. This yields the correct near-wall scalings, and quite satisfactory results for channel flow (see, e.g., Piomelli (1993)).

A substantial computational saving accrues if the viscous wall region is not resolved, and, indeed, this is the only feasible approach at high Reynolds number (see Section 13.4.5). In this case of LES-NWM, the effects of the unresolved motions are usually modelled through the use of boundary conditions similar to the wall functions used in turbulent-viscosity and Reynolds-stress models (Section 11.7.6). These conditions are applied either at the wall ($y = 0$) or at the first grid node away from the wall ($y = y_p$).

The impermeability condition, $\overline{U}_2 = 0$, is applied to the normal velocity. Boundary conditions on the tangential components of $\overline{U}$ are applied implicitly through the specification of the shear stresses as

$$\tau_{i2}^r(x,0,z) = u^*(x,z)\overline{U}_i(x,y_p,z), \quad \text{for } i = 1 \text{ and } 3. \tag{13.299}$$

The velocity scale u^* is obtained by assuming that, instantaneously, $\overline{U}$ satisfies the log law (Grötzbach 1981, Mason and Callen 1986), or a power law (Werner and Wengle 1989). A modification suggested by Piomelli, Ferziger, and Moin (1989) is to base the shear stress at x on the filtered velocity $\overline{U}_i(x + \Delta_s, y_p, z)$ some distance Δ_s downstream of x.

The assumptions embodied in the specification of u^* and the assumption that the shear stress is aligned with the tangential velocity are of course subject to criticism. In addition, the impermeability condition $\overline{U}_2 = 0$ is less innocuous than it may appear. The Reynolds stress $\langle u_i u_j \rangle$ can be decomposed into the resolved Reynolds stress R_{ij}^f (Eq. (13.185)) and the mean residual Reynolds stress $\langle \tau_{ij}^r \rangle$; and in LES ideally R_{ij}^f is the dominant contribution. However, the specification $(\overline{U}_2)_p = 0$ forces $(R_{i2}^f)_p$ to be zero, so that the shear stresses $\langle u_i u_2 \rangle$ arise entirely from the modelled boundary condition, Eq. (13.299).

For channel flow up to $\mathrm{Re}_\tau = 5,000$, Balaras, Benocci, and Piomelli (1995) obtained satisfactory results using the dynamic model combined with this type of near-wall treatment. It is generally recognized, however, that, for complex flows – with separation, reattachment, impingement, etc. – these wall treatments are the least satisfactory aspect of LES-NWM.

13.7 Discussion

In this section the LES methodology is discussed with respect to the criteria for appraising models described in Chapter 8; and then a final perspective on the simulation and modelling of turbulent flows is given.

13.7.1 An appraisal of LES

The level of description

Since the filtered velocity field provides a direct representation of the energy-containing motions, the level of description in LES is more than sufficient for most purposes. Because the resolved fields are filtered, some approximations are involved in extracting unfiltered one-point statistics such as $\langle U \rangle$ and $\langle u_i u_j \rangle$ (see Exercise 13.32); but in practice this is not a major concern. Compared with RANS, LES has the advantage of describing the unsteady, large-scale turbulent structures, and hence can be used to study phenomena such as unsteady aerodynamic loads on structures and the generation of sound.

Completeness

The mixing-length model is incomplete (because $\ell_m(x)$ has to be specified), whereas the k–ε and Reynolds-stress models are complete. Is LES complete? Or are flow-dependent specifications required? The answer hinges on the specification of the filter width Δ – in general as a function of position and time.

In applying LES to a complex flow, the usual practice is to specify the grid to be used in the numerical solution, and then (implicitly or explicitly) to take the filter width Δ to be proportional to the local grid size h. If the LES solution depends on the specification of the grid, then the model is incomplete. Experience shows that LES calculations are likely to depend on the grid if the energy-containing motions are poorly resolved (i.e., VLES). To be complete, this approach therefore requires prior knowledge of the flow so that the grid (and filter) can be appropriately specified to ensure that the bulk of the turbulence energy is resolved throughout the domain. The ideal numerical method for LES would include adaptive gridding to ensure automatically that the grid, and hence the filter, are everywhere sufficiently fine to resolve the energy-containing motions.

Cost and ease of use

The computational cost of LES requires careful consideration, especially for wall-bounded flows. For homogeneous isotropic turbulence, 40^3 modes are sufficient to resolve 80% of the energy, even at high Reynolds number (see Exercise 13.34). (Sometimes fewer modes (e.g., 32^3) are used so that there is less resolution of the spectrum either at high or at low wavenumbers.)

For free shear flows, the number of modes or grid points generally used is comparable. Hence, LES of such flows is quite practicable; and, in marked contrast to DNS, the computational cost is independent of the Reynolds number. On the other hand, compared with a Reynolds-stress model, LES is more expensive by approximately two orders of magnitude for each direction of statistical homogeneity.

For LES of wall-bounded flows, the resolution requirements are qualitatively different, because the size of the important near-wall motions scales with the viscous lengthscale δ_v (which decreases with the Reynolds number relative to the flow lengthscale δ). Chapman (1979) estimates that the cost of LES with near-wall resolution (LES-NWR) in aerodynamic applications increases with the Reynolds number as $Re^{1.8}$; that is, the cost increases by a factor of 60 for each decade increase in Reynolds number. (Essentially the same result is obtained in Exercise 13.29 on page 599.) Thus, while LES-NWR has been applied successfully to boundary layers and channel flows with low and moderate Reynolds numbers, its application to high-Reynolds-number flows such as those over aircrafts' wings, ships' hulls, and in the atmospheric boundary layer is infeasible.

The alternative is LES with near-wall modelling (LES-NWM), in which the near-wall energy-containing motions are not resolved. Then, depending on the implementation, the computational cost is either independent of the Reynolds number, or it increases weakly as $\ln Re$ (see Exercise 13.29).

Since LES – even for isotropic turbulence – requires a time-dependent three-dimensional computation, its cost penalty compared with a Reynolds-stress calculation is least for statistically three-dimensional and non-stationary flows (for which the Reynolds-stress-model calculation is also three-dimensional and unsteady). Hence, for example, LES is an attractive approach for the inherently three-dimensional and unsteady flow in spark-ignition engines (Haworth and Jansen 2000). For statistically stationary three-dimensional flows (e.g., Rodi *et al.* (1997)), the LES computation is generally performed for several flow-transit times so that time averaging can be used to estimate means. This need for time averaging increases the computational cost relative to that of a Reynolds-stress calculation.

For the three-dimensional flows generally encountered in engineering applications, the numerical task of solving the LES equation is substantially the same as that of solving the Navier–Stokes or turbulence-model equations.[4] Consequently the availability of codes to perform LES is comparable to that for turbulence models. Many commercial CFD codes include an 'LES option.' It should be appreciated, however, that on the grids generally employed, the calculations are VLES, and the spatial truncation errors may be substantial (compared with the modelled residual stresses).

The range of applicability

For constant-density inert flows, LES is generally applicable – although, as discussed above, computational cost dictates the use of LES-NWM rather than LES-NWR for high-Reynolds-number, wall-bounded flows. Compared with RANS, the higher level of description provided by LES increases its range of applicability to aeroacoustics and other phenomena associated with the unsteady turbulent motions.

Methodologies based on LES have also been developed for more complex turbulent-flow phenomena, such as high-speed compressible flows (see Hussaini (1998)), and reacting flows (e.g., Cook and Riley (1994) and Colucci *et al.* (1998)). Some of the important processes involved (e.g., molecular mixing and reaction) can occur predominantly on subgrid scales, and hence statistical modelling is required.

Accuracy

As with other turbulence approaches, the ideal way to assess the accuracies of various LES methodologies is to compare their performances over a broad range of test flows. Extraneous errors (see Fig. 8.2 on page 342) must be shown to be small; the test flows should be different than those used in the model's development; and the tests should not be performed by the model's developers. An additional requirement particular to LES is that the calculation should be performed for the filter width being varied

---

[4] However, see Boris *et al.* (1992) for a discussion of the numerical issues particular to LES.

over some range, to allow an assessment of the sensitivity of the calculation to the specification of the filter. Tests satisfying these requirements have not been performed, so an assessment of the accuracy of LES is subject to considerable uncertainty. Nevertheless, some useful comparative results are presented by Fureby *et al.* (1997) (for isotropic turbulence), by Vreman *et al.* (1997) (for the mixing layer), and by Rodi *et al.* (1997) (for two flows over square obstacles).

The mixing-layer comparisons of Vreman *et al.* (1997) demonstrate that quite accurate results can be obtained by using, for example, the dynamic mixed model (Eq. (13.274)). These and other calculations support the view that LES can be reasonably accurate for free shear flows.

For simple boundary-layer-type flows at moderate Reynolds number, there are numerous LES-NWR calculations based on the dynamic model that support a similar conclusion for these flows (e.g., Piomelli (1993)). However, it should be acknowledged that the number of different flows (used both for development and testing) is not large.

For more complex and higher-Reynolds-number flows, the limited evidence suggests that the accuracy of LES-NWM is less certain. For example, on the basis of a comparative study of various LES-NWM calculations of flow over obstacles, Rodi *et al.* (1997) conclude that the 'results are sensitive to the grid and numerical method' and that the wall treatments used are 'not reliable to be used with confidence in separated flows.'

13.7.2 Final perspectives

In the context of turbulent flows, a continuing major challenge to research is to develop methodologies (using the ever increasing available computer power) to calculate the flow and turbulence properties of practical relevance in engineering, atmospheric sciences, and elsewhere. This is the theme of Part II of this book.

There is a broad range of turbulent flow problems, varying in geometrical complexity, many involving additional physical and chemical processes, for which various levels of description and accuracy are sought. Consequently, it is valuable to have a broad range of approaches with various attributes, whose computational costs vary by as much as a factor of 10^6. Each of the approaches described in Chapters 10–13 has its place: no one approach will supplant all others.

Furthermore, there is a useful synergy among the approaches. While the computational cost of DNS precludes its use for high-Reynolds-number flows, it provides invaluable information for the development and testing of other approaches. RANS modelling can be expected to contribute to the major outstanding issue in LES – the modelling of the near wall region in high-Reynolds-number flows. Statistical approaches, such as PDF methods, can be used in conjunction with LES for the treatment of turbulent reactive flows, and other flows in which important processes occur on subgrid scales.

For some turbulent-flow problems, existing methodologies are acceptably accurate, and have an acceptable computational cost. For many others, substantial advances are required. Consequently, turbulent flows will remain a worthwhile and challenging area for research for some time to come.

Part three
Appendices

Appendix A
Cartesian tensors

From vector calculus we are familiar with *scalars* and *vectors*. A scalar has a single value, which is the same in any coordinate system. A vector has a magnitude and a direction, and (in any given coordinate system) it has three components.[1] With Cartesian tensors, we can represent not only scalar and vectors, but also quantities with more directions associated with them. Specifically, an *Nth-order tensor* ($N \geq 0$) has N directions associated with it, and (in a given Cartesian coordinate system) it has 3^N components. A zeroth-order tensor is a scalar, and a first-order tensor is a vector. Before defining higher-order tensors, we briefly review the representation of vectors in Cartesian coordinates.

A.1 Cartesian coordinates and vectors

Fluid flows (and other phenomena in classical mechanics) take place in the three-dimensional, Euclidean, *physical space*. As sketched in Fig. A.1, let E denote a Cartesian coordinate system in physical space. This is defined by the position of the origin O, and by the directions of the three mutually perpendicular axes. The unit vectors in the three coordinate directions are denoted by e_1, e_2, and e_3. We write e_i to refer to any one of these, with the understanding that the *suffix i* (or any other suffix) takes the value 1, 2, or 3.

The basic properties of the unit vectors e_i are succinctly expressed in terms of the *Kronecker delta* δ_{ij}. This is defined by

$$\delta_{ij} = \begin{cases} 1, & \text{if } i = j, \\ 0, & \text{if } i \neq j, \end{cases} \tag{A.1}$$

---

[1] We consider here three-dimensional space. The extension of tensors to spaces of different dimensions is straightforward.

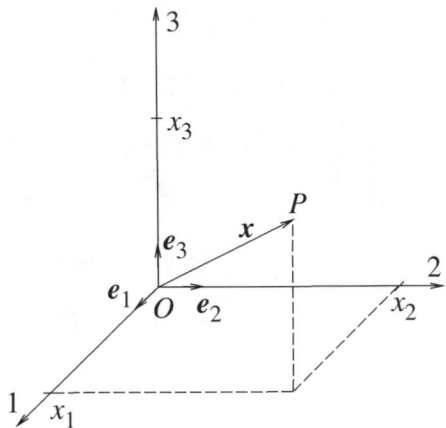

Fig. A.1. A sketch of the E coordinate system showing the origin O, the three orthonormal basis vectors, e_i, and the general point P with position $x = x_1e_1 + x_2e_2 + x_3e_3$.

or, in matrix notation,

$$
\begin{bmatrix} \delta_{11} & \delta_{12} & \delta_{13} \\ \delta_{21} & \delta_{22} & \delta_{23} \\ \delta_{31} & \delta_{32} & \delta_{33} \end{bmatrix} = \begin{bmatrix} 1 & 0 & 0 \\ 0 & 1 & 0 \\ 0 & 0 & 1 \end{bmatrix}.
\tag{A.2}
$$

It is evident from their construction that the vectors e_i have the *orthonormality* property:

$$
e_i \cdot e_j = \delta_{ij}.
\tag{A.3}
$$

For, if $i = j$, then the dot product is unity (since they are unit vectors), whereas, for $i \neq j$, the dot product is zero (since the vectors are orthogonal). Thus e_1, e_2, and e_3 form a set of *orthonormal basis vectors*.

Referring to Fig. A.1, x is the *position vector*, giving the position of the general point P relative to the origin O. This vector can be written

$$
x = e_1x_1 + e_2x_2 + e_3x_3,
\tag{A.4}
$$

where x_1, x_2, and x_3 are the *components* of x in the E coordinate system.

The summation convention

The notation of Cartesian tensors is considerably simplified by use of the Einstein summation convention. According to this convention, if a suffix is repeated (e.g., the suffix i in e_ix_i), then summation over all three values of the suffix ($i = 1$, 2 and 3) is implied. Hence, with this implied summation, Eq. (A.4) is written

$$
x = e_ix_i.
\tag{A.5}
$$

Note that the symbol ascribed to the *repeated* or *dummy* suffix is irrelevant (i.e., $e_i x_i = e_j x_j$).

In tensor equations, $+$, $-$, and $=$ signs separate different *expressions*. In a given expression, a suffix that is not repeated is called a *free* suffix. An important rule associated with the summation convention is that *a suffix cannot appear more than twice in an expression*.

EXERCISES _____

A.1 By explicitly expanding the summations, show the *substitution rule*

$$\delta_{ij} x_j = x_i. \tag{A.6}$$

Obtain the results

$$\delta_{ii} = 3, \tag{A.7}$$

$$\delta_{ij}\delta_{jk} = \delta_{ik}. \tag{A.8}$$

A.2 Show that the components of a vector can be obtained by

$$e_j \cdot x = e_j \cdot (e_i x_i) = x_j. \tag{A.9}$$

Let u and v be vectors with components (u_1, u_2, u_3) and (v_1, v_2, v_3). Show that the dot product is

$$u \cdot v = (e_i u_i) \cdot (e_j v_j) = u_i v_i = (e_i \cdot u)(e_i \cdot v). \tag{A.10}$$

Coordinate transformations

Tensors are defined in terms of their transformation properties. We now introduce a second Cartesian coordinate system (denoted by $\overline{E}$) and determine the transformation rules for the components of a vector.

The $\overline{E}$ system is obtained from E by any combination of rotations and reflections of the axes; an example is shown in Fig. A.2. It should be noted that, if there is an odd number of reflections, then the handednesses of E and $\overline{E}$ will be different. That is, if E is a right-handed system, and $\overline{E}$ is obtained by reflecting one axis, then $\overline{E}$ is a left-handed system. In Cartesian tensors, both right-handed and left-handed systems are valid, and there is no necessity to distinguish between them.

Let $\overline{e}_1$, $\overline{e}_2$, and $\overline{e}_3$ be the orthonormal basis vectors of the $\overline{E}$ system. The nine quantities defined by the dot products

$$a_{ij} \equiv e_i \cdot \overline{e}_j \tag{A.11}$$

are the *direction cosines*: a_{ij} is the cosine of the angle between the i axis in

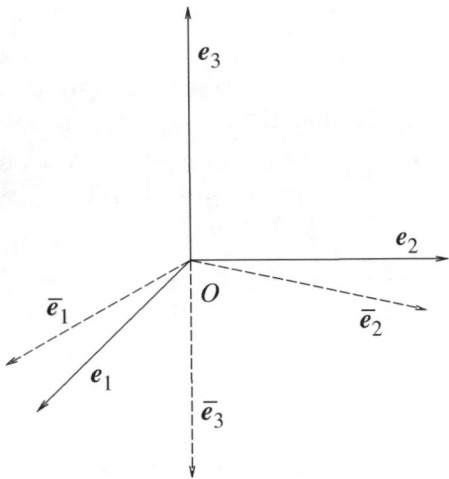

Fig. A.2. A sketch of the E (solid lines) and $\overline{E}$ (dashed lines) coordinate systems. In this particular example, $\overline{E}$ is obtained from E by a reflection of the $\overline{e}_3$ axis, and a rotation in the e_1–e_2 plane.

the E system and the j axis in the $\overline{E}$ system. Further, a_{ij} is the i component in the E system of $\overline{e}_j$ (see Eq. (A.9)). From this interpretation we obtain a fundamental property of the direction cosines: the dot product of $\overline{e}_j$ and $\overline{e}_k$ is (see Eq. (A.10))

$$\overline{e}_j \cdot \overline{e}_k = (e_i \cdot \overline{e}_j)(e_i \cdot \overline{e}_k)$$
$$= a_{ij}a_{ik}. \tag{A.12}$$

Hence, because the vectors $\overline{e}_j$ are orthonormal, we obtain

$$a_{ij}a_{ik} = \delta_{jk}, \tag{A.13}$$

and similarly it can be shown that

$$a_{ji}a_{ki} = \delta_{jk}. \tag{A.14}$$

The vector x is the same in the E and $\overline{E}$ systems, but its components (x_1, x_2, x_3) and $(\overline{x}_1, \overline{x}_2, \overline{x}_3)$ are different:

$$x = e_i x_i = \overline{e}_j \overline{x}_j. \tag{A.15}$$

The transformation rules are obtained by taking the dot product of this equation with e_k and $\overline{e}_k$:

$$x_k = a_{kj}\overline{x}_j, \tag{A.16}$$

$$\overline{x}_k = a_{ik}x_i. \tag{A.17}$$

Derivatives

The components of x are coordinates in the E system. We can therefore differentiate with respect to x_i functions defined in physical space. Some useful results are contained in the following exercises.

EXERCISE _____

A.3   Obtain the following results:

$$\frac{\partial x_j}{\partial x_i} = \delta_{ij}, \quad \frac{\partial x}{\partial x_i} = e_i, \quad \frac{\partial r}{\partial x_i} = \frac{x_i}{r}, \quad \text{where } r \equiv (x \cdot x)^{1/2},$$

$$\frac{\partial \bar{x}_j}{\partial x_i} = a_{ij}, \quad \frac{\partial x_j}{\partial \bar{x}_i} = a_{ji}, \quad \frac{\partial \bar{x}_i}{\partial x_j}\frac{\partial x_j}{\partial \bar{x}_k} = \delta_{ik}. \tag{A.18}$$

A.2 The definition of Cartesian tensors

A *zeroth-order tensor* is a scalar. It has $3^0 = 1$ component, which has the same value in every coordinate system. Examples are physical quantities such as density, temperature, and pressure, as well as dot products of vectors.

A *first-order tensor*, u,

$$u = e_i u_i = \bar{e}_j \bar{u}_j, \tag{A.19}$$

has $3^1 = 3$ components which transform by

$$\bar{u}_j = a_{ij} u_i, \tag{A.20}$$

(cf. Eq. (A.17)). As illustrated in the next two exercises, the way to determine whether a quantity of the form of Eq. (A.19) is a first-order tensor is to determine its transformation rule, and then to compare this rule with Eq. (A.20).

EXERCISE _____

A.4   Let $X(t)$ and $U(t) \equiv dX(t)/dt$ be the position and velocity of a particle. Determine the transformation rule between the components $U_i(t)$ and $\bar{U}_i(t)$, and hence show that velocity is a first-order tensor.

A.5   Let $\phi(x)$ be a scalar field. Determine the transformation rule between $g_i \equiv \partial\phi/\partial x_i$ and $\bar{g}_i \equiv \partial\phi/\partial \bar{x}_i$, and hence show that the scalar gradient

$$\nabla\phi = g = e_i g_i = \bar{e}_i \bar{g}_i \tag{A.21}$$

is a first-order tensor.

648 A Cartesian tensors

Second-order tensors

A second-order tensor, **b**,

$$\mathbf{b} = e_i e_j b_{ij} = \overline{e}_k \overline{e}_\ell \overline{b}_{k\ell}, \tag{A.22}$$

has $3^2 = 9$ components which transform by

$$\overline{b}_{k\ell} = a_{ik} a_{j\ell} b_{ij}. \tag{A.23}$$

There are several observations to be made about the expression $e_i e_j b_{ij}$ in Eq. (A.22). There is no operator (e.g., dot) between e_i and e_j, and the quantity $e_i e_j$ has no equivalent in standard vector notation. As always, the symbol used for the dummy suffixes is irrelevant, i.e.,

$$e_p e_q b_{pq} = e_i e_j b_{ij}, \tag{A.24}$$

but the ordering of the unit vectors is important, i.e.,

$$e_i e_j b_{ij} \neq e_j e_i b_{ij}. \tag{A.25}$$

In fact, the right-hand side of Eq. (A.25) defines a different second-order tensor, namely the transpose of **b**:

$$\mathbf{b}^{\mathrm{T}} = e_i e_j b_{ij}^{\mathrm{T}} = e_i e_j b_{ji} = e_j e_i b_{ij}. \tag{A.26}$$

To determine whether a quantity of the form of Eq. (A.22) is a second-order tensor, the transformation rule needs to be determined and compared with Eq. (A.23). For example, consider the quantities c_{ij} and $\overline{c}_{k\ell}$ defined by

$$c_{ij} \equiv u_i v_j, \quad \overline{c}_{k\ell} \equiv \overline{u}_k \overline{v}_\ell, \tag{A.27}$$

where **u** and **v** are vectors. Are these the components of a second-order tensor? To determine the answer, we need to determine the transformation rule for $\overline{c}_{k\ell}$, which we obtain from those of $\overline{u}_k$ and $\overline{v}_\ell$ (Eq. (A.19)):

$$\overline{c}_{k\ell} \equiv \overline{u}_k \overline{v}_\ell = (a_{ik} u_i)(a_{j\ell} v_j)$$
$$= a_{ik} a_{j\ell} c_{ij}. \tag{A.28}$$

This is indeed the transformation rule Eq. (A.23): thus $\mathbf{c} = e_i e_j c_{ij} = \overline{e}_k \overline{e}_\ell \overline{c}_{k\ell}$ is a second-order tensor.

EXERCISE

A.6 Show that the following are the components of second-order tensors:

$$\delta_{ij}, \quad \frac{\partial^2 \phi}{\partial x_i \partial x_j}, \quad \frac{\partial u_i}{\partial x_j},$$

where $\phi(x)$ is a scalar field and $u(x)$ is a vector field.

Tensors of higher order

Extension to higher-order tensors is straightforward. A *third-order tensor*

$$\mathbf{d} = e_i e_j e_k \, d_{ijk} = \overline{e}_\ell \overline{e}_m \overline{e}_n \, \overline{d}_{\ell mn} \tag{A.29}$$

has $3^3 = 27$ components, which transform by

$$\overline{d}_{\ell mn} = a_{i\ell} a_{jm} a_{kn} d_{ijk}. \tag{A.30}$$

An Nth-order tensor ($N \geq 0$) is of the form

$$\mathbf{f} = e_i e_j \ldots e_k f_{ij\ldots k} = \overline{e}_\ell \overline{e}_m \ldots \overline{e}_n \overline{f}_{\ell m\ldots n} \tag{A.31}$$

and has 3^N components, which transform by

$$\overline{f}_{\ell m\ldots n} = a_{i\ell} a_{jm} \ldots a_{kn} f_{ij\ldots k}. \tag{A.32}$$

A.3 Tensor operations

We now describe the operations that can be performed with tensors. If the operands are tensors, then so also is the result. It is left to the reader to demonstrate this fact.

Addition

Two tensors *of the same order* can be added or subtracted. For example, if $\mathbf{b}$ and $\mathbf{c}$ are second-order tensors, then their sum $\mathbf{s}$ is given by

$$
\begin{aligned}
\mathbf{s} &= e_i e_j s_{ij} \\
&= \mathbf{b} + \mathbf{c} = e_i e_j \, b_{ij} + e_i e_j c_{ij} \\
&= e_i e_j \, (b_{ij} + c_{ij}).
\end{aligned} \tag{A.33}
$$

Thus, in terms of components, we have, simply,

$$s_{ij} = b_{ij} + c_{ij}. \tag{A.34}$$

Tensor products

As an example, the tensor product of a first-order tensor $\boldsymbol{u}$ and the second-order tensor $\mathbf{b}$ is the third-order tensor $\mathbf{d}$:

$$
\begin{aligned}
\mathbf{d} &= e_i e_j e_k \, d_{ijk} \\
&= \boldsymbol{u}\mathbf{b} = (e_i u_i)(e_j e_k b_{jk}) \\
&= e_i e_j e_k \, u_i b_{jk}.
\end{aligned} \tag{A.35}
$$

Hence, in terms of components,

$$d_{ijk} = u_i b_{jk}. \tag{A.36}$$

In general, the tensor product of an Nth-order tensor and an Mth-order tensor is a tensor of order $N + M$.

The product of a scalar (i.e., a zeroth-order tensor) and an Nth-order tensor is an Nth-order tensor. In particular (taking the scalar to be -1) if b_{ij} is a tensor, so also is $-b_{ij}$, and hence subtraction is a valid operation, e.g.,

$$h_{ij} = b_{ij} - c_{ij}. \tag{A.37}$$

Equations (A.34), (A.36), and (A.37) are examples of *tensor equations* in component form. In these equations, the various expressions are separated by $+$, $-$, or $=$. In any tensor equation, each expression has the same free suffixes, e.g., i, j, and k in Eq. (A.36). The ordering of the suffixes is significant, but it does not need to be the same in each expression. So, for example, the equations

$$d_{ijk} = u_i b_{jk}, \tag{A.38}$$

$$d'_{ijk} = u_j b_{ik}, \tag{A.39}$$

$$d''_{ijk} = u_j b_{ki} \tag{A.40}$$

are all valid tensor equations, but they define different tensors $\mathbf{d}$, $\mathbf{d}'$, and $\mathbf{d}''$.

Contraction

Given the components d_{ijk} of a third-order tensor, consider

$$w_k \equiv d_{iik}. \tag{A.41}$$

Are w_1, w_2, and w_3 the components of a first-order tensor? In the $\overline{E}$ system we have

$$\overline{d}_{\ell mn} = a_{i\ell} a_{jm} a_{kn} d_{ijk}, \tag{A.42}$$

$$\overline{w}_n = \overline{d}_{\ell\ell n}. \tag{A.43}$$

Now in Eq. (A.42), setting m to ℓ, we obtain

$$\begin{aligned}
\overline{d}_{\ell\ell n} &= a_{i\ell} a_{j\ell} a_{kn} d_{ijk} \\
&= \delta_{ij} a_{kn} d_{ijk} \\
&= a_{kn} d_{iik}.
\end{aligned} \tag{A.44}$$

Thus we obtain

$$\overline{w}_n = a_{kn} w_k, \tag{A.45}$$

which is indeed the transformation rule for a first-order tensor.

The process of 'setting m to ℓ' is referred to as 'contracting m and ℓ.' In general, when any two indices of an Nth-order tensor ($N \geq 2$) are contracted, the result is a tensor of order $N - 2$. Note that d_{iik}, d_{iji}, and d_{ijj} are all valid (but different) contractions of d_{ijk}.

Inner products

If **b** and **c** are second- and third-order tensors, respectively, then their inner product is the third-order tensor **d**:

$$\begin{aligned}
\mathbf{d} &= \mathbf{b} \cdot \mathbf{c} \\
&= (e_i e_j b_{ij}) \cdot (e_k e_\ell e_m c_{k\ell m}) \\
&= e_i e_\ell e_m \delta_{jk} b_{ij} c_{k\ell m} \\
&= e_i e_\ell e_m b_{ij} c_{j\ell m}.
\end{aligned} \tag{A.46}$$

Thus, in terms of components,

$$d_{i\ell m} = b_{ij} c_{j\ell m}. \tag{A.47}$$

In general, taking the inner product of an Nth order tensor and an Mth-order tensor ($N, M \geq 1$) results in a tensor of order $N + M - 2$. The most familiar inner product is, of course, the dot product of two vectors.

Division

There is no tensor operation corresponding to division. (However, as discussed in Appendix B, a second-order tensor may have an inverse.)

Gradients

Gradients are obtained by applying the gradient operator

$$\nabla \equiv e_i \frac{\partial}{\partial x_i} \tag{A.48}$$

to a tensor field. For example, if $\mathbf{b}(x)$ is a second-order tensor field,

$$\mathbf{b}(x) = e_i e_j \, b_{ij}(x), \tag{A.49}$$

then the gradient of **b** is the third-order tensor **h** given by

$$\begin{aligned}
\mathbf{h} &= e_k e_i e_j \, h_{kij} \\
&= \nabla \mathbf{b} = e_k \frac{\partial}{\partial x_k} [e_i e_j \, b_{ij}(x)] \\
&= e_k e_i e_j \frac{\partial b_{ij}}{\partial x_k}.
\end{aligned} \tag{A.50}$$

Or, in terms of components,

$$h_{kij} = \frac{\partial b_{ij}}{\partial x_k}. \tag{A.51}$$

A shorthand notation that is sometimes useful is to write $b_{ij,k}$ for $\partial b_{ij}/\partial x_k$.

It may be seen from Eq. (A.50) that the gradient of $\mathbf{b}$ is similar to the tensor product of ∇ and $\mathbf{b}$. In general, the gradient of an Nth order tensor ($N \geq 0$) is a tensor of order $N + 1$. Repeated application of the gradient operator ∇ yields successively higher derivatives $- \partial^2 b_{ij}/(\partial x_k \, \partial x_\ell)$, for example.

Divergence

The divergence of the tensor $\mathbf{b}$ is the inner product of ∇ and $\mathbf{b}$:

$$\nabla \cdot \mathbf{b} = e_k \cdot \frac{\partial}{\partial x_k} [e_i e_j \, b_{ij}(\mathbf{x})]$$

$$= \delta_{ki} e_j \frac{\partial b_{ij}}{\partial x_k} = e_j \frac{\partial b_{ij}}{\partial x_i}. \tag{A.52}$$

In general the divergence of an Nth-order tensor ($N \geq 1$) is a tensor of order $N - 1$. In particular, the divergence of a vector is a scalar.

The Laplacian

The Laplacian is the scalar operator

$$\nabla^2 = \nabla \cdot \nabla = \left(e_i \frac{\partial}{\partial x_i} \right) \cdot \left(e_j \frac{\partial}{\partial x_j} \right)$$

$$= \delta_{ij} \frac{\partial^2}{\partial x_i \, \partial x_j} = \frac{\partial^2}{\partial x_i \, \partial x_i}. \tag{A.53}$$

Taylor Series

If $\mathbf{b}(\mathbf{x})$ is a smooth second-order tensor field, then the value of $\mathbf{b}$ at position $\mathbf{y} + \mathbf{r}$ can be obtained from $\mathbf{b}$ and its derivatives at $\mathbf{y}$ from a Taylor series. In terms of components the series is

$$b_{ij}(\mathbf{y} + \mathbf{r}) = b_{ij}(\mathbf{y}) + \left(\frac{\partial b_{ij}}{\partial x_k} \right)_y r_k + \frac{1}{2!} \left(\frac{\partial^2 b_{ij}}{\partial x_k \, \partial x_\ell} \right)_y r_k r_\ell$$

$$+ \frac{1}{3!} \left(\frac{\partial^3 b_{ij}}{\partial x_k \, \partial x_\ell \, \partial x_m} \right)_y r_k r_\ell r_m + \dots, \tag{A.54}$$

and similarly for tensors of different orders.

Gauss's theorem

This applies to (once continuously differentiable) tensor fields of any order. We use the second-order tensor $\mathbf{b}(\mathbf{x})$ as an example.

Let $\mathcal{A}$ be a piecewise smooth closed orientable surface that encloses a volume $\mathcal{V}$, and let $\mathbf{n}$ denote the outward-pointing unit normal on $\mathcal{A}$ (see Fig. 4.1 on page 85). Then,

$$\iiint\limits_{\mathcal{V}} \frac{\partial b_{ij}}{\partial x_k}\, \mathrm{d}V = \iint\limits_{\mathcal{A}} b_{ij}\, n_k\, \mathrm{d}A, \tag{A.55}$$

where $\mathrm{d}A$ and $\mathrm{d}V$ are elements of area and volume. If i and k are contracted, the result is the *divergence theorem*.

Notation

Equations involving tensors can be written either in *direct notation*, e.g., $\mathbf{b} = \mathbf{c}$, or in *suffix notation*, e.g., $b_{ij} = c_{ij}$. We follow the tradition in fluid mechanics of using suffix notation for tensors of order two or higher, and of using both notations interchangeably for first-order tensors (e.g., $\nabla \cdot \mathbf{u} = \partial u_i / \partial x_i$).

In using suffix notation we follow the normal practice of referring to 'the tensor b_{ij}'. However, it should be understood that, more correctly, b_{ij} is the i–j component of the tensor $\mathbf{b}$ in the E coordinate system.

EXERCISES

A.7 With $\mathbf{u}(\mathbf{x})$ being a vector field, and $\phi(\mathbf{x})$ and $\theta(\mathbf{x})$ being scalar fields, write in Cartesian tensor suffix notation

(a) $\mathbf{u} = \nabla\phi$,
(b) $\theta = \nabla \cdot \mathbf{u}$, and
(c) an equation, deduced from (a) and (b), relating ϕ to θ.

A.8 The quantities u_i, A_{ij}, B_{ijk}, and $C_{ijk\ell}$ are given tensors (not tensor fields). In each of (a)–(e), use every one of these tensors to complete a valid equation for the indicated quantity:

(a) a scalar: $\phi =$
(b) a first-order tensor: $v_j =$
(c) a second-order tensor: $T_{k\ell} =$
(d) a third-order tensor: $F_{pqr} =$
(e) a fourth-order tensor: $V_{ijk\ell} =$

(for example, a possible answer to (a) is $\phi = A_{ii} + u_i B_{ijj} + C_{iijj}$).

A.9 Let $\mathbf{u}(\mathbf{x}, t)$ be the velocity field, and let $\mathbf{n}$ be a given unit vector. Using Cartesian tensor suffix notation, write down expressions for

(a) the fluid acceleration,

(b) the component of acceleration in the direction of n, and

(c) the components of velocity in the plane perpendicular to n.

A.10   Consider two moving points A and B whose positions are $X^A(t)$ and $X^B(t)$, and whose velocities are $u^A(t)$ and $u^B(t)$. Let $s(t)$ denote the distance between A and B, and let $n(t)$ be the unit vector in the direction of $X^A(t) - X^B(t)$. Using Cartesian tensor suffix notation, obtain expressions for

(a) s in terms of X_i^A and X_i^B,

(b) n_i in terms of X_i^A, X_i^B and s,

(c) $\dfrac{ds}{dt}$ in terms of n_i, u_i^A and u_i^B, and

(d) $\dfrac{dn_i}{dt}$ in terms of n_i, u_i^A, u_i^B and s.

A.4 The vector cross product

In developing Cartesian tensors we have introduced many quantities that have direct equivalents in vector calculus – scalars, vector, dot products, gradient and divergence. However, for *Cartesian tensors there is no equivalent to the cross product or to the curl*. The considerations surrounding this observation are somewhat subtle, and often unappreciated or misunderstood. In the context of modelling turbulent flows it is an important issue, which we illustrate and clarify in this section.

The cross product can be written in suffix notation by using the *alternating symbol* (or Levi-Civita symbol) defined by

$$\varepsilon_{ijk} = \begin{cases} 1, & \text{if } (i,j,k) \text{ are cyclic,} \\ -1, & \text{if } (i,j,k) \text{ are anticyclic,} \\ 0, & \text{otherwise.} \end{cases} \tag{A.56}$$

Cyclic orderings are 123, 231, and 312; anticyclic orderings are 321, 132, and 213; otherwise two or more of the suffixes are the same. The cross product of two vectors u and v is then

$$r = u \times v = \det \begin{vmatrix} e_1 & e_2 & e_3 \\ u_1 & u_2 & u_3 \\ v_1 & v_2 & v_3 \end{vmatrix}$$

$$= \varepsilon_{ijk}\, e_i u_j v_k. \tag{A.57}$$

Thus the components are

$$r_i = \varepsilon_{ijk} u_j v_k, \tag{A.58}$$

and, for example,

$$r_1 = \varepsilon_{123} u_2 v_3 + \varepsilon_{132} u_3 v_2$$
$$= u_2 v_3 - u_3 v_2. \tag{A.59}$$

Similarly, if $U(x)$ is a vector field (e.g., velocity) then its curl, $\omega(x) = \nabla \times U$ (i.e., vorticity), is

$$\omega = e_i \varepsilon_{ijk} \frac{\partial U_k}{\partial x_j}, \tag{A.60}$$

with components

$$\omega_i = \varepsilon_{ijk} \frac{\partial U_k}{\partial x_j}. \tag{A.61}$$

Are r and ω first-order tensors? It is evident from Eqs. (A.58) and (A.61) that the answer depends upon whether or not ε_{ijk} is a third-order tensor. A simple example suffices to show that ε_{ijk} *is not a tensor*. Let the $\overline{E}$ system be obtained from the E system by a reflection of each axis, i.e., $\overline{e}_i = -e_i$. The direction cosines are then

$$a_{ij} = e_i \cdot \overline{e}_j = e_i \cdot (-e_j) = -\delta_{ij}. \tag{A.62}$$

Hence

$$a_{ip} a_{jq} a_{kr} \varepsilon_{ijk} = (-\delta_{ip})(-\delta_{jq})(-\delta_{kr})\varepsilon_{ijk}$$
$$= -\varepsilon_{pqr}, \tag{A.63}$$

which, because of the minus sign, is different than the transformation rule for a third-order tensor. (It can be shown that, if there is an even number of reflections (i.e., 0 or 2), then ε_{ijk} transforms as a tensor, but, if there is an odd number (i.e., 1 or 3), it transforms according to Eq. (A.63).)

We conclude, then, that the cross product $r = u \times v$ and the curl $\omega = \nabla \times U$ are not first-order tensors: they are more properly called *pseudovectors*.

The concept of *handedness* is central to the issues surrounding vectors and pseudovectors. In vector calculus we insist on using coordinate systems of fixed handedness – usually right-handed, as depicted in Fig. A.1. Objects, motions and flows can also exhibit handedness, the primary example being right-handed and left-handed screws.

Figure A.3 shows (in a right-handed coordinate system) a spinning arrow moving away from the observer toward a mirror. The velocity of the arrow is represented by the vector u, and the rate of rotation by the pseudovector ω.

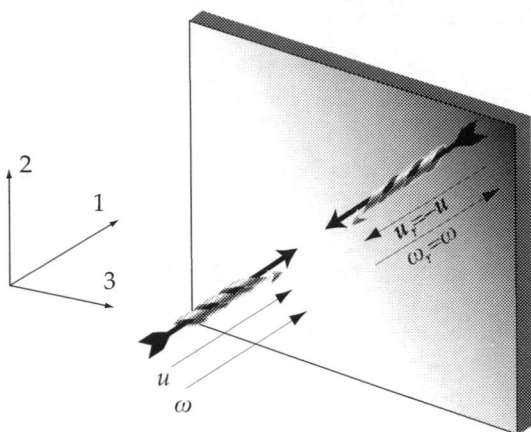

Fig. A.3. A sketch in a right-handed coordinate system of a spinning arrow (bottom left) moving toward a mirror, and its image in the mirror (top right). The velocity vector $\boldsymbol{u}$ changes direction but the rotation pseudovector $\boldsymbol{\omega}$ does not.

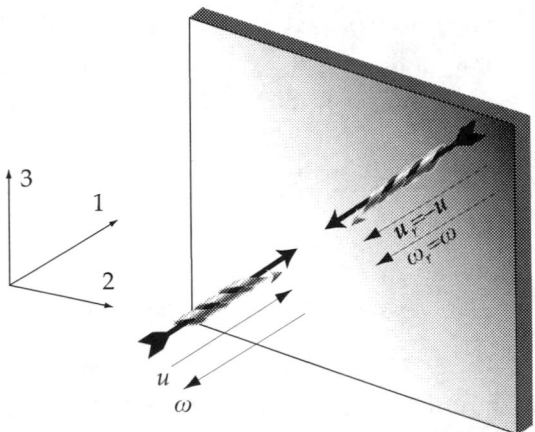

Fig. A.4. A sketch in a left-handed coordinate system of the same spinning arrow and mirror as those depicted in Fig. A.3. Note the direction of the rotation pseudovector $\boldsymbol{\omega}$.

Both $\boldsymbol{u}$ and $\boldsymbol{\omega}$ are parallel to the axis of the arrow, pointing from tail to head. The product $\boldsymbol{u} \cdot \boldsymbol{\omega}$ is positive, corresponding to motion as a right-handed screw. Also shown in Fig. A.3 is the reflection of the arrow in the mirror. The reflected image is of an arrow moving toward the observer with velocity $\boldsymbol{u}_r = -\boldsymbol{u}$, and rotating at the rate $\omega_r = \omega$. This illustrates the difference between the behavior of a vector and a pseudovector under a reflection. Note that, for the image arrow, $\boldsymbol{u}_r \cdot \boldsymbol{\omega}_r$ is negative and correspondingly the motion is as a left-handed screw.

Figure A.4 shows the same motion but in a left-handed coordinate system.

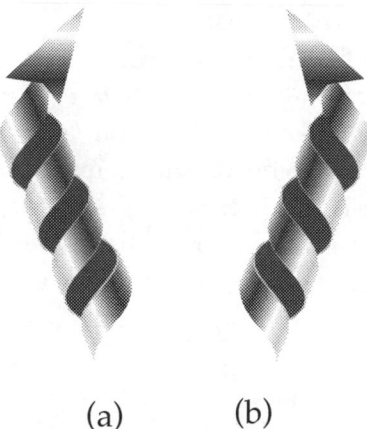

(a)   (b)

Fig. A.5. Sketches of helical motion (a) with positive helicity ($u \cdot \omega > 0$) and (b) with negative helicity ($u \cdot \omega < 0$).

The direction of ω is reversed, and the right-handed screw motion of the arrow now corresponds to a negative value of $u \cdot \omega$.

The reason that these are important considerations is that neither the laws of mechanics nor the properties of Newtonian fluids are biased toward right-handed or left-handed motion.[2] To illustrate the contrary behavior we consider the equation

$$\frac{\partial \omega}{\partial t} = u\omega \cdot \omega, \tag{A.64}$$

purporting to model some aspect of fluid behavior (with $u(x,t)$ and $\omega(x,t)$ being the velocity and vorticity fields). Apart from its lack of a physical basis, this equation is an atrocious model, because it is dimensionally incorrect, and it violates the principle of Galilean invariance (see Section 2.9). Pertinent to the present discussion is that the equation incorrectly biases the implied motion to a particular handedness. To see this, let a local right-handed coordinate system be chosen such that (at the moment and location of observation) the velocity is in the e_1 direction. Then, according to Eq. (A.64), the components of ω evolve by

$$\frac{\partial \omega_1}{\partial t} = |u||\omega|^2 \geq 0, \tag{A.65}$$

$$\frac{\partial \omega_2}{\partial t} = \frac{\partial \omega_3}{\partial t} = 0. \tag{A.66}$$

[2] A flow may be biased one way or the other, but this is brought about by the initial and boundary conditions, not by the governing equations.

It may be seen from Eq. (A.65) that the model implies that the component of ω in the direction of u always increases, and hence is biased toward positive values; or, equivalently, the *helicity* $u \cdot \omega$ is biased to positive values corresponding to right-handed screw motion, as depicted in Fig. A.5(a). If a left-handed coordinate system were used instead, then Eq. (A.64) would imply a bias toward left-handed screw motion.

Since both the laws of mechanics and the properties of Newtonian fluids are unbiased with respect to handedness, a biased equation such as Eq. (A.64) is fundamentally wrong. In constructing model equations to describe turbulent flows we want to ensure that there is no bias of handedness. This is simply achieved by demanding that equations be written in Cartesian tensor notation. For then, the implied motion is the same in right-handed and left-handed coordinate systems; so that, because of this symmetry, there can be no bias. The handedness symmetry of Cartesian tensors stems from the inclusion of reflections in their required transformation properties.

The conclusions are summarized as follows.

(i) The alternating symbol ε_{ijk} is not a third order tensor, and $r = u \times v$ and $\omega = \nabla \times u$ are not first-order tensors. They are pseudovectors that do not transform as tensors under a reflection of (an odd number of) coordinate axes.

(ii) The physical behavior implied by an equation written (legitimately) in Cartesian tensors is unbiased with respect to handedness, and is independent of the handedness of the coordinate system.

(iii) In contrast to (ii), an equation written in vector notation involving pseudovectors, or equivalently an equation written in suffix notation involving ε_{ijk} or pseudovectors, can incorrectly imply an asymmetry in the handedness in the physical behavior that depends on the handedness of the coordinate system.

While (in view of (iii) above) ε_{ijk} and pseudovectors are to be avoided in the construction of equations describing fluid behavior, they can legitimately and usefully be employed in other circumstances. In particular, vorticity and the vorticity equation are central to our understanding of many phenomena in fluid mechanics.

It may also be observed that an expression containing the alternation symbol an even number of times transforms like a tensor – because the minus signs cancel. Hence such expressions can be re-expressed as tensors.

A.11   Show the following:

$$\varepsilon_{ijk} = \varepsilon_{jki} = \varepsilon_{kij} = -\varepsilon_{jik}, \tag{A.67}$$

$$\varepsilon_{iik} = 0, \tag{A.68}$$

$$\varepsilon_{ijk}\varepsilon_{\ell mk} = \delta_{i\ell}\,\delta_{jm} - \delta_{im}\delta_{j\ell}, \tag{A.69}$$

$$\varepsilon_{ijk}\,\varepsilon_{ijk} = 6. \tag{A.70}$$

A.12   Use Eq. (A.61) to show that

$$\omega_i = -\varepsilon_{ijk}\frac{\partial U_j}{\partial x_k} = -\varepsilon_{ijk}\,\Omega_{jk}, \tag{A.71}$$

where the *rate-of-rotation tensor* is

$$\Omega_{ij} \equiv \frac{1}{2}\left(\frac{\partial U_i}{\partial x_j} - \frac{\partial U_j}{\partial x_i}\right), \tag{A.72}$$

and write explicit expressions for ω_1, ω_2 and ω_3.

A.13   Show that

$$\Omega_{ij} = -\tfrac{1}{2}\,\varepsilon_{ijk}\omega_k, \tag{A.73}$$

and rewrite this equation expressing each side explicitly as a 3×3 matrix.

A.14   Verify that

$$\varepsilon_{ijk}\varepsilon_{\ell mn} = \delta_{i\ell}\delta_{jm}\delta_{kn} + \delta_{im}\delta_{jn}\delta_{k\ell} + \delta_{in}\delta_{j\ell}\delta_{km}$$
$$- \delta_{in}\delta_{jm}\delta_{k\ell} - \delta_{im}\delta_{j\ell}\delta_{kn} - \delta_{i\ell}\delta_{jn}\delta_{km}. \tag{A.74}$$

A.5 A summary of Cartesian-tensor suffix notation

The definitions, rules, and operations involved with Cartesian tensors using suffix notation can be summarized thus.

1. In a tensor equation, *tensor expressions* are separated by +, −, or =. For example;

$$b_{ij} + c_{ij} = f_{ijkk}. \tag{A.75}$$

2. In a tensor expression, a suffix that appears once is a *free suffix* (e.g., i and j in Eq. (A.75)).

3. In a tensor expression, a suffix that appears twice is a *repeated suffix*

⠀

or a *dummy suffix* (e.g., the suffixes k in d_{ijkk}). The symbol used for a dummy suffix is immaterial, i.e., $d_{ijkk} = d_{ijpp}$.

4. *Summation convention*: a repeated suffix implies summation, i.e.,

$$d_{ijkk} = \sum_{k=1}^{3} d_{ijkk}. \tag{A.76}$$

5. In a tensor expression, a suffix cannot appear more than twice. For example, the expression f_{ijii} is invalid.

6. A tensor expression with N free suffixes is (or, more correctly, represents the components of) an Nth-order tensor. For example, each expression in Eq. (A.75) is a second-order tensor.

7. Each expression in a tensor equation must be a tensor of the same order, with the same free suffixes (not necessarily in the same order). Equation (A.75) is valid, whereas $b_{ij} = d_{ijk}$ and $b_{ij} = c_{ik}$ are both invalid.

8. The *Kronecker delta* δ_{ij} is defined by

$$\begin{aligned} \delta_{ij} &= 1, & \text{for } i = j, \\ &= 0, & \text{for } i \neq j. \end{aligned} \tag{A.77}$$

It is a second-order tensor. Note that $\delta_{ii} = 3$.

9. The *alternation symbol* ε_{ijk} in Eq. (A.56) is *NOT* a tensor.

10. *Addition*, e.g., $b_{ijk} = c_{ijk} + d_{ikj}$. Each tensor must be of the same order with the same free suffixes.

11. The *tensor product* of an Nth-order tensor and an Mth-order tensor is an $(N + M)$th-order tensor, e.g., $b_{ijk\ell m} = c_{ij}d_{k\ell m}$.

12. An Nth-order tensor ($N \geq 2$) can be *contracted* by changing two free suffixes into repeated suffixes. The result is a tensor of order $N - 2$. Different contractions of d_{ijk} are d_{iik}, d_{iji}, and d_{ijj}.

13. The *inner product* of an Nth-order tensor and an Mth-order tensor ($N \geq 1, M \geq 1$) is a tensor of order $N + M - 2$: e.g., $f_{ik\ell} = c_{ij}d_{jk\ell}$.

14. The *substitution rule* is that the inner product with the Kronecker delta is, for example,

$$\delta_{ij}c_{jk} = c_{ik}. \tag{A.78}$$

15. There is no tensor operation corresponding to *division*.

16. The *gradient of a tensor* is a tensor of one order higher, e.g., $d_{jk\ell} = \partial c_{k\ell}/\partial x_j$.

17. The *divergence* of an Nth-order tensor ($N \geq 1$) is a tensor of order $(N - 1)$, e.g., $v_k = \partial c_{jk}/\partial x_j$.

18. There are no tensor operations corresponding to the *vector cross product* or to the *curl*.

Appendix B

Properties of second-order tensors

In the study of turbulent flows, we encounter several second-order tensors, notably the velocity-gradient tensor $\partial U_i/\partial x_j$, the viscous-stress tensor τ_{ij}, and the Reynolds-stress tensor $\langle u_i u_j \rangle$. The purpose of this appendix is to review some of the properties of these tensors that are used in the text.

Matrix notation

The first observation is that the components b_{ij} of a second-order tensor **b** form a 3×3 matrix,

$$\mathbf{B} = \begin{bmatrix} b_{11} & b_{12} & b_{13} \\ b_{21} & b_{22} & b_{23} \\ b_{31} & b_{32} & b_{33} \end{bmatrix}, \tag{B.1}$$

and similarly the components u_i of a first-order tensor *u* form a column vector

$$U = \begin{bmatrix} u_1 \\ u_2 \\ u_3 \end{bmatrix}. \tag{B.2}$$

Consequently, the properties of matrices – that are familiar from the study of linear algebra – apply directly to second-order tensors.

Before proceeding, we clarify two aspects of notation and terminology. First, in this appendix we use lower case letters (e.g., **b** and *u*) for tensors, and upper-case letters for the corresponding matrices and column vectors (i.e., **B** and *U*). Second, the term 'vector' is used both in Cartesian tensors and in matrix analysis. When necessary the two different meanings are distinguished by using the terms 'first-order tensor' and 'column vector'. Table B.1 shows various operations expressed in the three notations being considered.

Table B.1. *Operations involving first- and second-order tensors expressed in various notations*

Operation	Cartesian tensor: direct notation	Cartesian tensor: suffix notation	Matrix notation
Inner product of two vectors	$\boldsymbol{u} \cdot \boldsymbol{v}$	$u_i v_i$	$U^T V$
Tensor product of two vectors	$\boldsymbol{uv}$	$u_i v_j$	$U V^T$
Inner product of second-order tensor and vector	$\mathbf{b} \cdot \boldsymbol{u}$	$b_{ij} u_j$	$\mathbf{B} U$
Inner product of transpose of second-order tensor and vector	$\mathbf{b}^T \cdot \boldsymbol{u}$	$b_{ji} u_j$	$\mathbf{B}^T U$
Inner product of two second-order tensors	$\mathbf{b} \cdot \mathbf{c}$	$b_{ij} c_{jk}$	$\mathbf{BC}$
Addition of two second-order tensors	$\mathbf{b} + \mathbf{c}$	$b_{ij} + c_{ij}$	$\mathbf{B} + \mathbf{C}$

Decomposition

The unique contraction b_{ii} of a second-order tensor corresponds to the trace of the matrix $\mathbf{B}$, trace($\mathbf{B}$), i.e., the sum of the diagonal components:

$$b_{ii} = \text{trace}(\mathbf{B}) = b_{11} + b_{22} + b_{33}. \tag{B.3}$$

The trace can be used to decompose the tensor into an *isotropic* part and a *deviatoric* part. By definition, an isotropic tensor has the same components in every coordinate system. There are no odd-order isotropic tensors, whereas isotropic second-order tensors are scalar multiples of the Kronecker delta. The isotropic part of $\mathbf{b}$ is

$$b_{ij}^{I} = \tfrac{1}{3} b_{\ell\ell} \, \delta_{ij}. \tag{B.4}$$

By definition, a deviatoric tensor has zero trace, so that the deviatoric part of $\mathbf{b}$ is

$$b_{ij}' \equiv b_{ij} - \tfrac{1}{3} b_{\ell\ell} \, \delta_{ij}. \tag{B.5}$$

Thus, $\mathbf{b}$ has the decomposition

$$b_{ij} = \tfrac{1}{3} b_{\ell\ell} \, \delta_{ij} + b_{ij}', \tag{B.6}$$

Or, in matrix notation, with $\mathbf{I}$ being the 3×3 unit matrix (corresponding to δ_{ij}), the decomposition is:

$$\mathbf{B} = \tfrac{1}{3} \, \text{trace}(\mathbf{B}) \, \mathbf{I} + \mathbf{B}', \tag{B.7}$$

with

$$\mathbf{B}' = \mathbf{B} - \tfrac{1}{3} \, \text{trace}(\mathbf{B}) \, \mathbf{I}. \tag{B.8}$$

The deviatoric part can be further decomposed into symmetric ($\mathbf{S}$) and antisymmetric ($\mathbf{R}$) parts:

$$s_{ij} = \tfrac{1}{2}(b'_{ij} + b'_{ji}) = s_{ji}, \tag{B.9}$$

$$r_{ij} = \tfrac{1}{2}(b'_{ij} - b'_{ji}) = -r_{ji}, \tag{B.10}$$

or, in matrix notation,

$$\mathbf{S} = \tfrac{1}{2}(\mathbf{B}' + \mathbf{B}'^{\mathrm{T}}) = \mathbf{S}^{\mathrm{T}}, \tag{B.11}$$

$$\mathbf{R} = \tfrac{1}{2}(\mathbf{B}' - \mathbf{B}'^{\mathrm{T}}) = -\mathbf{R}^{\mathrm{T}}. \tag{B.12}$$

Thus the second-order tensor $\mathbf{b}$ can be decomposed as the sum of an isotropic part $\tfrac{1}{3}b_{\ell\ell}\delta_{ij}$, a symmetric deviatoric part s_{ij}, and an antisymmetric deviatoric part r_{ij}:

$$b_{ij} = \tfrac{1}{3}b_{\ell\ell}\,\delta_{ij} + s_{ij} + r_{ij}. \tag{B.13}$$

EXERCISE _____

B.1   With s_{ij} being symmetric and r_{ij} being antisymmetric, use both suffix and matrix notation to show that

   (a) the diagonal components of r_{ij} are zero,
   (b) $s_{ik}s_{kj}$, $r_{ik}r_{kj}$, and $s_{ik}r_{kj} + s_{jk}r_{ki}$ are symmetric,
   (c) $s_{ik}r_{kj} - s_{jk}r_{ki}$ is antisymmetric, and
   (d) $s_{ij}r_{ji} = 0$.

Unitary transformations

The matrix $\mathbf{A}$ of the direction cosines a_{ij} is a *unitary* matrix. Recalling the definition $a_{ij} \equiv e_i \cdot \bar{e}_j$, we may observe that the jth column of $\mathbf{A}$ is the column vector corresponding to the components of $\bar{e}_j$ in the E coordinate system. Thus the columns of $\mathbf{A}$ are mutually orthogonal unit vectors. Hence, in accord with Eqs. (A.13) and (A.14), we obtain

$$\mathbf{A}^{\mathrm{T}}\mathbf{A} = \mathbf{A}\mathbf{A}^{\mathrm{T}} = \mathbf{I}. \tag{B.14}$$

The transformation rules for first- and second-order tensors (Eqs. (A.17) and (A.23)) correspond therefore to the unitary transformations

$$\overline{\mathbf{X}} = \mathbf{A}^{\mathrm{T}}\mathbf{X}, \tag{B.15}$$

$$\overline{\mathbf{B}} = \mathbf{A}^{\mathrm{T}}\mathbf{B}\mathbf{A}. \tag{B.16}$$

Principal axes

We now consider further the properties of *symmetric* (not necessarily deviatoric) second-order tensors. From linear algebra (see e.g., Franklin (1968), p. 100) we have the following important result. If $\mathbf{S}$ is a real symmetric matrix, then there exist unitary matrices $\tilde{\mathbf{A}}$ that diagonalize $\mathbf{S}$:

$$\tilde{\mathbf{A}}^{\mathrm{T}}\mathbf{S}\tilde{\mathbf{A}} = \boldsymbol{\Lambda} = \begin{bmatrix} \lambda_1 & 0 & 0 \\ 0 & \lambda_2 & 0 \\ 0 & 0 & \lambda_3 \end{bmatrix}. \tag{B.17}$$

The diagonal components λ_1, λ_2, and λ_3 (which are real) are the *eigenvalues* of $\mathbf{S}$, and the columns of $\tilde{\mathbf{A}}$ (denoted by $\tilde{\boldsymbol{e}}_1$, $\tilde{\boldsymbol{e}}_2$, and $\tilde{\boldsymbol{e}}_3$) are the corresponding *eigenvectors*. By their definition, the eigenvalues and eigenvectors satisfy the equation

$$\mathbf{S}\tilde{\boldsymbol{e}}_{(i)} = \lambda_{(i)}\tilde{\boldsymbol{e}}_{(i)}, \tag{B.18}$$

where bracketed suffixes are excluded from the summation convention. The eigenvectors $\tilde{\boldsymbol{e}}_1$, $\tilde{\boldsymbol{e}}_2$, and $\tilde{\boldsymbol{e}}_3$ provide an orthonormal basis for a special coordinate system – called the *principal axes* of $\mathbf{S}$.

If the eigenvalues are distinct, then the unit eigenvectors are determined up to their ordering and their sign. Hence the principal axes are determined up to reflections and $90°$ rotations of the axes. If two eigenvalues are equal, $\lambda_1 = \lambda_2 \neq \lambda_3$ say, then every vector orthogonal to $\tilde{\boldsymbol{e}}_3$ is an eigenvector. Hence, any coordinate system containing $\tilde{\boldsymbol{e}}_3$ as a basis vector provides principal axes of $\mathbf{S}$. If all three eigenvalues are equal, $\lambda_1 = \lambda_2 = \lambda_3 = \lambda$, then $\mathbf{S} = \lambda\mathbf{I}$, every vector is an eigenvector, and every coordinate system provides principal axes.

The distinction between diagonal and off-diagonal components is often made. For example, for the stress tensor, the diagonal components are the *normal stresses*, and the off-diagonal components are the *shear stresses*. The above result shows that this distinction is not intrinsic, but rather depends entirely on the coordinate system: in principal axes, the off-diagonal components are zero. The intrinsic distinction is between the isotropic part of the tensor and the anisotropic deviatoric part.

B.2   Given that the columns of $\tilde{\mathbf{A}}$ are the eigenvectors $\tilde{e}_1$, $\tilde{e}_2$, and $\tilde{e}_3$, obtain the results

$$\tilde{\mathbf{A}}^{\mathrm{T}}\tilde{e}_1 = \begin{bmatrix} 1 \\ 0 \\ 0 \end{bmatrix}, \tag{B.19}$$

$$\Lambda\tilde{\mathbf{A}}^{\mathrm{T}}\tilde{e}_1 = \begin{bmatrix} \lambda_1 \\ 0 \\ 0 \end{bmatrix}, \tag{B.20}$$

$$\tilde{\mathbf{A}}\Lambda\tilde{\mathbf{A}}^{\mathrm{T}}\tilde{e}_1 = \lambda_1\tilde{e}_1. \tag{B.21}$$

What is the significance of the last result?

B.3   From Eq. (B.17) obtain the results

$$\Lambda^2 \equiv \Lambda\Lambda = \begin{bmatrix} \lambda_1^2 & 0 & 0 \\ 0 & \lambda_2^2 & 0 \\ 0 & 0 & \lambda_3^2 \end{bmatrix}, \tag{B.22}$$

$$\mathbf{S} = \tilde{\mathbf{A}}\Lambda\tilde{\mathbf{A}}^{\mathrm{T}}, \tag{B.23}$$

$$\mathbf{S}^2 \equiv \mathbf{S}\mathbf{S} = \tilde{\mathbf{A}}\Lambda^2\tilde{\mathbf{A}}^{\mathrm{T}}. \tag{B.24}$$

Hence argue that the principal axes of $\mathbf{S}$ are also principal axes of $\mathbf{S}^2$, and (by induction) of $\mathbf{S}^3$, $\mathbf{S}^4$, ...

B.4   Let $v^{(\alpha)}$, $\alpha = 1, 2$, and 3 be mutually orthogonal vectors, and define the symmetric second-order tensor $\mathbf{s}$ by

$$s_{ij} = v_i^{(1)}v_j^{(1)} + v_i^{(2)}v_j^{(2)} + v_i^{(3)}v_j^{(3)}. \tag{B.25}$$

By considering $\mathbf{s}$ in principal axes, show that its eigenvalues are

$$\lambda_\alpha = v^{(\alpha)} \cdot v^{(\alpha)} \geq 0,$$

and (for $\lambda_\alpha \neq 0$) the corresponding unit eigenvectors are

$$\tilde{e}_\alpha = v^{(\alpha)}/(v^{(\alpha)} \cdot v^{(\alpha)})^{1/2}. \tag{B.26}$$

(Hence every symmetric positive semi-definite second-order tensor has the decomposition Eq. (B.25).)

B.5   Let f, g, and h be mutually orthogonal unit vectors. Use the results of Exercise B.4 to show that

$$f_i f_j + g_i g_j + h_i h_j = \delta_{ij}. \tag{B.27}$$

Invariants

For any second-order tensor **b**, powers are defined by

$$\mathbf{b}^2 \equiv \mathbf{b} \cdot \mathbf{b}, \quad \mathbf{b}^3 \equiv \mathbf{b} \cdot \mathbf{b} \cdot \mathbf{b}, \tag{B.28}$$

etc., corresponding to the matrix expressions

$$\mathbf{B}^2 = \mathbf{BB}, \quad \mathbf{B}^3 = \mathbf{BBB}, \tag{B.29}$$

etc. In terms of the components b_{ij}, Eq. (B.28) is

$$b_{ij}^2 = b_{ik}b_{kj}, \quad b_{ij}^3 = b_{ik}b_{k\ell}b_{\ell j}, \tag{B.30}$$

etc. Note that b_{ij}^2 denotes the i–j component of $\mathbf{b}^2$, which is different than the square of the i–j component of **b**, i.e., $(b_{ij})^2$. (The nine quantities $(b_{ij})^2$ are not the components of a tensor.)

Any scalar obtained from a tensor (e.g., b_{ii}, $b_{ii}^2 = b_{ij}b_{ji}$, $b_{ii}^3 = b_{ij}b_{jk}b_{ki}$) is called an *invariant*, because its value is the same in any coordinate system. For a symmetric second-order tensor **s** (with corresponding matrix **S**), three *principal invariants* are defined by

$$I_s = s_{ii} = \text{trace}(\mathbf{S}), \tag{B.31}$$

$$II_s = \tfrac{1}{2}[(s_{ii})^2 - s_{ii}^2] = \tfrac{1}{2}\{[\text{trace}(\mathbf{S})]^2 - \text{trace}(\mathbf{S}^2)\}, \tag{B.32}$$

$$III_s = \tfrac{1}{6}(s_{ii})^3 - \tfrac{1}{2}s_{ii}s_{jj}^2 + \tfrac{1}{3}s_{ii}^3 = \det(\mathbf{S}), \tag{B.33}$$

where det denotes the determinant.

Invariants are, by definition, the same in every coordinate system. On evaluating the principal invariants in principal axes, we obtain

$$I_s = \lambda_1 + \lambda_2 + \lambda_3, \tag{B.34}$$

$$II_s = \lambda_1\lambda_2 + \lambda_2\lambda_3 + \lambda_1\lambda_3, \tag{B.35}$$

$$III_s = \lambda_1\lambda_2\lambda_3. \tag{B.36}$$

It may be noted that I_s, II_s, and III_s are the simplest possible linear, quadratic, and cubic combinations of the eigenvalues that do not depend on their ordering.

The characteristic equation

The eigenvalue–eigenvector equation (Eq. (B.18)) can be written

$$(\mathbf{S} - \lambda\mathbf{I})\tilde{e} = 0, \tag{B.37}$$

and we know from the theory of linear equations that a non-trivial solution exists only if the determinant of $(\mathbf{S}-\lambda\mathbf{I})$ is zero. Evaluation of the determinant (Exercise B.6) leads to the cubic *characteristic equation*

$$\lambda^3 - I_s\lambda^2 + II_s\lambda - III_s = 0. \tag{B.38}$$

Hence the significance of the principal invariants.

EXERCISE _____

B.6   Evaluate the determinant

$$\begin{vmatrix} s_{11} - \lambda & s_{12} & s_{13} \\ s_{21} & s_{22} - \lambda & s_{23} \\ s_{31} & s_{32} & s_{33} - \lambda \end{vmatrix},$$

and hence verify Eq. (B.38). Evaluate $(\lambda_1 - \lambda)(\lambda_2 - \lambda)(\lambda_3 - \lambda)$ and discuss its significance.

The Cayley–Hamilton theorem

In the present context, the important *Cayley–Hamilton theorem* states that the matrix $\mathbf{S}$ satisfies its own characteristic equation:

$$\mathbf{S}^3 - I_s\mathbf{S}^2 + II_s\mathbf{S} - III_s\mathbf{I} = 0, \tag{B.39}$$

(see Exercise B.7). This equation shows that $\mathbf{S}^3$ can be expressed as a linear combination of $\mathbf{S}^2$, $\mathbf{S}$, and $\mathbf{I}$.

Further, by premultiplying the equation by $\mathbf{S}$, we find that $\mathbf{S}^4$ can also be expressed in the same way. Hence, by induction, $\mathbf{S}^n$ ($n \geq 3$) can be expressed as a linear combination of $\mathbf{S}^2$, $\mathbf{S}$, and $\mathbf{I}$, where the three coefficients of these matrices are invariants of $\mathbf{S}$. These observations are fundamental to *representation theorems*.

If all the eigenvalues λ_1, λ_2, and λ_3 are non-zero, then so also is $\det(\mathbf{S})$, and then $\mathbf{S}$ is *non-singular*. In this case there is an inverse $\mathbf{S}^{-1}$ such that $\mathbf{S}^{-1}\mathbf{S} = \mathbf{I}$. When it is transformed to principal axes, $\mathbf{S}$ becomes the diagonal matrix of eigenvalues (Eq. (B.17)), and then evidently $\mathbf{S}^{-1}$ is

$$\tilde{\mathbf{A}}^\mathrm{T}\mathbf{S}^{-1}\tilde{\mathbf{A}} = \mathbf{\Lambda}^{-1} = \begin{bmatrix} \lambda_1^{-1} & 0 & 0 \\ 0 & \lambda_2^{-1} & 0 \\ 0 & 0 & \lambda_3^{-1} \end{bmatrix}. \tag{B.40}$$

An explicit expression for S^{-1} is obtained by premultiplying Eq. (B.39) by S^{-1}. In terms of tensor components s_{ij}^{-1} of the inverse s^{-1}, the result is

$$s_{ij}^{-1} = (s_{ij}^2 - I_s s_{ij} + II_s \delta_{ij})/III_s. \tag{B.41}$$

On the other hand, if one or more of the eigenvalues is zero, then S is *singular*, and S^{-1} is undefined. The *rank deficiency* of S is equal to the number of zero eigenvalues.

The tensor s is *positive definite* if, for every non-zero vector v, the quantity

$$Q(v) \equiv s_{ij} v_i v_j \tag{B.42}$$

is strictly positive. Since $Q(v)$ is a scalar, its value is unaffected by a change of coordinates. In the principal axes of s, with $\tilde{v}_i$ being the components of v, Q is

$$Q(v) = \Lambda_{ij} \tilde{v}_i \tilde{v}_j = \lambda_1 \tilde{v}_1^2 + \lambda_2 \tilde{v}_2^2 + \lambda_3 \tilde{v}_3^2. \tag{B.43}$$

Clearly s is positive definite if, and only if, all of the eigenvalues are strictly positive. If the eigenvalues are non-negative ($\lambda_i \geq 0$) then $Q(v)$ is non-negative, and s is then *positive semi-definite*.

EXERCISES

B.7   Transform Eq. (B.39) to principal axes by premultiplying by $\tilde{A}^T$ and postmultiplying by $\tilde{A}$, expressing the result with terms of Λ. Compare the result with the characteristic equation (Eq. B.38) and hence verify the Cayley–Hamilton theorem.

B.8   For positive semi-definite S, define

$$\Lambda^{1/2} = \begin{bmatrix} \lambda_1^{1/2} & 0 & 0 \\ 0 & \lambda_2^{1/2} & 0 \\ 0 & 0 & \lambda_3^{1/2} \end{bmatrix}, \tag{B.44}$$

$$S^{1/2} = \tilde{A}\Lambda^{1/2}\tilde{A}^T. \tag{B.45}$$

Show that $s^{1/2}$ is a symmetric tensor with the property

$$s_{ik}^{1/2} s_{kj}^{1/2} = s_{ij}, \tag{B.46}$$

i.e., $s^{1/2}$ is the (symmetric) square root of s.

B.9   With s^{-1} being the inverse of the symmetric positive definite second-order tensor s, consider the equation

$$s_{ij}^{-1} x_i x_j = 1. \tag{B.47}$$

By considering principal axes, show that this equation defines an ellipsoid whose principal axes are aligned with the eigenvectors of **s**, and that the lengths of the axes are equal to the square roots of the eigenvalues of **s**.

Appendix C
Dirac delta functions

In the text, we make extensive use of the delta function $\delta(x)$ and related quantities such as the Heaviside function $H(x)$. In this appendix, the basic properties of these *generalized functions* are reviewed, and some particularly useful results are derived.

C.1 The definition of $\delta(x)$

The Dirac delta function has the properties

$$\delta(x) = \begin{cases} 0, & \text{for } x \neq 0, \\ \infty, & \text{for } x = 0, \end{cases} \tag{C.1}$$

$$\int_{-\infty}^{\infty} \delta(x)\,\mathrm{d}x = 1. \tag{C.2}$$

However, it should be appreciated that $\delta(x)$ is not a function in the usual sense, and the above equations do not provide an adequate definition. Instead, $\delta(x)$ is a *generalized function*. As such, its definition and properties depend on the integral

$$\int_{-\infty}^{\infty} \delta(x)g(x)\,\mathrm{d}x, \tag{C.3}$$

where $g(x)$ is any test function (with appropriate differentiability and behavior at infinity).

As a way to approach the Dirac delta function, consider the functions

$$D_n(x) \equiv \frac{n}{\sqrt{2\pi}} \exp(-\tfrac{1}{2}x^2 n^2), \tag{C.4}$$

for $n = 1, 2, 3, \ldots$. This is an example of a *delta sequence*. It may be recognized

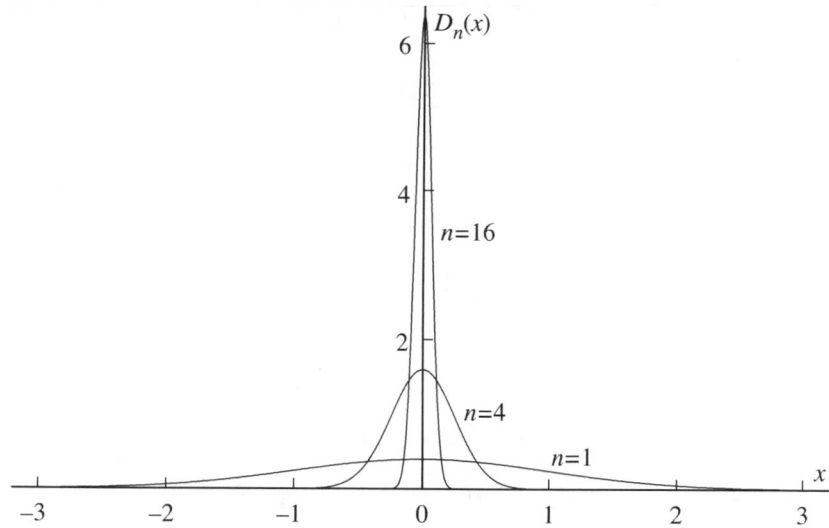

Fig. C.1. The functions $D_n(x)$ (Eq. (C.4)) for $n = 1, 4$, and 16.

that $D_n(x)$ is the Gaussian distribution with mean zero and standard deviation $1/n$ (see Eq. (3.41)). Thus we have

$$\int_{-\infty}^{\infty} D_n(x)\,\mathrm{d}x = 1, \tag{C.5}$$

and, more generally, for $m \geq 0$,

$$\int_{-\infty}^{\infty} x^m D_n(x)\,\mathrm{d}x = n^{-m}\hat{\mu}_m, \tag{C.6}$$

where $\{\hat{\mu}_0, \hat{\mu}_1, \hat{\mu}_2, \hat{\mu}_3, \hat{\mu}_4, \ldots\} = \{1, 0, 1, 0, 3, \ldots\}$ are the moments of the standardized Gaussian distribution. Figure C.1 shows $D_n(x)$ for several values of n.

For the test function $g(x)$, with the Taylor series

$$g(x) = g(0) + \sum_{m=1}^{\infty} g^{(m)}(0) x^m / m! \tag{C.7}$$

(where $g^{(m)}(x)$ denotes the mth derivative of $g(x)$), and for the integral of the form of Eq. (C.3), we obtain

$$\int_{-\infty}^{\infty} D_n(x)g(x)\,\mathrm{d}x = g(0) + \sum_{m=1}^{\infty} \frac{g^{(m)}(0)}{m!} \int_{-\infty}^{\infty} x^m D_n(x)\,\mathrm{d}x,$$

$$= g(0) + \sum_{m=1}^{\infty} g^{(m)}(0)\, n^{-m}\hat{\mu}_m / m!. \tag{C.8}$$

While the limit $\lim_{n\to\infty} D_n(x)$ does not exist, evidently we have

$$\lim_{n\to\infty} \int_{-\infty}^{\infty} D_n(x)g(x)\,\mathrm{d}x = g(0). \tag{C.9}$$

By definition, the Dirac delta function is the generalized function that has precisely this property:

$$\int_{-\infty}^{\infty} \delta(x)g(x)\,\mathrm{d}x = g(0). \tag{C.10}$$

C.2 Properties of $\delta(x)$

The basic properties of $\delta(x)$ are now itemized.

(i) By a simple change of variable, it follows from Eq. (C.10) that, for any constant a,

$$\int_{-\infty}^{\infty} \delta(x - a)g(x)\,\mathrm{d}x = g(a). \tag{C.11}$$

This is the *sifting property* of the delta function: $\delta(x - a)$ is referred to as a 'delta function at a,' and, from the function $g(x)$, the integral in Eq. (C.11) sifts out the particular value $g(a)$.

(ii) More generally, if the function $f(x)$ is continuous at a, then, for $\epsilon_1 > 0$ and $\epsilon_2 > 0$,

$$\int_{a-\epsilon_1}^{a+\epsilon_2} \delta(x - a)f(x)\,\mathrm{d}x = f(a). \tag{C.12}$$

This integral is not defined if ϵ_1 or ϵ_2 is zero, or if $f(x)$ is discontinuous at a.

(iii) Legitimate operations include multiplying by a function, e. g., $f(x)\delta(x - a)$ (provided that $f(x)$ is continuous at a), and addition, e.g., $f(x)\delta(x - a) + h(x)\delta(x - b)$. Illegitimate operations include multiplying two delta functions of the same variable, e.g., $\delta(x - a)\delta(x - b)$, and division by a delta function.

(iv) The only significance of an expression involving delta functions, e.g., $f(x)\delta(x - a)$, is the value obtained when it is multiplied by a test function $g(x)$ and integrated. For example, we have

$$\int_{-\infty}^{\infty} g(x)[f(x)\delta(x - a)]\,\mathrm{d}x = g(a)f(a), \tag{C.13}$$

and also

$$\int_{-\infty}^{\infty} g(x)[f(a)\delta(x - a)]\,\mathrm{d}x = g(a)f(a). \tag{C.14}$$

Consequently, the equation

$$f(x)\,\delta(x-a) = f(a)\delta(x-a) \tag{C.15}$$

is correct. (Recall that division by $\delta(x-a)$ is not legitimate, so there is no implication that $f(x)$ equals $f(a)$.)

(v) It is evident from the normalization condition $\int_{-\infty}^{\infty} \delta(x)\,\mathrm{d}x = 1$ that $\delta(x)$ is a *density* – as are PDFs. So, if x has dimensions of length, $\delta(x)$ has dimensions of inverse length. Under a change of independent variable, $\delta(x)$ transforms as a density. For example, consider $\delta[(x-a)/b]$ for $b > 0$. On setting $y = (x-a)/b$, we obtain

$$\int_{-\infty}^{\infty} g(x)\delta\left(\frac{x-a}{b}\right)\mathrm{d}x = \int_{-\infty}^{\infty} g(a+by)\delta(y)b\,\mathrm{d}y$$
$$= g(a)b$$
$$= \int_{-\infty}^{\infty} g(x)b\,\delta(x-a)\,\mathrm{d}x. \tag{C.16}$$

By comparing the first and last expressions in this equation, and by repeating the analysis for $b < 0$, we conclude that the transformation rule is (for $b \neq 0$)

$$\delta\left(\frac{x-a}{b}\right) = |b|\delta(x-a). \tag{C.17}$$

Note that, for $b = -1$, this equation shows the delta function to be symmetric:

$$\delta(x-a) = \delta(a-x). \tag{C.18}$$

C.3 Derivatives of $\delta(x)$

The derivatives of the delta sequence $D_n(x)$ (Eq. (C.4)) are defined by

$$D_n^{(1)}(x) = \frac{\mathrm{d}}{\mathrm{d}x}D_n(x). \tag{C.19}$$

It is shown in Fig. C.2 for three values of n. On integrating by parts, we obtain

$$\int_{-\infty}^{\infty} D_n^{(1)}(x)g(x)\,\mathrm{d}x = \int_{-\infty}^{\infty} \frac{\mathrm{d}D_n(x)}{\mathrm{d}x}g(x)\,\mathrm{d}x$$
$$= [D_n(x)g(x)]_{-\infty}^{\infty} - \int_{-\infty}^{\infty} D_n(x)\frac{\mathrm{d}g(x)}{\mathrm{d}x}\,\mathrm{d}x$$
$$= -\int_{-\infty}^{\infty} D_n(x)g^{(1)}(x)\,\mathrm{d}x. \tag{C.20}$$

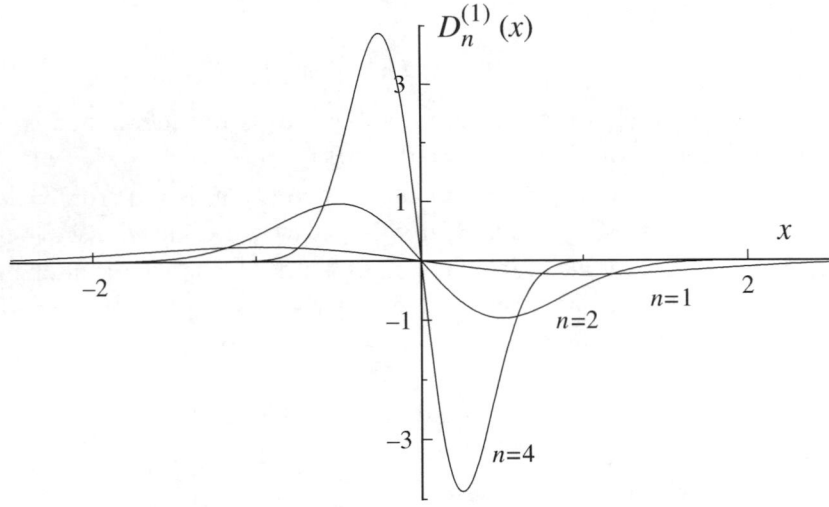

Fig. C.2. The functions $D_n^{(1)}(x)$ (Eq. (C.19)) for $n = 1, 2$, and 4.

(The test functions satisfy the weak conditions required for $D_n(x)g(x)$ to vanish at infinity.) The derivative of the delta function $\delta^{(1)}(x)$ is then defined such that

$$\int_{-\infty}^{\infty} \delta^{(1)}(x)g(x)\,dx = \lim_{n\to\infty} \int_{-\infty}^{\infty} D_n^{(1)}(x)g(x)\,dx$$

$$= \int_{-\infty}^{\infty} -\delta(x)g^{(1)}(x)\,dx$$

$$= -g^{(1)}(0). \tag{C.21}$$

It may be observed then that $-\delta^{(1)}(x-a)$ sifts the derivative at a, i.e., $g^{(1)}(a)$; and also that $\delta^{(1)}$ is anti-symmetric:

$$\delta^{(1)}(x-a) = -\delta^{(1)}(a-x) \tag{C.22}$$

(see also Fig. C.2).

Higher derivatives are defined similarly. For the mth derivative we obtain

$$\int_{-\infty}^{\infty} \delta^{(m)}(x)g(x)\,dx = \lim_{n\to\infty} \int_{-\infty}^{\infty} D_n^{(m)}(x)g(x)\,dx = (-1)^m g^{(m)}(0). \tag{C.23}$$

The transformation rule is (for $b \neq 0$)

$$\delta^{(m)}\left(\frac{x-a}{b}\right) = b^m |b| \delta^{(m)}(x-a) \tag{C.24}$$

(see Exercise C.1), and hence

$$\int_{-\infty}^{\infty} \delta^{(m)}\left(\frac{x-a}{b}\right) g(x)\,dx = (-b)^m |b| g^{(m)}(a). \tag{C.25}$$

EXERCISE _____

C.1   Following the method employed in Eq. (C.16), show that the transformation rule for $\delta^{(1)}$ is (for $b \neq 0$)

$$\delta^{(1)}\left(\frac{x-a}{b}\right) = b|b|\delta^{(1)}(x-a). \tag{C.26}$$

_____

C.4 Taylor series

A possibly surprising observation is that $\delta(x)$ has a Taylor-series expansion. The Taylor series for the test function $g(x)$ is

$$g(h) = g(0) + \sum_{m=1}^{\infty} \frac{h^m}{m!} g^{(m)}(0). \tag{C.27}$$

In terms of delta functions, the left-hand side is

$$g(h) = \int_{-\infty}^{\infty} \delta(x-h)g(x)\,dx, \tag{C.28}$$

while the right-hand side is

$$g(0) + \sum_{m=1}^{\infty} \frac{h^m}{m!} g^{(m)}(0) = \int_{-\infty}^{\infty} \left(\delta(x) + \sum_{m=1}^{\infty} \frac{(-h)^m}{m!} \delta^{(m)}(x) \right) g(x)\,dx. \tag{C.29}$$

Hence we obtain

$$\delta(x-h) = \delta(x) + \sum_{m=1}^{\infty} \frac{(-h)^m}{m!} \delta^{(m)}(x). \tag{C.30}$$

C.5 The Heaviside function

Corresponding to the delta sequence $D_n(x)$ (Eq. (C.4)), the sequence $S_n(x)$ is defined by

$$S_n(x) = \int_{-\infty}^{x} D_n(y)\,dy, \tag{C.31}$$

and is shown in Fig. C.3. Evidently, as n tends to infinity, $S_n(x)$ tends to the unit step function with

$$\lim_{n\to\infty} S_n(x) = \begin{cases} 0, & \text{for } x < 0, \\ 1, & \text{for } x > 0. \end{cases} \tag{C.32}$$

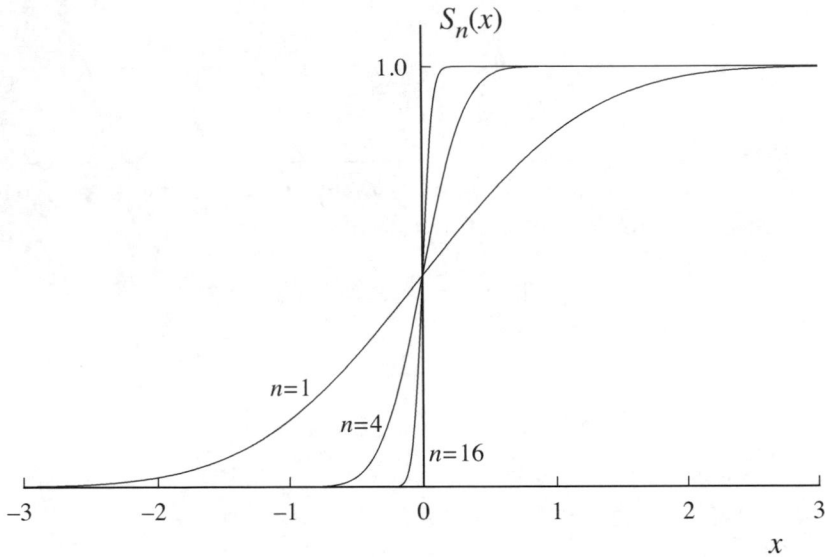

Fig. C.3. The functions $S_n(x)$ (Eq. (C.31)) for $n = 1, 4$, and 16.

The Heaviside function $H(x)$ is defined as the generalized function with the property

$$\int_{-\infty}^{\infty} H(x)g(x)\,dx = \lim_{n\to\infty} \int_{-\infty}^{\infty} S_n(x)g(x)\,dx = \int_{0}^{\infty} g(x)\,dx. \qquad (C.33)$$

The derivative of $H(x)$ is $\delta(x)$; and, conversely, $H(x)$ is the integral $\int_{-\infty}^{x} \delta(y)\,dy$. This is evident from the above definitions, or, alternatively, from

$$\frac{d}{dy} \int_{-\infty}^{\infty} H(x-y)g(x)\,dx = \int_{-\infty}^{\infty} -H^{(1)}(x-y)g(x)\,dx$$

$$= \frac{d}{dy} \int_{y}^{\infty} g(x)\,dx = -g(y). \qquad (C.34)$$

EXERCISES

C.2   Show that the transformation rule for the Heaviside function is

$$\begin{aligned} H[(x-a)/b] &= H(x-a), && \text{for } b > 0, \\ &= 1 - H(x-a), && \text{for } b < 0. \end{aligned} \qquad (C.35)$$

C.3   Consider the sequence of functions

$$\bar{S}_n(x) \equiv \exp(-e^{-nx}). \qquad (C.36)$$

How do $S_n(x)$ and $\overline{S}_n(x)$ differ as n tends to infinity? Show that

$$\lim_{n \to \infty} \int_{-\infty}^{\infty} \overline{S}_n(x)g(x)\,\mathrm{d}x = \int_{-\infty}^{\infty} H(x)g(x)\,\mathrm{d}x. \qquad \text{(C.37)}$$

C.6 Multiple dimensions

Let y by a point in three-dimensional Euclidean space. The three-dimensional delta function at y is written $\delta(x - y)$, and is just the product of the three one-dimensional delta functions:

$$\delta(x - y) = \delta(x_1 - y_1)\delta(x_2 - y_2)\delta(x_3 - y_3). \qquad \text{(C.38)}$$

It has the sifting property

$$\iiint_{-\infty}^{\infty} g(x)\delta(x - y)\,\mathrm{d}x = g(y). \qquad \text{(C.39)}$$

where $\mathrm{d}x$ is written for $\mathrm{d}x_1\,\mathrm{d}x_2\,\mathrm{d}x_3$. (Note that it is legitimate to form the product of delta functions of different variables, e.g., $\delta(x_1 - a)\delta(x_2 - b)$, but not of delta functions of the same variable, e.g., $\delta(x - a)\delta(x - b)$.)

Appendix D
Fourier transforms

The purpose of this appendix is to provide definitions and a summary of the properties of the Fourier transforms used elsewhere in this book. For further explanations and results, the reader may refer to standard texts, e.g., Bracewell (1965), Lighthill (1970), and Priestley (1981).

Definition

Given a function $f(t)$, its Fourier transform is

$$g(\omega) = \mathcal{F}\{f(t)\} \equiv \frac{1}{2\pi} \int_{-\infty}^{\infty} f(t)e^{-i\omega t}\,\mathrm{d}t, \qquad (\text{D.1})$$

and the inverse transform is

$$f(t) = \mathcal{F}^{-1}\{g(\omega)\} = \int_{-\infty}^{\infty} g(\omega)e^{i\omega t}\,\mathrm{d}\omega. \qquad (\text{D.2})$$

For $f(t)$ and $g(\omega)$ to form a Fourier-transform pair it is necessary that the above integrals converge, at least as generalized functions. The transforms shown here are between the time domain (t) and the frequency domain (ω); the corresponding formulae for transforms between physical space (x) and wavenumber space (κ) are obvious. Some useful Fourier-transform pairs are given in Table D.1 on page 679, and a comprehensive compilation is provided by Erdélyi, Oberhettinger, and Tricomi (1954).

There is not a unique convention for the definition of Fourier transforms. In some definitions the negative exponent $-i\omega t$ appears in the inverse transform, and the factor of 2π can be split in different ways between $\mathcal{F}$ and $\mathcal{F}^{-1}$. The convention used here is the same as that used by Batchelor (1953), Monin and Yaglom (1975), and Tennekes and Lumley (1972).

Table D.1. *Fourier-transform pairs (a, b, and v are real constants with b > 0 and v > $-\frac{1}{2}$)*

$f(t)$	$g(\omega)$
1	$\delta(\omega)$
$\delta(t-a)$	$\dfrac{1}{2\pi}e^{-i\omega a}$
$\delta^{(n)}(t-a)$	$\dfrac{(i\omega)^n}{2\pi}e^{-i\omega a}$
$e^{-b\lvert t\rvert}$	$\dfrac{b}{\pi(b^2+\omega^2)}$
$\dfrac{1}{b\sqrt{2\pi}}e^{-t^2/(2b^2)}$	$\dfrac{1}{2\pi}e^{-b^2\omega^2/2}$
$H(b-\lvert t\rvert)$	$\dfrac{\sin(b\omega)}{\pi\omega}$
$(b^2+t^2)^{-(v+1/2)}$	$\dfrac{2\sqrt{\pi}}{\Gamma(v+\frac{1}{2})}\left(\dfrac{\lvert\omega\rvert}{2b}\right)^v K_v\left(\dfrac{\lvert\omega\rvert}{b}\right)$

Derivatives

The Fourier transforms of derivatives are

$$\mathcal{F}\left\{\frac{d^n f(t)}{dt^n}\right\} = (i\omega)^n g(\omega), \tag{D.3}$$

$$\mathcal{F}^{-1}\left\{\frac{d^n g(\omega)}{d\omega^n}\right\} = (-it)^n f(t). \tag{D.4}$$

The cosine transform

If $f(t)$ is real, then $g(\omega)$ has conjugate symmetry:

$$g(\omega) = g^*(-\omega), \quad \text{for } f(t) \text{ real}, \tag{D.5}$$

as may be seen by taking the complex conjugate of Eq. (D.2). If $f(t)$ is real and even (i.e., $f(t) = f(-t)$), then Eq. (D.1) can be rewritten

$$g(\omega) = \frac{1}{2\pi}\int_{-\infty}^{\infty} f(t)\cos(\omega t)\,dt$$

$$= \frac{1}{\pi}\int_{0}^{\infty} f(t)\cos(\omega t)\,dt, \tag{D.6}$$

showing that $g(\omega)$ is also real and even. The inverse transform is

$$f(t) = 2 \int_0^\infty g(\omega) \cos(\omega t) \, d\omega. \tag{D.7}$$

Equations (D.6) and (D.7) define the *cosine Fourier transform* and its inverse.

In considering spectra, it is sometimes convenient to consider *twice* the Fourier transform of a real even function $f(t)$, i.e.,

$$\begin{aligned} \bar{g}(\omega) &\equiv 2g(\omega) \\ &= \frac{2}{\pi} \int_0^\infty f(t) \cos(\omega t) \, dt, \end{aligned} \tag{D.8}$$

so that the inversion formula

$$f(t) = \int_0^\infty \bar{g}(\omega) \cos(\omega t) \, d\omega, \tag{D.9}$$

does not contain a factor of 2 (cf. Eq. (D.7)).

The delta function

The Fourier transform of the delta function $\delta(t - a)$ is (from Eq. (D.1) and invoking the sifting property Eq. (C.11))

$$\mathcal{F}\{\delta(t - a)\} = \frac{1}{2\pi} e^{-i\omega a}, \tag{D.10}$$

and, in particular,

$$\mathcal{F}\{\delta(t)\} = \frac{1}{2\pi}. \tag{D.11}$$

Setting $g(\omega) = (1/2\pi)e^{-i\omega a}$, the inversion formula (Eq. (D.2)) yields

$$\delta(t - a) = \int_{-\infty}^\infty \frac{1}{2\pi} e^{i\omega(t-a)} \, d\omega. \tag{D.12}$$

This is a remarkable and valuable result. However, since the integral in Eq. (D.12) is clearly divergent, it – like $\delta(t - a)$ – must be viewed as a generalized function. That is, with $G(t)$ being a test function, Eq. (D.12) has the meaning

$$\begin{aligned} \int_{-\infty}^\infty G(t)\delta(t - a) \, dt &= \int_{-\infty}^\infty \int_{-\infty}^\infty \frac{1}{2\pi} G(t) e^{i\omega(t-a)} \, d\omega \, dt \\ &= G(a). \end{aligned} \tag{D.13}$$

Further explanation is provided by Lighthill (1970) and Butkov (1968).

Convolution

Given two functions $f_a(t)$ and $f_b(t)$ (both of which have Fourier transforms) their convolution is defined by

$$h(t) \equiv \int_{-\infty}^{\infty} f_a(t-s)f_b(s)\,ds. \tag{D.14}$$

With the substitution $r = t - s$, the Fourier transform of the convolution is

$$
\begin{aligned}
\mathcal{F}\{h(t)\} &= \frac{1}{2\pi} \int_{-\infty}^{\infty} e^{-i\omega t} \int_{-\infty}^{\infty} f_a(t-s)f_b(s)\,ds\,dt \\
&= \frac{1}{2\pi} \int_{-\infty}^{\infty}\int_{-\infty}^{\infty} e^{-i\omega(r+s)} f_a(r)f_b(s)\,ds\,dr \\
&= \frac{1}{2\pi} \int_{-\infty}^{\infty} e^{-i\omega r} f_a(r)\,dr \int_{-\infty}^{\infty} e^{-i\omega s} f_b(s)\,ds \\
&= 2\pi \mathcal{F}\{f_a(t)\}\mathcal{F}\{f_b(t)\}.
\end{aligned} \tag{D.15}
$$

That is, the Fourier transform of the convolution is equal to the product of 2π and the Fourier transforms of the functions.

Parseval's theorems

We consider the integral of the product of two functions $f_a(t)$ and $f_b(t)$ that have Fourier transforms $g_a(\omega)$ and $g_b(\omega)$. By writing f_a and f_b as inverse Fourier transforms, we obtain

$$
\int_{-\infty}^{\infty} f_a(t)f_b(t)\,dt = \int_{-\infty}^{\infty}\int_{-\infty}^{\infty} g_a(\omega)e^{i\omega t}\,d\omega \int_{-\infty}^{\infty} g_b(\omega')e^{i\omega' t}\,d\omega'\,dt \tag{D.16}
$$

$$
= \int_{-\infty}^{\infty}\int_{-\infty}^{\infty}\int_{-\infty}^{\infty} g_a(\omega)g_b(-\omega'')e^{i(\omega-\omega'')t}\,d\omega\,d\omega''\,dt. \tag{D.17}
$$

The integral of the exponential term over all t yields $2\pi\delta(\omega - \omega'')$, see Eq. (D.12), so that the integration of all ω'' is readily performed, producing

$$
\int_{-\infty}^{\infty} f_a(t)f_b(t)\,dt = 2\pi \int_{-\infty}^{\infty} g_a(\omega)g_b(-\omega)\,d\omega. \tag{D.18}
$$

This is *Parseval's second theorem*.

For the case in which f_a and f_b are the same function (i.e., $f_a = f_b = f$ and correspondingly $g_a = g_b = g$), Eq. (D.18) becomes *Parseval's first theorem*:

$$
\int_{-\infty}^{\infty} f(t)^2\,dt = 2\pi \int_{-\infty}^{\infty} g(\omega)g(-\omega)\,d\omega. \tag{D.19}
$$

If $f(t)$ is real, this can be re-expressed as

$$\int_{-\infty}^{\infty} f(t)^2 \, dt = 2\pi \int_{-\infty}^{\infty} g(\omega)g^*(\omega) \, d\omega$$

$$= 4\pi \int_{0}^{\infty} g(\omega)g^*(\omega) \, d\omega. \tag{D.20}$$

EXERCISES

D.1   With $f(t)$ being a differentiable function with Fourier transform $g(\omega)$, obtain the following results:

$$f(0) = \int_{-\infty}^{\infty} g(\omega) \, d\omega, \tag{D.21}$$

$$\int_{-\infty}^{\infty} f(t) \, dt = 2\pi g(0), \tag{D.22}$$

$$\int_{-\infty}^{\infty} \left(\frac{d^n f}{dt^n} \right)^2 \, dt = 2\pi \int_{-\infty}^{\infty} \omega^{2n} g(\omega)g(-\omega) \, d\omega. \tag{D.23}$$

Re-express the right-hand sides for the case of $f(t)$ being real.

D.2   Let $f_a(t)$ be the zero-mean Gaussian distribution with standard deviation a, i.e.,

$$f_a(t) = \mathcal{N}(t; 0, a^2) \equiv \frac{1}{a\sqrt{2\pi}} \, e^{-t^2/(2a^2)}, \tag{D.24}$$

and let $f_b(t) = \mathcal{N}(t; 0, b^2)$, where a and b are positive constants. Show that the convolution of f_a and f_b is $\mathcal{N}(t; 0, a^2 + b^2)$.

Appendix E
Spectral representation of stationary random processes

The purpose of this appendix is to show the connections among a statistically-stationary random process $U(t)$, its spectral representation (in terms of Fourier modes), its frequency spectrum $E(\omega)$, and its autocorrelation function $R(s)$.

A statistically stationary random process has a constant variance, and hence does not decay to zero as $|t|$ tends to infinity. As a consequence, the Fourier transform of $U(t)$ does not exist. This fact causes significant technical difficulties, which can be overcome only with more elaborate mathematical tools than are appropriate here. We circumvent this difficulty by first developing the ideas for periodic functions, and then extending the results to the non-periodic functions of interest.

E.1 Fourier series

We start by considering a non-random real process $U(t)$ in the time interval $0 \le t < T$. The process is continued periodically by defining

$$U(t + NT) = U(t), \tag{E.1}$$

for all non-zero integer N.

The time average of $U(t)$ over the period is defined by

$$\langle U(t) \rangle_T \equiv \frac{1}{T} \int_0^T U(t) \, dt, \tag{E.2}$$

and time averages of other quantities are defined in a similar way. The fluctuation in $U(t)$ is defined by

$$u(t) = U(t) - \langle U(t) \rangle_T, \tag{E.3}$$

and clearly its time average, $\langle u(t) \rangle_T$, is zero.

683

For each integer n, the frequency ω_n is defined by

$$\omega_n = 2\pi n/T. \tag{E.4}$$

We consider both positive and negative n, and observe that

$$\omega_{-n} = -\omega_n. \tag{E.5}$$

The nth complex *Fourier mode* is

$$
\begin{aligned}
e^{i\omega_n t} &= \cos(\omega_n t) + i\sin(\omega_n t) \\
&= \cos(2\pi n t/T) + i\sin(2\pi n t/T).
\end{aligned} \tag{E.6}
$$

Its time average is

$$
\begin{aligned}
\langle e^{i\omega_n t} \rangle_T &= 1, \quad \text{for } n = 0, \\
&= 0, \quad \text{for } n \neq 0, \\
&= \delta_{n0},
\end{aligned} \tag{E.7}
$$

and hence the modes satisfy the orthogonality condition

$$\langle e^{i\omega_n t} e^{-i\omega_m t} \rangle_T = \langle e^{i(\omega_n - \omega_m)t} \rangle_T = \delta_{nm}. \tag{E.8}$$

The process $u(t)$ can be expressed as a Fourier series,

$$u(t) = \sum_{n=-\infty}^{\infty} (a_n + ib_n)e^{i\omega_n t} = \sum_{n=-\infty}^{\infty} c_n e^{i\omega_n t}, \tag{E.9}$$

where $\{a_n, b_n\}$ are real, and $\{c_n\}$ are the complex *Fourier coefficients*. Since the time average $\langle u(t) \rangle_T$ is zero, it follows from Eq. (E.7) that c_0 is also zero. Expanded in sines and cosines, Eq. (E.9) is

$$
\begin{aligned}
u(t) = \sum_{n=1}^{\infty} [(a_n + a_{-n}) + i(b_n + b_{-n})]\cos(\omega_n t) \\
+ \sum_{n=1}^{\infty} [i(a_n - a_{-n}) - (b_n - b_{-n})]\sin(\omega_n t).
\end{aligned} \tag{E.10}
$$

Since $u(t)$ is real, c_n satisfies *conjugate symmetry*,

$$c_n = c_{-n}^*, \tag{E.11}$$

(i.e., $a_n = a_{-n}$ and $b_n = -b_{-n}$) so that the imaginary terms on the right-hand side of Eq. (E.10) vanish. Thus the Fourier series (Eq. (E.10)) becomes

$$u(t) = 2\sum_{n=1}^{\infty} [a_n \cos(\omega_n t) - b_n \sin(\omega_n t)], \tag{E.12}$$

which can also be written

$$u(t) = 2 \sum_{n=1}^{\infty} |c_n| \cos(\omega_n t + \theta_n), \qquad (E.13)$$

where the *amplitude* of the nth Fourier mode is

$$|c_n| = (c_n c_n^*)^{1/2} = (a_n^2 + b_n^2)^{1/2}, \qquad (E.14)$$

and its *phase* is

$$\theta_n = \tan^{-1}(b_n/a_n). \qquad (E.15)$$

An explicit expression for the Fourier coefficients is obtained by multiplying Eq. (E.9) by the −mth mode and averaging:

$$\langle e^{-i\omega_m t} u(t) \rangle_T = \left\langle \sum_{n=-\infty}^{\infty} c_n e^{i\omega_n t} e^{-i\omega_m t} \right\rangle_T$$

$$= \sum_{n=-\infty}^{\infty} c_n \delta_{nm} = c_m. \qquad (E.16)$$

It is convenient to introduce the operator $\mathcal{F}_{\omega_n}\{ \ \}$ defined by

$$\mathcal{F}_{\omega_n}\{u(t)\} \equiv \langle u(t) e^{-i\omega_n t} \rangle_T = \frac{1}{T} \int_0^T u(t) e^{-i\omega_n t} \, dt, \qquad (E.17)$$

so that Eq. (E.16) can be written

$$\mathcal{F}_{\omega_n}\{u(t)\} = c_n. \qquad (E.18)$$

Thus, the operator $\mathcal{F}_{\omega_n}\{ \ \}$ determines the Fourier coefficient of the mode with frequency ω_n.

Equation (E.9) is the spectral representation of $u(t)$, giving $u(t)$ as the sum of discrete Fourier modes $e^{i\omega_n t}$, weighted with Fourier coefficients, c_n. With the extension to the non-periodic case in mind, the spectral representation can also be written

$$u(t) = \int_{-\infty}^{\infty} z(\omega) e^{i\omega t} \, d\omega, \qquad (E.19)$$

where

$$z(\omega) \equiv \sum_{n=-\infty}^{\infty} c_n \delta(\omega - \omega_n), \qquad (E.20)$$

with ω being the continuous frequency. The integral in Eq. (E.19) is an inverse Fourier transform (cf. Eq. (D.2)), and hence $z(\omega)$ is identified as the Fourier transform of $u(t)$. (See also Exercise E.1.)

E.2 Periodic random processes

We now consider $u(t)$ to be a statistically stationary, periodic random process. All the results obtained above are valid for each realization of the process. In particular, the Fourier coefficients c_n are given by Eq. (E.16). However, since $u(t)$ is random, the Fourier coefficients c_n are random variables. We now show that the means $\langle c_n \rangle$ are zero, and that the coefficients corresponding to different frequencies are uncorrelated.

The mean of Eq. (E.9) is

$$\langle u(t) \rangle = \sum_{n=-\infty}^{\infty} \langle c_n \rangle e^{i\omega_n t}. \tag{E.21}$$

Recall that c_0 is zero, and that for $n \neq 0$, the stationarity condition – that $\langle u(t) \rangle$ be independent of t – evidently implies that $\langle c_n \rangle$ is zero.

The covariance of the Fourier modes is

$$
\begin{aligned}
\langle c_n c_m \rangle &= \langle \langle e^{-i\omega_n t} u(t) \rangle_T \langle e^{-i\omega_m t} u(t) \rangle_T \rangle \\
&= \frac{1}{T^2} \int_0^T \int_0^T e^{-i\omega_n t} e^{-i\omega_m t'} \langle u(t) u(t') \rangle \, dt' \, dt \\
&= \frac{1}{T} \int_0^T e^{-i(\omega_n + \omega_m)t} \left(\frac{1}{T} \int_{-t}^{T-t} e^{-i\omega_m s} R(s) \, ds \right) dt \\
&= \delta_{n(-m)} \mathcal{F}_{\omega_m}\{R(s)\}.
\end{aligned}
\tag{E.22}
$$

The third line follows from the substitution $t' = t + s$, and from the definition of the autocovariance

$$R(s) \equiv \langle u(t) u(t+s) \rangle, \tag{E.23}$$

which (because of stationarity) is independent of t. The integrand $e^{-i\omega_m s} R(s)$ is periodic in s, with period T, so the integral in large parentheses is independent of t. The last line then follows from Eqs. (E.8) and (E.17).

It is immediately evident from Eq. (E.22) that the covariance $\langle c_n c_m \rangle$ is zero unless m equals $-n$: that is, *Fourier coefficients corresponding to different frequencies are uncorrelated.* For $m = -n$, Eq. (E.22) becomes

$$\langle c_n c_{-n} \rangle = \langle c_n c_n^* \rangle = \langle |c_n|^2 \rangle = \mathcal{F}_{\omega_n}\{R(s)\}. \tag{E.24}$$

Thus the variances $\langle |c_n|^2 \rangle$ are the Fourier coefficients of $R(s)$, which can therefore be expressed as

$$R(s) = \sum_{n=-\infty}^{\infty} \langle c_n c_n^* \rangle e^{i\omega_n s} = 2 \sum_{n=1}^{\infty} \langle |c_n|^2 \rangle \cos(\omega_n s). \tag{E.25}$$

It may be observed that $R(s)$ is real and an even function of s, and that it depends only on the amplitudes $|c_n|$ independent of the phases θ_n.

Again with the extension to the non-periodic case in mind, we define the *frequency spectrum* by

$$\check{E}(\omega) = \sum_{n=-\infty}^{\infty} \langle c_n c_n^* \rangle \delta(\omega - \omega_n), \tag{E.26}$$

so that the autocovariance can be written

$$R(s) = \sum_{n=-\infty}^{\infty} \langle c_n c_n^* \rangle e^{i\omega_n s} = \int_{-\infty}^{\infty} \check{E}(\omega) e^{i\omega s} \, d\omega \tag{E.27}$$

(cf. Eq. E.19).

It may be seen from its definition that $\check{E}(\omega)$ is a real, even function of ω (i.e., $\check{E}(\omega) = \check{E}(-\omega)$). It is convenient, then, to define the (alternative) frequency spectrum by

$$E(\omega) = 2\check{E}(\omega), \quad \text{for } \omega \geq 0, \tag{E.28}$$

and to rewrite Eq. (E.27) as the inverse cosine transform (Eq. (D.9))

$$R(s) = \int_0^{\infty} E(\omega) \cos(\omega s) \, d\omega. \tag{E.29}$$

Setting $s = 0$ in the above equation, we obtain

$$R(0) = \langle u(t)^2 \rangle = \sum_{n=-\infty}^{\infty} \langle c_n c_n^* \rangle = \int_{-\infty}^{\infty} \check{E}(\omega) \, d\omega$$

$$= \int_0^{\infty} E(\omega) \, d\omega. \tag{E.30}$$

Consequently, $\langle c_n c_n^* \rangle$ represents the contribution to the variance from the nth mode, and similarly

$$\int_{\omega_a}^{\omega_b} E(\omega) \, d\omega$$

is the contribution to $\langle u(t)^2 \rangle$ from the frequency range $\omega_a \leq |\omega| < \omega_b$. It is clear from Eq. (E.26) that, like $R(s)$, the spectrum $E(\omega)$ is independent of the phases.

It may be observed that Eq. (E.27) identifies $R(s)$ as the inverse Fourier transform of $\check{E}(\omega)$ (cf. Eq. (D.2)). Hence, as may be verified directly from Eq. (E.25), $\check{E}(\omega)$ is the Fourier transform of $R(s)$:

$$\check{E}(\omega) = \frac{1}{2\pi} \int_{-\infty}^{\infty} R(s) e^{-i\omega s} \, ds. \tag{E.31}$$

Similarly, $E(\omega)$ is *twice* the Fourier transform of $R(s)$:

$$E(\omega) = \frac{2}{\pi} \int_0^\infty R(s) \cos(\omega s) \, ds. \tag{E.32}$$

Having identified $R(s)$ and $\check{E}(\omega)$ as a Fourier-transform pair, we now take Eq. (E.31) as the definition of $\check{E}(\omega)$ (rather than Eq. (E.26)).

The spectrum $\check{E}(\omega)$ can also be expressed in terms of the Fourier transform $z(\omega)$, defined in Eq. (E.20). Consider the infinitesimal interval $(\omega, \omega + d\omega)$, which contains either zero or one of the discrete frequencies ω_n. If it contains none of the discrete frequencies, then

$$z(\omega) \, d\omega = 0, \quad \check{E}(\omega) \, d\omega = 0. \tag{E.33}$$

On the other hand, if it contains the discrete frequency ω_n, then

$$z(\omega) \, d\omega = c_n, \quad \check{E}(\omega) \, d\omega = \langle c_n c_n^* \rangle. \tag{E.34}$$

Thus, in general,

$$\check{E}(\omega) \, d\omega = \langle z(\omega) z(\omega)^* \rangle \, d\omega^2. \tag{E.35}$$

The essential properties of the spectrum are that

 (i) $\check{E}(\omega)$ is non-negative ($\check{E}(\omega) \geq 0$) (see Eq. (E.35));
 (ii) $\check{E}(\omega)$ is real (because $R(s)$ is even, i.e., $R(s) = R(-s)$); and
 (iii) $\check{E}(\omega)$ is even, i.e., $\check{E}(\omega) = \check{E}(-\omega)$ (because $R(s)$ is real).

Table E.1 provides a summary of the relationships among $u(t)$, c_n, $z(\omega)$, $R(s)$, and $\check{E}(\omega)$.

EXERCISES

E.1   By taking the Fourier transform of Eq. (E.9), show that $z(\omega)$ given by Eq. (E.20) is the Fourier transform of $u(t)$. (Hint: see Eq. (D.12).)

E.2   Show that the Fourier coefficients $c_n = a_n + ib_n$ of a statistically stationary, periodic random process satisfy

$$\langle c_n^2 \rangle = 0, \quad \langle a_n^2 \rangle = \langle b_n^2 \rangle, \quad \langle a_n b_n \rangle = 0, \tag{E.36}$$

$$\langle a_n a_m \rangle = \langle a_n b_m \rangle = \langle b_n b_m \rangle = 0, \quad \text{for } n \neq m. \tag{E.37}$$

Table E.1. *Spectral properties of periodic and non-periodic statistically stationary random processes*

	Periodic	Non-periodic
Autocovariance	$R(s) \equiv \langle u(t)u(t+s) \rangle$, periodic	$R(s) \equiv \langle u(t)u(t+s) \rangle$, $R(\pm\infty) = 0$
Spectrum	$\check{E}(\omega) \equiv \dfrac{1}{2\pi} \displaystyle\int_{-\infty}^{\infty} R(s)e^{-i\omega s}\, \mathrm{d}s$, discrete	$\check{E}(\omega) \equiv \dfrac{1}{2\pi} \displaystyle\int_{-\infty}^{\infty} R(s)e^{-i\omega s}\, \mathrm{d}s$, continuous
	$E(\omega) = 2\check{E}(\omega)$	$E(\omega) = 2\check{E}(\omega)$
	$= \dfrac{2}{\pi} \displaystyle\int_{0}^{\infty} R(s)\cos(\omega s)\, \mathrm{d}s$	$= \dfrac{2}{\pi} \displaystyle\int_{0}^{\infty} R(s)\cos(\omega s)\, \mathrm{d}s$
	$R(s) = \displaystyle\int_{-\infty}^{\infty} \check{E}(\omega)e^{i\omega s}\, \mathrm{d}\omega$	$R(s) = \displaystyle\int_{-\infty}^{\infty} \check{E}(\omega)e^{i\omega s}\, \mathrm{d}\omega$
	$= \displaystyle\int_{0}^{\infty} E(\omega)\cos(\omega s)\, \mathrm{d}\omega$	$= \displaystyle\int_{0}^{\infty} E(\omega)\cos(\omega s)\, \mathrm{d}\omega$
Fourier coefficient	$c_n = \langle e^{-i\omega_n t} u(t) \rangle_T$ $= \mathcal{F}_{\omega_n}\{u(t)\}$	
Fourier transform	$z(\omega) = \displaystyle\sum_{n=-\infty}^{\infty} c_n \delta(\omega - \omega_n)$	
Spectral representation	$u(t) = \displaystyle\int_{-\infty}^{\infty} e^{i\omega t} z(\omega)\, \mathrm{d}\omega$ $= \displaystyle\sum_{n=-\infty}^{\infty} e^{i\omega_n t} c_n$	$u(t) = \displaystyle\int_{-\infty}^{\infty} e^{i\omega t}\, \mathrm{d}Z(\omega)$
Spectrum	$\check{E}(\omega)\, \mathrm{d}\omega = \langle z(\omega)z(\omega)^* \rangle\, \mathrm{d}\omega^2$	$\check{E}(\omega)\, \mathrm{d}\omega = \langle \mathrm{d}Z(\omega)\, \mathrm{d}Z(\omega)^* \rangle$

E.3 Non-periodic random processes

We now consider $u(t)$ to be a non-periodic, statistically stationary random process. Instead of being periodic, the autocovariance $R(s)$ decays to zero as $|s|$ tends to infinity. Just as in the periodic case, the spectrum $\check{E}(\omega)$ is defined to be the Fourier transform of $R(s)$, but now $\check{E}(\omega)$ is a continuous function of ω, rather than being composed of delta functions.

In an approximate sense, the non-periodic case can be viewed as the periodic case in the limit as the period T tends to infinity. The difference between adjacent discrete frequencies is

$$\Delta\omega \equiv \omega_{n+1} - \omega_n = 2\pi/T, \tag{E.38}$$

which tends to zero as T tends to infinity. Consequently, within any given frequency range ($\omega_a \leq \omega < \omega_b$), the number of discrete frequencies ($\approx (\omega_b - \omega_a)/\Delta\omega$) tends to infinity, so that (in an approximate sense) the spectrum becomes continuous in ω.

Mathematically rigorous treatments of the non-periodic case are given by Monin and Yaglom (1975) and Priestley (1981). Briefly, while the non-periodic process $u(t)$ does not have a Fourier transform, it does have a spectral representation in terms of the Fourier–Stieltjes integral

$$u(t) = \int_{-\infty}^{\infty} e^{i\omega t}\,dZ(\omega), \tag{E.39}$$

where $Z(\omega)$ is a non-differentiable complex random function. It may be observed that $dZ(\omega)$ (in the non-periodic case) corresponds to $z(\omega)\,d\omega$ (in the periodic case, Eq. (E.20)). The spectrum for the non-periodic case corresponding to Eq. (E.35) for the periodic case is

$$\check{E}(\omega)\,d\omega = \langle\,dZ(\omega)\,dZ(\omega)^*\rangle. \tag{E.40}$$

In Table E.1 the spectral representations for the periodic and non-periodic cases are compared.

E.4 Derivatives of the process

For the periodic case, the process $u(t)$ has the spectral representation Eq. (E.9). On differentiating with respect to time, we obtain the spectral representation of du/dt:

$$\frac{du(t)}{dt} = \sum_{n=-\infty}^{\infty} i\omega_n c_n e^{i\omega_n t}. \tag{E.41}$$

Similarly, the spectral representation of the kth derivative is

$$u^{(k)}(t) \equiv \frac{d^k u(t)}{dt^k} = \sum_{n=-\infty}^{\infty} (i\omega_n)^k c_n e^{i\omega_n t}. \tag{E.42}$$

Because $u^{(k)}(t)$ is determined by the Fourier coefficients of $u(t)$ (i.e., c_n in Eq. (E.42)), the autocovariance and spectrum of $u^{(k)}(t)$ are determined by $R(s)$ and $E(\omega)$.

It follows from the same procedure as that which leads to Eq. (E.25) that the autocorrelation of $u^{(k)}(t)$ is

$$R_k(s) \equiv \langle u^{(k)}(t)u^{(k)}(t+s)\rangle$$

$$= \sum_{n=-\infty}^{\infty} \omega_n^{2k} \langle c_n c_n^* \rangle e^{i\omega_n s}. \tag{E.43}$$

By comparing this with the $(2k)$th derivative of Eq. (E.25),

$$\frac{d^{2k}R(s)}{ds^{2k}} = (-1)^k \sum_{n=-\infty}^{\infty} \omega_n^{2k} \langle c_n c_n^* \rangle e^{i\omega_n s}, \tag{E.44}$$

we obtain the result

$$R_k(s) = (-1)^k \frac{d^{2k}R(s)}{ds^{2k}}. \tag{E.45}$$

The spectrum of $u^{(k)}(t)$ is (cf. Eq. (E.26))

$$\check{E}_k(\omega) \equiv \sum_{n=-\infty}^{\infty} \omega_n^{2k} \langle c_n c_n^* \rangle \delta(\omega - \omega_n),$$

$$= \omega^{2k} \sum_{n=-\infty}^{\infty} \langle c_n c_n^* \rangle \delta(\omega - \omega_n)$$

$$= \omega^{2k} \check{E}(\omega), \tag{E.46}$$

a result that can, alternatively, be obtained by taking the Fourier transform of Eq. (E.45).

In summary, for the kth derivative $u^{(k)}(t)$ of the process $u(t)$, the autocovariance $R_k(s)$ is given by Eq. (E.45), while the spectrum is

$$\check{E}_k(\omega) = \omega^{2k} \check{E}(\omega). \tag{E.47}$$

These two results apply both to the periodic and to non-periodic cases.

Appendix F

The discrete Fourier transform

We consider a periodic function $u(t)$, with period T, sampled at N equally spaced times within the period, where N is an even integer. On the basis of these samples, the *discrete Fourier transform* defines N Fourier coefficients (related to the coefficients of the Fourier series) and thus provides a discrete spectral representation of $u(t)$.

The *fast Fourier transform* (FFT) is an efficient implementation of the discrete Fourier transform. In numerical methods, the FFT and its inverse can be used to transform between time and frequency domains, and between physical space and wavenumber space. In DNS and LES of flows with one or more directions of statistical homogeneity, pseudo-spectral methods are generally used (in the homogeneous directions), with FFTs being used extensively to transform between physical and wavenumber spaces.

The time interval Δt is defined by

$$\Delta t \equiv \frac{T}{N}, \tag{F.1}$$

the sampling times are

$$t_j \equiv j\Delta t, \quad \text{for } j = 0, 1, \ldots, N - 1, \tag{F.2}$$

and the samples are denoted by

$$u_j \equiv u(t_j). \tag{F.3}$$

The complex coefficients $\tilde{c}_k$ of the discrete Fourier transform are then defined for $1 - \frac{1}{2}N \le k \le \frac{1}{2}N$ by

$$\tilde{c}_k \equiv \frac{1}{N} \sum_{j=0}^{N-1} u_j e^{-i\omega_k t_j} = \frac{1}{N} \sum_{j=0}^{N-1} u_j e^{-2\pi i j k/N}, \tag{F.4}$$

where (as with Fourier series) the frequency ω_k is defined by

$$\omega_k = \frac{2\pi k}{T}.$$ (F.5)

As demonstrated below, the inverse transform is

$$u_\ell = \sum_{k=1-\frac{1}{2}N}^{\frac{1}{2}N} \tilde{c}_k e^{i\omega_k t_\ell} = \sum_{k=1-\frac{1}{2}N}^{\frac{1}{2}N} \tilde{c}_k e^{2\pi i k\ell/N}.$$ (F.6)

In order to confirm the form of the inverse transform, we consider the quantity

$$\mathcal{I}_{j,N} \equiv \frac{1}{N} \sum_{k=1-\frac{1}{2}N}^{\frac{1}{2}N} e^{2\pi i jk/N}.$$ (F.7)

Viewed in the complex plane, $\mathcal{I}_{j,N}$ is the centroid of the N points $e^{2\pi i jk/N}$ for the N values of k. For j being zero or an integer multiple of N, each point is located at $(1,0)$, so $\mathcal{I}_{j,N}$ is unity. For j not being an integer multiple of N, the points are distributed symmetrically about the origin, so $\mathcal{I}_{j,N}$ is zero. Thus

$$\mathcal{I}_{j,N} = \begin{cases} 1, & \text{for } j/N \text{ integer,} \\ 0, & \text{otherwise.} \end{cases}$$ (F.8)

With this result, the right-hand side of Eq. (F.6) can be written

$$\sum_{k=1-\frac{1}{2}N}^{\frac{1}{2}N} \tilde{c}_k e^{2\pi i k\ell/N} = \sum_{k=1-\frac{1}{2}N}^{\frac{1}{2}N} \frac{1}{N} \sum_{j=0}^{N-1} u_j e^{-2\pi i jk/N} e^{2\pi i k\ell/N}$$

$$= \sum_{j=0}^{N-1} u_j \frac{1}{N} \sum_{k=1-\frac{1}{2}N}^{\frac{1}{2}N} e^{2\pi i k(\ell-j)/N}$$

$$= \sum_{j=0}^{N-1} u_j \mathcal{I}_{(\ell-j),N} = u_\ell.$$ (F.9)

In the final sum, the only non-zero contribution is for $j = \ell$. This verifies the inverse transform, Eq. (F.6).

It is informative to study the relationship between the coefficients of the discrete Fourier transform $\tilde{c}_k$ and those of the Fourier series c_k. From the definitions of these quantities (Eqs. (F.6) and (E.9)) we have

$$u_\ell = \sum_{k=1-\frac{1}{2}N}^{\frac{1}{2}N} \tilde{c}_k e^{i\omega_k t_\ell} = \sum_{k=-\infty}^{\infty} c_k e^{i\omega_k t_\ell}. \tag{F.10}$$

Before considering the general case, we consider the simpler situation in which the Fourier coefficients c_k are zero for all modes with $|\omega_k| \geq \omega_{max}$, where ω_{max} is the highest frequency represented in the discrete Fourier transform,

$$\omega_{max} \equiv \frac{\pi}{\Delta t} = \omega_{N/2}. \tag{F.11}$$

In this case, the sums in Eq. (F.10) are both effectively from $-(\frac{1}{2}N - 1)$ to $(\frac{1}{2}N - 1)$, so the coefficients $\tilde{c}_k$ and c_k are identical.

For the general case, we need to consider frequencies higher than ω_{max}. For k in the range $-(\frac{1}{2}N - 1) \leq k \leq \frac{1}{2}N$, and for a non-zero integer m, the $(k + mN)$th mode has the frequency

$$\omega_{k+mN} = \omega_k + 2m\omega_{max}, \tag{F.12}$$

with

$$|\omega_{k+mN}| \geq \omega_{max}. \tag{F.13}$$

At the sampling times t_j, the $(k + mN)$th mode is indistinguishable from the kth mode, since

$$e^{i\omega_{k+mN}t_j} = e^{2\pi i j(k+mN)/N} = e^{2\pi ijk/N} = e^{i\omega_k t_j}. \tag{F.14}$$

The $(k + mN)$th mode is said to be *aliased* to the kth mode.

The coefficients $\tilde{c}_k$ can be determined from their definition (Eq. (F.4)) with the Fourier series substituted for u_j:

$$\tilde{c}_k = \frac{1}{N} \sum_{j=0}^{N-1} \left(\sum_{n=-\infty}^{\infty} c_n e^{2\pi inj} \right) e^{-2\pi ijk/N}$$

$$= \sum_{n=-\infty}^{\infty} \mathcal{I}_{n-k,N} c_n = \sum_{m=-\infty}^{\infty} c_{k+mN}. \tag{F.15}$$

Thus the coefficient $\tilde{c}_k$ is the sum of the Fourier coefficients of all the modes that are aliased to the kth mode.

In view of conjugate symmetry, the N complex coefficient $\tilde{c}_k$ can be expressed in terms of N real numbers (e.g., $\Re\{\tilde{c}_k\}$ for $k = 0, 1, \ldots, \frac{1}{2}N$ and $\Im\{\tilde{c}_k\}$ for $k = 1, 2, \ldots, \frac{1}{2}N-1$, see Exercise F.1). The discrete Fourier transform and its inverse provide a one-to-one mapping between u_j and $\tilde{c}_k$. On the order of N^2 operations are required in order to evaluate $\tilde{c}_k$ directly from the sum in Eq. (F.4). However, the same result can be obtained in on the order of $N \log N$ operations by using the *fast Fourier transform* (FFT) (see, e.g., Brigham (1974)). Thus, for periodic data sampled at sufficiently small time intervals, the FFT is an efficient tool for evaluating Fourier coefficients, spectra, autocorrelations (as inverse Fourier transforms of spectra), convolutions, and derivatives as

$$\frac{d^n u(t)}{dt^n} = \mathcal{F}^{-1}\{(i\omega)^n \mathcal{F}\{u(t)\}\}. \tag{F.16}$$

As with the Fourier transform, there are various definitions of the discrete Fourier transform. The definition used here makes the most direct connection with Fourier series. In numerical implementations, the alternative definition given in Exercise F.2 is usually used.

EXERCISES

F.1   Show that, for real $u(t)$, the coefficients $\tilde{c}_k$ satisfy

$$\tilde{c}_k = \tilde{c}_{-k}^*, \quad \text{for } |k| < \tfrac{1}{2}N, \tag{F.17}$$

and that $\tilde{c}_0$ and $\tilde{c}_{\frac{1}{2}N}$ are real.
      Show that

$$\cos(\omega_{\max} t_j) = (-1)^j, \tag{F.18}$$

$$\sin(\omega_{\max} t_j) = 0. \tag{F.19}$$

F.2   An alternative definition of the discrete Fourier transform is

$$\bar{c}_k = \sum_{j=0}^{N-1} u_j e^{-2\pi i jk/N}, \quad \text{for } k = 0, 1, \ldots, N-1. \tag{F.20}$$

Show that the inverse is

$$u_\ell = \frac{1}{N} \sum_{k=0}^{N-1} \bar{c}_k e^{2\pi i k\ell/N}. \tag{F.21}$$

What is the relationship between the coefficients $\tilde{c}_k$ and $\bar{c}_k$?

Appendix G
Power-law spectra

In the study of turbulence, considerable attention is paid to the shape of spectra at high frequency (or large wavenumber). The purpose of this appendix is to show the relationships among a power-law spectrum ($E(\omega) \sim \omega^{-p}$, for large ω), the underlying random process $u(t)$, and the second-order structure function $D(s)$.

We consider a statistically stationary process $u(t)$ with finite variance $\langle u^2 \rangle$ and integral timescale $\bar{\tau}$. The autocorrelation function

$$R(s) \equiv \langle u(t)u(t+s) \rangle \tag{G.1}$$

and half the frequency spectrum form a Fourier-cosine-transform pair:

$$R(s) = \int_0^\infty E(\omega) \cos(\omega s) \, d\omega, \tag{G.2}$$

$$E(\omega) = \frac{2}{\pi} \int_0^\infty R(s) \cos(\omega s) \, ds. \tag{G.3}$$

The third quantity of interest is the second-order structure function

$$
\begin{aligned}
D(s) &\equiv \langle [u(t+s) - u(t)]^2 \rangle \\
&= 2[R(0) - R(s)] \\
&= 2 \int_0^\infty [1 - \cos(\omega s)] E(\omega) \, d\omega.
\end{aligned}
\tag{G.4}
$$

By definition, a power-law spectrum varies as

$$E(\omega) \sim \omega^{-p}, \quad \text{for large } \omega, \tag{G.5}$$

whereas a power-law structure function varies as

$$D(s) \sim s^q, \quad \text{for small } s. \tag{G.6}$$

The aim here is to understand the significances of particular values of p and q and the connection between them.

The first observation – obtained by setting $s = 0$ in Eq. (G.2) – is that the variance is

$$\langle u^2 \rangle = R(0) = \int_0^\infty E(\omega) \, d\omega. \tag{G.7}$$

By assumption $\langle u^2 \rangle$ is finite. Hence, if $E(\omega)$ is a power-law spectrum, the requirement that the integral converges dictates $p > 1$.

A sequence of similar results stems from the spectra of the derivatives of $u(t)$. Suppose that the nth derivative of $u(t)$ exists, and denote it by

$$u^{(n)}(t) = \frac{d^n u(t)}{dt^n}. \tag{G.8}$$

The autocorrelation of $u^{(n)}(t)$ is

$$
\begin{aligned}
R_n(s) &\equiv \langle u^{(n)}(t) u^{(n)}(t+s) \rangle \\
&= (-1)^n \frac{d^{2n} R(s)}{ds^{2n}}
\end{aligned}
\tag{G.9}
$$

(see Appendix E, Eq. (E.45)), and its frequency spectrum is

$$E_n(\omega) = \omega^{2n} E(\omega) \tag{G.10}$$

(see Eq. (E.47)). Hence we obtain

$$
\begin{aligned}
\left\langle \left(\frac{d^n u}{dt^n} \right)^2 \right\rangle &= R_n(0) = \int_0^\infty E_n(\omega) \, d\omega \\
&= \int_0^\infty \omega^{2n} E(\omega) \, d\omega.
\end{aligned}
\tag{G.11}
$$

The left-hand side is finite if $u(t)$ is differentiable n times (in a mean-square sense). Then, if $E(\omega)$ is a power-law spectrum, the requirement that the integral in Eq. (G.11) converges dictates

$$p > 2n + 1. \tag{G.12}$$

For an infinitely differentiable process – such as the velocity evolving by the Navier–Stokes equations – it follows from Eq. (G.12) that (for large ω) the spectrum decays more rapidly than any power of ω: it may instead decay as $\exp(-\omega)$ or $\exp(-\omega^2)$, for example. Nevertheless, over a significant range of frequencies ($\omega_l < \omega < \omega_h$, say) a power-law spectrum may occur, with exponential decay beyond ω_h.

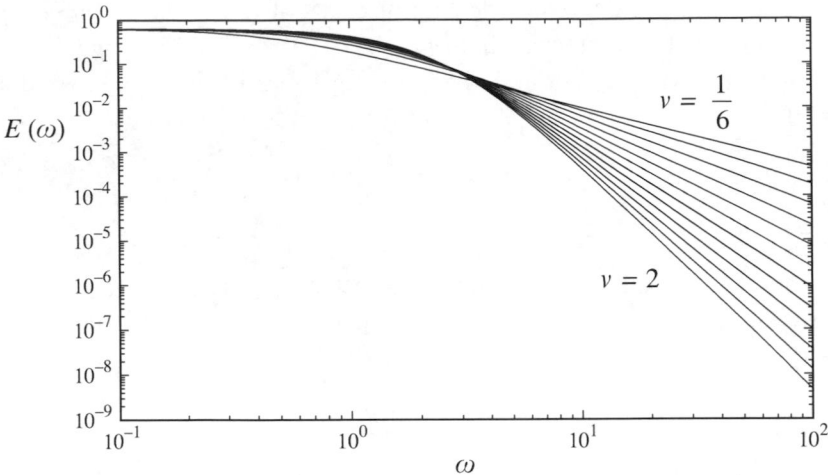

$E(\omega)$

ω

Fig. G.1. Non-dimensional power-law spectra $E(\omega)$: Eq. (G.14) for $v = \frac{1}{6}, \frac{1}{3}, \dots, 1\frac{5}{6}, 2$.

If the process $u(t)$ is at least once continuously differentiable, then, for small s, the structure function is

$$D(s) \approx \left\langle \left(\frac{du}{dt} \right)^2 \right\rangle s^2, \tag{G.13}$$

i.e., a power law (Eq. (G.6)) with $q = 2$.

It is instructive to examine the non-dimensional power-law spectrum

$$E(\omega) = \frac{2}{\pi} \left(\frac{\alpha^2}{\alpha^2 + \omega^2} \right)^{(1+2v)/2}, \tag{G.14}$$

for various values of the positive parameter v. The non-dimensional integral timescale is unity, i.e.,

$$\int_0^\infty R(s)\,ds = \frac{\pi}{2} E(0) = 1, \tag{G.15}$$

while (for given v) α is specified (see Eq. (G.18) below) so that the variance is unity. Figure G.1 shows $E(\omega)$ for v between $\frac{1}{6}$ and 2. For large ω, the straight lines on the log–log plot clearly show the power-law behavior with

$$p = 1 + 2v. \tag{G.16}$$

The corresponding autocorrelation (obtained as the inverse transform of Eq. (G.14)) is

$$R(s) = \frac{2\alpha}{\sqrt{\pi}\,\Gamma(v + \frac{1}{2})} (\tfrac{1}{2} s\alpha)^v K_v(s\alpha), \tag{G.17}$$

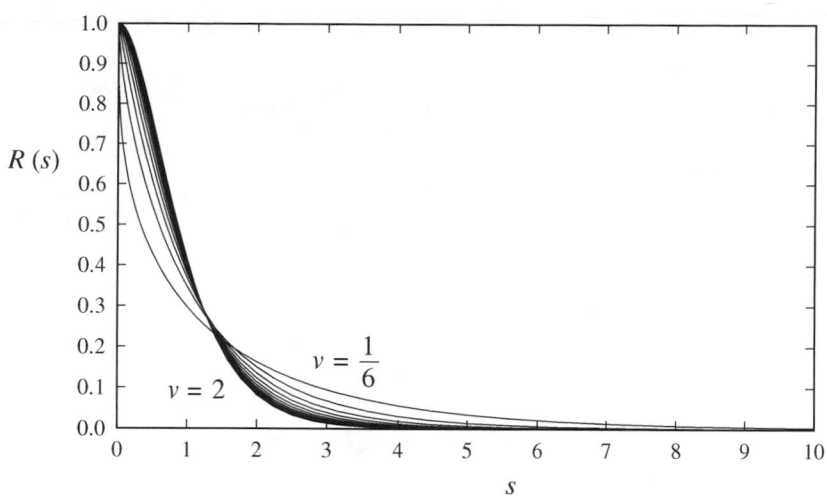

Fig. G.2. Autocorrelation functions $R(s)$, Eq. (G.19), for $v = \frac{1}{6}, \frac{1}{3}, \ldots, 1\frac{5}{6}, 2$.

where K_v is the modified Bessel function of the second kind. The normalization condition $\langle u^2 \rangle = R(0) = 1$ yields

$$\alpha = \sqrt{\pi}\,\Gamma(v + \tfrac{1}{2})/\Gamma(v),\tag{G.18}$$

so that Eq. (G.17) can be rewritten

$$R(s) = \frac{2}{\Gamma(v)}(\tfrac{1}{2}s\alpha)^v K_v(s\alpha),\tag{G.19}$$

These autocorrelations are shown in Fig. G.2. (For $v = \frac{1}{2}$, the autocorrelation given by Eq. (G.19) is simply $R(s) = \exp(-|s|)$.)

The expression for the autocorrelation is far from revealing. However, the expansion for $K_v(s\alpha)$ (for small argument) leads to very informative expressions for the structure function (for small s):

$$D(s) = \begin{cases} 2\dfrac{\Gamma(1-v)}{\Gamma(1+v)}(\tfrac{1}{2}\alpha s)^{2v}\ldots, & \text{for } v < 1, \\[2mm] 2(v-1)(\tfrac{1}{2}\alpha s)^2\ldots, & \text{for } v > 1. \end{cases}\tag{G.20}$$

Hence the structure function varies as a power-law with exponent

$$q = \begin{cases} 2v, & \text{for } v < 1, \\ 2, & \text{for } v > 1. \end{cases}\tag{G.21}$$

This behavior is evident in Fig. G.3.

The conclusions to be drawn from the expressions for the power-law exponents p and q (Eqs. (G.16) and (G.21)) are straightforward. The case

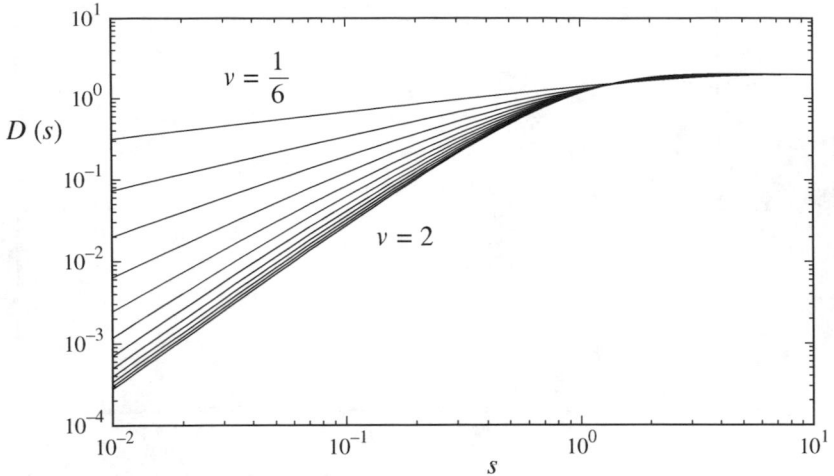

Fig. G.3. Second-order structure functions $D(s)$, Eq. (G.20), for $v = \frac{1}{6}, \frac{1}{3}, \ldots, 1\frac{5}{6}, 2$. Observe that, for $v > 1$ and small s, all the structure functions vary as s^2.

Table G.1. *The relationships among the spectral exponent p, the structure-function exponent q, and the differentiability of the underlying process u(t) for the power-law spectrum Eq. (G.14)*

Parameter in Eq. (G.14)	Spectrum	Structure function	Process $u(t)$
	$E(\omega) \sim \omega^{-p}$	$D(s) \sim s^q$	$\left\langle \left(\dfrac{d^n u}{dt^n}\right)^2 \right\rangle < \infty$
v	p	q	n
$\frac{1}{3}$	$\frac{5}{3}$	$\frac{2}{3}$	0
$\frac{1}{2}$	2	1	0
> 1	> 3	2	≥ 1
> 2	> 5	2	≥ 2

$v > 1$ corresponds to an underlying process that is at least once continuously differentiable, for which p is greater than 3 and q is 2. The case $0 < v < 1$ corresponds to a non-differentiable process $u(t)$, and the power laws are connected by

$$p = q + 1. \tag{G.22}$$

Table G.1 displays these results for particular cases.

For $v < 1$, for large ω the power-law spectrum is

$$E(\omega) \approx C_1 \omega^{-(1+q)}, \tag{G.23}$$

while for small s the structure function is

$$D(s) \approx C_2 s^q. \tag{G.24}$$

These relations stem from Eqs. (G.14) and (G.20), from which C_1 and C_2 (which depend upon q) can be deduced. It is a matter of algebra (see Exercise G.1) to show that C_1 and C_2 are related by

$$\frac{C_1}{C_2} = \frac{1}{\pi}\Gamma(1 + q)\sin\left(\frac{\pi q}{2}\right). \tag{G.25}$$

For the particular case $q = \frac{2}{3}$, this ratio is 0.2489; or, to an excellent approximation,

$$(C_1/C_2)_{q=2/3} \approx \tfrac{1}{4}. \tag{G.26}$$

Although we have considered a specific example of a power-law spectrum (i.e., Eq. (G.14)), the conclusions drawn are general. If a spectrum exhibits the power-law behavior $E(\omega) \approx C_1 \omega^{-p}$ over a significant range of frequencies, then there is corresponding power-law behavior $D(s) \approx C_2 s^q$ for the structure function with $q = \min(p - 1, 2)$. (This assertion can be verified by analysis of Eq. (G.4), see Monin and Yaglom (1975).) For $q < 2$, C_1 and C_2 are related by Eq. (G.25).

EXERCISE _____

G.1   Identify C_1 and C_2 in Eqs. (G.23) and (G.24). With the use of the following properties of the gamma function:

$$\Gamma(1 + v) = v\Gamma(v), \tag{G.27}$$

$$\Gamma(v)\Gamma(1 - v) = \pi/\sin(\pi v), \tag{G.28}$$

$$\Gamma(v)\Gamma(v + \tfrac{1}{2}) = (2\pi)^{\frac{1}{2}} 2^{(1/2-2v)}\Gamma(2v), \tag{G.29}$$

verify Eq. (G.25).

Appendix H
Derivation of Eulerian PDF equations

In this appendix, the transport equation for the Eulerian PDF of velocity $f(V;x,t)$ is derived from the Navier–Stokes equations. This derivation is based on the properties of the *fine-grained PDF*; some other techniques are described by Dopazo (1994).

The fine-grained PDF

For a given realization of the flow, the (one-point, one-time, Eulerian) fine-grained PDF of velocity is defined by

$$f'(V;x,t) \equiv \delta(U(x,t) - V) = \prod_{i=1}^{3} \delta(U_i(x,t) - V_i). \qquad \text{(H.1)}$$

At each point x and time t, f' is a three-dimensional delta function in velocity space located at $V = U(x,t)$. (Appendix C reviews the properties of delta functions.)

The fine-grained PDF is very useful in obtaining and manipulating PDF equations, because of the following two properties:

$$\langle f'(V;x,t) \rangle = f(V;x,t), \qquad \text{(H.2)}$$

$$\langle \phi(x,t)f'(V;x,t) \rangle = \langle \phi(x,t)|U(x,t) = V \rangle f(V;x,t). \qquad \text{(H.3)}$$

The first of these relations is obtained from the general expression for a mean $\langle Q(U(x,t)) \rangle$, Eq. (12.2) on page 464, by replacing the integration variable V by V', making the substitution $Q(U(x,t)) = \delta(U(x,t) - V)$, and correspondingly $Q(V') = \delta(V' - V)$:

$$\langle f'(V;x,t) \rangle = \langle \delta(U(x,t) - V) \rangle$$

$$= \int \delta(V' - V)f(V';x,t)\,dV'$$

$$= f(V;x,t). \qquad \text{(H.4)}$$

The last step follows from the sifting property of delta functions (Eq. (C.39)). The second relation, Eq. (H.3), is obtained in an analogous way by

$$
\begin{aligned}
\langle \phi(x,t) f'(V;x,t) \rangle &= \langle \phi(x,t) \delta(U(x,t) - V) \rangle \\
&= \iint_{-\infty}^{\infty} \psi \delta(V' - V) f_{U\phi}(V', \psi; x, t) \, dV' \, d\psi \\
&= \int_{-\infty}^{\infty} \psi f_{U\phi}(V, \psi; x, t) \, d\psi \\
&= f(V;x,t) \int_{-\infty}^{\infty} \psi f_{\phi|U}(\psi | V; x, t) \, d\psi \\
&= f(V;x,t) \langle \phi(x,t) | U(x,t) = V \rangle.
\end{aligned}
\tag{H.5}
$$

These five steps follow from the definition of f', Eq. (H.1); the definition of an unconditional mean of a function of U and ϕ; the sifting property; the substitution of Eq. (12.3); and the definition of a conditional mean, Eq. (12.6).

Derivatives of the fine-grained PDF

The temporal and spatial derivatives of the fine-grained PDF $f'(V;x,t)$ are required in the derivation of the PDF transport equation. As a simpler preliminary, consider a scalar-valued differentiable process $u(t)$ with fine-grained PDF,

$$
f_u'(v;t) = \delta(u(t) - v),
\tag{H.6}
$$

where v is the sample-space variable. The derivative of the delta function $\delta(v - a)$ (with a being a constant) is denoted by $\delta^{(1)}(v - a)$ (Eq. (C.21)), and it is an odd function (Eq. (C.22)):

$$
\frac{d}{dv} \delta(v - a) = \frac{d}{dv} \delta(a - v) = \delta^{(1)}(v - a) = -\delta^{(1)}(a - v).
\tag{H.7}
$$

Thus, differentiating Eq. (H.6) with respect to t (using the chain rule), we obtain

$$
\begin{aligned}
\frac{\partial}{\partial t} f_u'(v;t) &= \delta^{(1)}(u(t) - v) \frac{du(t)}{dt} = -\delta^{(1)}(v - u(t)) \frac{du(t)}{dt} \\
&= -\frac{\partial f_u'(v;t)}{\partial v} \frac{du(t)}{dt} = -\frac{\partial}{\partial v} \left(f_u'(v;t) \frac{du(t)}{dt} \right).
\end{aligned}
\tag{H.8}
$$

The last step follows because $u(t)$ is independent of v.

For the fine-grained PDF of velocity, the same procedure yields the required derivatives:

$$\frac{\partial}{\partial t} f'(V;x,t) = -\frac{\partial f'(V;x,t)}{\partial V_i} \frac{\partial U_i(x,t)}{\partial t}, \tag{H.9}$$

$$\frac{\partial}{\partial x_i} f'(V;x,t) = -\frac{\partial f'(V;x,t)}{\partial V_j} \frac{\partial U_j(x,t)}{\partial x_i}. \tag{H.10}$$

A final result required in the derivation that follows is

$$U_i(x,t)\frac{\partial}{\partial x_i} f'(V;x,t) = \frac{\partial}{\partial x_i}\left[U_i(x,t)f'(V;x,t)\right]$$

$$= \frac{\partial}{\partial x_i}\left[V_i f'(V;x,t)\right] = V_i \frac{\partial}{\partial x_i} f'(V;x,t). \tag{H.11}$$

The first step relies on incompressibility ($\nabla \cdot U = 0$), the second follows from the sifting property (Eq. (C.15)), and the third step follows because V_i is an independent variable.

The PDF transport equation

With the results already obtained, the derivation of the transport equation for the PDF $f(V;x,t)$ is straightforward. The substantial derivative of $f'(V;x,t)$ is obtained from Eqs. (H.9)–(H.11), is

$$\frac{Df'}{Dt} = \frac{\partial f'}{\partial t} + V_i \frac{\partial f'}{\partial x_i} = -\frac{\partial}{\partial V_i}\left(f'\frac{DU_i}{Dt}\right), \tag{H.12}$$

and the mean of this equation yields

$$\frac{\partial f}{\partial t} + V_i \frac{\partial f}{\partial x_i} = -\frac{\partial}{\partial V_i}\left(f\left\langle \frac{DU_i}{Dt}\middle| V\right\rangle\right). \tag{H.13}$$

In Eq. (H.11) it is assumed that the velocity field is incompressible; but otherwise, this PDF equation (Eq. (H.13)) is quite general, and contains no physics. The physics enters when the Navier–Stokes equations (Eq. (2.35) on page 17) are used to substitute for DU_i/Dt:

$$\frac{\partial f}{\partial t} + V_i \frac{\partial f}{\partial x_i} = -\frac{\partial}{\partial V_i}\left(f\left\langle \nu \nabla^2 U_i - \frac{1}{\rho}\frac{\partial p}{\partial x_i}\middle| V\right\rangle\right). \tag{H.14}$$

The Reynolds decomposition of pressure ($p = \langle p \rangle + p'$) leads to

$$\left\langle \frac{\partial p}{\partial x_i}\middle| V\right\rangle = \frac{\partial \langle p \rangle}{\partial x_i} + \left\langle \frac{\partial p'}{\partial x_i}\middle| V\right\rangle. \tag{H.15}$$

Note that $\partial\langle p\rangle/\partial x_i$ is non-random, and so is unaffected by the mean and conditional-mean operations; i.e.,

$$\frac{\partial\langle p\rangle}{\partial x_i} = \left\langle\frac{\partial\langle p\rangle}{\partial x_i}\right\rangle = \left\langle\frac{\partial\langle p\rangle}{\partial x_i}\bigg|V\right\rangle. \tag{H.16}$$

The substitution of the decomposition Eq. (H.15) into Eq. (H.14) yields the PDF equation given in the text, Eq. (12.9).

The following exercises treat the derivation of the transport equation for the PDF $g(v;x,t)$ of the fluctuating velocity, and also develop various decompositions of the terms arising in the equations for f and g.

EXERCISES

H.1   By following the same procedure as that used to derive the transport equation for $f(V;x,t)$, Eq. (H.14), show that the equation for $g(v;x,t)$ (the PDF of the fluctuating velocity) is

$$\frac{\partial g}{\partial t} + (\langle U_i\rangle + v_i)\frac{\partial g}{\partial x_i} = -\frac{\partial}{\partial v_i}\left(g\left\langle\frac{\mathrm{D}u_i}{\mathrm{D}t}\bigg|v\right\rangle\right). \tag{H.17}$$

Use Eq. (5.138) on page 126 to substitute for $\mathrm{D}u_i/\mathrm{D}t$, and hence obtain the result

$$\frac{\partial g}{\partial t} + (\langle U_i\rangle + v_i)\frac{\partial g}{\partial x_i} = v_i\frac{\partial\langle U_j\rangle}{\partial x_i}\frac{\partial g}{\partial v_j} - \frac{\partial\langle u_i u_j\rangle}{\partial x_i}\frac{\partial g}{\partial v_j}$$
$$-\frac{\partial}{\partial v_i}\left(g\left\langle v\,\nabla^2 u_i - \frac{1}{\rho}\frac{\partial p'}{\partial x_i}\bigg|v\right\rangle\right). \tag{H.18}$$

H.2   From the relationship

$$g(v;x,t) = f(\langle U\rangle + v;x,t) \tag{H.19}$$

between the PDFs of $u(x,t)$ and $U(x,t)$ show that their evolution equations are related by

$$\frac{\partial g}{\partial t} + (\langle U_i\rangle + v_i)\frac{\partial g}{\partial x_i}$$
$$= \frac{\partial f}{\partial t} + (\langle U_i\rangle + v_i)\frac{\partial f}{\partial x_i} + \frac{\partial g}{\partial v_i}\left(\frac{\partial\langle U_i\rangle}{\partial t} + (\langle U_j\rangle + v_j)\frac{\partial\langle U_i\rangle}{\partial x_j}\right)$$
$$= v_j\frac{\partial g}{\partial v_i}\frac{\partial\langle U_i\rangle}{\partial x_j} - \frac{\partial}{\partial v_j}\left[g\left(\left\langle\frac{\mathrm{D}U_j}{\mathrm{D}t}\bigg|V\right\rangle - \frac{\bar{\mathrm{D}}\langle U_j\rangle}{\bar{\mathrm{D}}t}\right)\right]. \tag{H.20}$$

Show that this is equivalent to Eq. (H.18).

H.3   With $g'(v; x, t)$ being the fine-grained PDF $\delta(u(x, t) - v)$, obtain the result

$$g' \frac{\partial p'}{\partial x_i} = \frac{\partial}{\partial x_i}(g'p') + p' \frac{\partial g'}{\partial v_j} \frac{\partial u_j}{\partial x_i}. \tag{H.21}$$

Hence show that the term in $\partial p'/\partial x_j$ in Eq. (H.18) can be re-expressed as

$$\frac{\partial}{\partial v_i} \left(g \left\langle \frac{1}{\rho} \frac{\partial p'}{\partial x_i} \middle| v \right\rangle \right) = \frac{\partial^2}{\partial x_i \, \partial v_i} \left(g \left\langle \frac{p'}{\rho} \middle| v \right\rangle \right) + \frac{1}{2} \frac{\partial^2}{\partial v_i \, \partial v_j} \left[g \mathcal{R}_{ij}^c(v) \right], \tag{H.22}$$

where the *conditional pressure–rate-of-strain tensor* $\mathcal{R}_{ij}^c(v, x, t)$ is defined by Eq. (12.20).

H.4   Obtain the relation

$$\nabla^2 g' = \frac{\partial^2 g'}{\partial v_j \, \partial v_k} \frac{\partial u_j}{\partial x_i} \frac{\partial u_k}{\partial x_i} - \frac{\partial g'}{\partial v_j} \nabla^2 u_j, \tag{H.23}$$

and hence show that the viscous term in Eq. (H.18) can be re-expressed as

$$-\frac{\partial}{\partial v_i} (g \langle \nu \nabla^2 u_i | v \rangle) = \nu \nabla^2 g - \frac{1}{2} \frac{\partial^2}{\partial v_i \, \partial v_j} \left[g \, \varepsilon_{ij}^c(v) \right], \tag{H.24}$$

where the *conditional dissipation tensor* $\varepsilon_{ij}^c(v, x, t)$ is defined by Eq. (12.21).

H.5   By using the techniques of Exercises H.3 and H.4, show that the PDF transport equation, Eq. (H.14), can be re-expressed as

$$\frac{\partial f}{\partial t} + V_i \frac{\partial f}{\partial x_i} = \nu \nabla^2 f + \frac{1}{\rho} \frac{\partial \langle p \rangle}{\partial x_i} \frac{\partial f}{\partial V_i} + \frac{\partial^2}{\partial x_i \, \partial V_i} \left(f \left\langle \frac{p'}{\rho} \middle| V \right\rangle \right)$$
$$+ \frac{1}{2} \frac{\partial^2}{\partial V_i \, \partial V_j} \left[f \left\langle \frac{p'}{\rho} \left(\frac{\partial U_i}{\partial x_j} + \frac{\partial U_j}{\partial x_i} \right) - 2\nu \frac{\partial U_i}{\partial x_k} \frac{\partial U_j}{\partial x_k} \middle| V \right\rangle \right]. \tag{H.25}$$

Appendix I
Characteristic functions

For the random variable U, the characteristic function is defined by

$$\Psi(s) \equiv \langle e^{iUs} \rangle. \tag{I.1}$$

It is a non-dimensional complex function of the real variable s (which has the dimensions of U^{-1}). Like the PDF of U, $f(V)$, the characteristic function fully characterizes the random variable U.

In some circumstances, results can be obtained more simply in terms of the characteristic function than in terms of the PDF. These circumstances include calculation of moments, transformation of random variables, sums of independent random variables, normal and joint-normal distributions, and solutions to model PDF equations. The basic properties of characteristic functions used in the text are now given.

The relationship to the PDF

Using the definition of the mean as an integral over the PDF (Eq. (3.20) on page 41), the equation (Eq. (I.1)) defining the characteristic function can be rewritten

$$\Psi(s) = \int_{-\infty}^{\infty} f(V) e^{iVs} \, dV. \tag{I.2}$$

This may be recognized as an inverse Fourier transform (cf. Eq. (D.2) on page 678) so that $f(V)$ and $\Psi(s)$ form a Fourier-transform pair:

$$f(V) = \mathcal{F}\{\Psi(s)\} = \frac{1}{2\pi} \int_{-\infty}^{\infty} \Psi(s) e^{-iVs} \, ds, \tag{I.3}$$

$$\Psi(s) = \mathcal{F}^{-1}\{f(V)\}. \tag{I.4}$$

Because $f(V)$ is real, $\Psi(s)$ has conjugate symmetry: $\Psi(s) = \Psi^*(-s)$.

The behavior at the origin

Setting s to zero in Eq. (I.1) yields the important result

$$\Psi(0) = 1, \tag{I.5}$$

which corresponds to the normalization condition for the PDF. The kth derivative of $\Psi(s)$ is

$$\Psi^{(k)}(s) \equiv \frac{d^k \Psi(s)}{ds^k} = i^k \langle U^k e^{iUs} \rangle. \tag{I.6}$$

Thus, the moments of U (about the origin) are given by

$$\langle U^k \rangle = (-i)^k \Psi^{(k)}(0). \tag{I.7}$$

Linear transformations

With a and b being constants, the random variable

$$\widetilde{U} \equiv a + bU \tag{I.8}$$

has the characteristic function

$$\widetilde{\Psi}(s) = \langle e^{i\widetilde{U}s} \rangle = \langle e^{ias + ibUs} \rangle$$
$$= e^{ias} \Psi(bs). \tag{I.9}$$

Sums of independent random variables

If U_1 and U_2 are *independent* random variables with characteristic functions $\Psi_1(s)$ and $\Psi_2(s)$, then the characteristic function of their sum $\widetilde{U} = U_1 + U_2$ is

$$\widetilde{\Psi}(s) = \langle e^{i(U_1 + U_2)s} \rangle = \langle e^{iU_1 s} e^{iU_2 s} \rangle$$
$$= \langle e^{iU_1 s} \rangle \langle e^{iU_2 s} \rangle = \Psi_1(s) \Psi_2(s), \tag{I.10}$$

i.e., the product of the characteristic functions.

The normal distribution

If U is a standardized normal random variable, so that its PDF is

$$f(V) = \frac{1}{\sqrt{2\pi}} e^{-\frac{1}{2}V^2}, \tag{I.11}$$

then taking the inverse Fourier transform (see Table D.1 on page 679) yields

$$\Psi(s) = e^{-s^2/2}. \tag{I.12}$$

More generally, if U is normal with mean μ and variance σ^2, then Eqs. (I.9) and (I.12) yield

$$\Psi(s) = e^{i\mu s - \sigma^2 s^2/2}. \tag{I.13}$$

The fine-grained characteristic function

Just as the fine-grained PDF $f'(V) \equiv \delta(U - V)$ is defined so that its mean is the PDF $f(V)$, so also the fine-grained characteristic function is defined by

$$\Psi'(s) \equiv e^{isU}, \tag{I.14}$$

so that its mean is $\Psi(s)$. Consistently, $f'(V)$ and $\Psi'(s)$ form a Fourier-transform pair. With Q being a second random variable, we have (see Eq. (H.5))

$$\langle Qf'(V) \rangle = \langle Q|V \rangle f(V). \tag{I.15}$$

Correspondingly, for the fine-grained characteristic function we have

$$\begin{aligned}
\langle Q\Psi'(s) \rangle &= \int_{-\infty}^{\infty} \langle Q\Psi'(s)|V \rangle f(V) \, dV \\
&= \int_{-\infty}^{\infty} \langle Q|V \rangle f(V) e^{isV} \, dV \\
&= \mathcal{F}^{-1}\{\langle Q|V \rangle f(V)\}. \tag{I.16}
\end{aligned}$$

Thus $\langle Qf' \rangle$ and $\langle Q\Psi' \rangle$ also form a Fourier-transform pair.

Summary

Table I.1 summarizes several of the useful properties of the characteristic function, and its relationship to the PDF.

EXERCISES _____

I.1   On the basis of the Fourier transforms Eqs. (I.3) and (I.4), verify the relations

$$\mathcal{F}^{-1}\left\{\frac{d^2 f(V)}{dV^2}\right\} = -s^2 \Psi(s), \tag{I.17}$$

$$\mathcal{F}^{-1}\left\{\frac{d}{dV}(fV)\right\} = -s\frac{d\Psi(s)}{ds}. \tag{I.18}$$

I.2   Use Eqs. (I.7) and (I.12) to determine the first six moments of the standardized normal distribution.

I.3   Let $U_1, U_2, \ldots, U_N$ be independent, identically distributed, standardized random variables. Show that

$$X_N \equiv \frac{1}{\sqrt{N}} \sum_{i=1}^{N} U_i, \tag{I.19}$$

Table I.1. *Relationships between characteristic functions and probability density functions (the quantities on the right-hand side are the Fourier transforms of those on the left-hand side)*

Characteristic function $\Psi(s) = \mathcal{F}^{-1}\{f(V)\}$	Probability density function $f(V) = \mathcal{F}\{\Psi(s)\}$
$\Psi'(s) = e^{isU}$	$f'(V) = \delta(U - V)$
$\begin{aligned}\Psi(s) &= \langle \Psi'(s)\rangle \\ &= \int_{-\infty}^{\infty} f(V)e^{isV}\,\mathrm{d}V\end{aligned}$	$\begin{aligned}f(V) &= \langle f'(V)\rangle \\ &= \frac{1}{2\pi}\int_{-\infty}^{\infty}\Psi(s)e^{-isV}\,\mathrm{d}s\end{aligned}$
$(-i)^n \dfrac{\mathrm{d}^n\Psi(s)}{\mathrm{d}s^n}$	$V^n f(V)$
$(-is)^n\Psi(s)$	$\dfrac{\mathrm{d}^n f(V)}{\mathrm{d}V^n}$
$-s\dfrac{\mathrm{d}\Psi(s)}{\mathrm{d}s}$	$\dfrac{\mathrm{d}}{\mathrm{d}V}[Vf(V)]$
$\langle Q\Psi'(s)\rangle$	$\langle Qf'(V)\rangle = \langle Q|V\rangle f(V)$
$(-is)^n\langle Q\Psi'(s)\rangle$	$\dfrac{\mathrm{d}^n}{\mathrm{d}V^n}[\langle Q|V\rangle f(V)]$
$\exp(i\mu s - \tfrac{1}{2}\sigma^2 s^2)$	$\dfrac{1}{\sigma\sqrt{2\pi}}\exp(-\tfrac{1}{2}[V-\mu]^2/\sigma^2)$

is a standardized random variable. Show that the characteristic function of X_N, $\Psi_N(s)$, is related to that of U_i, $\Psi_u(s)$, by

$$\Psi_N(s) = \Psi_u(N^{-1/2}s)^N, \tag{I.20}$$

and therefore

$$\ln\Psi_N(s) = N\ln\left(1 - \frac{1}{2}\frac{s^2}{N} + \frac{1}{3!}\Psi_u^{(3)}(0)\frac{s^3}{N^{\frac{3}{2}}} + \dots\right). \tag{I.21}$$

Hence obtain the result

$$\lim_{N\to\infty}\Psi_N(s) = e^{-s^2/2}, \tag{I.22}$$

that is, X_N tends to a standardized normal random variable. (This is a simple version of the *central-limit theorem*.)

I.4   Let U and Q be random variables, with $f(V)$ and $\Psi'(s)$ being the

PDF and fine-grained characteristic function of U, respectively. Show that the characteristic function of $U + Q$ is

$$\langle e^{is(U+Q)} \rangle = \sum_{n=0}^{\infty} \frac{(is)^n}{n!} \langle Q^n \Psi'(s) \rangle, \tag{I.23}$$

and correspondingly that the PDF of $U + Q$ is

$$\langle \delta(U + Q - V) \rangle = \sum_{n=0}^{\infty} \frac{(-1)^n}{n!} \frac{d^n}{dV^n} [f(V) \langle Q^n | V \rangle]. \tag{I.24}$$

Joint random variables

Let $U = \{U_1, U_2, \ldots, U_D\}$ be a set of D random variables. The corresponding characteristic function is defined by

$$\Psi(s) \equiv \langle e^{iU_j s_j} \rangle, \tag{I.25}$$

where $s = \{s_1, s_2, \ldots, s_D\}$ are the independent variables.

Let A be a constant D-vector, and let $\mathbf{B}$ be a constant $D \times D$ matrix, which are used to define the linear transformation

$$\widetilde{U} = A + \mathbf{B}U. \tag{I.26}$$

Then the characteristic function $\widetilde{\Psi}(s)$ of $\widetilde{U}$ is

$$\widetilde{\Psi}(s) = \langle e^{i\widetilde{U}_j s_j} \rangle = e^{iA_j s_j} \langle e^{iB_{jk} U_k s_j} \rangle = e^{iA^{\mathrm{T}}s} \Psi(\mathbf{B}^{\mathrm{T}}s). \tag{I.27}$$

The characteristic function $\Psi_{(j)}(s)$ of the single random variable U_j can be obtained from $\Psi(s)$ by setting $s_k = 0, k \neq j$ (as is evident from Eq. (I.25)).

If the D random variables are mutually independent then

$$\Psi(s) \equiv \left\langle \prod_{j=1}^{D} e^{iU_{(j)} s_{(j)}} \right\rangle = \prod_{j=1}^{D} \langle e^{iU_{(j)} s_{(j)}} \rangle = \prod_{j=1}^{D} \Psi_{(j)}(s_{(j)}), \tag{I.28}$$

where suffixes in brackets are excluded from the summation convention. It follows from Eqs. (I.12) and (I.28) that the characteristic function of D jointly normal standardized random variables U is

$$\Psi(s) = e^{-s_j s_j/2} = e^{-s^{\mathrm{T}}s/2}. \tag{I.29}$$

If these random variables are subjected to the linear transformation Eq. (I.26), then the resulting random vector $\widetilde{U}$ has mean

$$\langle \widetilde{U} \rangle = A, \tag{I.30}$$

covariance matrix

$$\mathbf{C} = \mathbf{B}\mathbf{B}^{\mathrm{T}}, \tag{I.31}$$

and characteristic function (from Eqs. (I.27) and (I.29))

$$\begin{aligned}
\widetilde{\Psi}(s) &= \exp(iA_j s_j - \tfrac{1}{2}C_{jk}s_j s_k) \\
&= \exp(i\boldsymbol{A}^{\mathrm{T}}s - \tfrac{1}{2}s^{\mathrm{T}}\mathbf{C}\,s).
\end{aligned} \tag{I.32}$$

In fact, this equation is usually taken as the definition of jointly normal random variables.

Appendix J
Diffusion processes

This appendix provides a brief introduction to diffusion processes and to some of the mathematical techniques used in their description and analysis. More comprehensive and rigorous accounts are provided by Gardiner (1985), Gillespie (1992), Arnold (1974), and Karlin and Taylor (1981).

A diffusion process is a particular kind of stochastic process. It is a continuous-time Markov process with continuous sample paths (and other properties described below).

Markov processes

Let $U(t)$ for $t \geq t_0$ be a stochastic process with one-time PDF $f(V;t)$. We introduce N times $t_1 < t_2 < \ldots < t_N$, (with $t_1 > t_0$), and consider the PDF of $U(t_N)$ conditioned on $U(t)$ at the earlier times $\{U(t_{N-1}), U(t_{N-2}), \ldots, U(t_1)\}$, which is denoted by

$$f_{N-1}(V_N;t_N|V_{N-1},t_{N-1}, V_{N-2},t_{N-2},\ldots, V_1,t_1).$$

The PDF of $U(t)$ conditioned on a single past time is denoted by, for example,

$$f_1(V_N;t_N|V_{N-1},t_{N-1}).$$

By definition, if $U(t)$ is a Markov process then these conditional PDFs are equal:

$$f_{N-1}(V_N;t_N|V_{N-1},t_{N-1},V_{N-2},t_{N-2},\ldots, V_1,t_1) = f_1(V_N;t_N|V_{N-1},t_{N-1}). \quad \text{(J.1)}$$

This means that, given $U(t_{N-1}) = V_{N-1}$, knowledge of the previous values $U(t_{N-2}), U(t_{N-3}),\ldots, U(t_1)$ provides no further information about the future value $U(t_N)$.

The Chapman–Kolmogorov equation

For any process, from the definition of conditional PDFs, we have

$$f_1(V_3;t_3|V_1,t_1) = \int_{-\infty}^{\infty} f_2(V_3;t_3|V_2,t_2,V_1,t_1)f_1(V_2;t_2|V_1,t_1)\,dV_2, \qquad \text{(J.2)}$$

(see Exercise J.1). For a Markov process, Eq. (J.1) can be used to replace f_2 by $f_1(V_3;t_3|V_2,t_2)$ which leads to the *Chapman–Kolmogorov equation*

$$f_1(V_3;t_3|V_1,t_1) = \int_{-\infty}^{\infty} f_1(V_3;t_3|V_2,t_2)f_1(V_2;t_2|V_1,t_1)\,dV_2. \qquad \text{(J.3)}$$

Increments

A useful concept is the *increment* in a process: the increment in a positive time interval h is defined by

$$\Delta_h U(t) \equiv U(t+h) - U(t). \qquad \text{(J.4)}$$

It is important to note that h is positive and that the increment is defined forward in time. A process can be considered as a sum of its increments, e.g.,

$$U(t_N) = U(t_0) + \Delta_{t_1-t_0} U(t_0) + \Delta_{t_2-t_1} U(t_1) + \ldots + \Delta_{t_N-t_{N-1}} U(t_{N-1}). \qquad \text{(J.5)}$$

The PDF of the increment $\Delta_h U(t)$, conditional on $U(t) = V$, is denoted by $g(\hat{V};h,V,t)$. If h is taken to be $t_3 - t_2$, then $U(t_2)$ can be re-expressed as

$$U(t_2) = U(t_3) - \Delta_h U(t_2), \qquad \text{(J.6)}$$

and the first conditional PDF on the right-hand side of Eq. (J.3) is

$$f_1(V_3;t_2+h|V_3 - \hat{V}, t_2) = g(\hat{V};h,V_3 - \hat{V}, t_2). \qquad \text{(J.7)}$$

Thus the Chapman–Kolmogorov equation can be rewritten as

$$f_1(V;t_2+h|V_1,t_1) = \int_{-\infty}^{\infty} g(\hat{V};h,V-\hat{V},t_2)f_1(V-\hat{V};t_2|V_1,t_1)\,d\hat{V}. \qquad \text{(J.8)}$$

Diffusion processes

There are qualitatively different kinds of continuous-time Markov processes, which are distinguished from each other by the behaviors of their increments $\Delta_h U(t)$ in the limit as h tends to zero. One defining property of a diffusion process is that its sample paths are continuous. More precisely, for every $\epsilon > 0$,

$$\lim_{h\downarrow 0} \frac{1}{h} P\{|\Delta_h U(t)| > \epsilon | U(t) = V\} = 0. \qquad \text{(J.9)}$$

If they exist, the *infinitesimal parameters* of a process are defined by

$$B_n(V,t) \equiv \lim_{h \downarrow 0} \frac{1}{h} \langle [\Delta_h U(t)]^n \,|\, U(t) = V \rangle,$$

$$= \lim_{h \downarrow 0} \frac{1}{h} \int_{-\infty}^{\infty} \hat{V}^n g(\hat{V}; h, V, t) \, d\hat{V}, \tag{J.10}$$

for $n = 1, 2, \ldots$. In addition to Eq. (J.9), the defining properties of a diffusion process are that the *drift coefficient*,

$$a(V,t) \equiv B_1(V,t), \tag{J.11}$$

and the *diffusion coefficient*,

$$b(V,t)^2 \equiv B_2(V,t), \tag{J.12}$$

exist, and that the remaining infinitesimal parameters are zero:

$$B_n(V,t) = 0, \quad \text{for } n \geq 3. \tag{J.13}$$

A differentiable deterministic process governed by the ordinary differential equation

$$\frac{dU(t)}{dt} = a(U(t), t) \tag{J.14}$$

is a degenerate diffusion process, with drift $a(V,t)$ and diffusion coefficient $b(V,t)^2 = 0$. A non-degenerate diffusion process (i.e., $b(V,t) > 0$) is clearly nowhere differentiable, for the fact that $\langle [\Delta_h U(t)]^2 / h \rangle$ tends to a positive limit implies that $\langle [\Delta_h U(t)/h]^2 \rangle$ tends to infinity.

The Kramers–Moyal equation

In the Chapman–Kolmogorov equation (Eq. (J.8)), both g and f_1 on the right-hand side involve the argument $V - \hat{V}$. Expanding these quantities in a Taylor series about V yields

$$f_1(V; t_2 + h | V_1, t_1) = f_1(V; t_2 | V_1, t_1)$$

$$+ \int_{-\infty}^{\infty} \sum_{n=1}^{\infty} \frac{(-\hat{V})^n}{n!} \frac{\partial^n}{\partial V^n} \left[g(\hat{V}; h, V, t_2) f_1(V; t_2 \,|\, V_1, t_1) \right] d\hat{V}. \tag{J.15}$$

By dividing by h, taking the limit $h \to 0$, and using Eq. (J.10), we obtain the *Kramers–Moyal equation*

$$\frac{\partial}{\partial t} f_1(V; t | V_1, t_1) = \sum_{n=1}^{\infty} \frac{(-1)^n}{n!} \frac{\partial^n}{\partial V^n} [B_n(V,t) f_1(V; t | V_1, t_1)]. \tag{J.16}$$

This equation applies to processes for which the parameters $B_n(V, t)$ exist, and for $t \geq t_1$. The appropriate initial condition is

$$f_1(V; t_1 | V_1, t_1) = \delta(V - V_1). \tag{J.17}$$

The Fokker–Planck equation

For a diffusion process, all of the parameters B_n are zero, except for the drift, $B_1 = a$, and the diffusion, $B_2 = b^2$. In this case, Eq. (J.16) reduces to the *Fokker–Planck* or *forward Kolmogorov* equation:

$$\frac{\partial}{\partial t} f_1(V; t | V_1, t_1) = -\frac{\partial}{\partial V} [a(V, t) f_1(V; t | V_1, t_1)]$$

$$+ \frac{1}{2} \frac{\partial^2}{\partial V^2} [b(V, t)^2 f_1(V; t | V_1, t_1)]. \tag{J.18}$$

This equation determines the evolution of the conditional PDF.

The corresponding equation for the marginal PDF $f(V; t)$ is obtained by multiplying by $f(V_1; t_1)$ and integrating over V_1. Since, in Eq. (J.18), only f_1 has any dependence on V_1, the result is simply

$$\frac{\partial}{\partial t} f(V; t) = -\frac{\partial}{\partial V} [a(V, t) f(V; t)] + \frac{1}{2} \frac{\partial^2}{\partial V^2} [b(V, t)^2 f(V; t)]. \tag{J.19}$$

For the deterministic process governed by the ordinary differential equation Eq. (J.14), the diffusion coefficient is zero, and hence the last terms in Eqs. (J.18) and (J.19) vanish. The resulting equations are called the *Liouville equations*.

The stationary distribution

If the coefficients a and b are independent of time, it is possible for the diffusion process to be statistically stationary. In this case, the Fokker–Planck equation (Eq. (J.19)) reduces to

$$0 = -\frac{d}{dV} [a(V) f(V)] + \frac{1}{2} \frac{d^2}{dV^2} [b(V)^2 f(V)], \tag{J.20}$$

which has the solution

$$f(V) = \frac{C}{b(V)^2} \exp\left(\int_{V_o}^{V} \frac{2a(V')}{b(V')^2} \, dV' \right), \tag{J.21}$$

where the lower limit V_o can be chosen for convenience, and the constant C is determined by the normalization condition. (If the integral of the right-hand side of Eq. (J.21) over all V does not converge, then $U(t)$ does not have a stationary distribution.)

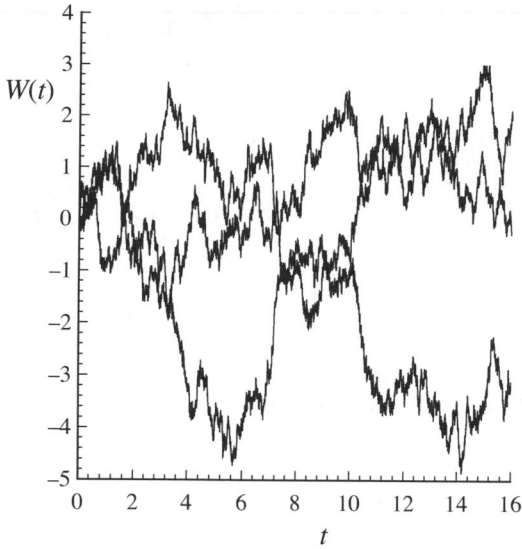

Fig. J.1. Three sample paths of the Wiener process.

The Wiener process

The most fundamental diffusion process, from which all others can be derived, is the Wiener process, denoted by $W(t)$. This is defined (for $t \geq 0$) by the initial condition $W(0) = 0$, and by the specification of the drift and diffusion coefficients,

$$a(V,t) = 0, \quad b(V,t)^2 = 1. \tag{J.22}$$

Some sample paths of $W(t)$ are shown in Fig. J.1.

As may readily be verified, the solution to the Fokker–Planck equation (Eq. (J.18)) with $a = 0$ and $b^2 = 1$ from the initial condition Eq. (J.17) is

$$f_1(V;t|V_1,t_1) = \frac{1}{\sqrt{2\pi(t-t_1)}} \exp\left(-\frac{\frac{1}{2}(V-V_1)^2}{t-t_1}\right), \tag{J.23}$$

i.e., a normal distribution with mean V_1 and variance $t - t_1$. Thus, for all $h > 0$, the increment $\Delta_h W(t)$ is normally distributed with mean zero and variance h:

$$\Delta_h W(t) \overset{\mathrm{D}}{=} \mathcal{N}(0,h). \tag{J.24}$$

(The symbol $\overset{\mathrm{D}}{=}$ is read 'is equal in distribution to,' and $\mathcal{N}(\mu, \sigma^2)$ denotes the normal with mean μ and variance σ^2, Eq. (3.41).)

Some important properties (not all independent) of Wiener-process increments are

(i) $\langle W(t_2) - W(t_1) \rangle = 0$,

(ii) $\langle [W(t_2) - W(t_1)]^2 \rangle = \text{var}\,[W(t_2) - W(t_1)] = t_2 - t_1$,

(iii) $W(t_2) - W(t_1) \overset{D}{=} \mathcal{N}(0, t_2 - t_1)$,

(iv) $h^{-1/2}\Delta_h W(t) \overset{D}{=} \mathcal{N}(0,1)$,

(v) $W(t_2) - W(t_1)$ is independent of $W(t)$ for $t \le t_1$,

(vi) $\langle [W(t_3) - W(t_2)][W(t_2) - W(t_1)] \rangle = 0$ – increments in non-overlapping time intervals are independent,

(vii) $\langle [W(t_4) - W(t_2)][W(t_3) - W(t_1)] \rangle = t_3 - t_2$ – the covariance of increments equals the duration of the overlap of the time intervals,

(viii) $\sum_{n=1}^{N} \langle [W(t_n) - W(t_{n-1})]^2 \rangle = t_N - t_0$, and

(ix) $W(t)$ is a Gaussian process: the joint PDF of $W(t_1), W(t_2), \ldots, W(t_N)$ is a joint normal.

Several other interesting properties of the Wiener process are deduced in Exercise J.2.

Stochastic differential equations

Because diffusion processes are not differentiable, the standard tools of differential calculus cannot be applied. Instead of differential calculus, the appropriate method is the Ito calculus; and, instead of being described by ordinary differential equations, diffusion processes are described by stochastic differential equations.

The *infinitesimal increment* of the process $U(t)$ is defined by

$$dU(t) \equiv U(t + dt) - U(t), \qquad (\text{J}.25)$$

where dt is a positive infinitesimal time interval. For the Wiener process in particular, we have

$$dW(t) = W(t + dt) - W(t) \overset{D}{=} \mathcal{N}(0, dt). \qquad (\text{J}.26)$$

Now consider the process $U(t)$ defined by the initial condition $U(t_0) = U_0$, and by the increment

$$dU(t) = a[U(t), t]\,dt + b[U(t), t]\,dW(t), \qquad (\text{J}.27)$$

for given functions $a(V, t)$ and $b(V, t)$. It is readily verified that the process $U(t)$ defined by this stochastic differential equation is a diffusion process; and, as implied by the notation, the drift and diffusion coefficients are $a(V, t)$ and $b(V, t)^2$.

A random variable is fully characterized by its PDF; and two random variables with the same PDF are statistically identical. Similarly, a diffusion process is fully characterized by its drift and diffusion coefficients; and two diffusion processes with the same coefficients are statistically identical. Thus the stochastic differential equation Eq. (J.27) provides a general expression for a diffusion process.

The stochastic differential equation Eq. (J.27) shows that the infinitesimal increment of a diffusion process is Gaussian, i.e.,

$$dU(t) = \mathcal{N}(a[U(t), t] \, dt, \, b[U(t), t]^2 \, dt). \tag{J.28}$$

This Gaussianity is not a defining property of diffusion processes, but rather a deduction from their definition.

White noise

Prior to the development of the theory of stochastic differential equations, diffusion processes were commonly expressed as ordinary differential equations involving white noise. On dividing Eq. (J.27) by dt we obtain

$$\frac{dU(t)}{dt} = a[U(t), t] + b[U(t), t] \dot{W}(t), \tag{J.29}$$

where the white noise $\dot{W}(t)$ is $dW(t)/dt$. Since neither dU/dt nor dW/dt exists, this equation cannot be interpreted in the usual way. Consequently it is preferable not to use the concept of white noise, but instead to express diffusion processes as stochastic differential equations.

The evolution of moments

Equations for the evolution of the unconditional moments $\langle U(t)^n \rangle$ can be derived from the Fokker–Planck equation (Eq. (J.19)), or from the stochastic differential equation (Eq. (J.27)). The latter approach is instructive.

Taking the mean of Eq. (J.25) and substituting Eq. (J.27), we obtain

$$\langle U(t + dt) \rangle - \langle U(t) \rangle = \langle dU(t) \rangle$$
$$= \langle a[U(t), t] \rangle dt + \langle b[U(t), t] \, dW(t) \rangle. \tag{J.30}$$

Now $dW(t)$ has zero mean, and it is independent of $U(t')$ for $t' \leq t$. Thus the last term vanishes, leading to

$$\frac{d}{dt} \langle U(t) \rangle = \langle a[U(t), t] \rangle. \tag{J.31}$$

Similarly, the mean of the square of $U(t + dt)$ is

$$\langle U(t + dt)^2 \rangle = \langle [U(t) + dU(t)]^2 \rangle$$
$$= \langle U(t)^2 \rangle + 2 \langle U(t) \, dU(t) \rangle + \langle dU(t)^2 \rangle. \tag{J.32}$$

The cross-term is $2\langle Ua \rangle \, dt$, while the final term is

$$
\begin{aligned}
\langle dU(t)^2 \rangle &= \langle a(U[t],t)^2 \rangle \, dt^2 + \langle b(U[t],t)^2 \, dW(t)^2 \rangle \\
&= \langle b(U[t],t)^2 \rangle \langle dW(t)^2 \rangle + o(dt) \\
&= \langle b(U[t],t)^2 \rangle \, dt + o(dt),
\end{aligned}
\tag{J.33}
$$

where $o(h)$ denotes a quantity such that

$$
\lim_{h \downarrow 0} \frac{o(h)}{h} = 0,
\tag{J.34}
$$

(e.g., $h^{1+\epsilon} = o(h)$, for all $\epsilon > 0$). Thus the mean square of $U(t)$ evolves by

$$
\frac{d}{dt} \langle U(t)^2 \rangle = 2 \langle U(t)a[U(t),t] \rangle + \langle b[U(t),t]^2 \rangle.
\tag{J.35}
$$

Notice that, for a differentiable process $\langle dU(t)^2 \rangle / dt$ is zero, but, for a diffusion process, it is $\langle b^2 \rangle$, and this leads to the final term in Eq. (J.35).

The Ornstein–Uhlenbeck (OU) process

The OU process is the simplest statistically stationary diffusion process. It is defined by the linear drift coefficient

$$
a(V,t) = -\frac{V}{T},
\tag{J.36}
$$

the constant diffusion coefficient

$$
b(V,t)^2 = \frac{2\sigma^2}{T},
\tag{J.37}
$$

and the initial condition

$$
U(0) \overset{D}{=} \mathcal{N}(0,\sigma^2),
\tag{J.38}
$$

where T is a positive timescale, and σ is a constant. The corresponding stochastic differential equation is the *Langevin equation*

$$
dU(t) = -U(t)\frac{dt}{T} + \left(\frac{2\sigma^2}{T} \right)^{1/2} dW(t).
\tag{J.39}
$$

For the OU process, the Fokker–Planck equation (Eq. (J.18)) for the PDF of $U(t)$ conditional on $U(t_1) = V_1$ (for $t > t_1$), $f_1(V;t|V_1,t_1)$, is

$$
\frac{\partial f_1}{\partial t} = \frac{1}{T}\frac{\partial}{\partial V}(Vf_1) + \frac{\sigma^2}{T}\frac{\partial^2 f_1}{\partial V^2}.
\tag{J.40}
$$

With the deterministic initial condition Eq. (J.17), the solution to this equation is the normal distribution

$$
f_1(V;t|V_1,t_1) = \mathcal{N}\left[V_1 e^{-(t-t_1)/T}, \ \sigma^2 \left(1 - e^{-2(t-t_1)/T} \right) \right],
\tag{J.41}
$$

(see Exercise J.4). This solution, which fully characterizes the process, shows that the conditional mean $\langle U(t)|V_1\rangle$ decays from V_1 to 0 on the timescale T; while the conditional variance increases from 0 to σ^2 on the timescale $\frac{1}{2}T$. At large times, the conditional PDF tends to the stationary distribution $\mathcal{N}(0, \sigma^2)$.

An important deduction from Eq. (J.41) is that the OU process is a *Gaussian process*. A consequence of the Markov property is that the joint PDF of $U(t)$ at the $N+1$ times $\{t_0 = 0, t_1, t_2, \ldots, t_N\}$ can be written as the product of the marginal PDF at t_0, and the N conditional PDFs $f_1(V_n; t_n \mid V_{n-1}, t_{n-1})$ for $n = 1, 2, \ldots, N$. Since each of these PDFs is normal (see Eqs. (J.38) and (J.41)), the $(N+1)$-time joint PDF is joint normal, satisfying the definition of a Gaussian process.

Since the mean $\langle U(t)\rangle$ is zero, the autocovariance (for $s \geq 0$) is

$$R(s) = \langle U(t_1 + s)U(t_1)\rangle. \tag{J.42}$$

This is readily evaluated from the conditional mean $\langle U(t_1 + s)|V_1\rangle = V_1 e^{-s/T}$ by

$$\begin{aligned} R(s) &= \langle\langle U(t_1 + s)|U(t_1)\rangle\, U(t_1)\rangle \\ &= \langle U(t_1)^2\rangle e^{-s/T} = \sigma^2 e^{-s/T}. \end{aligned} \tag{J.43}$$

For any statistically stationary process, the autocovariance $R(s)$ and the autocorrelation function $\rho(s)$ are even functions. Hence, for the OU process, we have

$$R(s) = \sigma^2 e^{-|s|/T}, \tag{J.44}$$

$$\rho(s) = e^{-|s|/T}. \tag{J.45}$$

Notice that the integral timescale defined by $\int_0^\infty \rho(s)\,ds$ is T.

The second-order structure function $D_2(s)$ and the frequency spectrum $E(\omega)$ contain the same information as does the autocovariance $R(s)$. For the OU process, the structure function is

$$\begin{aligned} D_2(s) &\equiv \langle[U(t+s) - U(t)]^2\rangle = 2[\sigma^2 - R(s)] \\ &= 2\sigma^2(1 - e^{-|s|/T}) = \frac{2\sigma^2}{T}|s| + \mathcal{O}(s^2), \end{aligned} \tag{J.46}$$

while the spectrum (twice the Fourier transform of $R(s)$) is

$$E(\omega) = \frac{(2/\pi)\sigma^2 T}{1 + T^2\omega^2}. \tag{J.47}$$

The lack of differentiability of $U(t)$ manifests itself in the discontinuous slope of $\rho(s)$ at the origin, and in the variations $D_2(s) \sim s$ (for small s) and $E(\omega) \sim \omega^{-2}$ (for large ω).

In summary; the Ornstein–Uhlenbeck process (which is generated by the Langevin equation, Eq. (J.39)) is a statistically stationary Gaussian process. As such, it is fully characterized by its mean (which is zero), the variance σ^2, and the autocorrelation function $\rho(s) = e^{-|s|/T}$.

The Ito transformation

Consider the process $q(t)$ defined by

$$q(t) = Q[U(t)], \tag{J.48}$$

where $U(t)$ is a diffusion process (with drift $a(V,t)$ and diffusion $b(V,t)^2$), and $Q(V)$ is a differentiable function, with derivatives $Q'(V)$, $Q''(V)$, etc. The infinitesimal increment in q is

$$dq(t) = Q[U(t) + dU(t)] - Q[U(t)]. \tag{J.49}$$

By expanding $Q(U + dU)$ in a Taylor series about $U(t)$, and substituting Eq. (J.27) for dU, we obtain

$$\begin{aligned} dq(t) &= Q'[U(t)]\,dU + \tfrac{1}{2}Q''[U(t)]\,dU^2 + o(dt) \\ &= (Q'a + \tfrac{1}{2}Q''b^2)\,dt + Q'b\,dW + o(dt). \end{aligned} \tag{J.50}$$

Thus, $q(t)$ is itself a diffusion process, with stochastic differential equation

$$dq(t) = a_q[q(t), t]\,dt + b_q[q(t), t]\,dW(t), \tag{J.51}$$

where the coefficients are

$$a_q = Q'a + \tfrac{1}{2}Q''b^2, \tag{J.52}$$

$$b_q = Q'b. \tag{J.53}$$

These relations form the *Ito transformation*. The essential difference – compared with the transformation rule for ordinary differential equations – is the additional drift $\tfrac{1}{2}Q''b^2$.

Vector-valued diffusion processes

The development presented above for the scalar-valued diffusion process extends straightforwardly to the vector-valued process $U(t) = \{U_1(t), U_2(t), \ldots, U_D(t)\}$. Only the principal results are given here.

The drift coefficient is the vector

$$a(V, t) = \lim_{h \downarrow 0} \frac{1}{h} \langle [\Delta_h U(t)] | U(t) = V \rangle, \tag{J.54}$$

while the diffusion coefficient is the $D \times D$ matrix **B**, with elements

$$B_{ij}(V,t) = \lim_{h \downarrow 0} \frac{1}{h} \langle \Delta_h U_i(t) \, \Delta_h U_j | U(t) = V \rangle. \qquad (\mathrm{J.55})$$

It follows from its definition that **B** is symmetric positive semi-definite.

The Fokker–Planck equation for the PDF of $U(t), f(V;t)$, and also for the conditional PDF, is

$$\frac{\partial f}{\partial t} = -\frac{\partial}{\partial V_i}(a_i f) + \frac{1}{2} \frac{\partial^2}{\partial V_i \partial V_j}(B_{ij} f), \qquad (\mathrm{J.56})$$

cf. Eq. (J.19).

The vector-valued Wiener process $W(t) = \{W_1(t), W_2(t), \ldots, W_D(t)\}$ is simply composed of the independent scalar processes $W_i(t)$. The increment $dW(t)$ is a joint normal, with zero mean, and covariance

$$\langle dW_i \, dW_j \rangle = dt \, \delta_{ij}. \qquad (\mathrm{J.57})$$

The process is statistically isotropic: if **A** is a unitary matrix, then $\hat{W}(t) \equiv AW(t)$ is also a vector-valued Wiener process.

The stochastic differential equation for $U(t)$ is written

$$dU_i(t) = a_i[U(t), t] \, dt + b_{ij}[U(t), t] \, dW_j(t), \qquad (\mathrm{J.58})$$

where (for consistency with Eq. (J.55)) the coefficients b_{ij} satisfy

$$b_{ik} b_{jk} = B_{ij}. \qquad (\mathrm{J.59})$$

Notice that the non-symmetric matrix **b** is not uniquely determined by the symmetric matrix **B**. Two possible choices of **b** are the symmetric square root of **B** and the lower triangular matrix given by the Cholesky decomposition of **B**. All choices of **b** (consistent with Eq. (J.59)) result in statistically identical diffusion processes.

EXERCISES

J.1   Consider three random variables U_1, U_2, and U_3. Conditional PDFs are defined from the joint PDFs by, for example,

$$f_{3|1}(V_3|V_1) = f_{13}(V_1, V_3)/f_1(V_1), \qquad (\mathrm{J.60})$$

$$f_{3|12}(V_3|V_1, V_2) = f_{123}(V_1, V_2, V_3)/f_{12}(V_1, V_2). \qquad (\mathrm{J.61})$$

Obtain the result

$$\int_{-\infty}^{\infty} \frac{f_{123}(V_1, V_2, V_3)}{f_{12}(V_1, V_2)} \frac{f_{12}(V_1, V_2)}{f_1(V_1)} \, dV_2 = f_{3|1}(V_3|V_1), \qquad (\mathrm{J.62})$$

and hence verify Eq. (J.2).

J.2   Let the time interval $(0, T)$ be divided into M equal sub-intervals of duration $h = T/M$. For the discrete times nh $(n = 0, 1, \ldots, M)$, the process $\tilde{W}$ is defined by $\tilde{W}(0) = 0$ and

$$\tilde{W}(nh) = \sum_{i=1}^{n} h^{1/2}\xi_i, \quad \text{for } n \geq 1, \tag{J.63}$$

where $\xi_1, \xi_2, \ldots, \xi_M$ are independent standardized normal random variables.

(a) Show that $\tilde{W}$ is statistically identical to a Wiener process sampled at the same times.

(b) Let S_M denote the sum of the squares of the increments of the Wiener process

$$S_M \equiv \sum_{n=0}^{M-1} [\Delta_h W(nh)]^2. \tag{J.64}$$

Obtain the results $\langle S_M \rangle = T$ and $\text{var}(S_M) = 2hT = 2T^2/M$. Hence argue that the random variable S_M has the non-random limit $S_\infty = T$.

(c) Consider a plot of $\tilde{W}$ against t, in which successive values are connected by straight line segments. Show that the expected length of each line segment exceeds $\sqrt{2h/\pi}$, and that the expectation of the sum of the lengths exceeds $\sqrt{2MT/\pi}$. Hence argue that, in every positive interval, the sample path of a Wiener process has infinite arc length.

(d) What is the joint PDF $f_n(V, \hat{V}; h)$ of $\tilde{W}(nh)$ and $\Delta_h \tilde{W}(nh)$? For $n \geq 1$, consider the event C_n defined by $\tilde{W}(nh)$ and $\tilde{W}[(n+1)h]$ having opposite signs. What region of the $V - \hat{V}$ sample space corresponds to the event C_n? Show that the probability of C_n is

$$P(C_n) = \frac{1}{\pi} \tan^{-1}\left(\frac{1}{\sqrt{n}}\right) \geq \frac{1}{4\sqrt{n}}. \tag{J.65}$$

(Hint: consider the simple transformation that transforms f_n to a standardized joint normal.) The expected number of times that $\tilde{W}$ changes sign is $N_M = \sum_{n=1}^{M-1} P(C_n)$. How does N_M behave as M increases? Hence argue that, in any positive interval (t_1, t_2), the Wiener process takes every value between $W(t_1)$ and $W(t_2)$ an infinite number of times.

(e) For any positive constant c, show that, in the transformed time $\hat{t} = ct$, the process

$$\hat{W}(\hat{t}) \equiv \frac{1}{\sqrt{c}} W(ct) \tag{J.66}$$

is a Wiener process.

J.3   For the diffusion process defined by Eq. (J.27), which of the following quantities are correlated: $W(t)$, $dW(t)$, $U(t)$, $dU(t)$, and $U(t+dt)$?

J.4   Let $\Psi(s,t)$ (for $t \geq t_1$) be the characteristic function of the OU process from the deterministic initial condition $U(t_1) = V_1$; so that its Fourier transform is the conditional PDF $f_1(V;t|V_1,t_1)$. From the Fokker–Planck equation for f_1 (Eq. (J.40)), show that $\Psi(s,t)$ evolves by

$$\left(\frac{\partial}{\partial t} + \frac{s}{T} \frac{\partial}{\partial s} \right) \Psi(s,t) = -\frac{\sigma^2 s^2}{T} \Psi(s,t). \tag{J.67}$$

Use the method of characteristics to obtain the solution

$$\Psi(s,t) = \Psi(se^{-(t-t_1)/T}, t_1) \exp(-\tfrac{1}{2} s^2 \Sigma(t)^2), \tag{J.68}$$

where

$$\Sigma(t)^2 \equiv \sigma^2 (1 - e^{-2(t-t_1)/T}). \tag{J.69}$$

Show that, with the initial condition $\Psi(s,t_1) = e^{isV_1}$, this solution corresponds to the normal distribution with mean

$$\mu(t) = V_1 e^{-(t-t_1)/T} \tag{J.70}$$

and variance $\Sigma(t)^2$.

J.5   Let each component of the velocity $U(t) = \{U_1(t), U_2(t), U_3(t)\}$ evolve by an independent Langevin equation, i.e.,

$$dU = -U \frac{dt}{T} + \left(\frac{2\sigma^2}{T} \right)^{1/2} dW. \tag{J.71}$$

What is the joint PDF of U? The speed $q(t)$ is defined by

$$q(t) = [U_i(t)U_i(t)]^{1/2}. \tag{J.72}$$

Show that $q(t)$ evolves by the stochastic differential equation

$$dq = \left(\frac{2\sigma^2}{q} - q \right) \frac{dt}{T} + \left(\frac{2\sigma^2}{T} \right)^{1/2} dW. \tag{J.73}$$

Show that the stationary distribution of q is

$$f(v) = \sqrt{\frac{2}{\pi}} \frac{v^2}{\sigma^3} e^{-v^2/(2\sigma^2)}. \tag{J.74}$$

J.6   Let $\Psi(s, t)$ be the characteristic function of the general vector-valued diffusion process given by Eq. (J.58). By expanding $\langle \exp(is.[U(t) + dU(t)]) \rangle$ (see Eq. (I.23)), obtain the result

$$\Psi(s, t + dt) = \Psi(s, t) + is_j \langle a_j e^{is \cdot U(t)} \rangle \, dt - \tfrac{1}{2} s_j s_k \langle b_{j\ell} b_{k\ell} e^{is \cdot U(t)} \rangle \, dt + o(dt), \tag{J.75}$$

where the coefficients are, for example, $a_j = a_j[U(t), t]$. Show that the Fourier transform of this equation (when it is divided by dt) is the Fokker–Planck equation, Eq. (J.56).

Bibliography

Adrian, R. J. (1990). Stochastic estimation of sub-grid scale motions. *Appl. Mech. Rev. 43*, S214–S218.

Akselvoll, K. and P. Moin (1996). Large-eddy simulation of turbulent confined coannular jets. *J. Fluid Mech. 315*, 387–411.

Anand, M. S., A. T. Hsu, and S. B. Pope (1997). Calculations of swirl combustors using joint velocity-scalar probability density function method. *AIAA J. 35*, 1143–1150.

Anand, M. S. and S. B. Pope (1985). Diffusion behind a line source in grid turbulence. In L. J. S. Bradbury *et al.* (Eds.), *Turbulent Shear Flows 4*, pp. 46–61. Berlin: Springer-Verlag.

Anand, M. S., S. B. Pope, and H. C. Mongia (1993). PDF calculations of swirling flows. Paper 93-0106, AIAA.

Anselmet, F., Y. Gagne, E. J. Hopfinger, and R. A. Antonia (1984). High-order velocity structure functions in turbulent shear flows. *J. Fluid Mech. 140*, 63–89.

Antonia, R. A., M. Teitel, J. Kim, and L. W. B. Browne (1992). Low-Reynolds-number effects in a fully developed turbulent channel flow. *J. Fluid Mech. 236*, 579–605.

Arnal, D. and R. Michel (1990). *Laminar–Turbulent Transition*. Heidelberg: Springer-Verlag.

Arnold, L. (1974). *Stochastic Differential Equations: Theory and Applications*. New York: John Wiley.

Aronson, D., A. V. Johansson, and L. Löfdahl (1997). Shear-free turbulence near a wall. *J. Fluid Mech. 338*, 363–385.

Aubry, N., P. Holmes, J. L. Lumley, and E. Stone (1988). The dynamics of coherent structures in the wall region of a turbulent boundary layer. *J. Fluid Mech. 192*, 115–173.

Bakewell, H. P. and J. L. Lumley (1967). Viscous sublayer and adjacent wall region in turbulent pipe flow. *Phys. Fluids 10*, 1880–1889.

Balaras, E., C. Benocci, and U. Piomelli (1995). Finite-difference computations of high Reynolds number flows using the dynamic subgrid-scale model. *Theor. Comput. Fluid Dyn. 7*, 207–216.

Baldwin, B. S. and T. J. Barth (1990). A one-equation turbulence transport model for high Reynolds number wall-bounded flows. TM 102847, NASA.

Baldwin, B. S. and H. Lomax (1978). Thin-layer approximation and algebraic model for separated turbulent flows. Paper 78-257, AIAA.

Bardina, J., J. Ferziger, and W. C. Reynolds (1983). Improved turbulence models based on large-eddy simulation of homogeneous, incompressible, turbulent flows. Mechanical Engineering Department TF-19, Stanford University.

Bardina, J., J. H. Ferziger, and W. C. Reynolds (1980). Improved subgrid models for large eddy simulation. Paper 80-1357, AIAA.

Bardina, J., J. H. Ferziger, and R. S. Rogallo (1985). Effect of rotation on isotropic turbulence: computation and modelling. *J. Fluid Mech. 154*, 321–336.

Barenblatt, G. I. (1993). Scaling laws for fully developed turbulent shear flows. Part 1. Basic hypotheses and analysis. *J. Fluid Mech. 248*, 513–520.

Barenblatt, G. I. and A. J. Chorin (1998). Scaling of the intermediate region in wall-bounded turbulence: the power law. *Phys. Fluids 10*, 1043–1044.

Barenblatt, G. I. and A. S. Monin (1979). Similarity laws for turbulent stratified flows. *Arch. Ration. Mech. Anal. 70*, 307–317.

Batchelor, G. K. (1953). *The Theory of Homogeneous Turbulence*. Cambridge: Cambridge University Press.

Batchelor, G. K. (1967). *An Introduction to Fluid Dynamics*. Cambridge: Cambridge University Press.

Batchelor, G. K. and I. Proudman (1954). The effect of rapid distortion of a fluid in turbulent motion. *Q. J. Appl. Math. 7*, 83–103.

Batchelor, G. K. and A. A. Townsend (1948). Decay of turbulence in the final period. *Proc. R. Soc. London Ser. A 194*, 527–543.

Batchelor, G. K. and A. A. Townsend (1949). The nature of turbulent motion at large wave-numbers. *Proc. R. Soc. London Ser. A 199*, 238–255.

Béguier, C., I. Dekeyser, and B. E. Launder (1978). Ratio of scalar and velocity dissipation time scales in shear flow turbulence. *Phys. Fluids 21*, 307–310.

Behnia, M., S. Parneix, and P. A. Durbin (1998). Prediction of heat transfer in an axisymmetric turbulent jet impinging on a flat plate. *Int. J. Heat Mass Transfer 41*, 1845–1855.

Belin, F., J. Maurer, P. Tabeling, and H. Willaime (1997). Velocity gradient distributions in fully developed turbulence: an experimental study. *Phys. Fluids 9*, 3843–3850.

Bell, J. H. and R. D. Mehta (1990). Development of a two-stream mixing layer from tripped and untripped boundary layers. *AIAA J. 28*, 2034–2042.

Berkooz, G., P. Holmes, and J. L. Lumley (1993). The proper orthogonal decomposition in the analysis of turbulent flows. *Annu. Rev. Fluid. Mech. 25*, 539–575.

Birdsall, C. K. and D. Fuss (1969). Clouds-in-clouds, clouds-in-cells physics for many-body plasma simulation. *J. Comput. Phys. 3*, 494–511.

Blackburn, H. M., N. N. Mansour, and B. J. Cantwell (1996). Topology of fine-scale motions in turbulent channel flow. *J. Fluid Mech. 310*, 269–292.

Blackwelder, R. F. and J. H. Haritonidis (1983). Scaling of the bursting frequency in turbulent boundary layers. *J. Fluid Mech. 132*, 87–103.

Blackwelder, R. F. and R. E. Kaplan (1976). On the wall structure of the turbulent boundary layer. *J. Fluid Mech. 76*, 89–112.

Blackwelder, R. F. and L. S. G. Kovasznay (1972). Time scales and correlations in a turbulent boundary layer. *Phys. Fluids 15*, 1545–1554.

Blasius, H. (1908). Grenzschichten in Flüssigkeiten mit kleiner Reibung. *Z. Math. Phys. 56*, 1–37.

Boris, J. P., F. F. Grinstein, E. S. Oran, and R. L. Kolbe (1992). New insights into large eddy simulation. *Fluid Dyn. Res. 10*, 199–228.

Borue, V. and S. A. Orszag (1996). Numerical study of three-dimensional Kolmogorov flow at high Reynolds numbers. *J. Fluid Mech. 306*, 293–323.

Bracewell, R. N. (1965). *The Fourier Transform and its Applications*. New York: McGraw Hill.

Bradbury, L. J. S. (1965). The structure of a self-preserving turbulent plane jet. *J. Fluid Mech. 23*, 31–64.

Bradshaw, P. (1967). The structure of equilibrium boundary layers. *J. Fluid Mech. 29*, 625–645.

Bradshaw, P. (1973). Effects of streamline curvature on turbulent flow. AGAR-Dograph 169, AGARD, Paris.

Bradshaw, P. (1987). Turbulent secondary flows. *Annu. Rev. Fluid. Mech. 19*, 53–74.

Bradshaw, P. and G. P. Huang (1995). The law of the wall in turbulent flow. *Proc. R. Soc. London Ser. A 451*, 165–188.

Bradshaw, P., B. E. Launder, and J. L. Lumley (1996). Collaborative testing of turbulence models. *J. Fluids Eng. Trans. ASME 118*, 243–247.

Brigham, E. O. (1974). *The Fast Fourier Transform*. Englewood Cliffs, NJ: Prentice-Hall.

Brown, G. L. and A. Roshko (1974). On density effects and large structure in turbulent mixing layers. *J. Fluid Mech. 64*, 775–816.

Brumley, B. (1984). Turbulence measurements near the free surface in stirred tank experiments. In W. Brutsaert and G. H. Jirka (Eds.), *Gas Transfer at Water Surfaces*, pp. 83–92. Dordrecht: Reidel.

Buell, J. C. and N. N. Mansour (1989). Asymmetric effects in three-dimensional spatially-developing mixing layers. In *Proc. Seventh Symp. Turbulent Shear Flows*, Stanford University, pp. 9.2.1–9.2.6.

Burgers, J. M. (1940). Application of a model system to illustrate some points of the statistical theory of free turbulence. *Proc. Koninklijke Nederlandse Academie van Wetenschappen XLIII*, 2–12.

Bushnell, D. M. and C. B. McGinley (1989). Turbulence control in wall flows. *Annu. Rev. Fluid. Mech. 21*, 1–20.

Butkov, E. (1968). *Mathematical Physics*. Reading MA: Addison-Wesley.

Cannon, S. and F. Champagne (1991). Large-scale structures in wakes behind axisymmetric bodies. In *Proc. Eighth Symp. Turbulent Shear Flows*, Munich, pp. 6.5.1–6.5.6.

Cannon, S., F. Champagne, and A. Glezer (1993). Observations of large-scale structures in wakes behind axisymmetric bodies. *Exp. Fluids 14*, 447–450.

Cantwell, B. J. (1981). Organized motion in turbulent flow. *Annu. Rev. Fluid. Mech. 13*, 457–515.

Carmody, T. (1964). Establishment of the wake behind a disk. *ASME J. Basic Eng. 86*, 869–882.

Cazalbou, J. B., P. R. Spalart, and P. Bradshaw (1994). On the behavior of two-equation models at the edge of a turbulent region. *Phys. Fluids 6*, 1797–1804.

Champagne, F. H., Y. H. Pao, and I. J. Wygnanski (1976). On the two-dimensional mixing region. *J. Fluid Mech. 74*, 209–250.

Chapman, D. R. (1979). Computational aerodynamics development and outlook. *AIAA J. 17*, 1293–1313.

Chapman, S. and T. G. Cowling (1970). *The Mathematical Theory of Non-Uniform Gases* (3rd ed.). Cambridge: Cambridge University Press.

Chasnov, J. R. (1991). Simulation of the Kolmogorov inertial subrange using an improved subgrid model. *Phys. Fluids A 3*, 188–200.

Chasnov, J. R. (1993). Computation of the Loitsianski integral in decaying isotropic turbulence. *Phys. Fluids 5*, 2579–2581.

Chasnov, J. R. (1995). The decay of axisymmetric homogeneous turbulence. *Phys. Fluids 7*, 600–605.

Chen, H., S. Chen, and R. H. Kraichnan (1989). Probability distribution of a stochastically advected scalar field. *Phys. Rev. Lett. 63*, 2657–2660.

Chen, J.-Y., R. W. Dibble, and R. W. Bilger (1990). PDF modeling of turbulent nonpremixed $CO/H_2/N_2$ jet flames with reduced mechanisms. In *Twenty-Third Symp. (Int.) on Combustion*, pp. 775–780. Pittsburgh: The Combustion Institute.

Chen, S., G. D. Doolen, R. H. Kraichnan, and Z.-S. She (1993). On statistical correlations between velocity increments and locally averaged dissipation in homogeneous turbulence. *Phys. Fluids A 5*, 458–463.

Chevray, R. (1968). The turbulent wake of a body of revolution. *ASME J. Basic Eng. 90*, 275–284.

Choi, K.-S. (Ed.) (1991). *Recent Developments in Turbulence Management*. Dordrecht: Kluwer.

Chollet, J. P. (1984). Two-point closures as a subgrid scale modelling for large eddy simulations. In F. Durst and B. E. Launder (Eds.), *Turbulent Shear Flows IV*, pp. 62–72. Heidelberg: Springer-Verlag.

Chollet, J.-P. and M. Lesieur (1981). Parameterization of small scales of three-dimensional isotropic turbulence utilizing spectral closures. *J. Atmos. Sci. 38*, 2747–2757.

Chou, P. Y. (1945). On velocity correlations and the solution of the equations of turbulent fluctuation. *Quart. Appl. Math. 3*, 38–54.

Chung, M. K. and S. K. Kim (1995). A nonlinear return-to-isotropy model with Reynolds number and anisotropy dependency. *Phys. Fluids 7*, 1425–1436.

Clark, R. A., J. H. Ferziger, and W. C. Reynolds (1979). Evaluation of subgrid-scale models using an accurately simulated turbulent flow. *J. Fluid Mech. 91*, 1–16.

Coleman, H. W. and F. Stern (1997). Uncertainties and CFD code validation. *J. Fluids Eng. Trans. ASME 119*, 795–807.

Coles, D. (1956). The law of the wake in the turbulent boundary layer. *J. Fluid Mech. 1*, 191–226.

Colucci, P. J., F. A. Jaberi, P. Givi, and S. B. Pope (1998). Filtered density function for large eddy simulation of turbulent reacting flows. *Phys. Fluids 10*, 499–515.

Comte-Bellot, G. and S. Corrsin (1966). The use of a contraction to improve the isotropy of grid-generated turbulence. *J. Fluid Mech. 25*, 657–682.

Comte-Bellot, G. and S. Corrsin (1971). Simple Eulerian time correlations of full- and narrow-band velocity signals in grid-generated isotropic turbulence. *J. Fluid Mech. 48*, 273–337.

Cook, A. W. and J. J. Riley (1994). A subgrid model for equilibrium chemistry in turbulent flows. *Phys. Fluids 6*, 2868–2870.

Corino, E. R. and R. S. Brodkey (1969). A visual investigation of the wall region in turbulent flow. *J. Fluid Mech. 37*, 1–30.

Corrsin, S. (1943). Investigation of flow in an axially symmetrical heated jet of air. Technical Report W-94, NACA.

Corrsin, S. (1958). On local isotropy in turbulent shear flow. R&M 58B11, NACA.

Corrsin, S. and A. L. Kistler (1954). The free-stream boundaries of turbulent flows. Technical Note 3133, NACA.

Craft, T. J. and B. E. Launder (1996). A Reynolds stress closure designed for complex geometries. *Int. J. Heat Fluid Flow 17*, 245–254.

Craft, T. J., B. E. Launder, and K. Suga (1996). Development and application of a cubic eddy-viscosity model of turbulence. *Int. J. Heat Fluid Flow 17*, 108–115.

Craya, A. (1958). Contribution à l'analyse de la turbulence associée à des vitesses moyennes. Publications scientifiques et techniques 345, Ministère de l'air, France.

Crow, S. C. (1968). Viscoelastic properties of fine-grained incompressible turbulence. *J. Fluid Mech. 33*, 1–20.

Curl, R. L. (1963). Dispersed phase mixing: I. Theory and effects of simple reactors. *AIChE J. 9*, 175–181.

Dahm, W. J. A. and P. E. Dimotakis (1990). Mixing at large Schmidt number in the self-similar far field of turbulent jets. *J. Fluid Mech. 217*, 299–330.

Daly, B. J. and F. H. Harlow (1970). Transport equations in turbulence. *Phys. Fluids 13*, 2634–2649.

Davidov, B. I. (1961). On the statistical dynamics of an incompressible turbulent fluid. *Dokl. Akad. Nauk S.S.S.R. 136*, 47–50.

de Souza, F. A., V. D. Nguyen, and S. Tavoularis (1995). The structure of highly sheared turbulence. *J. Fluid Mech. 303*, 155–167.

Dean, R. B. (1978). Reynolds number dependence of skin friction and other bulk flow variables in two-dimensional rectangular duct flow. *J. Fluids Eng. Trans. ASME 100*, 215–223.

Deardorff, J. W. (1970). A numerical study of three-dimensional turbulent channel flow at large Reynolds numbers. *J. Fluid Mech. 41*, 453–480.

Deardorff, J. W. (1974). Three-dimensional numerical study of the height and mean structure of a heated planetary boundary layer. *Boundary-Layer Meteorol. 7*, 81–106.

Deardorff, J. W. (1980). Stratocumulus-capped mixed layers derived from a three-dimensional model. *Boundary-Layer Meteorol. 18*, 495–527.

Demuren, A. O., M. M. Rogers, P. Durbin, and S. K. Lele (1996). On modeling pressure diffusion in non-homogeneous shear flows. In *Studying Turbulence Using Numerical Simulation Databases VI: Proceedings of the 1996 Summer Program*, pp. 63–72. Center for Turbulence Research, Stanford University.

Dimotakis, P. E. (1991). Turbulent free shear layer mixing and combustion. In S. N. B. Murthy and E. T. Curran (Eds.), *High-Speed Flight Propulsion Systems*, pp. 265–340. Washington: AIAA.

Domaradzki, J. A. (1992). Nonlocal triad interactions and the dissipation range of isotropic turbulence. *Phys. Fluids 4*, 2037–2045.

Domaradzki, J. A. and R. S. Rogallo (1990). Local energy transfer and nonlocal interactions in homogeneous, isotropic turbulence. *Phys. Fluids A 2*, 413–426.

Dopazo, C. (1994). Recent developments in PDF methods. In P. A. Libby and F. A. Williams (Eds.), *Turbulent Reacting Flows*, Chapter 7, pp. 375–474. London: Academic Press.

Dopazo, C. and E. E. O'Brien (1974). An approach to the autoignition of a turbulent mixture. *Acta Astronaut. 1*, 1239–1266.

Dowling, D. R. and P. E. Dimotakis (1990). Similarity of the concentration field of gas-phase turbulent jets. *J. Fluid Mech. 218*, 109–141.

Dreeben, T. D. and S. B. Pope (1997a). Probability density function and Reynolds-stress modeling of near-wall turbulent flows. *Phys. Fluids 9*, 154–163.

Dreeben, T. D. and S. B. Pope (1997b). Wall-function treatment in PDF methods for turbulent flows. *Phys. Fluids 9*, 2692–2703.

Dreeben, T. D. and S. B. Pope (1998). PDF/Monte Carlo simulation of near-wall turbulent flows. *J. Fluid Mech. 357*, 141–166.

Durbin, P. A. (1991). Near-wall turbulence closure modeling without damping functions. *Theor. Comput. Fluid Dyn. 3*, 1–13.

Durbin, P. A. (1993). A Reynolds stress model for near-wall turbulence. *J. Fluid Mech. 249*, 465–498.

Durbin, P. A. (1995). Separated flow computations with the k–ε–v^2 model. *AIAA J. 33*, 659–664.

Durbin, P. A. and C. G. Speziale (1994). Realizability of second-moment closure via stochastic analysis. *J. Fluid Mech. 280*, 395–407.

Durst, F., A. Melling, and J. H. Whitelaw (1974). Low Reynolds number flow over a plane symmetric sudden expansion. *J. Fluid Mech. 64*, 111–128.

Durst, F. and F. Schmitt (1985). Experimental studies of high Reynolds number backward-facing step flow. In *Proc. Fifth Symp. Turbulent Shear Flows*, Cornell University, pp. 5.19–5.24.

Erdélyi, A., F. Oberhettinger, and F. G. Tricomi (1954). *Tables of Integral Transforms*, Volume I. New York: McGraw-Hill.

Escudier, M. P. (1966). The distribution of mixing length in turbulent flows near walls. Heat Transfer Section TWF/TN/12, Imperial College, London.

Eswaran, V. and S. B. Pope (1988). Direct numerical simulations of the turbulent mixing of a passive scalar. *Phys. Fluids 31*, 506–520.

Eubank, R. L. (1988). *Spline Smoothing and Nonparametric Regression*. New York: Marcel Dekker.

Fabris, G. (1979). Conditional sampling study of the turbulent wake of a cylinder. Part 1. *J. Fluid Mech. 94*, 673–709.

Falco, R. E. (1974). Some comments on turbulent boundary layer structure inferred from the movements of a passive contaminant. Paper 74-99, AIAA.

Falco, R. E. (1977). Coherent motions in the outer region of turbulent boundary layers. *Phys. Fluids 20*, S124–S132.

Fiedler, H. and M. R. Head (1966). Intermittency measurements in the turbulent boundary layer. *J. Fluid Mech. 25*, 719–735.

Fischer, H. B., E. J. List, R. C. Y. Koh, J. Imberger, and N. H. Brooks (1979). *Mixing in Inland and Coastal Waters*. New York: Academic Press.

Foster, I. (1995). *Designing and Building Parallel Programs*. Addison-Wesley.

Fox, R. O. (1992). The Fokker–Planck closure for turbulent molecular mixing: passive scalars. *Phys. Fluids A 4*, 1230–1244.

Fox, R. O. (1994). Improved Fokker–Planck model for the joint scalar, scalar gradient PDF. *Phys. Fluids 6*, 334–348.

Fox, R. O. (1996). On velocity-conditioned scalar mixing in homogeneous turbulence. *Phys. Fluids 8*, 2678–2691.

Fox, R. O. (1997). The Lagrangian spectral relaxation model of the scalar dissipation in homogeneous turbulence. *Phys. Fluids 9*, 2364–2386.

Franklin, J. N. (1968). *Matrix theory*. Englewood Cliffs, N.J.: Prentice-Hall.

Frederiksen, R. D., W. J. A. Dahm, and D. R. Dowling (1997). Experimental assessment of fractal scale similarity in turbulent flows. Part 2. Higher-dimensional intersections and non-fractal inclusions. *J. Fluid Mech. 338*, 89–126.

Fu, S., B. E. Launder, and D. P. Tselepidakis (1987). Accommodating the effects of high strain rates in modelling the pressure–strain correlation. Mechanical Engineering Department Report TFD/87/5, UMIST.

Fureby, C., G. Tabor, H. G. Weller, and A. D. Gosman (1997). A comparative study of subgrid scale models in homogeneous isotropic turbulence. *Phys. Fluids 9*, 1416–1429.

Gad-el-Hak, M. and P. R. Bandyopadhyay (1994). Reynolds number effects in wall-bounded turbulent flows. *Appl. Mech. Rev. 47*, 307–365.

Galperin, B. and S. A. Orszag (1993). *Large Eddy Simulation of Complex Engineering and Geophysical Flows*. Cambridge: Cambridge University Press.

Gao, F. and E. E. O'Brien (1991). A mapping closure for multispecies Fickian diffusion. *Phys. Fluids A 3*, 956–959.

Gao, F. and E. E. O'Brien (1993). A large-eddy simulation scheme for turbulent reacting flows. *Phys. Fluids A 5*, 1282–1284.

Gardiner, C. W. (1985). *Handbook of Stochastic Methods for Physics, Chemistry and the Natural Sciences* (2nd ed.). Berlin: Springer-Verlag.

Gatski, T. B. and C. G. Speziale (1993). On explicit algebraic stress models for complex turbulent flows. *J. Fluid Mech. 254*, 59–78.

George, W. K. (1989). The self-preservation of turbulent flows and its relation to initial conditions and coherent structures. In W. K. George and R. Arndt (Eds.), *Advances in Turbulence*, pp. 39–73. New York: Hemisphere.

George, W. K., L. Castillo, and P. Knecht (1996). The zero pressure-gradient turbulent boundary layer. TRL 153, State University of New York at Buffalo.

George, W. K. and H. J. Hussein (1991). Locally axisymmetric turbulence. *J. Fluid Mech. 233*, 1–23.

Germano, M. (1986). A proposal for a redefinition of the turbulent stresses in the filtered Navier–Stokes equations. *Phys. Fluids 29*, 2323–2324.

Germano, M. (1992). Turbulence: the filtering approach. *J. Fluid Mech. 238*, 325–336.

Germano, M., U. Piomelli, P. Moin, and W. H. Cabot (1991). A dynamic subgrid-scale eddy viscosity model. *Phys. Fluids A 3*, 1760–1765.

Ghosal, S., T. S. Lund, P. Moin, and K. Akselvoll (1995). A dynamic localization model for large-eddy simulation of turbulent flows. *J. Fluid Mech. 286*, 229–255.

Ghosal, S. and P. Moin (1995). The basic equations for the large eddy simulation of turbulent flows in complex geometry. *J. Comput. Phys. 118*, 24–37.

Ghosal, S. and M. M. Rogers (1997). A numerical study of self-similarity in a turbulent plane wake using large-eddy simulation. *Phys. Fluids 9*, 1729–1739.

Gibson, M. M. and B. E. Launder (1978). Ground effects on pressure fluctuations in the atmospheric boundary layer. *J. Fluid Mech. 86*, 491–511.

Gicquel, L. Y. M., F. A. Jaberi, P. Givi, and S. B. Pope (1998). Filtered density function of velocity for large eddy simulation of turbulent flows. *Bull. Am. Phys. Soc. 43*, 2120.

Gillespie, D. T. (1992). *Markov Processes: An Introduction for Physical Scientists*. Boston: Academic Press.

Gillis, J. C. and J. P. Johnston (1983). Turbulent boundary-layer flow and structure on a convex wall and its redevelopment on a flat wall. *J. Fluid Mech. 135*, 123–153.

Girimaji, S. S. (1996). Fully explicit and self-consistent algebraic Reynolds stress model. *Theor. Comput. Fluid Dyn. 8*, 387–402.

Girimaji, S. S. (1997). A Galilean invariant explicit algebraic Reynolds stress model for turbulent curved flows. *Phys. Fluids 9*, 1067–1077.

Gleick, J. (1988). *Chaos: Making a New Science*. New York: Penguin.

Godin, P., D. W. Zingg, and T. E. Nelson (1997). High-lift aerodynamic computations with one- and two-equation turbulence models. *AIAA J. 35*, 237–243.

Grötzbach, G. (1981). Numerical simulation of turbulent temperature fluctuations in liquid metals. *Int. J. Heat Mass Transfer 24*, 475–490.

Guckenheimer, J. and P. Holmes (1983). *Nonlinear Oscillations, Dynamical Systems, and Bifurcations of Vector Fields*. New York: Springer-Verlag.

Gutmark, E. and I. Wygnanski (1976). The planar turbulent jet. *J. Fluid Mech. 73*, 465–495.

Hanjalić, K. (1970). *Two-dimensional Asymmetric Turbulent Flow in Ducts*. Ph. D. thesis, University of London.

Hanjalić, K. (1994). Advanced turbulence closure models: a view of current status and future prospects. *J. Heat Fluid Flow 15*, 178–203.

Hanjalić, K. and S. Jakirlić (1993). A model of stress dissipation in second-moment closures. *Appl. Sci. Res. 51*, 513–518.

Hanjalić, K., S. Jakirlić, and I. Hadžić (1997). Expanding the limits of equilibrium second-moment turbulence closures. *Fluid Dyn. Res. 20*, 25–41.

Hanjalić, K. and B. E. Launder (1972). A Reynolds stress model of turbulence and its application to thin shear flows. *J. Fluid Mech. 52*, 609–638.

Hanjalić, K. and B. E. Launder (1976). Contribution towards a Reynolds-stress closure for low-Reynolds-number turbulence. *J. Fluid Mech. 74*, 593–610.

Hanjalić, K. and B. E. Launder (1980). Sensitizing the dissipation equation to irrotational strains. *J. Fluids Eng. Trans. ASME 102*, 34–40.

Hanjalić, K., B. E. Launder, and R. Schiestel (1980). Multiple-time-scale concepts in turbulent transport modelling. In L. J. S. Bradbury *et al.* (Eds.), *Turbulent Shear Flows 2*, pp. 36–49. Berlin: Springer-Verlag.

Härdle, W. (1990). *Applied Nonparametric Regression*. Cambridge: Cambridge University Press.

Harlow, F. H. and P. I. Nakayama (1968). Transport of turbulence energy decay rate. University of California Report LA-3854, Los Alamos Science Laboratory.

Haworth, D. C. and S. H. El Tahry (1991). Probability density function approach for multidimensional turbulent flow calculations with application to in-cylinder flows in reciprocating engines. *AIAA J. 29*, 208–218.

Haworth, D. C. and K. Jansen (2000). Large-eddy simulation on unstructured deforming meshes: toward reciprocating IC engines. *Computers and Fluids 29*, 493–524.

Haworth, D. C. and S. B. Pope (1986). A generalized Langevin model for turbulent flows. *Phys. Fluids 29*, 387–405.

Haworth, D. C. and S. B. Pope (1987). A PDF modelling study of self-similar turbulent free shear flow. *Phys. Fluids 30*, 1026–1044.

Head, M. R. and P. Bandyopadhyay (1981). New aspects of turbulent boundary-layer structure. *J. Fluid Mech. 107*, 297–338.

Heisenberg, W. (1948). On the theory of statistical and isotropic turbulence. *Proc. R. Soc. London Ser. A 195*, 402–406.

Heskestad, G. (1965). Hot-wire measurements in a plane turbulent jet. *J. Appl. Mech. 32*, 721–734.

Hinze, J. O. (1975). *Turbulence* (2nd ed.). New York: McGraw-Hill.

Hockney, R. W. and J. W. Eastwood (1998). *Computer Simulation Using Particles.* Bristol: Adam Hilger.

Hoffmann, P. H., K. C. Muck, and P. Bradshaw (1985). The effect of concave surface curvature on turbulent boundary layers. *J. Fluid Mech. 161*, 371–403.

Holmes, P., J. L. Lumley, and G. Berkooz (1996). *Turbulence, Coherent Structures, Dynamical Systems, and Symmetry.* New York: Cambridge University Press.

Hsu, A. T., M. S. Anand, and M. K. Razdan (1990). An assessment of PDF versus finite-volume methods for turbulent reacting flow calculations. Paper 96-0523, AIAA.

Huffman, G. D. and P. Bradshaw (1972). A note on von Kármán's constant in low Reynolds number turbulent flows. *J. Fluid Mech. 53*, 45–60.

Hunt, J. C. R. (1973). A theory of turbulent flow round two-dimensional bluff bodies. *J. Fluid Mech. 61*, 625–706.

Hunt, J. C. R. (1985). Turbulent diffusion from sources in complex flows. *Annu. Rev. Fluid. Mech. 17*, 447–485.

Hunt, J. C. R. and D. J. Carruthers (1990). Rapid distortion theory and the 'problems' of turbulence. *J. Fluid Mech. 212*, 497–532.

Hunt, J. C. R. and J. M. R. Graham (1978). Free-stream turbulence near plane boundaries. *J. Fluid Mech. 84*, 209–235.

Hussaini, M. Y. (1998). On large-eddy simulation of compressible flows. Paper 98-2802, AIAA.

Hussein, H. J., S. Capp, and W. K. George (1994). Velocity measurements in a high-Reynolds-number, momentum-conserving, axisymmetric, turbulent jet. *J. Fluid Mech. 258*, 31–75.

Janicka, J., W. Kolbe, and W. Kollmann (1977). Closure of the transport equation for the probability density function of turbulent scalar fields. *J. Non-Equilib. Thermodyn. 4*, 47–66.

Jayesh and S. B. Pope (1995). Stochastic model for turbulent frequency. FDA 95-05, Cornell University.

Jeong, J., F. Hussain, W. Schoppa, and J. Kim (1997). Coherent structures near the wall in a turbulent channel flow. *J. Fluid Mech. 332*, 185–214.

Jiménez, J. and P. Moin (1991). The minimal flow unit in near-wall turbulence. *J. Fluid Mech. 225*, 213–240.

Jiménez, J. and R. D. Moser (1998). LES: where are we and what can we expect? Paper 98-2891, AIAA.

Jones, W. P. (1994). Turbulence modelling and numerical solution methods for variable density and combusting flows. In P. A. Libby and F. A. Williams (Eds.), *Turbulent Reacting Flows*, Chapter 6, pp. 309–374. London: Academic Press.

Jones, W. P. and M. Kakhi (1997). Application of the transported PDF approach to hydrocarbon turbulent jet diffusion flames. *Combust. Sci. Technol. 129*, 393–430.

Jones, W. P. and B. E. Launder (1972). The prediction of laminarization with a two-equation model of turbulence. *Int. J. Heat Mass Transfer 15*, 301–314.

Jones, W. P. and P. Musonge (1988). Closure of the Reynolds stress and scalar flux equations. *Phys. Fluids 31*, 3589–3604.

Jovic, S. and D. M. Driver (1994). Backward-facing step measurements at low Reynolds number, $Re_h = 5000$. TM 108807, NASA.

Juneja, A. and S. B. Pope (1996). A DNS study of turbulent mixing of two passive scalars. *Phys. Fluids 8*, 2161–2184.

Kachanov, Y. S. (1994). Physical mechanisms of laminar-boundary-layer transition. *Annu. Rev. Fluid. Mech. 26*, 411–482.

Karlin, S. and H. M. Taylor (1981). *A Second Course in Stochastic Processes*. New York: Academic Press.

Kassinos, S. C. and W. C. Reynolds (1994). A structure-based model for the rapid distortion of homogeneous turbulence. TF 61, Stanford University.

Kebede, W., B. E. Launder, and B. A. Younis (1985). Large-amplitude periodic pipe flow: a second-moment closure study. In F. Durst (Ed.), *Proc. Fifth Symp. Turbulent Shear Flows*, Ithaca, pp. 16.23–16.29. Cornell University.

Kellogg, O. D. (1967). *Foundations of Potential Theory*. Berlin: Springer-Verlag.

Kellogg, R. M. and S. Corrsin (1980). Evolution of a spectrally local disturbance in grid-generated, nearly isotropic turbulence. *J. Fluid Mech. 96*, 641–669.

Kim, H. T., S. J. Kline, and W. C. Reynolds (1971). The production of turbulence near a smooth wall in a turbulent boundary layer. *J. Fluid Mech. 50*, 133–160.

Kim, J., P. Moin, and R. Moser (1987). Turbulence statistics in fully developed channel flow at low Reynolds number. *J. Fluid Mech. 177*, 133–166.

Klebanoff, P. S. (1954). Characteristics of turbulence in a boundary layer with zero pressure gradient. TN 3178, NACA.

Kline, S. J., B. J. Cantwell, and G. M. Lilley (1982). Proceedings of the 1980-81 AFOSR-HTTM-Stanford conference on complex turbulent flows. Technical report, Stanford University.

Kline, S. J., W. C. Reynolds, F. A. Schraub, and P. W. Runstadler (1967). The structure of turbulent boundary layers. *J. Fluid Mech. 30*, 741–773.

Kline, S. J. and S. K. Robinson (1990). Quasi-coherent structures in the turbulent boundary layer: part I. Status report on a community-wide summary of the data. In S. J. Kline and N. H. Afgan (Eds.), *Near-wall Turbulence*, pp. 200–217. New York: Hemisphere.

Knight, D., G. Zhou, N. Okong'o, and V. Shukla (1998). Compressible large eddy simulation using unstructured grids. Paper 98-0535, AIAA.

Kolmogorov, A. N. (1941a). Dissipation of energy in locally isotropic turbulence. *Dokl. Akad. Nauk SSSR 32*, 19–21 [in Russian].

Kolmogorov, A. N. (1941b). The local structure of turbulence in incompressible viscous fluid for very large Reynolds numbers. *Dokl. Akad. Nauk SSSR 30*, 299–303 [in Russian].

Kolmogorov, A. N. (1942). The equations of turbulent motion in an incompressible fluid. *Izvestia Acad. Sci., USSR; Phys. 6*, 56–58 [in Russian].

Kolmogorov, A. N. (1962). A refinement of previous hypotheses concerning the local structure of turbulence in a viscous incompressible fluid at high Reynolds number. *J. Fluid Mech. 13*, 82–85.

Kolmogorov, A. N. (1991). The local structure of turbulence in incompressible viscous fluid for very large Reynolds numbers. *Proc. R. Soc. London Ser. A 434*, 9–13.

Kovasznay, L. S. G., V. Kibens, and R. F. Blackwelder (1970). Large-scale motion in the intermittent region of a turbulent boundary layer. *J. Fluid Mech. 41*, 283–325.

Kraichnan, R. H. (1959). The structure of isotropic turbulence at very high Reynolds numbers. *J. Fluid Mech. 5*, 497–543.

Kraichnan, R. H. (1976). Eddy viscosity in two and three dimensions. *J. Atmos. Sci. 33*, 1521–1536.

Kral, L. D., M. Mani, and J. A. Ladd (1996). Application of turbulence models for aerodynamic and propulsion flowfields. *AIAA J. 34*, 2291–2298.

Kuznetsov, V. R. and V. A. Sabel'nikov (1990). *Turbulence and Combustion*. New York: Hemisphere.

Lai, Y. G. and R. M. C. So (1990). On near-wall turbulent flow modelling. *J. Fluid Mech. 221*, 641–673.

Langevin, P. (1908). Sur la théorie du mouvement Brownien. *Comptes Rendus Acad. Sci., Paris 146*, 530–533.

LaRue, J. C. and P. A. Libby (1974). Temperature fluctuations in the plane turbulent wake. *Phys. Fluids 17*, 1956–1967.

LaRue, J. C. and P. A. Libby (1976). Statistical properties of the interface in the turbulent wake of a heated cylinder. *Phys. Fluids 19*, 1864–1875.

Launder, B. E. (1986). Low-Reynolds-number turbulence near walls. Mechanical Engineering Department. Report TFD/86/4, UMIST.

Launder, B. E. (1990). Phenomenological modelling: present... and future? In J. L. Lumley (Ed.), *Whither Turbulence? Turbulence at the Crossroads*, pp. 439–485. Berlin: Springer-Verlag.

Launder, B. E. (1996). An introduction to single-point closure methodology. In T. B. Gatski, M. Y. Hussaini, and J. L. Lumley (Eds.), *Simulation and Modeling of Turbulent Flows*, Chapter 6, pp. 243–310. New York: Oxford University Press.

Launder, B. E., G. J. Reece, and W. Rodi (1975). Progress in the development of a Reynolds-stress turbulence closure. *J. Fluid Mech. 68*, 537–566.

Launder, B. E. and W. C. Reynolds (1983). Asymptotic near-wall stress dissipation rates in a turbulent flow. *Phys. Fluids 26*, 1157–1158.

Launder, B. E. and B. I. Sharma (1974). Application of the energy-dissipation model of turbulence to the calculation of flow near a spinning disc. *Lett. Heat Mass Transf. 1*, 131–138.

Launder, B. E. and D. B. Spalding (1972). *Mathematical Models of Turbulence*. London: Academic Press.

Le, H., P. Moin, and J. Kim (1997). Direct numerical simulation of turbulent flow over a backward-facing step. *J. Fluid Mech. 330*, 349–374.

Lee, M. J., J. Kim, and P. Moin (1990). Structure of turbulence at high shear rate. *J. Fluid Mech. 216*, 561–583.

Lee, M. J. and W. C. Reynolds (1985). Numerical experiments on the structure of homogeneous turbulence. Technical Report TF-24, Stanford University.

Leonard, A. (1974). Energy cascade in large eddy simulation of turbulent fluid flow. *Adv. Geophys. 18A*, 237–248.

Lesieur, M. (1990). *Turbulence in Fluids* (2nd ed.). Dordrecht: Kluwer.

Lesieur, M. and O. Métais (1996). New trends in large-eddy simulations of turbulence. *Annu. Rev. Fluid. Mech. 28*, 45–82.

Leslie, D. C. and G. L. Quarini (1979). The application of turbulence theory to the formulation of subgrid modelling procedures. *J. Fluid Mech. 91*, 65–91.

Lighthill, M. J. (1970). *Introduction to Fourier Analysis and Generalized Functions*. Cambridge: Cambridge University Press.

Lilly, D. K. (1967). The representation of small-scale turbulence in numerical simulation experiments. In H. H. Goldstine (Ed.), *Proc. IBM Scientific Computing Symp. on Environmental Sciences*, pp. 195–210. Yorktown Heights, NY: IBM.

Lilly, D. K. (1992). A proposed modification of the Germano subgrid-scale closure method. *Phys. Fluids A 4*, 633–635.

Liu, S., C. Meneveau, and J. Katz (1994). On the properties of similarity subgrid-scale models as deduced from measurements in a turbulent jet. *J. Fluid Mech. 275*, 83–119.

Loitsyanskii, L. G. (1939). Some basic laws for isotropic turbulent flow. *Trudy Tsentr. Aero.-Giedrodin. Inst. 440*, 3–23 [in Russian].

Long, R. R. and T.-C. Chen (1981). Experimental evidence for the existence of the 'mesolayer' in turbulent systems. *J. Fluid Mech. 105*, 19–59.

Lorenz, E. N. (1963). Deterministic non-periodic flow. *J. Atmos. Sci. 20*, 130–141.

Lumley, J. L. (1965). Interpretation of time spectra measured in high-intensity shear flows. *Phys. Fluids 8*, 1056–1062.

Lumley, J. L. (1967a). Similarity and the turbulent energy spectrum. *Phys. Fluids 10*, 855–858.

Lumley, J. L. (1967b). The structure of inhomogeneous turbulence. In A. M. Monin and V. I. Tatarski (Eds.), *Atmospheric Turbulence and Wave Propagation*, pp. 166–178. Moscow: Nauka.

Lumley, J. L. (1975). Pressure–strain correlation. *Phys. Fluids 18*, 750.

Lumley, J. L. (1978). Computational modeling of turbulent flows. *Adv. Appl. Mech. 18*, 123–176.

Lumley, J. L. (1992). Some comments on turbulence. *Phys. Fluids A 4*, 203–211.

Lundgren, T. S. (1969). Model equation for nonhomogeneous turbulence. *Phys. Fluids 12*, 485–497.

Luo, J. and B. Lakshminarayana (1997). Prediction of strongly curved turbulent duct flows with Reynolds stress model. *AIAA J. 35*, 91–98.

Makita, H. (1991). Realization of a large-scale turbulence field in a small wind tunnel. *Fluid Dyn. Res. 8*, 53–64.

Mandelbrot, B. B. (1974). Intermittent turbulence in self-similar cascades: divergence of high moments and dimension of the carrier. *J. Fluid Mech. 62*, 331–358.

Mansour, N. N., J. Kim, and P. Moin (1988). Reynolds-stress and dissipation-rate budgets in a turbulent channel flow. *J. Fluid Mech. 194*, 15–44.

Marušić, I., A. K. M. Uddin, and A. E. Perry (1997). Similarity law for the streamwise turbulence intensity in zero-pressure-gradient turbulent boundary layers. *Phys. Fluids 9*, 3718–3726.

Mason, P. J. (1989). Large-eddy simulation of the convective atmospheric boundary layer. *J. Atmos. Sci. 46*, 1492–1516.

Mason, P. J. (1994). Large-eddy simulation: a critical review of the technique. *Quart. J. R. Meteorol. Soc. 120*, 1–26.

Mason, P. J. and N. S. Callen (1986). On the magnitude of the subgrid-scale eddy coefficient in large-eddy simulations of turbulent channel flow. *J. Fluid Mech. 162*, 439–462.

Mason, P. J. and D. J. Thomson (1992). Stochastic backscatter in large-eddy simulations of boundary layers. *J. Fluid Mech. 242*, 51–78.

McComb, W. D. (1990). *The Physics of Fluid Turbulence*. Oxford: Oxford University Press.

McMillan, O. J. and J. H. Ferziger (1979). Direct testing of subgrid-scale models. *AIAA J. 17*, 1340–1346.

Melling, A. and J. H. Whitelaw (1976). Turbulent flow in a rectangular duct. *J. Fluid Mech. 78*, 289–315.

Mellor, G. L. and H. J. Herring (1973). A survey of the mean turbulent field closure models. *AIAA J. 11*, 590–599.

Meneguzzi, M., H. Politano, A. Pouquet, and M. Zolver (1996). A sparse-mode spectral method for the simulation of turbulent flows. *J. Comput. Phys. 123*, 32–44.

Meneveau, C. and T. S. Lund (1997). The dynamic Smagorinsky model and scale-dependent coefficients in the viscous range of turbulence. *Phys. Fluids 9*, 3932–3934.

Meneveau, C., T. S. Lund, and W. H. Cabot (1996). A Lagrangian dynamic subgrid-scale model of turbulence. *J. Fluid Mech. 319*, 353–385.

Meneveau, C. and K. R. Sreenivasan (1991). The multifractal nature of turbulent energy dissipation. *J. Fluid Mech. 224*, 429–484.

Menter, F. (1994). Two-equation eddy-viscosity turbulence models for engineering applications. *AIAA J. 32*, 1598–1605.

Métais, O. and M. Lesieur (1992). Spectral large-eddy simulation of isotropic and stably stratified turbulence. *J. Fluid Mech. 239*, 157–194.

Miller, P. L. (1991). *Mixing in High Schmidt Number Turbulent Jets*. Ph. D. thesis, California Institute of Technology.

Millikan, C. B. (1938). A critical discussion of turbulent flows in channels and circular tubes. In J. P. Den Hartog and H. Peters (Eds.), *Proc. 5th Int. Congr. Applied Mechanics*, New York, pp. 386–392. New York: Wiley.

Minier, J.-P. and J. Pozorski (1995). Analysis of a PDF model in a mixing layer case. In *Proc. Tenth Symp. Turbulent Shear Flows*, Pennsylvania State University, pp. 26.25–26.30.

Mohamed, M. S. and J. C. LaRue (1990). The decay power law in grid-generated turbulence. *J. Fluid Mech. 219*, 195–214.

Moin, P. and J. Kim (1982). Numerical investigation of turbulent channel flow. *J. Fluid Mech. 118*, 341–377.

Moin, P. and K. Mahesh (1998). Direct numerical simulation: a tool for turbulence research. *Annu. Rev. Fluid. Mech. 30*, 539–578.

Moin, P. and R. D. Moser (1989). Characteristic-eddy decomposition of turbulence in a channel. *J. Fluid Mech. 200*, 471–509.

Monaghan, J. J. (1992). Smoothed particle hydrodynamics. *Annu. Rev. Astron. Astrophys. 30*, 543–574.

Monin, A. S. and A. M. Yaglom (1971). *Statistical Fluid Mechanics: Mechanics of Turbulence*, Volume 1. Cambridge, MA: MIT Press.

Monin, A. S. and A. M. Yaglom (1975). *Statistical Fluid Mechanics: Mechanics of Turbulence*, Volume 2. Cambridge, MA: MIT Press.

Moon, F. C. (1992). *Chaotic and fractal dynamics : an introduction for applied scientists and engineers*. New York: Wiley.

Moser, R. D., J. Kim, and N. N. Mansour (1999). Direct numerical simulation of turbulent channel flow up to $Re_\tau = 590$. *Phys. Fluids 11*, 943–946.

Muck, K. C., P. H. Hoffmann, and P. Bradshaw (1985). The effect of convex surface curvature on turbulent boundary layers. *J. Fluid Mech. 161*, 347–369.

Mungal, M. G. and D. Hollingsworth (1989). Organized motion in a very high Reynolds number jet. *Phys. Fluids A 1*, 1615–1623.

Murlis, J., H. M. Tsai, and P. Bradshaw (1982). The structure of turbulent boundary layers at low Reynolds numbers. *J. Fluid Mech. 122*, 13–56.

Mydlarski, L. and Z. Warhaft (1996). On the onset of high-Reynolds-number grid-generated wind tunnel turbulence. *J. Fluid Mech. 320*, 331–368.

Mydlarski, L. and Z. Warhaft (1998). Passive scalar statistics in high Péclet number grid turbulence. *J. Fluid Mech. 358*, 135–175.

Naot, D., A. Shavit, and M. Wolfshtein (1970). Interactions between components of the turbulent velocity correlation tensor due to pressure fluctuations. *Israel J. Technol. 8*, 259–269.

Narasimha, R. and K. R. Sreenivasan (1979). Relaminarization of fluid flows. *Adv. Appl. Mech. 19*, 222–309.

Nee, V. W. and L. S. G. Kovasznay (1969). Simple phenomenological theory of turbulent shear flows. *Phys. Fluids 12*, 473–484.

Nelkin, M. (1994). Universality and scaling in fully developed turbulence. *Adv. Phys. 43*, 143–181.

Nilsen, V. and G. Kosály (1999). Differential diffusion in turbulent reacting flows. *Combust. Flame 117*, 493–513.

Novikov, E. A. (1990). The effects of intermittency on statistical characteristics of turbulence and scale similarity of breakdown coefficients. *Phys. Fluids A 2*, 814–820.

Obukhov, A. M. (1941). The spectral energy distribution in a turbulent flow. *Dokl. Akad. Nauk SSSR 32*, 22–24 [in Russian].

Obukhov, A. M. (1962). Some specific features of atmospheric turbulence. *J. Fluid Mech. 13*, 77–81.

Okong'o, N. and D. Knight (1998). Compressible large eddy simulation using unstructured grids: channel and boundary layer flows. Paper 98-3315, AIAA.

Orszag, S. A. and G. S. Patterson (1972). Numerical simulation of three-dimensional homogeneous isotropic turbulence. *Phys. Rev. Lett. 28*, 76–79.

Orszag, S. A., I. Staroselsky, W. S. Flannery, and Y. Zhang (1996). Introduction to renormalization group modeling of turbulence. In T. B. Gatski, M. Y. Hussaini, and J. L. Lumley (Eds.), *Simulation and Modeling of Turbulent Flows*, Chapter 4, pp. 155–183. New York: Oxford University Press.

Oster, D. and I. Wygnanski (1982). The forced mixing layer between parallel streams. *J. Fluid Mech. 123*, 91–130.

Overholt, M. R. and S. B. Pope (1996). Direct numerical simulation of a passive scalar with imposed mean gradient in isotropic turbulence. *Phys. Fluids 8*, 3128–3148.

Panchapakesan, N. R. and J. L. Lumley (1993a). Turbulence measurements in axisymmetric jets of air and helium. Part 1. Air jet. *J. Fluid Mech. 246*, 197–223.

Panchapakesan, N. R. and J. L. Lumley (1993b). Turbulence measurements in axisymmetric jets of air and helium. Part 2. Helium jet. *J. Fluid Mech. 246*, 225–247.

Panchev, S. (1971). *Random Functions and Turbulence*. Oxford: Pergamon Press.

Panton, R. L. (1984). *Incompressible Flow*. New York: Wiley.

Pao, Y.-H. (1965). Structure of turbulent velocity and scalar fields at large wavenumbers. *Phys. Fluids 8*, 1063–1075.

Parneix, S., P. A. Durbin, and M. Behnia (1998). Computation of 3-D turbulent boundary layers using the V2F model. *Flow, Turbulence Combust. 60*, 19–46.

Parneix, S., D. Laurence, and P. A. Durbin (1998). A procedure for using DNS databases. *J. Fluids Eng. Trans. ASME 120*, 40–47.

Patel, V. C. and M. R. Head (1969). Some observations on skin friction and velocity profiles in fully developed pipe and channel flows. *J. Fluid Mech. 38*, 181–201.

Patel, V. C., W. Rodi, and G. Scheuerer (1985). Turbulence models for near-wall and low Reynolds number flows: a review. *AIAA J. 23*, 1308–1319.

Patel, V. C. and F. Sotiropoulos (1997). Longitudinal curvature effects in turbulent boundary layers. *Proc. Aerospace Sci. 33,* 1–70.

Pennisi, S. (1992). On third order tensor-valued isotropic functions. *Int. J. Eng. Sci. 30,* 679–692.

Pennisi, S. and M. Trovato (1987). On the irreducibility of Professor G. F. Smith's representations for isotropic functions. *Int. J. Eng. Sci. 25,* 1059–1065.

Perot, B. and P. Moin (1995). Shear-free turbulent boundary layers. Part 1. Physical insights into near-wall turbulence. *J. Fluid Mech. 295,* 199–227.

Perry, A. E. and M. S. Chong (1982). On the mechanism of wall turbulence. *J. Fluid Mech. 119,* 173–217.

Perry, A. E., S. Henbest, and M. S. Chong (1986). A theoretical and experimental study of wall turbulence. *J. Fluid Mech. 165,* 163–199.

Perry, A. E. and I. Marušić (1995). A wall–wake model for the turbulence structure of boundary layers. Part I. Extension of the attached eddy hypothesis. *J. Fluid Mech. 298,* 361–388.

Peterson, V. L., J. Kim, T. L. Holst, G. S. Deiwert, D. M. Cooper, A. B. Watson, and F. R. Bailey (1989). Supercomputer requirements for selected disciplines important to aerospace. *Proc. IEEE 77,* 1038–1055.

Phillips, O. M. (1955). The irrotational motion outside a free turbulent boundary. *Proc. Cambr. Phil. Soc. 51,* 220–229.

Piomelli, U. (1993). High Reynolds number calculations using the dynamic subgrid-scale stress model. *Phys. Fluids A 5,* 1484–1490.

Piomelli, U. and J. R. Chasnov (1996). Large-eddy simulations: theory and applications. In M. Hallbäck, D. S. Henningson, A. V. Johansson, and P. H. Alfredsson (Eds.), *Turbulence and Transition Modelling,* Chapter 7, pp. 269–336. Dordrecht: Kluwer.

Piomelli, U., J. Ferziger, and P. Moin (1989). New approximate boundary conditions for large eddy simulations of wall-bounded flows. *Phys. Fluids A 6,* 1061–1068.

Piomelli, U. and J. Liu (1995). Large-eddy simulation of rotating channel flow using a localized dynamic model. *Phys. Fluids 7,* 839–848.

Piomelli, U., P. Moin, and J. H. Ferziger (1988). Model consistency in the large eddy simulation of turbulent channel flows. *Phys. Fluids 31,* 1884–1891.

Piomelli, U., Y. Yu, and R. J. Adrian (1996). Subgrid-scale energy transfer and near-wall turbulence structure. *Phys. Fluids 8,* 215–224.

Pope, S. B. (1975). A more general effective-viscosity hypothesis. *J. Fluid Mech. 72,* 331–340.

Pope, S. B. (1978). An explanation of the turbulent round-jet/plane-jet anomaly. *AIAA J. 16,* 279–281.

Pope, S. B. (1981a). A Monte Carlo method for the PDF equations of turbulent reactive flow. *Combust. Sci. Technol. 25,* 159–174.

Pope, S. B. (1981b). Transport equation for the joint probability density function of velocity and scalars in turbulent flow. *Phys. Fluids 24,* 588–596.

Pope, S. B. (1983a). Consistent modeling of scalars in turbulent flows. *Phys. Fluids 26,* 404–408.

Pope, S. B. (1983b). A Lagrangian two-time probability density function equation for inhomogeneous turbulent flows. *Phys. Fluids 26,* 3448–3450.

Pope, S. B. (1985). PDF methods for turbulent reactive flows. *Prog. Energy Combust. Sci. 11,* 119–192.

Pope, S. B. (1990). Computations of turbulent combustion: progress and challenges. In *Twenty-Third Symp. (Int.) on Combustion*, pp. 591–612. Pittsburgh: The Combustion Institute.

Pope, S. B. (1991a). Application of the velocity-dissipation probability density function model to inhomogeneous turbulent flows. *Phys. Fluids A 3*, 1947–1957.

Pope, S. B. (1991b). Mapping closures for turbulent mixing and reaction. *Theor. Comput. Fluid Dyn. 2*, 255–270.

Pope, S. B. (1994a). Lagrangian PDF methods for turbulent flows. *Annu. Rev. Fluid. Mech. 26*, 23–63.

Pope, S. B. (1994b). On the relationship between stochastic Lagrangian models of turbulence and second-moment closures. *Phys. Fluids 6*, 973–985.

Pope, S. B. (1998a). The implied filter in large-eddy simulations of turbulence. *Bull. Am. Phys. Soc. 43*, 2087–2088.

Pope, S. B. (1998b). The vanishing effect of molecular diffusivity on turbulent dispersion: implications for turbulent mixing and the scalar flux. *J. Fluid Mech. 359*, 299–312.

Pope, S. B. and Y. L. Chen (1990). The velocity-dissipation probability density function model for turbulent flows. *Phys. Fluids A 2*, 1437–1449.

Porté-Agel, F., C. Meneveau, and M. B. Parlange (2000). A scale-dependent dynamic model for large-eddy simulation: application to the atmospheric boundary layer. *J. Fluid Mech. 415*, 261–284.

Prandtl, L. (1925). Bericht über die Entstehung der Turbulenz. *Z. Angew. Math. Mech. 5*, 136–139.

Prandtl, L. (1933). Attaining a steady air stream in wind tunnels. TM 726, NACA.

Prandtl, L. (1945). Über ein neues Formelsystem für die ausgebildete Turbulenz. *Nachr. Akad. Wiss. Göttingen Math-Phys. K1*, 6–19.

Priestley, M. B. (1981). *Spectral Analysis and Time Series*. London: Academic Press.

Rayleigh, Lord (1916). On the dynamics of revolving fluids. *Proc. R. Soc. London Ser. A 93*, 148–154.

Reynolds, O. (1883). An experimental investigation of the circumstances which determine whether the motion of water shall be direct or sinuous, and the law of resistance in parallel channels. *Philos. Trans. R. Soc. London Ser. A 174*, 935–982.

Reynolds, O. (1894). On the dynamical theory of incompressible viscous flows and the determination of the criterion. *Philos. Trans. R. Soc. London Ser. A 186*, 123–161.

Reynolds, W. C. (1987). Fundamentals of turbulence for turbulence modeling and simulation. In *Special Course on Modern Theoretical and Experimental Approaches to Turbulent Flow Structure and its Modelling*, AGARD-R-755, Neuilly-sur-Seine, pp. 1–65.

Reynolds, W. C. (1990). The potential and limitations of direct and large eddy simulations. In J. L. Lumley (Ed.), *Whither Turbulence? Turbulence at the Crossroads*, pp. 313–343. Berlin: Springer-Verlag.

Reynolds, W. C. and S. C. Kassinos (1995). A one-point modelling of rapidly deformed homogeneous turbulence. *Proc. R. Soc. London Ser. A 451*, 87–104.

Richardson, L. F. (1922). *Weather Prediction by Numerical Process*. Cambridge: Cambridge University Press.

Ristorcelli, J. R., J. L. Lumley, and R. Abid (1995). A rapid-pressure covariance representation consistent with the Taylor–Proudman theorem materially frame indifferent in the 2-D limit. *J. Fluid Mech. 292*, 111–152.

Robertson, H. P. (1940). The invariant theory of isotropic turbulence. *Proc. Cambr. Phil. Soc. 36*, 209–223.

Robinson, S. K. (1991). Coherent motions in the turbulent boundary layer. *Annu. Rev. Fluid. Mech. 23*, 601–639.

Rodi, W. (1972). *The Prediction of Free Turbulent Boundary Layers Using a Two-equation Model of Turbulence.* Ph. D. thesis, Imperial College, London.

Rodi, W., J. H. Ferziger, M. Breuer, and M. Pourquié (1997). Status of large eddy simulation: results of a workshop. *J. Fluids Eng. Trans. ASME 119*, 248–262.

Rodi, W. and N. N. Mansour (1993). Low Reynolds number k–ε modelling with the aid of direct numerical simulation data. *J. Fluid Mech. 250*, 509–529.

Rodi, W. and G. Scheuerer (1986). Scrutinizing the k–ε turbulence model under adverse pressure gradient conditions. *J. Fluids Eng. Trans. ASME 108*, 174–179.

Roekaerts, D. (1991). Use of a Monte Carlo PDF method in a study of the influence of turbulent fluctuations on selectivity in a jet-stirred reactor. *Appl. Sci. Res. 48*, 271–300.

Rogallo, R. S. (1981). Numerical experiments in homogeneous turbulence. Technical Report TM81315, NASA.

Rogers, M. M. and P. Moin (1987). The structure of the vorticity field in homogeneous turbulent flows. *J. Fluid Mech. 176*, 33–66.

Rogers, M. M. and R. D. Moser (1994). Direct simulation of a self-similar turbulent mixing layer. *Phys. Fluids 6*, 903–923.

Rotta, J. C. (1951). Statistische Theorie nichthomogener Turbulenz. *Z. Phys. 129*, 547–572.

Rubinstein, R. and J. M. Barton (1990). Nonlinear Reynolds stress models and the renormalization group. *Phys. Fluids A 2*, 1472–1476.

Saddoughi, S. G. and S. V. Veeravalli (1994). Local isotropy in turbulent boundary layers at high Reynolds number. *J. Fluid Mech. 268*, 333–372.

Saffman, P. G. (1960). On the effect of the molecular diffusivity in turbulent diffusion. *J. Fluid Mech. 8*, 273–283.

Saffman, P. G. (1967). The large-scale structure of homogeneous turbulence. *J. Fluid Mech. 27*, 581–593.

Saffman, P. G. (1970). A model for inhomogeneous turbulent flow. *Proc. R. Soc. London Ser. A 317*, 417–433.

Sarkar, S. and C. G. Speziale (1990). A simple nonlinear model for the return to isotropy in turbulence. *Phys. Fluids A 2*, 84–93.

Sato, Y. and K. Yamamoto (1987). Lagrangian measurement of fluid-particle motion in an isotropic turbulent field. *J. Fluid Mech. 175*, 183–199.

Savill, A. M. (1987). Recent developments in rapid-distortion theory. *Annu. Rev. Fluid. Mech. 19*, 531–575.

Saxena, V. and S. B. Pope (1998). PDF calculations of major and minor species in a turbulent piloted jet flame. In *Twenty-Seventh Symp. (Int.) on Combustion*, pp. 1081–1086. Pittsburgh: The Combustion Institute.

Schlichting, H. (1933). Laminare Strahlenausbreitung. *Z. Angew. Math. Mech. 13*, 260–263.

Schlichting, H. (1979). *Boundary-Layer Theory* (7th ed.). New York: McGraw-Hill.

Schmidt, H. and U. Schumann (1989). Coherent structure of the covective bound-
    ary layer derived from large-eddy simulations. *J. Fluid Mech. 200*, 511–562.
Schneider, W. (1985). Decay of momentum in submerged jets. *J. Fluid Mech. 154*,
    91–110.
Schoppa, W. and F. Hussain (1998). A large-scale control strategy for drag
    reduction in turbulent boundary layers. *Phys. Fluids 10*, 1049–1051.
Schumann, U. (1975). Subgrid scale model for finite difference simulations of
    turbulent flows in plane channels and annuli. *J. Comput. Phys. 18*, 376–404.
Schumann, U. (1977). Realizability of Reynolds-stress turbulence models. *Phys.
    Fluids 20*, 721–725.
Schwarz, W. R. and P. Bradshaw (1994). Term-by-term tests of stress-transport
    turbulence models in a three-dimensional boundary layer. *Phys. Fluids 6*,
    986–998.
Scotti, A., C. Meneveau, and M. Fatica (1997). Dynamic Smagorinsky model on
    anisotropic grids. *Phys. Fluids 9*, 1856–1858.
Scotti, A., C. Meneveau, and D. K. Lilly (1993). Generalized Smagorinsky model
    for anisotropic grids. *Phys. Fluids A 5*, 2306–2308.
Shih, T.-H. and J. L. Lumley (1985). Modeling of pressure correlation terms in
    Reynolds stress and scalar flux equations. FDA 85-3, Cornell University.
Shir, C. C. (1973). A preliminary study of atmospheric turbulent flow in the
    idealized planetary boundary layer. *J. Atmos. Sci. 30*, 1327–1339.
Simpson, R. L. (1989). Turbulent boundary-layer separation. *Annu. Rev. Fluid.
    Mech. 21*, 205–34.
Sirovich, L., K. S. Ball, and L. R. Keefe (1990). Plane waves and structures in
    turbulent channel flow. *Phys. Fluids A 2*, 2217–2226.
Smagorinsky, J. (1963). General circulation experiments with the primitive equa-
    tions: I. The basic equations. *Mon. Weather Rev. 91*, 99–164.
Smith, A. M. O. and T. Cebeci (1967). Numerical solution of the turbulent
    boundary layer equations. Report DAC 33735, Douglas Aircraft Division.
Smith, C. R. and S. P. Metzler (1983). The characteristics of low-speed streaks in
    the near wall region of a turbulent boundary layer. *J. Fluid Mech. 129*, 27–54.
Smith, L. M. and W. C. Reynolds (1992). On the Yakhot–Orszag renormalization
    group method for deriving turbulence statistics and models. *Phys. Fluids A 4*,
    364–390.
Smith, L. M. and S. L. Woodruff (1998). Renormalization-group analysis of
    turbulence. *Annu. Rev. Fluid. Mech. 30*, 275–310.
Spalart, P. R. (1988). Direct simulation of a turbulent boundary layer up to
    $R_\theta = 1410$. *J. Fluid Mech. 187*, 61–98.
Spalart, P. R. and S. R. Allmaras (1994). A one-equation turbulence model for
    aerodynamic flows. *Recherche Aérospatiale 1*, 5–21.
Spalart, P. R. and J. H. Watmuff (1993). Experimental and numerical study of a
    turbulent boundary layer with pressure gradients. *J. Fluid Mech. 249*, 337–371.
Spencer, A. J. M. (1971). Theory of invariants. In A. C. Eringen (Ed.), *Continuum
    Physics*, Volume 1, pp. 240–352. New York: Academic.
Speziale, C. G. (1981). Some interesting properties of two-dimensional turbulence.
    Phys. Fluids 24, 1425–1427.
Speziale, C. G. (1985). Galilean invariance of subgrid-scale stress models in
    large-eddy simulation of turbulence. *J. Fluid Mech. 156*, 55–62.
Speziale, C. G. (1987). On nonlinear k–l and k–ε models of turbulence. *J. Fluid
    Mech. 178*, 459–475.

Speziale, C. G. (1996). Modeling of turbulent transport equations. In T. B. Gatski, M. Y. Hussaini, and J. L. Lumley (Eds.), *Simulation and Modeling of Turbulent Flows*, Chapter 5, pp. 185–242. New York: Oxford University Press.

Speziale, C. G., R. Abid, and E. C. Anderson (1992). Critical evaluation of two-equation models for near-wall turbulence. *AIAA J. 30*, 324–331.

Speziale, C. G., S. Sarkar, and T. B. Gatski (1991). Modelling the pressure–strain correlation of turbulence: an invariant dynamical systems approach. *J. Fluid Mech. 227*, 245–272.

Sreenivasan, K. R. (1989). The turbulent boundary layer. In M. Gad-el-Hak (Ed.), *Frontiers in Experimental Fluid Mechanics*, Chapter 4, pp. 159–209. Berlin: Springer-Verlag.

Sreenivasan, K. R. (1991). Fractals and multifractals in fluid turbulence. *Annu. Rev. Fluid. Mech. 23*, 539–600.

Sreenivasan, K. R. (1995). On the universality of the Kolmogorov constant. *Phys. Fluids 7*, 2778–2784.

Sreenivasan, K. R. and P. Kailasnath (1993). An update on the intermittency exponent in turbulence. *Phys. Fluids A 5*, 512–514.

Sreenivasan, K. R., R. Ramshankar, and C. Meneveau (1989). Mixing, entrainment and fractal dimensions of surfaces in turbulent flows. *Proc. R. Soc. London Ser. A 421*, 79–108.

Stapountzis, H., B. L. Sawford, J. C. R. Hunt, and R. E. Britter (1986). Structure of the temperature field downwind of a line source in grid turbulence. *J. Fluid Mech. 165*, 401–424.

Stolovitzky, G., P. Kailasnath, and K. R. Sreenivasan (1995). Refined similarity hypotheses for passive scalars mixed by turbulence. *J. Fluid Mech. 297*, 275–291.

Subramaniam, S. and S. B. Pope (1998). A mixing model for turbulent reactive flows based on Euclidean minimum spanning trees. *Combust. Flame 115*, 487–514.

Sullivan, P. P., J. C. McWilliams, and C.-H. Moeng (1994). A subgrid-scale model for large-eddy simulation of planetary boundary-layer flows. *Boundary-Layer Meteorol. 71*, 247–276.

Tamura, T. and K. Kuwahara (1989). Numerical analysis on aerodynamic characteristics of an inclined square cylinder. Paper 89-1805, AIAA.

Taulbee, D. B. (1992). An improved algebraic Reynolds stress model and corresponding nonlinear stress model. *Phys. Fluids A 4*, 2555–2561.

Tavoularis, S. and S. Corrsin (1981). Experiments in nearly homogeneous turbulent shear flow with a uniform mean temperature gradient. Part 1. *J. Fluid Mech. 104*, 311–347.

Tavoularis, S. and S. Corrsin (1985). Effects of shear on the turbulent diffusivity tensor. *Int. J. Heat Mass Transfer 28*, 265–276.

Tavoularis, S. and U. Karnik (1989). Further experiments on the evolution of turbulent stresses and scales in uniformly sheared turbulence. *J. Fluid Mech. 204*, 457–478.

Taylor, G. I. (1921). Diffusion by continuous movements. *Proc. London Math. Soc. 20*, 196–212.

Taylor, G. I. (1935a). Statistical theory of turbulence: Parts I–III. *Proc. R. Soc. London Ser. A 151*, 421–464.

Taylor, G. I. (1935b). Turbulence in a contracting stream. *Z. Angew. Math. Mech. 15*, 91–96.

Taylor, G. I. (1938). The spectrum of turbulence. *Proc. R. Soc. London Ser. A 164*, 476–490.

Taylor, G. I. and G. K. Batchelor (1949). The effect of wire gauze on small disturbances in a uniform stream. *Quart. J. Appl. Math. 2*, 1–26.

Tennekes, H. and J. L. Lumley (1972). *A First Course in Turbulence*. Cambridge, MA: MIT Press.

Theodorsen, T. (1952). Mechanism of turbulence. In *Proc. Second Midwestern Conf. on Fluid Mechanics*, Ohio. Ohio State University.

Thomas, N. H. and P. E. Hancock (1977). Grid turbulence near a moving wall. *J. Fluid Mech. 82*, 481–496.

Tong, C. and Z. Warhaft (1995). Passive scalar dispersion and mixing in a turbulent jet. *J. Fluid Mech. 292*, 1–38.

Townsend, A. A. (1976). *The Structure of Turbulent Shear Flow* (2nd ed.). Cambridge: Cambridge University Press.

Tritton, D. J. (1988). *Physical Fluid Dynamics* (2nd ed.). Oxford: Oxford University Press.

Tsai, K. and R. O. Fox (1996). PDF modeling of turbulent mixing effects on initiator efficiency in a tubular LDPE reactor. *AIChE J. 42*, 2926–2940.

Tucker, H. J. (1970). The distortion of turbulence by irrotational strain. Mechanical Engineering 70-7, McGill University.

Uberoi, M. S. (1956). Effect of wind-tunnel contraction on free-stream turbulence. *J. Aerospace Sci. 23*, 754–764.

Uberoi, M. S. and P. Freymuth (1970). Turbulent energy balance and spectra of the axisymmetric wake. *Phys. Fluids 13*, 2205–2210.

Valiño, L. and C. Dopazo (1991). A binomial Langevin model for turbulent mixing. *Phys. Fluids A 3*, 3034–3037.

Van Atta, C. W. and R. A. Antonia (1980). Reynolds number dependence of skewness and flatness factors of turbulent velocity derivatives. *Phys. Fluids 23*, 252–257.

Van Atta, C. W. and W. Y. Chen (1970). Structure functions of turbulence in the atmospheric boundary layer over the ocean. *J. Fluid Mech. 44*, 145–159.

van Driest, E. R. (1956). On turbulent flow near a wall. *J. Aerospace Sci. 23*, 1007–1011.

Van Dyke, M. (1982). *An Album of Fluid Motion*. Stanford, CA: Parabolic Press.

Van Slooten, P. R., Jayesh, and S. B. Pope (1998). Advances in PDF modeling for inhomogeneous turbulent flows. *Phys. Fluids 10*, 246–265.

Van Slooten, P. R. and S. B. Pope (1997). PDF modeling of inhomogeneous turbulence with exact representation of rapid distortions. *Phys. Fluids 9*, 1085–1105.

Van Slooten, P. R. and S. B. Pope (1999). Application of PDF modeling to swirling and nonswirling turbulent jets. *Flow, Turbulence Combust. 62*, 295–333.

Vasilyev, O. V., T. S. Lund, and P. Moin (1998). A general class of commutative filters for LES in complex geometries. *J. Comput. Phys. 146*, 82–104.

Villermaux, J. and J. C. Devillon (1972). In *Proc. Second Int. Symp. on Chemical Reaction Engineering*, New York: Elsevier.

Vincenti, W. G. and C. H. Kruger Jr (1965). *Introduction to Physical Gas Dynamics*. Malabar, FL: Krieger.

Voke, P. R. (1996). Subgrid-scale modelling at low mesh Reynolds numbers. *Theor. Comput. Fluid Dyn. 8*, 131–143.

von Kármán, T. (1930). Mechanische Ähnlichkeit und Turbulenz. In *Proc. Third Int. Congr. Applied Mechanics*, Stockholm, pp. 85–105.

von Kármán, T. (1948). Progress in the statistical theory of turbulence. *Proc. Natl. Acad. Sci. USA 34*, 530–539.

von Kármán, T. and L. Howarth (1938). On the statistical theory of isotropic turbulence. *Proc. R. Soc. London Ser. A 164*, 192–215.

Vreman, B., B. Geurts, and H. Kuerten (1997). Large-eddy simulation of the turbulent mixing layer. *J. Fluid Mech. 339*, 357–390.

Waleffe, F. (1997). On a self-sustaining process in shear flows. *Phys. Fluids 9*, 883–900.

Wallace, J. M., H. Eckelmann, and R. S. Brodkey (1972). The wall region in turbulent shear flow. *J. Fluid Mech. 54*, 39–48.

Warhaft, Z. (1980). An experimental study of the effect of uniform strain on thermal fluctuations in grid-generated turbulence. *J. Fluid Mech. 99*, 545–573.

Warhaft, Z. (1984). The interference of thermal fields from line sources in grid turbulence. *J. Fluid Mech. 144*, 363–387.

Weber, R., B. M. Visser, and F. Boysan (1990). Assessment of turbulence modeling for engineering prediction of swirling vortices in the near burner zone. *Int. J. Heat Fluid Flow 11*, 225–235.

Wei, T. and W. W. Willmarth (1989). Reynolds-number effects on the structure of a turbulent channel flow. *J. Fluid Mech. 204*, 57–95.

Welton, W. C. (1998). Two-dimensional PDF/SPH simulations of compressible turbulent flows. *J. Comput. Phys. 139*, 410–443.

Welton, W. C. and S. B. Pope (1997). PDF model calculations of compressible turbulent flows using smoothed particle hydrodynamics. *J. Comput. Phys. 134*, 150–168.

Werner, H. and H. Wengle (1989). Large-eddy simulation of the flow over a square rib in a channel. In *Proc. Seventh Symp. Turbulent Shear Flows*, pp. 10.2.1–10.2.6. Stanford University.

Wilcox, D. C. (1988). Multiscale model for turbulent flows. *AIAA J. 26*, 1311–1320.

Wilcox, D. C. (1993). *Turbulence modeling for CFD*. La Cañada, CA: DCW Industries.

Willmarth, W. W. and S. S. Lu (1972). Structure of the Reynolds stress near the wall. *J. Fluid Mech. 55*, 65–92.

Wizman, V., D. Laurence, M. Kanniche, P. Durbin, and A. Demuren (1996). Modeling near-wall effects in second-moment closures by elliptic relaxation. *Int. J. Heat Fluid Flow 17*, 255–266.

Wong, V. C. (1992). A proposed statistical-dynamic closure method for the linear or non-linear subgrid-scale stresses. *Phys. Fluids A 4*, 1080–1082.

Wygnanski, I., F. Champagne, and B. Marasli (1986). On the large-scale structures in two-dimensional, small-deficit, turbulent wakes. *J. Fluid Mech. 168*, 31–71.

Wygnanski, I. and H. Fiedler (1969). Some measurements in the self-preserving jet. *J. Fluid Mech. 38*, 577–612.

Wygnanski, I. and H. E. Fiedler (1970). The two-dimensional mixing region. *J. Fluid Mech. 41*, 327–361.

Yakhot, V. and S. A. Orszag (1986). Renormalization group analysis of turbulence. I. Basic theory. *J. Sci. Comput. 1*, 3–51.

Yeung, P. K. (1998). Correlations and conditional statistics in differential diffusion: scalars with uniform mean gradients. *Phys. Fluids 10*, 2621–2635.

Yeung, P. K. and S. B. Pope (1989). Lagrangian statistics from direct numerical simulations of isotropic turbulence. *J. Fluid Mech. 207*, 531–586.

Yoshizawa, A. (1984). Statistical analysis of the deviation of the Reynolds stress from its eddy-viscosity representation. *Phys. Fluids 27*, 1377–1387.

Zagarola, M. V., A. E. Perry, and A. J. Smits (1997). Log laws or power laws: the scaling in the overlap region. *Phys. Fluids 9*, 2094–2100.

Zagarola, M. V. and A. J. Smits (1997). Scaling of the mean velocity profile for turbulent pipe flow. *Phys. Rev. Lett. 78*, 239–242.

Zang, Y., R. L. Street, and J. R. Koseff (1993). A dynamic mixed subgrid-scale model and its application to turbulent recirculating flows. *Phys. Fluids A 5*, 3186–3196.

Zeman, O. and J. L. Lumley (1976). Modeling buoyancy driven mixed layers. *J. Atmos. Sci. 33*, 1974–1988.

Zhou, Y. (1993). Interacting scales and energy transfer in isotropic turbulence. *Phys. Fluids 5*, 2511–2524.

Author index

Abid, R., 383, 427
Adrian, R. J., 615
Akselvoll, K., 560, 625, 626, 631
Allmaras, S. R., 385
Anand, M. S., 504, 509, 514, 547, 556
Anderson, E. C., 383
Anselmet, F., 257
Antonia, R. A., 256, 257, 262, 285
Arnal, D., 300
Arnold, L., 715
Aronson, D., 433
Aubry, N., 331

Bailey, F. R., 339
Bakewell, H. P., 325, 330
Balaras, E., 636
Baldwin, B. S., 308, 367, 385
Ball, K. S., 330
Bandyopadhyay, P. R., 308, 324–327
Bardina, J., 346, 382, 628, 629
Barenblatt, G. I., 308–310
Barth, T. J., 385
Barton, J. M., 454, 456
Batchelor, G. K., 7, 10, 11, 110, 148, 203, 205, 259, 406, 680
Béguier, C., 165
Behnia, M., 448, 461
Belin, F., 258
Bell, J. H., 143, 144, 395
Benocci, C., 636
Berkooz, G., xviii, 325, 326, 328, 330, 331
Bilger, R. W., 547
Birdsall, C. K., 525
Blackburn, H. M., 328
Blackwelder, R. F., 325, 327, 328
Blasius, H., 301
Boris, J. P., 633, 638
Borue, V., 352
Boysan, F., 364
Bracewell, R. N., 680
Bradbury, L. J. S., 135
Bradshaw, P., 305, 321, 322, 327, 365, 379, 384, 432, 461, 462

Breuer, M., 638, 639
Brigham, E. O., 697
Britter, R. E., 503
Brodkey, R. S., 324, 325
Brooks, N. H., 494
Brown, G. L., 178
Browne, L. W. B., 285
Brumley, B., 433
Buell, J. C., 355
Burgers, J. M., 634
Bushnell, D. M., 332
Butkov, E., 682

Cabot, W. H., 593, 620, 625, 626
Callen, N. S., 635
Cannon, S., 153, 179, 181
Cantwell, B. J., 324, 328, 461
Capp, S., 6, 97, 98, 100, 102, 105–108, 119, 127
Carmody, T., 153
Carruthers, D. J., 415, 421, 422
Castillo, L., 308, 310
Cazalbou, J. B., 379, 384
Cebeci, T., 308, 367
Champagne, F. H., 140–142, 149, 153, 179, 181, 510
Chapman, D. R., 599, 600, 637
Chapman, S., 361
Chasnov, J. R., 205, 222, 560, 612, 614
Chen, H., 551
Chen, J.-Y., 547
Chen, S., 352, 551
Chen, T.-C., 308
Chen, W. Y., 258
Chen, Y. L., 507, 509
Chevray, R., 153
Choi, K.-S., 332
Chollet, J.-P., 611
Chong, M. S., 326
Chorin, A. J., 310
Chou, P. Y., 390
Chung, M. K., 398, 399, 402, 403
Clark, R. A., 602, 629
Coleman, H. W., 343

749

Subject index

Boldface page numbers indicate principal references or definitions.

Italic page numbers refer to tables or figures.